Advances in Intelligent Systems and Computing

Volume 686

About this Series

The series "Advances in Intelligent Systems and Computing" contains publications on theory, applications, and design methods of Intelligent Systems and Intelligent Computing. Virtually all disciplines such as engineering, natural sciences, computer and information science, ICT, economics, business, e-commerce, environment, healthcare, life science are covered. The list of topics spans all the areas of modern intelligent systems and computing.

The publications within "Advances in Intelligent Systems and Computing" are primarily textbooks and proceedings of important conferences, symposia and congresses. They cover significant recent developments in the field, both of a foundational and applicable character. An important characteristic feature of the series is the short publication time and world-wide distribution. This permits a rapid and broad dissemination of research results.

More information about this series at http://www.springer.com/series/11156

Fatos Xhafa · Srikanta Patnaik
Albert Y. Zomaya
Editors

Advances in Intelligent Systems and Interactive Applications

Proceedings of the 2nd International Conference on Intelligent and Interactive Systems and Applications (IISA2017)

Springer

Editors
Fatos Xhafa
Technical University of Catalonia
Barcelona
Spain

Srikanta Patnaik
Department of Computer Science and Engineering, Faculty of Engineering and Technology
SOA University
Bhubaneswar, Odisha
India

Albert Y. Zomaya
School of Information Technologies
University of Sydney
Sydney, NSW
Australia

ISSN 2194-5357 ISSN 2194-5365 (electronic)
Advances in Intelligent Systems and Computing
ISBN 978-3-319-69095-7 ISBN 978-3-319-69096-4 (eBook)
https://doi.org/10.1007/978-3-319-69096-4

Library of Congress Control Number: 2017956074

Printed on acid-free paper

This Springer imprint is published by Springer Nature
The registered company is Springer International Publishing AG
The registered company address is: Gewerbestrasse 11, 6330 Cham, Switzerland

Preface

As we all know, Intelligent-interactive Systems (IIS) are systems that interact with human beings, media, or virtual agents in intelligent computing environments. The emergence of Big Data and Internet of Things opened new opportunities in both academic and industrial research for successful design and development of Intelligent-interactive Systems. These systems are usually embodied with human capabilities such as to reason, learn, perceive, take decisions, make plans, and perform actions in a wide range of applications such as smart navigation of automobiles so as to avoid traffic congestion, intelligent transportations, smart light, virtual environments, mood-based music systems, and many more. The development of these Intelligent-interactive Systems involves fabricating and controlling machines to perform tasks usually need to be done by human beings.

Since research on Intelligent-interactive Systems deals with the pressing need of designing innovative systems requiring new approaches to current computing architectures and interactions between systems and their peripherals, it explicitly demands knowledge acquisition in many cross-disciplinary subjects such as algorithmic foundations, computational and mathematical modeling fundamentals, artificial and cognitive intelligence foundations, human–computer interaction, software architecture, and developmental concepts and logics of embedded systems.

Some of the significant underlying questions include the following: How huge amounts of data are to be collected and how can they be further analyzed for enhancing the interaction between intelligent systems and users, what are the various issues and challenges faced during technical design and realization of Intelligent-interactive Systems?

This volume explores how novel interactive systems can intelligently face various challenges and limitations previously encountered by human beings using different machine learning algorithms along with analysis of recent trends.

However, designing and development of Intelligent-interactive Systems are quite hard to implement due to variations in abilities of human being, preferences, and limitations. The current researchers are focusing on two major areas: artificial intelligence and human–computer interaction. Further, these two areas are classified into application-specific sub-areas such as computer vision, speech recognition or processing, Web-based systems, data visualization. Some of the widely adopted research areas include information retrieval, recommender systems, intelligent learning systems, natural language processing, multimodal interactions, ubiquitous computing, personalized applications, and semantic-based applications.

Although many researchers are working on various issues of the above-mentioned areas, still lots of challenges and problems arise since it is still in an immature stage. One of the common issues is how both human and artificial intelligence can be combined together to attain efficient results. Apart from these, there are issues relating to users' preferences and privacy protection, which have to be addressed by various intelligent algorithms.

The volume further discusses current research issues and sheds light on future research directions by including various recent and original research articles systematically addressing real-world problems, proposing novel interfaces such as virtual environments, wearable interfaces, and novel interaction models such as touch screens, speech-enabled systems, establishing new standards and metrics for evaluations of model effectiveness and so on. The current research scenario of Intelligent-interactive Systems involves understanding the mechanism lying behind human interactions and how that can be embedded into various applications.

Various dimensions of the IIS includes intelligent user interfaces, context awareness and adaptability, human–computer interaction, wearable technologies, smart solutions for real-time problems, smart navigation, data-driven social analytics, mobile robotics, virtual communication environments, face and gesture analysis, and crowdsourcing.

This volume contains 125 contributions from diverse areas of Intelligent-interactive Systems (IIS), which has been categorized into seven sections, namely (i) Autonomous Systems; (ii) Pattern Recognition and Vision Systems; (iii) E-Enabled Systems. (iv) Mobile Computing and Intelligent Networking; (v) Internet & Cloud Computing; (vi) Intelligent Systems and (vii) Various Applications.

(I) **Autonomous Systems**: This is one of the established areas of interactive intelligence system typically consisting of learning, reasoning, decision making which support the system's primary function. There are nine (9) contributions consisting of various algorithms, models, and learning techniques.

(II) **Pattern Recognition and Vision Systems**: This is one of the primary functions of any Intelligent-interactive Systems. There are forty-five (45) contributions comprised in this section covering the developments in this area of deep learning to binocular stereovision to 3D vision.

(III) **E-Enabled Systems**: This is one of the essential areas of Intelligent-interactive Systems, as many interactive systems are now designed through Internet. It covers information navigation and retrieval, designing intelligent learning environments, and model-based user interface design. There are twelve (12) contributions covered in this section.

(IV) **Mobile and Wireless Communication**: This area is one of the leading areas of IIS, which covers ubiquitous or mobile computing and networking. This section covers twenty-three (23) contributions.

(V) **Internet and Cloud Computing**: It is one of the essential areas of IIS, which caters to enhance communication between the system and users, in a way which may not be closely related to the system's main function. This is commonly found in the areas of multimodal interaction, natural language processing, embodied conversational agents, computer graphics, and accessible computing. In this section, there are four (4) contributions consisting of micro-blogging, user satisfaction modeling to the design and construction of graphical cloud computing platform.

(VI) **Intelligent Systems**: This is the nervous system of IIS, and many researchers are engaged in this area of research. This section contains 11 contributions.

(VII) **Applications**: Applications of IIS in various domains are covered in the last section which consists of 21 contributions.

Acknowledgements

The contributions covered in this proceeding are the outcome of the contributions from more than hundred researchers. We are thankful to the authors and paper contributors of this volume.

We are thankful to the editor in chief of the Springer Book series on **"Advances in Intelligent Systems and Computing"** Prof Janusz Kacprzyk for his support to bring out the second volume of the conference, i.e., IISA-2017. It is noteworthy to mention here that constant support from the editor in chief and the members of the publishing house makes the conference fruitful for the second edition.

We would like to appreciate the encouragement and support of Dr. Thomas Ditzinger, executive editor, and his Springer publishing team.

We are thankful to our friend Prof. Madjid Tavana, Professor and Lindback Distinguished Chair of Business Analytics of La Salle University, USA, for his keynote address. We are also thankful to the experts and reviewers who have worked for this volume despite the veil of their anonymity.

We look forward to your valued contribution and supports to next editions of the International Conference on Intelligent and Interactive Systems and Applications.

We are sure that the readers shall get immense benefit and knowledge from the second volume of the area of Intelligent and Interactive Systems and Applications.

Fatos Xhafa
Srikanta Patnaik
Albert Y. Zomaya

(IV) Mobile and Wireless Communication. [illegible]

(V) Internet and Cloud Computing. [illegible]

Contents

Pattern Recognition and Vision Systems

Others

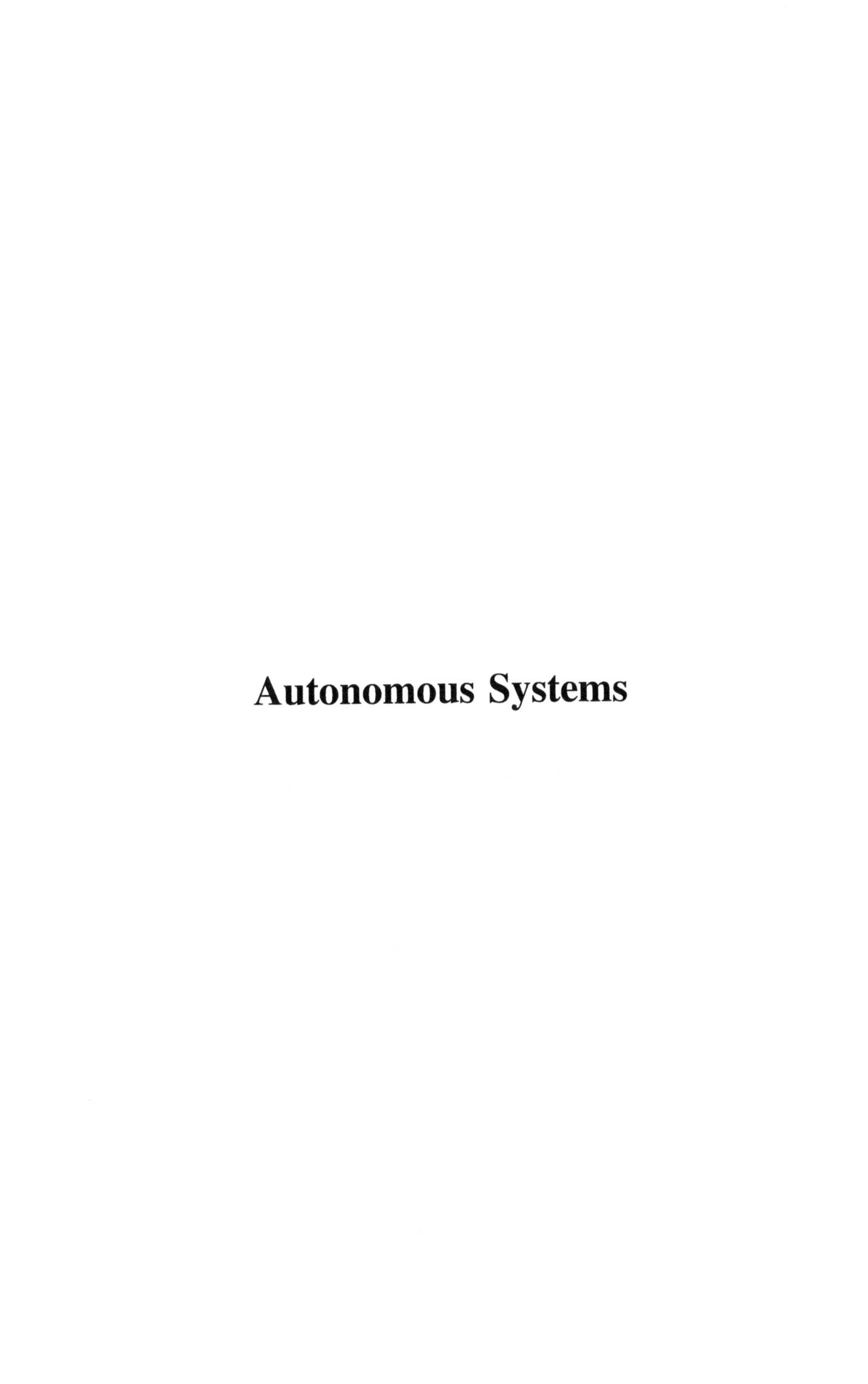

Autonomous Systems

Prediction of Rolling Force Based on a Fusion of Extreme Learning Machine and Self Learning Model of Rolling Force

AZiGuLi[1,2], Can Cui[1,2], Yonghong Xie[1,2](✉), Shuang Ha[1,2], and Xiaochen Wang[3]

[1] School of Computer and Communication Engineering, University of Science and Technology Beijing, Beijing 100083, China
894547148@qq.com

[2] Beijing Key Laboratory of Knowledge Engineering for Materials Science, Beijing 100083, China

[3] School of Electronic Engineering, Xidian University, Xi'an 710071, China

Abstract. Aiming at the rolling force model of hot strip rolling mill, the forecasting technique based on ELM (extreme learning machine) is presented in this paper. Initially, the variables associated with control rolling are identified by the analysis of the traditional formula of rolling force, in order to ensure the effectiveness of the model, and then apply ELM network to predict models. Production data is applied to train and test the above network, and compare with the modified calculated value of rolling force, which got from the self-learning model of rolling force. The results show that the thickness can be predicted more rapidly and exactly, which can meet the actual demand of production, when this model and the rolling force learning model are integrated.

Keywords: ELM · Rolling force model · Self learning

1 Introduction

Today, hot rolling production has more strict requirements on the dimension of steel strip and the precision of plate-shape, in particular for preset of finishing block. And the precision of finishing block preset be the center of the control model of finishing block, rolling force involves a series of domain of operation, such as device parameter, rolling parameter of adjustment and rolling enforcement [1]. And rolling force is related to the thickness model control, strip shape model control and temperature model control, therefore the control of rolling force is key to hot rolling production. The traditional rolling force is counted according to the mathematical model, and the certain physical quantities of mathematical model is counted according to the experiment many times, which make the accuracy of forecast not high enough.

And ELM Algorithm have absolute advantages in such sectors as precision and predicting speed. Therefore, paper will apply the advantage of ELM to rolling force prediction. Rolling force in the every pass set different values according to the different

F. Xhafa et al. (eds.), *Advances in Intelligent Systems and Interactive Applications*, Advances in Intelligent Systems and Computing 686,
https://doi.org/10.1007/978-3-319-69096-4_1

rolled piece parameters, because the multiple passes are set on rolling force. For precision rolling, it need to set the rolling force of seven pass, so precision rolling need to predict the rolling force of seven pass, in turn, according to ELM [2].

2 ELM Algorithm

ELM Algorithm is a simple and effective single invisible layer feed back network SLFNs learning algorithm, which is firstly proposed by Prof. G.B. Huang [3] in 2004. The learning rate of the tradition feed-forward [4] neural network far less than the actual demand, which has been caused by two reasons. The reason one is that the slowly learning algorithm, which based on gradient, has been widely applied to the training of neural network, the all network parameters must be repeat adjusted for each iteration. One the other hand, SVN and transformed smallest squared SVN has applied to binary classification due to its powerful classification capability, and the traditional SVN can not be used directly in regression and multi-class classification, which the problem has not been solved though having proposed different SVN variant.

ELM turned to broad SLFNs, and hidden layers of ELM need not to adjust. Parameters of hidden nodes is not only independent of training data, but also is independent of each others [5]. The standard feed forward neural network having hidden nodes has the ability to the separation and common approximate. ELM features high efficiency learning, and generalization ability is better than traditional feed forward neural network [4].

Taking the typical structure as an example, as shown in Fig. 1, the network is comprised of three parts: input layer, hidden layer and output layer.

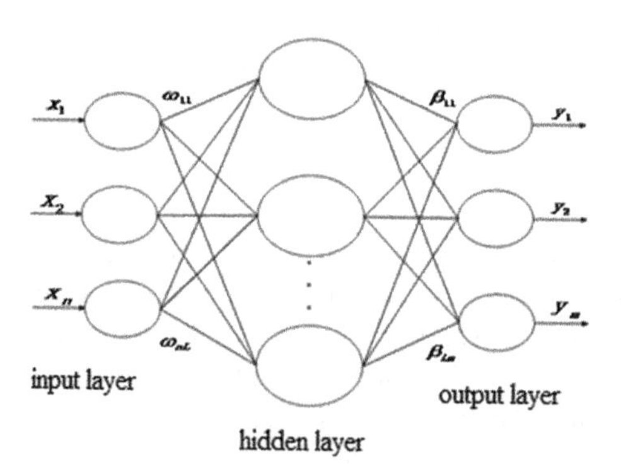

Fig. 1. Typical structure of SLFNs

Figure 1, in the model, there are n input variable, it means that nerve cell number of output layer is n. This indicates that the input vector of sample is $x_1, x_2, \ldots, x_n$, the variable of output sample is m, this indicates that the output vector of sample is $y_1, y_2, \ldots, y_n$. The connection of neighboring layers is determined by joining factor ω and β, and the linked weights matrix are formed by these factors [5].

Hidden layers nodes of n and standard SLFNs math expression of active function g(x) is as followed:

$$\sum_{i=1}^{N'} \beta_i g(\omega_i \cdot x_j + b_i) = o_j, j = 1, \ldots, N \tag{2.1}$$

$\omega_i \cdot x_j$ inner product of vectors and $x_j \cdot \omega_i$
ω_i weight matrix of hidden layer i and input layer,$\omega_i = [\omega_{i1}, \omega_{i2}, \ldots, \omega_{in}]^T$,
β_i weight matrix of hidden layer i and output layer, $\beta_i = [\beta_{i1}, \beta_{i2}, \ldots, \beta_{im}]^T$,
b_i threshold value of hidden layer i,

When formula is employed to approximate the number N of samples in zero error, that can be expressed as $\sum_{i=1}^{N'} \|o_j - t_j\| = 0$ [5]. Existing ω_i, β_i and b_i makes formula become

$$\sum_{i=1}^{N'} \beta_i g(\omega_i \cdot x_j + b_i) = t_j, j = 1, \ldots, N \tag{2.2}$$

Sample formula can be abbreviated:

$$H\beta = T \tag{2.3}$$

H, β and T can be expressed as two formula:

$$H(\omega_1, \omega_2, \ldots \omega_{N'}; b_1, b_2, \ldots b_{N'}; x_1, x_2, \ldots x_N) = \begin{bmatrix} g(\omega_1 x_1 + b_1) & g(\omega_2 x_1 + b_2) & \ldots & g(\omega_1 x_1 + b_{N'}) \\ g(\omega_1 x_2 + b_1) & g(\omega_2 x_2 + b_2) & \ldots & g(\omega_1 x_2 + b_{N'}) \\ \vdots & \vdots & \ldots & \vdots \\ g(\omega_1 x_N + b_1) & g(\omega_2 x_N + b_2) & \ldots & g(\omega_1 x_N + b_{N'}) \end{bmatrix}_{N \times N'} \tag{2.4}$$

$$\beta = \begin{bmatrix} \beta_1^T \\ \vdots \\ \beta_{N'}^T \end{bmatrix}_{N' \times m}, T = \begin{bmatrix} t_1^T \\ \vdots \\ t_N^T \end{bmatrix}_{N \times m} \tag{2.5}$$

H is output matrix of hidden layer [3], the number of hidden layer should be much less than the number of sample in most cases, it means that when $N' \ll N$, H is rectangle matrix [5]. When learning algorithm, H can be calculated and can be made stable through given the any value of ω and b, and parameter β should be determined. Through formula (2.3), the least-square solution $\beta^{'}$ can be get:

$$\left\|H\beta^{'} - T\right\| = \min\|H\beta - T\| \tag{2.6}$$

The minimum norm and the least-square solution can be get according to the formula (2.3):

$$\beta^{'} = H + T \tag{2.7}$$

A few important property can be get according to the generalized inverse matrix:

Least training error. The least-square solution of formula (2.3) is $\beta^{'}$ from formula (2.7). Although almost of learning algorithm hope to reach least training error, most of it can not be done because of local minimum value and infinite iteration algorithm.

Weight minimum norm and optimal generation ability. $\beta^{'}$ from formula (2.7) is minimal norm of the least-square solution. For SLFNs, magnitude of weight is very important, the smaller weight, the more powerful generation ability. It means $\beta^{'}$ has more better generation ability.

The minimal norm $\beta^{'}$ of the least-square solution from (2.3) is unique.

The sample is trained by ELM [3], according to the random weight ω and threshold b, output matrix H can be get according to the set of hidden layers nodes and activation function, corresponding prediction models can be get according to the formula of feed forward network. Mainly ELM algorithm including:

Input weight ω, b are given randomly, and corresponding hidden layers nodes are get according to the real cases.

Select activation function g(X), calculate the hidden layer output matrix H [2].

Calculate the output weight $\beta^{'}, \beta^{'} = H + T$.

For ELM, input weight ω and threshold b can be generated at random, the nodes of hidden layers are adjusted correspondingly to reach optimal output conclusion. This study shows that compared with traditional BP network, ELM has the excellence of faster learning speed, better generalization ability, and avoid the traditional algorithm problems, including locally optimal tragedy, sensitive selection of learning rate and so on [2].

So far, ELM has obtained approval by most researchers, and has been used in so many fields, such as biology identification technology, image-processing, signal processing, big data analysis, robot and internet. ELM could be used in the following respects:

ELM can be trained in hidden layers feed forward network of widespread type, the one layer of ELM [4] has been trained by hidden layer, which used in characteristics learning, cluster, regression and classify. The entire network can be seen as no adjusting ELM.

ELM can move into many local regions of multi hidden layer feed forward network, and work together with other study architecture or model.

Each hidden layer nodes of ELM are sub-networks formed by some nodes, and hidden layer can form local receptive domain [3].

In each hidden layer, the connection from the input layer to hidden layer node can be full connection, and random local connection can be generated by the connection according to the different continuous probability distribution. For the network [5], if a limited number of hidden layer nodes and the connection have changed, the power of the network still can make stable.

Compared with present mainstream president algorithm, the most advantage of ELM is that the hidden neurons of the entire learning machine can not adjust, and have high efficiency. Some algorithms of neural network have been applied to the steel industry, like BP, RBF and Hopfield, but for ELM, there is a little application in steel production industry for now. In this paper, ELM is applied to the steel rolling production, and together with other methods, hot continuous rolling exit thickness and rolling force are used to model for guiding steel production [5].

3 Rolling Force Mathematical Model

Rolling force can be calculated through some combination of the mathematical model and empirical data, therefore, there are a lot of formula for rolling force. For now, the correlative formula of rolling force has focus on the SIMS formula of OROWAN, and researchers use the formula to calculate result in different rolling force condition, the get different deformation formula, the original rolling force formula of SIMS follows [5]:

$$P = Bl_c'Q_pKK_T \tag{3.1}$$

P rolling force, KN,

l_c' flattened roll and horizontal projection length of strip contact arc, mm,

Q_p the impact influence of stress state is caused by friction on contact arc,

K decide the inner components of mental material and the deformation condition, mental deformation resistance about the deformation temperature, deformation speed and the deformation degree,

K_T coefficient of front and back tensions on the influence of rolling force,

B bandwidth, mm

By some of these formulas and existed relationship among all variables, corresponding physical makeup of the test samples are found by the database of Beihai steel production. Because set-up calculation of Beihai steel production, modified calculation and adaptive learning in steel production, that some of data need be discarded and the measurement data need be chosen as samples, when samples are chosen. And according to the corresponding formula of rolling force, variables are chosen such as roll speed, roll rotate, forward slide coefficient, back slide coefficient, rolling force measured value, FET measured temperature, FDT measured temperature, exit thickness gauge measured value (deviation), exit width gauge, roll gap measured value, bend-roller force measured value, roll speed measured value, median width slab, entry thickness measured value, entry temperature, exit temperature and roll gap.

4 Rolling Force Model Self Learning of Finishing Mill

The corresponding technology parameters for the rolling force model of finishing mill is different depending on the site environment and the equipment situation [6]. In order to ensure the shape and thick of the production, the computer of the model need be optimized based on the data of the actual rolling. The rolling force method is modeled according to the analysis of the deformation resistance and the coefficient of stress state. And the initial parameters are set, AGC control is used for model, which can achieve continuous self learning, to adjust rolling force. As compared with self learning of the other parameters, the biggest difference is that rolling force can not compare the measured rolling force of the N pass steel with the original estimate rolling force. Because bounce is a remarkable feature of mil, when measured rolling force become the ups and downs at the change, because of the elastic deformation, the value of actual thickness is more or less than the value of the original prediction thickness. So the value set in rolling force and the measured value of rolling force can be get in two different technology parameters, which can not be compared. And the self learning model used the measured rolling force to correct the model parameters, rolling force model recalculated the rolling force depending on correction factor, this step calculation is called correction calculation. So compared with the measured value and the value of rolling force got from the correction calculation, the feature of the model can be reflected.

Through the above process of the rolling force model, in production, initially, when slab is required to roll through the first finishing mill, the measured value of rolling force in the first pass can be get, and the parameters of rolling force model are corrected by the model according to the measured value of rolling force in the first pass, then, the value of rolling force is recalculated by the original model, and the parameters of the production line in the second pass are set by the value of rolling force which was got according to the correction calculation [8], the second roll is finished. When the second roll is finished, the rolling force are corrected to calculate by repeating the step of the first roll, finally, the value of the rolling force is controlled by repeating the self learning in order to the guide the steel production.

5 Forecasting Experiment and Result Based on Rolling Force of ELM

According to the above the finishing mill principle, initially, through the formula of the rolling force, affecting factors about the rolling force are found and regarded as the input variable in order to predict for the rolling force in the first pass, the measured value of rolling force in the first pass which is got in the steel production and the corresponding affecting data about the rolling force are regarded as the input variable into the next ELM model in order to predict the rolling force in the second pass, by analogy, when predicting the rolling force in the seventh pass, the prediction rolling force in the top six pass is got by the model in order to guide the production, so the measured value of rolling force in the top six pass which is got in the steel production and the corresponding affecting data about the rolling force are regarded as the input variable into the next ELM [6] model in order to predict the rolling force in the seventh pass, finally, the all of the prediction data in the seven passes are got, the multiple ELM models superposition are realized.

First of all, divide the 522 filtrated groups of data which is related to the rolling force into training set and test set, with 472 group as the training set, the remaining 50 group as a test set, to predict the first pass rolling force of the finishing mill. Due to the number of nodes in hidden layer can produce certain influence on prediction accuracy, so many experiments are needed to determine the optimal number of nodes. The conclusion can get through the experiment, when the number of hidden layer nodes is 80, network performance achieves the best. The activation function is "sig" function. Based on the data, the study found some dimensions of the data which is related to the rolling force are small, while some are very large. If different dimensions are not processed, the rate of convergence of the network will be reduced. First, process the input data as normalization, put the rolling force data to be quantitative without the outline, and then carries on the training and obtain the comparison between the actual rolling force with predict rolling force model.

From the picture the most of points of first time prediction of rolling force and actual rolling force are close, occasionally there are some larger errors. The reason of these errors is the rolling force value is bigger, so the influence of the amplitude is bigger. For the traditional rolling force model, the comparison with the measured rolling force can not be done by setting calculation of rolling force. Correct calculation of rolling force and the measured rolling force should be compared [7]. Therefore corrected average standard deviation of the calculated rolling force and ELM prediction model can be seen in the Fig. 2.

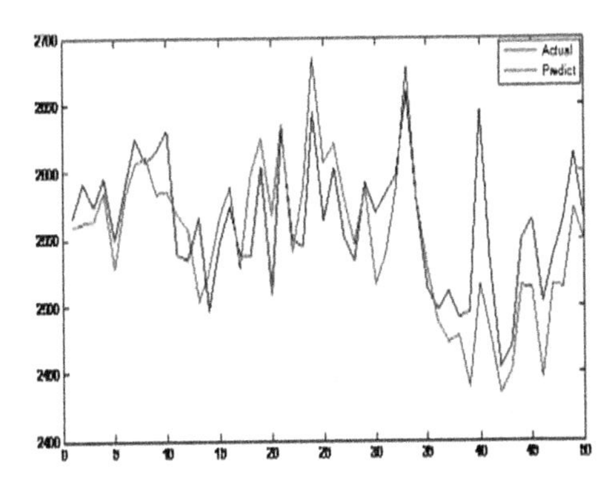

Fig. 2. The predicted value of the rolling force and the actual value of the rolling force for ELM

The first mill prediction precision of rolling force is slightly higher than the precision of rolling force obtained by modified computing. Because correction calculation of rolling force is for the guidelines to rolling mill, so the predicted rolling force can be used in place of the correct calculation of rolling force to guide production. Because the order of magnitude of rolling force unit is big, the experiment will increase the numerical calculation, so orders of magnitude unit need to be downsized, and then according to the calculation results restore the order of magnitude of the corresponding unit. Make the first mill rolling force production actual value from ELM model with other variables as the input of the second mill rolling force prediction model [7], the actual value of rolling force by second mill again make the third mill modeling and so on to get seven mills of ELM superposition model. The predicted results compare with the measured values and get the average standard deviations, and compare with average standard deviation fixed calculated (Table 1).

Table 1. The error value of first pass rolling

	The correction calculate of rolling force	The prediction of rolling force
RMSE (10^3)	0.083703207	0.037146671

The first mill prediction precision of rolling force is slightly higher than the precision of rolling force obtained by modified computing [8]. Because correction calculation of rolling force is for the guidelines to rolling mill, so the predicted rolling force can be used in place of the correct calculation of rolling force to guide production [8]. Because the order of magnitude of rolling force unit is big, the experiment will increase the numerical calculation, so orders of magnitude unit need to be downsized, and then according to the calculation results restore the order of magnitude of the corresponding unit. Make the first mill rolling force production actual value from ELM model with other variables as the input of the second mill rolling force prediction model [5], the actual value of rolling force by second mill again make the third mill modeling and so on to get seven mills of ELM superposition model [4]. The predicted results compare with the measured values and get the average standard deviations, and compare with average standard deviation fixed calculated (Table 2).

Table 2. The error value of each pass rolling force for precision rolling

Pass	The prediction of rolling force		The correction calculation of rolling force	
	Time (s)	RMSE (10^3)	Time (s)	RMSE (10^3)
1	0.004	0.03708315	–	0.08370321
2	0.001	0.03686208	–	0.05952837
3	0.001	0.05461632	–	0.04789749
4	0.001	0.03943099	–	0.02624345
5	0.001	0.04748756	–	0.03547987
6	0.001	0.03010246	–	0.03407099
7	0.001	0.02245528	–	0.07047431

6 Conclusion

From the above experimental data, the use of ELM for continuous prediction of seven passes in the results, the first and second mills of the rolling force prediction accuracy compared with the correction precision of the calculation model is higher; the third to the fifth mills of prediction precision deviation is bigger [8], the deflection prediction accuracy is lower than those of correction calculation model, the effect is poorer. Mills of 6 and 7 of the predicted results compared with the correction results found that 6 and 7 of the finishing passes by using machine learning to predict the effect of is better than the effect of the modified calculation.

If all seven passes of finishing mill group through ELM rolling force prediction model, then in the third to the fifth mills due to big errors may lead the final exported steel does not meet the actual requirements. So according to the above experimental results to improve the model, through the ELM and self learning of system integration method to predict the rolling force, build the model as shown in Fig. 3. Two mills before can use ELM overlay to predict the rolling force, and according to the predicted rolling force to guide production [9]. The third to the fifth mills on-line correction can through the system of self learning, Then the sixth, seven mills again apply ELM rolling force prediction model.

Through improving model to forecast the rolling force, the time of forecast can be greatly shortened and solve the real-time demanding of rolling technology problem, and get the strong robustness. High accuracy model will has a broad application prospect.

References

1. Sun, Y.K.: Model of Control Hot Strip Mill, pp. 124–163. Metallurgical Industry Press, Beijng (2002)
2. Huang, G.B., Zhu, Q.Y., Siew, C.K.: Extreme learning machine: theory and applications. Neuro-Computing **70**, 489–501 (2006)
3. Huang, G.B., Zhu, Q.Y., Siew, C.K.: Extreme learning machine: a new learning scheme of feed-forward neural networks. In: Proceedings of International Joint Conference on Neural Networks (IJCNN 2004), pp. 985–990 (2004)
4. Lonen, J., Kamarainen, J.-K., Lampinen, J.: Differential evolution training algorithm for feed-forward neural networks. Neural Process. Lett. **7**(1), 93–105 (2003)
5. Huang, G.B., Chen, L., Siew, C.K.: Universal approximation using incremental constructive feed-forward networks with random hidden nodes. IEEE Trans. Neural Netw. **17**(4), 879–892 (2006)
6. Huang, G.B., Chen, L.: Convex incremental extreme learning machine. Neurocomputing **70**, 3056–3062 (2007)
7. Xie, J., Jiang, S., Xie, W., et al.: An efficient global K-means clustering algorithm. J. Comput. **6**(2), 271–279 (2011)
8. Al-Zoubi, M.B., Hudaib, A., Huneiti, A., et al.: New efficient strategy to accelerate k-means clustering algorithm. Am. J. Appl. Sci. **5**(9), 1247–1250 (2008)
9. Gaemperle, R., Mueller, S.D., Koumoutsakos, P.: A parameter study for differential evolution. In: Grmela, A., Mastorakis, N.E. (eds.) Advances in Intelligent Systems, Fuzzy Systems, Evolutionary Computation, pp. 293–298. WSEAS Press (2002)

User Behavior Profiles Establishment in Electric Power Industry

Wei Song[1], Gang Liu[1], Di Luo[2(✉)], and Di Gao[1]

[1] State Grid Jibei Electric Power Science Research Institute, Nanjing, China
[2] School of Computer and Communication Engineering, University of Science & Technology Beijing, Beijing, China
braveld@163.com

Abstract. The big data of electric power industry contains the information of users' values, credits, and behavior preferences. On the basis of the data, electric power companies can provide personal services as well as increasing profits. In this paper, we proposed a labeling system based on the clustering algorithms and Gradient Boost Decision Tree (GBDT) algorithm to establish user behavior profiles for State Grid Group of China, including basic information labels, behavior labels, behavior description labels, behavior prediction, and user classification. The experimental results showed that the approach can describe the behavior features of users in the electric power industry effectively.

Keywords: Behavior profile · Behavior prediction · Clustering · Gradient boost decision tree · Electric power industry

1 Introduction

There are many kinds of behaviors in electric power industry including electricity consuming, paying, capacity changing, illegally electricity using, etc. Although the electric power companies have a lot of users' behavior data, they are numerical and hard to understand directly. Some rules are needed to turn the initial numerical data to semantic labels which could easily illustrate the users' feature. Here are some of the labels: 'Zhangjiakou', 'Low monthly consumption', 'capacity change frequently', 'Recover from capacity suspend every 7 months', 'High risk of arrearage', 'Low user value'. These labels are called user profiles. Profiling the users is the basis of providing the personal services, and its applications in the electric power industry are various. First, companies can select users by labels, which are more detailed and semantic. Also, it's easy to recognize the occurrence of the user behaviors so that companies can easily monitor the user behaviors and give advise. At the same time, as the companies handle the feature of the user behaviors, precision marketing will be reasonable and the companies can arrange business better. What the most important is, according to the user profiles of electric power industry, the companies can get the semantic credit report of the users. User profile is commonly used in IT industry, but we can not build the user profiles in the electricity industry using the same approach in the IT industry

F. Xhafa et al. (eds.), *Advances in Intelligent Systems and Interactive Applications*, Advances in Intelligent Systems and Computing 686,
https://doi.org/10.1007/978-3-319-69096-4_2

due to the great differences between the two industries. The electricity industry needs its own labeling systems and new methods to generate the semantic labels. The essence of user profiles is to label the user features, which depends on label generation techniques.

2 Related Work

User profiles are very useful in the IT, telecom, financial industry. Commonly applied techniques such as frequent pattern mining, probability matrix, etc., are used to build the mobile user behavior profiles based on the data of the users' frequent behaviors, data of regular stops and moving speeds [1]. Some labels are built base on the expert experience. These labels represent the long and short term preference of users, which can be used in the collaborative filtering algorithm to recommend web page [2]. Collaborative filtering matrix is an effective method to create user label from the recording history. Ali et al. built a user profile system named "TiVo" based on the collaborative filter matrix technique [3]. Dynamic Bayes algorithm was employed to build a user's profile by online education resource [4]. However, there is few related work to build the user profile of electricity users. Sparse encoding model was applied to create user label of abnormal electricity consumption [5]. Logistic regression algorithm was well employed to create the label showing the risk of arrearage in electric power industry [6]. Back propagation neural network model was used to predict the possibility of stealing electricity [7].

3 User Label System of Electric Power Industry

Electric user behavior profile contains static basic information and dynamic behavior information. The static basic information is stable, while dynamic behavior information is changing over time. This paper proposed a new hierarchical label system for electric power industry, including basic information label, behavior label, behavior description label, behavior prediction and user classification label.

3.1 Behavior Label

Most of the data are numerical. To show what a user did directly, the semantic label can be an effective way. The paper combines the expert experience and the clustering algorithm to achieve the discretization of the data, dividing the data into several comprehensible labels. That is, turn the numerical data to semantic label T.

3.2 Behavior Description Label

The behavior description label shows the temporal features of the user behavior and user preference. Each behavior can be described from its occurrence frequency, average variation, coverage rate, standard deviation, average time interval, periodicity, period preference. create Ratio is the coverage rate of the behavior *at* of the user u, and we use

$sum(at_j, u)_{ET-ST}$ as the sum of the behavior in the period T. Therefore, the formula is as follow:

$$\text{CreateRatio} = \frac{sum(at_j, u)_{ET-ST}}{\sum_{i=0}^{n} sum(at_i, u)_{ET-ST}} \tag{1}$$

Periodicity is for estimating if this behavior has periodic regular. If not, we use 0 to represent it, otherwise, we use *d* to show the periodic time interval. We use *sum()* as the sum of the occurrence of d_j, and the *period(at, u)* is calculated as follow:

$$period(at, u) = \begin{cases} d_i \ \exists d_i \in (d_1, d_2, \ldots, d_n), \ sum(d_i, u) \geq 0.6 \sum_{i=0}^{n} sum(d_i, u) \\ 0 \ \forall d_i \in (d_1, d_2, \ldots, d_n), \ sum(d_i, u) \leq 0.6 \sum_{i=0}^{n} sum(d_i, u) \end{cases} \tag{2}$$

The paper define *TF* as the period preference. If the number of the occurrence of the behavior in the one period of time is more than 60% of the total number of the occurrence of the behavior, we will label the user has periodical preference, and the label name will be the period.

Distribute the one dimension temporal feature such as occurrence frequency into several levels to describe the different frequency of the behaviors. Besides, distribute occurrence frequency, coverage rate, average time interval into several levels, and let the experts name each level so we can get the labels of the behavior preference. The labels of the period preference can be created as the name of the time period.

3.3 Behavior Prediction and User Classification Label

The electricity company will be able to arrange the product plan more reasonable by predicting the future behaviors of users. The user classification label is for dividing users into different groups, likes 'high-value user', 'low-value user', 'high credit', 'bad credit'. The clustering algorithm is always used to do that, and it performs well. The behavior prediction labels are based on various classification algorithm and GBDT algorithm is the first choice. Using the clustering algorithm and GBDT algorithm, the paper classifies users into different groups which represent different risk of being behind in payment and labels them.

4 Label Generation Technique

4.1 Label Generation Technique Based on Clustering Algorithm

k-means cluster algorithm can't be used directly in the situation [8]. While generating the label, we have to deal with the outliers instead of just ridding off them. The algorithm based on the idea of the Alex and Alessandro can perform well in such situation. The algorithm has its basis in the assumptions that cluster centers are surrounded by neighbors with lower local density and that they are at a relatively large distance from any points with a higher local density [9].

We use S to represent the N size dataset that waiting to be classified, X for each data point in this dataset, and it defined as $S = \{X_i\}_{i=1}^{N}$. Distances between the data point X_i and X_i are defined as δ. For each X, calculate its density ρ_i and distance δ_i. First, we have to define a cutoff distance d_c, which is given by experience. ρ_i is defined as follow:

$$\rho_{\mathrm{i}} = \sum\nolimits_{i,j \in I_S} \phi\left(d_{ij} - d_c\right) \tag{3}$$

where

$$\phi(\mathrm{x}) = \begin{cases} 1, & x<0 \\ 0, & x \geq 0 \end{cases} \tag{4}$$

Assume that $\{q_i\}_{i=1}^{N}$ is the index of $\{p_i\}_{i=1}^{N}$ in descending order, which is $\rho_{q1} \geq \rho_{q2} \geq \ldots \geq \rho_{qN}$. So δ_{qi} is defined as:

$$\delta_{qi} = \begin{cases} \min\left(d_{qiqj}\right), & i \geq 2, j<i \\ \max\left(\delta_{qj}\right), & i = 1, j \geq 2 \end{cases} \tag{5}$$

Assume that there are n_c $(n_c > 0)$ cluster centers in the dataset S. $\{m_i\}_{i=1}^{N}$ is the serial number of the data point corresponding to each of the cluster center. $d_{\max} = \max\left\{d_{ij}\right\}$ is the biggest distance between two data point in S. $\{n_i\}_{i=1}^{N}$ represents the number of the data point whose density is bigger than X_i and is the closest with X_i. The steps of the algorithm are as follow:

Step one: Initialization and pretreatment

(1) Given t to determine the cutoff distance, where $t \in (0, 1)$;
(2) Compute the distance d_{ij} between each two data points, where $i > j$, $i, j \in I_S$;
(3) According to 2), M is the number of the distances between each two data points in dataset S and $M = 0.5(N-1)N$. Sort them in ascending order, making $d_1 \leq d_2 \leq \ldots \leq d_M$. So we can get $d_c = d_{f(Mt)}$, *f(Mt)* is the function that rounds the *Mt;*
(4) According to the formula (1), compute the density of each data point to get $\{p_i\}_{i=1}^{N}$, and sort them in descending order to get the index $\{p_i\}_{i=1}^{N}$;
(5) According to the formula (3), compute the distance between data points, and get $\{n_i\}_{i=1}^{N}$.

Step two: Determine the cluster centers

The cluster centers must fit two conditions: (1) cluster centers are surrounded by neighbors with lower local density; (2) cluster centers are at a relatively large distance from any points with a higher local density. For each data point, the bigger ρ_i and δ_i it has, it has more chance to be the cluster centers. Consider both of the ρ_i and δ_i, we can get γ_i defined as follow:

$$\gamma_i = \rho_i \delta_i \ , \ i \in I_S \tag{8}$$

Apparently, the bigger γ_i it has, it is much likely to be the cluster centers. Draw the figure of the γ_i in the descending order. There must be some points that have lager γ_i and can be easily identified. Make them as the cluster centers.

Step three: Classify the rest of the data points

For the rest of the data points, they are sorted in the descending order of their ρ_i. So compare each of them to the different cluster centers, and classify them into the cluster that has the smallest distance.

5 Experiment and Analysis

The paper selects the real data of large-scale industry users from State Grid Group of China, which contains 183662 monthly electricity consumption data, 23562 electricity inspection data, 13839 capacity changing data, ranging from 2014.1 to 2016.10. Besides, there are also 7008 arrearage data ranging from 2016.1 to 2016.10.

5.1 Generate the Behavior Label

Create behavior labels according to monthly consumption data, and get the 'Extreme high electricity consumption', 'High electricity consumption', 'Medium electricity consumption', 'Low electricity consumption' and 'Extreme low electricity consumption' labels. The monthly consumption of large-scale industry users has a wide distribution. A small number of extreme large-scale steel plants have much higher monthly consumption than the others, which become the outliers. Therefore, we rid this part of data into the 'High electricity consumption' level and label the behavior. The paper set t = 0.05 to get the best result. And Fig. 1 is what we got.

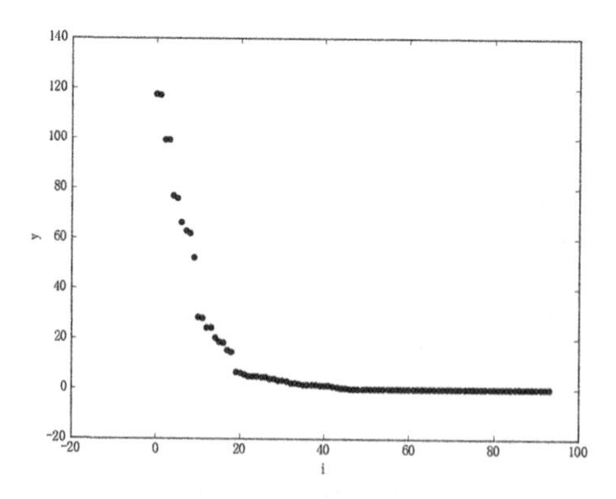

Fig. 1. γ_i in descending order of each data point

From the Fig. 1, we can see there are 20 points on the figure that are much high than the others. These points will be selected to be the cluster centers, and after that, the cluster centers will be classified into the rest of the four levels to get the other labels.

5.2 Generate the Behavior Description Label

The paper is using the monthly consuming behavior and capacity changing behavior as an example in this section. According to the formula in Sect. 3, we can get Table 1.

Table 1. The behavior feature of monthly consuming and capacity changing

Behavior	Frequency	Coverage rate	Average time interval	Standard deviation	Periodicity	Period preference
No consumption	17	0.57	1.5 month	1.36	1 month	Former half of year
Low electricity consumption	7	0.23	2.1 months	2.06	No periodicity	Latter half of year
Extreme low electricity consumption	4	0.13	2.75 months	3.05	No periodicity	0
Capacity suspend	5	0.56	6 months	3.36	No periodicity	Beginning of years
Recovery from capacity suspend	4	0.44	7 months.	0	7 months	Latter half of year

From the table above, classify them with cluster algorithm to get the behavior description label: ‘No consumption frequently’, ‘Low electricity consumption’, ‘Capacity suspend happens at the beginning of the year’, ‘Recover from capacity suspend every 7 months’.

5.3 Generate the Behavior Prediction Label and User Classification Label

The paper uses RFM model [12] combined with expert experience to define the risk of the arrearage. The paper uses nine months arrearage data as training data ranging from 2016.1 to 2016.9 and classifies the RFM value into different clusters using the clustering algorithm above and gets 17 clusters. Giving the R, F and M the same weight, classify the 17 clusters into 4 levels according to the expert experience. Finally, we can label users with ‘High risk in arrearage’, ‘Medium risk in arrearage’, ‘Low risk in arrearage’, ‘No risk in arrearage’.

Build the predicting model for risk of the arrearage based on GBDT algorithm according to the monthly consumption record, contract breaking record, capacity changing record. The result is displayed using the confusion matrix as follow:

GBDT	No	Low	Medium	High
No	363	0	0	0
Low	0	589	126	3
Medium	0	196	588	23
High	0	3	32	70

The paper compares it with the same training and testing data using the random forest (RF) algorithm. Compare the two different algorithms with the total accuracy, the recognition rate of the 'No risk in arrearage', 'Low risk in arrearage', 'Medium risk in arrearage', 'High risk in arrearage' respectively. The comparison is showed as Table 2.

Table 2. The comparison between GBDT and RF

Algorithm	Total	No	Low	Medium	High
GBDT	0.807	1	0.820	0.729	0.667
RF	0.803	0.997	0.822	0.719	0.634

We can know that both algorithms have almost the same accuracy in predicting the risk of the arrearage. However, the GBDT has better performance in recognizing the higher risk users.

6 Conclusion

The paper proposes a method to build the user behavior profiles of the electric power industry. The user behavior profile can recognize and describe behaviors, make predictions and classify users into different groups. Through the experiments, the user behavior profile that we build can describe the user perfectly. However, there are some disadvantages. Some of the labels need human involvement and are not generated quantitatively. The predicting model of the behavior may update after a period of time. In the future, we are going the build the user profile with more detailed data to fit the need of providing personal services.

References

1. Huang, W., Xu, S., Wu, J. et al.: The profile construction of the mobile user. J. Mod. Inf. **10** (2016)
2. Liu, Y.: A Generation Method of Mobile Users' Tags baesd on Time Features. Dalian University of Technology (2015)
3. Ali, K., Van Stam, W.: TiVo: making show recommendations using a distributed collaborative filtering architecture// Tenth ACM SIGKDD International Conference on Knowledge Discovery and Data Mining, Seattle, Washington, USA, pp. 394–401 (Aug 2004)
4. Slimani, H., Faddouli, N.E., Bennani, S., et al.: Models of digital educational resources indexing and dynamic user Profile evolution. Int. J. Emerg. Technol. Learn. **11**(1), 26 (2016)

5. Li, Z., Lujun, Z., Weiguo, G.: Application of sparse coding in detection for abnormal electricity consumption behaviors. Power Syst. Technol. **11**, 3182–3188 (2015)
6. Xu, R., Nd, W.D.: Survey of clustering algorithms. IEEE Trans. Neural Netw. **16**(3), 645–678 (2005)
7. Chen, X., Li, C., Xiaoxiao, L., et al.: Research on electricity demand forecasting based on ABC-BP neural network. Comput. Meas. Control **22**(3), 912–914 (2014)
8. Kojima, K.: Proceedings of the fifth Berkeley symposium on mathematical statistics and probability. Berkeley Symposium on Mathematical Statistics & Probability (1969)
9. Rodriguez, A., Laio, A.: Clustering by fast search and find of density peaks. Science **344**, 1492 (2014). doi:10.1126/science.1242072
10. Deng, X.: Research of O2O E-commerce Recommendation Model on Gradient Boosting Regression Trees. Anhui University of Science & Technology (2016)
11. Friedman, J.H.: Greedy function approximation: A gradient boosting machine. Ann. Stat. **29**(5), 1189–1232 (2000)
12. Fader, P.S., Hardie, B.G.S.: RFM and CLV: Using Iso-Value Curves for Customer Base Analysis. J. Mark. Res. XLII(4), 415–430 (2005)

The Optimal Pan for Baking Brownie

Haiyan Li[1], Tangyu Wang[1], Xiaoyi Zhou[2], Guo Lei[1(✉)], Pengfei Yu[1], Yaqun Huang[1], and Jun Wu[1]

[1] School of Information Science and Engineering, Yunnan University, Kunming, China
guolei@ynu.edu.cn
[2] School of Business, Renmin University, Beijing, China

Abstract. In order to maximize the pan number placed in the oven and the even distribution of heat in a pan when baking brownie, we propose two mathematical models, named as "edge cutting model" and "weight evaluation model", to get the solutions. The "edge cutting model" mathematically proves how the heat diffusion distributes over the outer edge of the pans with various geometric shapes by cutting the edge into small components and then deducing the functional relation between the heat distribution respect to the ratio of the surface area to the mass of each component. And we prove why the heat concentrates in the four corners of the rectangular pan and why the heat evenly distributes on the outer edge of the round pan. Further, we simulate the heat distribution results of different shapes of the pan when it is placed in the oven by using the software of the ANSYS. And the results are consistent with the theory analysis, which demonstrate the feasibility of the proposed model.

Keywords: Edge cutting model · Weight evaluation model · Even distributuion of heat

1 Introduction

Nothing warms the soul like a warm brownie. Brownies are definitely America's favorite bar cookie. However, it is not easy to bake a perfect brownie. Brownie is often overcooked at the 4 corners and the edges for a rectangle pan. Interestingly, compared the rectangular pan with the circle pan, we find the outer edge of brownie is not overcooked by the circle pan. In our daily life, we prefer to cook a brownie tasted well without being overcooked. Thus, we want to know the distribution of heat across the outer edge of the pan. In addition, the primary goal of the model is to solve the following problems: To gain the maximum number of the pans that can make full use of the oven space.

In order to maximize the pan number placed in the oven and the even distribution of heat in a pan when baking brownie, we propose two mathematical models, named as "edge cutting model" and "weight evaluation model", to get the solutions. The "edge cutting model" mathematically proves how the heat diffusion distributes over the outer edge of the pans with various geometric shapes by cutting the edge into small components and then deducing the functional relation between the heat distribution respect

F. Xhafa et al. (eds.), *Advances in Intelligent Systems and Interactive Applications*, Advances in Intelligent Systems and Computing 686,
https://doi.org/10.1007/978-3-319-69096-4_3

to the ratio of the surface area to the mass of each component. And we prove why the heat concentrates in the four corners of the rectangular pan and why the heat evenly distributes on the outer edge of the round pan. Further, we simulate the heat distribution results of different shapes of the pan when it is placed in the oven by using the software of the ANSYS. And the results are consistent with the theory analysis, which demonstrate the feasibility of the proposed model.

2 The Distribution of Heat for Pan

2.1 Finite Element Analysis

As for different shape pan, since the distribution of heat in a pan varies with the shape of the pan, we originally propose mode named as *Edge Cutting Model* to mathematically prove the heat at the corner is greater than that at the edges. In order to get the distribution of heat at any place on the pan, we must know the way how heat is transferred in the pan. The heat transfer is mainly classified into three mechanisms according to physical definition, including thermal conduction, thermal convection and thermal radiation.

(1) Thermal conduction: On a microscopic scale, heat conduction occurs as hot, rapidly moving or vibrating atoms and molecules interact with neighboring atoms and molecules, transferring some of their energy (heat) to these neighboring particles [1].
(2) Thermal Convection: Convective heat transfer, or convection, is the transfer of heat from one place to another by the movement of fluids, a process that is essentially the transfer of heat via mass transfer [2].
(3) Thermal radiation: Thermal radiation is energy emitted by matter as electromagnetic waves due to the pool of thermal energy that all matter possesses that has a temperature above absolute zero [2].

The oven actually consists of the box, electrical heating element, thermostat, timing device and the power adjustment switch. In addition, we know the oven heat the product by using electrical heating element which can produce radiation heat [3]. Furthermore, we assume that the distribution of heat is even in the whole space of the oven. Since the pan is heated mainly by the air temperature in the oven, we mainly consider the heat transfer as thermal convection.

We know the thermal conductivity varies with different material according to physical knowledge, which metal can transfer heat better than non-metallic material. The thermal conductivity is associated with the physical structure and the chemical structure [4]. Hence, we assume that the pan discussed is made of by the same material, which is assumed to be steel. According to physical definition, for steel, the thermal conductivity is $50\mathrm{W/m \cdot K}$, the specific heat capacity is 460 J/kg °C, the heat transfer coefficient is $125\mathrm{W/m^2 \cdot K}$, the density is $7850\,\mathrm{kg/m^3}$.

Differential equation is widely used to solve the problems of thermodynamics. However, it is difficult to get the exact result by solving the differential equation. Hence, in our real life, such as baking brownie, we cannot use the differential functions to get the maximum even heat distribution of a pan. Thus, in order to solve the differential equation of thermodynamic field in a oven, we use finite element method to gain an approximate solution [5]. According to the First Law of Thermodynamics, we know that the amount of energy gained by the system must be exactly equal to the amount of energy lost during an interaction between a system and its surroundings [6].

Here we use transient heat analysis to compute the temperature field of an oven changing with time. For transient heat analysis, the load is changed with time. We propose to use ANSYS software to simulate the situation with various initial conditions.

ANSYS software can be used to analyze the problems of heat field, electric field, magnetic field and the sound field. We get results which can visually show how the temperature changes with time. We first simulate the heat field with the following predefined requirements by ANSYS software.

- The material is steel.
- This steel's width is 2 m, length is 2 m and thickness is 0.1 m.
- The initial temperature is 20 °C.

The oven's temperature is 1120 °C [7].

2.2 Edge Cutting Model

At first, we obtain the temperature field of an oven by using ANSYS software emulation, which the simulation condition is shown in the Table 1 and the result is shown in Fig. 1.

Table 1. Simulation condition

Material	Steel
Thermal conductivity	$50\mathrm{W/m \cdot K}$
Specific heat capacity	$460\,\mathrm{J/kg}$ °C
Heat exchange coefficient	$125\mathrm{W/m^2 \cdot K}$
Density	$7850\mathrm{kg/m^3}$
Length and width	2 m
Thickness	0.1 m
Initial temperature	20 °C
Heating temperature	1120 °C

We can conclude that the temperature is the highest at the corners and it decreases along the edges (and to a lesser extent at the edges), which remain unchanged before encountering the next corner, according to the simulation result so it can be directly used to explain why the four corners of a rectangular pan are overcooked. Thereafter, we abstract phenomenon to find a mathematical model to prove why heat is concentrated in the four corners of a rectangular pan.

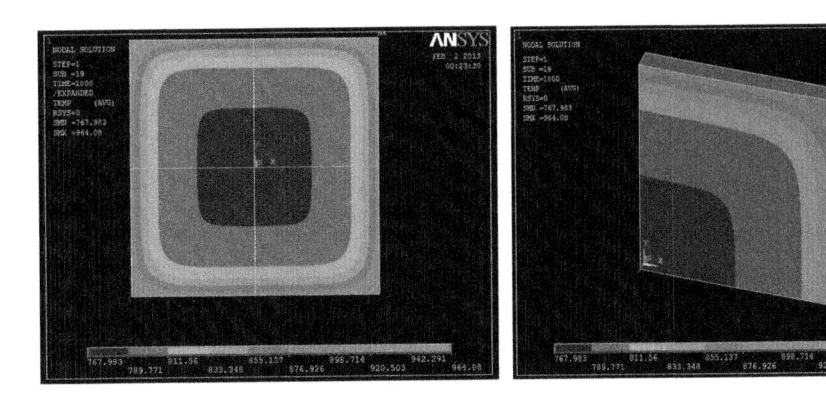

Fig. 1. The ANSYS software emulation result of the square

Our model is to divide the edge of a pan into finite small units. The heat at each part can be represented by the ratio of the surface of the unit to its mass, that is S/M.

We began to consider from the corners of the pan. The size of the contact area to the mass of per unit mass is used to measure the heat extent of each unit.

ABCD is the outer edge of a pan. Now we take ab side as example to analyze and the ab edge is cut into small units with fixed intervals. The ratio of the surface area to the mass of each unit, represented as S/M, is worked out and shown in Table 2.

Table 2. The S, M, S/M of each unit

编号	S	M	S/M
S1	$\frac{a}{2}(d(2r-\mathrm{d})+2\mathrm{rh})$	$\frac{a}{2}d(2r-\mathrm{d})\mathrm{h}\rho$	$\frac{1}{h\rho}+\frac{2r}{d(2r-\mathrm{d})\rho}$
S2	$\frac{a}{2}(d(2r-\mathrm{d})+2\mathrm{rh})$	$\frac{a}{2}d(2r-\mathrm{d})\mathrm{h}\rho$	$\frac{1}{h\rho}+\frac{2r}{d(2r-\mathrm{d})\rho}$
S3	$\frac{a}{2}(d(2r-\mathrm{d})+2\mathrm{rh})$	$\frac{a}{2}d(2r-\mathrm{d})\mathrm{h}\rho$	$\frac{1}{h\rho}+\frac{2r}{d(2r-\mathrm{d})\rho}$
….	….	….	….
Sn	$\frac{a}{2}(d(2r-\mathrm{d})+2\mathrm{rh})$	$\frac{a}{2}d(2r-\mathrm{d})\mathrm{h}\rho$	$\frac{1}{h\rho}+\frac{2r}{d(2r-\mathrm{d})\rho}$

The trend graph of S/M is shown in the Fig. 2. The curve sharply degrades originating from the start point and then remains unchanged, indicating the mathematical calculation is consistent with the simulation result.

The outer edge of a round pan is divided into small unit with the same interval, shown in Fig. 3, and we get the result of S/M by calculating. We can use MATLAB to plot the graph. It shows in Fig. 4.

The S/M of all units are the same, so we conclude that heat is distributed evenly at the edge of the circle. Then, the result of ANSYS simulation is shown in Fig. 5, which is consistent with the calculation by our mathematical model, demonstrating the feasibility of our model.

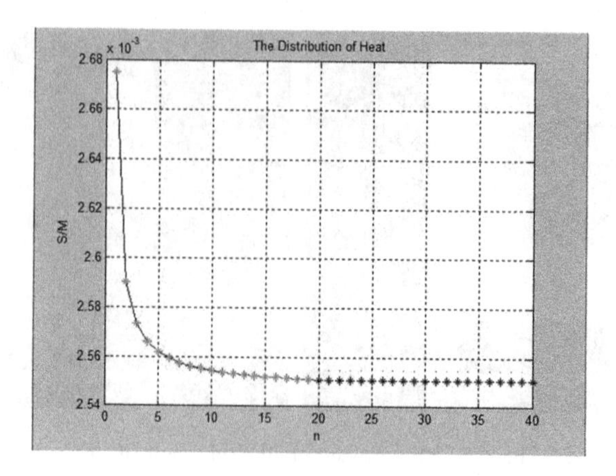

Fig. 2. The trends of S/M from the corner to the center of the edge

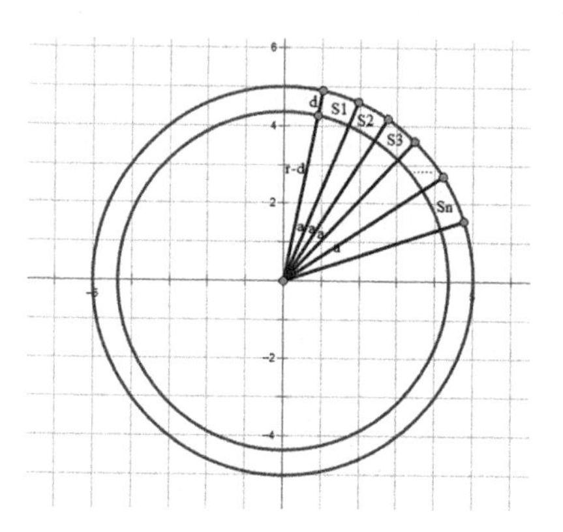

Fig. 3. Cutting the edge uniform of circle

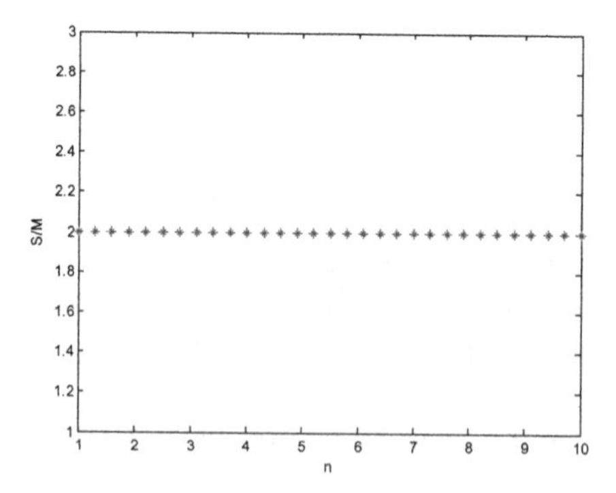

Fig. 4. The trends of S/M for the circle

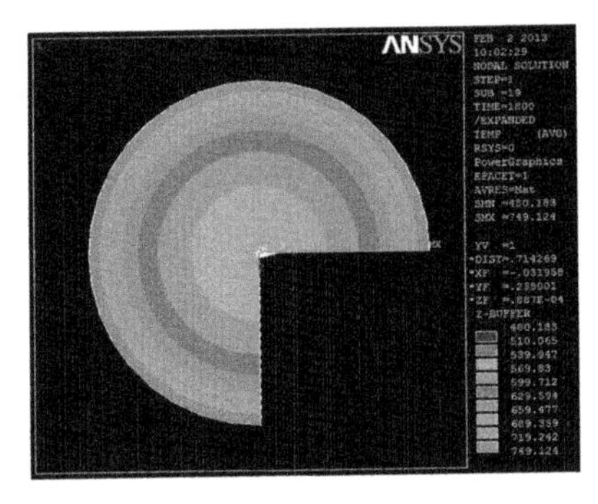

Fig. 5. The ANSYS software emulation result of the square

The proposed model can be extended to illustrate the distribution of heat across the outer edge of a pan with different shapes.

We now consider a general situation, which the edges of a pan with arbitrary shape is into small unit and then we calculate the S/M of each unit. The greater the value of S/M, the higher heat concentrated in this unit, shown in Fig. 6.

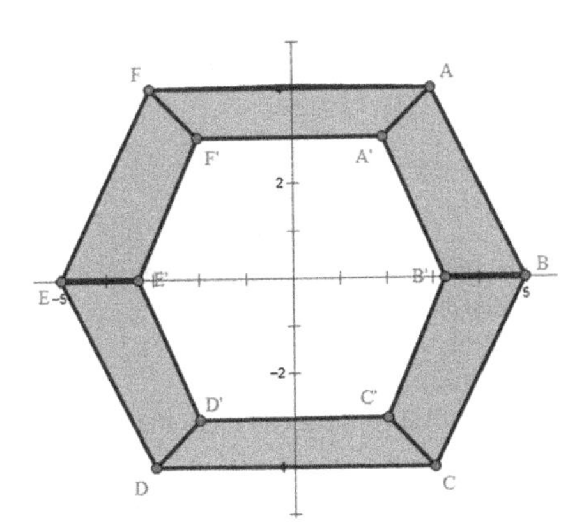

Fig. 6. The edge of the regular hexagon

For example, the distribution of heat across the outer edge of a regular hexagon is calculated and the result is shown in Fig. 7. The six highest points corresponds to the six corners.

With the increase in the number of edges, the heat distribution tends to be more uniform and the polygon edge eventually becomes an arc which the heat distribution is even.

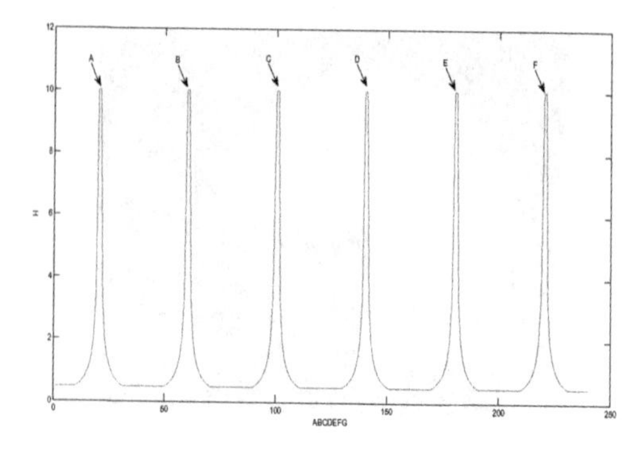

Fig. 7. The trends of S/M of the edge

3 Conclusion

By the edge cutting model, we can get the functional relation of the heat distribution with respect to the surface area and the mass of the pan. The heat distribution shows negative correlations to the surface area and the mass of the pan.

Acknowledgements This work was supported by the Grant (61561050), (61462094), (61661050) from the Natural Science Foundation of China.

References

1. Moaveni, S.: Finite element analysis theory and application with ansys, 3rd edn, pp. 1–5. Prentice Hall, New Jersey (2007)
2. http://en.wikipedia.org/wiki/Heat_transfer
3. http://baike.baidu.com/view/1017221.htm
4. Eckert, E.R.G., Drake, R.M.: Analysis of heat and mass transfer, Qing Hang translate, pp. 31–33. Science Press, Beijing (1983)
5. Kong, X.Q.: Finite element method apply in heat transfer, 3rd edn, pp. xiii–xiv. Science Press, Beijing (1998)
6. http://www.sfu.ca/~mbahrami/ENSC%20388/Notes/First%20Law%20of%20Thermodynamics_Closed%20Systems.pdf
7. Xin, W.T.: ANSYS 13.0 the thermodynamic finite element analysis from entry to the master, 2nd edn, pp. 114–123. Machinery Industry Press, Beijing (2011)

Intelligent Steering Control Based the Mathematical Motion Models of Collision Avoidance for Fishing Vessel

Renqiang Wang[1(✉)], Yuelin Zhao[2], Keyin Miao[1,2], and Jianming Sun[1,2]

[1] College of Navigation Technology, Jiangsu Maritime Institute, Nanjing, China
wangrenqiang2009@126.com
[2] College of Navigation, Dalian Maritime University, Dalian, China

Abstract. To solve the problem of safety avoidance of fishing boat in open water, the mathematical motion models of collision avoidance is established on the basis of geometry of collision avoidance. Taking into account that alteration of course alone is the most commonly used action to avoid collision in sufficient sea-room. Therefore, the intelligent steering control based on Radial Basis Function (RBF) neural networks is proposed and added in the above models. The simulation results verify the effectiveness of the models.

Keywords: Intelligent steering control · Collision avoidance · Mathematical motion models · Neural network

1 Introduction

In recent years, the proportion of collision between merchant ships and fishing vessels in the waters of the East China Sea has been high, therefore, fisheries safety issues remain grim, and the main reason include the defects existed in fishing vessel itself, as well as human error of crew in merchant ships and fishing vessel [1]. To this end, in literature [2], from the merchant point of view, relevant strategies are put forward so as to prevent collisions between merchant ships and fishing vessels; Specific avoidance methods are put forward in literature [3] so that the situation can be improved from the two aspects of merchant ships and fishing vessels; In literature [4], a strategy is proposed to solve the collision problem of merchant ships and fishing vessels in accordance with the “International Regulations for Preventing Collisions at Sea” and good seamanship; The strategy of collision avoidance for fishing vessel is put forward according to the fishing way of fishing vessels in reference [5], on the basis of analyzing the reason of collisions between merchant ships and fishing boats; In reference [6], different methods of collision avoidance action against specific fishing vessels are carried out based on analysis of the specific situation of fishing operations.

However, methods of collision avoidance action for fishing vessels are merely qualitatively analyzed in the above-mentioned references, but the specific avoidance action plans are not proposed. To this end, on the basis of the analysis of the

F. Xhafa et al. (eds.), *Advances in Intelligent Systems and Interactive Applications*, Advances in Intelligent Systems and Computing 686,
https://doi.org/10.1007/978-3-319-69096-4_4

characteristics of fishing vessels, an intelligent steering control based the mathematical motion models of collision avoidance for fishing vessel was investigated in the paper based on the idea of reference [7–9]. And the results can be used for reference of the seafarers.

2 Intelligent Steering Control Based the Mathematical Motion Models of Collision Avoidance for Fishing Vessel

2.1 Merchant Ship Maneuvering Motion Equation

When sailing at sea, merchant vessel always expresses hysterical quality and nonlinear characteristics due to be affected by wind, wave and other conditions. Therefore, the Bech ship maneuvering nonlinear motion equations [10] which is shown as below was adopted in this paper.

$$\begin{cases} T_1T_2\dddot{\varphi} + (T_1+T_2)\ddot{\varphi} + KH(\dot{\varphi}) = K(T_3\dot{\delta}+\delta) + \omega_0 \\ \dot{V} + a_{vv}V^2 + a_{rr}\dot{\varphi}^2 + a_{\delta\delta}V^2\delta^2 = a_{nn}n^2 + a_{nv}nV \end{cases} \tag{1}$$

Where, T_1, T_2, T_3 and K are the ship manipulative capability parameters; $H(\dot{\varphi}) = \alpha_1\dot{\varphi} + \alpha_2\dot{\varphi}^3$, φ is ship course; δ is rudder angle, $\dot{\delta}$ is rudder speed; n is rotational speed of the main engine; V is ship speed; a_{vv}, a_{rr}, $a_{\delta\delta}$, a_{nn} and a_{nv} are coefficients; ω_0 is external disturbance.

2.2 Merchant Ship Intelligent Steering Control

The formula (2) as following is transformed by formula (1) so as to facilitate the design of intelligent steering control algorithm [11–13], which can be investigated by back-stepping method [14, 15] combined with sliding-mode control theory.

$$\begin{cases} \dot{x}_1 = x_2 \\ \dot{x}_2 = x_3 \\ \dot{x}_3 = f(x) + bu + \omega \\ y = x_1 \end{cases} \tag{2}$$

Where, $x_1 = \varphi$, $x_2 = \dot{x}_1 = r$, $x_3 = \dot{x}_2 = \dot{r}$, $\omega = -\mathrm{b}(\alpha_1x_2 + \alpha_2x_2^3) - \mathrm{a}x_3 + \mathrm{b}u + \omega_0$, $u = T_3\dot{\delta} + \delta$. ω is the total uncertainty, $a = (T_1+T_2)/T_1T_2$, $b = K/T_1T_2$, α_1 and α_2 are parameter, which can be estimated by taking advantage of adaptive method [13, 14].

Step1, sliding mode surface is constructed.

$$\begin{cases} z_1 = x_1 - \varphi_r \\ z_2 = x_2 - \sigma_1(z_1) \\ z_3 = x_3 - \sigma_2(z_1, z_2) \end{cases} \tag{3}$$

Where, φ_r is expected course, σ_1 and σ_2 are virtual stabilization function, z_1 is the deviation of course.

Step 2, the asymptotic stability of control system is proved by Lyapunov theory with backstepping method.

The first Lyapunov function is constructed as following.

$$V_1(z_1) = \frac{1}{2}z_1^2 + \frac{1}{2}z_2^2 + \frac{1}{2}z_3^2 \quad (4)$$

According to the sliding mode surface formula, the following settings are made:

$$\begin{cases} \sigma_1(z_1) = -k_1 z_1 + \dot{\varphi}_r \\ \sigma_2(z_1, z_2) = -k_2 z_2 - z_1 + \dot{\sigma}_1(z_1) \end{cases} \quad (5)$$

It can be inferred the formula (6) by formula (3).

$$\dot{z}_3 = \dot{x}_3 - \dot{\sigma}_2(z_1, z_2) = f(x) + cu + d - \dot{\sigma}_2(z_1, z_2) \quad (6)$$

It can be inferred the formula (7) by combing formula (4) and (6).

$$\dot{V}_3(z_1, z_2, z_3) = -k_1 z_1^2 - k_2 z_2^2 + z_3 \dot{z}_3 \quad (7)$$

Step 3, utilizing RBF neural network to compensate the disturbances so as to make estimation error $\tilde{\omega} \to 0$. It is consideration that RBF network has a universal approximation properties [13], therefore, the RBF network is used to approach the interference ω, and to achieve effective compensation. The RBF network algorithm is:

$$\begin{cases} h_j = \exp\left(\frac{\|\mathbf{x}-\mathbf{c}_j\|^2}{2b_j^2}\right) \\ d = \mathbf{W}^{*\mathrm{T}}\mathbf{h}(\mathbf{x}) + \delta \end{cases} \quad (8)$$

Where, $\mathbf{x}$ is the network input, i represents one of the first input network input layer, j is a input of hidden layer of the network, $\mathbf{h} = [h_j]^{\mathrm{T}}$ is high Gaussian function, $\mathbf{W}^*$ is the ideal weighting value, δ is ideal approximation error of neural network, and there is $\delta \leq \delta_{\max}$, $\hat{\omega}$ is the network output, and there is $\hat{\omega} = \hat{\mathbf{W}}^{\mathrm{T}}\mathbf{h}(x)$, $\hat{\mathbf{W}}$ is the estimation weights of the neural network.

In this paper, the network layer to the system is the input switching surfacez_3, and $\tilde{\mathbf{W}} = \hat{\mathbf{W}} - \mathbf{W}^*$ is taken into account, there is,

$$\omega - \hat{\omega} = \mathbf{W}^{*\mathrm{T}}\mathbf{h} + \delta - \hat{\mathbf{W}}^{\mathrm{T}}\mathbf{h} = -\tilde{\mathbf{W}}^{\mathrm{T}}\mathbf{h} + \delta \quad (9)$$

Step 4, construct global Lyapunov function,

$$V = V_3(z_1, z_2, z_3) + \frac{1}{2\gamma}\tilde{w}^T\tilde{w} \tag{10}$$

And, it can be inferred the formula (11) by derivation of formula (10),

$$\begin{aligned}\dot{V} &= \dot{V}_3(z_1, z_2, z_3) + \frac{1}{\gamma}\tilde{w}^T\dot{\hat{w}} \\ &= \frac{1}{\gamma}\tilde{w}^T\dot{\hat{w}} + -k_1 z_1^2 - k_2 z_2^2 + z_3[z_2 + f(x) + cu + \omega - \dot{\sigma}_2(z_1, z_2)]\end{aligned} \tag{11}$$

The following compensation control law is considered,

$$u = \frac{1}{c}[-f(x) - z_2 - \hat{\omega} + \dot{\sigma}_2(z_1, z_2) - \varepsilon\,\mathrm{sgn}(z_3) - \varepsilon\tanh(z_3) - k_3 z_3] \tag{12}$$

Therefore, it can be inferred the formula (13) by combing formula (11) and (12),

$$\dot{V} \le \delta_{\max} z_3 - \varepsilon|z_3| + \tilde{w}^T(\frac{1}{\gamma}\dot{\hat{w}} - z_3\mathbf{h}(z_3)) \tag{13}$$

Making the following settings,

$$\begin{cases} \varepsilon \ge \delta_{\max} \\ \dot{\hat{w}} = \gamma z_3 \mathbf{h}(z_3) \end{cases} \tag{14}$$

And, it can be inferred the formula (15) by combing formula (13) and (14),

$$\dot{V} \le \delta z_3 - \varepsilon|z_3| \le 0 \tag{15}$$

Therefore, the entire control system is globally asymptotically stable.

2.3 Acquisition of Collision Avoidance Elements

The acquisition of collision avoidance elements are obtained based on geometric plotting triangle through establishing coordinate system XOY, which is shown as Fig. 1.

In the coordinate system XOY, X axis stands for north direction, Y axis stands for east direction, and the merchant ship set as the origin of coordinate, where, V_o and φ_o respectively represents initial speed and course of merchant ship, V_L and φ_L respectively represents initial speed and course of fishing boat, α_0 and R_0 respectively represents initial relative bearing and relative distance.

According to the coordinate system mentioned above, it can be inferred the initial position which is noticed as formula (16) of fishing boat.

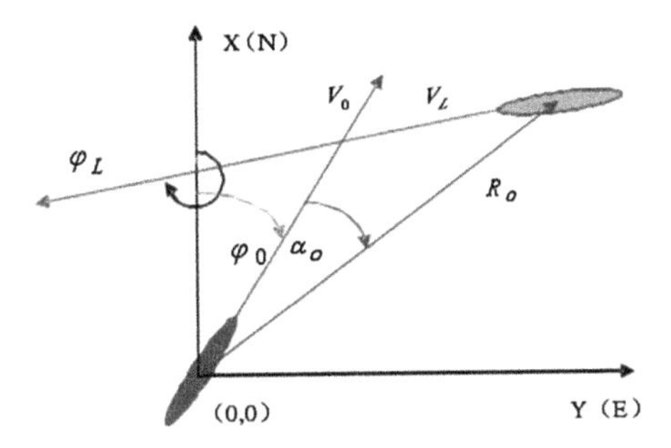

Fig. 1. Coordinate system of true motion of two ships

$$\begin{cases} X_{LO} = R_0 \cdot \cos(\varphi_0 + \alpha_0) \\ Y_{LO} = R_0 \cdot \sin(\varphi_0 + \alpha_0) \end{cases} \tag{16}$$

Where, an assumption is that two ships are in sight of one another. Therefore, the merchant ship is give-way vessel on the basis of the International Regulations for Preventing Collisions at Sea. In view of that, merchant ship usually alters her course by steering to avoid collision for fishing boat in sufficient sea-room. An unknown steering angle C_m is proposed, which will be executed, and new course φ of merchant ship is satisfied the demand of avoidance. And the φ is described as follows.

$$\varphi = \varphi \pm C_m \tag{17}$$

In the expression, the symbol "+" is expressed alter course to starboard side, the symbol "–" is expressed alter course to port side.

And the speed V of merchant ship is less than initial speed V_0 during the turning process. Therefore, the instantaneous displacement $X(t)$ & $Y(t)$ of merchant vessel and the instantaneous relative displacement X_R & Y_R of fishing ship are described as following.

$$\begin{cases} X(t) = \int_0^t V \cdot \cos\phi dt \\ Y(t) = \int_0^t V \cdot \sin\phi dt \end{cases} \tag{18}$$

$$\begin{cases} X_R = X_{LO} + \int_0^t (V_L \cdot \cos\phi_L - V \cdot \cos\phi)dt \\ Y_R = Y_{LO} + \int_0^t (V_L \cdot \sin\phi_L - V \cdot \sin\phi)dt \end{cases} \tag{19}$$

In view of that, the instantaneous relative bearing, distance, relative course and speed of fishing ship are shown as follows.

$$\begin{cases} \alpha(t) = arctg\dfrac{Y_R}{X_R} \\ R(t) = \sqrt{X_R^2 + Y_R^2} \\ \phi_R = arctg\dfrac{V_L \cdot \cos\phi_L - V \cdot \cos\phi}{V_L \cdot \sin\phi_L - V \cdot \sin\phi} \\ V_R(t) = \sqrt{V_L^2 + V^2 - 2V_L \cdot V \cdot \cos(\phi - \phi_L)} \end{cases} \tag{20}$$

It can be inferred the DCPA and TCPA shown as formulas (17) of fishing ship by combining formulas (9) to (16) on the basis of geometry of collision avoidance.

$$\begin{cases} DCPA = R(t) \cdot |\sin(\phi_R - \alpha(t) - 180)| \\ TCPA = R(t) \cdot \cos(\phi_R - \alpha(t) - 180)/V_R(t) \end{cases} \tag{21}$$

3 Application

The simulation of collision avoidance for fishing ship is carried out by applying the above mathematical motion models of collision avoidance. The safe passing distance of two ships is 0.7 nautical miles in paper [3–6].

3.1 Simulation Object and Conditions Setting

Effectiveness of the mathematical motion models of collision avoidance for fishing vessel is verified through simulation of training vessel "YULONG". It is known that when ship's speed is 13 knots the related parameters of ship mathematical model [10] are: $a_{vv} = 1.3586 \times 10^{-4}$, $a_{\delta\delta} = 1.6658 \times 10^{-3}$, $a_{nv} = 5.9167 \times 10^{-4}$, $a_{nn} = 1.4042 \times 10^{-2}$, $a_{rr} = 101.52$, $T_1 = 0.0425$, $T_2 = 23.901$, $T_3 = 10.064$, $K = 7.927$,$\alpha_1 = 1$, $\alpha_2 = 30$.

Assumption is that the speed and course of merchant vessel "YULONG" is 13 knots and 000 degrees, the speed and course of fishing boat is 5 knots and 150 degrees, relative azimuth is 350 degrees, and relative distance is two nautical miles.

3.2 Simulation Results and Analysis

The relationship of DCPA and steering size is shown in Table 1. And the relative distance and true motion trajectory of the two ships are shown in Figs. 2 and 3.

There are two conclusions are drawn from Table 1. The first is that it is too early for the action taken to avoid collision to expand XTD (cross track distance). The second is that an optimal collision avoidance action is invested, which is that the distance of two ships is 1.2 nautical miles and the steering size is 60 degrees.

Table 1. Relationship of DCPA and steering size

DCPA/XTD (*n mile*)		Steering size (°)				
		40	50	60	70	80
Distance (*n mile*)	1.2		0.60/0.48	0.70/0.51	0.73/0.56	0.77/0.60
	1.3		0.64/0.51	0.72/0.56	0.78/0.58	
	1.4	0.60/0.50	0.71/0.54	0.80/0.60		
	1.5	0.66/0.55	0.78/0.61	0.88/0.68		

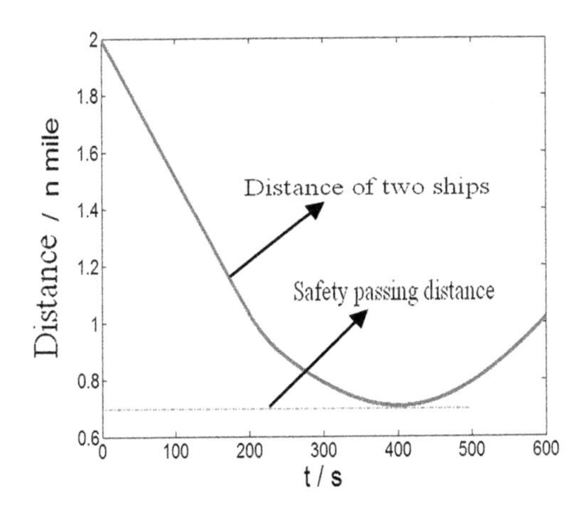

Fig. 2. The output of relative distance

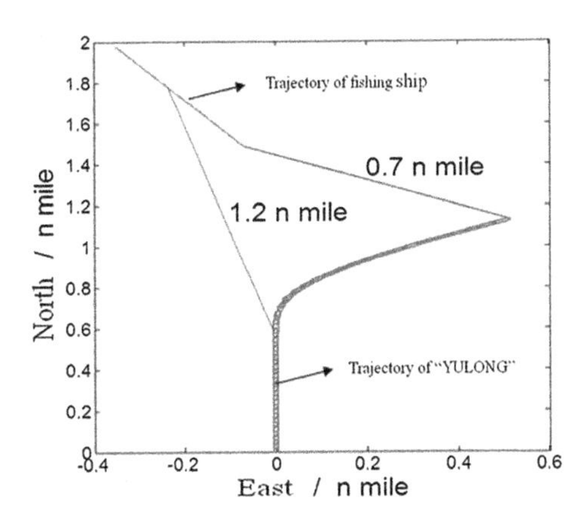

Fig. 3. True motion trajectory of vessels

4 Conclusion

In accordance with the International Regulations for Preventing Collisions at Sea, the merchant ships should give way for fishing boats in a positive way with in ample time. Therefore, intelligent steering control based the mathematical motion models of collision avoidance for fishing vessel is established on the basis of geometry of collision avoidance. The simulation verifies effectiveness of the models. While, it is noteworthy that the model is valid and available only if the fishing boat maintains its course and speed, and that the navigable waters are open waters. At the same time, the model does not consider the impact of external environment.

Acknowledgements This work is supported by Foundation of Jiangsu Nautical Institute under Grant No. 2015B09 and Jiangsu Maritime Institute under Grant No. XR1501 & No.2015KJZD-01.

References

1. http://www.chinanews.com/gn.shtml
2. Tao, Zhou: Methods of navigation and collision avoidance for coastal fishing zone in China. Word Shipping **25**(5), 24–26 (2010)
3. Wang, Y.: A brief analysis of collision avoidance for fishing vessels by merchant vessels in coastal water. China Water Transport **9**(9), 18–19 (2009)
4. Sun, Y., Chen, L., Chang, H.: Study on the safety measures of ships' avoidance to fishing, Proceedings of Maritime Technology Practice, 12, 83–87 (2009)
5. Li, J.: Security measures of avoidance for fishing vessels by merchant at sea, Proceedings of Ships Avoidance and Safe Operation, 4, 59–63 (2007)
6. Liu, K., Li, L., Liu, B.: Study on methods of evading fishing during ship sailing, Proceedings of Practical Research on Ship Navigation, 10, 107–111 (2010)
7. Wang, R., Zhao, Y.: A study on dynamic mathematical model to avoid collision by ship steering. J. Dalian Maritime University **40**(2), 17–20 (2014)
8. Yan, Y.: Study on mathematical model for ship collision avoidance dynamic system. J. Dalian Maritime University **21**(2), 30–36 (1995)
9. Fang, C., Deng, H.: The ship's mathematic motion models of altering course to avoid collision based on the optimization of genetic algorithm, Proceedings of the 33rd Chinese Control Conference, pp. 6364–6367 (2014)
10. Jia, X., Yang Y.: Ship motion mathematical model, 978-7-5632-1137-1/4/$6.00 © Dalian Maritime University Press, Dalian (1999)
11. Wang R., Gong, J., Zhao, Y.: Neural network sliding mode control under new reaching law and application, Proceedings of Advanced Information Technology, Electronic and Automation Control Conference, 911–917 (2015)
12. Xin, W., Zhengjiang, L., Tieshan, L., Weilin, L.: Neural network-based adaptive control for a ship course discrete-time nonlinear system. J. Harbin Engineering Univ. **37**(1), 42–48 (2016)
13. Lin, F.J., Shen, P.H., Hsu, S.P.: Adaptive backstepping sliding mode control for linear induction motor drive. IEEE Proc. Electr. Power Appl. **149**(3), 184–194 (2002)

14. Wang, Y., Guo, C., Sun, F., Guo, D.: Dynamic neural-fuzzified adaptive control of ship course with parametric modelling uncertainties. Int. J. Model. Ident. Control **13**(8), 251–258 (2015)
15. Hu, G.S., Zhou, Y.B., Xiao, H.R.: Application of fuzzy neural network in ship course control. Appl. Mech. Mater. **13**(10), 309–315 (2012)

Dynamical Analysis of Fractional-Order Hyper-chaotic System

Junqing Feng(✉) and Guohong Liang

School of Science, Air Force Engineering University, 710051 Xi'an, China
fjq1983@163.com

Abstract. The purpose of this paper is to study the dynamical behavior of fractional order hyper-chaotic complex systems based on the bifurcation theorem. The variation of the system parameters and fractional order can induce the bifurcation by the simulation results.

Keywords: Fractional-order system · System order · System parameters

1 Introduction

The fractional calculus has been widely concerned in mathematics. However, At the beginning of its development, it has been paid much attention in the field of pure mathematics, Famous mathematician Leibniz and Euler give their initial understanding of fractional calculus [1]; In 1819, based on the Gamma function, Lacroix gave the first definition of fractional calculus, And a simple fractional calculus [2]. This has inspired mathematicians to study fractional calculus, However, due to the lack of a reasonable physical interpretation of fractional calculus, The lack of application background, in the field of application of slow development.

Since the first discovery of chaotic attractor in Lorenz numerical experiments, People put forward many chaotic systems, such as Chen system, *Lü* system, Duffing system, VanderPol system [3–6]. Early research focused on low dimensional chaotic systems. Because the hyperchaotic system has at least two positive Lyapunov exponents, And it contains more abundant and more complex dynamic behaviors than the low dimensional chaotic system, so it is more suitable for the design of digital cryptography and secure communication. It is a fractional calculus operator that can describe the dynamical behavior of chaotic system more accurately. Initial sensitivity and pseudo randomness are common properties of fractional order chaotic systems. In addition, it also has some properties of fractional order systems, such as the ability to reflect the historical information of the system, strong historical memory and so on. Therefore, the study of fractional order chaotic systems has extensive theoretical significance and practical value.

F. Xhafa et al. (eds.), *Advances in Intelligent Systems and Interactive Applications*, Advances in Intelligent Systems and Computing 686,
https://doi.org/10.1007/978-3-319-69096-4_5

2 Related Work

2.1 Fractional-Order Hyper-chaotic Complex System

In the literature [7], the author puts forward the integer orderhyper-chaotic system and makes a detailed analysis. Based on this, this paper presents a fractional order hyper-chaotic system with complex variables:

$$\begin{cases} D_*^{q_1} y_1 = a(y_1 - y_2) + y_4 \\ D_*^{q_2} y_2 = by_2 - y_1 y_3 + y_3 \\ D_*^{q_3} y_3 = \frac{1}{2}(\bar{y}_1 y_2 + y_1 \bar{y}_2) - cy_3 \\ D_*^{q_4} y_4 = \frac{1}{2}(\bar{y}_1 y_2 + y_1 \bar{y}_2) - dy_4 \end{cases} \tag{1}$$

The $y = (y_1, y_2, y_3, y_4)^{\mathrm{T}}$ is state variable, $y_1 = x_1 + ix_2, y_2 = x_3 + ix_4$ is complex state variable, $y_3 = x_5, y_4 = x_6$ is real state variable, $i = \sqrt{-1}$, state variables can be divided into imaginary and real parts, according to the linear property of the Caputo differential operator, the system (1) can be written as follows:

$$\begin{cases} D_*^{q_1} x_1 = a(x_2 - x_1) + x_6 \\ D_*^{q_1} x_2 = a(x_3 - x_2) + x_6 \\ D_*^{q_2} x_3 = bx_4 - x_2 x_5 + x_6 \\ D_*^{q_2} x_4 = bx_4 - x_1 x_5 + x_6 \\ D_*^{q_3} x_5 = x_1 x_3 + x_2 x_4 - cx_5 \\ D_*^{q_4} x_6 = x_1 x_3 + x_2 x_4 - dx_6 \end{cases} \tag{2}$$

2.2 Dynamic Behavior Analysis of the System

2.2.1 System Equilibrium Point

The equilibrium point of the system can be obtained by the following formula:

$$\begin{cases} D_*^{q_1} x_1 = a(x_3 - x_1) + x_6 = 0 \\ D_*^{q_1} x_2 = a(x_1 - x_2) = 0 \\ D_*^{q_2} x_3 = bx_3 - x_1 x_4 + x_6 = 0 \\ D_*^{q_2} x_4 = bx_4 - x_2 x_3 = 0 \\ D_*^{q_3} x_5 = x_1 x_3 + x_2 x_5 - cx_5 = 0 \\ D_*^{q_4} x_6 = x_1 x_2 + x_2 x_3 - dx_6 = 0 \end{cases} \tag{3}$$

According to the second equations of equation set (3), we have $x_1 = x_2$, by fourth equations, we have $x_2(b - x_5) = 0$, then $x_2 = 0$ or $x_5 = b$. if $x_5 = b$, then $a = b$, $x_6 = \frac{cx_5}{d} = \frac{cb}{d}$. Thus the equilibrium points of the system (3) are distributed in the circle which center for $(\frac{c}{2d}, 0)$ and radius of $r = \left(\sqrt{4bcd^2 + c^2}\right)/2d$, the equation of the circle can be written as:

$$(x_1 - \frac{c}{2d})^2 + x_2^2 = (\frac{\sqrt{4bcd^2 + c^2}}{2d})^2 \tag{4}$$

Let $x_1 - \frac{c}{2d} = r\cos\theta$, $x_2 = x_4 = r\sin\theta$, The balance point is as follows:

$$E_\theta = (r\cos\theta + \frac{c}{2b}, r\sin\theta, r\cos\theta - \frac{c}{2b}, r\sin\theta, b, \frac{bc}{d})$$

for $\theta \in [0, 2\pi]$.
If $x_2 = 0$, the equilibrium points of the system (3) are (0, 0, 0, 0, 0, 0), and there are two isolated unstable points:

$$(s_1, 0, \frac{das_1}{da + s_1}, 0, \frac{das_1^2}{c(da + s_1)}, \frac{as_1^2}{(da + s_1)}),$$
$$(s_2, 0, \frac{das_2}{da + s_2}, 0, \frac{das_2^2}{c(da + s_2)}, \frac{as_2^2}{(da + s_2)}),$$

For

$$s_1 = \frac{c}{2d} + \frac{1}{2}\sqrt{(\frac{c}{d})^2 + 4bc}, s_2 = \frac{c}{2d} - \frac{1}{2}\sqrt{(\frac{c}{d})^2 + 4bc}$$

At the same time, the Jacobi matrix E_0 of the system (3) is

$$J_{E_0} = \begin{bmatrix} -a & 0 & a & 1 & 0 & 1 \\ 0 & -a & 0 & a & 0 & 0 \\ 0 & 0 & b & 0 & a & 1 \\ 0 & 0 & 0 & b & 0 & a \\ 0 & 0 & 0 & 0 & -c & 0 \\ 0 & 0 & 0 & 0 & 0 & -d \end{bmatrix} \tag{5}$$

The characteristic equation is $(\lambda + d)(\lambda + c)(\lambda - b)^2(\lambda + a)^2 = 0$, and it's characteristic value $\lambda_1 = -c, \lambda_2 = -d, \lambda_3 = \lambda_4 = b, \lambda_5 = \lambda_6 = -a$, the equilibrium point (0, 0, 0, 0, 0, 0) is stable, if b have negative eigenvalue, c, a, d is positive, otherwise the system (3) is unstable.

2.2.2 The Influence of System Order Variation on the System

We let $q_1 = q_2 = q_3 = q_4 = \alpha = 0.15$, and selection system parameter $(a, b, c, d) = (23, 20, 4, 1)$, The fractional order system is an equal order system. When$q_1 = q_2 = q_3 = q_4 = \alpha = 0.80$, the fractional order hyperchaotic complex system has abalancepoint. if has $\alpha = 0.87$, the attractor of the system is shown in Fig. 1a. If we continued increase $\alpha = 0.90$, the attractor of the system is shown in Fig. 1b, It can be seen that the attractor with order 0.87 is completely different.

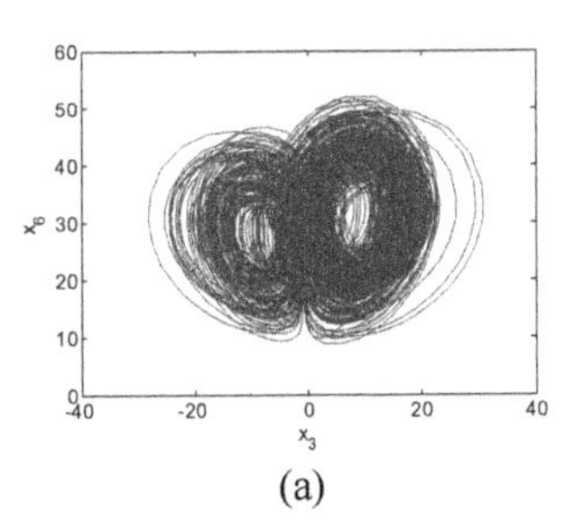

(a)

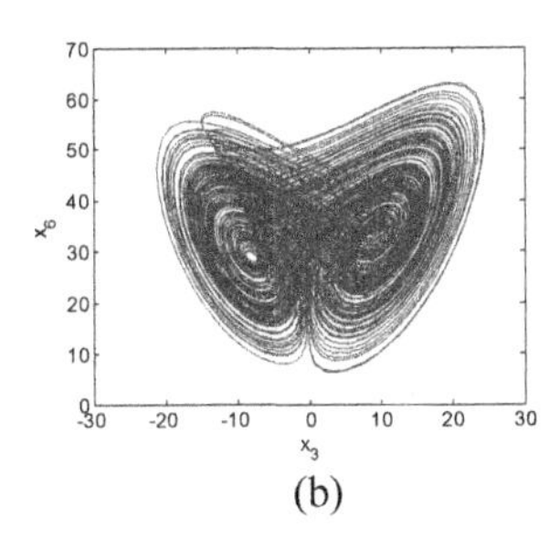

(b)

Fig. 1.

Because the fractional order system is very complex, we only consider some special classes,

Selected $q_2 = q_3 = q_4 = 0.80$, when the fractional order $q_1 \in [0.81, 0.99]$, the fractional order q_1 of the system is shown in Fig. 2. When the order $q_1 = 0.86$, the fractional order hyperchaotic complex passes through the fork type bifurcation into the chaotic state. When selecting fractional order $q_1 = q_3 = q_4 = 0.92$, $q_2 \in [0.80, 0.95]$, the dynamic behavior of the system is shown in Fig. 3, As we can see in Fig. 3, the fractional order hyperchaotic complex system is chaotic in the range $q_2 \in [0.897, 1]$, when $q_2 \in (0.80, 0.997)$, there is a period doubling window, when $q_2 < 0.82$, the system is a fixed point.

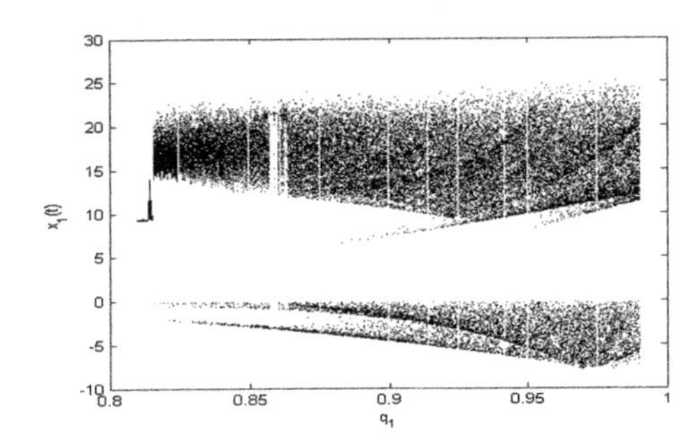

Fig. 2. Dynamical behavior of complex system when order $q_1 \in (0.80, 0.90)$

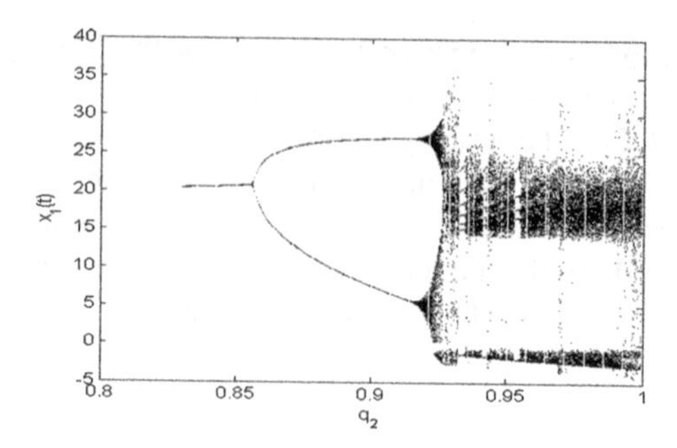

Fig. 3. Dynamical behavior of complex system when order $q_2 \in (0.80, 0.99)$

We do a lot of numerical simulation, for fractional order $q_3 = q_4$, $q_3 \in [0.80, 1]$, $q_4 \in [0.80, 1]$, fractional order hyperchaotic complex system has no obvious bifurcation behavior in this range.

2.2.3 The Influence of System Parameter Change on the System

First, we select the order of the system $q_1 = q_2 = q_3 = q_4 = 0.91$, the dynamic behavior of the system is further analyzed by means of bifurcation diagrams and phase diagrams.

Second, let the system parameters $(b, c, d) = (20, 5, 4)$, let parameters increase from 31.5 to 33, the bifurcation diagram of the system with parameters is shown in Fig. 4, it can be seen that the system starts from 31.5, and the system enters chaos through a series of period doubling bifurcations. When the system parameters starts from 33.5, then the tangent bifurcation occurs and system enters the chaotic state. after that, a series of period doubling bifurcations occurred,

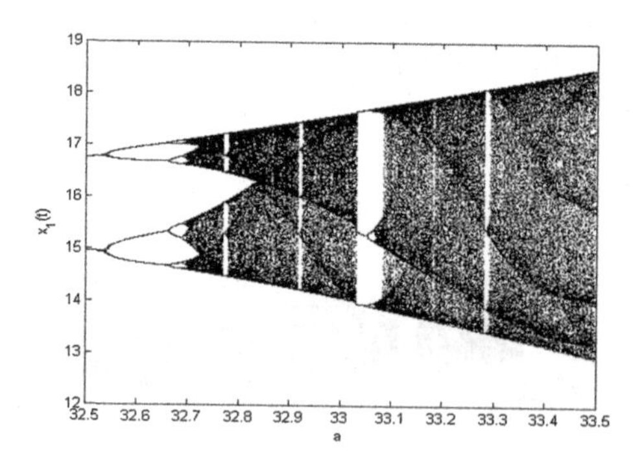

Fig. 4. Dynamical behavior of complex system when parameter $a \in (32.5, 33.5)$

Third, when the system parameters are selected by $(a, c, d) = (40, 5, 4)$, the system increases from 16 to 20, and a series of periodic doubling bifurcations occur. at the same time, the phase diagram of the system in three-dimensional phase space is obtained, the system can be seen in the case when the parameter is increased to 19.25,

When the parameter b > 19.25, the system goes through a series of period doubling bifurcation into chaos, the phase diagram can be seen, the system has a cycle, cycle three and chaotic attractor.

Finally, the parameters of the system are changed with the change of parameters, the bifurcation diagram of the system with parameter d can be described, when the parameter $d = 0.3$ takes a tangent bifurcation, then the system enters a period doubling window, when the parameter $d = 1.12$ appears two order period doubling bifurcation, and then goes into chaos.

3 Conclusion

The existence and stability of the equilibrium points of the system are studied in this paper, and then the dynamic behavior of the system is discussed in terms of the order of the fractional order system and the bifurcation of the system parameters.

References

1. Petráš, I.: Fractional-Order Nonlinear Systems: Modeling, Analysis and Simulation. Higher Education Press, Beijing (2011)
2. Podlubny, I.: Fractional Differential Equations. Academic Press, San Diego (1999)
3. Caputo, M.: Linear models of dissipation whose Q is almost frequency independent-II. Geophys. J. R. Astron. Soc.**13**(5), 529–539 (1967)
4. Hartley, T.T., Lorenzo, C.F., Qammer, H.K.: Chaos in a fractional order Chua's system. IEEE Trans. Circ. Syst. I **42**(8), 485–490 (1995)
5. Wu, X.J., Shen, S.L.: Chaos in the fractional-order Lorenz system. Int. J Comput. Math. **86** (7), 1274–1282 (2009)
6. Li, C.G., Chen, G.: Chaos in the fractional order Chen system and its control. Chaos Solitons Fractals **22**(4), 549–554 (2004)
7. Mahmoud, G.M., Mahmoud, E.E., Ahmed, M.E.: On the hyper-chaotic complex Lü system. Nonlinear Dyn. **58**(4), 725–738 (2009)

Evaluating the Performance of the Logistics Parks: A State-of-the-Art Review

Yingxiu Qi[1], Yan Sun[2], and Maoxiang Lang[1](✉)

[1] School of Traffic and Transportation, Beijing Jiaotong University, Beijing 100044, China
mxlang@bjtu.edu.cn

[2] School of Management Science and Engineering, Shandong University of Finance and Economics, Jinan 250014, Shandong Province, China

Abstract. The remarkable development of the logistics industry motivates the optimization researches on the logistics park planning. After the construction period of the logistics parks when the relative location and layout problems are highlighted by the scholars, the performance evaluation on the logistics parks during their operation period is of great significance. On one hand, the well-performed logistics parks can be determined and stand out from their peers through performance evaluation, and can hence provide a benchmark for others to learn and further make progress. On the other hand, logistics parks can analyze their SWOT based on such evaluation and realize their sustainable development. Consequently, the performance evaluation on the logistics parks has been attached great importance in recent years. In this study, we present a systematical review on the logistics park performance evaluation problem from two aspects, including evaluation index system and quantitative evaluation methods (e.g., AHP, TOPSIS and DEA, etc.). We wish this review can help the readers clearly understand the research progress of this problem and also draw more colleagues to this research field.

Keywords: Performance evaluation · Logistics park · Evaluation index · Evaluation model · Literature review

1 Introduction

Nowadays, under the background of globalization, international trade is developing significantly, which also remarkably motivates the commodity circulation [1]. The modern logistics industry is an essential component in the circulation system, and has also become an important support for the economic development. On one hand, companies spare no effort to fully utilize and integrate social logistics resources to lower their logistics costs. On the other hand, local governments also promote the development of modern logistics industry as an important strategy to support sustainable economic development, improve the investment environment, and enhance social and economic benefits [2]. Consequently, as the most important nodes in the logistics network, the logistics parks in China is enjoying a prosperous construction period.

F. Xhafa et al. (eds.), *Advances in Intelligent Systems and Interactive Applications*, Advances in Intelligent Systems and Computing 686,
https://doi.org/10.1007/978-3-319-69096-4_6

The logistics park is a comprehensive facility that consists of logistics management systems, advanced information systems and cooperative freight transportation systems, etc. [3]. Logistics parks play an important role in the freight transport system, including organizing freight transportation, synergizing the participating organizations [4] and promoting regional economic development [2]. With the fierce market competition and the various customer demands, logistics enterprises began to cluster in the logistics parks together with some manufacturing enterprises, so that they can effectively share resources (including logistics infrastructures, information and customer orders, etc.), learn from each other, and finally improve their service quality, e.g., reducing the customer order cycle and increasing the additional value to the final products [5–8].

However, negative effects, e.g., congestion, pollution, noise and vibration brought by freight transportation considerably affect the livability [9] and the sustainability of the cities where logistics parks locate, which have been worried about by the public, government and the industry [10]. Worse still, the increasing logistics demand aggravates the negative effects. To solve this paradox, first of all, we should clearly understand the performance situations of the logistics parks from various aspects, which motivates the raise of the performance evaluation problem on logistics parks that aim at making objective judgment on the performance of the logistics parks by using reasonable index system and quantitative evaluation models. After such evaluation, we can further analyze the SWOT of the logistics parks and find the directions to realize their sustainable operation and development.

Performance evaluation on logistics parks now become one of the hottest areas of logistic research fields. Faced with this highlighted problem, this study gives it a systematical review by analyzing and summarizing its current research progress. The aim of this study is to help readers understand this problem as well as draw more attention to it. The remaining sections are organized as follows. Section 2 makes a review of the evaluation indexes and the principle for selecting indicators. The review of evaluation models are presented in Sect. 3. Finally, Sect. 4 presents the conclusions.

2 Review on the Evaluation Index System

2.1 Establishment of the Evaluation Index System

Majority of the current studies classify the evaluation indexes into three main categories, including the (1) *state of the logistics park*, (2) *operation state of the settled enterprises* and (3) *social and environmental contribution of the logistics park*. As for the third category, the transportation/logistics system is a human–machine–environment system. People and environment are the key factors in such system [11]. The logistics activities will not only influence the society (including the social economy and people related factors, such as employment and income) but also impact the environment (including pollution and energy consumption). Therefore, besides the contribution to the social economy, contribution to the people and the environment should also be formulated as the indexes to evaluate the logistics park performance. A comprehensive evaluation index system that combines the accomplishments of various relative studies is shown as Table 1.

Table 1. Evaluation index system of the logistics park performance

Criteria	Sub-criteria	Definition of sub-criteria	References
State of the logistics park	Impact zone	• The utilization area of the park; • Market scale	[4, 5, 12]
	Facilities	• Storage facilities; • Transportation facilities; • Handling facilities; • Facility capacities	[5, 8, 12]
	Transportation accessibility	• Distances between the park and highway entrances/exits, ports, airports, railway stations; • Number of such nodes	[5, 12–16]
	Finance	• Annual revenues; • Economic output–input ratio; • Government investment; • Operational activity cost; • Net cash flow; • Total capital input	[4, 5, 8, 12, 14, 15, 17, 18]
	Information level	• Information facilities; • Information service type; • Applied informational technology	[4, 5, 8, 13–15, 19, 20]
	Service level	• Response time; • Billing of orders without errors; • Number of customer complaints; • Rate of service satisfaction	
	External environment	• Land cost; • Labor cost; • Regional primary, secondary, and tertiary industry values; • Average GDP	[13, 16]
Operation state of the settled enterprises	Enterprise scale	• Number of settled enterprises; • Development level	[5]
	Enterprise activities	• Re-financing activities; • Land ownership; • Logistics properties	[5]
Social and environmental contribution of the logistics park	Economic contribution	• Annual revenues; • GDP growth of hinterland	[5, 15]
	Society contribution	• Current number of employees; • Income of the employees; • Educational level of employees; • Urban traffic pressure relief	[5, 17, 18]
	Environmental contribution	• New clear resource utilization; • Logistics park greenbelt coverage rate; • Carbon dioxide or noxious gas emissions; • Waste release; • Total energy consumption; • Ratio of eco-environment protection investment to GDP	[4, 5, 14, 15, 17, 18]

2.2 Discussion on the Evaluation Index System

The evaluation index system is built based on the experience of the experts and the scholars. Following aspects should be taken into account when establishing the evaluation index system.

(1) The evaluation index system should consist both qualitative indexes and quantitative indexes.
(2) The indexes should defined as definitely as possible in order to gain reliable data to further conduce evaluation process.
(3) The indexes should present the developing directions to the logistics parks, for example, the utilization of clear energies.
(4) The index should keep evolving, due to the continuous development of the society that motivates the emerging demands for e-commerce, clean energy vehicles, drones and driverless vehicles, etc. [21].

During the data processing, following three issues should be considered.

(1) Part of the data are attained from questionnaires. How to design comprehensive but objective and concise questionnaires to avoid misunderstanding and to gain reliable data is quite important.
(2) The assignment of weights to the indexes should reduce subjectivity in order to gain objective and fair evaluation results.
(3) The correlation analysis on the various evaluation indexes is required to avoid the information duplication.

3 Review on the Evaluation Models

3.1 Analysis on the Evaluation Models

The evaluation models can be classified into two categories, including qualitative methods and quantitative methods. Delphi method, Gray Relational Analysis (GRA), Analytic Hierarchy Process (AHP) and Analytic Network Process (ANP), etc. belong to the former category, while Data Envelopment Analysis (DEA) and Technique for Order of Preference by Similarity to Ideal Solution (TOPSIS), etc. belong to the latter. The advantages, disadvantages and corresponding applicability of the widely used evaluation models are summarized in Table 2.

3.2 Discussion on the Evaluation Models

To the best of our knowledge, majority of the relevant studies only used one evaluation model to generate the evaluation results. However, all these models have their own superiorities and none of them are perfect in dealing with the performance evaluation problem. Currently, there exist quite a few studies that explore the combination of AHP and Fuzzy Synthetic Evaluation to solve the problem. In our opinion, how to further develop combined models to gain more objective and fair evaluation results is one of the study directions of the performance evaluation on the logistics parks.

Table 2. Review on the evaluation models

Models	Advantages	Disadvantages	Applicability	References
Delphi method	• Can be easily operated; • Fully utilize the expert experience	• Involve significant subjectivity	• Evaluation problems significantly relied on the preferences of the decision makers	[14, 21]
AHP & ANP	• Combine both quantitative and qualitative analysis; • Feasible to deal with the multi-objective evaluation problems	• Involve obvious subjectivity; • Incapable to address the system with many hierarchies and indexes	• Evaluation problems involving both objective data and subjective opinions of the decision makers	[8, 20, 22]
Neural networks	• Good self-learning capability; • Good fault tolerance; • High computational efficiency	• Need large numbers of samples	• Complex systematic problem, such as nonlinear evaluation problems	[16, 23, 24]
DEA	• No need to know the structure of data; • No need to assign subjective weights to the indexes; • Can analyze the weaknesses of objects	• The model is complicated; • The types of the objects should be identical	• Complex system with large-scale input and output	[25]
TOPSIS	• The model is simple; • Can compare the relative good or bad among the evaluation objects	• Can only give the internal ranking of objects; • The evaluation indexes should be monotonic	• Multi-object evaluation problem	[26]
Fuzzy synthetic evaluation	• Demanded information is not very large; • Feasible to deal with the fuzzy system	• Will lose some information; • Cause error due to the duplication of information	• Many indexes are difficult to be described by definite values	[20, 27, 28]
Gray relational analysis	• Can analysis the magnitude relationships among subsystems	• Need the best or worst sequence	• Dynamic process analysis	[29, 30]

4 Conclusions

In this study, we present a systematical review on the evaluation performance problem on the logistics parks. We summarize and classify the various indexes that appear on the relevant studies and give an evaluation index system as comprehensively as possible, which can guide scholars to understand and select the indexes to build their own systems in their relevant studies. Meanwhile, we analyze different evaluation models, identify and summarize their advantages and disadvantages, which can help scholars select suitable models to solve their problems. Moreover, based on our summary, scholars can effectively develop combined models to avoid the disadvantages of the single model and to gain better evaluation results.

Acknowledgements This study was supported by the National Natural Science Foundation Project of China (No. 71390332-3) and the Scientific Research Developing Project of the China Railway Corporation (No. 2014X009-B)

References

1. Sun, Y., Lang, M., Wang, D.: Optimization models and solution algorithms for freight routing planning problem in the multi-modal transportation networks: a review of the state-of-the-art. Open Civ. Eng. J. **9**, 714–723 (2012)
2. Zheng, W., Sun, Y.: Model confirmation of logistics park land size based on classification of Chinese national standard. Procedia-Social Behav. Sci. **43**, 799–804 (2012)
3. Tang, J., Tang, L., Wang, X.: Solution method for the location planning problem of logistics park with variable capacity. Comp. Oper. Res. **40**(1), 406–417 (2013)
4. Silva, R.M.D., Senna, E.T.P., Júnior, L., Fontes, O., Senna, L.A.D.S.: A framework of performance indicators used in the governance of logistics platforms: the multiple-case study. J. Transp. Lit. **9**(1), 5–9 (2015)
5. Notteboom, T.E.R.J.: Port Regionalization: Towards a new phase in port development. Marit. Pol. Mgmt. **3**(32), 297–313 (2005)
6. McCalla, R.J., Slack, B., Comtois, C.: Intermodal freight terminals: locality and industrial linkages. Can. Geogr. **45**(3), 404–413 (2011)
7. Silva, R.M.S.E.: Logistics platform: A framework based on systematic review of the literature. In: 22nd International Conference on Production Research (2013)
8. Zhang, J., Tan, W.: Research on the performance evaluation of logistics enterprise based on the analytic hierarchy process. Energ. Procedia **14**, 1618–1623 (2012)
9. Boerkamps, J., Van Binsbergen, A.: GoodTrip-A new approach for modelling and evaluation of urban goods distribution. In Urban Transport Conference, 2nd KFB Research Conference (1999)
10. Agrawal, S., Singh, R.K., Murtaza, Q.: A literature review and perspectives in reverse logistics. Resour. Conserv. Recycl. **97**, 76–92 (2015)
11. Sun, Y., Lang, M., Wang, D.: Bi-objective modelling for hazardous materials road-rail multimodal routing problem with railway schedule-based space-time constraints. Int. J. Environ. Res. Public Health **13**(8), 1–31 (2016)
12. Suksri, J., Raicu, R., Yue, W.L.: Towards sustainable urban freight distribution-a proposed evaluation framework. In Proceedings of 2012 Australasian Transport Research Forum, Perth, Australia, 26–28 September (2012)
13. Yue, H., Yue, W., Long, X.: Engineering evaluation system of logistics park capability. Syst. Eng. Procedia **2**, 295–299 (2011)
14. Abdulrahman, M.D., Gunasekaran, A., Subramanian, N.: Critical barriers in implementing reverse logistics in the Chinese manufacturing sectors. Int. J. Prod. Econ. **147**, 460–471 (2014)
15. Shaik, M.N., Abdul-Kader, W.: Comprehensive performance measurement and causal-effect decision making model for reverse logistics enterprise. Comput. Ind. Eng. **68**, 87–103 (2014)

16. Sun, Y., Lang, M., Wang, D.: Optimization and integration method for railway freight stations based on a hybrid neural network model. Comput. Model. New Technol. **18**(11), 1233–1241 (2014)
17. Liu, M.F., Zhang, J.: Appraisal on logistics park niche fitness based on gray relation analysis. In Conference on Systems Science, Management Science and System Dynamic (2007)
18. da Silva, R.M., Senna, E.T., Senna, L.A.: Governance of logistics platforms: the use of a survey for building a framework of performance indicators. Afr. J. Bus. Manage. **8**(10), 350–365 (2014)
19. Fasanghari, M.: Assessing the impact of information technology on supply chain management. In 2008 International Symposium on Electronic Commerce and Security, 726–730 (2008)
20. Jun, W.: A fuzzy evaluation model of the performance evaluation for the reverse logistics management. In 2009 WRI World Congress on Computer Science and Information Engineering. **1**, 724–727 (2009)
21. Lagorio, A., Pinto, R., Pinto, R., Golini, R., Golini, R.: Research in urban logistics: a systematic literature review. Int. J. Phys. Distrib. Logist. Manag. **46**(10), 908–931 (2016)
22. Kunadhamraks, P., Hanaoka, S.: Evaluating the logistics performance of intermodal transportation in Thailand. Asia Pac. J. Market. Logist. **24**(1), 323–342 (2008)
23. Sun, Y., Lang, M., Wang, D.: BP neural network based optimization for China railway freight transport network. Adv. Mater. Res. **1037**, 404–410 (2014)
24. Sun, Y., Lang, M., Wang, D., Liu, L.: A PSO-GRNN model for railway freight volume prediction: empirical study from China. J. Ind. Eng. Manage. **7**(2), 413–433 (2014)
25. Hu, Q.: Research on the planning of logistics park and analysis of assessment system. Master Thesis, Dalian Maritime University, China (2007)
26. Wang, X.: Research on service innovation performance evaluation of modern logistics enterprises based on entropy-dual point method. Doctoral Thesis, Jilin University, China (2013)
27. Wang, D., Chen, J.: Analysis on logistics park performance appraisal based on fuzzy synthetic evaluation method. Logist. Sci. Tech. **10**, 79–82 (2009)
28. Peng, J., Song, W.: Analysis on service performance appraisal of logistics enterprise based on fuzzy integral method. Railway Transp. Econ. **38**(7), 17–21 (2016)
29. Liu, M., Zhang, J.: Appraisal on logistics park niche fitness based on gray relation analysis. In Conference on Systems Science, Management Science and System Dynamic, 1393–1398 (2007)
30. Cui, H.: Study on the operating performance evaluation of multi-service logistics parks. Master Thesis, Beijing Jiaotong University, China (2015)

A Tabu Search Algorithm for Loading Containers on Double-Stack Cars

Zepeng Wang[1], Maoxiang Lang[1(✉)], Xuesong Zhou[2], Jay Przybyla[3], and Yan Sun[4]

[1] School of Traffic and Transportation, Beijing Jiaotong University, 100044 Beijing, People's Republic of China
mxlang@bjtu.edu.cn
[2] School of Sustainable Engineering and the Built Environment, Arizona State University, AZ 85281 Tempe, USA
[3] Department of Civil and Environmental Engineering, University of Utah, 84112 Salt Lake City, UT, USA
[4] School of Management Science and Engineering, Shandong University of Finance and Economics, 250014 Jinan, Shandong Province, China

Abstract. This study explores a multi-objective optimization for loading containers on double-stack cars that considers safety issues including lowering the center-of-gravity height and balancing the wheelsets' load of the loaded cars. To effectively solve this problem, a Tabu Search algorithm with 2-opt and Tabu list techniques is designed to obtain the close-to-optimal solutions to the problem. In this study, two experimental cases are presented to demonstrate the efficiency of the proposed algorithm and its computational accuracy by comparing with the exact solution strategy proposed in our previous study. The experimental results indicate that the Tabu Search algorithm can obtain the optimal solutions to the double-stack car loading problem more efficiently.

Keywords: Multi-objective optimization · Double-stack cars · Container loading problem · Center-of-gravity height · Tabu search algorithm

1 Introduction

The growing international trade has created both opportunities and challenges for railroad system all over the world. On one hand, it provides railroad transport with an adequate source of goods, which brings large profit for railroad enterprises and stimulates the development of the railroad system. On the other hand, it proposes a great challenge for railroad transport operation under the condition of limited transport resources, which leads to the fact that capacities of many rail terminals are inadequate.

Faced with this challenge, many researchers have switched their study emphasis to an efficient and economic freight transport mode, namely *Containerization*. As a kind of containerized transport modes, double-stack container trains have widely been operated in North America and Australia after introduction in 1984 from west coast ports of United States. The containers are allowed to be stacked two high (double-stacked) on the platform of the double-stack car so that the containers can be

F. Xhafa et al. (eds.), *Advances in Intelligent Systems and Interactive Applications*, Advances in Intelligent Systems and Computing 686,
https://doi.org/10.1007/978-3-319-69096-4_7

transported more efficiently than the regular mode. However, as for double-stack cars, their center of gravity are higher than single-stack ones and the center-of-gravity height varies from car to car after they get loaded, which raises higher safety concerns with the stability of loading containers on double-stack cars.

Many studies have been done to try and solve the problem of loading containers on rail cars. In the early period, heuristic algorithms were designed based on a set of given loading rules [1–3], while the optimization strategies drew less attention. The double-stack car loading problem is known as an NP hard problem. Heuristic algorithms have been proved to be the most efficient approaches to obtain the close-to-optimal solution of the similar problems in many relevant studies [4], e.g., the genetic algorithm [5], the adaptive link adjustment evolutionary algorithm [6], the memetic algorithm [7], and the column generation algorithm [8].

Loading containers on cargo ships or in airplanes has similar safety issue to the loading problem explored in this study. In the academic field, almost all the recent studies have formulated the center-of-gravity constraint and load-balancing constraint in their optimization models [9–12]. Only a few studies adopted a simplified function to represent the center-of-gravity height as the optimization objective [13]. Based on the analysis above, in order to strengthen the safety performance of loading containers on double stack cars and further guarantee the their operational safety, from the multi-objective optimization point of view, in our optimization strategy, lowering the center of gravity and improving load balancing of the loaded cars are set as the second and third objectives, and maximizing the loaded containers on the cars is the first one.

The rest sections of this study are organized as follows. First of all, we define the problem of loading containers on double-stack cars in Sect. 2 and propose a set of optimization strategies on the double-stack car loading problem in the same section. Then in Sect. 3, we design a Tabu Search algorithm with 2-opt and Tabu list techniques to obtain the close-to-optimal solutions to the problem. In Sect. 4, two experimental cases are used to verify the efficiency of the proposed algorithm and its computational accuracy in solving the double-stack car loading problem by comparing with the exact solution strategy proposed in our previous study. In Sect. 5, we draw the conclusions of this study.

2 Problem Description

2.1 Background Information

In China, the double-stack cars are all manufactured with the same configurations, with the maximal payload of 78 tons, the tare weight of 22 tons, the length of 19,466 mm, the platform length of 12,300 mm, the height of platform from rail surface of 290 mm, and the empty car's center-of-gravity height from the rail surface of 650 mm.

Typically in China, 20 ft ISO container and 40 ft ISO container can be loaded on the double-stack cars, where ISO is short for International Organization for Standardization. One 20 ft ISO container can be represented by 1 TEU, and 2 TEUs for one 40 ft ISO container.

According to *Regulations on Railroad Double-Stack Container Transport* [14] issued by the former Ministry of Railway of the People's Republic of China in 2007, three feasible loading patterns can be adopted to load containers on double-stack cars in China, which can be seen in Fig. 1.

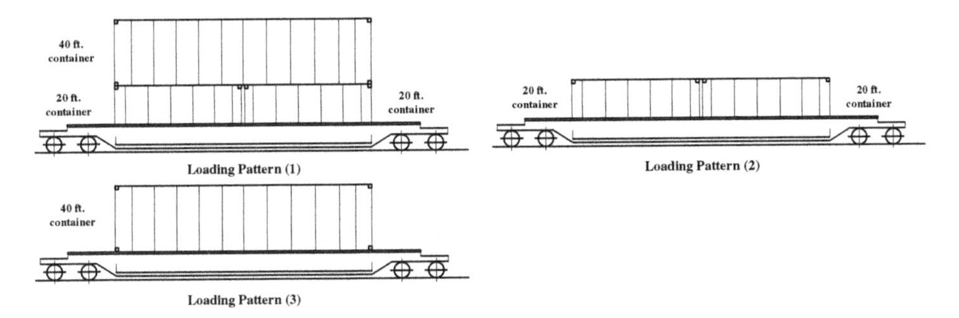

Fig. 1. Feasible double-stack car loading patterns

2.2 Optimization Strategy

The double-stack car loading problem aims at planning best loading schemes to load a set of 20 ft containers and a set of 40 ft containers on a set of double-stack rail cars by using the loaded patterns shown as Fig. 1. As claimed in Sect. 1, the objectives of the optimization strategy and their priorities based on the lexicographic goal programming approach [15] are separately:

(1) **Economic goal:** maximizing the TEUs loaded on the double-stack cars in order to improve the double-stack transportation economy. This objective is of the highest priority.
(2) **First safety goal:** minimizing the maximal center-of-gravity heights of the cars. As the most important factor that influences the operational safety of the double-stack transportation, we grant it as the second priority.
(3) **Secondary safety goal:** minimizing the maximal weight difference of two 20 ft containers in Loading Patterns (1) and (2) in order to improve the load balance. This objective is grant the third priority.

According to Reference [14], the constraints of double-stack car loading problem are as follows.

(1) **Loading pattern constraint**, i.e., the loading pattern of a car should be only selected from the three feasible loading patterns, otherwise the car will be kept empty.
(2) **Payload constraint** that ensures the gross weight of the containers loaded on each car should not exceed 78 tons.
(3) **Center-of-gravity height constraint** that ensures the center-of-gravity height of each car should not exceed 2400 mm.
(4) **Load balancing constraint** that ensures the weight difference between the two 20 ft containers on each car that adopts Loading Patterns (1) and (2) should not exceed 10 tons.

3 Tabu Search Algorithm for the Container Loading Problem

3.1 Introduction to the Algorithm

Tabu Search algorithm is a modified iterative local search algorithm. Compared the general local search algorithms, Tabu Search algorithm allows a degradation of the objective so that local optimal solutions can be avoided. Moreover, the algorithm will forbid the recently examined local optimal solutions to be repetitively searched by inserting them into a Tabu list that is constantly updated. Due to its advantages above, Tabu Search algorithm has been widely utilized in the combinational optimization field. Figure 2 indicates the framework of the algorithm [16].

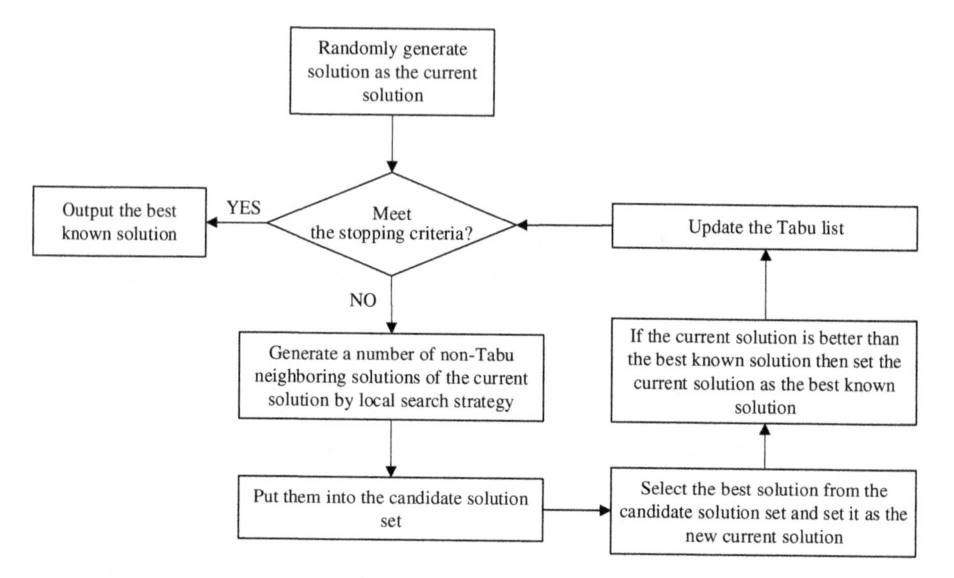

Fig. 2. Framework of the Tabu Search algorithm

3.2 Tabu Search Algorithm Design

(1) **Representation of the solutions**

In this study, we use a 20 ft container list and a 40 ft container list to represent a solution to the problem. A loading scheme can be encoded to a solution by loading the 20 ft containers and 40 ft containers sequentially on the cars based on the loading pattern constraint.

Take loading eight 20 ft containers (1, 2, 3, 4, 5, 6, 7, 8) and five 40 ft containers (A, B, C, D, E) on six cars (No. 1, 2, 3, 4, 5, 6) as example, the representation of a solution and corresponding loading scheme are illustrated by Fig. 3.

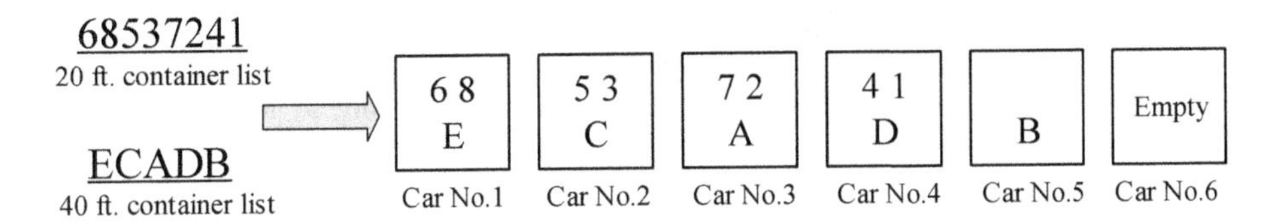

Fig. 3. Diagram of the representation of a solution

(2) **Evaluation of the solutions**

As for a double-stack car loading problem with K cars, we use S to represent one of its solution. First, we calculate the total gross weight of the loaded containers, the center-of-gravity height, and the weight difference of the kth car for $k = 1, 2, \ldots, K$. Then we check if solution S satisfies the constraints presented in Sect. 2.2. If S satisfies all the constraints, the corresponding loading scheme is feasible, and we use $F_k = 1$ to represent its loading feasibility, otherwise infeasible and there is $F_k = 0$. The three objective values of solution S can be determined by Eqs. 1, 2, and 3 where $TEUs_k$, H_k and Wd_k separately represent the loaded TEUs, the center-of-gravity height and the weight difference of the kth car. The three objective values will be further used to evaluate the quality of solution.

$$TEUs = \sum_{k=1}^{K} (TEUs_k \cdot F_k) \tag{1}$$

$$H = \max_{k=1}^{K} (H_k \cdot F_k) \tag{2}$$

$$Wd = \max_{k=1}^{K} (Wd_k \cdot F_k) \tag{3}$$

(3) **Local search strategy**

The local search is as follows. First, we randomly select two 20 ft containers and two 40 ft containers in the current solution and separately switch their positions. After the selection and switch operations, we can then get a new neighboring solution. Figure 4 gives an example to illustrate the local search strategy.

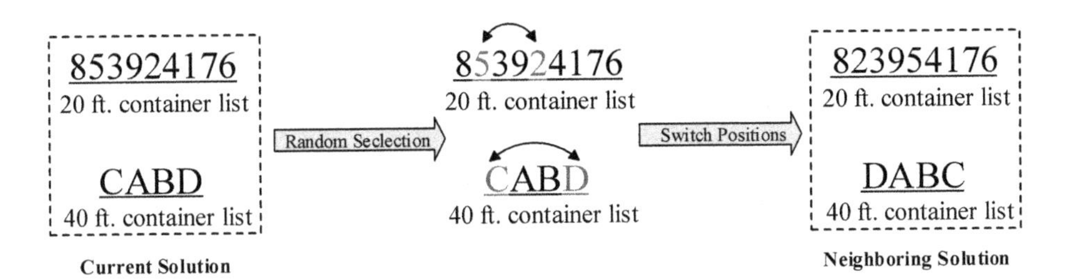

Fig. 4. Diagram of the local search strategy

(4) **Determining the candidate solution set**
First a certain number of non-Tabu neighboring solutions can be randomly generated from the current solution to the problem according to the local search strategy. Then we put them into the candidate solution set and further evaluate their quality.

(5) **Determining the Tabu solutions**
After examination, we can get the best solution from the candidate solution set. Then we add the best solution to the end of the Tabu list. Meanwhile, the first solution on the Tabu list gets released. The length of the Tabu list is determined by the problem scale.

(6) **Stopping criterion**
After the iteration researches up to a certain number of generations determined by the designer, the Tabu Search algorithm will stop.

4 Numerical Cases

To demonstrate the feasibility of the proposed optimization strategies and the Tabu Search algorithm, we design two experimental cases shown as Table 1.

Table 1. Numerical cases

Case No.	Number of 40 ft containers	Number of 20 ft containers	Number of double-stack cars
1	5	12	5
2	12	20	10

First, we randomly generate the container weights according to the practical distributions. Then we use an exact solution strategy that combines lexicographic goal programming approach and linear-fractional programming technique to solve the two small-scale experimental cases above and obtain their optimal solutions. For detailed information on this solution strategy and corresponding optimization model, readers can refer to our previous study [17]. This solution strategy is performed by the mathematical programming software GAMS. In the two experimental cases, the length of the Tabu list is set to 5. The experimental results are presented in Table 2.

Table 2. Optimization results of the two approaches

Case No.	Approach	Optimization objectives			CPU times (s)
		TEUs	H_{max} (m)	Wd_{max} (mm)	
1	GAMS	20	2226.4	3.71	801
	Tabu search algorithm	20	2226.4	3.71	0.015
2	GAMS	40	2182.5	3.59	8960
	Tabu search algorithm	40	2182.5	3.59	1.0

Table 2 indicates that the two approaches can obtain the same solutions of the two experimental cases, while GAMS takes approximate 13 min to 2.5 h to solve the two double-stack car loading problems, which is very time consuming. By comparison, the efficiency of the Tabu Search algorithm and its computational accuracy in solving the double-stack loading problems are hence verified.

5 Conclusions

In this study, we propose a multi-objective optimization strategies for loading containers on double-stack cars, in which lowering the center-of-gravity height and improving the load balance of the loaded cars are set as the second and third optimization objectives in order to improve the operational safety of the double-stack cars.

Then a Tabu Search algorithm is designed to solve the double-stack car loading problem. The algorithm mainly contains two steps, including randomly generating initial solutions and improving the solutions by the 2-opt and Tabu list techniques. In the case study, the computational efficiency of the Tabu Search and its computational accuracy are verified by solving two experimental cases and comparing with the previous designed exact solution strategy that combines lexicographic goal programming approach and linear-fractional programming technique.

Acknowledgements. This work was financially supported by the National Natural Science Foundation of China under Grant No. 71390332-3.

References

1. Jahren, C.T., Rolle, S.S.: A computerized assignment algorithm for loading intermodal containers to double-stack railcars. Final Report, University of Washington, 1994
2. Pacanovsky, D.L., Jahren, R.N., Newman, R.R., Howland, D.: A decision support system to load containers to double-stack rail cars. Civ. Eng. Environ. Syst. **11**(4), 247–261 (1995)
3. Jahren, C.T., Rolle, S.S., Surgeon, L.E., Palmer, R.N., Newman, R.R., Howland, D.L.: Automatic assignment algorithms for loading double-stack railcars. Transp. Res. Record **1511**, 10–18 (1995)
4. Raidl, G.R., Kodydek, G.: Genetic algorithms for the multiple container packing problem. Lect. Notes Comput. Sci. **1498**, 875–884 (1998)
5. Soak, S.M., Lee, S.W., Jeon, M.G.: The improved adaptive link adjustment evolutionary algorithm for the multiple container packing problem. Appl. Intell. **33**(2), 144–158 (2010)
6. Soak, S.M., Lee, S.W.: A memetic algorithm for the quadratic multiple container packing problem. Appl. Intell. **36**(1), 119–135 (2012)
7. Zhu, W.B., Huang, W.L., Lim, A.: A prototype column generation strategy for the multiple container loading problem. Eur. J. Oper. Res. **233**(1), 27–39 (2012)
8. Mongeau, M., Bes, C.: Optimization of aircraft container loading. IEEE Trans. Aerosp. Electron. Syst. **39**(1), 140–150 (2003)
9. Imai, A., Sasaki, K., Nishimura, E., Papadimitriou, S.: Multi-objective simultaneous stowage and load planning for a container ship with container rehandle in yard stacks. Eur. J. Oper. Res. **171**(2), 373–389 (2006)

10. Chen, B.S., Yang, Z., Lv, Y.J., Meng, C.: Air arm transferring airlift loading plan optimization. In Proceedings of the 2nd International Conference on Transportation Engineering, 2448–2453 (2009)
11. Pacino, D., Delgado, A., Jensen, R.M., Bebbington, T.: An accurate model for seaworthy container vessel stowage planning with ballast tanks. Lect. Notes Comput. Sci. **7555**, 17–32 (2012)
12. Dahmani, N., Krichen, S.: On solving the bi-objective aircraft cargo loading problem. In 2013 5th International Conference on Modeling, Simulation and Applied Optimization, 535–540 (2013)
13. Hu, W.B., Hu, Z.B., Shi, L., Luo, P., Song, W.: Combinatorial optimization and strategy for ship stowage and loading schedule of container terminal. J. Comput. **7**(8), 2078–2092 (2012)
14. Ministry of Railway of the People's Republic of China: regulations on railroad Double-stack container transport (2007)
15. Sun, Y., Lang, M., Wang, D.: Optimization models and solution algorithms for freight routing planning problem in the multi-modal transportation networks: a review of the state-of-the-art. Open Civ. Eng. J. **9**, 714–723 (2012)
16. Lang, M.X., Wang, Y.L., Zhou, X.S.: A two-stage algorithm for a dynamic multi-trip vehicle scheduling problem. in Proceedings of 2010 WASE International Conference on Information Engineering, 188–191 (2010)
17. Lang, M.X., Zhou, X.S., Sun, Y.: Multi-objective optimization for double stack train loading problem. J. Transp. Syst. Eng. Inf. Technol. **15**(6), 94–100, 106 (2015)

Optimization Model Under Grouping Batch for Prefabricated Components Production Cost

Chun Guang Chang(✉) and Yu Zhang

School of Management, Shenyang Jianzhu University, 110168 Shenyang, China
ccg7788@sohu.com

Abstract. During the prefabricated component (PC) production stage, for many kinds of PCs, production technologies are not the same. It is an important reason for high cost of PC production that reasonable group batch plan is lacked. By group technology and planning theory, an optimization model of grouping batch production for a variety of PCs is established. The model regards the minimum of the cost of production of PCs as the objective function. The size of the bottom mold, the type of embedded parts and connectors and the number of molds in the production process of PCs are constrains. By comparative analysis of application instance, the production cost under grouping batch for PCs is significantly reduced. It proves that the optimization model has a reference for production cost control under grouping batch for PCs.

Keywords: Prefabricated component (PC) · Production cost · Grouping batch · Optimization model

1 Introduction

Due to the prefabricated buildings in our country are in the initial stage, the demand for prefabricated components is low and the domestic prefabricated component (PC) plant did not form economy scale. There are many types of PCs, but the demand of each kind of PCs is often less. At the same time, the technical level in our country is generally not high, and the traditional production plan is still used in PC plants. So the production efficiency is reduced, the serious waste problem of resources is gradually increased. The increase in the cost under grouping batch for PCs will cause a large change in the production cost of the PCs.

On grouping batch optimization aspect, a linear programming model to determine operational production planning is set up in [1]. By chaotic particle swarm optimization algorithm, the integration of process planning and scheduling is solved in [2]. A three-level discrete-time algorithm which uses nonlinear models and integrates planning and detailed scheduling is introduced in [3]. On cost control aspect, an adaptive control with optimization methodology for optimizing the production cost subjected to quality constraints in high-performance milling operations of hardened steel is studied in [4]. A total cost minimization control scheme for biological wastewater treatment process is proposed in [5]. Above research provides an important theoretical basis for the production cost control under grouping batch for PCs.

F. Xhafa et al. (eds.), *Advances in Intelligent Systems and Interactive Applications*, Advances in Intelligent Systems and Computing 686,
https://doi.org/10.1007/978-3-319-69096-4_8

However, the research on the grouping batch for PCs is still less, the scholars paid more attentions on the construction technology and method. Therefore, this paper combines the theory of linear programming with the principle of grouping technology under grouping batch for PCs. Focusing on the reasonable division of production batches of PCs so that bottlenecks in production scheduling link of PCs restricting the production cost can be solved.

2 Problem Description and Variables Defining

2.1 Problem Description

To meet the requirements of the construction period and the quality of PCs, the production of a certain amount of PCs should be completed in the PC plant. During the production stage, it is a typical production cost optimization problem to minimize the production cost of PCs by dividing into production batches in a reasonable way. Namely, the production batches of PCs are divided according to the PC size, the groups of molds, the number of embedded parts and connectors, the size of the bottom mold, the technical level of the workers in the assembling mold, demolding and so on in a reasonable way. Eventually, it will lead to the optimization of labor costs, mold amortization fees and other costs under grouping batch PCs.

2.2 Variables Defining

i-serial number of PC type;
m-total number of PC types;
j -serial number of the production batch;
n-total number of production batches;
b_j'-upper limit of the short side length of the PC allowed to be produced by the bottom mold at the jth batch;
b_j-lower limit of the short side length of the PC allowed to be produced by the bottom mold at the $j$thbatch;
b_i-the short side length of the PC i;
f_j-upper limit of the number of groups that mold workers can assemble or demold on the jth batch production line;
f_j'-lower limit of the number of groups that mold workers can assemble or demold on the jth batch production line;
f_i-the number of groups that mold workers required for PC i;
q_j-upper limit of the number of embedded parts and connectors types that workers can assemble on the jth batch production line;
q_j'-lower limit of the number of embedded parts and connectors types that workers can assemble on the jth batch production line;
q_i-the number of embedded parts and connectors types required for PC i;
T_i-duration of PC i;
t_i-production time of PC i;

t_{j0}-start time of the jth batch;
c_{ij}-production cost of the PC i in the jth batch;
x_{ij} -0-1variable for judging whether component i is produced on the jth batch;
Where, $x_{ij} = \begin{cases} 1 & \text{PC } i \text{ is produced on the } j\text{th batch} \\ 0 & \text{else} \end{cases}$

3 Model Establishing

Based on above analysis and variables defining, the production cost optimization model for production scheduling of PCs is established as follows:

$$\text{Min } Z = \sum_{i=1}^{m}\sum_{j=1}^{n} x_{ij}c_{ij} \tag{1}$$

$$\text{s.t.} \sum_{j=1}^{n} x_{ij} = 1 \quad i = 1, 2 \ldots m \tag{2}$$

$$\sum_{i=1}^{m} x_{ij} \geq 1 \quad j = 1, 2 \ldots n \tag{3}$$

$$t_{j0} + \sum_{i=1}^{k} x_{ij} * t_i \leq T_k \quad k = 1, 2 \ldots m \; j = 1, 2, 3 \ldots n \tag{4}$$

$$t_{j0} + \sum_{i=1}^{m} x_{ij} * t_i \leq t_{j+1,0} \quad j = 1, 2, 3 \ldots n \tag{5}$$

$$b'_j x_{ij} \leq b_i x_{ij} \leq b_j x_{ij} \tag{6}$$

$$f'_j x_{ij} \leq f_i x_{ij} \leq f_j x_{ij} \tag{7}$$

$$q'_j x_{ij} \leq q_i x_{ij} \leq q_j x_{ij} \tag{8}$$

$$x_{ij} = 1 \; or \; 0 \quad i = 1, 2 \ldots m \; j = 1, 2, 3 \ldots n \tag{9}$$

Where, formula (1) is the objective function, and it indicates minimizing the production cost of PCs. Formulas from formula (2) to formula (9) are constraints. Formula (2) indicates that one PC can only be produced in one batch. Formula (3) indicates that each batch of production lines must produce at least one PC, production line is not allowed to be free. Formula (4) indicates the when PC k is produced in batch j, it will meet the requirements of its production period. Formula (5) indicates that only when the production of one batch is all over, can the next batch be put into production. Formula (6) indicates that when PC i is produced in batch j, it will meet the module size requirements on the production line of batch j. Formula (7) indicates that the number of mold groups of PC meets the requirements of mold workers production capacity on the production line of batch j. Formula (8) shows the number of embedded parts and connectors types of component i will meet the requirements of embedded parts installation capacity on the production line of batch j. Formula (9) introduces 0–1 variables.

4 Application Instance and Comparative Analysis

4.1 Application Instance

The prefabricated factory will provide a batch of PCs for one construction project. Five typical PCs are selected as the research objects, i.e., m = 5, the specific technics parameters are shown in Table 1. Combining with the actual situation of the PCs plant, the components will be divided into three batches to be produced, i.e., n = 3. According to the actual situation of the PC plant, the size of bottom mold on the first batch of production line is 12 m * 4 m * 0.2 m. The upper and lower limits of the number of embedded part and connector types are determined by the ratio of the number of skilled workers and general workers in each batch of embedded parts. So, the number of mold group F1 on the first production batch is 25–35. The method that identifies the upper and lower limits of the mold groups is the same. The technics parameters are represented in detail as Table 2.

Table 1. Typical component parameter table

Serial number of component	Size b (m)	Number of embedded parts and connectors	Number of mold group m	Production time t (d)	Duration T (d)
1	4.7 * 3.4 * 0.2	5	6	8	30
2	2.9 * 2.8 * 0.2	4	7	7	30
3	5.9 * 2.2 * 0.2	9	8	6	43
4	0.61 * 1.2 * 0.08	10	9	9	50
5	4.2 * 2.1 * 0.2	8	4	5	52

In the prefabricated plant, they are put into production in the order of the first, the second, and the third installments. Only when the production of this batch is all over, can the next batch be put into production. The PCs in the same batch are put into production in the order of the serial number of the components from 1 to 5.

The optimization model under grouping batch for PCs production cost is solved, and the optimization solutions can be seen in Table 3.

4.2 Comparative Analysis

To verify the application effect of this model, the optimized production scheme is compared with the traditional production scheme as shown in Tables 4 and 5.

Table 2. Production process parameters of each batch

Production batch	The size of the mold b	The number of embedded and connectors F	The number of mold group M	Start time t_0
I	12 × 4 m	25–35	10–15	0
II	9 × 4 m	30–45	13–20	15
III	7 × 3 m	40–60	15–22	37

Table 3. Optimized results.

Production batch \ Serial number of component	1	2	3	4	5	Min Z
I	1	1	0	0	0	17708
II	0	0	0	1	1	
III	0	0	1	0	0	

Note: The number "1" in the table indicates that the component is produced on the production batch; the number "0" indicates that the component is not produced on the production batch.

Table 4. Traditional production scheduling program

Production batch	I	II			III	Total
Serial number of PC	3	1	2	4	5	——
Labor costs Y_{ij}	100	80	345	137	122	846
Mold amortization fees P_{ij}	583	1374	1534	357	783	4631
Production cost C_{ij}	3802	3929	3390	3407	3858	18386

Table 5. Optimized production batch program and effect

Production batch	I		II		III	Total
Serial number of PC	1	2	4	5	3	——
Labor costs Y_{ij}	220	100	145	260	131	856
ΔY	+140	−245	+8	+138	+31	+10
Mold amortization fees P_{ij}	983	1689	357	430	430	3889
ΔP	−391	155	0	−353	−153	−742
Production costs C_{ij}	3678	3300	3407	3643	3680	17708
ΔC	−251	−90	8	−215	−122	−732
$\Delta C/C_{ij}$ (%)	−6.60	−2.29	0.24	−6.31	−3.16	−3.98

Respectively, on the labor cost and mold amortization fee, the production cost of PCs changes. Taking PC 1 for example, the amortization cost of the bottom mold is reduced by about 391 ¥, the labor cost is increased by 140 ¥, and the total cost of the PC is reduced by 251 ¥, So the production cost reduction rate of the PC 1 is 6.60%.

The total cost of labor in the whole production process increased by 10 ¥. The total cost of molds decreased by 742 ¥. The total production cost decreased by 732 ¥. So it shows that the model has achieved the purpose of optimizing and reducing PCs production cost under grouping batch.

In the above example, there are only five typical prefabricated members selected as the research object, the number is small, so the optimization effect is not very obviously. If there are many types of PCs on a large scale of production, the model will bring considerable economic benefits to the prefabricated plant and have important practical value.

5 Conclusion

This paper is based on the group batch production for PCs, and the idea and principles of group technology are introduced. Under the premise of guaranteeing the construction period and product quality, the production cost of the PCs is optimized by optimizing the production batches of the PCs. It can achieve the purpose of controlling the production cost of the PCs. By applying this model to an instance, it can be seen that the optimized group batch plan is superior than the traditional one. It has strong application and reference through the comparative analysis. With the decreasing of PCs production cost, the prefabricated building cost will be reduced, and the prefabricated building can be widely accepted by the consumers in China. A new era for real estate enterprise will come into being.

Acknowledgements. This work was supported by the National Natural Science Foundation of China (51678375); The Natural Science Foundation of Liaoning Province (2015020603; 201602604); Liaoning provincial social science planning fund (L15BJY018) and Shenyang Scientific and Technological Planning (F15-198-5-15).

References

1. Zhang, Y.: A study of linear programming modeling and optimization on operational production scheduling for batch/continuous mixed production in M company. Manag. Sci. Eng. **5**(4), 237–247 (2016)
2. Milica, P., Najdan, V., Marko, M., Zoran, M.: Integration of process planning and scheduling using chaotic particle swarm optimization algorithm. Expert Syst. Appl. **64**, 569–588 (2016)
3. Pedro, A., Castillo, C., Vladimir, M.: Inventory pinch based, multiscale models for integrated planning and scheduling-part II: gasoline blend scheduling. AIChE J. **60**(7), 2475–2497 (2014)
4. Jorge, A.S., José, V.A.-N., Hector, R.S., Federico, G.-E.: Adaptive control optimisation system for minimising production cost in hard milling operations. Int. J. Comput. Integr. Manuf. **27**(4), 348–360 (2014)
5. Yamanaka, O., Obara, T., Yamamoto, K.: Total cost minimization control scheme for biological wastewater treatment process and its evaluation based on the COST benchmark process. Water Sci. Technol. **53**(4–5), 203–214 (2006)

Air Quality Evaluation System Based on Stacked Auto-Encoder

Yuxuan Zhuang, Liang Chen(✉), and Xiaojie Guo

School of Mechanical and Electric Engineering, Soochow University, 215021 Suzhou Jiangsu, China
chenl@suda.edu.cn

Abstract. Air quality reflects the extent of air pollution, and it is evaluated based on the concentration of air pollutants. Air quality level is traditionally assessed by mathematical formula, which cannot precisely represent the level of air pollution in some circumstances. This paper introduces a novel Air Quality Evaluation system based on Stacked Auto-Encoder (named AQES-SAE) to tackle these problems. The data of air pollutant concentrations and the air quality index are collected and clustered with improved K-means algorithm. The labeled clusters as training samples are then transferred to a deep neural networks and trained on the SAE model, which can be used as the classifier in the air quality evaluation. The proposed method is compared with traditional mathematical formula, and it is verified that the AQES-SAE system is prior to the traditional method because it can identify air quality level more accurately and more reasonably.

Keywords: SAE · Air quality level · Deep neural networks · K-means

1 Introduction

As the changing of environment, people pay more attention to air quality index (AQI), which is used to evaluate the air quality [1]. Unifying the air pollutant concentrations into a single numerical form, grading characterization of air pollution concentration and air quality status is suitable for short-term air quality status and trends [2]. At present, the international community generally uses these forms of index to issue air quality information and make environmental regulatory decision.

Presently, people are trying to find better methods for air quality prediction and assessment [3–5]. For example, BP network has a widespread application in air pollution forecast, which can be used for the assessment of air quality levels [6]. However, with the breakout of large amount of data, it is insufficient for a shallow model to extract features among input pollutant variables and give accurate air quality level, so air quality assessment can not be made effectively by traditional methods.

To tackle big data problem, deep learning has excellent function in terms of computational elements and parameters required to represent unknown functions [7]. The greedy layer-wise unsupervised training strategy mostly helps the optimization, by initializing weights in a region near a good local minimum, thus SAE can correctly extract features and cluster data [8, 9], and recent studies in machine learning have

F. Xhafa et al. (eds.), *Advances in Intelligent Systems and Interactive Applications*, Advances in Intelligent Systems and Computing 686,
https://doi.org/10.1007/978-3-319-69096-4_9

shown that a deep or hierarchical architecture is useful to find highly non-linear and complex patterns in data [10, 11].

In this paper, We introduces a novel Air Quality Evaluation system based on Stacked Auto-Encoder (named AQES-SAE). This method call be regarded as all extension of our previous work [12], as the latter laid the theoretical foundation for the present study.

A categorization database has created by K-means algorithm. The labeled clusters as training samples are transferred to a deep neural networks and trained on the SAE model. Through the pre-training and fine-tuning, the model can accurately and quickly identify air quality level. Examples of typical samples are presented to validated the superiority of proposed method, which is of great significance for air quality assessment.

2 The Methods

2.1 Introduction to Traditional Method

In a traditional way, mathematical formulas are used to assess air quality levels, AQI is used to indicate the extent of air pollution. Table 1 shows air pollutant concentration limits in each small range, which is needed for the following formula.

Table 1. Air pollutant concentration limits

IAQI	Air pollutant concentration limits					
	PM2.5	PM10	SO_2	NO_2	CO	O_3
0	0	0	0	0	0	0
50	35	50	50	40	2	100
100	75	150	150	80	4	160
150	115	250	475	180	14	215
200	150	350	800	280	24	265
300	250	420	1600	565	36	800
400	350	500	2100	750	48	1000
500	500	600	2620	940	60	1200

According to the actual concentration of pollutants, Individual Air Quality Index (IAQI) separately can be computed by the following formula (1) and formula (2). The maximum value of the IAQI for each pollutant is chosen as AQI, which is based on a major decision of air quality level.

$$IAQI_p = \frac{IAQI_{Hi} - IAQI_{Lo}}{BP_{Hi} - BP_{Lo}}(C_p - BP_{Lo}) + IAQI_{Lo} \tag{1}$$

$$AQI = \max\{IAQI_1, IAQI_2, IAQI_3, \ldots, IAQI_n\} \tag{2}$$

For the above formula, $IAQIp$ refers to $IAQI$ of p, and C_p is concentration of p, BP_{Hi} refers to minimum concentration in the vicinity, and BP_{Lo} refers to maximum concentration in the vicinity. $IAQI_{Hi}$ refers to the corresponding highest $IAQI$, and $IAQI_{Lo}$ refers to the corresponding lowest $IAQI$, n refers to types of pollutants.

2.2 AQES-SAE System

Figure 1 shows the whole process of AQES-SAE. The labeled samples are processed with a deep neural network, which consists of several auto-encoders. The number of layers depends on the task complexity. All the parameters of the neural network need to be pre-trained and fine-tuned according to the supervising criterion, thus the learning performance can be greatly improved. Trained model can be used to identify features and determine the label, which is helpful for us to learn better about air quality comparing with the the traditional mathematical method.

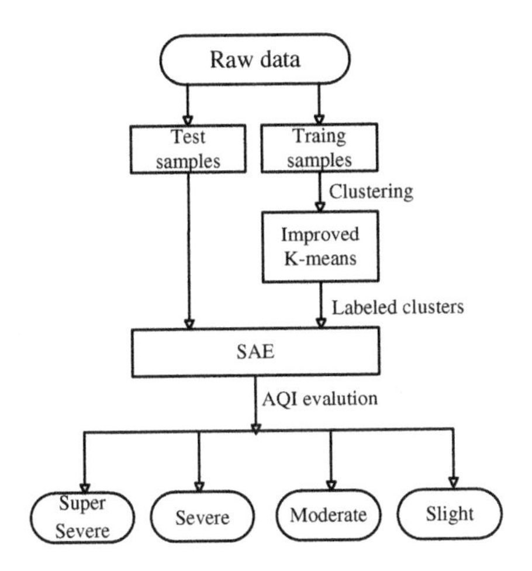

Fig. 1. Flow chart of AQES-SAE

The labeled samples are processed with SAE, which consists of several auto-encoders. The number of SAE layers depends on several factors, including the requirements of the speed and accuracy, the complexity of the task. Furthermore, all the parameters of the neural network need to be pre-trained and fine-tuned according to the supervising criterion, thus the learning performance can be greatly improved. Trained model can be used to identify features and determine the label, which is helpful for us to learn better about air quality comparing with the the traditional mathematical method.

3 The Experiments

3.1 Classification with Improved K-means Algorithm

K-means algorithm is a typical distance-based clustering algorithm, the adjustment rule of iterative operation is obtained by using the method of extremum, a more appropriate initial point calculated by the hierarchical method. A more accurate distance calculation method is chosen, so that the coefficient is minimal for the association between each sample. The detail process can be find on our previous work on [12].

According to our previous paper [12], we can precisely cluster the samples of air quality records in the past 1200 days consisting of concentrations of PM2.5, PM10, SO_2, NO_2, CO, O_3.

3.2 Identification with the SAE Model

The AQES-SAE model used in this paper is briefly shown in Fig. 2. There are six nodes of input layer representing six features (PM2.5, PM10, SO_2, NO_2, CO, O_3). In order to get better effect, the three hidden layers with one hundred nodes are used in the SAE model, each nodes has a activation function in the three hidden layers. Four nodes of output layer show the different probability of each air quality level, the maximum of which is regarded as air quality level.

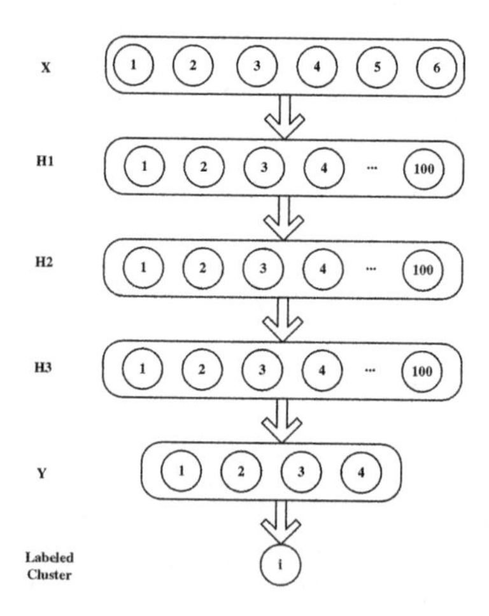

Fig. 2. The SAE model in the AQES-SAE system

It is of the necessity and importance to adjust parameters so that SAE model has characteristics of good effect, high speed and correct identification. In this model, learning rate of pre-training is 0.001 and epochs of training is 800. Learning rate of fine-tuning is 0.8 and epochs of fine-tuning is 12, and batch sizes is 5.

4 The Results

4.1 Clustering Analysis

Figure 3 shows the change trend of verification errors. In the beginning, validation errors drops very fast, dropping from 59.39 to 7.88% during 100 epochs. After 800 iterations, the final validation error can reduce to 3.07%. Its trend is stable and non-oscillatory, and this model effectively depicts the random fluctuation and non-linearity of the data. This indicates that SAE is valuable for popularized application, a satisfying result is obtained with the proposed SAE model.

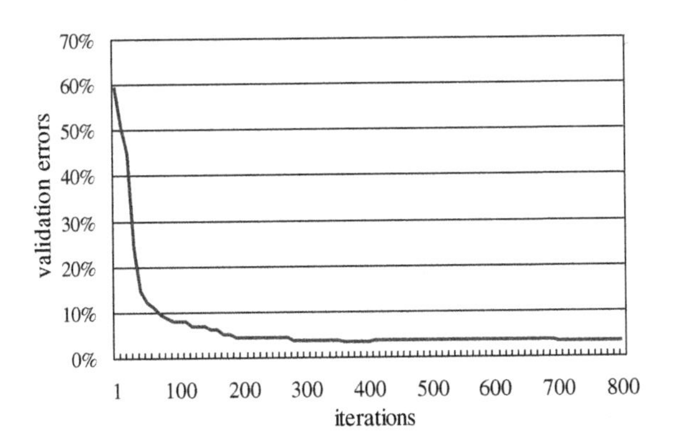

Fig. 3. The change of validation error caption

4.2 Superiority of the AQES-SAE System

According to the data in the Table 2, which shows two sets of examples, different methods are used to identify air quality levels. PM10 is the same value in the Table 2, however, other elements are dramatically different between those two sets of data.

Table 2. Evaluation results

Labeled cluster	AQI	PM2.5	PM10	SO_2	NO_2	CO	O_3
D	104	88	158	54	88	0.8	72
C	104	36	158	14	45	0.6	34

In the traditional mathematical method, which is PM10 as a basis for evaluation, air quality levels are severe according to division of grade standards of air quality levels.

In the AQES-SAE, these two samples can be distinguished, one is super severe, the other is moderate, because the non-linear structure of SAE can give each index an appropriate weight more intelligently. Thus, AQES-SAE is better than mathematical formula, for it can identify air quality level accurately and be more reasonable.

5 Summary

Due to people's increasing attention to air quality, there have been great interests in assessing air quality level. In order to improve the accuracy and rationality of identifying air quality levels, we cluster data with K-means algorithm precisely to give labels to the samples, then, a deep neural network is built based on SAE. Thus an AQES-SAE is proposed, Thanks to the two-step learning scheme of greedy layer-wise pre-training and the fine-tuning in deep learning, the air quality level is identified precisely, which is helpful for air quality monitoring system to improve precision.

References

1. Carvalho, V.S.B., Freitas, E.D., Martins, L.D., et al.: Air quality status and trends over the Metropolitan Area of São Paulo Brazil as a result of emission control policies. Environ. Sci. Policy **47**(9), 68–79 (2015)
2. Cogliani, E.: Air pollution forecast in cities by an air pollution index highly correlated with meteorological variables. Atmos. Environ. **35**(16), 2871–2877 (2001)
3. Chen, Q., Shao, Y.: The application of improved BP neural network algorithm in urban air quality prediction: evidence from China, In: IEEE Pacific-Asia Workshop on Computational Intelligence and Industrial Application, pp. 160–163. IEEE Computer Society, Washington (2008)
4. Gorai, A.K., Kanchan, Upadhyay A., et al.: Design of fuzzy synthetic evaluation model for air quality assessment. Environ. Syst. Decisions. **34**(3), 456–469 (2014)
5. Salama, K., Alrashed, Y., Alghamdi, F., et al.: Assessment of the air quality levels in the King Abdul Aziz Port in Dammam. Int. J. Med. Sci. Public Health 5(2), 282–286 (2015)
6. Bai, H.M., Shen, R.P., Shi, H.D., et al.: Forecasting model of air pollution index based on BP neural network. Environ. Sci. Technol. 36, 186–189 (2013)
7. Bengio, Y.: Learning deep architectures for AI, Found Trends Machine Learn. 2(1):1–127 (2009)
8. Bengio, Y., Lamblin, P., Popovici, D., et al.: Greedy layer-wise training of deep networks. In: International Conference on Neural Information Processing Systems, pp. 153–160. MIT Press (2006)
9. Bengio, Y., Lamblin, P., Popovici, D., Larochelle, H.: Greedy layer-wise training of deep networks. In: Advances in Neural Information Processing Systems, pp. 153–160 (2007)
10. Sun, Z.J.: Marginal fisher feature extraction algorithm based on deep learning. Dianzi Yu Xinxi Xuebao/ J. Electron. Inf. Technol. **35**(4), 805–811 (2013)
11. Vincent, P., Larochelle, H., Bengio, Y., Manzagol, P.A.: Extracting and composing robust features with denoising autoencoders. In: Proceedings of the 25th International Conference on Machine Learning, pp. 1096–1103 (2008)
12. Guo, X., Chen, L., Zhou, H., et al.: An improved K-means algorithm and its application in the evaluation of air quality levels. In: Chinese Control and Decision Conference, pp. 3324–3329 (2015)

Pattern Recognition and Vision Systems

Urban-Rural Difference of the Coupling Between Social-Economic Development and Landscape Pattern in Chengdu Plain

Zhang Huabing[1(✉)], Lu Dapei[1], Zhen Yan[2], Han Shuang[1], and Chen Hongquan[1]

[1] College of Urban and Planning, Yancheng Teacher's University, Yancheng 224051, China
jszhbing@163.com

[2] The Third Surveying and Mapping Engineering Institute of Sichuan Province, Chengdu 610500, China

Abstract. Taking Chengdu plain as the research object, using 2014 TM image and the data of social economic development, from the perspective of urban and suburban, studied the differences of the coupling relationships between landscape pattern and economic development in the studying area, the results are as follows: the coupling degree show that the coupling degree Chengdu plain is 0.411, the coupling coordination degree is 0.332, which is high strength and low coordination. The coupling degree is 0.498 and the coupling coordination degree is 0.436 in the urban, which is high strength and normal coordination. The coupling degree is 0.342 and the coupling coordination degree is 0.325 in the suburban, which is low strength and low coordination. The coupling differences in urban and rural between the economic development and landscape pattern is because the result by natural conditions and human activities, the human economic activity is the leading cause of urban and rural landscape pattern differences.

Keywords: Landscape pattern · Social-economic · Coupling model · Chengdu plain

1 Introduction

Regional landscape pattern is the result of natural factors and human activities. In the process of urbanization, different levels of human activities in urban and rural areas have different impact on the region. Urban and rural areas usually take human as subject, developing from the transformation of the surrounding environment continuously and purposefully, they are the concentration of population, social, economic and cultural activities [1]. They differ in the ways and intensity of their transformation and use of the environment. As the results, cities and villages show different characteristics in social-economic development, landscape structure and function. Therefore, it is necessary to understand the relationship between regional economic development and ecological environment.

F. Xhafa et al. (eds.), *Advances in Intelligent Systems and Interactive Applications*, Advances in Intelligent Systems and Computing 686,
https://doi.org/10.1007/978-3-319-69096-4_10

In recent years, a series of quantitative researches have been carried out on urban landscape pattern, especially many achievements from the use of quantitative methods, which combined with remote sensing technology, carrying out multiply-temporal landscape pattern dynamic changes and simulation [2]. Researches also noted that economic development and ecological landscape are two aspects of mutual influence in the process of urbanization. Many studies have shown that there is a certain relationship between social-economic development and the ecological environment. However, from a quantitative perspective on the coupling between the two is still relatively scarce. Therefore, this paper takes the Chengdu Plain in China as a case study area, and applies the coupling degree model to analyze the coupling differences between the landscape pattern and the economic development system from the perspective of the large urban and suburban counties.

2 Materials and Methods

2.1 Study Area and Data Source

Chengdu plain to the north of Qinling Mountains, southern to Daba Mountains, YunGui Plateau, western to Qionglai and Longshan mountain, Eastern to Longquan mountain. The southwestern of Chengdu plain is the Mingjiang River system, northeastern is the Tuojiang River system. The annual average temperature is 16.2–18.6 °C and the annual rainfall is in the 873–1265 mm. We select key development zone in Chengdu plain as the research object, which GDP was RMB 10 056.6 billion, and the per capita GDP was RMB 58 460 in 2014. The three production structure is 10.5:61.3:28.2. The end of 2014, the total population was 12107364, the population density was 989 people per square kilometer. In order to compare the difference between urban and rural, further to divide the study area for two parts by Chengdu Beltway. One is in Chengdu City, the other is rural areas.

The landscape indexes are from TM images, social and economic development data from the Statistical Yearbook 2014 of Chengdu city and Deyang city.

2.2 Coupling Model

The coupling degree can reflect the intensity, the coupling effect and the degree of the interaction among the elements or systems in the system, which determines the trend of the system from disorder to order [3]. The degree of coupling can reflect the extent of the intensity and contribution to the landscape pattern and economic development in the study area [4].

$$C = \frac{[F(x) \times G(y)]^k}{[\alpha F(x) + \beta G(y)]^{2k}} \tag{1}$$

$$T = \alpha F(x) + \beta G(y) \tag{2}$$

$$D = \sqrt{C \times T} \tag{3}$$

where C is Coupling degree, k is the adjustment coefficient, the value of it is 2–5. α and β are weight coefficients. In the article, the value of k is set to 2, and the value of α and β is 0.5. T, D refer to comprehensive harmonic degree and coupling coordination degree. System indicators are selected as follows (Table 1).

Table 1. Social-economic and ecological environment index systems in Chengdu plain

Target layer	Criteria layer	Indicator layer
Urbanization	Social economic	Population density
		Per person GDP
		Non-agricultural proportion
Ecological environment	Landscape pattern	Average fractal dimension
		Average patch area

3 Results

3.1 Urban-Rural Differences in Landscape Pattern

The proportion of cultivated land is 33.536%, the proportion of forest land is 30.35%, and construction land is 15.981%. The proportion of wasteland is the lowest, only 0.952%, which indicates that the land use degree of Chengdu Plain is high.

In the urban area, the proportion of construction land is largest, the next is cultivated land, the third is garden land, which proportion are 31.078%, 23.554% and 11.982%, the proportion of wasteland is only 0.091%. Which indicates that the human activities are strong. In the rural area, cultivated land occupies the largest area, forest land is the second, next is construction land, which proportion are 36.198%, 35.249% and 11.955%. It shows that the proportion of cultivated land is higher than that of urban area, and the proportion of construction land is lower than that of rural area. Mainly because of the development of the two and the three industry in the urban area, the population density is large, which leading to the proportion of construction land is large. The proportion of garden in urban is higher than in rural, mainly urban take landscape architecture for important. The proportion of forest land in the rural is much higher than that in the urban area, and forestry belongs to the primary industry and is suitable for development in the suburbs.

The Shannon diversity index in the rural is higher than the urban, indicating that the ecological diversity of the rural is abundant and the urban species diversity was scarce. This may be due to strong human activities in the urban area, which make a large impact on the ecological. The lower the fractal dimension, the more regular the landscape patch shape and the more intense human activities. It can be seen from the table that the fractal dimension of the urban area is lower than that of the rural. The average patch area in the urban area is obviously lower than that in the rural, which indicates that the urban landscape has become more broken due to strong human activities.

3.2 Urban-Rural Coupling Differences Between the Landscape Pattern and the Social-economic Development

The coupling value is between 0 and 1, and the higher the value, the higher the coupling, indicating a benign resonance coupling between the systems, which tends to a more ordered structure [5]. The coupling value is between 0 and 0.3, the coupling between landscape pattern and economic development is at a low level. The value is between 0.3 and 0.5, the coupling is in the antagonistic stage. The urban development of this stage has surpassed the infection point of S-type urban development curve and entered the rapid development stage. The value is in the range of 0.5–0.8, the landscape pattern and economic development enter the run-in stage. The value between 0.8 and 1, the level of urban development not only in quantity has been greatly developed, the quality has also been greatly improved. The coordination degree is between 0 and 1, the coordination degree is between 0 and 0.4, which is the low-level coordination. The degree is between 0.4 and 0.5, which is moderate coordination. The degree is between 0.5 and 0.8, for a high degree of coordination. The degree is between 0.8 and 1, for the extreme coordination of the coupling (Table 2).

Table 2. The coupling degree and coupling coordination between landscape pattern and economic development in the study area

Region	Coupling degree	Coupling coordination degree	Coupling intensity and coordination degree
Urban areas	0.498	0.436	High-intensity coordination
Rural areas	0.342	0.325	Low intensity low coordination
Overall	0.411	0.332	High intensity and low coordination

It can be seen from the results that the coupling value of the whole study area is 0.411 and the coupling coordination value is 0.332, showing high-intensity and low-coupling, which indicates that the economic development and landscape pattern of Chengdu Plain are in antagonistic stage, the level of urbanization is high, the ecological environment is destroyed, and the landscape pattern is disturbed by human influence. The coupling degree of Chengdu urban area is 0.498, and the coupling degree is 0.436. The coupling and coupling stage of high-intensity shows that the urban landscape pattern and economic development are in antagonistic stage. The urbanization development has entered a rapid development stage, while the ecological environment is destructed. The coupling degree of rural is 0.342 and the coupling coordination value is 0.325, which shows low intensity and low coupling. It shows that the rural landscape pattern and economic development are at the low level coupling stage, the development level of urbanization is low, human disturbance effect is small, the landscape diversity is rich, the landscape pattern is destroyed for a little by the urbanization.

Natural factors of the study area are the basis for the existence of the differences in landscape pattern and have important impacts on the economic development such as Jianyang, Chongzhou, Dayi County, Qionglai City, due to superior natural conditions, farmland and other natural landscape account for a higher proportion. Urbanization development and human activities are the important driving factors. Chengdu city has a great location, economic foundation is great and population concentrated to the central is obvious. In the process of urbanization, the population size affects the landscape pattern change. When the population density of the urban area is large, the greater the proportion of non-agricultural industries, the higher the level of economic development, the greater the value of per person GDP, man-made landscape is much more, the landscape shape is more regular. But the rapid development of urbanization had bring much negative effects on the ecological environment, such as water environment, soil environment pollution and biodiversity loss.

Acknowledgments This study were founded by Geographic National Condition Monitoring Engineering Research Center of Sichuan Province of China (No.GC201518) and Natural Science Research Major Projects of colleges and universities in Jiangsu province (14KJA170006).

References

1. Xiaojiang, Qiu: Fracture pattern of urban landscape and the relationship between urbanization and economic development. Acta. Ecol. Sin. **24**(3), 57–61 (2012)
2. Yang, Q., Lihuan, H., Guihua, D., et al.: Analysis of urban landscape pattern and dynamic simulation method. Anhui Agric. Sci. **41**(21), 8979–8981 (2013)
3. Yaobin, L.: Analysis of coupling degree between urbanization and ecological environment in China. J. Nat. Resour. **20**(1), 105–112 (2005)
4. Yaobin, L., Rendong, L., Xuefeng, S.: Genetic Correlation Analysis of Regional Urbanization and Eco-environment Coupling in China. Acta. Geogr. Sin. **60**(2), 237–247 (2005)
5. Simayi, Z., Chen, S., Aimurela, D., et al.: Hotan urbanization and ecological environment coupling between. Northwest. Norm. Univ. **51**(1), 97–102 (2015)

A New Model and Algorithm for Clustering

Guohong Liang[1,2(✉)], Ying Li[2], and Junqing Feng[1]

[1] School of Science, Air Force Enigeering University, 710051 Xi'an, China
liangguohong321@163.com
[2] School of Computer Science, Northwestern Polytechnical University, 710129 Xi'an, China

Abstract. Clustering is a unsupervised pattern recognition method and an important research content in data mining and artificial intelligence. A new mathematical model of clustering is proposed using graph theory. It is proved theoretically that the model is a submodular function. According to it, a corresponding algorithm is proposed.

Keywords: Clustering · Model · Submodular function · Algorithm

1 Introduction

The research of cluster analysis has a long history. The importance of cluster analysis and its cross-cutting characteristics with other research directions have been recognized by people for a long history. Clustering is one of the important research directions of data mining and pattern recognition, It plays an extremely important role in identifying the internal structure of data. Clustering is mainly used in speech recognition, character recognition, etc. The clustering algorithm in machine learning is applied to image segmentation and machine vision, In addition, clustering is applied to the statistical science, clustering analysis also plays an important role in biology, psychology, archaeology, geology, geography and marketing research. At present, many kinds of clustering algorithms have been successfully applied in many fields [1]. Two problems are usually solved in clustering research and application process [2]. One is how to divide a given data set so that the results are optimal. Second, how many classes of the data sets are most appropriate. Among them, the first problem is solved by clustering algorithm, the second problem is clustering validity problem [3]. Although in some applications, Number of cluster can be estimated through the user experience and domain knowledge. But in general, Number of cluster is not known in advance, Determining the best number of cluster is a difficult job [4].

Spectral clustering is a popular clustering algorithm in recent years, the spectral clustering algorithm is based on the spectral graph partition theory. Its essence is to transform clustering problem into the optimal partition problem, is a point to clustering algorithm, easy to achieve and very effective, can effectively find complex shape and uneven density, not easy to fall into local advantages, the distribution of data without any hypothesis, has a good application prospect. Adjacency matrix is a typical representation of graph in graph theory, used to represent the adjacency information of

F. Xhafa et al. (eds.), *Advances in Intelligent Systems and Interactive Applications*, Advances in Intelligent Systems and Computing 686,
https://doi.org/10.1007/978-3-319-69096-4_11

vertices in the graph, and defines the connectivity between vertices in the graph. The graph clustering algorithm regards the similarity matrix of image set as the adjacency matrix of a weighted undirected graph. If the data points x_i are a vertex v_i in the graph, the edge e_{ij} represents the relationship between the vertex i and the vertices j, and the weight w_{ij} of the edge is the similarity between two points, the undirected graph $G = (V, E)$ based on similarity is a collection of vertices V, a collection of edges E. Therefore, the task of clustering based on graph becomes the graph divided into disjoint subgraphs according to some criteria. The simplest objective function in graph clustering is to minimize the cost of cutting. The cutting cost is defined as the sum of all edges of the cut to be cut into different subgraph (partitions). But this objective function often leads to very uneven division, Therefore, in order to improve this problem, some new target functions such as proportional cutting, normalized cut and min-max cutting are proposed [5]. Min-max excision attempts at the minimum cost and maximizes the sum of the weights within a partition. If there is a certain degree coverage (this phenomenon is very common in text), the effect of minimal maximal cutting will be better [6] because the minimum maximum cutting objective function tends to balance the clustering, and has certain anti-interference to the imbalance of clustering results.

2 Related Work

2.1 New Model of Clustering

Given data set $X = \{x_1, x_2, \ldots, x_n\}$, where $x_i \in R^d$ and X true class is K, We want to cluster the data set, first build a similar connection diagram of the data set $G = (X, E)$, inside E is the collection of all edges of the connection, $E = \{(i,j)|W_{ij} \neq 0\}$, W is the corresponding similar graph matrix of the connected graph, where W_{ij} is similarity between x_i, x_j.

To set the data set X cluster K class, we want to look for one of its subgraphs from a similar diagram $G^* = (X, E_{G^*})$, where $E_{G^*} \subseteq E$, which contains a connected K subgraph, and the intra-class compact of the data is similarity and the inter-class size similarity. Finally, the data corresponding to all the vertices in the graph is aggregated into a class.

2.2 Compact Homogenizing of Class

2.2.1 The Establishment of Similar Graph Matrix

The edge set E of the graph G is take out the edge of the subgraph H of the edge E_H of the partially associated phase vertices. The edge set of the graph is take out the edge of the subgraph of the edge of the partially associated phase vertices, for the correlation degree of vertices remains unchanged, the vertice x_i of the graph G, when removing some edges associated with the vertex $\{(i,k) \in E|k \neq i\}$. The weights $\{W_{ik}|k \neq i\}$ of these removed edges are Again to weight. W_{ii},

$$W_{ii}^{H} = W_{ii} + \sum_{\{k|(i,k)\in E\setminus E_H\}} W_{ik} = W_i - \sum_{\{k|(i,k)\in E_H, i\neq k\}} W_{ik}$$

where the degree of the vertex x_i in the graph G is $W_i = \sum_{k=1}^{n} W_{ik}$. Defining the weight of the edge$\{(i,k) \in E|k \neq i\}$ is $W_{ik}^{H} = W_{ik}$, if $(i,j) \notin E_H$, $W_{ij}^{H} = 0$ $W_i^H = W_i = \sum_{k=1}^{n} W_{ik}$, indicates that the vertex x_i of G is in the same degree as in the subgraph H.

2.2.2 The Establishment of Compact Homogenizing Between Class

Definition: The probability distribution of the edge of any vertex x_i in a subgraph H is:

$$p_{ij}(H) = \frac{W_{ij}^{H}}{W_i^{H}} = \begin{cases} \frac{W_{ij}}{W_i}, & \text{if } i \neq j \text{ and } (i,j) \in E_H; \\ 0, & \text{if } i \neq j \text{ and } (i,j) \notin E_H; \\ 1 - \frac{\sum_{\{k|(i,k)\in H\}} W_{ik}}{W_i}, & \text{if } i = j. \end{cases}$$

Therefore, the distribution uniformity of the edges of the vertices x_i of the subgraph H is represented as [7]:

$$L_i(H) = - \sum_{\{j|(i,j)\in E_H\}} p_{ij}(H) \log p_{ij}(H).$$

Vector $\lambda = (\lambda_1, \lambda_2, \ldots, \lambda_n)$, where $\lambda_i = \frac{W_i}{W_T}, i = \{1, 2, \ldots, n\}, W_T = \sum_{j=1}^{n} W_j$

When λ_i is larger, the distribution of the vertices x_i of subgraph H in the neighborhood is more compact. For vertice x_i, definition:

$$B_i(H) = \lambda_i L_i(H) = -\lambda_i \sum_{\{j|(i,j)\in E_H\}} p_{ij}(H) \log p_{ij}(H).$$

Using $B_i(H)$ explains the compact uniformity of the vertices x_i in the subgraph H.

Definition:

$$B(H) = \sum_i B_i(H) = -\sum_i \lambda_i \sum_{\{j|(i,j)\in E_H\}} p_{ij}(H) \log p_{ij}(H).$$

If the subgraph H is composed of K_H connected subgraph $H_1, H_2, \ldots, H_{K_H}$, $B(H)$ can be written $\sum_{t=1}^{K_H} B(H_t)$, where H_t is the collection of edge of connected subgraphs,

$$B(H_t) = -\sum_i \lambda_i \sum_{\{j|(i,j)\in E_H\}} p_{ij}(H) \log p_{ij}(H).$$

When the function of $B(H)$ is bigger, the corresponding function $B(H_t)$ of each subgraph H_t is larger, the compact uniformity of the vertices of the graph H_t is better, so the compact uniformity of the data distribution can be described in the mean sense.

2.3 Size Similarity Between Class

The uniformity of the size distribution is called the size similarity of the subgraph between connected subgraph $H_1, H_2, \ldots, H_{K_H}$.

Let $V_H = \{V_{,1} V_2, \ldots, V_{K_H}\}$, V_i is set of vertices of connected subgraph H_i, X_H is a class random variable, The probability distribution is:

$$p(X_H = i) = \frac{|V_i|}{|X|}, i = \{1, 2, \ldots, K_H\}.$$

So size similarity of graph H denotes:

$$B(X_H) = -\sum_i p(X_H = i) \log p(X_H = i)$$

Without knowing the number of true categories, In order to ensure the size of the claim and make the number of cluster as small as possible, further correction $H(X_H)$, get a function which is better size similarity of the new subgraph H:

$$\begin{aligned} f(H) &= B(X_H) - K_A \\ &= -\sum_i p(X_H = i) \log p(X_H = i) - K_H \end{aligned}$$ <!endaligned>

2.4 Cluster Model

In order to reflect the clustering effect as much as possible, that is compact homogenizing between class and size similarity between class, The optimization model is as follows:

$$\begin{aligned} \max \quad & B(H) + \delta f(H) \\ s.t. \quad & E_H \subseteq E, K_H = K. \end{aligned} \quad (1)$$ <!endaligned>

Where $\delta \geq 0$, For convenience $F(H) = B(H) + \delta f(H)$.

To solve the model (1) easily, Make reasonable improvements to it. We defines two sets:

$$C_1 = \{H | K_H = K, E_H \subseteq E\},$$
$$C_2 = \{H | K_H > K, E_H \subseteq E\}$$

In fact, $B(H)$ and $f(H)$ are all [8], so linear combination of $B(H)$ and $f(H)$ is also monotonically increasing,

$$\max_{H \in C_2} F(H) < \max_{H \in C_1} F(H)$$

Hence, the model (1) is equivalent to mode:

$$\begin{aligned} &\max \; F(H) \\ &s.t.\, E_H \subseteq E, K_H \geq K \end{aligned}$$

The model is improved to:
$$\begin{aligned} &\max \; F(H) \\ &s.t.\, E_H \subseteq E, K_H \geq K \textit{ and } \text{H without connection loop} \end{aligned}$$

In fact, if $I_K = \{H|E_H \subseteq E, K_H \geq K \textit{ and } \text{H without connection loop}\}$. Then $M = (E, I_K)$ is matroid. The model is to solve the monotonically increasing submodular optimization problem of matroid constraints. The solution of such problems has been discussed [9].

3 Algorithm

Step 1 Put $S \leftarrow \Phi$.
Step 2 For each $i = 1, 2, \ldots, |E|$ do the following (2.1) and (2.2).
(2.1) If there exists no element $e \in E - S$ such that $S \cup \{e\} \in D$ and $x(S \cup \{e\}) = f(S \cup \{e\})$, then stop($x$ is not an extreme base)[10].
Otherwise find one such element $e \in E - S$ and put $S \leftarrow S \cup \{e\}$ and $e_i \leftarrow e$.
(2.2) Put $T \leftarrow S$.
For each $j = 1, 2, \cdots, i - 1$ do the following $(*)$.
If $x(T - \{e_{i-j}\}) = I(T - \{e_{i-j}\})$, then put $T \leftarrow T - \{e_{i-j}\}$.
Put dep $(x, e_i) \leftarrow T$.

References

1. Omran, M.G.H., Engelbrecht, A.P., Salman, A.: An overview of clustering methods[J]. Intell. Data Anal. **11**(6), 583–605 (2007)
2. Giancarlo, R., Utro, F.: Algorithmic paradigms for stability-based cluster validity and model selection statistical methods, with applications to microarray data analysis. Theoret. Comput. Sci. **428**(4), 58–70 (2012)
3. Bezdek, J.C.: Cluster validity with fuzzy sets. J. Cybern. **3**(3), 58–74 (1974)
4. Liang, J.Y., Zhao, X.W., Li, D.Y., et al.: Determining the number of clusters using information entropy for mixed data. Pattern Recognit. **45**(6), 2251–2265 (2012)
5. Luxburg, Ulrike: Max Plank. A tutorial on spectral clustering. Stat. Comput. **17**(4), 395–416 (2007)

6. Ding, C., He, X., Zh, H., et al.: A Min-Max cut algorithm for graph partioning and data clustering. Proceedings of ICDM, 107–114 (2001)
7. Hua, Le: Application of Spectral Clustering and Entropy in Clustering, pp. 25–29. Zhejiang University, Hangzhou (2014)
8. Liu, M.Y., Tuzel, O., Ramalingam, S., Chellappa, R.: Entropy rate superpixel segmentation. IEEE Conf. Computer vision and pattern recognition (2011)
9. Calinescu, G., Chekuri, C., Pal, M., Vondrak, J.: Maximizing a submodular set function subject to a matroid constraint. Integer programming and combinatorial optimization. Springer, New york (2007)
10. Fujishige, S.: Submodular functions and optimization, vol. 65 (1991)

Gibbs Phenomenon for Bi-orthogonal Wavelets

Jie Zhou(✉) and Hongchan Zheng

Department of Applied Mathematics, Northwestern Polytechnical University, Xi'an, Shaanxi 710072, People's Republic of China
zhjie@mail.nwpu.edu.cn

Abstract. In this paper, we consider that a Gibbs phenomenon exists for the biorthogonal wavelets expansions of a discontinuous function. Firstly, we study properties of the biorthogonal wavelets kernel. Based on these properties of the kernel, we discuss existence of this phenomenon for the biorthogonal wavelets expansions of a discontinuous function. Also, we present the necessary condition of this phenomenon existence for the biorthogonal wavelets expansions of a discontinuous function.

Keywords: Biorthogonal dual basis · Gibbs phenomenon · Scaling function

1 Introduction

It is well known that given arbitrary continuous and periodic functions, the Fourier series can be used very effectively to approximate these functions. However, when we approximate a discontinuous function by using the Fourier series, the overshoot around the discontinuous point can be occured. This phenomenon is the famous Gibbs phenomenon. It was pointed out by Gibbs [1], and gave the mathematical descriptions of this phenomenon in 1899.

A similar phenomenon exists for wavelets expansions. In 1991, Richards [2] showed a Gibbs phenomenon for periodic spline functions. Based on wavelet's multi-resolution analysis and the related knowledges of Daubechies wavelets, the Gibbs phenomenon for Daubechies wavelets can be presented in [3], and gave the conditions of this phenomenon existence for wavelets expansions. In the same time, for a wide class of wavelets, Shim and Volkmer [4] discussed the existence of this phenomenon. By using related knowledges of wavelet transforms, Karanikas [5] discussed that the phenomenon existed in continuous and discrete wavelet analysis. Ruch and Fleet [6] studied this phenomenon for scaling vectors, and illustrated that avoiding this phenomenon via positive scaling vectors. Shim and kim [7] discussed this phenomenon for sampling series in wavelet subspaces. For the Gibbs phenomenon of the periodic wavelet frame series, Zhang [8] gave analysis of integral representations of partial sums, and studied convergence of the periodic wavelet frame series.

However, in many applications, orthogonal MRAS are harder to be found, The Riesz or biorthogonal basis are usually considered. So, we require the maps to be oblique projections. The orthogonal requirement can be replaced by biorthogonal conditions. In this paper, we will consider that whether exist a Gibbs phenomenon for non-orthogonal wavelets expansions. This motivates us to study the Gibbs

F. Xhafa et al. (eds.), *Advances in Intelligent Systems and Interactive Applications*, Advances in Intelligent Systems and Computing 686,
https://doi.org/10.1007/978-3-319-69096-4_12

phenomenon for the dual basis. We generalize the result of [4] to the biorthogonal scaling functions, and analyze the existence of this phenomenon.

2 Preliminaries

To study the Gibbs phenomenon for the dual basis, we first give some basic definitions and results for Gibbs phenomenon.

We first must study this phenomenon for Fourier series. Considering a analytic and periodic function $g(x)$, let S_n be the nth partial sum of the Fourier expansion of the function $g(x)$. For any fixed x_1, we have $\lim_{n\to\infty} S_n(x_1) = g(x_1)$. However, whenever $g(x)$ has a jump discontinuity x_m, for any fixed × approach the discontinuity x_m, we have $\lim_{n\to\infty} S_n(x) \neq g(x_m)$. The overshoot or undershoot is called the Gibbs phenomenon. For more details for Fourier series, see [1]. Next, we consider this phenomenon for the biorthogonal wavelets expansions.

Assume that we have two MRAs $\{V_n\}$ and $\{\tilde{V}_n\}, n \in Z$ of the square integrable function $L_2(R)$, and generated by biorthogonal scaling functions $\phi(x)$ and $\tilde{\phi}(x)$.

For any function $g(x) \in L_2(R)$ admits the wavelet expansions

$$g(x) = \sum_{n,j} \langle g, \tilde{\phi}_{n,j} \rangle \phi_{n,j} = \sum_{n,j} \langle g, \phi_{n,j} \rangle \tilde{\phi}_{n,j} \tag{1}$$

The projection Q_n and $\tilde{Q}_n$ from $L_2(R)$ into $\{V_n\}$ and $\{\tilde{V}_n\}$ are given by

$$Q_n g = \sum_{n\to\infty} \langle g, \tilde{\phi}_{n,j} \rangle \phi_{n,j}, \quad \tilde{Q}_n g = \sum_{n\to\infty} \langle g, \phi_{n,j} \rangle \tilde{\phi}_{n,j}.$$

The projections are now oblique projections, instead of orthogonal projection. Next, we analyze the convergence of a partial sum of wavelet expansions (1). The convergence plays a central role in the research of the Gibbs phenomenon. Similar to the results of orthogonal wavelet expansions, we have that oblique projections $Q_n g$ converges to $g(x)$ as $n\to\infty$. Similarly, $\tilde{Q}_n g$ converges to $g(x)$ as $n\to\infty$. For more detail about the convergence wavelet expansion, see [10]. Now, we can use $Q_n g$ as the partial sum of biorthogonal expansions considering the Gibbs phenomenon. It is possible to give integral representations of $Q_n g$, we have

$$Q_n g(x) = \int_{-\infty}^{+\infty} 2^n \, d(2^n x, 2^n y) \, g(y) \, dy, \tag{2}$$

where the kernel

$$d(x, y) = \sum_{j\in Z} \phi(x - j)\tilde{\phi}(y - j). \tag{3}$$

The kernel is very important roles for Gibbs phenomenon. Some questions of Gibbs phenomenon can be given by analyzing their kernel. Next, we state some properties of the kernel $d(x, y)$ in following lemmas.

Lemma 2.1 Let $\phi(x), \tilde{\phi}(y)$ be continuous scaling functions, and satisfy

$$|\phi(x)| \leq M(1+|x|)^{-2}, \qquad \tilde{\phi}(y) \leq M'(1+|y|)^{-2}, \tag{4}$$

for $x, y \in R$, and constants M, M'. Then the kernel $d(x, y)$ is continuous and satisfies

$$|d(x, y)| \leq J(1+|x-y|)^{-2}, \qquad \text{for all } x, y \in R, \tag{5}$$

where J is a constant.

Proof According to the definition of the kernel, it converges locally uniformly, since $\phi(x), \tilde{\phi}(y)$ are continuous, so $d(x, y)$ is continuous too. Assume $|x+y| \leq 1,$ *and* $x \geq y.$ If $j \geq 0$, we have $d(x+1, y+1) = d(x, y) = d(y, x)$, then

$$\begin{aligned} |d(x,y)| &= |\sum_{j \in Z} \phi(x-j)\tilde{\phi}(y-j)| \leq |\sum_{j \in Z} \phi(x-j)||\tilde{\phi}(y-j)| \\ &\leq \sum_{j \in Z} M(1+|x-j|)^{-2} M'(1+|y-j|)^{-2}. \end{aligned}$$

Since $(1+|x-j|)\,(1+|y-j|) \geq \frac{1}{4}\,(1+x-y)(1+|x-y-2j|)$, we have

$$\begin{aligned} |d(x,y)| \leq MM' 4^2 (1+(x-y))^{-2} &\sum_{j \geq 0} (1+|x-y-2j|)^{-2} \\ &+ \sum_{j<0} (1+|x-y-2j|)^{-2}. \end{aligned}$$

According to the sum $\sum_{j \in Z} (1+|z-2j|)^{-2}$ is bounded, this result can be completes proved.

Lemma 2.2 [10] Let $d_n(x, y) = 2^n d(2^n x, 2^n y)$ be the kernel of V_n, then $\{d_n(x, y)\}$ satisfying

(1) There exist a constant $b > 0$, such that $\int_{-\infty}^{+\infty} |d_n(x, y)| dx \leq b, \quad y \in R, n \in N.$
(2) There exist a constant $b > 0$, such that $\int_{y-b}^{y+b} |d_n(x, y)| dx \to 1$, uniformly on compact subsets of R, as $n \to \infty$.
(3) For each $r > 0$, $\sup_{|x-y| \geq r} |d_n(x, y)| \to 0 \quad as\ n \to \infty.$

If the kernel $d_n(x, y)$ satisfying (1), (2), (3) of Lemma 2.2, we called that it is a quasi-positive delta kernel sequence. For more the details and proof, we can see [10].

Lemma 2.3 Let $g(x) \in L_2(R)$ be a real valued continuous function, Let $d_n(x, y)$ be a quasi-positive delta sequence, then $\lim_{n \to \infty} Q_n g(y) = \int_{-\infty}^{+\infty} d_n(x, y) g(x) dx = g(y)$.

3 Gibbs Phenomenon of Biorthogonal Wavelet Expansion

Based on the properties of the kernel functions, in this section, we will discuss the Gibbs phenomenon for the biorthogonal wavelet expansion. First, we give the following precise definition.

Definition 3.1 [4] Let $g : R \to R$ be a square integrable bounded function with a jump discontinuity at 0, the limits $\lim_{x\to 0^+} g(x) = g(0+)$ *and* $\lim_{x\to 0^-} g(x) = g(0-)$ exist and are different. Without loss of generality we assume $g(0+) > g(0-)$. For the given scaling function $\phi(x), \tilde{\phi}(x)$ and an oblique projection $Q_n g(x)$, we say that the biorthogonal wavelet expansion of $g(x)$ exhibit a Gibbs phenomenon at 0, if there exist a positive sequence $\{x_n\}$ with $\lim_{n\to\infty} x_n = 0$ such that $\lim_{n\to\infty} Q_n g(x_n) > g(0+)$, or if there exist a negative sequence $\{y_n\}$ with $\lim_{n\to\infty} y_n = 0$ such that $\lim_{n\to\infty} Q_n g(y_n) < g(0-)$.

Theorem 3.1 Let $\phi(x)$ be a continuous compactly supported scaling function, and has a dual biorthogonal basis $\tilde{\phi}(x)$. The function $g(x)$ satisfy the conditions of the Definition 3.1. Set $g(0+) - g(0-) = 2h > 0$ *for* $h \in N$.

(1) If for some $a > 0$, $\int_{-\infty}^{0} |d_n(t,a)|dt < 0$, then $\lim_{n\to\infty} Q_n g(\frac{a}{n}) > g(0+)$.

(2) If for some $a < 0$, $\int_{-\infty}^{0} |d_n(t,a)|dt > 0$, then $\lim_{n\to\infty} Q_n g(\frac{a}{n}) < g(0-)$.

Proof We consider the biorthogonal wavelet expansions of this function $g(x) = \begin{cases} 1 & 0 < x \le 1, \\ -1 & -1 \le x < 0. \end{cases}$

Since the oblique projection $Q_n g$ converges to $g(x)$ as $n \to \infty$, for the given scaling function, we know that the oblique projection $Q_n g$ is too a partial sum of the biorthogonal wavelet expansion of this function. By (2, 3), we have

$$d_n(2^{-n}a, y) = 2^n \sum_{j\in Z} \phi(a-j)\tilde{\phi}(2^n y - j) = 2^n d(a, 2^n y).$$

$$Q_n g(2^{-n}a) = \sum_{j\in Z} \int_{-\infty}^{+\infty} g(y) 2^n d(a, 2^n y)dy = \int_0^1 2^n d(a, 2^n y)dy - \int_{-1}^{0} 2^n d(a, 2^n y)dy$$
$$= \int_0^{2^n} d(a,t)dt - \int_{-2^n}^{0} d(a,t)dt.$$

Since $\int_{-\infty}^{+\infty} d(a,t)dt = 1$, we have $Q_n g(2^{-n}a) = 1 - 2\int_{-\infty}^{0} d(a,t)dt + o(1)$. For some $a > 0$, if $\int_{-\infty}^{0} d(a,t)dt = -r < 0$, then we has $Q_n g(2^{-n}a) = 1 + 2r + o(1)$.

Hence, $\lim_{n\to\infty} Q_n g(2^{-n}a) = 1 + 2r > 1 = g(0+)$. (1) follows. Similarly, we can get (2)

Corollary 3.1 Let $g(x)$ be a square integrable bounded function with a jump discontinuity at 0. Let $\phi(x)$ be a real valued function on R, and has a dual biorthogonal basis $\tilde{\phi}(x)$. Set the biorthogonal wavelet expansion of $g(x)$ exhibit a Gibbs phenomenon at 0, then we have

$$\int_{-\infty}^{0} d(a,y)dy < 0,\ for\, a > 0, or \int_{-\infty}^{0} d(a,y)dy > 0,\ for\, a < 0.$$

4 Conclusion

In this paper, we discuss the Gibbs phenomenon for the biorthogonal wavelets expansions of a discontinuous function. Based on MRA, we generalize the results of the orthogonal basis to the biorthogonal basis, and discuss the existence of this phenomenon. First, we gave properties of the kernel, based in these properties of the kernel, we proved the existence of this phenomenon for the biorthogonal wavelets expansions of a discontinuous function. Moreover, we obtained the necessary condition of the existence of this phenomenon. In the future, we will focus on the existence of this phenomenon for vector wavelets. On the other hand, by considering the mask of scaling functions, we will study the conditions of the existence of Gibbs phenomenon.

Acknowledgments. This work is supported by Natural Science Basic Research Plan in Shaanxi Province of China (Program No, 2016, JM6065)

References

1. Gibbs, J.W.: Letter to the editor. Nat. (London) **59**, 606 (1899)
2. Richards, F.B.: A Gibbs phenomenon for spline functions. J. Approx. Theory. **66**, 334–351 (1991)
3. Lelly, S.: Gibbs phenonmenon for wavelets. Appl. Comput. Harmoninc Anal. **3**, 72–81 (1996)
4. Shim, H., Volkmer, H.: On the Gibbs phenomenon for wavelet expansions. J. Approx. Theory. **84**, 74–95 (1996)
5. Karanikas, C.: Gibbs phenomenon in wavelet analysis. Result. Math. **34**, 330–341 (1998)
6. Ruch, David K., Van Fleet, Patrick J.: Gibbs'phenomenon for nonnegative compactly supported scaling vectors. J. Math. Appl. **304**, 370–382 (2005)
7. Hong, T.S., Hong, O.K.: On Gibb's phenomenon for samping series in wavelet subspaces. App. Anal. **61**, 97–109 (1996)
8. Zhang, Z.W.: Convergence and Gibbs phenomenon of periodic wavelets frame series. Rocky Mt. J. Math. **39**, 4 (2009)
9. Gilbert, G.W.: Pointwise convergence of wavelet expansions. J. Approximation Theory. **80**, 108–118 (1995)

α-Convergence Theory

Chun-Qiu Ji[1], Zhen-Guo Xu[2(✉)], and Yi Wang[2]

[1] School of Mathematical Sciences, Mudanjiang Normal University, Mudanjiang 157000, China
[2] National Science and Technology Infrastructure Center, Beijing 100862, China
zhenguoxu@126.com

Abstract. In this paper, we give α-convergence theory of nets, ideals and filters by means of the concept of α-closed L-sets. Its applications are presented.

Keywords: L-topological space · α-closed remote sets · α-convergence

1 Introduction

In this paper, we give the α-convergence theory of ideals, filters, nets based on the idea of [2].

An element k in L is called co-prime if k' is a prime element. An element k in L is called prime if $k \geq l \wedge h$ implies $k \geq l$ or $k \geq h$. An L-topological space (or L-ts for short) is a pair (X, τ) where τ is a subfamily of L^X which contains $\underline{0}$, $\underline{1}$ and is closed for any suprema and finite infima. τ is called an L-topology on X. Each member of τ is called an open L-set and its complement is called a closed L-set.

Definition 1.1. [6]. Suppose (X, τ) be an L-ts and $B \in L^X$. If $B \leq \mathrm{int}(cl(\mathrm{int}(B)))$, then B is called α-open; B is called α-closed if B' is α-open.

Definition 1.2. [6]. Suppose (X_1, τ) and (X_2, σ) be two L-tss, map $f : X \to Y$. If for every $B \in \alpha O(Y)$, we have $f_L^{\leftarrow}(B) \in \alpha O(X)$, then we say that f is α-irresolute.

Definition 1.3. [6]. Suppose $K \in L^X$. Define:

(1) $bb_\alpha(K) = \cap\{\Lambda | \Lambda \geq K, \Lambda \text{ is } \alpha\text{-closed}\}$;
(2) $nb_\alpha(K) = \cup\{\Pi | \Pi \leq K, \Pi \text{ is } \alpha\text{-open}\}$.

$nb_\alpha(K)$ and $bb_\alpha(K)$ are called α-nb and α-bb of K.

2 α-Adherence Dots and α-Accumulation Dots

Definition 2.1. Suppose (Y, ς) be an L-ts, $y_\lambda \in F(L^Y)$, $Q \in L^Y$. If $y_\lambda \not\leq Q$, then we say that Q is a remote set of y_λ. Define $\eta_\alpha(y_\lambda) = \{Q | Q \text{ is a } \alpha\text{-closed remote set of } y_\lambda\}$.

Definition 2.2. Suppose (Y, ς) be an L-ts, $G \in L^Y$, $y_\lambda, y_\mu \in F(L^Y)$. If $G \not\leq Q$ for each $Q \in \eta_\alpha(y_\lambda)$, then we say that x_λ is an α-adherence dot of G.

F. Xhafa et al. (eds.), *Advances in Intelligent Systems and Interactive Applications*, Advances in Intelligent Systems and Computing 686,
https://doi.org/10.1007/978-3-319-69096-4_13

An α-adherence dot y_λ of G is called an α-accumulation dot of G if $y_\lambda \not\leq G$ or $y_\lambda \leq G$ implies that for each dot y_μ with $y_\lambda \leq y_\mu \leq G$ such that $G \not\leq y_\mu \vee P$. Define $G^{d_\alpha} = \cup\{$ all α-accumulation dots of $G\}$, G^{d_α} is called the α-derived set of G.

Theorem 2.3. Suppose (Y, ς) be an L-ts, $G \in L^Y$ and $y_\lambda \in F(L^Y)$. Then

(1) y_λ is an α-adherence dot of G if and only if $y_\lambda \leq bb_\alpha(G)$;
(2) $cl_\alpha(G) = \cup\{$all α-adherence dots of $G\}$;
(3) $cl_\alpha(G) = G \vee G^{d_\alpha}$;
(4) $cl_\alpha(G^{d_\alpha}) \leq cl_\alpha(G)$.

Proof. Omitted.

Theorem 2.4. Suppose (Y, ς) be an L-ts $\Delta \in L^Y$. So G is α-closed iff for every dot $\beta_\lambda \leq \Delta$, have $Q \in \eta_\alpha(\beta_\lambda)$ ensure $\Delta \leq Q$.

Proof. We only prove the sufficiency. Assuming for each dot $\beta_\lambda \leq \Delta$, there exists Q belong to $\eta_\alpha(\beta_\lambda)$ such that Δ less than or equal to Q, i.e., have $Q \in \eta_\alpha(\beta_\lambda)$ ensure $cl_\alpha(\Delta) \leq Q$. Then $\beta_\lambda \leq cl_\alpha(\Delta)$. Hence above statement implies that $\beta_\lambda \leq \Delta$ push onto $\beta_\lambda \leq cl_\alpha(\Delta)$. So $G \geq cl_\alpha(\Delta)$. Therefore Δ is α-closed.

3 α-Convergence of Nets

In this section, we shall discuss α-convergence for nets.

Definition 3.1. Suppose (Y, ς) be an L-ts, $G \in L^Y$, $y_\lambda \in F(L^Y)$ and W be a net. So

(1) If for every $Q \in \eta_\alpha(y_\lambda)$, $W(n) \leq Q$ is final true, then we say that y_λ is an α-limit dot of W, remember to $W \xrightarrow{\alpha} y_\lambda$;
(2) If for every $Q \in \eta_\alpha(y_\lambda)$, $W(n) \leq Q$ is often true, then we say that y_λ is an α-cluster dot of W, remember to $W \overset{\alpha}{\infty} y_\lambda$.

Define $\lim_\alpha W = \vee\{y_\lambda | y_\lambda$ is an α-limit dot of $W\}$; $ad_\alpha W = \vee\{y_\lambda |\ y_\lambda$ is an α-cluster dot of $W\}$. Obviously $\lim_\alpha(W) \leq ad_\alpha(W)$.

Theorem 3.2. Suppose (Y, ς) be an L-ts, $\beta_\lambda \in F(L^Y)$, $W = \{W(a) | a \in Z\}$ a net in L^X.

(1) Let $K = \{K(a) | a \in Z\}$ and for every $a \in Z$, $W(a) \leq K(a)$. If $W \xrightarrow{\alpha} \beta_\lambda$, then $K \xrightarrow{\alpha} \beta_\lambda$;
(2) Let $K = \{K(a) | a \in Z\}$ be and for every $a \in Z$, $W(a) \leq K(a)$. If $W \overset{\alpha}{\infty} \beta_\lambda$, then $K \overset{\alpha}{\infty} \beta_\lambda$;
(3) If $W \xrightarrow{\alpha} \beta_\lambda$ and $\beta_\mu \leq \beta_\lambda$, then $W \xrightarrow{\alpha} \beta_\mu$;
(4) If $W \overset{\alpha}{\infty} \beta_\lambda$ and $\beta_\mu \leq \beta_\lambda$, then $W \overset{\alpha}{\infty} \beta_\mu$.

Theorem 3.3. Suppose (Y, ς) be an L-ts, $\beta_\lambda \in F(L^Y)$, $W = \{W(a)|a \in Z\}$ a net in L^X. So

(1) $W \overset{\alpha}{\to} \beta_\lambda$ iff $\beta_\lambda \leq \lim_\alpha W$;
(2) $W \overset{\alpha}{\infty} \beta_\lambda$ iff $\beta_\lambda \leq \mathrm{ad}_\alpha W$.

Proof.

(1) The necessity is obvious. Suppose that $y_\lambda \leq \lim_\alpha W$ and $Q \in \eta_\alpha(y_\lambda)$. Then $\lim_\alpha W \leq Q$. By $\lim_\alpha W$, we obtain an **α**-limit dot δ of W insure $\delta \leq Q$, id est, $Q \in \eta_\alpha(\delta)$. By δ is an **α**-limit dot of W, we obtain W is eventually not in Q, So $W \overset{\alpha}{\to} y_\lambda$.
(2) Similar to the proof of (1), we can prove (2).

Theorem 3.4. Suppose $\beta_\lambda \in F(L^Y)$, $W = \{W(a)|a \in Z\}$ a net in L^X. If W has a subnet K satisfy $K \overset{\alpha}{\to} \beta_\lambda$, then $W \overset{\alpha}{\infty} \beta_\lambda$.

Proof. Omitted.

Theorem 3.5. Suppose $W = \{W(a)|a \in Z\}$ a net in L^X. Then $\lim_\alpha W$ is **α**-closed and $ad_\alpha W$ too is.

Proof. Let $\beta_\lambda \leq cl_\alpha(\lim_\alpha W)$. Then for each $Q \in \eta_\alpha(\beta_\lambda)$. So have a dot δ satisfy $\delta \leq \lim_\alpha W$, $\delta \leq Q$. therefore $Q \in \eta_\alpha(\delta)$. By Theorem 3.4 $W \overset{\alpha}{\to} \delta$. Hence $W \leq Q$ is eventually true. Thus $\beta_\lambda \leq \lim_\alpha W$. This implies that $\lim_\alpha W$ are **α**-closed. Similarly $ad_\alpha W$ are **α**-closed.

Theorem 3.6. Suppose $\Delta \in L^X$, $\beta_\lambda \in F(L^X)$.Suppose that in Δ exists a net $W = \{W(a)|a \in Z\}$ ensure $W \overset{\alpha}{\infty} \beta_\lambda$, we have $\beta_\lambda \leq bb_\alpha(\Delta)$.

Theorem 3.7. Suppose $\Delta \in I^X$ $(I = [0, 1])$, $y_\lambda \in F(I^X)$ in Δ. Suppose that in $\Delta - \beta_\lambda$ exists a net $W = \{W(a)|a \in Z\}$ such that $W \overset{\alpha}{\to} \beta_\lambda$, then β_λ is an **α**-accumulation dot of Δ.

4 α-Convergence of Ideals

Definition 4.1. Suppose (Y, ς) be an L-ts, I be an ideal in L^Y and $y_\lambda \in F(L^Y)$.

(1) If $\eta_\alpha(y_\lambda) \subset I$, then we say that y_λ is **α**-limit dot of I, remember to $I \overset{\alpha}{\to} y_\lambda$;
(2) If for every $G \in I$ and every $\Lambda \in \eta_\alpha(y_\lambda)$, it follows that $G \vee \Lambda \neq 1$, then we say that y_λ is **α**-cluster dot of I, remember to $I \overset{\alpha}{\infty} y_\lambda$.

Define $\lim_\alpha I = \vee\{y_\lambda | y_\lambda$ is an **α**-limit dot of $I\}$; $ad_\alpha I = \vee\{y_\lambda | \ y_\lambda$is an **α**-cluster dot of $I\}$. Obviously $\lim_\alpha I \leq ad_\alpha I$.

Theorem 4.2. Suppose I and S be two ideals in L^X and $I \subset S$ and $y_\lambda, y_\mu \in F(L^X)$. So

(1) $I \overset{\alpha}{\rightarrow} y_\lambda$ implies $S \overset{\alpha}{\rightarrow} y_\lambda$;
(2) $S \overset{\alpha}{\infty} y_\lambda$ implies $I \overset{\alpha}{\infty} y_\lambda$;
(3) If $I \overset{\alpha}{\rightarrow} y_\lambda$ and $y_\mu \leq y_\lambda$, then $I \overset{\alpha}{\rightarrow} y_\mu$;
(4) If $I \overset{\alpha}{\infty} y_\lambda$ and $y_\mu \leq y_\lambda$, then $I \overset{\alpha}{\infty} y_\mu$.

Theorem 4.3. Suppose $\Delta \in L^X$, $y_\lambda \in F(L^Y)$. If have an idea I in L^Y ensure $G \not\in I$, $I \overset{\alpha}{\rightarrow} y_\lambda$, then $y_\lambda \leq bb_\alpha(G)$.

Proof. Suppose that $I \overset{\alpha}{\rightarrow} y_\lambda$ and $G \not\in I$. Let $Q \in \eta_\alpha(y_\lambda)$, by $\eta_\alpha(y_\lambda) \subset I$ and I is a lower set, we know that $G \not\leq Q$, so y_λ is an α-adherence dot of G, therefore $y_\lambda \leq bb_\alpha(G)$.

The followings Theorems 4.4–4.6 can be proved.

Theorem 4.4. Suppose $G \in I^X(I = [0,1])$, $y_\lambda \in F(I^X)$ in Δ and $y_\lambda \leq \Delta$. If have an ideal I in I^X ensure $\Delta - y_1 \notin I$, $I \overset{\alpha}{\rightarrow} y_\lambda$, therefore y_λ is an α-accumulation dot of G.

Theorem 4.5. Suppose (Y, ς) be an L-ts, I be an ideal in L^Y and $y_\lambda \in F(L^Y)$. So

(1) $I \overset{\alpha}{\rightarrow} y_\lambda \Leftrightarrow y_\lambda \leq \lim_\alpha I$;
(2) $I \overset{\alpha}{\infty} y_\lambda \Leftrightarrow y_\lambda \leq \mathrm{ad}_\alpha I$.

Theorem 4.6. Suppose I be an ideal in L^X. Then $\lim_\alpha I$, $ad_\alpha I$ are α-closed.

5 α-Convergence of Filters

Definition 5.1. Suppose (Y, ς) be an L-ts, $y_\lambda \in F(L^Y)$, $Q \in L^X$. Q is said a quasi set of y_λ if $y_\lambda \not\leq Q'$, in this case, we also say that y_λ quasicoincides with Q and it denote by $y_\lambda \hat{q} Q$. A quasi set Q of y_λ is called a α-open quasi set of y_λ if Q is α-open. The set of all α-open quasi sets of $y_\lambda$is denoted by $\Omega_\alpha(y_\lambda)$.

Remark 5.2. From the above definition, we know that if $\Sigma, \Upsilon \in L^{\tilde{X}}, \Sigma \leq \Upsilon$, $y_\lambda \in F(L^{\tilde{X}})$ and $x_\lambda \hat{q} \Sigma$, then $x_\lambda \hat{q} \Upsilon$.

Definition 5.3. Suppose (Y, ς) be an L-ts, Γ be a proper filter in $F(L^Y)$.

(1) If for every $U \in \Omega_\alpha(\delta)$ and every $A \in \Gamma$, it follows that $U \wedge A \neq \underline{0}$, then we say that δ is an α-cluster dot of Γ, remember to $\Gamma \overset{\alpha}{\infty} \delta$, we also called Γ α-accumulates to δ.
(2) If $\Omega_\alpha(\delta) \subset P$, then we say that δ is an α-limit dot of Γ, remember to $\Gamma \overset{\alpha}{\infty} \delta$.

Define: $\mathrm{ad}_\alpha \Gamma = \cup\{$ all α-cluster dots of $\Gamma\}$ and $\lim_\alpha \Gamma = \cup\{$ all α-limit dots of $\Gamma\}$.

Theorem 5.4. Suppose (Y, ς) be an L-ts, Γ be a proper filter in L^Y and $\delta \in F(L^Y)$. So

(1) $\Gamma \overset{\alpha}{\to} \delta \Rightarrow \Gamma \overset{\alpha}{\infty} \delta$;
(2) $\lim_\alpha \Gamma \leq ad_\alpha \Gamma$;
(3) $\Gamma \overset{\alpha}{\to} \delta,\ \lambda \leq \delta \Rightarrow \Gamma \overset{\alpha}{\to} \lambda$;
(4) $\Gamma \overset{\alpha}{\infty} \delta,\ \lambda \leq \delta \Rightarrow \Gamma \overset{\alpha}{\infty} \lambda$;
(5) $\Gamma \overset{\alpha}{\to} \delta \Leftrightarrow \delta \leq \lim_\alpha \Gamma$;
(6) $\Gamma \overset{\alpha}{\infty} \delta \Leftrightarrow \delta \leq \mathrm{ad}_\alpha \Gamma$.

Definition 5.5. Suppose (Y, ς) be an L-ts, Γ, Φ be two proper filters in L^Y and $\delta \in F(L^Y)$. Say Φ is finer than Γ or say Γ is coarser than Φ, if $\Gamma \subset \Phi$.

Theorem 5.6. Suppose (Y, ς) be an L-ts, Γ, Φ be two proper filters in L^Y and $\delta \in F(L^Y)$. So

(1) $\mathrm{ad}_\alpha \Phi \leq \mathrm{ad}_\alpha \Gamma$;
(2) $\lim_\alpha \Gamma \leq \lim_\alpha \Phi$;
(3) If $\Phi \overset{\alpha}{\infty} e$, then $\Gamma \overset{\alpha}{\infty} e$;
(4) If $\Gamma \overset{\alpha}{\to} e$, then $\Phi \overset{\alpha}{\to} e$.

Proof. Omitted because it is simple.

6 Relations Among Nets, Ideals, Filters

In this section, we discuss relations among nets, ideals and filters.

Theorem 6.1. Suppose (Y, ς) be an L-ts, I be an ideal in L^Y, W be a net in L^Y. So

(1) $\lim_\alpha I = \lim_\alpha W(I)$;
(2) $ad_\alpha I = ad_\alpha W(I)$;
(3) $\lim_\alpha W = \lim_\alpha I(W)$;
(4) $ad_\alpha W \leq ad_\alpha I(W)$.

Proof. Let $\lambda \leq \lim_\alpha I$. Then $I \overset{\alpha}{\to} \lambda$, so $Q \in I$ for each $Q \in \eta_\alpha(\lambda)$. Hence $(\lambda, Q) \in Z(I)$. Suppose $(\delta, \Delta) \in Z(I)$, $(\delta, \Delta) \geq (\lambda, Q)$, therefore we have $W(I)(\delta, \Delta) = \delta \leq \Delta \geq Q$. So $W(I)$ is not eventually in Q for every $Q \in \eta_\alpha(\lambda)$, i.e., $W(I) \overset{\alpha}{\to} \lambda$. Conversely, let $\lambda \leq \lim_\alpha W(I)$. Then $W(I) \overset{\alpha}{\to} \lambda$. So for every $Q \in \eta_\alpha(\lambda)$ there exists $(\delta, \Delta) \in Z(I)$ such that $W(I)(c, H) = c \leq Q$ whenever $(c, H) \geq (\delta, \Delta)$ and $(c, H) \in W(I)$. In particular, take $H = \Delta$, we know that $c \leq \Delta$ implies $c \not\leq P$, i.e., $c \leq Q$ implies $c \leq \Delta$. Hence $Q \leq \Delta$. Because I is a down set, $\Delta \in I$, so $Q \in I$. Hence $\eta_\alpha(\lambda) \subset I$. Hence $I \overset{\alpha}{\to} \lambda$. From Theorem 4.5 we obtain $b \leq \lim_\alpha I$. Thus (1) holds.

Theorem 6.2. Suppose (Y, ς) be an L-ts, W be a net in L^Y, Γ be a proper filter in L^Y, $\delta \in F(L^Y)$. So

(1) $W \stackrel{\alpha}{\rightarrow} \delta \Leftrightarrow \Gamma(W) \stackrel{\alpha}{\rightarrow} \delta$;
(2) $\Gamma \stackrel{\alpha}{\rightarrow} \delta \Leftrightarrow W(\Gamma) \stackrel{\alpha}{\rightarrow} \delta$
(3) $\Gamma \stackrel{\alpha}{\infty} \delta \Leftrightarrow W(\Gamma) \stackrel{\alpha}{\infty} \delta$;
(4) $W \stackrel{\alpha}{\infty} \delta$ implies $\Gamma(W) \stackrel{\alpha}{\infty} \delta$.

7 Applications of α-Convergence Theory of Nets

We can prove the followings theorems.

Theorem 7.1. Suppose (X_1, τ) and (X_2, σ) be two L-tss. A mapping f is α-irresolute if f $cl_\alpha(f_L^{\leftarrow}(B)) \in \eta_\alpha(x_\lambda)$ for each $B \in \eta_\alpha(f_L^{\rightarrow}(x_\lambda))$, where $x_\lambda \in F(L^{X_1})$.

Theorem 7.2. Suppose (X_1, τ) and (X_2, σ) be two L-tss. f is a mapping and it is α-irresolute. If a net $W \stackrel{\alpha}{\rightarrow} x_\lambda$ in L^{X_1}, then $f_L^{\rightarrow}(W) \stackrel{\alpha}{\rightarrow} f_L^{\rightarrow}(x_\lambda)$ in L^{X_2}.

Theorem 7.3. Suppose (X_1, τ) and (X_2, σ) be two L-tss. A mapping f is α-irresolute, then

(1) $f_L^{\rightarrow}(\lim_\alpha W) \leq \lim_\alpha f_L^{\rightarrow}(W)$ for every net W in L^{X_1};
(2) $\lim_\alpha f_L^{\leftarrow}(K) \leq f_L^{\leftarrow}(\lim_\alpha K)$ for every net K in L^{X_2}.

References

1. Pu, B.-M., Liu, Y.-M.: Fuzzy topology I. Neighborhood structure of a fuzzy dot and Moore-Smith convergence. J. Math. Anal. Appl. **76**, 571–599 (1980)
2. Wang, G.-J.: Theory of L-Fuzzy Topological Spaces. Shaanxi Normal University Press, Xi'an (1988). (in Chinese)
3. Bai, S.-Z.: Q-convergence of fuzzy nets and weak separation axioms in fuzzy lattices. Fuzzy Sets Syst. **88**, 379–386 (1997)
4. Shi, F.-G., Zheng, C.-Y.: O-convergence of fuzzy nets and its applications. Fuzzy Sets Syst. **140**, 499–507 (2003)
5. Chen, S.-L., Cheng, J.-S.: θ-convergence of nets of L-fuzzy sets and its applications. Fuzzy Sets Syst. **86**, 235–240 (1997)
6. Singal, M.K.: Fuzzy alpha-sets and alpha-continuous maps. Fuzzy Sets Syst. **48**, 383–390 (1992)

Feature Selection Optimization Based on Atomic Set and Genetic Algorithm in Software Product Line

Zhijuan Zhan[1(✉)], Weilin Luo[2], Zonghao Guo[2], and Yumei Liu[2]

[1] Science and Technology on Avionics Integration Laboratory, China National Aeronautical Radio Electronics Research Institute, Shanghai, China
zzjaldm@aliyun.com

[2] Department of Computer Science and Technology, Nanjing University of Aeronautics and Astronautics, NangJing, JiangSu, China

Abstract. Software product line (SPL) engineering is an effective method to improve the software development process in terms of development costs and time-to-market by using comprehensive software reuse technology. The feature model is a demand model that describes the common and variability of software product family and the relationship between features in SPL engineering. The difficulty of product configuration based on the feature model is how to choose the optimal combination of features from the complex feature model to satisfy the constraints. In order to achieve the problem of constrained feature selection optimization, we propose a method based on atomic set and a genetic algorithm to optimize feature selection. Firstly, the feature model is optimized by using the atomic set algorithm. Then, the whole constraints of the model are modeled as the evaluation function of the effective and invalid configuration in the genetic algorithm. Finally, by the genetic operations of combining the effective configuration and the invalid configuration, such as crossover, selection and mutation, it selects the best effective configuration for output.

Keywords: Feature model · Atomic set · Genetic algorithm · Optimized feature selection

1 Introduction

Software product line (SPL) [1] engineering is a software development process, which improves the reuse rate of components and the efficiency of software development. In SPLE, Feature Model [2] is a requirement model describing variability, commonality as well as the relationships among features in software product lines. It guides the user to select valid product configurations satisfying the constraints in the process of configuration. In SPLE, product configuration is a complicated process. So, how to find an optimized product configuration is a hot topic of research at home and abroad. Genetic algorithm [3] is proposed as a global heuristic search algorithm by reference to biological evolution process. It is an effective method [4, 5] to optimize the feature selection process with resource constraints by using a genetic algorithm.

F. Xhafa et al. (eds.), *Advances in Intelligent Systems and Interactive Applications*, Advances in Intelligent Systems and Computing 686,
https://doi.org/10.1007/978-3-319-69096-4_14

In view of the above problems, we propose a method to optimize feature selection based on atomic set and genetic algorithm. Firstly, This method uses the atomic set to optimize the feature model. Secondly, based on atomic set model, we model the integrity constraints of the feature model which is employed as a fitness function to divide the valid configuration and invalid configuration. Thirdly, through crossover between a valid configuration and a invalid configuration, adaptive selection as well as mutation operation, it will converge to the optimal solution rapidly with the combination of integrity constraints function of the feature model and the crossover and mutation operator of genetic algorithm. Finally, we find an optimized feature selection as the optimal solution that minimizes or maximizes an objective function.

The remainder of the paper is structured as follows: Sect. 2 introduces the related theoretical knowledge. Section 3 presents the atomic set computation. Section 4 details the implementation process of the genetic algorithm for optimized feature selection. In Sect. 5, the experiment and analysis of the different scale feature model are presented. The related work is discussed in Sect. 6. Finally, we summary our conclusion and future work in Sect. 7.

2 Basic Concept

2.1 Feature Models

Feature model is the requirement model which is widely used for describing the variability and commonality in software product lines as well as the relationships among features. Feature mode is represented by a tree structure called feature diagram. In feature diagram [6], there is only a root node usually indicating a domain system. Typical node represents a feature identified by feature name. Edges represent top-down decomposition relation of feature.

Figure 1 refers to feature diagram in literature [7]. The feature diagram depicts the variability and commonality in mobile phones product lines.

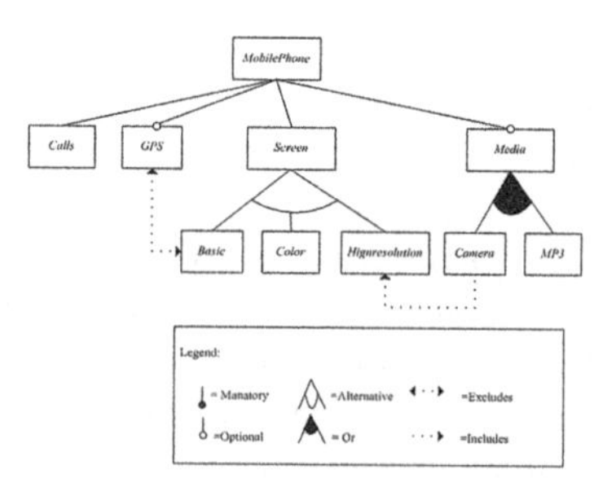

Fig. 1. Mobile phone feature model

2.2 Atomic Sets

The use of atomic sets [8] for the automated analysis of feature models was proposed by Zhang et al. in 2004. An atomic set represents a group of features (at least one) that can be treated as a unit during the analysis of a feature model. Those features can never appear in a product separately. For a given feature model, The intuitive idea of atomic sets is that mandatory features and their parent features can be considered as a whole. Taking the feature model of Fig. 1 as an example, the areas delimited by dashed lines illustrate the atomic sets (Fig. 2).

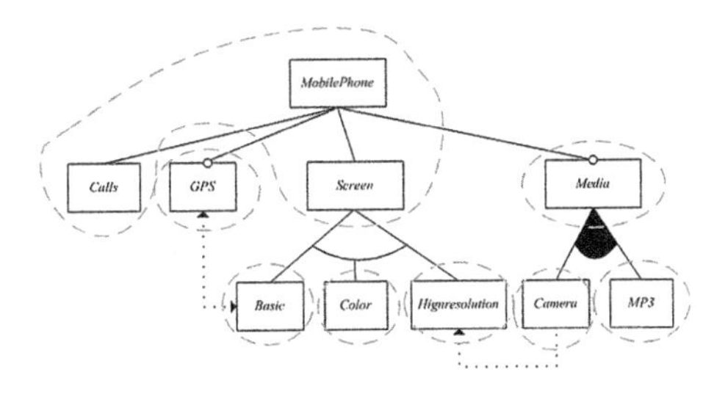

Fig. 2. Atomic set model

3 Atomic Sets Computation

For using atomic set as the basic unit in the analysis of feature model, Algorithm 1 depicts the pseudo-code of atomic sets computation.

Algorithm 1 Computing the atomic set model

input: feature model fm
output: atomic set

```
1:  Feature root = fm.getRoot()
2:  AtomicSet as = new AtomicSet("AS-0")
3:  as.addFeature(root)
4:  as.setRoot(true)
5:  as.setRoot(true)
6:  return as
```

After atomic set algorithm is implemented to feature model in Fig. 1, the atomic set model shown in Fig. 3 can be got.

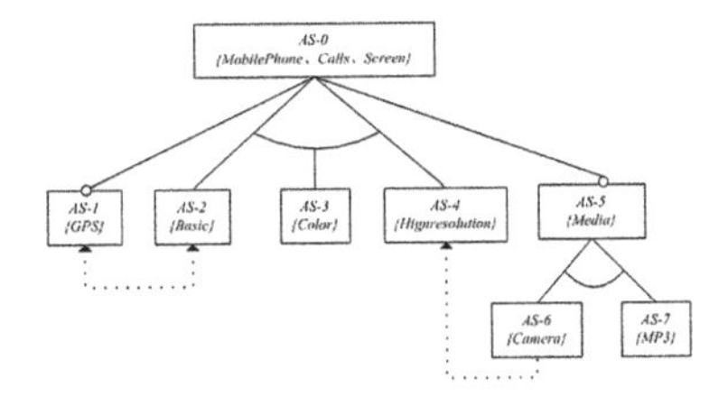

Fig. 3. Mobile phone atomic set model

4 Genetic Algorithm for Feature Selection Optimization

4.1 Initial Population

In initializing population, the individual represents a group of feature selection. For a feature model with N features, each individual is a randomly generated string with N binary digit, such as 10101100. Each binary digit represents a feature and a value of 1 for a digit indicates that the feature is selected (0 is unselected).

4.2 Fitness Function

The fitness function in genetic algorithm also called evaluation function is a metric of evaluating individual quality in population. In all research on feature selection optimization with resource constraint, the value of resource consumption is undoubtedly a fitness function. Usually, the computation of resource consumption is given by domain experts.

4.3 Crossover Operation

In genetic algorithm, the crossover is a key operation to maintain the population diversity. Crossover operation is that two parent individuals swap genes to generate new offspring according to crossover probability and specified crossover rule. This paper uses uniform crossover in a valid configuration set Pop_{valid} and invalid configuration set $Pop_{invalid}$ in order to generate more valid configurations.

4.4 Selection Operation

Selection operation implements that the individual with high fitness is selected from the population into the next generation. Because of the crossover operation performed in a valid product configuration and invalid product configuration, the invalid product configuration should be reserved in the process of the selection operation. Suppose that the size of population is popsize and the proportion of invalid configuration accounted for the population is ε. Based on a certain proportion, the algorithm selects the valid product configuration and the invalid product configuration according to individual incorrect value. After selection, the total number of valid product configuration and invalid product configuration is popsize.

4.5 Mutation Operation

In genetic algorithm, Crossover operator can improve the search ability. However, it may lead to the situation of the local convergence or premature convergence. Mutation operation produces gene mutation to a gene in parent individual.

4.6 The Flow Diagram of Genetic Algorithm

According to the analysis of the genetic algorithm, the Fig. 4 shows the operation process of the algorithm.

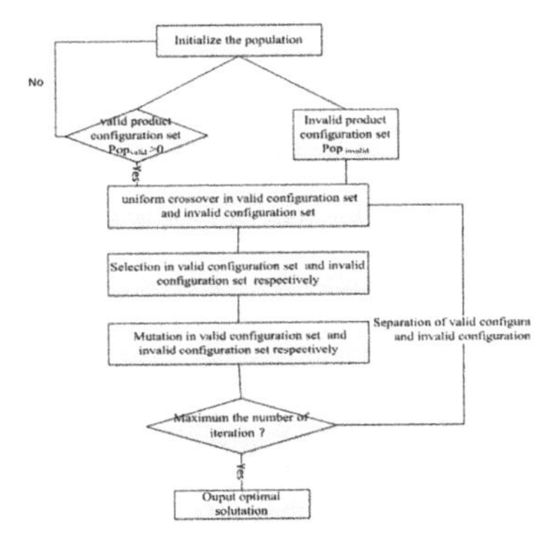

Fig. 4. The flow diagram of genetic algorithm

5 Experiment and Analysis

5.1 Test Model

For verifying the efficiency of the algorithm, this paper adopts different scale feature models as test model. Table 1 presents the number of feature, the number of atomic set, the number of constraint rule and reference information about two feature models. In addition, in order to optimize the feature model with resource constraints, we extends the feature model by adding cost property to each feature where the value of cost is typical of double and generated randomly between 5.0 and 15.0.

Table 1. Test model

Feature model	Feature	Atomic set	Constraints	Reference
Web portal	43	35	63	Mendoncal et al. [10]
E-shop	290	206	419	Lau [11]

Table 1 illustrates that after using atomic set to optimize the feature model, the number of variables is decreased by 19 and 29% respectively on the corresponding feature model Web Portal and E-Shop. The number of variables is reduced from 68 to 56% compared with the constraint rules on the Web Portal model. The number of variables is reduced from 69 to 49% compared with the constraint rules on the E-Shop model.

5.2 Experiment Result and Analysis

5.2.1 The Simulation Results of Atomic Set

The size of the initial population is 30. The number of iterations is 100. The crossover probability is 1. The mutation rate is 0.1. After the program is run ten times iteratively, the optimal solution is indicated in the Table 2. Optimization efficiency is around 40–60%.

Table 2. The simulation results of atomic set

Feature model	Method	T_{init} (s)	T_{GA} (s)	$V_{incorrect}$
Web portal	Atomic set	1.16	12.68	0
	Non atomic set	4.91	19.25	2
E-shop	Atomic set	9.39	40.11	32
	Non atomic set	13.48	111.56	44

5.2.2 The Simulation Results of Genetic Algorithm

To increase the search space of valid configuration, the size of the initial population is 100. The number of iteration is 100. The crossover probability is 1.0. The mutation rate is 0.1 in a valid configuration set. The mutation rate is 0.2 in an invalid configuration set. After the program is run iteratively after ten times, the optimal solution can be obtained from a valid configuration set shown in Table 3.

Table 3. The simulation results of genetic algorithm

Feature model	T_{GAFES} (s)	T_{GA} (s)	$\frac{T_{GA}}{T_{GAFES}}$ (%)
Web portal	15.91	13.84	13
E-shop	147.30	94.94	35

In order to compare effectiveness of the algorithm, the genetic algorithm in this paper is compared with GAFES algorithm proposed by Guo et al. [4] in time efficiency where the optimal time of Web Portal model is compared with the optimal time of 50-sized model in GAFES algorithm and the optimal time of E-Shop is compared with the optimal time of 200-sized model.

6 Related Work

White et al. [12] present a method called Filtered Cartesian Flattening which maps the problem of feature selection according to several constraints to a multi-dimensional multi-choice knapsack problem and offers an approximate optimal product configuration in the literature. Chen et al. [13] propose a method which uses the selectivity of feature to optimize product configuration. The method firstly calculates the selectivity of each feature. According to selectivity of feature, it helps users to select the optimal feature sequence. Sayyad et al. [5] propose a method which implements multi-object optimized feature selection according to user's requirement. The method concludes that IBEA (Indicator-Based Evolutionary Algorithm) has a higher efficiency than other multi-object genetic algorithm for feature selection optimization.

7 Conclusion

We propose an approach to optimize feature selection based on atomic set and genetic algorithm. Firstly, the approach uses atomic set to optimize the feature model. To a certain extent, the operation can reduce the number of variables on the feature model. At the same time, the property of all nodes are converted to optional property. Then in the process of optimizing feature selection by using genetic algorithm, the individual can be divided into valid product configuration and invalid product configuration through modeling the integrity constraints of the feature model. The crossover and mutation operator of genetic algorithm makes this approach possess both the balance of global and local search ability and speed up the convergence to the optimal solution. The experimental results show that the approach is valid. In future work, we will adopt a more complex example to improve this approach. At the same time, we will try to use other genetic algorithm to optimize feature selection with resource constraints and user's requirement and realize the multi-objective optimization.

Acknowledgments. The work of the article Product Configuration Based on Feature Model is based on contribution, ideas, and inspiration from my tutor and friends, and the support of Aviation Science Fund Project (20155552047), National Basic Research Program of China (973) (No. 2014CB744904).

References

1. Pohl, K., Böckle, G., van der Linden, F.J.: Software Product Line Engineering: Foundations, Principles and Techniques. Springer Publishing Company, Incorporated (2010)
2. Schobbens, P.Y., Heymans, P., Trigaux, J.C., et al.: Generic semantics of feature diagrams. Comput. Netw. **51**(2), 456–479 (2007)
3. Zhou, M., Sun, S.D.: The Principle and Application of Genetic Algorithm. National Defence Industry Press, Beijing (1999)
4. Guo, J., White, J., Wang, G., et al.: A genetic algorithm for optimized feature selection with resource constraints in software product lines. J. Sys. Softw. **84**(12), 2208–2221 (2011)
5. Sayyad, A.S., Menzies, T., Ammar, H.: On the value of user preferences in search-based software engineering: a case study in software product lines. In: Proceedings of the 35th International Conference on Software Engineering (ICSE), pp. 492–501. IEEE Computer Society (2013)
6. Schobbens, P., Heymans, P., Trigaux, J.: Feature diagrams: a survey and a formal semantics. In: Proceedings of 14th IEEE International Conference on Requirements Engineering, pp. 139–148. IEEE computer society, Washington (2006)

7. Benavides, D., Segura, S., Ruiz-Cortés, A.: Automated analysis of feature models 20 years later: a literature review. Inf. Sys. **35**(6), 615–636 (2010)
8. Segura, S.: Automated analysis of feature models using atomic sets. In: Proceedings of the First Workshop on Analyses of Software Product Lines (ASPL 2008), pp. 201–207. Limerick, Ireland (2008)
9. Mendonca, M., Branco, M., Cowan, D.: S.P.L.O.T. - Software Product Lines Online Tools. In: Proceedings of OOPSLA, USA (2009)
10. Mendonca, M., Bartolomei, T., Cowan, D.: Decision-making coordination in collaborative product configuration. In: Proceedings in ACM Symposium on Applied Computing, USA, pp. 108–113 (2008)
11. Lau, S.Q.: Domain analysis of e-commerce systems using feature-based model templates. Electrical and Computer Engineering, University of Waterloo, Canada (2006)
12. White, J., Schmidt, D.C.: Optimizing and automating product-line variant selection for mobile devices. In: Proceedings of the 11th International Software Product Line Conference, pp. 129–140 (2007)
13. Chen, S., Erwig, M.: Optimizing the product derivation process. In: Proceedings of the 15th Software Product Line International Conference, pp. 35–44. IEEE (2011)

Product Configuration based on Feature Model

Zhijuan Zhan[1(✉)], Yunjiao Zhan[2], Mingyu Huang[2], and Yumei Liu[2]

[1] Science and Technology on Avionics Integration Laboratory, China National Aeronautical Radio Electronics Research Institute, Shanghai, China
zzjaldm@aliyun.com

[2] Department of Computer Science and Technology, Nanjing University of Aeronautics and Astronautics, Nangjing, Jiangsu, China

Abstract. The approaches of product configuration based on feature are about how to select features from a feature model based on specific domain requirements and stakeholders' goals. Although the literature on this topic has gained most importance in Academic and industrial fields, only little effort is dedicated to compare and analyze them. In order to address this shortcoming and to provide a basis for more structured research on feature modeling in the future, we first build a framework model to describe model structure, dependency management, automated support, configuration approaches and so on shared in the approaches family. Secondly, we understand and classify different configuration method based on the framework. Meanwhile we analyze the commonalities and variabilities among different approaches.

Keywords: Software product line · Feature model · Product configuration

1 Introduction

In recent years, software developers are desperately seeking a kind of new technology to improve the reusability of software under the pressure of the diversity of user requirements and the high cost of development and maintenance. Software product line engineering (SPLE) has become an increasingly prominent approach for software development [1]. Software product line engineering sees similar systems in a specific field as a whole and deals with commonalities and variabilities between the systems. By maximizing the platform reuse and batch customization, SPLE achieves products family development. Industrial experience shows that SPLE can significantly improve software development productivity, quality and reduce time to market.

The software product line consists of two processes: domain engineering and application engineering. In domain engineering, the most important of the process is scope, commonality and variability analysis. In application engineering phase, we select features from the feature model to meet the specific needs of the specific software product. From the above two processes, we can know that the product configuration is the purpose of feature modeling.

Based on the specific goals, requirements and limitations of the stakeholders, and the integrity constraints of the feature model, the product configuration based on feature

F. Xhafa et al. (eds.), *Advances in Intelligent Systems and Interactive Applications*, Advances in Intelligent Systems and Computing 686,
https://doi.org/10.1007/978-3-319-69096-4_15

model is about selecting the best set of desirable features from the feature model. A significant amount of research exists on product configuration technology.

So, we introduce a method called review based on feature model. The main idea regarding of the review process and its structure is detailed below.

1. Construct feature model to capture all the commonality and variability in the domain.
2. Present an analysis about the commonality and variability among the approaches.
3. Discuss the results obtained and describe some challenges.

The remainder of this paper is structured as follows: Section 2 simply presents feature modeling in the configuration process. Section 3 presents the automated techniques used for analyzing feature model. Section 4 describes the characteristic of configuration approaches. Section 5 covers related work. Section 6 concludes the paper.

2 Configuration Architecture Model

The feature-based configuration approaches have commonalities and variabilities. In this paper, we build a feature model to describe the commonalities and variabilities among the configuration approaches. On the basis, we analyze the product configuration research by analyzing the feature model.

Figure 1 depicts a feature model. Illustrates how features are used to specify the approaches for software product line configuration. Root feature PCA (Product Configuration Approaches) represents the domain of configuration approaches. According to the feature model, all approaches must include feature FM (Feature Modeling), AS (Automated Support) and FS (Feature Selection) which represents the necessary parts in the configuration process.

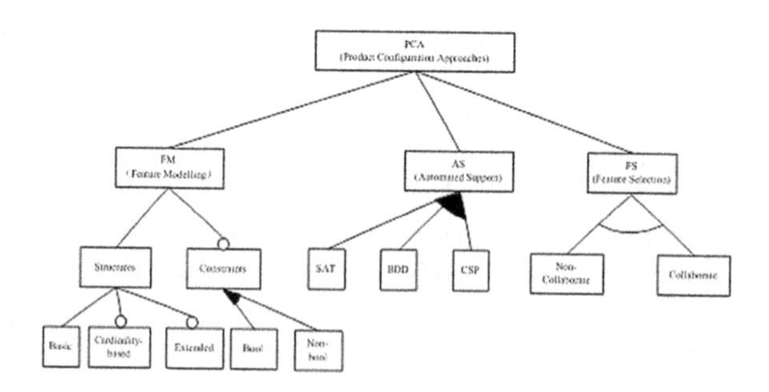

Fig. 1. Product configuration methods product line feature model

In automated support process, feature SAT, BDD and CSP are the three main tools for automated analysis feature model in software product line. The relationship among SAT, BDD and CSP is OR decomposition mode which means all the approaches may include one of them or all of them.

3 Feature Modeling

The feature model is the basis of the product configuration method, and its complexity determines the characteristics of the configuration method. Feature model usually consists of a tree like feature graph and cross-tree constraints. Feature graph's nodes denote features and edges represent top-down hierarchical decomposition of features. Cross-tree constraints are represented in text added to feature model.

3.1 Feature Diagram

3.1.1 Basic Feature Diagram

The basic feature model notation proposes four relations, namely: mandatory, optional, alternative and or-relation. The relationships group as following:

- Mandatory: Mandatory's child is included in all products in which its parent feature appears.
- Alternative: only one of the child feature can be selected when its parent feature is part of the product.
- Optional: Optional's child can be optionally included in all products in which its parent feature appears.
- Or: one or more of them can be included in the products in which their parent feature appears.

3.1.2 Cardinality-Based Feature Diagram

Feature models allow specification of cardinality relations in the form of [n..m] for feature cardinality and [n..m] for feature group cardinality, which enforce that at least n and at most m features in a group must be present in any product containing their parent.

3.2 Constraints

In variability management, feature diagram is limited to describe the relationships between parent feature and child feature. Below, we will review the constraints classified into Boolean constraints and non-Boolean constraints.

3.2.1 Boolean Constraints

Boolean constraints are the constraints can be expressed by boolean expressions. The boolean constraints expression can use classical operators: !, &&, ||, $\rightarrow$, $\leftrightarrow$ for boolean values. Classical feature model cross-tree constraints keywords *requires* and *excludes* can also be used as boolean expressions whose form is $A \rightarrow B$ and $A \rightarrow \neg B$ respectively.

3.2.2 Non-Boolean Constraints

Product configuration uses complex constraints to propagate configuration choices made by the stakeholders. Boolean constraints can't express some complex requirements of user effectively. Such constraints are called non-boolean constraints. For example, Czarnecket et al. [2] have used joint probability distribution to describe the constraints among features.

4 Automated Support

Due to the complexity of feature model, it is a big challenge to find a valid configuration. We will introduce the characteristics of three automated analysis tools in the software product line. Acknowledgments

4.1 SAT

Boolean satisfiability problem [3] is about determining whether a given propositional formula is satisfiable. The formulas are usually specified in Conjunctive Normal Form (CNF). Mapping a feature model into a propositional formula is the key step when using SAT solver to analyze feature models. Benavides et al. [4] concludes the mapping relation of basic feature model to propositional logic.

4.2 BDD

Similarly, Binary Decision Diagrams(BDD) [5] is also about deciding whether a formula is satisfiable or not. Different to SAT, BDD translates the propositional formula into a graph representation. BDD's advantage is that the data structure can provide efficient algorithm for counting the number of the possible solutions.

4.3 CSP

A CSP is solved by finding states (values for variables) in which all constraints are satisfiable. Benavides et al. [5, 6] first proposed the usage constraint programming for analysis on feature model. Benavides et al. also proposed the support for the analysis of extended feature model and the operation of optimization.

In conclusion, each of the above automated support tools has it's scope of application and disadvantages. Benavides et al. [7] developed a tool called FAMA that integrated SAT, BDD and CSP for analyzing the different operations based on the characteristics of tools.

5 Feature Selection

Feature model configuration is also a feature selection process. The selection of the best set of features for a product would be based on specific goals, requirements and limitation of stakeholders, as well as the integrity constraints of feature model. The configuration process can be classified as non-collaborative configuration and collaborative configuration according to the number and role of participants.

5.1 Non-collaborative Configuration

The key point of the non-collaborative configuration process is analyzing and modeling the requirements of the stakeholders which can be classified into hard requirements and soft requirements. Soft requirement is a fuzzy concept which is helpful for making correct decision. Therefore, the difficulty is how to capture and model the soft requirements in configuration process.

White et al. [8] proposed an mobile devices automating product-line variant selection. To address the challenge of downloading the software and browsing the website on demand, the approach first captures the requirement of the product line architecture (PLA) and then constructs a custom variant from PLA for a device.

Bartholdt et al. [9] proposed an integrated quality modeling based on feature modeling in software product lines (IQ-SPLE). The approach supports quantitatively and qualitatively model the quality.

5.2 Collaborative Configuration

Compared to non-collaborative configuration process, the collaborative configuration can effectively address the large and complex software product line configuration problem.

Hubaux et al. [10] ration. It tailors a large and complex feature model into several views according to the profile of stakeholders. Hubaux et al. [11] developed a tool set for feature-based configuration workflow. In order to avoid the conflict among the different views, the workflow is used to drive the configuration of the views.

Similarly, Czarnecki et al. [12] proposed multi-staged configuration based on feature model. Each stage corresponds to a stakeholder and the later stage is based on the former. Finally, an ultimate product is configured through integrating the configuration in different stage. The difference is the way to select feature. The former selects the desired feature and eliminated the undesired. The later excludes the irrelevant features through assigning some values to features.

6 Conclusion

In this paper, we revise the state of the art on the feature model configuration process as a product line. Accordingly, a feature model is constructed to describe the communalities and variabilities among the families of the methods. We then analyses the feature model to provide an overview of the diversity of the methods. We also provide information about the research status and prospect of the field.

In future work, we will improve the review method based on feature modeling to better support the analysis of similar methods in specific domain. On the basis, we will compare the differences and commonalities of feature modeling and decision modeling.

Acknowledgments. The work of the article Product Configuration Based on Feature Model is based on contribution, ideas, and inspiration from my tutor and friends, and the support of Aviation Science Fund Project (20155552047) and National Basic Research Program of China (973) (No. 2014CB744904).

References

1. Pohl, K., Böckle, G., Linden, F.J.V.D.: Software product line engineering: foundations, principles, and techniques. In: Proceedings of the First International Workshop on Formal Methods in Software Product Line Engineering, vol. 49, no. 12, pp. 29–32 (2005)
2. Cook, S.A.: The complexity of theorem-proving procedures. In: ACM Symposium on Theory of Computing, Shaker Heights, Ohio, USA, 3–5 May 1971
3. Benavides, D., Segura, S., Ruiz-Cortés, A.: Automated analysis of feature models 20 years later: a literature review ☆. Inf. Syst. **35**(6), 615–636 (2010)
4. Tsang, E.P.K.: Foundations of constraint satisfaction. In: DBLP, pp. 188–224 (1993)
5. Benavides, D., Cortés, R.A., Trinidad, P.: Coping with automatic reasoning on software product lines (2004)
6. Benavides, D., Trinidad, P., Cortés, A.R.: Using constraint programming to reason on feature models. In: International Conference on Software Engineering and Knowledge Engineering, pp. 677–682 (2005)
7. Trinidad, P., Ruiz-Cortés, A., Benavides, D.: Automated analysis of stateful feature models (2013)
8. White, J., Schmidt, D.C., Wuchner, E., et al.: Automating product-line variant selection for mobile devices. In: International Software Product Line Conference. SPLC 2007, pp. 129–140 (2007)
9. Bartholdt, J., Medak, M., Oberhauser, R.: Integrating quality modeling with feature modeling in software product lines. In: International Conference on Software Engineering Advances, pp. 365–370 (2009)
10. Hubaux, A., Heymans, P., Schobbens, P.Y., et al.: Supporting multiple perspectives in feature-based configuration. Softw. Syst. Model. **12**(3), 641–663 (2013)
11. Abbasi, E.K., Hubaux, A., Heymans, P.: A toolset for feature-based configuration workflows. In: International Software Product Line Conference, pp. 65–69 (2011)
12. Czarnecki, K., Helsen, S., Eisenecker, U.: Staged configuration through specialization and multilevel configuration of feature models. Softw. Process Improv. Pract. **10**(2), 143–169 (2010)

Research on Fast Browsing for Massive Image

Fang Wang[1,2(✉)], Ying Peng[1], and Xiaoya Lu[1]

[1] College of Computer Science and Technology, Southwest University for Nationalities, Chengdu, China
wangfang@swun.cn

[2] Computer System Key Laboratory of the National Council for Nationalities, Southwest University for Nationalities, Chengdu, China

Abstract. The development and application of image data has the characteristics of high resolution, large data volume and so on. Research on how to take advantage of MapReduce to distributed processing efficient and fast is one of the focuses and hotspots in the field of massive image data management. To solve the above problems, combing efficient distributed programming and running frame provided by MapReduce model and image pyramid algorithm, proposing and realizing a distributed model for massive image data. Experiment expresses that this model is good performance in massive image's browsing and rooming.

Keywords: Massive image · Distributed processing · MapReduce · Image pyramid

1 Introduction

The image data is an important data source of Geographic Information System (GIS), it contains rich information of resources and the environment. Under the support of GIS, it can be applied in complex and comprehensive information analysis of the geology, geophysics and geochemistry, earth biology and military and so on. With the progress of science and technology, the extensive application of the computer, the rapid growth of data source, a sharp increase in the amount of data, data processing scale has reached the level of PB, such as transaction data of the Alibaba and eBay online trading sites, real-time monitoring data of video monitoring system, instant messaging log data of the Tencent, data of astronomical observation. However, with the rapid growth of data size, whether the existing methods of data management has ability to cope with challenges become crucial, how to efficient store, organize, process, manage and distribute the image data and shorten the time from acquisition and processing to using reach minute level even second level has become an urgent problem to be solved.

High resolution (high spatial resolution, high spectral resolution, high time resolution) remote sensing image has become effective means of field monitoring, remote sensing vegetation, water and marine remote sensing, agricultural remote sensing, atmospheric research, environmental monitoring, geological hazard investigation, application of high resolution remote sensing image has been covering all aspects of

F. Xhafa et al. (eds.), *Advances in Intelligent Systems and Interactive Applications*, Advances in Intelligent Systems and Computing 686,
https://doi.org/10.1007/978-3-319-69096-4_16

earth and related sciences. With the resolution of image data becomes more and more high, and the amount of data is more and more big, the current research mainly focuses on how to browse, display and update the large amount of data quickly and efficiently. One of the most successful methods to solve these problems is to resample the image, cut the blocks, and establish the image pyramid model, which can save the storage space for the relatively high efficiency in the fast display. The pyramid model has been widely used in the field of image display and so on. However, the current pyramid model is mainly limited to serial mode, and there is still much space to development in distributed computing.

2 Distributed Computing and Image Pyramid

2.1 Distributed Computing

Distributed computing is a computing method which sharing information between two or more software which can not only run in the same computer but also run on multiple computers connected with network. It works in the environment which is composed of multiple hosts on the basis of data storage, data analysis and computation. Compared with the single computer system, the main difference between them is the scale of the problem, including the size of data processing and storage. Distributed computing environment has the characteristics of resource sharing, high transparency, high cost performance, high reliability and high flexibility.

2.2 MapReduce

Distributed computing environment has the characteristics of resource sharing, high transparency, high cost performance, high reliability and high flexibility. However, there are many problems in the distributed computing environment, such as the complexity of programming, the failure of nodes and the network blocking. The MapReduce parallel programming model which proposed by Google solves this problem. The MapReduce is an architecture of software which is proposed by Google, it mainly be applied to parallel computing on large data set. Map and Reduce are two important concepts and core idea of it. Programmers who are not sophisticated in distributed parallel programming are able to run their program in distributed system.

Each Map function parallel processes the divided data and produces different outputs from different inputs. And each Reduce function parallel computing data themselves. A synchronous barrier is needed before Reduce because that Reduce function is not able to begin until all Map functions have finished. And in this stage, the intermediate results of Map are aggregated and shuffled so that final result can be got by Reduce by computing and summarizing all the outputs of itself efficiently.

2.3 Image Pyramid

Image pyramid is the structure whose shape likes a pyramid. This structure is composed of a set of image whose resolving power from coarse to fine and data form small to large recording to the user's needs in the same spatial reference. It is usually used in image coding and gradual image transporting and it is typical layered data structure which is proper for multi-resolution organization of raster data and image data. At the bottom of image pyramid, there is image's high resolution representation of image to be processed. The higher the number layer, the lower the size and the lower resolving power.

3 MapReduce Model for Massive Image

3.1 Data Blocking Prepared for MapReduce

The first important problem in parallel is how to divide computing task or computing data and then be able to compute these sub tasks or data blocks at the same time. Task is not able to be split or there is dependency relationship in data. In other words, the best method to improve processing speed is parallel computing if a big data can be split to the data blocks with same computing process and there is no dependency between them. So the first step in this model is to divides the big image into the same N unit whatever in space form or spectral form. In this step that all the big data can be processed parallel. As shown in Fig. 1, the one size of spatial block is N rows multiply all columns and a size of spectral block is always sample multiply the image band.

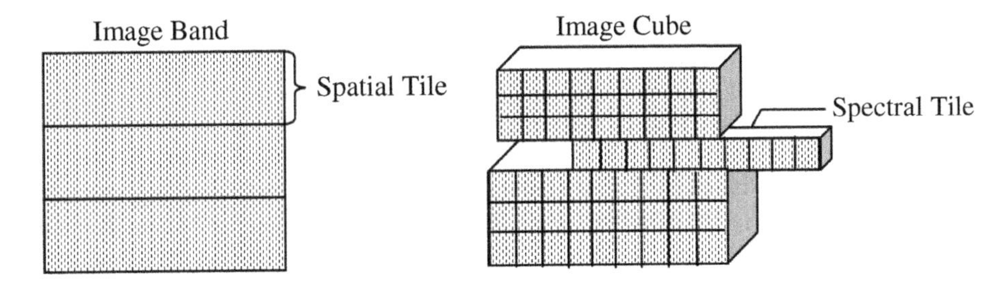

Fig. 1. Data blocking

3.2 MapReduce Image Model

As shown in Fig. 2, each Map function parallel processes the sub-block. It builds image pyramid for each block in the same rule until reaches the peak of pyramid, and writes on the result of each layers. After all Map functions have finished, each Reduce function parallel get appropriate layer's data. At last, spelling up these new blocks into a new image which is more small than original but loading fast.

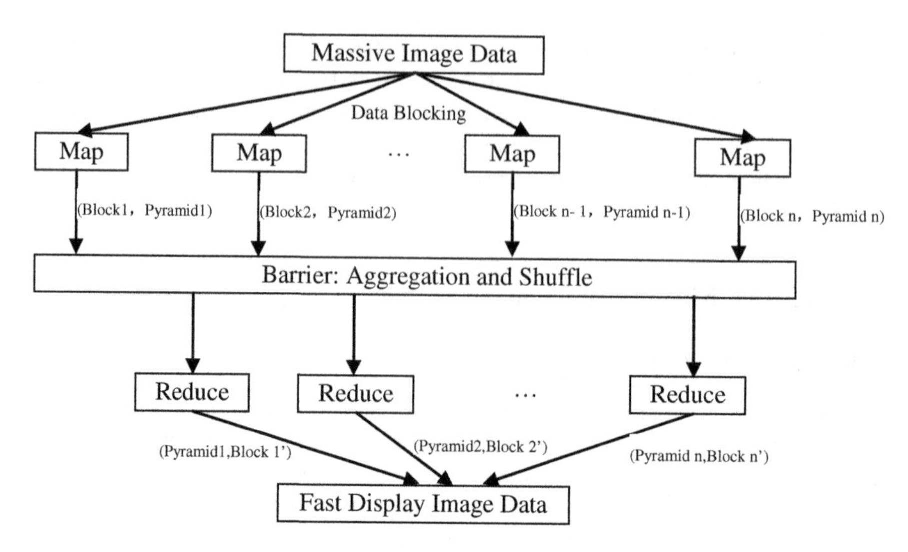

Fig. 2. MapReduce model for massive image

In Map, it takes the sub-block as the bottom of image pyramid that is zero layer and blocking it again to get zero layer matrix. Firstly, it builds image pyramid in two steps: At first, doing Gauss kernel convolution for the i layer of image. At second, removing all even rows in the matrix, then the i + 1 layer image is get. The formula is as follow:

$$G_o(i,j) = \sum_{m=-2}^{2} \sum_{n=-2}^{2} s(m,n) G_{o-1}(2i+m, 2j+n) \tag{1}$$

It smooth the sub-block by Gauss smoothing filter before down sampling, the size of upper layer image is 1/4 of lower layer. Repeat he steps above, the Map will get a whole pyramid. In the Formula (1), $s(m,n)$ is the Gauss convolution kernel whose length is 5.

In Reduce, it collects all the results of image pyramid of sub-block, and spells them together to generate a new image which is based on the original massive image.

4 Experiment

4.1 Experimental Data and Equipment

The hardware is a HP workstation whose CPU model is intel 7, memory size is 128 GB, hard-disk's capacity is 10TB, speed is 7200 RPM and the software is Microsoft Window10 OS. In order to validate the model, we select 5 m resolving power of 16 bit multispectral and panchromatic image data (origin from the resource one 02C satellite) of certain area as experimental data to construct a MapReduce Model. The initial size of the image is 48 GB. And we divide image into 48 sub-blocks, and construct a 10 layers image pyramid for each sub-block.

4.2 Experimental Result

The response speed of original image is 28 s, and the output image's response speed is 4 s which reduces 7 times. The output image's resolving power can match 2 m. Figure 3 is the original image and Fig. 4 is the output image for browsing.

Fig. 3. Original image

Fig. 4. Output image

Meanwhile, we realize zoom function for browsing image by using appropriate lay of the Gaussian pyramid. Experiment display that browsing is fluent whether internal zooming or switching between different layers.

Acknowledgments. This work was financially supported by the Fundamental Research Funds for the Central Universities, Southwest University for Nationalities (No. 2015NZYQN71).

References

1. Li, Y., Zhang, H., Li, S.: Parallel processing of long strip remote sensing image based on pipeline with data distribution strategy optimization. Comput. Appl. Softw. **33**(11), 117–121 (2016)
2. Yang, L., Tian, S.: On identifying water body in remote sensing images based on distributed computing. **33**(6), 138–140 (2016)
3. Li, X., Xing, J., Liu, D., Wang, H., Liu, W.: Distributed processing practice of massive GIS data based on HBase. Big Data **2**(3), 73–82 (2016)
4. Zhang, H., Wang, X., Cao, J., Ma, Y., Guo, Y., Wang, M.:. Comput. Eng. Appl. **52**(22), 22–25 (2016)
5. Dong, X., Deng, C., Yuan, S., Wu, Z., Zhang, Z.: Distributed differential evolution algorithm based on MapReduce model. J. Chin. Comput. Syst. **37**(12), 2695–2701 (2016)
6. Qiu, L., Du, Z., Xie, J., Qiu, Z., Xu, W., Zhang, Y.: A real-time, visualization method of high resolution remote sensing images bigfiles. Geomatics Inf. Sci. Wuhan Univ. **41**(8) (2016)
7. Li, Z., Qiang, Y.: Medical CI image enhancement algorithm based on laplacian pyramid and wavelet transform. Comput. Sci. **43**(1), 300–303 (2016)
8. Wang, Q., Nie, R., Jin, X., Zhou, D., He, K., Yu, J.: Image fusion algorithm using LP transformation and PCNN-SML. Comput. Sci. **43**(6A), 122–124 (2016)
9. Liu, P., Gong, J.: Parallel construction of global pyramid for large remote sensing image. Geomatics Inf. Sci. Wuhan Univ. **41**(1), 117–122 (2016)
10. Li, S., Yu, H., Han, J., Hei, B.: Design and Implementation of efficient visualization management system for massive remote sensing images based on three-dimensional globe. Remote Sens. Technol. Appl. **31**(1), 170–176 (2016)
11. Yang, J., Liao, Z., Feng, C.: Survey on big data storage framework and algorithm. J. Comput. Appl. **36**(9), 2465–2471 (2016)
12. Gou, W., Zhai, Q., Yu, S., Han, P., Liu, X.: The explore base on Ziyuan-3 satellite imaging in the DSM data production. Geomatics Spat. Inf. Technol. **39**(1), 141–143 (2016)

RVM for Recognition of QRS Complexes in Electrocardiogram

Lu Bing[1(✉)], Xiaolei Han[1], and Wen Si[2]

[1] School of Information and Computer Science, Shanghai Business School, 201400, Shanghai, People's Republic of China
betty20006@163.com

[2] Department of Rehabilitation, Huashan Hospital, Fudan University, 200040 Shanghai, People's Republic of China

Abstract. QRS complex is with significant use in electrocardiogram (ECG) components. This paper uses a special algorithm to recognize QRS complexes in the ECG, which is based on relevance vector machines (RVM). RVM is used as a classifier to detect QRS areas. Performance of the adopted approach is validated by cross-validation. The experimental result has showed that, the RVM method has better classification effect, along with fewer parameter settings and fewer kernel functions than that of support vector machine (SVM). It can perform a satisfying result under the poor condition of the ECG signal.

Keywords: ECG · RVM · QRS recognition · SVM

1 Introduction

Electrocardiogram (ECG) is a vital medical tool to provide vivid information on condition of the cardiac affairs [1]. Studying ECG is quite important in the found of heart diseases. As displayed in Fig. 1, ECG is consist of P, QRS and T-wave in a whole cardiac cycle, where, the QRS complex is by ventricular depolarization in the cardiac circle, and is a key waveform. If we find the positions of the QRS complexes, the other parts of ECG such as P-waves, T-waves and ST segment can also be found, in order to analyze the whole cardiac period.

By now, many QRS recognition algorithms have been proposed. Many pattern recognition and classification methods based on the artificial intelligence have been developed for the analysis of problems in bioinformatics [2]. As a widely used learning algorithm, support vector machine (SVM) has been applied for classification and regression in many aspects [3]. Though SVM classifier has outcome some successful results and performance, there are still many disadvantages [4]. For instance, the number of support vectors become larger when the training data grows. The training procedure requires to estimate the regularizing parameter C. What is more, the kernel functions have to meet the Mercer's condition.

The relevance vector machine (RVM) is originated from Bayesian area of the sparse learning problem, and enjoys a satisfying generalization performance [5]. RVM can be applied to a lot of areas, where it proved itself the effectiveness of the probability

F. Xhafa et al. (eds.), *Advances in Intelligent Systems and Interactive Applications*, Advances in Intelligent Systems and Computing 686,
https://doi.org/10.1007/978-3-319-69096-4_17

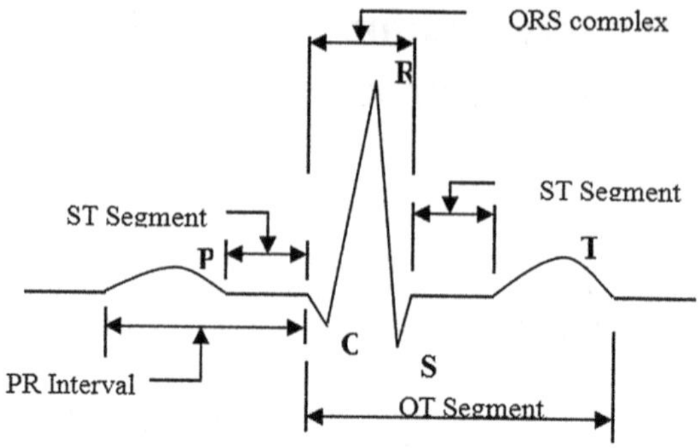

Fig. 1. ECG signal

output. The appearance of RVM saves some shortcomings of SVM, including the Mercer kernels and the regularizing parameter C. The most important advantage is to lead to reduction of computational complexity [4, 5]. In this paper, RVM is applied to recognize the QRS complexes both in training and testing period.

2 RVM Classifier

We have known that for RVM, the kernel functions include the polynomial kernel and the Gaussian kernel: $K(x, x_i) = (1 + x \cdot x_i)^d$ and $K(x, x_i) = exp\left(-x - x_i^2/\zeta^2\right)$, where, d and ζ are kernel parameters [5, 6]. For a two-class problem, we note the sigmoid function: $\sigma(\mathrm{y}) = 1/(1 + \mathrm{e}^{-\mathrm{y}})$, hence we have the following function:

$$\mathrm{P}(\mathrm{t}|\mathrm{w}) = \prod_{\mathrm{i}=1}^{\mathrm{n}} \sigma\{\mathrm{y}(\mathrm{x_i}; \mathrm{w})\}^{\mathrm{t_i}} [1 - \sigma\{\mathrm{y}(\mathrm{x_i}; \mathrm{w})\}]^{1-\mathrm{t_i}} \tag{1}$$

where, x is a vector as the input, t is scalar represent for its class label, $\mathrm{t_i} \in \{0, 1\}$. The above function can be re-transferred and result in the following abstract functions [7]:

$$\mathrm{logp}\{\mathrm{w}|\mathrm{t}, \alpha\}|_{W_{MP}} = -\left(\Phi^{\mathrm{T}}\mathrm{B}\Phi + \mathrm{A}\right) \tag{2}$$

where B is a diagonal matrix, and Φ represents for a complex matrix. We gain the right part of the above function as a single form as:

$$\sum = \left(\Phi^{\mathrm{T}}\mathrm{B}\Phi + \mathrm{A}\right)^{-1} \tag{3}$$

By this method, we adopt MacKay thought [8] to update the inner parameters α_i by

$$\alpha_i^{new} = (1 - \alpha_i \Sigma_{ii})/w_{MP}^2 \tag{4}$$

where Σ_{ii} is the ith element of the diagonal covariance. Samples with $w_i \neq 0$ are called relevance vectors.

3 QRS Detection

The ECG wave is inevitably interfered by all kinds of noises, including power frequency noise, baseline wander, and high frequency noise, etc. It is of vital importance of carrying on reprocessing before the QRS complex detection. In our study, we apply wavelet transformation and band-pass filter to suppress interference in order to gain relatively pure training samples.

Because slope of the QRS area is larger than that in the non-QRS area, slope is seen as an important feature. We calculate every single max value of slope on sampling point of the reprocessed ECG signal to form and enhance the QRS part.

In the experiment, we use 8 slopes, and regard them as the input to the RVM method. Then, on training period, a threshold of 8 points is transferred within the training parts. Later, on testing period, we select a group of 8 maximum slopes to use as the input vector of the RVM.

4 Implementation and Performance Evaluation

The data set of the QRS recognition is selected: Single-lead ECG records, with 500 Hz sampling in 15s.

In RVM, there are two kernel functions: Gaussian kernel and polynomial kernel. A key parameter needs to consider, which is kernel parameter ζ or d. For SVM, we also adopt it to act as a classifier for this special problem. Parameters include ζ or d, along with regularizing parameter C. In this experiment, we use ten-fold cross validation to perform and select the best kernels and margin-loss tradeoff C [9, 10].

In this section, we evaluate the classification results by our method. The following index are used to test the performance:

Classification accuracy (C_A): Right classification results in testing data;

P_{RV} : Number of the support vectors of the RVM;

P_{SV} : Number of the support vectors of the SVM;

In this part, we use the leave-one-out cross validation to measure the classification performance. The algorithm is to gain the excellent performance. The results of the two methods are shown in Tables 1 and 2, respectively. We may see from the tables that the support vectors of the RVM is less than that of SVM.

Figures 2 and 3 show selected QRS recognition by RVM. Firstly, the preprocessor removes noises and baseline wanders. We can find that in this case, the maximum of the QRS area is not very larger than that of other waves, such as the P- and T-waves. However, all the QRS areas are successfully recognized using RVM, which shows its effectiveness. However, RVM could not detect one QRS complex due to lower amplitude of slope.

Table. 1. Results of 4-fold cross validation by RVM

Kernels	ζ	d	Correction (%)	Number of support vectors (%)
Gaussian	**3**	-	**93.9**	**2.97**
Gaussian	5	-	91.4	3.35
Polynomial	-	1	92.2	3.13
Polynomial	-	**2**	**92.6**	**3.77**
Polynomial	-	3	89.1	3.96

Table. 2. Results of 10-fold cross validation by SVM

Kernels	ζ	d	C	Correction (%)	Number of support vectors (%)
Gaussian	**4**	-	**1**	**89.8**	**36.6**
Gaussian	5	-	1	89.1	38.6
Gaussian	3	-	3	88.2	32.4
Polynomial	-	1	1	84.6	51.1
Polynomial	-	3	1	85.1	49.1
Polynomial	-	**1**	**3**	**87.2**	**42.1**
Polynomial	-	2	3	85.6	53.7

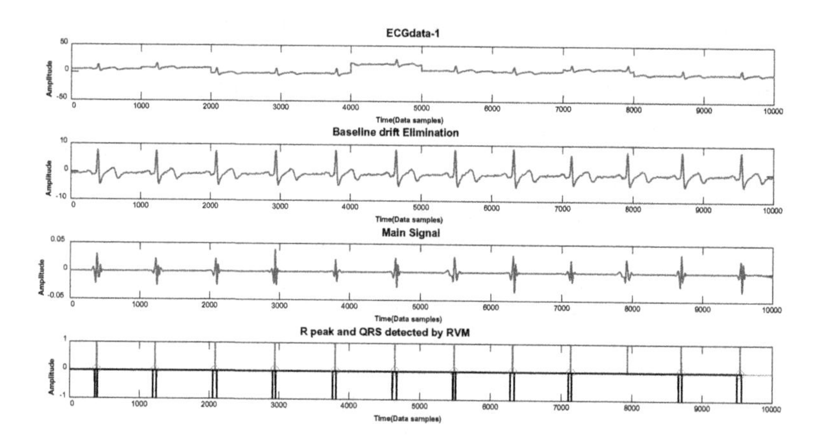

Fig. 2. QRS detection for ECG data-1

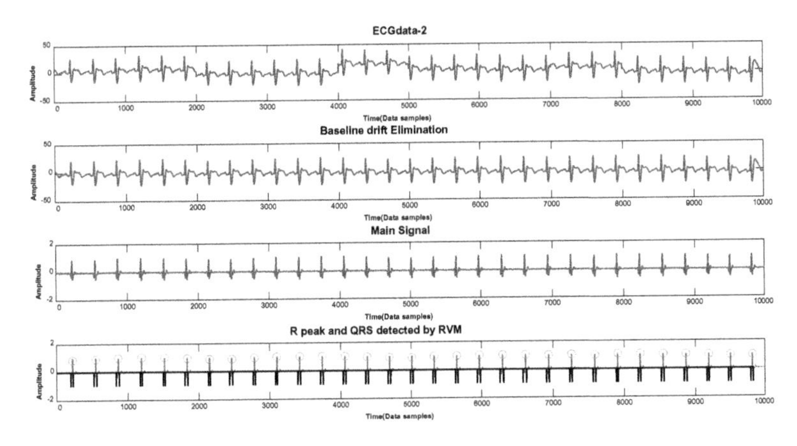

Fig. 3. QRS detection for ECG data-2

The result shows that the RVM enjoys much less number of RVs than support vectors of the SVM, which may accelerate the convergence speed. At the same time, the performance of RVM is better than then the latter.

5 Conclusions

In this paper, QRS recognition method adopting RVM is used and applied to the ECG signals. The results have shown that the RVM enjoys satisfying recognition result, and has outcome many short comes of SVM. However, computation of posterior weight covariance matrix may need more time for RVM to train samples. But all in all, RVM can be seen as a promising method for signals such as ECG/EEG' processing.

Acknowledgments. This research was partially supported by the Shanghai Business School "Phosphor" Science Foundation (Grant No. 16-11051), the Training Foundation for The Excellent Youth Teachers of Shanghai Education Committee (Grant No. ZZsxy15008), the Shanghai Business School Scholars Programme (Grant No. 16-11051), the Shanghai Natural Science Fund (Grant No. 14ZR1429800), and the Experimental Course Reform Project for Commercial Application of Internet of Things.

References

1. He, Bin: Modeling and Imaging of Bioelectrical Activity Principles and Applications, pp. 161–180. Kluwer Academic/Plenum Publishers, The Netherlands (2004)
2. Rani, R., Chouhan, V.S., Sinha, H.P.: Automated detection of QRS complex in ECG signal using wavelet transform. IJCSNS Int. J. Comput. Sci. Netw. Secur. **15**(1), 1 (2016)
3. Brudzewski, K., Osowski, S., Markiewicz, T.: Classification of milk by means of an electronic nose and SVM neural network. Sens. Actuators B **98**, 291–298 (2004)
4. Tipping, M.E.: Sparse Bayesian learning and the relevance vector machine. J. Mach. Learn. Res. **1**, 211–244 (2001)

5. Tipping, M.E.: The relevance vector machine. In: Advances in Neural Information Processing Systems, vol. 12, pp. 652–658. MIT Press, Cambridge (2000)
6. Oikonomopoulos, I., Patras, M. Pantic: Spatiotemporal salient points for visual recognition of human actions. IEEE Trans. Syst. Man Cybern. B **36**, 710–719 (2005)
7. Krishnapuram, B., Carin, L.: Sparse multinomial logistic regression: fast algorithms and generalization bounds. IEEE Trans. Pattern Anal. **27**, 957–968 (2005)
8. Mackay, D.J.C.: The evidence framework applied to classification networks. Neural Comput. **4**, 720–736 (1992)
9. Nabney, I.T.: Efficient training of RBF networks for classification. In: Proceedings of the 9th International Conference on Artificial Neural Networks, pp. 210–215 (1999)
10. Hsu, C.W., Chang, C.C., Lin, C.J.: A practical guide to support vector classification, Technical report (2003)

The Architecture of the RFID-Based Intelligent Parking System

Xiao Xiao(✉) and Qingquan Zou

Research and Advanced Technology, SAIC MOTOR Corporation Limited, Shanghai, China
xiaoxiao@saicmotor.com

Abstract. The increasing amount of the automobile has led to many problems. How to carry out parking management work efficiently and reliably has become a vital part of the smart city development. In this paper, an intelligent parking system based on RFID technology is proposed, which identifies a vehicle by a reader and a tag connected to the car and achieves the management functions like non-conduct and cashless charging. The system could effectively deal with many problems of the traditional charging work such as inefficiencies, imbalance of resource allocation and loss of fare.

Keywords: RFID · Intelligent parking system · Parking management · Non-conduct charging

1 Introduction

With rapid economic development in China, the number of the automobile is increasing at a very fast rate, which has led to many parking problems such as inefficiency of management, imbalance of resources, loss of fare. The parking requirement on safety, comfort, efficiency is in urgent need of a highly automatic, convenient and intelligent parking system [1].

RFID (radio frequency identification) as a new technology identifies and tracks goods, which consists of four parts: the e-tags, the readers, the middleware and the application. Generally, we called the application and the middleware application system [2].

The RFID-based intelligent parking system is designed in this paper, which identifies a vehicle by a reader and a tag connected to the car and achieves the management functions like automatic fare collection. The system could effectively deal with many problems of the traditional parking system.

2 The Research Status of the Intelligent Parking System

The research of the intelligent parking system began earlier in Germany and Japan with more sophisticated techniques and better functions. In 1971, the world's first intelligent parking system was established in Aachen, Germany. In 1973, Japan had its first intelligent parking system in Kashiwa which could provide users parking guidance

F. Xhafa et al. (eds.), *Advances in Intelligent Systems and Interactive Applications*, Advances in Intelligent Systems and Computing 686,
https://doi.org/10.1007/978-3-319-69096-4_18

information such as the location, total parking number and the present use of the parking lot. And in 1995, the intelligent parking system had been applied in traffic management of over 40 cities in Japan. As the first successful case, the intelligent parking system of the Shinjuku area in Tokyo was established based on the comprehensive consideration of the related factors like traffic control and safety monitoring system [3].

In the theoretical research, Russell G. Thompson had a research on the optimization of the parking search model, parking guidance system and information display [4]. D. Teodorovi proposed a parking reservation system based on integer programming and fuzzy logic method [5]. V.W.S. Tang put forward an intelligent parking system based on low-cost WiFi technology which could enable data sharing.

The intelligent parking system develops rapidly in the theoretical research and applications. However, it still has some shortcomings such as imperfect working process, imbalance of resource allocation, loss of fare and lack of global insight. And some subsystems such as vehicle identification system, information publishing system, safety detecting system and user management system should be brought together to form a comparatively complete system. To solve the problem above, RFID technology is applied to the intelligent parking system with the overall configuration and the work flow studied in this paper.

3 The Overall Configuration of the System

With the application of the high frequency RFID technology and the digital video technology, the intelligent parking system can capture and record the car in and out information quickly and efficiently over from a distance without affecting traffic efficiency. The system also surfaces powerful information processing and query capabilities to ensure safety, efficient and orderly of vehicle management.

The readers, digital video devices, LED display devices and voice devices which are connected in the network are installed in the entrance of the parking areas when the system is used. All the cars in need of parking are required to have passive e-tags which bind personal information such as drives' license numbers and identification card numbers, license plate numbers, parking areas, EC balance. When the car drives to the entrance of the parking areas, the system can perceive the car, read the information of the tag and transmits the information to data processing center. The center records and analyzes the information with the assistance of the internet so as to judge the direction of the car and figure out the traffic flow over some time in this parking area. With the analysis of the traffic flow in this area, the design of entrances will be more reasonable and information service will be provided in the case of emergency.

In order to achieve effective management and monitoring, the entrance road of the parking areas is required to have mounting brackets on which there are RFID readers and surveillance cameras. And the cars in need of parking are required to have e-tags, the schematic diagram as shown in Fig. 1.

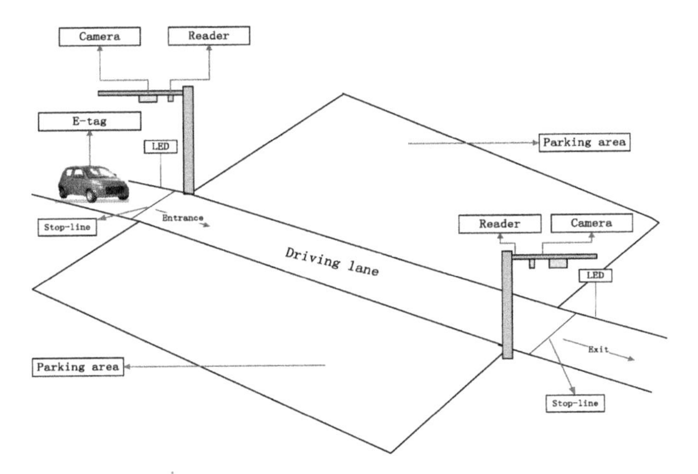

Fig. 1. The schematic diagram of intelligent parking system

3.1 The Basic Flow of the Intelligent Parking System

The system considers the parking processing as the base. Because of the distinct advantages over other technologies such as non-conduct identification, non-cash transactions, RFID technology is applied to the intelligent fee colleting management system. The prerequisite has a legal e-tag in which the unique ID number is connected to user information such as ID card number, license plate number, EC balance. The users can generally be divided into two kinds: with or without the e-tag. When a car stops at the stop line in the entrance of the parking areas, the reader receives the information from the e-tag of the car and transmits the information to the data processing center. The center judges the legitimacy of the tag and assigns a parking space to the user if the tag is legal, the basic flow of the intelligent parking system as shown in Fig. 2.

3.1.1 The Parking Process of the Vehicle with Legal e-tag

When a car with legal e-tag would enter the parking area, it should stop at the stop-line at the entrance firstly. The RFID reader detects the car and records the relevant information such as user profiles, vehicle images and arrival time. After the analysis of the data processing center, the user is told to have the permission to park by automatic broadcasting, with the parking space number and user information shown in the LED screen. After hearing the voice and confirming the information, the user drives into the parking area slowly and finds the assigned parking space.

When the car with legal e-tags would leave the parking area, it should stop at the stop-line at the exit firstly. The RFID reader detects the car and records the relevant information such as user profiles, vehicle images and departure time. After the analysis of the data processing center, the user is charged by total parking time (departure time-arrival time). With the assigned parking space number, user information, parking fees, current balances and total parking time shown in the LED screen. After hearing the voice and confirming the information, the user leaves the parking area slowly, which marks the end of the parking process.

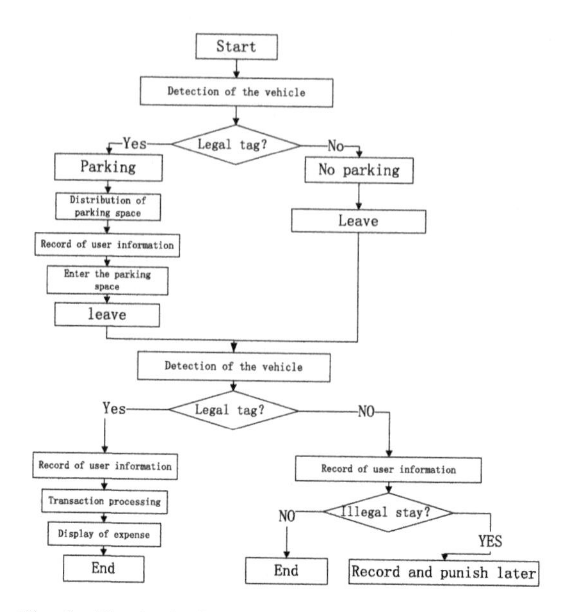

Fig. 2. The basic flow of the intelligent parking system

3.1.2 The Parking Process of the Vehicle Without Legal e-tag

When a car without legal e-tag would enter the parking area, it should stop at the stop-line at the entrance firstly. The RFID reader detects the car and records the relevant information such as user profiles, vehicle images and arrival time. The data processing center analyzes the information and finds that the e-tag of the car is illegal (such as information mismatch, incomplete information and null information). The user is told not have the permission to park by automatic broadcasting and required to leave via the driving lane within the stipulated time with “illegal e-tag” shown in the LED screen. After hearing the voice and confirming the information, the user leaves within the driving area slowly.

When the car without legal e-tags would leave the parking area, it should stop at the stop-line at the exit firstly. The RFID reader detects the car and records the relevant information such as user profiles, vehicle images and departure time. The data processing center analyzes the information and finds that the e-tag of the car is illegal and judges whether the user stays beyond the time limit. In that case, the system records the user profiles, vehicle images and arrival time. The authorities will comply with relevant regulations.

3.1.3 The Parking Process of the Vehicle Without Enough Electronic Balance

When the car without legal e-tags would leave the parking area, the procedures are similar to those with legal e-tags. The system records the information like user profiles, vehicle images and arrival time.

When the car would leave the parking area, the system records the information and the data processing center figures out the parking fee by total parking time (departure time-arrival time). If the user's electric balance is not enough, he is told that his charge is overdue with the assigned parking space number, user information, parking fees, total parking time and the amount owing shown in the LED screen. If the user has not paid all fees owed, he will be punished.

3.2 The Design of the Intelligent Parking Management System

According to the analysis of the function demand of the system, external entities can be determined. And according to the external entities and system functions, the parking management system can be designed in general, the structure of the intelligent parking management system as shown in Fig. 3. The data processing center is composed of four subsystems: vehicle identification system, information publishing system, safety detecting system, user management system. The vehicle identification system is responsible for the collection and the analysis of user information. The information publishing system is responsible for the display of user information and the voice prompts to users. The safety detecting system is responsible for the monitoring of parking areas and alarming in time when emergencies occur. The user management system is responsible for the summary and management of user information, the accounting and charging in parking process.

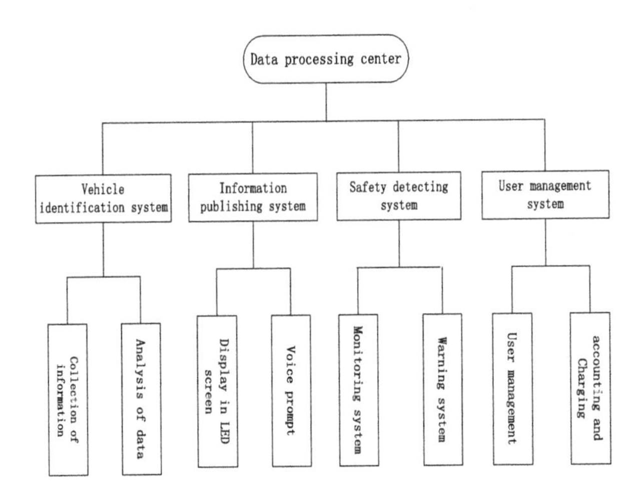

Fig. 3. The structure of the intelligent parking management system

3.3 The Basic Requirements for the Intelligent Parking System

- Reliable: All the equipment can be reliable and work well in any possible weather, environmental and traffic conditions. The system can deal with extreme cases and the percentage of charge errors and loss of expense should be controlled within the limits.
- Safe: The high safety factor of the system avoids the occurrence of accidents.

- Convenient: The process should be quick and easy so that users could drive through a toll booth in a short time.
- Good compatible: The establishment of the intelligent parking system requires that different systems built in various parking areas can be interconnected and various equipment must meet certain standards. In the same time, the information system not only manages user information but also provides other information like current traffic conditions, current weather and future climate.
- Economical: The cost of the system construction and operation should be kept as low as possible under the preconditions of function satisfaction to promote the popularity of the intelligent parking system.
- Good privacy: The design of the system should prevent the misuse of user information such as the leakage of user profiles.

4 Summary

As the new technology, RFID is being vastly applied in many high-tech industries and plays a more important role in intelligent traffic fields. This paper introduces the RFID technology, analyzes the framework architecture of the intelligent parking system and puts forward a new scheme to the solution of parking problems and the efficient use of resources and the improvement of paring efficiency.

However, many problems such as the plan of reasonable charge rate for parking remain to be solved. Further research and analysis will be conducted later.

References

1. Zhang, S.: Intelligent parking management system. J. Hunan Eng. Coll. **13**(2), 16–17 (2013)
2. You, Z., Li, S., et al.: The Theory and Application of RFID. Publishing House of Electronics Industry, Beijing (2004)
3. Zhou, Z., Chen, J., Wang, W., et al.: Theory trend researchom urban parking guidance and information system. J. Trans. Eng. Inf. **4**(1), 5–10 (2006)
4. Lo Lacono, L., Wang, J.: Web service layer security. Netw. Secur. **10**(2), 10–13 (2008)
5. Feng, Y.H.: Convergence theorems for fuzzy random variables and fuzzy martingales. Fuzzy Sets Syst. **10**, 435–441 (1999)

A Review of Cognitive Psychology Applied in Robotics

Huang Qin[1,2], Tao Yun[1,2], Wang Yujin[3(✉)], Wu Changlin[3], and Yu Lianqing[4]

[1] Department of Education and Management, Yunnan Normal University, Kunming 650092, Yunnan, China
[2] Key Laboratory of Educational Informatization for Nationalities (YNNU), Ministry of Education, Kunming 650092, China
[3] School of Mechanical Science and Engineering, Huazhong University of Science and Technology, Wuhan 430074, Hubei, China
wangyujin@hust.edu.cn
[4] School of Mechanical Engineering and Automation, Wuhan Textile University, Wuhan 430073, Hubei, China

Abstract. The application of robot is expanded unceasingly followed by the development of artificial intelligence, such as household and nursing, education and other services. The human–machine interactive mode changes fundamentally as the relation within them is closer, while emotional interaction is the prevailing trend. And thus there is a need of the guidance of cognitive psychology in robotics. This paper reviews the robotic emotion recognition, and then introduces the cognitive theory applied in robotics, including robot autism intervention, robot education and household service robot.

Keywords: Cognition · Emotion recognition · Autism · Robot education · Services

1 Introduction

Cognitive psychology emerged and spread in the western psychological field since the mid-1950s, in which cognitive process, such as perception, memory, attention, recognition, learning, emotion, planning, reasoning and problem solving, etc., is handled with information processing [1]. Contemporary cognitive psychology has the features of situational and embodiment, that it emphasizes to study cognition in real life as the cognition is activities of living body in real-time environment [2].

Human beings are disposed as information processing systems in cognitive psychology, while the cognitive process includes sensory process, memory process, reaction process and control process [3]. The basic idea of cognitive psychology is that the brain is similar to a computer, and thus, cognitive process is deemed as the representation and operation process of the computer. Based on this viewpoint, artificial intelligence springs up through the coalition of cognitive psychology and computer science, with the aim to imitate human intelligence and further to design robots with the ability to respond in anthropomorphic intelligence style. Robotics is a hot topic in the

F. Xhafa et al. (eds.), *Advances in Intelligent Systems and Interactive Applications*, Advances in Intelligent Systems and Computing 686,
https://doi.org/10.1007/978-3-319-69096-4_19

artificial intelligence field as it provides application instances for simulating human thinking. The expanding application of robot, such as household, nursing, education and other services, encounters with more complex work environment, flexible assignment, and more in-depth human–machine interaction. Therefore, it requires the theoretical support of the cognitive psychology which can provide the human logical thinking mode.

2 Robot Affective Computing and Interaction

The relationship between human and robot in the service fields is more like the interaction of two intelligent systems. In-depth interaction and further emotional interaction require that service robots should have the ability of emotion cognition which is an important role in the reasoning, learning and memory of human beings. Consequently, the aim of robot affective computing is to imitate human emotion process through information method by the signal recognition of physiological characteristics and behavior characteristics. In this way, the emotional interaction based on affective computing provides a natural and harmonious interactive mode since that the robot has the ability of emotional recognition, understanding and expression, which are also the main research contents of the affective computing.

2.1 Emotion Recognition

The methods of the recognition of human emotional states in robotics include facial expression recognition, physiological signal recognition, speech emotion recognition, and also the poses recognition.

Facial expression is a kind of information representation of nonverbal communication, which is adequately represented by the movements of facial muscles. The facial expression recognition is a necessary condition for social interaction, and is directly affected by the attention and cognitive time. Many researchers have studied on how to recognize the expression using artificial intelligence. Mase [4] proposed a recognition method that using optical flow to track the muscle movements rather than feature points matching, and then got accuracy rate of 80% in the experiment. Li Hai-bo put forward a pattern recognition method which combined computer graphics and visual servo [5]. Trevor [6] proposed an approach to synthesize facial expression with neural network and interpolation algorithms.

Speech emotion recognition is a problem of pattern recognition since that the signals generated by different tones have the different structure characteristics of time, amplitude, fundamental frequency and formant. To define each sub-network of as a type of basic emotion, Nicholson [7] used the output of the topologies to represent a speech emotion composed with different types, and then got an accuracy of 73% for the training set but lower than 60% for beyond. Park et al. simulated the nonlinear kinetic behavior of brain using a dynamic recurrent neural network composed with an input node, two hidden layer nodes and four output nodes, thus it can recognize four kinds of speech emotion and with an accuracy of 77% on average [8].

Physiological signal recognition is to recognize emotional situation by the detection of the physiological signals. Schwartz [9] verified that positive emotions are more likely to arouse the activities of zygomaticus, while negative arouse the corrugator. And then, Picard [10] estimated the types of emotion except intensity with an accuracy rate of 81%. Andrea [11] identified six types of emotion through feature extraction and pattern recognition for the collected signals of pulse, electromyography, skin conductance and dermal thermometry from subjects. Wagner [12] collected signals of respiration, skin conductance and blood pressure from the subjects stimulated by music, and then the original signals with noise were analyzed by Fisher mapping and pattern recognition.

Pose especially for gesture is another aspect of emotion recognition, since it also implies emotional information. Camurri et al. [13] designed a system to acquire the posture data, which were mapped as four basic emotions after multi-level signal vector analysis. Institute of Computing Technology, Chinese Academy of Science has developed a system based on data glove for sign language recognition and synthesis, which is very useful for the interaction of deaf-mutes.

2.2 Affective Modeling

The core issue of whether the robot has the anthropomorphic emotion is to build an appropriate emotion model. Many scholars have studied on affective modeling, including HMM, OCC, Cathexis, Kismet, personality-mood-emotion and Euclidean space emotion model.

The HMM model proposed by Picard includes a Markov process and a random process, which are used to describe the state transition, and the relation of the states and observation sequence, respectively [14]. The OCC model proposed by Ortony, Clore and Collins [15] defines 22 kinds of emotion states generated with the extensible rule. Cathexis proposed by Velbsquez is a distributed model of emotion generation [16], which is composed with emotion proto-specialists. Each emotion proto-specialist represents a kind of emotion, and then four types of sensor are used to activate the emotion. Breazeal designed an emotional robot called Kismet [17] with the ability to respond to external stimulus according to the emotional point in its 3-dimension emotion space represented by awake, posture and valence. Kshirsagar [18] proposed a personality-mood-emotion layered emotion model, in which the mood is perceived as a transformational psychological state situated between personality and emotion. Euclidean space emotion model is proposed by Wei Zhehua [19], which involves three axes representing three basic emotional states. Therefore, each emotional state could be shown in the form of 3-dimension coordinates, and then the emotional intensity is described as the coordinate value. There are 27 emotional states in the Euclidean space model caused by dispersing the axes into discrete values, such as 0, 0.5 and 1. The transfer among the states is described as a Markov process.

2.3 Emotion Expression

To give robot the ability of emotion expression is a major research topic in robotics and there are many examples around the world.

In the 1990s, Cynthia Breazeal worked in Artificial Intelligence Laboratory of MIT designed a robot called Kismet, which is a humanoid robot with six types of facial expression. The emotional robot WE-3RV is developed in Waseda University, which has nine types of basic emotion in its 3-dimension emotion space. The robot SAYA, which is developed by Tokyo Polytechnic University, generates facial expression based on pneumatic artificial muscles. NAO is humanoid robot with a yearling emotional and behavioral learning mode, thus, it can analyze human emotion and interact with people bilaterally through the observation of expression and poses. Professor Hiroshi of Osaka University designed a series of female humanoid robot Repliee Q1, Repliee Q2 and Geminoid TMF.

Professor Wu of Harbin Institute of Technology designed a humanoid head robot called H & F robot-I with the capacity of showing eight basic emotions. Sequentially, he developed the robot H & F robot-II and H & F robot-III with the performances of facial movements coordinated and synchronized mouth shapes with voice recognition [20]. Zhi-liang Wang [21] carried out the study on artificial emotion and multi-information fusion interaction, then developed humanoid head robot with the characteristics of man-machine mutual attention and dynamic emotion regulation. Xian-xin Ke developed an expression robot SHFR-1 using AT89S52 and seven servo motors [22]. In addition, "Hundred Wise Star" and "Tong Tong" are also the facial expression robots developed by Harbin Institute of Technology and Institute of Automation, Chinese Academy of Sciences, respectively.

The facial expression robots proposed recently are shown in Fig. 1.

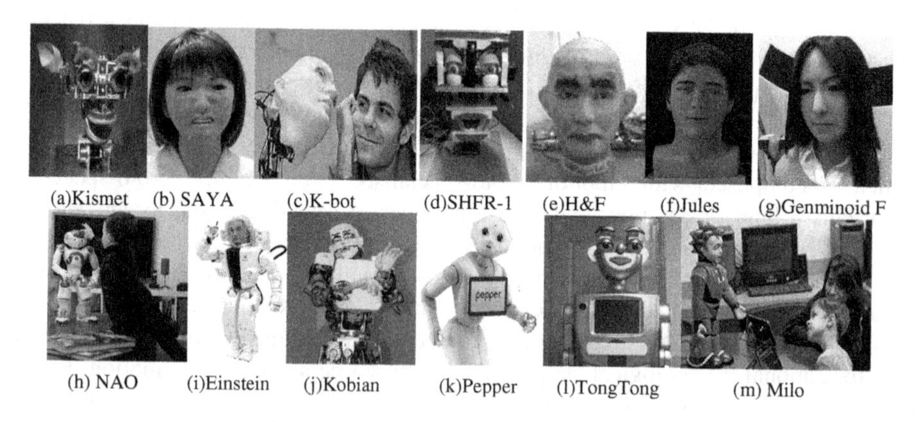

Fig. 1 Plots of facial expression robots

3 Application of Robot Cognition

3.1 Robot Autism Intervention

Autism is a developmental disorder diseased in the early child with the symptoms of communicating obstacle, language development disorders, narrow interests and repetitive stereotyped behaviors. The three symptoms of autism are explained by the defect of theory of mind (TOM), deficient central coherence and deficiency of executive function in cognitive psychology, respectively [23]. Serious lack of emotion recognition, which is inseparable from social interaction and the development capacity of emotion and behavior, causes autistic children missing the facial expression cognition, such as angry, joy, fear, sad, surprise and disgust.

The rehabilitation of autism has an important significance as the incidence becomes higher recently, but there is no significant breakthrough in drug therapy approaches. Medical practice researches show that early intervention training is an effective means of autism rehabilitation; however, the difficulty of intervention is to communicate with autistic children who are incapable of expression cognition. The research of Langdell shows that non-social stimuli, such as nonliving objects, are easier to be recognized by autism children, rather than social stimuli, like human faces [24]. The study on cognition process of autism in cognitive psychology declares that it is possible for early intervention using external stimuli, such as robots and toys. The interactivity of robots causes the attention and curiosity of autism children, and then improves the capacities of facial expression cognition and social interaction [25]. More importantly, the simple facial expression and body langue of the humanoid robots simplify the communication and make the possible of interaction with autism children who fear about the rich facial expression and body langue of human.

Many researchers have studied on robot autism intervention therapy, such as NAO, Milo, and Lego Mindstorms. The University of North Carolina and the school of psychology of university of Notre Dame bring the robot NAO into autism treatment firstly. In the study of interaction, Tapus, Shamsuddin, Dong-fan Chen et al. found that autism children are most interest in the robot NAO, and the social behavior is more frequently than with human, thus, they confirmed that robots can help autism children to reduce the stereotyped behavior, improve the communication behavior, and then, social interaction [26, 27]. Another humanoid robot named Milo, developed by RoboKind Company, is also used in autism intervention therapy.

As a new method of autism therapy, robot intervention has clear effects on facial expression recognition, language communication, and social interaction. Moreover, robot can help physician to do the repeated work as the rehabilitation specialists are scarce recently, due to the advantage of repeatedly of the robots. The role of robot in autism intervention will be more significant as the close combination of artificial intelligence with cognitive psychology.

3.2 Robot Education

Learning is a cognition process, including reception, conversion, storage and extraction of the input information. Constructivism considers the learning as a process of knowledge construction composed with the new knowledge construction and experience recombination, and the appearing and solving of cognitive conflict between the two component parts improve the development of cognition. According to the education situation model proposed by Bernd Weidenmann, the influence factors of learning include learner, teacher, media and environment, and the relationships among them are shown as Fig. 2. Using robot as the learning media conforms to the practical, situational and embodiment of cognition theory, and also conforms to the interactive cognitivism philosophy which emphasizes cognition comes from interaction. Robot education motivates the interest, thus, improves the learning efficiency, innovation ability, operational ability, teamwork, logical thinking, independent learning and social interaction [28]. The research on robot education includes robotic education theory, robotic teaching and competition.

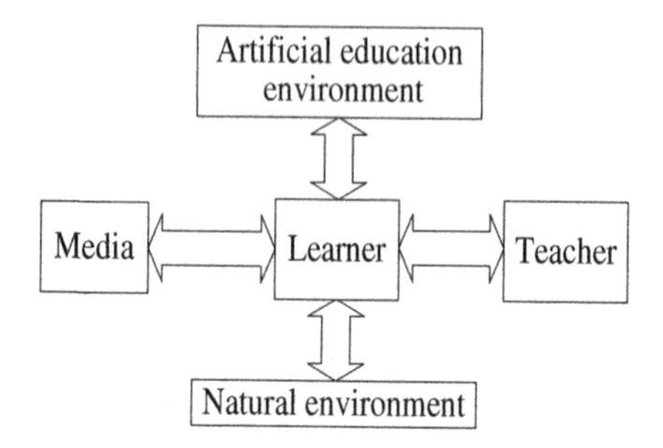

Fig. 2 Education situation model

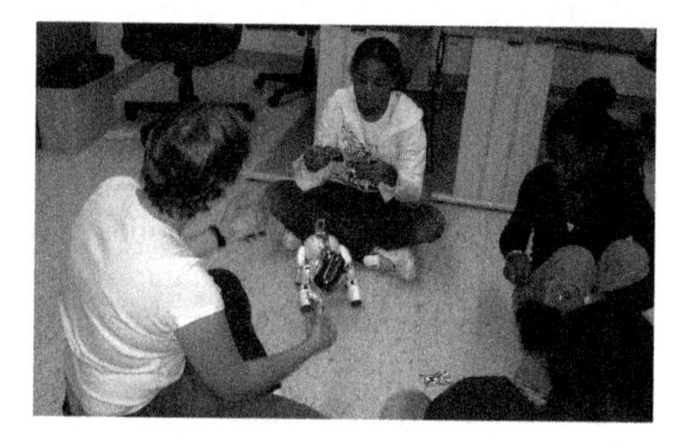

Fig. 3 Students work with robots [31]

In the aspect of theoretical research, constructivism learning, project-based learning, problem-based learning and self-directed learning are the general theoretical guidance of robot education. Bersetal et al. [29] consider that the four principles of constructivism learning are more appropriate for robot education. LEGO Education proposed a 4C model of instruction in which the learning process is divided into 4 phases that are Connect, Construct, Contemplate, and Continue. The four phases form a spiral development enabling learner to connect experience, construct new task, contemplate operation process and continue new challenge. Mauch thinks that robot education can overcome the repeated practice and passive acceptance existing in the traditional education [30]. Currently, robot teaching is a common phenomenon. For example, the Center for Distributed Robotics at the University of Minnesota established an annual robotics day camp to inspire both women and minorities to pursue careers in technology as shown as Fig. 3 [31]. On the other hand, robot competitions are most prevalent over the world, including RoboCup, Federation of International Robot-soccer

Association (FIRA), First Lego League (FLL), International Robot Olympiad (IRO), THE Trinity College Fire-Fighting Home Robot Contest (TCFFHRC), China's National Robot Competition, China Adolescent Robotics Competition (CARC), etc.

3.3 Household and Nursing Robot

The world faces a severe demographic challenge as the rapidly-growing elderly population. The increasing aged population and changes in family structure lead to the weakness of the family support. The lacking of company and communication of the aged could result in depression and autism. Besides, the aged population generally shows degradation of cognition and physiological function [32], such as movement disorders, visual and hearing impairments, language communication disorder, degradation of memory and fluid intelligence, etc. A new mode of the providing for the aged is necessary, because of the gradually weakening in traditional aged mode provided by family caused by the great number of elderly population.

To use robots for household and nursing is an effective solution, because that they fulfill various needs of the aged and disabled according to Maslow's theory of hierarchy of human needs. In 1984, Engelberger developed a home nursing robot named Help Mate with the functions of obstacle avoidance and autonomous navigation [33]. Riken Institute developed an intelligent care robot RI-MAN with vision, smell and auditory senses, and can help nurses to take care of the elderly [34]. In 2006, Mitsubishi Heavy Industries Ltd. marketed a household robot named "AR", which can wash clothes and mop the floor. In 2008, National Chiao Tung University developed a household robot called ROLA with the abilities of speech recognition and natural conversation. Due to the vision-servo, ROLA recognize the faces and indoor environment, thus, avoid obstacle, and locate user's location. Besides, it also receives the physiological signals monitored by wearable devices, and then sounds an alarm if necessary. Tokai Rubber Industries has partnered with Riken Institute to develop a nursing support robot named RIBA II with a rubber covered exterior. This robot can transfer patients safely from one place to another, due to the usage of series sensors. Toyota developed a Human Support Robot (HSR) with the aim to help the disabled and the aged people [35]. Additionally, the robots Pepper and JIBO also have great potential in household. With the rapid development of network, wearable devices and smart home, and also the pressures of the aged providing and nursing, household and nursing robot will occur new opportunities.

4 Conclusion and Prospect

The application of robot will draw much close to real life as the advancements in technology, especially for the machine vision and wearable devices. Then the combination of robotics with cognitive psychology will become closer increasingly; therefore, bring in higher cognition compatibility within them. The cognitive psychology provides theoretical support for the spontaneous and harmonious interaction between robot and human by means of study on the cognitive structure and cognitive process.

The intelligent level of the robot is far less than human beings at present, due to the lacking of intentionality, semantics and natural language understanding. Among them, natural language understanding is the most important goal of artificial intelligence which needs a long-term study, caused by the complexity and non-normalization. The robotics must combine more closely with computer science, neurosciences, cognitive psychology, philosophy of mind and linguistics to tackle these key problems.

Acknowledgments. Support from the National Natural Science Foundation of China, Grant No. 51275363 and Grant No. 31660282 toward this paper is gratefully appreciated.

References

1. Ye, Hao-Sheng: Embodied cognition: a new approach in cognitive psychology. Adv. Psychol. Sci. **18**(5), 705–710 (2010)
2. Anderson, M.L.: Embodied cognition: a field guide. Artif. Intell. **149**(1), 91–130 (2003)
3. Glass, A.L., Holyock, K.J.: Cognition, p. 5. Random House, New York (1986)
4. Mase, K.: Recognition of facial expression for optical flow. IEICE Trans. Spec. Issue Comput. Vision Appl. **E74**(10): 3474–3483 (1991)
5. Li, H.-B., Roivainen, P., Forheimer, R.: 3D motion estimation in model based facial Image coding. IEEE Trans. Pattern Anal. Mach. Intell. **15**, 545–555 (1993)
6. Terzopoulos, Demetri: Physically-based facial modeling analysis and animation. J. Vis. Comput. Anim. **1**(2), 73–80 (1990)
7. Nicholson, J., Takahashi, K., Nakatsu, R.: Emotion recognition in speech using neural networks. Neural Comput. Appl. **9**, 290–296 (2000)
8. Park, C.H., Sim, K.B.: Emotion recognition and acoustic analysis from speech signal. In: International Joint Conference on Neural Networks (IJCNN2003), vol. 4, pp. 2594–2598, IEEE, New York (2003)
9. Sehwartz, G.E., Fair, P.L., Salt, P., et al.: Facial muscle pattening to affective image indepressed and nondepressed subjects. Science **192**, 489–491 (1976)
10. Picard, R.W., Vyzas, E., Healey, J.: Towards machine emotional intelligence: analysis of affective physical state. In: IEEE Trans Pattern Analysis and Machine Intelligence, pp. 119–175, IEEE, New York (2001)
11. Haag, A., Goronzy, S., et al.: Emotion recognition using biosensors: first steps towards automatic system. In: Proceedings of the Kloster Irsee Tutorial and Research Workshop on Affective Dialogue Systems, pp. 36–48, Springer, Germany (2004)
12. Wagner, J., Kim, J., Andre, E.: From physiological signals to emotions: implementing an comparing selected methods for feature extraction and classification. In: Proceedings of IEEE International Conference on Multimedia and Expo. Paperback, pp. 940–943. IEEE Computer Society Press, Wiley (2005)
13. Camurri, A., Volpe, G., De Poli, G., et al.: Communicating expressiveness and affect in multimodal interactive systems. IEEE Multimedia **12**(1), 43–53 (2005)
14. Picard, R.W.: Affective Computing. MIT Press, USA (1997)
15. Ortony, A., Clore, G.L., Colins, A.: The Cognitive Structure of Emotions. Cambridge University Press, Cambridge, UK (1988)
16. Velásquez, J.D.: Modeling emotions and other motivations in synthetic agents. In: Proceedings of the Fourteenth National Conference on Artificial Intelligence and Ninth Innovative Applications of Artificial Intelligence Conference, Menlo Park (1997)

17. Breazeal, C.: A motivational system for regulating human-robot interaction. In: Proceedings of the National Conference on Artificial Intelligence, pp. 54–61. Madison, WI (1998)
18. Kshirsagar, S.: A multilayer personality model. In: SMARTGRAPH'02: Proceedings of the 2nd International Symposium on Smart Graphics, pp. 107–115. New York (2002)
19. Wang, Z.: Artificial psychology—a most accessible science research to human brain. J. Univ. Offence Technol. Beijing (2000)
20. Meng, Q., Wu, W., Zhong, Y., et al.: Research and experiment of lip coordination with speech for the humanoid head robot-"H&Frobot-III". In: Humanoids 2008, Ieee-Ras International Conference on Humanoid Robots, pp. 603–608. IEEE (2008)
21. Wang, W., Wang, Z., Zheng, S., et al.: Joint attention for a kind of facial expression robots with the same structure. Robot **34**(3), 265–274 (2012) (in Chinese)
22. Tang, W.B., Xian-Xin, K.E., Bai, Y., et al.: Control system design of a facial robot. J. Mach. Des. (2010) (in Chinese)
23. Klin, A., Jones, W., Schultz, R., et al.: Defining and quantifying the social phenotype in autism. Am. J. Psychiatry **159**(6), 895–908 (2002)
24. Zhou, N., Fang, J.: An Experimental study of the characteristics of emotion cognition in autistic children. Psychol. Sci. **3**, 25–28 (2003). (in Chinese)
25. Cabibihan, J.J., Javed, H., Ang, M., et al.: Why robots? A survey on the roles and benefits of social robots in the therapy of children with autism. Int. J. Soc. Rob. **5**(4), 593–618 (2013)
26. Tapus, A., Peca, A., Aly, A., et al.: Children with autism social engagement in interaction with Nao, an imitative robot: a series of single case experiments. Interact. Stud. **13**(3), 315–347 (2012)
27. Shamsuddin, S., Yussof, H., Ismail, L.I., et al.: Initial response in HRI—a case study on evaluation of child with autism spectrum disorders interacting with a humanoid robot Nao. Procedia Eng. **41**(41), 1448–1455 (2012)
28. Nourbakhsh, I., Crowley, K., et al.: The robotic autonomy mobile robotics course: robot design, curriculum design and educational assessment. Auton. Rob. **18**(l), 103–127 (2005)
29. Bers, M., Ponte, I., Juelich, C.: Teachers as designers: integrating robotics in early childhood education. Inf. Technol. Child. Educ. Annu. **14**, 123–145 (2002)
30. Maueh, E.: Using technological innovation to improve the problem solving skills of middle school students. Clearing House **74**(4), 211–213 (2001)
31. Cannon, K., Lapoint, M.A., Bird, N., et al.: Using robots to raise interest in technology among underrepresented groups. IEEE Robot. Autom. Mag. **14**(2), 73–81 (2007)
32. Meng, Q., Lee, M.H.: Design issues for the elderly. Adv. Eng. Inform. **5**(20), 171–186 (2006)
33. Engelberger, J.: Helpmate G, a service robot with experience. Ind. Robot **22**(2), 101–104 (1998)
34. Onishi, M., Takagi, K., Luo, Z., et al.: A Soft Human-Interactive Robot RI-MAN. In: IEEE/RSJ International Conference on Intelligent Robots and Systems. p. 1. New York, IEEE (2006)
35. Hashimoto, K., Saito, F., Yamamoto, T., et al.: A field study of the human support robot in the home environment. In: Advanced Robotics and ITS Social Impacts, pp. 143–150, New York, IEEE (2013)

Robot Vision Navigation Based on Improved ORB Algorithm

Sun-Wen He[1(✉)], Zhang-Guo Wei[1,2], and Lu-Qiu Hong[2]

[1] Shanghai University of Electric Power, Shanghai, China
402094040@qq.com
[2] Shanghai HRSTEK Co., Ltd., Shanghai, China

Abstract. This paper presents a visual navigation method based on Improved ORB mobile robot. In view of the fact that ORB algorithm does not have scale invariance at feature point matching, an improved ORB algorithm is proposed based on the idea of SIFT algorithm. Firstly, the multi-scale space of the image is generated, and the stable extremum is detected in the multi-scale space, so that the extracted feature points have the scale invariant information. Then, the feature points are described by the ORB descriptor to generate the binary invariant descriptor. Using Improved ORB to extract the ORB features of the input scene, and combined with the robot odometer information to achieve robot navigation. The experimental results show that the method can accomplish the navigation task of the robot well, and it has some robustness to the dynamic information in the environment.

Keywords: Mobile robots · Feature point matching · Scale invariance · Improved ORB · Visual navigation

1 Introduction

At present, the robot navigation methods are the following: inertial navigation, magnetic navigation, satellite navigation such as GPS, visual navigation and so on [1]. Among these methods, visual navigation has become an important direction in the field of intelligent robot navigation because of its abundant information and versatility of environmental cognition.

Image feature matching is very critical in the issue of visual navigation. Once the matching error, it is possible to cause the robot localization failure, so the robustness of the image feature matching algorithm is high. At present, the main image feature matching algorithms are [2]: SIFT, SURF and ORB. Among these algorithms, SIFT and SURF are widely used, and their matching accuracy is high, which is applicable to many kinds of image transformations, but the operation speed is slow. ORB [3] is one of the most rapid image feature matching algorithms. Its feature point detection and descriptive extraction are based on FAST corner and BRIEF description algorithm respectively, and add rotation invariance. However, the ORB does not have scale invariance, and the feature point matching effect is poor when the image scale changes. Therefore, this paper combines the SIFT algorithm and improves the ORB to solve the problem that ORB does not have scale invariance.

F. Xhafa et al. (eds.), *Advances in Intelligent Systems and Interactive Applications*, Advances in Intelligent Systems and Computing 686,
https://doi.org/10.1007/978-3-319-69096-4_20

2 Improved ORB Algorithm

The original ORB matching algorithm mainly includes the following steps [4]: Firstly, the Oriented FAST detector is used to detect the feature points on two consecutive images. Then the binary descriptive vector of the feature points is generated by the Rotated BRIEF descriptor. Finally the matching of the feature points of the image uses the Hamming distance ratio criterion to get the final feature points matching.

The fundamental reason why the ORB does not have the scale invariance is that the feature point detected by FAST does not contain the scale invariant information, so that in the subsequent ORB step, although the direction of the feature point can be introduced to obtain the rotation invariance, it can not make the descriptor have the scale invariant [5]. Considering that SIFT is using multi-scale space theory to extract stable feature points, then the SIFT descriptor is used to describe and match the feature points, which makes the SIFT algorithm have good scale invariance [6]. Therefore, based on the idea of SIFT algorithm, the multi-scale space of SIFT algorithm is used to extract feature points with scale invariance. Then the feature points are described by ORB to generate the ORB descriptor, and the feature points are matched by using the Hamming distance. This not only solves the problem that ORB does not have scale invariance, but also retains the advantages of fast speed of ORB descriptor and other advantages such as rotation invariance.

2.1 Extreme Points in Scale Space Detection

Lindeberg et al. have proved [7] that under some reasonable constraints, the Gaussian function is the only smooth kernel function of the scaling kernel. The dimension of the image $F(x, y)$ is defined as a function $L(x, y, \sigma)$ (σ represents the scale factor) which consists of the Gaussian function $G(x, y, \sigma)$ and the image $F(x, y)$:

$$L(x, y, \sigma) = G(x, y, \sigma)F(x, y) \tag{1}$$

$$G(x, y, \sigma) = \frac{1}{2\pi\sigma^2}\exp(-\frac{x^2 + y^2}{2\sigma^2}) \tag{2}$$

In order to effectively detect the stable feature points in the scale space, the extreme points are found in the space $D(x, y, \sigma)$ obtained by the Gaussian difference function and the image convolution, and the local extreme points are used as the candidate feature points of the scale space:

$$D(x, y, \sigma) = [G(x, y, k\sigma) - G(x, y, \sigma)]F(x, y) = L(x, y, k\sigma) - L(x, y, \sigma) \tag{3}$$

In Eq. (3), the adjacent two scales are separated by a constant k.

2.2 Extract Stable Feature Points

After obtaining the extreme points in the scale space, we need to select these extreme points to remove the unstable points, so as to enhance the stability and anti-noise ability of the feature points matching [8]. For the extreme points obtained on a certain scale, the 2 dimensional function of the 3 dimension is used to find the position of the extreme point on the original image, and the extreme point of the low contrast is removed. At first, $D(x, y, \sigma)$ was carried out at a certain extreme point by Taylor expansion:

$$D(X) = D + \frac{\partial D^T}{\partial X} X + \frac{1}{2} X^T \frac{\partial^2 D}{\partial X^2} X \tag{4}$$

For the partial derivative of formula (4), and the partial derivative is zero, we can get the extreme point $\dot{X}$:

$$\dot{X} = -\frac{\partial^2 D^{-1}}{\partial X^2} \frac{\partial D}{\partial X} \tag{5}$$

The formula (5) is brought into the formula (4) to obtain:

$$D(\dot{X}) = D + \frac{1}{2} \frac{\partial D^T}{\partial X} \dot{X} \tag{6}$$

If the absolute value of $D(\dot{X})$ is less than 0.03, the corresponding extreme point is removed to filter the extreme points of low contrast. In order to obtain a stable extreme point, it is necessary to remove the extreme points on the edge, which can be achieved by calculating the principal curvature ratio. First calculate the threshold to be detected Heather matrix:

$$H = \begin{pmatrix} D_{xx} & D_{xy} \\ D_{xy} & D_{yy} \end{pmatrix} \tag{7}$$

Suppose α is H the largest eigenvalue, β is the smallest eigenvalue, the matrix properties can be seen:

$$\begin{cases} Trace(H) = D_{xx} + D_{yy} = \alpha + \beta \\ Det(H) = D_{xx} D_{yy} - (D_{xy}^2) = \alpha\beta \end{cases} \tag{8}$$

Since the eigenvalues of H are proportional to the principal curvature of D, only the ratio of eigenvalues is concerned. Suppose γ is the ratio of the largest and smallest eigenvalues, namely $\alpha = \gamma\beta$. It can be obtained:

$$\frac{Trace(H)^2}{Det(H)} = \frac{(\alpha + \beta)^2}{\alpha\beta} = \frac{(\alpha\beta + \beta)^2}{\gamma\beta^2} = \frac{(r+1)^2}{r} \tag{9}$$

The result of Eq. (9) is only related to the ratio of two eigenvalues, but not to the specific eigenvalue. The right term of Eq. (9) increases with γ. So to check the ratio of the main curvature is less than a certain threshold γ, as long as check whether the following holds:

$$\frac{Trace(H)^2}{Det(H)} < \frac{(r+1)^2}{r} \tag{10}$$

The literature [8] indicates the threshold, that is, the extreme point of the principal curvature ratio greater than $\gamma = 8$ is the point on the edge, which is filtered out.

3 Visual Navigation Method

Mobile robot navigation methods including two parts, the path of learning and state positioning and navigation. The first part: the path of learning, in an unknown environment, a pre-specified path for robot learning, through continuous testing environment and record the environmental characteristics of the visual system, while recording the odometer information to calculate the corresponding behavior model, the environmental characteristics and behavior patterns as state nodes to construct the topological map. The second part is the status of location and navigation path: state node robot comparing the original study, namely the environment characteristics of the visual detection system and the preservation of the state, locate and select the corresponding mode of navigation behavior. Figure 1 is the navigation method for mobile robot.

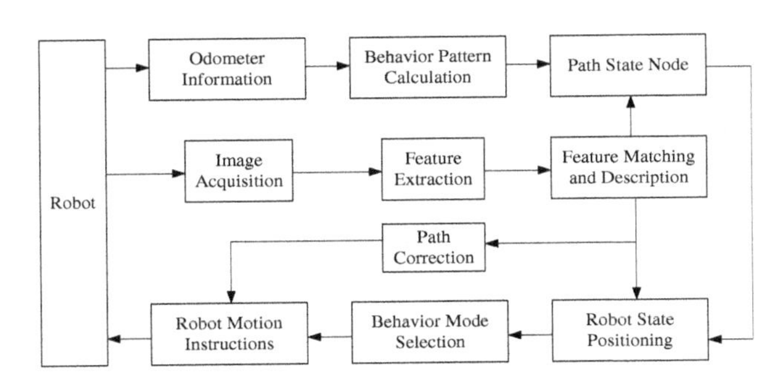

Fig. 1 Mobile robot vision navigation method flow chart

4 Experimental Results

In order to verify that the Imoroved ORB algorithm can effectively overcome the defects that the ORB does not have the scale invariance, the contrast experiment is carried out on the scale of the test image. The experimental results of the image are shown in Fig. 2. It can be seen from Fig. 2a, ORB due to the use of FAST corner detection so extracted more feature points, but also because there is no scale invariance, making the feature point of the match is more messy; When using Imoroved ORB to match, the more satisfactory experimental results are obtained, as shown in Fig. 2b. It can be found that the Imoroved ORB algorithm has a good matching effect when the target has a large scale change, which overcomes the defects that the ORB does not have scale invariance.

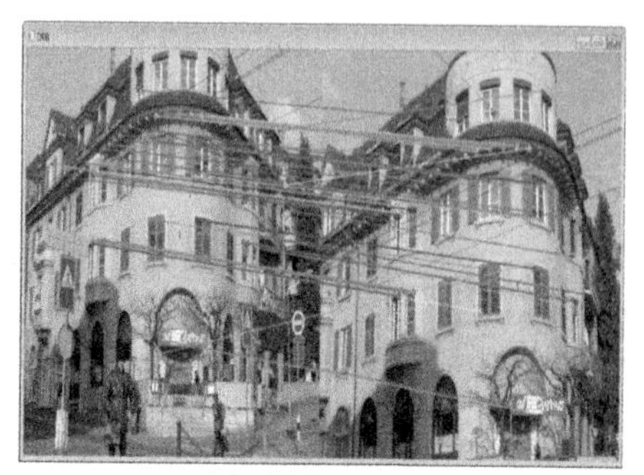

(a) ORB match effect

(b) Imoroved ORB match effect

Fig. 2 **a** ORB match effect. **b** Imoroved ORB match effect

In order to verify the practicability of Improved ORB algorithm, the matching time and accuracy of each algorithm are compared. The statistical results are shown in Tables 1 and 2. The following conclusions are obtained: (1) The average matching time of Improved ORB is about 17.4 ms longer than ORB, which can be roughly equivalent, and the average matching speed is about 61.9 times faster than that of SIFT, which shows the superiority of Improved ORB. (2) When the image scale changes, Improved ORB can effectively and accurately match the feature points, and the average matching accuracy is about 92.9%, which is about 71.2% higher than that of ORB.

Table 1 Comparison of matching time

	Matching Algorithm		
	Improved ORB	ORB	SIFT
1	50.3	30.6	3238.6
2	43.1	26.7	2896.5
3	50.7	31.2	3268.9
4	63.4	33.1	3498.7
5	15.3	9.8	783.2
6	49.2	30.1	3354.4
7	53.5	32.6	3471.2
8	20.4	12.3	864.5
Average value	43.2	25.8	2672

Table 2 Comparison of matching accuracy with the changes of image scale

	Matching algorithm		
	ORB	ORB	SIFT
1	94.2	27.5	97.6
2	93.1	19 .2	95.3
3	92.6	16.9	94.1
4	95.8	31.1	98.5
5	93.4	20.2	96.1
6	91.8	17.8	92.8
7	88.5	13.6	91.3
8	93.7	27.2	95.9
Average value	92.9	21.7	95.2

In this paper, using Kinect to get the input scene image sequence to verify the validity of the proposed method. The image acquisition frequency is 30 FPS, the number of feature points is 500, and the image similarity threshold is less than 0.5 when the path is learned. The image matching rate threshold is above 0.8 when the state is located. The total length of the path of the robot walking in the indoor environment is 6.9 m, the path is inverted 'L' shape. In the process of constructing the map, the robot walks only once, and a total of 1200 frames are collected, and the state nodes are learned, as shown in Fig. 3. The robot trajectory obtained after navigation is compared with the learning path. As shown in Fig. 4, the robot navigation error is 0.06 m, it shows a stable navigation capability.

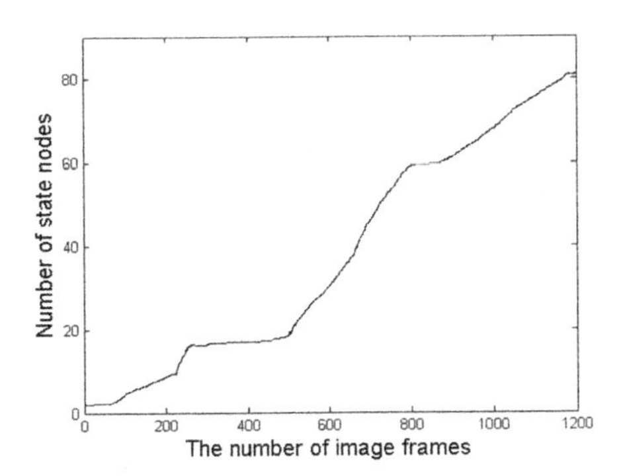

Fig. 3 State node generation

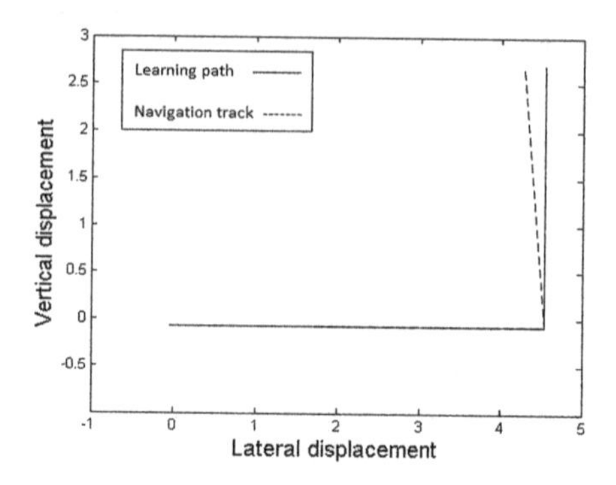

Fig. 4 Robot navigation track

5 Conclusion

The Improved ORB algorithm proposed in this paper solves the problem that ORB does not have scale invariance, while retaining the advantages of ORB fast. Then the experiments on mobile robot vision navigation methods based on ORB features and odometer information confirmed the feasibility of the algorithm.

References

1. Fuentes-Pacheco, J., Ruiz-Ascencio, J., Rendón-Mancha, J.M.: Visual simultaneous localization and mapping: a survey. Artif. Intell. Rev. **43**(1), 55–81 (2015)
2. Gauglitz, S., Höllerer, T., Turk, M.: Evaluation of interest point detectors and feature descriptors for visual tracking. Int. J. Comput. Vision **94**(3), 335–360 (2011)
3. Rublee E, Rabaud V, Konolige K, et al.: ORB: an efficient alternative to SIFT or SURF. In: IEEE International Conference on Computer Vision. pp. 2564–2571. IEEE, New York, (2011)
4. Mur-Artal, R., Montiel, J.M.M., Tardós, J.D.: ORB-SLAM: a versatile and accurate monocular SLAM system. IEEE Trans. Rob. **31**(5), 1147–1163 (2015)
5. Calonder M, Lepetit V, Strecha C, et al.: BRIEF: binary robust independent elementary features. In: European Conference on Computer Vision, pp. 778–792. Springer, Berlin (2010)
6. Rublee E, Rabaud V, Konolige K, et al.: ORB: an efficient alternative to SIFT or SURF. In: International Conference on Computer Vision, pp. 2564–2571. IEEE Computer Society, Washington, D.C. (2011)
7. Lowe, D.G.: Distinctive image features from scale-invariant keypoints. Int. J. Comput. Vision **60**(2), 91–110 (2004)
8. Rosten, E., Drummond, T.: Machine learning for high-speed corner detection. In: European Conference on Computer Vision, pp. 430–443. Springer, Berlin (2006)

An Improved Multi-label Relief Feature Selection Algorithm for Unbalanced Datasets

Yonghong Xie[1,2], Daole Li[1,2], Dezheng Zhang[1,2](✉), and Ha Shuang[1,2]

[1] School of Computer and Communication Engineering, University of Science and Technology Beijing, Beijing 100083, China
daoleli@126.com

[2] Beijing Key Laboratory of Knowledge Engineering for Materials Science, Beijing 100083, China

Abstract. RelieF algorithm is a series of feature selection methods, including the first proposed ReliffF algorithm and later extended ReliefF algorithm. The core idea is to give greater weight to the features which have great contribution to the classification problem. The algorithm is simple and has high efficiency. However, it is not ideal to apply the Relief algorithm directly to unbalanced multi-label datasets. Based on the Relief algorithm, this paper proposes an improved multi-label ReliefF feature selection algorithm for unbalanced datasets, called UBML-ReliefF algorithm, which effectively eliminates the problem of datasets unbalance. Finally, we use the stroke cases to do experiments, and the experimental results show that the multi-label classification effect in UBML-ReliefF algorithm is better than the original ReliefF algorithm.

Keywords: Feature selection · Unbalanced datasets · Multi-label classification

1 Introduction

Feature selection is sometimes called attribute selection, and its task is to select a feature subset [1] from all features to make the classifier has better performance. In the actual classification applications, there are many features which are not related to the classification problem. If the original data is used as the classification features, it will easily cause dimensional disaster, making the classification model become more complex and it will take more time to train the model.

The quality of the classification system depends on whether the selected features can well reflect the classification problem. Feature selection is to select $d < p$ features from the input p features to make the evaluation model to be best. In general, the feature set contains more or less features that do not contribute to the classification problem, which are also called redundant features and have a significant negative impact on learning problem. The results of the existing studies show that the size of training samples required for most classification systems is exponentially increasing with the unuseful features. So, feature selection has a significant effect on the design of

F. Xhafa et al. (eds.), *Advances in Intelligent Systems and Interactive Applications*, Advances in Intelligent Systems and Computing 686,
https://doi.org/10.1007/978-3-319-69096-4_21

classification systems in different conditions. Selecting good features can not only reduce the dimension of the feature space, speed up the efficiency of the algorithm, but also help to select appropriate features to get better classification results. So how to select features plays an important role in the design of classification systems and it is also a hot topic in the field of machine learning.

2 ReliefF Algorithm

There are series of algorithms in ReliefF [2–7]. The core idea of this series of algorithms is that good features should make similar samples close, leaving different classes of samples away. The ReliefF algorithm was first proposed by Kira [8, 9], which is only designed for binary classification. Kononenko has improved the feature selection algorithm on the basis of ReliefF, which is suitable for multi-class classification problem [10]. The basic idea of ReliefF is to assign the weights to each feature in the feature set of the classification samples, and then iterate to update the weights. Secondly, the feature subsets are selected according to the weight of feature, which makes the good features discrete different samples and aggregates similar samples.

Suppose there are K category tags in a given single tag datasets, the training datasets is defined as $D = \{(x_1, y_1), (x_2, y_2), \ldots, (x_n, y_n)\}$, where $x_i \in R^p$ and $y_i \in R^k$ respectively represent feature space and class label space of classified samples. If the sample x_i belongs to class k, then $y_i(k) = 1$, otherwise $y_i(k) = 0$. So the $p \times n$ feature matrix $X = [x_1, x_2, \ldots, x_n]$and the $K \times n$ label matrix $Y = [y_1, y_2, \ldots, y_n]$ constitute the classification sample D.

ReliefF algorithm is shown in the following Algorithm 1. The value of parameters m and k are set according to the sample size. $class(R_i)$ is the class label set of the Classification sample R_i. $M_j(C)$ represents the $j(j = 1, 2, \ldots, k)$ class of samples of class C target. $P(C)$ is the probability of class C. $diff(A, I_1, I_2)$ represents the distance between the sample I_1 and I_2 on the feature A.

Algorithm 1
$ReliefF(D, m, k)$

ReliefF(D,m,k)

Input: Training datasets D, the number of iterations m, the nearest neighbor number k;

Output: Predicted feature weight vector W;

Introduction: Calculate the eigenvector W of datasets D.

---

(1) *Initialize the feature weight vector W(A)=0.0, A=1,2,…,p;*

(2) *For i=1:m*

(3) *Randomly select a sample from D as* R_i;

(4) *Find the k nearest neighbor* H_j *with sample* R_i;

(5) *For each class* $C \neq class(R_i)$, *find k nearest neighbors* $M_j(C)$ *which are different from sample* R_i;

(6) *For* $A = 1:p$

Update each feature weight;

$$W[A] = W[A] - \frac{\sum_{j=1}^{k} diff(A, R_i, H_j)}{mk} + \frac{\sum_{c \neq class(Ri)} \left[\frac{P(C)}{1 - P(class(R_i))} \sum_{j=1}^{k} diff(A, R_i, M_j(C)) \right]}{mk}$$

(7) *End for*

(8) *End for*

(9) *Output W*

As shown in step (6), for a dimension feature A, if the distance $diff(A, R_i, H_j)$ between the two samples from the same category on A is smaller, or the distance $diff(A, R_i, H_j)$ between the two samples from the different category on A is larger, it is indicated that the feature A is more suitable for classification and its weight $W[A]$ is greater.

The weight of a feature is determined by the above algorithm. The greater the weight of a feature is, the more distinct the feature is. So a threshold can be used to select a new feature subset in order to achieve the purpose of reducing the dimension of features.

As shown in the *ReliefF* algorithm, the step (4) and step (5) are both to find the nearest neighbors based on the truth that each sample only belongs to one kind of category. It did not consider the sample can belong to multiple categories, and the features weight updating formula of the sixth step is not considered multi class label contribution, so the *ReliefF* algorithm is only suitable for single-label datasets, it is not suitable for multi-label classification.

3 UBML-ReliefF Algorithm

In fact, much data can't be clearly divided into a number of independent categories, each feature data may belong to multiple categories, such as a sunny sunrise picture can belong to the sea, the sun, the sky or other different categories.

Taking into account the fact that the number of different categories in original classification data is different (i.e., the original data is unbalanced), during multi-iteration of the randomly selected sample, the larger the proportion of the sample is, the greater probability that the sample is selected. Random selection is unfair to a small number of samples. So the result of feature selection with the ReliefF algorithm is not very good. Furthermore, ReliefF algorithm doesn't take into account the situation of multiclass and the relationship between class label and feature. In order to overcome the defects of ReliefF algorithm, an improved feature selection algorithm-UBML-ReliefF was proposed. It is shown in the following Algorithm 2.

Algorithm 2
UBML- ReliefF(D, m, k)

ReliefF(D,m,k)

Input: Training datasets D, the number of iterations m, the nearest neighbor number k;
Output: Predicted feature weight vector W;
Introduction: Calculate the eigenvector W of datasets D.

---

(1) *Initialize the feature weight vector W(A)=0.0, A=1,2,...,p;*
(2) *For i=1:m*
(3) *Select a class from D, then randomly select a sample R_i from the selected class, $class(R_i) = h = (h_1, h_2, \ldots, h_T)$;*
(4) *For t=1:T*
(5) *Find k nearest neighbors H_j that belong to class h_t as well as sample R_i;*
(6) *For each class $C \neq h_t$, find k nearest neighbors $M_j(C)$ which are different from sample R_i;*
(7) *For $A=1:p$*
Update each feature weight;

$$W[A] = W[A] + \frac{\frac{1}{P(h_t)}}{\sum_{C \in class(R_i)} \frac{1}{P(C)}} \{ - \frac{\sum_{j=1}^{k} diff(A, R_i, H_j)}{mk} + \frac{\sum_{C \neq class(R_i)} [\frac{P(C)}{1 - P(class(R_i))} \sum_{j=1}^{k} diff(A, R_i, M_j(C))]}{mk} \};$$

(8) *End for*
(9) *End for*
(10) *End for*
(11) *Output W*

ReliefF algorithm firstly needs to randomly select a sample from the datasets as a training sample, but in this random way it is likely to select an unrepresentative sample, and that may influence the accuracy of classification results. UBML-ReliefF algorithm first randomly selects a class,then randomly select a sample R_i from the selected class.

When looking for the nearest neighbor, first find the T class labels h_i that the sample has, where $h = (h_1, h_2, \ldots, h_T)$. Then for each class label $h_i (i = 1, 2, \ldots, T)$, finding k nearest neighbors $\mathrm{H_j}$ that belong to class h_i as well as sample R_i and for each class $C \neq h_t$, find k nearest neighbors $M_j(C)$ which are different from sample R_i. That solves the problem that the algorithm cannot deal with multiple classes simultaneously. Secondly, it is assumed that the class label of the sample has equal contribution to it, then add the contribution value to the feature weight update formula, $\frac{1}{P_{h_t}} / \sum_{C \in \mathrm{class}(\mathrm{R_i})} \frac{1}{P_C}$ is the contribution value of each class label in sample R_i. Adding the contribution value to the feature weight update formula eliminates the effect of the unbalanced number of randomly selected sample labels. Other parameters are same as Algorithm 1. The experimental results of UBML-ReliefF algorithm will be given in the next section.

4 Experiment

4.1 Datasets

This experiment adopts more than 40 thousand medical records from the China national 10th Five-year Plan, and extracts 1023 stroke cases. The purpose is to predict the main treatment of stroke. After the decomposition of stroke syndrome, there are six kinds of Syndrome Elements: Qi Deficiency (QD), Blood Stasis(BS), Yin Deficiency (YID), Yang Deficiency (YAD), Phlegm Turbidity (PT) and Blood Deficiency (BD). The vector model of stroke case is shown in Table 1, obviously, the prediction of stroke syndrome and classification problem belongs to the multi label classification problem.

As shown in Table 2, the number of Qi Deficiency (QD) and Blood Stasis (BS) is more than others, so the number of Symptom Element is unbalanced. If we use all features to make pre prediction, the accuracy will be not very good. So, in this paper, we use the multi-label feature selection algorithm which is suitable for unbalanced medical datasets (Table 2).

4.2 Performance Evaluation Index

In this paper, we use five popular evaluation indexes in the experiment of multi-label classification. Suppose that X is a D dimensional input space, which is a possible label set in the sample space, $Y = \{1, 2, \ldots, Q\}$ is a possible set of Q labels for classification sample space. So a multi-label datasets can be represented as $D = \{(x_i, Y_i)(1 \leq i \leq p)\}$, where $x_i \in X$ is a feature sample, $y_i \in Y$ is a label set of x_i.

The function $f(x_i, y)$ return a value, which represents the confidence of each possible label of X belonging to Y. $rank_f(x_i, y)$ returns the sort of $f(x_i, y)$. For the test datasets $\Gamma = (x_i, Y_i | 1 \leq i \leq m)$, we use the following five performance evaluation indexes.

Table 1. Medical datasets of stroke cases

Prescription properties									Syndrome elements					
Id	Medicine nature	Sweet	…	Bitter	Small intestine	Liver	…	Large intestine	QD	BS	YID	YAD	PT	BD
1	0.683	0.126	…	0.513	0.0	0.459	…	0.0	1	0	1	1	0	0
2	0.417	0.395	…	0.256	0.0	0.6	…	0.0	0	1	0	1	1	0
…	…	…	…	…	…	…	…	…	…	…	…	…	…	…
1023	0.753	0.3	…	0.687	0.0	0.437	…	0.0	0	0	1	0	0	0

Table 2. The number of each syndrome element

Syndrome element	QD	BS	YID	YAD	PT	BD
Frequency	536	428	357	278	219	55

(1) *Average precision*: The formula of Average precision is shown in formula (1).

$$avgprec_D(f) = \frac{1}{n}\sum_{i=1}^{n}\frac{1}{|y_i|}\sum_{y \in y_i}\frac{|\{y^{'}|rank_f(x_i, y') \leq rank_f(x_i, y), y' \in Y\}|}{rank_f(x_i, y)} \tag{1}$$

The average accuracy reflects the average accuracy of the predicted labels. The greater $avgprec_D(f)$ is, the better performance it will have.

(2) *Coverage*: The formula of Coverage is shown in formula (2).

$$cov_D(f) = \frac{1}{n}\sum_{i=1}^{n} \max_{y \in y_i} rank_f(x_i, y) - 1 \tag{2}$$

The smaller $cov_D(f)$ is, the better performance it will have.

(3) *Hamming Loss*: The formula of Hamming Loss is shown in formula (3).

$$hloss_D(f) = \frac{1}{n}\sum_{i=1}^{n}\frac{1}{m}|h(x_i)\Delta y_i| \tag{3}$$

The smaller $hloss_D(f)$ is, the better classification performance system will have. When $hloss_D(f) = 0$, system will has the best performance.

(4) *one* $-$ *error*: The one-error indicator is used to indicate the number of times that the highest ranked label is not in the sample label set. Its formula is shown in (4).

$$one - error_D(f) = \frac{1}{n}\sum_{i=1}^{n}\left|\left|\left[\arg\max_{y \in Y} f(x_i, y) \notin y\right]\right|\right| \tag{4}$$

The smaller $one - error_D(f)$ is, the better classification performance system will have. When $one - error_D(f) = 0$, system will have the best performance.

(5) *Ranking Loss*: The Ranking Loss index indicates that the sort error occurs in the sort sequence of class labels of the sample. Its formula is shown in (5).

$$rloss_D(f) = \frac{1}{n}\sum_{i=1}^{n}\frac{1}{|y_i||\overline{y_i}|}|\{(y_1, y_2) : f(x_i, y_1) \leq f(x_i, y_2), (y_1, y_2) \in y_i \times \overline{y_i})\}| \tag{5}$$

$\overline{y_i}$ represents the complement of y_i. The smaller $rloss_D(f)$ is, the better classification performance system will have. When $rloss_D(f) = 0$, system will has the best performance.

4.3 Experiment and Results Analysis

In order to demonstrate the performance of UBML-ReliefF algorithm, we choose BSVM [11], ML-KNN [12], Rank-SVM [13], CLR [14], ECC [15] and LEAD[16] as classification algorithm. They respectively use ReliefF and UBML-ReliefF algorithm to select features. In order to make the experimental results to be more convinced, we use 10-folder cross validation. The classification results of the feature subsets selected by the ReliefF and UBML-ReliefF algorithm are respectively shown in Table 3 and Table 4. Figure 1 shows the comparison of the experimental indicators.

Table 3. Classification results based on ReliefF feature selection algorithm

	Average precision	Coverage	Hamming loss	One-error	Ranking loss
ML-KNN	0.647	0.329	0.278	0.383	0.238
BSVM	0.638	0.279	0.234	0.333	0.246
Rank-SVM	0.559	0.326	0.338	0.501	0.212
CLR	0.68	0.291	0.281	0.352	0.194
ECC	0.686	0.297	0.255	0.289	0.203
LEAD	0.711	0.282	0.27	0.314	0.188

Table 4. Classification results based on UBML-ReliefF feature selection algorithm

	Average precision	Coverage	Hamming loss	One-error	Ranking loss
ML-KNN	0.752	0.211	0.161	0.276	0.172
BSVM	0.764	0.28	0.167	0.267	0.139
Rank-SVM	0.725	0.256	0.233	0.353	0.223
CLR	0.787	0.186	0.173	0.311	0.144
ECC	0.823	0.194	0.157	0.216	0.195
LEAD	0.815	0.175	0.152	0.305	0.136

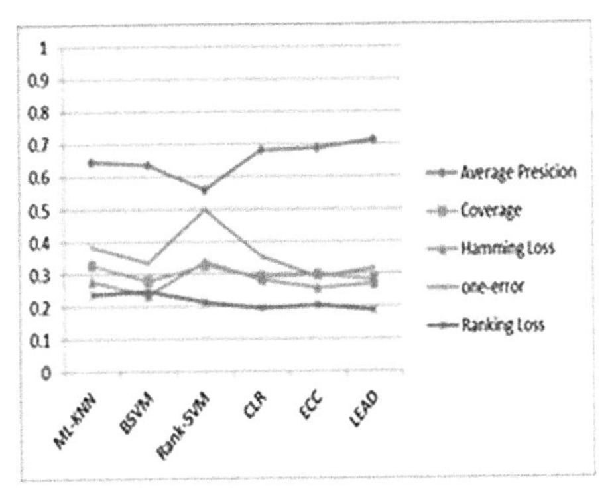

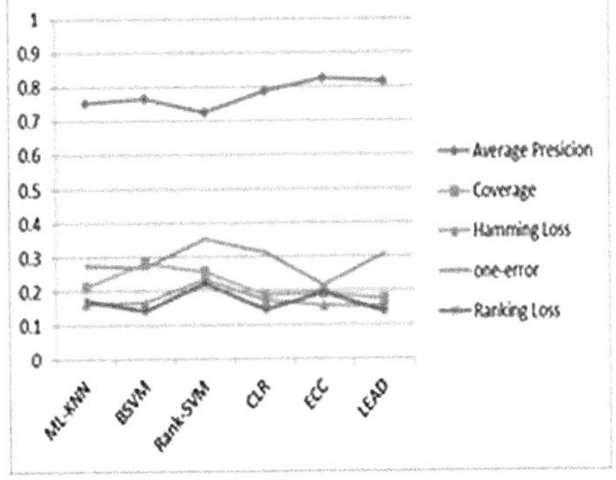

Fig. 1. Classification results after feature selection(Left ReliefF, Right UBML-ReliefF)

5 Conclusion

In this paper, an improved feature selection algorithm—UBML-ReliefF is proposed for the unbalanced multi-label datasets. It can effectively eliminate the problem of poor classification results caused by unbalanced datasets. We adopt more than 40 thousand medical records from the China national 10th Five-year Plan and extracts 1023 stroke cases to do the experiment. After the feature selection by ReliefF and UBML-ReliefF algorithm, the selected feature subsets are combined with other common multi-label classification algorithms. And after 10-folder cross validation, we prove that the UBML-ReliefF algorithm has better performance than the original ReliefF algorithm on the unbalanced data classification.

Acknowledgements. This work is supported by the National Natural Science Foundation of China (No. 61175048) and we would like to express our gratitude to Wan Yifei for he help us conduct experiments.

References

1. Yao, X., Wang, X.D., Zhang, Y.X., et al.: Summary of feature selection algorithms. Control Decis. **27**(2), 161–166 (2012)
2. Jia, H., Sleep, E.E.G.: Staging studies based on relief algorithm. Lect. Notes Electr. Eng. **163**, 935–940 (2013)
3. Gore, S., Govindaraju, V.: Feature selection using cooperative game theory and relief algorithm recent trends, advances and solutions. Knowledge, information and creativity support systems. Springer International Publishing, USA (2016)
4. Gao, L., Li, T., Yao, L., et al.: Research and application of data mining feature selection based on relief algorithm. J. Softw. **9**(2) (2014)
5. Lei, P., Wang, J., et al.: Study of emotion recognition under anxiety based on physiological signals by relief method. Chin. J. Med. Instrum. **3**, 186–189 (2014)
6. Xiang, Z., Zhao Hong, D., et al.: Maximum entropy relief feature weighting. J. Comput. Res. Dev. **48**(6), 1038–1048 (2011)
7. Li, X.L.: The study and application of feature selection algorithms based on relief. Dalian University of Technology (2013)
8. Xu, Y., Chen, L.: Term-frequency based feature selection methods for text categorization. In: Genetic and Evolutionary Computing (ICGEC), 2010 Fourth International Conference on. IEEE, pp. 280–283 (2010)
9. Inza, I., Larrañaga, P., Blanco, R., et al.: Filter versus wrapper gene selection approaches in DNA microarray domains. Artif Intell. Med. **31**(2), 91–103 (2004)
10. Zhou, Z.H., Zhang, M.L., Huang, S.J., et al.: Multi-instance multi-label learning. Artif. Intell. **176**(1), 2291–2320 (2012)
11. Kong, X., Ng, M.K., Zhou, Z.H.: Transductive multilabel learning via label set propagation. IEEE Trans. Knowl. Data Eng. **25**(3), 704–719 (2013)
12. Zhang, M.L., Zhou, Z.H.: ML-KNN: A lazy learning approach to multi-label learning. Pattern Recogn. **40**(7), 2038–2048 (2007)
13. Elisseeff, A., Weston, J.: A kernel method for multi-labelled classification. In: Dietterich TG, Becker S, Ghahramani Z, (eds.) Proceedings of the Advances in Neural Information Processing Systems 14, pp. 681–687. MIT Press, Cambridge (2002)

14. Fürnkranz, J., Hüllermeier, E., Mencía, E.L., et al.: Multilabel classification via calibrated label ranking. Mach. Learn. **73**(2), 133–153 (2008)
15. Read, J., Pfahringer, B., Holmes, G., Frank, E.: Classifier chains for multi-label classification. Mach. Learn. **85**(3), 333–359 (2011)
16. Zhang, M.L., Zhang, K.: Multi-Label learning by exploiting label dependency. In: Proceedings of the 16th ACM SIGKDD International Conference on Knowledge Discovery and Data Mining, pp. 999–1008, ACM Press, New York (2010)

Area Topic Model

Hongchen Guo[1], Liang Zhang[2], and Zhiqiang Li[2(✉)]

[1] Network Information Technology Centre, Beijing Institute of Technology, Beijing, China
[2] School of Software, Beijing Institute of Technology, Beijing, China
lizq@bit.edu.cn

Abstract. Location-based topic modeling is an emerging domain to capture topical trends over the area dimension, which makes it possible to conduct further analysis on the various preference of users in different areas, such as investigating users' social opinion tendencies, generating personalized recommendation and so on. In this paper, we proposed a novel topic model called Area-LDA to discover the latent area-specific topic. Previous works which based on pre-discretized or post-hoc analysis of area topics had no ability to help generate topic for a document. Different from previous works, our model extends the original LDA by associating each document with area factor and introducing a new area distribution over topics into the model. Therefore, for each generated document, the distribution over topics is influenced by both word co-occurrences of the document and word co-occurrences of the corresponding area. We present the experimental analysis over the real-world dataset and the results demonstrate the effectiveness of the proposed method to mine interpretable topic trends on different areas.

Keywords: Topic model · LDA · Location-based service

1 Introduction

As a statistical model, topic model is used to find latent topics in a series of documents. LDA [1] model is a classical hierarchical Bayesian model and has a large volume of variants on the basis of specific task, such as for categorization [2, 3], for customer satisfaction survey [4], for social network discovery and analysis [5–7], etc. In these applications, probabilistic graphical model structures are discreetly-designed to capture implied patterns.

In this paper, we put an eye on the relations between areas and topics. Different areas have varied impact on topic of the documents collected in these areas. For example, if collecting articles that people read every day in Beijing, we can find that people focus on both technological and economic news. Furthermore, we take two areas of articles, ZhongGuanCun (Technology Center) and GuoMao (International Trade Center), to do a deep analysis, it is obvious that people in ZhongGuanCun are more concerned about technology than people in GuoMao, and economy is more important in the eyes of people in GuoMao than those in ZhongGuanCun. This is topical preference over area.

F. Xhafa et al. (eds.), *Advances in Intelligent Systems and Interactive Applications*, Advances in Intelligent Systems and Computing 686,
https://doi.org/10.1007/978-3-319-69096-4_22

Previous works have made some efforts to discovery relations between areas and topics by pre-discredited or post-hoc analysis. These methods have explicit weakness that without considering the area factor will loss the regional characteristics, that means it missed the opportunity for area to improve topic discovery.

Therefore, in this paper, we propose a novel topic model called Area-LDA that integrates area factor into original LDA so as to influence topic choice. Area-LDA parameterizes a pre-area topic distribution. With a pre-document topic distribution, LDA-style model is responsible for choosing every topic. When a prominent word co-occurrence pattern appears in an area (actually it appears in an area's documents), Area-LDA will capture this pattern and adjust the distribution of document-topic and area-topic so as to match the topic reflected in the co-occurring words. In the end, when model training is completed, we can not only obtain the topic distribution of every document but also that of a specific area.

The contributions of this paper can be summarized as:

(1) We extend the original LDA by integrating the area information implied in the documents into the model and propose Area-LDA Model.
(2) Area-LDA Model can capture the topic of documents from the perspective of area. In other words, it can reflect the distribution of topics in certain areas.
(3) This paper applies Area-LDA Model in real-world data set. The model can accurately reflect the different topics focused by different areas.

2 Related Works

In recent years, research and application of topic model are mainly in specific tasks, such as assignment and classification, emotional analysis, public opinion supervision, etc. In this area, there is a huge space for application. In this section, we review the state of the art algorithms related to this paper.

As early as 2007, Wang [8] proposed LATM topic model through integrating geographic knowledge with LDA model. In this model, every word in documents was associated to a location. Therefore, from LATM, the relationship between words and location can be obtained. The Geographic Topic Model (GTM) [9] is a truly spatial topic model that generates spatial regions associated with topics. GTM is a cascading topic model that selects words conditioned by base topics that are further conditioned on spatial regions.

In addition, topic model with geographical in social network is a hot research direction. Hong [10] developed a model to find geographical topics in Twitter, which were based on geographic location as well as user information on the basis of the assumption that individual users of Twitter would tend to be geographically localized. Yin [11] combined speech on social network and GPS information. From three perspectives, position drive model, text drive model and the combination of the two models, the author analyzed the relationship between topic and location, thereby finding that topics of speech in different areas differ apparently. It was thus proved the hypothesis that geographical distributions can help modeling topics, while topics provide important cues to group different geographical regions. Sizov [12] proposed

GeoFolk framework to combine text and spatial information together to construct better algorithms for content management, retrieval, and sharing in social media. To make use of spatial information, GeoFolk assumes that each topic generates latitude and longitude from two topic-specific Gaussian distributions.

Different from the geo-topic model mention above, this paper does not simply regard area information as given text characteristic. First, Area-LDA regards documents as attached to a certain area. In other words, all words of documents in the same area have the same area information. Second, in Area-LDA, area information is not only used as a region identifier of a certain document, but also contributes an area-topic distribution, thereby influencing the topic generation of documents in this area. In the model referred to in Duan's article [13], location part of that model is quite similar to Area-LDA. Difference lies in that Area-LDA puts documents into the area instead of placing them individually in the course of generating documents. Such an arrangement is more realistic, and corresponds with the fact that we view topic distribution from the perspective of "area".

3 Area-LDA Topic Model

Area-LDA integrates area factor into original LDA so as to influence topic choice. Area-LDA parameterizes a pre-area topic distribution, with a pre-document topic distribution, LDA-style model is responsible for choosing every topic. Finally, we can not only obtain the topic distribution of every document but also that of a specific area.

3.1 Overall Design of Area-LDA

We present Area-LDA based on basic LDA topic model by introducing area-topic distribution. Its main intuition is that people in different areas interest in diverse topics. For example, the topical trends in boy's dormitory are mainly articles on sport and technology, while those in girl's dormitory are shopping, SNS and so on. Area-topic distribution is to capture the topical trends of a certain area in order to influence the computing of "document-topic" distribution. The graphical model representation of Area-LDA is shown in Fig. 1.

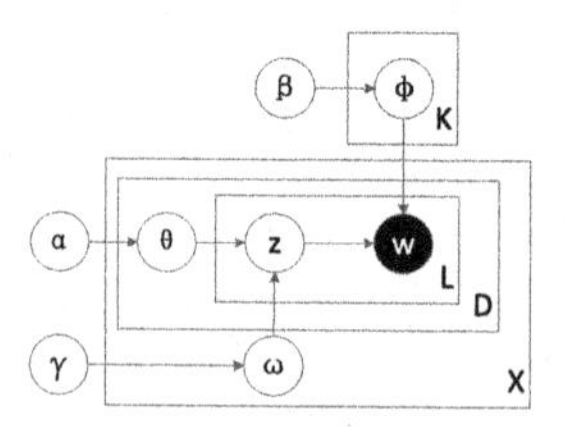

Fig. 1. The graphical model representation of Area LDA

θ is a document-topic distribution (multinomial distribution). Suppose that there are N potential topics in a document, θ represents the degree of correlation between this document and N potential topics. Similarly, φ is also multinomial distribution sampled from Dirichlet distribution, while β is its hyper-parameter. φ is topic-word distribution. If a topic is constituted by N words, φ represents the degree of importance of N words to one topic. ω is area-topic multinomial distribution and γ is its hyper-parameter. If there are N potential topic in an area, ω describes the distribution of N potential topics in this area. Detail information of notations is shown in Table 1.

Table 1. Notations description

Symbol	Description	Symbol	Descrption
X	Area number	α	Hyperparameters of doc-topic, topic-word, area-topic Dirichlet distribution respectively
D	Document number	β	
L	Length of document	γ	
K	Topic number	$\Phi = \{\varphi_{k,v}\}$	Matrix of topic-word, doc-topic, area-topic distribution respectively
V	Total number of words	$\Theta = \{\theta_{d,k}\}$	
$W = \{w_i\}$	W is a doc and w_i is a word in W	$\Omega = \{\omega_{x,k}\}$	
$Z = \{z_i\}$	Topic of a word in a doc		

Every element in matrix Φ, Θ and Ω indicate the probability of word v in topic k, the probability of topic k in document d and the probability of topic k in area x respectively. From the probability graphical representation, the generative process of Area-LDA is as follows:

(1) Draw a topic k from multinomial $\phi_k \sim \mathrm{Dir}(\beta)$;
(2) Draw an area x from multinomial $\omega_x \sim \mathrm{Dir}(\gamma)$;
(3) For every document d in area x:
(a) Draw the distribution of topic in document $\theta_d \sim \mathrm{Dir}(\alpha)$;
(b) For the place of every word in document d:
(1) According to θ_d and ω_x, choose a topic $z \sim \mathrm{Mult}(\theta_d)$ and $\mathrm{Mult}(\omega_x)$;
(2) According to z, choose a word $v \sim \mathrm{Mult}(\phi_z)$.

Therefore, the joint probability of word and topic of Area-LDA is defined as follows:

$$p(Z,W|\alpha,\beta,\gamma) = p(W|Z,\beta)p(Z|up\alpha,\gamma) \tag{1}$$

Symmetrical hyper-parameters are applied in the equation. That is the value of hyper-parameter in every Dirichlet distribution is the same.

Common solving algorithms of LDA model are variational inference, Gibbs Sampling and Expectation propagation. The final update equation of Gibbs sampling as follow:

$$p(z_i = k|Z^{-i}, W, \alpha, \beta, \gamma) = N_{k,v} + \beta - 1\left(\sum_{v=1}^{V} N_{k,v} + \beta\right) - 1 . \frac{M_{d,k} + \alpha - 1}{\left(\sum_{k=1}^{K} M_{d,k} + \alpha\right) - 1} \cdot \frac{A_{x,k} + \gamma - 1}{\left(\sum_{k=1}^{K} A_{x,k} + \gamma\right) - 1} \quad (2)$$

3.2 Process of Estimator

According to Area-LDA's definition of $\phi_{k,v}, \theta_{d,k}, \omega_{d,k}$:

$$f_{k,v} = \frac{N_{k,v} + \beta}{\sum_{v=1}^{V} (N_{k,v} + \beta)}, \theta_{d,k} = \frac{M_{d,k} + \alpha}{\sum_{k=1}^{K} (M_{d,k} + \alpha)}, \omega_{d,k} = \frac{A_{x,k} + \gamma}{\sum_{k=1}^{K} (A_{x,k} + \gamma)} \quad (3)$$

In this definition, two-dimensional array NZW[z,w] represents Φ distribution matrix; NMZ[m,z] means Θ distribution matrix; NXZ[x,z] refers to Ω distribution matrix. NZ[z] stands for $\sum_{v=1}^{V} N_{z,v}$, NM[d] for $\sum_{k=1}^{K} M_{d,k}$, and NX[x] for $\sum_{k=1}^{K} A_{x,k}$. The pseudo-code of the algorithm is shown in Table 2.

Table 2. Estimator process pseudo-code

```
// Estimator
while iterations
    foreach document m in M
        foreach position n in document m
            NZW[z_{m,n}, w_{m,n}] --; NMZ[m, z_{m,n}] --; NXZ[m.area, z_{m,n}] --;
            NZ[z_{m,n}] --; NM[m] --; NX[m.area] --;
            sample topic z_{m,n} according to Gibbs updating rule
            NZW[z_{m,n}, w_{m,n}] ++; NMZ[m, z_{m,n}] ++; NXZ[m.area, z_{m,n}] ++;
            NZ[z_{m,n}] ++; NM[m] ++; NX[m.area] ++;
        end
    end
end
```

In order to fix the problem, we applies an experiential method here. It is undeniable that experiential method is not verified by rigorous mathematical theories. But the change is proved to be effective. Hence modification is made on the third item of wheel function equation, with sigmoid function introduced to lower it impact.

Former estimator wheel function equation

$$p(z_i = k|Z^{-i}, W,\alpha,\beta,\gamma) = \frac{N_{k,v} + \beta - 1}{\left(\sum_{v=1}^{V} N_{k,v} + \beta\right) - 1} \times \frac{M_{d,k} + \alpha - 1}{\left(\sum_{k=1}^{K} M_{d,k} + \alpha\right) - 1} \times \frac{A_{x,k} + \gamma - 1}{\left(\sum_{k=1}^{K} A_{x,k} + \gamma\right) - 1} \quad (4)$$

After modification:

$$p(z_i = k|Z^{-i}, W,\alpha,\beta,\gamma) = \frac{N_{k,v} + \beta - 1}{\left(\sum_{v=1}^{V} N_{k,v} + \beta\right) - 1} \cdot \frac{M_{d,k} + \alpha - 1}{\left(\sum_{k=1}^{K} M_{d,k} + \alpha\right) - 1} \cdot \left(\frac{1}{1 + e^{\frac{A_{x,k} - 1}{\sum_{k=1}^{K} A_{x,k} - 1} - \frac{1}{K}}} + 0.5\right) \quad (5)$$

Given that sigmoid function is monotonically increasing, the modified wheel function still maintains its positive correlation with the former equation. Besides, sigmoid function maps the value field to [0.5, 1.5], at the same digit level as the two items of the equation, thereby lowering the influence of area-topic and successfully removing the error bias of Gibbs sampling.

4 Experiments

4.1 Dataset

To evaluate the performance of Area-LDA model, we use the web log of BIT network on July 02, 2014, each of which consists of request-time, inner ip, and request-url. The inner IP could indicate the location of online users; request-url could be converted into text. So we choose four buildings, including girl's and boy's dormitory of undergraduate and graduate students.

In the preprocessing, we convert IP into campus building tag. And, we fetch the HTML document of urls. Then, we do data cleaning on it, including cleaning html tag, deleting Non-Chinese characters, word tokenizing, cleaning the stopwords. And last, we deleting the high-frequency words and low-frequency words in the dataset.

The statistic of the dataset is shown in Table 3.

Table 3. The statistic of dataset

Area	Docs	Words number / doc	Avg words number / doc
No. 1 building	4055	936745	236.6466
No. 2 building	3017	771685	255.7789
No. 5 building	4000	794733	186.0834
No. 6 building	3062	706679	234.6388
Total	14134	3209842	227.1007

4.2 Experiment Settings

In our experiment, we train our model based on JGibbsLDA, and add area-topic distribution and a hyper-parameter γ. In multinomial sampling implementation, multinomial parameters not only affected by document-topic distribution and topic-word distribution, but also affected by area-topic distribution. Also, we modify file input and output format code.

Topic number is set based on specific data environment. In this experiment, we apply the method of personal experience to select topics. The number of topics is set to be [10, 20, 30, 40, 50, 60, 70, 80]. It is found in experiment that large value generates more detailed topic clustering. While, small value is likely to confuse the topic meaning. After being evaluated by artificial evaluation training model, the number ranging 30 to 50 is found to have good effect. In the second training, the topic number is set to be [30, 35, 40, 45, 50]. Training results show that the effect of model is best when the topic amount is 35. Due to the training outcome, and the weblog data scene, this experiment determines topic amount to be 35.

When applying Area-LDA model, this experiment endows classic value to the two hyper-parameters: alpha = 50/T, beta = 0.1. But for the gamma hyper-parameter in the area-topic distribution, this experiment controls variable to determine the best value. In the experiment, gamma hyper-parameter is used to train topic model with values from 0.1 to 0.9 (0.1, 0.2, 0.3, 0.4, 0.5, 0.6, 0.7, 0.8, 0.9). In the course of research finding analysis, the author found that gamma hyper-parameter is insensitive to the result. The gamma was thus set to be 0.5. In the particular application discussed in this paper, we find that the sensitivity to hyper-parameters is not very strong. Thus, we use fixed symmetric Dirichlet distributions ($\alpha = 50/T$, $\beta = 0.1$, $\gamma = 0.5$) in our experiments.

4.3 Experimental Restuls

Qualitative Analysis of Topical words. A number of topics are captured from the text of we blog. Due to space limitations, we just list some significant topics represented by top 10 words in Table 4, it is concluded that the meaning of most words are related to daily Internet surfing.

Table 4. Top 10 words of topic, we translate Chinese to English

Topic 4	Author time landlord forum reply super feeling bbs myself like
Topic 6	Beauty job hunting feinimoshu guest sohu live show tv position
Topic 11	original latest wechat novel follow role cartoon price limit patrol
Topic 14	World cup football match minute germany brazil player argentina team
Topic 16	Technic method design system analysis problem require manage application control
Topic 18	Automobile internet gold education technology stock football millitary sports digital
Topic 21	Breast woman doctor fashion female sex health picture hospital usage
Topic 25	Product market company investment enterprise trade bank stock economic relative
Topic 26	Channel shanghai network television news tv guangdong expand show comprehensive
Topic 28	Price color category purchase selected description item size detail sale
Topic 30	Hotel city minute travel beijing walking service hour booking review
Topic 34	Anchorman broadcast reward dubbing review comment rank hot story type

Among the 35 topics, topics like Topic 4 (BBS), Topic 6 (Job TV Show), Topic 14 (Football sport), Topic 16 (Engineering), Topic 25 (Investment), Topic 28 (Online Shopping) and Topic 30 (Travel) are worth to be considered. These topics differ in that they are captured in different area. The differences of some topics are caused by the special occasion on 2 July 2014.

Quantitative Analysis of Area-Topic Distribution. From the perspective of area, we make cross comparison of popularity among different topics. The Ω distribution of topic model is topic-area distribution, named as P(topic|area). In different areas, topics is shown in Fig. 2.

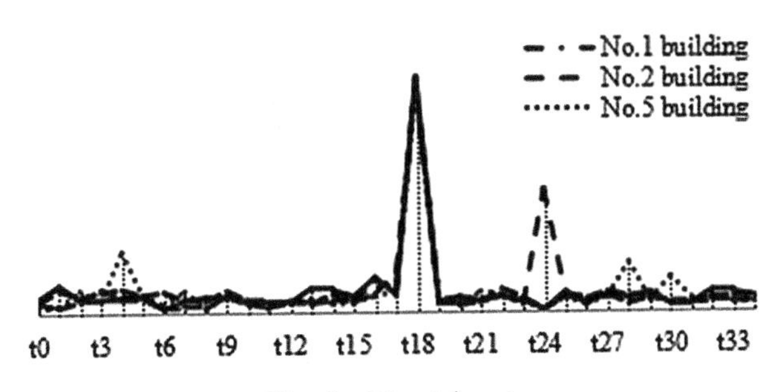

Fig. 2. P(topic|area)

From the figure, we can find out that Topic 18, namely portal website, is highly popular in all the four dormitories. As a collection and distribution center of Internet information, portal website is also a main entrance of netiziens to obtain online news of all sorts, which rightly reflects its importance. In the figure, the dot line represents women's dormitory, in which popularity of Topic 4 (BBS), Topic 28 (Online Shopping) and Topic 30 (Travel) is rather high. Of particular note, there is a summit in Topic 25 (Investment) because of a special occasion. After tracking the original data, it is found that the topic comes from http://kp.cngold.org/, a website designed to introduce investment. On 2 July 2014, a student living in No. 2 dormitory was keen on content related to stock and investment. Analysis on the original data reveals that the calculation result of this experimental model is fundamentally right.

Besides an explanation of the popularity of various topics in different areas, namely the analysis of P(topic|area), decision makers also need cross comparison among the four areas from the perspective of topics. To this end, P(area|topic) should be calculated.

$$P(\text{area}|\text{topic}) \propto P(\text{topic}|\text{area}) * P(\text{area}) \tag{6}$$

P(topic|area) can obtained from Ω distribution. P(area) is worked out through calculating the rate of the document amount of a certain area to the total amount. At last, the calculated result P(aera|topic) should be normalized.

In Fig. 3, the popularity of key topics in various areas is made cross comparison. The x-coordinate is the notation of topics, while y-coordinate is the degree of popularity of a certain area compared with other areas. Topics with distinct differences are selected to display.

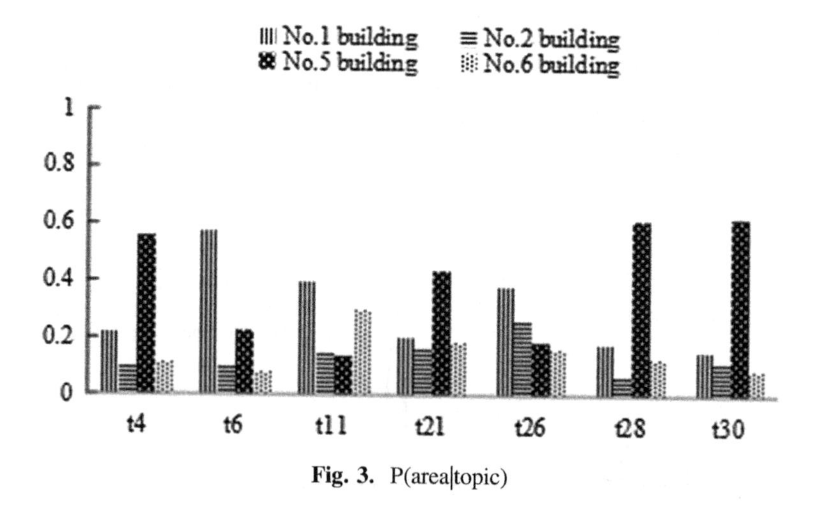

Fig. 3. P(area|topic)

As shown in Fig. 3, the black underground white point column represents No. 5 women's dormitory, and other columns stands for men's dormitories. It is obvious that Topic 4 (BBS), Topic 21 (Women's Health), Topic 28 (Online Shopping) and Topic 30 (Travel) are far more popular in women's dormitory than other areas. According to the topic list, Topic 4 (BBS) is related to Tianya BBS, Topic 21 (Women's Health) is woman's health, Topic 28 (Online Shopping) is shopping and Topic 30 (Travel) is travel and tourism. It can be concluded that the weblog recognized by topic model is substantially identical to the popular topics of network users in women's dormitory. The result is acceptable.

Topics like Topic 6 (Job), Topic 11 (Novel), Topic 26 (PPTV) are more popular in No. 1 men's dormitory of undergraduate students than the other three men's dormitory of graduate students, in which popular topics are job-seeking programs like Only You, websites for novel or comic and online television programs. But in No. 1 dormitory for undergraduate students, the key topics are online entertainment programs.

In this section, brief analysis on the calculation of model leads to the conclusion that experiment results are substantially identical to campus users' utilization of network, thereby revealing their online behaviors and habits. Experiment result shows that Area-LDA could discover area's topics tendency effectively. So the model has more promise for future use.

5 Conclusions

This paper proposes Area-LDA topic model based on the hypothesis that area influences the underlying meaning of documents. Through analyzing research results, the author has confirmed that the model proposed in this paper can accurately and effectively capture the meaningful underlying topics in network weblog.

Although this paper has made some progress, the author should continue to further investigation in order to solve the problems: (1) Due to the limitation of data environment, this paper has not fully mined and captured the underlying topics in the weblog of campus network. The Area-LDA topic model proposed in this paper should be trained with more data in the future. (2) The number of topics in this paper is selected in a rather subjective method. In ongoing study, more researchers should be invited to decide a proper number of topics.

In the model proposed in this paper, hyper-parameters like alpha, beta and gamma are decided based on experience. Nevertheless, automatic optimization parameter can make the model more proximate to the real distribution of data, and in turn improve the quality of topics.

References

1. Blei, D.M., Ng, A.Y., Jordan, M.I.: Latent dirichlet allocation[J]. J Mach. Learn. Res. **3**, 993–1022 (2003)
2. Lee, Y.S., Lo, R., Chen, C.Y., et al.: News topics categorization using latent Dirichlet allocation and sparse representation classifier. In: IEEE International Conference on Consumer Electronics—Taiwan (ICCE-TW), pp. 136–137 (2015)
3. Rasiwasia, N., Vasconcelos, N.: Latent Dirichlet allocation models for image classification. IEEE Trans. Pattern Anal. Mach. Intell. **35**(11), 2665–2679 (2013)
4. Tirunillai, S., Tellis, G.J.: Extracting dimensions of consumer satisfaction with quality from online chatter: strategic brand analysis of big data using latent Dirichlet allocation. Available at SSRN 2408855 (2014)
5. Zhang, H., Qiu, B., Giles, C.L., et al.: An LDA-based community structure discovery approach for large-scale social networks ISI 200 (2007)

6. Cha, Y., Cho, J.: Social-network analysis using topic models. In: Proceedings of the 35th International ACM SIGIR, pp. 565–574 (2012)
7. McCallum, A., Corrada-Emmanuel, A., Wang, X.: Topic and role discovery in social networks. Computer Science Department Faculty Publication Series, vol. 3 (2005)
8. Wang, C., Wang, J., Xie, X., et al.: Mining geographic knowledge using location aware topic model. In: Proceedings of the 4th ACM Workshop on Geographical Information Retrieval, pp. 65–70. ACM (2007)
9. Eisenstein, J., O'Connor, B., Smith, N.A., Xing, E.P.: A latent variable model for geographic lexical variation. In Proceedings of the 2010 Conference on Empirical Methods in Natural Language Processing, EMNLP '2010, pp. 1277–1287. Association for Computational Linguistics, Stroudsburg (2010)
10. Hong, L., Ahmed, A., Gurumurthy, S., Smola, A.J., Tsioutsiouliklis, K.: Discovering geographical topics in the twitter stream. In: Mille, A., Gandon, F.L., Misselis, J., Rabinovich, M., Staab, S. (eds.) WWW, pp. 769–778. ACM (2012)
11. Yin, Z., Cao, L., Han, J., et al.: Geographical topic discovery and comparison. In: Proceedings of the 20th International Conference on World Wide Web, pp. 247–256. ACM (2011)
12. Geofolk S.S.: latent spatial semantics in web 2.0 social media. In: Proceedings of the third acm international conference on web search and data mining, pp. 281–290. ACM (2010)
13. Duan, L., Guo, W., Zhu, X., Hu, B.: Constructing spatio-temporal topic model for microblog topic retrieving. Geomat. Inf. Sci. Wuhan Univ. **2**, 018 (2014)

Predicting Popularity of Topic Based on Similarity Relation and Co-occurrence Relation

Lu Deng(✉), Qiang Liu, Jing Xu, Jiuming Huang, Bin Zhou, and Yan Jia

College of Computer, National University of Defense Technology, Changsha 410073, China
denglu@nudt.edu.cn

Abstract. Interaction behaviors of users on different platforms on the Internet make user-generated content spread widely and become popular. How to model and predict the popularity of topic concerned by users is vital for many fields. Aiming at the problem of topic popularity prediction on microblog platform, a popularity prediction method based on similar topics and co-occurrence topics is proposed. The method is further evaluated with the Sina Weibo dataset. The experimental results show that our method can have relatively better performance in predicting topic popularity than the baseline methods.

Keywords: Popularity prediction · Similarity relation · Co-occurrence relation

1 Introduction

With the rapid development of the Internet, more and more application platforms are full of people's daily life. For a topic, how to judge its popularity based on the information in its early rapidly and accurately is significant in capture and analysis of public opinion, which results in more and more attention of researchers.

There are some researches in popularity prediction of online content such as Reservoir Computing model [1], Stochastic Model [2] of user behavior, Biology-inspired Survival Analysis Method [3] and so on. Kong et al. [4] discussed relevant dynamic factors having effect on post popularity and proposed a prediction method combining local characteristics and dynamic factors based on evolution information early. Wan et al. [5] proposed a prediction method of message popularity based on propagation simulation. Xiong et al. [6] proposed a method for quantitative description of topic popularity which takes into account user attributes and content attributes. Nie et al. [7] proposed a prediction method based on KNN method, which predicts the development trend of topic according to the popularity of its k most similar topics. Wang et al. [8] proposed a prediction method based on KNN method and LDA model. Wang et al. [9] used the concepts of knowledge instead of terms in bag-of-words model. He et al. [10] proposed a modified topic model considering topic interests of posted and reposted messages by the disparity between the words posted and reposted messages.

F. Xhafa et al. (eds.), *Advances in Intelligent Systems and Interactive Applications*, Advances in Intelligent Systems and Computing 686,
https://doi.org/10.1007/978-3-319-69096-4_23

All of the above methods did not consider co-occurrence information and there is a strong logical relationship between the topics that appears in the same document, which also plays a key role in predicting the popularity of topic. Therefore, this paper puts forward a method of predicting popularity of topics based on similar relation and co-occurrence relation at the same time.

2 The SCW Method for Prediction of Topic Popularity

2.1 Definition

Definition 1 (*Similarity Relation*) For two topics t_a and t_b, if the similarity degree in semantics meet the requirement, topic t_a and topic t_b will be called that they have similarity relation.

Definition 2 (*Co-occurrence Relation*) For two topics t_a and t_b, if there is a document belonging to these two topics at the same time and their weight probabilities to the document meet the requirement, topic t_a and topic t_b will be called that they have co-occurrence relation.

2.2 The Description of Topic Popularity

The different social platforms have their own characteristics. Compared with other measures, the number of retweeting for a topic is more suitable to describe the real spread of topic in Sina Weibo. Based on consideration mentioned above, we give the definition of popularity prediction in this paper: for a certain tweet c, posting time is set to 0 and the number of retweeting at time t is represented as $R(c, t)$ which means the population of tweet c at time t. For a topic z, it includes many tweets while posting time of its first tweet is set to 0 and the count of retweeting from tweets belonging to topic z at time t is represented as $R(z, t)$ which means the popularity of topic z at time t. The method treats evolution situation of topic z at time t_i as input and predicts the popularity level of topic z at time $t_p (t_p > t_i)$.

$$q = R(z,\ t_i)/R(z,\ t_p) \tag{1}$$

The threshold values are preplanned to measure the popularity of topics. There are four popularity levels which is based on the ratio from Eq. 1. If $q \le p_1$, it means that there will be lots of tweets and retweets produced, which implies that the topic will be popular amazingly for some time to come (level L_1); If $p_1 < q \le p_2$, it means that there will be a number of tweets and retweets produced, which implies that the topic will be popular highly for some time to come (level L_2); If $p_2 < q \le p_3$, it means that there will be some tweets and retweets produced, which implies that the topic will be popular partly for some time to come (level L_3); If $q > p_3$, it means that there will be few tweets and retweets produced, which implies that the topic will be popular hardly for some time to come (level L_4).

2.3 The Analysis of Similarity Relation

How to judge similarity relation between two topics is the key point for popularity prediction. The technology of KL divergence is taken to measure the similarity. KL divergence (Kullback–Leibler divergence) called relative entropy is a method to evaluate the difference degree between two probability distributions. The analysis of similarity relation can be achieved by the following steps:

(1) Similarity degree of topics. For topic t_a and topic t_b, their corresponding distributions are T_a and T_b. The similarity degree is calculate based on KL-divergence as shown in Eq. 2. The closer the semantics of two topics is, the more similar their distributions are, which means the closer the corresponding KL-divergence value to zero is.

$$D(T_a||T_b) = \sum_{i=1}^{n} T_a(i)\log(T_a(i)/T_b(i)) \tag{2}$$

(2) Similarity vector for each topic. For a topic t_a, its similarity degree with another topic t_b can be calculated by Eq. 2. Equation 3 can be used to calculate the similarity vector for topic t_a. Where t_i represents the ith topic and w_{s-ai} represents its corresponding weight in similarity. Due to the situation that the smaller KL divergence value is, the more similar two topics are, the way of reciprocal is adopted to calculate the weight of each topic to topic t_a.

$$\text{Similarity}(t_a) = \{(t_1, w_{s-a1}), (t_2, w_{s-a2}), \ldots, (t_k, w_{s-ak})\} \tag{3}$$

2.4 The Analysis of Co-occurrence Relation

If there is a document that belongs to two topics simultaneously and the weights meet the requirement, the number of co-occurrence of two topics plus one, which means the two topics are co-occurrence relation. The analysis of co-occurrence relation can be achieved by the following steps:

(1) Representative topics of each document. For each document, topics are sorted descending by their probabilities to the document. The top three topics [11] are regarded as representative topics. For document d, its corresponding set of representative topics represents as S(d) containing three topics.
(2) Co-occurrence of two topics. Boolean function φ is defined to judge if two topics are representative topics to document d at the same time as Eq. 4. For topic t_a and topic t_b, the number of co-occurrence can be calculated as Eq. 5 where D is the corpus.

$$\varphi(t_a, t_b, d) = \begin{cases} 1 & \text{if } t_a \text{ in } T(d) \text{ and } t_b \text{ in } T(d) \\ 0 & \text{if } t_a \text{ out of } T(d) \text{ or } t_b \text{out of } T(d) \end{cases} \tag{4}$$

$$\mathrm{Cor}(t_a, t_b) = \sum_{d \text{ in } D} (t_a, t_b, d) \tag{5}$$

(3) Co-occurrence vector for each topic. For a topic t_a, its co-occurrence vector can be represented as Eq. 6 where w_{c-ai} represents its corresponding weight in co-occurrence. w_{c-ab} can be calculated by the ratio of the count of co-occurrence with topic t_b and the count of co-occurrence with all the other topics.

$$\mathrm{Cooccurrence}(t_a) = \{(t_1, w_{c-a1}), (t_2, w_{c-a2}), \ldots, (t_m, w_{c-am})\} \tag{6}$$

2.5 The Method SCW to Predict Topic Popularity

According to the mind of KNN method, popularity of a topic can be predicted based on the popularities of its K most similar topics and its top k co-occurrence count topics. With the consideration mentioned above, SCW (Similarity Co-occurrence Weighting) method is proposed. It includes the following steps:

(1) Popularity level of a topic based on its similarity relation. For a topic t_a, the similar degrees that it is similar with other topics can be represented as Eq. 3. Its K most similar topics are chosen and represented as $S_{sim}(t_a)$. The probabilities that topic t_a belongs to four popularity levels can be calculated based on the popularity levels of elements in $S_{sim}(t_a)$ and their similarities with topic t_a as Eq. 7.

$$p_{sim}(t_a, L_i) = \sum_{t_j \in S_{sim}(t_a)} w_{s-kj} * \delta(L(t_j) = L_i) \tag{7}$$

Where the values of i can be 1, 2, 3, 4, which corresponding to four levels L_1, L_2, L_3, L_4 and $L(t_j)$ is a function that is used to judge the popularity level that topic t_j belongs to. $\delta()$ is a boolean function as shown in Eq. 8.

$$\delta(x) = \begin{cases} 1 \text{ if } x \text{ is true} \\ 0 \text{if } x \text{ is false} \end{cases} \tag{8}$$

(2) Popularity level of a topic based on its co-occurrence relation. For a topic t_a, the co-occurrence degrees that its co-occurrence status with other topics can be represented as Eq. 6. Its top k co-occurrence weight topics are chosen and represented as $S_{cor}(t_a)$. The probabilities that topic t_a belongs to four popularity levels can be calculated based on the popularity levels of elements in $S_{cor}(t_a)$ and their co-occurrence degrees with topic t_a as Eq. 9.

$$p_{cor}(t_a, L_i) = \sum_{t_j \in S_{cor}(t_a)} w_{c-kj} * \delta(L(t_j) = L_i) \tag{9}$$

(3) Final popularity level of a topic. For a topic t_a, the final popularity level can be predicted based on the steps one and two as shown in Eq. 10. The status

corresponding to maximum probability is chosen to be the final popularity level for topic t_a.

$$p(t_a, L_i) = \gamma * p_{sim}(t_a, L_i) + (1 - \gamma) * p_{cor}(t_a, L_i) \quad (10)$$

3 An Empirical Study

3.1 Data Set and Experiment Setting

The data on Sina Weibo is used to do the experiments. We crawl the text information on Sina Weibo from October 1, 2012 to October 31, 2012 through API interface with retweeting and comment information. Filter information and 137,266,573 texts are used for the experiment after preprocessing.

The method proposed by Steinbach [12] is adopted to do the evaluation, that is, the precision, recall rate and F measure. The evaluation of popularity each level L_i is calculated and the final precision P, recall rate R and F measure F of method is the average of four popularity levels.

The SCW method combines the consideration of similarity and co-occurrence with the mind of KNN method. In the experiment, the K value of KNN method is set to five, time t_r is set to 25 h and t_t is set to 40 h. For popularity levels, the parameters p_1, p_2,p_3 are set to 0.2, 0.4 and 0.7 respectively.

There are three variables in LDA model: parameters α, β and the number of topics k. According to the literature [13], we know that the best choice is that $\alpha = 50/k$ and $\beta = 0.01$. The evaluation standard of perplexity used widely in statistical language model is selected to choose the best value of k. Perplexity is a standard measuring fit degree of model and the lower the value is, he better the model fit. In the experiment, we can see the change of perplexity by taking different values of k as shown in Fig. 1. Perplexity reaches the minimum value when the number of topics is 1000, which means that when the k = 1000, the LDA model has a better performance. Therefore, k is set to 1000, $\alpha = 0.05$ and $\beta = 0.01$. We assign the training set and the test set with the ratio of 4:1, that is, the final training data set has 800 topics and the test data set has 200 topics.

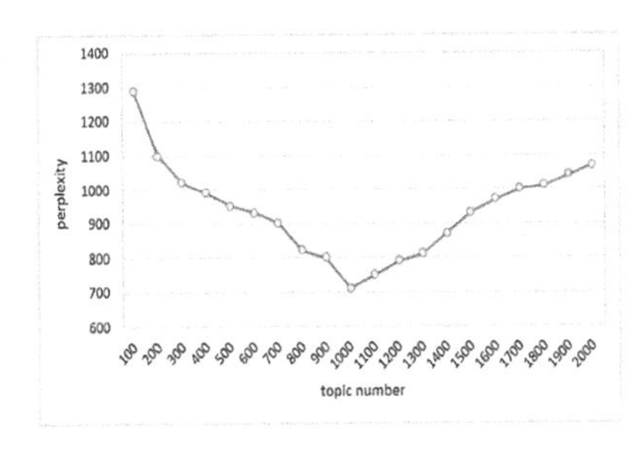

Fig. 1. Perplexity with different values of topic number k

Parameter γ is used to measure the effects of similarity relation and co-occurrence relation and a better value of parameter γ can improve performance of method to some extent. Here the training data is used to do the experiment. Due to the different values of γ from 0.1 to 0.8, count and observe the changes of P, R and F-measure to choice the best value for γ. The experiment results are shows as Fig. 2.

The values of P, R and F-measure all change from low to high and then of a gradual decrease. The P and F-measure are both maximums when $\gamma = 0.4$ while R is relatively high. So $\gamma = 0.4$ is selected for the experiment.

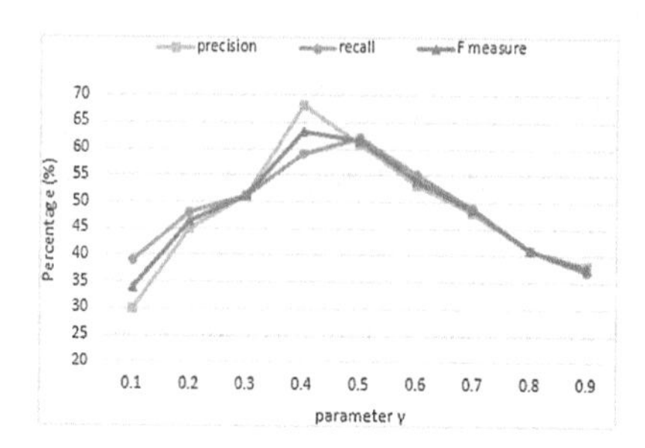

Fig. 2. Precision, recall and F-measure with different values of linear parameter γ

3.2 Result Analysis

To test the performance of method SCW, the methods shown below are chosen to do the comparison. And the parameters are the same as mentioned above.

1. Method based on similarity relation (KNN-S): parameter γ is set to 1, that is, popularity prediction of a topic is only affected by popularity levels of its K most similar topics.
2. Method based on co-occurrence relation (KNN-C): parameter γ is set to 0, that is, popularity prediction of a topic is only affected by popularity levels of its top k co-occurrence count topics.
3. S-H model: The ratio μ of retweet quantities at time t_t and time t_r is obtained by training. With current problem setting and hypothesis of S-H model, the classification result can be obtained as following way: if $\mu \leq p_1$, all samples will be classified to popular hardly (level L_4); if $p_1 < \mu \leq p_2$, all the samples will be classified to popular partly (level L_3); if $p_2 < \mu \leq p_3$, all the samples will be classified to popular highly (level L_2); If $\mu > p_3$, all samples are classified to popular amazingly (level L_1).

The results are shown as Fig. 3. To S-H model, all the test data were classified to the same popularity level because of that they shared the unique linear factor μ under the current problem definition, which greatly affects the performance of the model. To KNN-S method, it is the results of similarity. Hence it has a better performance than S-H model since it does not have to force classification. To KNN-C method, it is based on co-occurrence and its performance is better than KNN-S method, which shows that co-occurrence information has a greater impact on popularity than similarity information in some ways. The SCW method has a better popularity prediction effect than the other baseline methods as far as current parameter setting. The analysis results show that many topics can be both similar topics and co-occurrence topics for a certain topic at the same time. These topics should have greater impact on the popularity prediction, which is in keeping with our method.

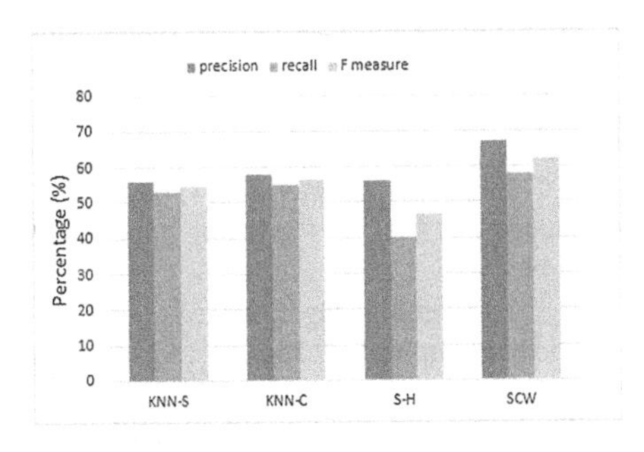

Fig. 3. Precision, recall and F-measure of different models

4 Conclusions

In the research of social media, the analysis and prediction of topic popularity can mine concerns and change trends of topics, which is easy to grasp, supervise and guide public opinions timely and accurately. Based on the idea of KNN method, this paper adopted retweet volume of topic in microblog to measure popularity measure, divided the popularity into four levels, combined similarity information with co-occurrence information and predicted topic popularity level according to early situation of the topic. The experiment results show that SCW method which is based on similarity relation and co-occurrence relation has a better performance than baseline methods and can predict popularity level of the topic after a period of time. How to integrate influence regional characteristics and other factors into the method and improve the performance of the method will be our focus in the future.

Acknowledgements. The work described in this paper is partially supported by National Key Fundamental Research and Development Program (No. 2013CB329601, No. 2013CB329602, No. 2013CB329604) and National Natural Science Foundation of China (No. 61502517, No. 61372191, No. 61572492, No. 61502517), 863 Program of China (Grant No. 2012AA01A401, 2012AA01A402, 2012AA013002), Project funded by China Postdoctoral Science Foundation (2013M542560, 2015T81129).

References

1. Wu, T., Timmers, M., Vleeschauwer, D.D., Leekwijck, W.V.: On the use of reservoir computing in popularity prediction. In: Evolving Internet (INTERNET), 2010 Second International Conference on, 19–24, 2010
2. Lerman, K., Hogg, T.: Using a model of social dynamics to predict popularity of news. In: Proceedings of the 19th international conference on World Wide Web, 621–630, 2010
3. Lee, J.G., Moon, S., Salamatian, K.: An approach to model and predict the popularity of online contents with explanatory factors. In: Web Intelligence and Intelligent Agent Technology (WI-IAT), 2010 IEEE/WIC/ACM International Conference on, 623–630, 2010
4. Kong, Q.C., Mao, W.J.: Predicting popularity of forum threads based on dynamic evolution. J. Softw. **25**(12), 2767–2776 (2014)
5. Wan, S.X., Guo, J.F., Lan, Y.Y., Cheng, X.Q.: Tweet popularity prediction based on propagation simulation. J. Chinese Info. Process. **28**(3), 68–74 (2014)
6. Xiong, X.B., Zhou, G., Huang, Y.Z., Ma, J.: Predicting popularity of tweets on Sina Weibo. J. Info. Eng. Univ. **13**(4), 496–502 (2012)
7. Nie, R.L., Chen, L., Wang, Y.Q., et al.: Algorithm for prediction of new topic's hotness using the K-nearest neighbors. Comput. Sci. **39**(6A), 257–260 (2012)
8. Wang, W.J., Chen, L., Wang, Y.Q., et al.: Algorithm for prediction of post's hotness using K-nearest neighbors and latent dirichlet allocation. J. Sichuan Univ. **51**(3), 467–473 (2014)
9. Xiang, W., Jia, Y., Chen, R.H., et al.: Improving text categorization with semantic knowledge in wikipedia. IEICE Trans. Info. Sys. **E96-D**(12), 2786–2794 (2013). (Dec)
10. Li, H., Yan, J., Weihong, H., et al.: Mining user interest in microblogs with a user-topic model. Comm. China **11**(8), 131–144 (2014)
11. Zhang, J., Li, F.: Context-based topic evolution and topic relations extraction. J. Chinese Info. Process. **29**(2), 179–189
12. Steinbach, M., Kapypis, G., Kumar, V.: A comparison of document c1ustering techniques. Proceedings of KDD Workshop on Text Mining, 109–111, 2000
13. David, B., et al.: Latent dirichlct allocation. J. Mach. Learn. Res. **3**, 993–1002 (2003)

A New Bayesian-Based Method for Privacy-Preserving Data Mining

Guang Li(✉)

School of Electronic and Control Engineering, Chang'an University, Xi'an, Shaanxi 710064, P. R. China
liguangteacher@126.com

Abstract. Recently, data mining developed fast and attracted a lot of attention. When using data mining in real world, privacy protection is an important problem. Over the past ten years, many researchers study this problem and propose a lot of PPDM (privacy preserving data mining) methods. These methods can complete data mining task when protecting privacy. This paper gives a new Bayesian-based PPDM method, which is designed for classification. This method is a data perturbation method and is algorithm-independent, which means the perturbed data can be used by normal classification methods directly. Experiments show that comparing with existing methods, this new method perform better for protecting privacy, when they keeping data utility both.

Keywords: Privacy preserving · Classification · Bayesian decision theory · Data perturbation

1 Introduction

Data mining is the process to extract valuable patterns from large data set. Recently, data mining developed fast. It attracted a lot of attention and has been applied in many fields. When using data mining in real world, privacy protection is an important problem. In the traditional data mining methods, the issue of privacy protection is not considered. They assume the data can be gotten directly. But in fact, in many fields, especially as medicine, financial, homeland security and so on, this assumption does not hold. For legal and social requirements, privacy protection is very important and the data can not be gotten directly in many cases. And the traditional data mining methods are not suitable for these cases.

Many researchers have studied the privacy protection problem for data mining. They proposed many methods to solve this problem [1, 2]. One important part of these methods is privacy-preserving data mining (PPDM) method. The PPDM methods both consider data mining and privacy protection. They always can complete data mining without gotten the original data directly. According to the differences of basic idea, PPDM methods can be divided to two categories. One category is the data perturbation-based methods [3–5]. The other category is using secure multi-party computation (SMC) [6, 7]. In the first category methods, when doing data mining, the

F. Xhafa et al. (eds.), *Advances in Intelligent Systems and Interactive Applications*, Advances in Intelligent Systems and Computing 686,
https://doi.org/10.1007/978-3-319-69096-4_24

original data can not be gotten. The executors of data mining only can get a perturbed data set and only can do data mining on this data set. If they can finish data mining task by using this data set and can not get the privacy information from it, the PPDM method is successful. In the second category methods, the application scenario is set to be distributed databases which have many nodes. Every node has part of the data set but data mining should be executed on the whole data set. So, every node can not finish data mining alone and they must co-operate. This kind of method designed SMC-based protocols to exchange information among nodes. These protocols can support the completion of data mining task. And every node should not get any privacy data from the information exchanging procession.

A new Bayesian-based method is proposed by this paper for PPDM. This method is designed for classification and is based on data perturbation strategy. From experiments, it can be seen that this method perform well. It can protect privacy well when keeping data utility. Additionally, it is an algorithm-independent method. So the normal classification method can be applied without modification on the perturbed data. It is an advantage for data mining application.

2 Bayesian Decision Theory

In Bayesian decision theory [8], it assume that there are some category: W_1, W_2, …, W_m. For a new sample x, whose category is not known, its category is predicted by using probability $P(W_i|x)$. The category maximizing this probability will be given to sample x. When actually calculating of this probability, Bayes' formula is often used. Then, it will be calculated as Eq. (1).

$$P(W_i|x) = \frac{P(x|W_i)P(W_i)}{\sum_{j=1}^{m} P(x|W_j)P(W_j)} \tag{1}$$

The Bayesian decision theory is a basic idea to solve classification problem. A lot of classification methods have been derived from it. KNN (k nearest neighbors) method is one of them.

KNN method is used to deal with the situation that the prior parameterized knowledge about the underlying probability structure is not known. In this situation, the probability $P(W_i|x)$ has to be estimated by using nonparametric method from training samples alone.

In KNN method, k-nearest-neighbor estimation is used. For the new sample x, a cell of volume V around x and capture k samples is set up. If assuming that the probability of sample occurrence is uniform in this cell, the probability $P(x, W_i)$ can be estimated as Eq. (2).

$$P(x, W_i) = \frac{k_i/n}{V} \tag{2}$$

k_i is the number of samples belonged to category W_i in this cell, and n is the total number of samples.

Then, the probability $P(W_i|x)$ is estimated as Eq. (3). And the category of x is predicted as the most frequent category in the k nearest training sample from x.

$$P(W_i|x) = \frac{k_i}{k} \tag{3}$$

3 The New Method

This paper's new PPDM method is a Bayesian decision theory-based method and is designed for classification. This new method is a data perturbation method. It will open a perturbed data set. And the original data will be hided.

This method believe that the categories of new samples will be predicted according to the probability $P(W_i|x)$ in the Bayesian decision theory as Sect. 2 shown above. The category maximizes this probability will be given to the new sample. If data perturbation never change the category maximizes probability $P(W_i|x)$, than it will never change the prediction category in Bayesian decision theory and should keep the data utility. If using the k-nearest-neighbor estimation, it should keep the most frequent category in the k nearest training sample from new samples not changed.

In detail, new sample x is generated randomly. The k nearest neighbor training samples for x will be selected out. If B_1 and B_2 are the most and least frequent category appeared in these k samples respectively, x will be given B_1 in Bayesian decision theory. The data is perturbed by add the average of samples belong to B_1 in these k samples and delete one sample belong to B_2. Obviously, after perturbation, B_1 is still the most frequent category appeared in these k samples and so the data utility is kept. This method does not request B_1 and B_2 must be two different categories. If there is only one category in these k samples, B_1 and B_2 will be the same. This data perturbation process will repeated n times. The workflow of this method is shown in Fig. 1.

Input: $D = \{x_1, x_2, \ldots, x_n\}$, which is the original data. The parameters k and n.
Output: The perturbed sample set T.

procedure **NewMethod** (D, k, n)
$T = D$
for i in 1:n
  generate sample x randomly
  Dx is the set of the k nearest neighbor samples in D for x
  B_1 is the most frequent category appeared in Dx
  B_2 is the least frequent category appeared in Dx
  s is the average of samples belong to B_1 in Dx
  t is one of the samples belong to B_2 in Dx selected randomly
  $T = T - t + s$
end for
return (T)
end procedure

Fig. 1. The detailed process of the new algorithm proposed in this paper

It needs attention that this new method never considers the risk. If considering the risk, Bayesian decision theory will give the category minimize the expected loss to the new sample. And it is not guarantee to be the category maximize probability $P(W_i|x)$. It is easy to change this new method to fit the situation considering the risk.

This paper use experiments to confirm that this new method perform well. When keeping data utility, it can protect privacy well. On the other hand, obviously, this new method is an algorithm-independent method. So the normal classification method can be applied without modification on the perturbed data. It is an advantage for application.

4 Experiments

For experiments, this paper used two data sets: the PID (Pima Indians Diabetes) and the WBC (Wisconsin Breast Cancer). These two data sets are all real-life data, and are all collected by UCI (University of California Irvine).

To compared with the new method, this paper use singular value decomposition (SVD)-based PPDM methods [9, 10]. The methods based on SVD have two forms. The first one is the basic form, which is called BSVD (basic SVD) method. The second one is the improved form, which is called SSVD (sparsified SVD) method.

For measuring the privacy protection, this paper used five metrics: CK, CP, RK, RP and VD. They are all used in the SVD-based methods. For better privacy protection, CK and RK should be smaller, and CP, RP and VD should be larger.

$$E = R_p - (1 - e)R_o \tag{4}$$

For measuring the data utility, this paper calculated the error E, which is defined as Eq. (4). In experiments, classifiers are trained on original trained data and perturbed trained data both. And then, classifiers are used to predict original tested samples. In Eq. (4), R_o is prediction accuracy for classifier trained by original data, and R_p is that by perturbed data. Three kinds classifier are trained. They are support vector machine (SVM), decision tree (j48 in WEKA [11]) and the nearest neighbor. In Eq. (4), e is a constant and is set to be 0.02. If error E not less than zero, the data utility is kept.

The new method has two parameters: k and n. It used k-nearest-neighbor estimation and n is the number of iterations. Figures 2 and 3 show the utility measures of the new method proposed by this paper for different values of k and n respectively. Figure 2 let $n = 1500$ and Fig. 3 let $k = 27$. As mentioned above, this paper selects three kinds of classification methods to test data utility. For every one, it can get an error E, which is defined as Eq. (4). In Figs. 2 and 3, the Y-axis E_m is the minimum value of these three errors. It can be seen that data utility is kept well for the new method. Different values of parameters have little effect for the data utility.

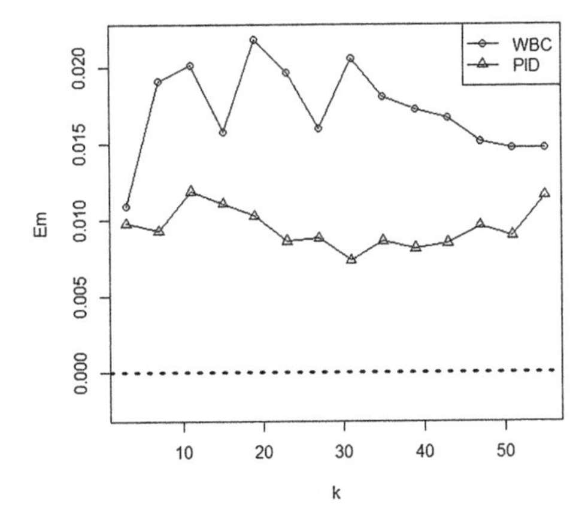

Fig. 2. The error E of the new algorithm with parameter k's different values.

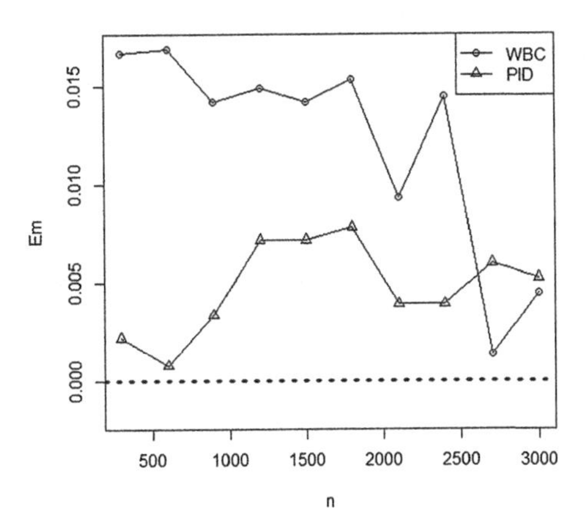

Fig. 3. The error E of the new algorithm with parameter n's different values.

Table 1 includes the comparing method and the proposed method's privacy protection measures. In Table 1, D1 is the WBC data set and D2 is the PID data set. SVD1 is the BSVD method and SVD2 is the SSVD method. In the new method, $n = 1500$ and $k = 27$, which is about at the middle position of range in previous experiment. For SVD-based method, the parameters' values are selected to guarantee data utility and optimize privacy measures. It can be seen that if comparing with the SVD-based method, privacy is protected better by the new proposed method, when data utility is kept.

Table 1. Privacy protection measures for SVD-based methods and the proposed method

Data	Algorithm	CK	CP	RK	RP	VD
D1	New method	0.7	0.3	0.015	37.6	0.36
D1	SVD1	0.8	0.3	0.019	31.9	0.11
D1	SVD2	0.8	0.3	0.015	37.3	0.25
D2	New method	0.94	0.06	0.012	60.9	0.22
D2	SVD1	1	0	0.126	48.3	0.01
D2	SVD2	1	0	0.064	56.2	0.03

5 Conclusions

To solve the privacy protection problem in data mining, a Bayesian-based PPDM method, which is designed for classification, is proposed in this paper. This new method is one of method using data perturbation scheme. The basic idea of it is that if the original and the perturbed data can give same category prediction for new samples under Bayesian decision theory, it can keep data utility and can be used for privacy preserving classification. Through experiments, it can confirm that the new proposed method perform well. Compared with existing methods, it can protect privacy better when keeping data utility. Additionally, it is an algorithm-independent method, which is an advantage for application.

Acknowledgements. This paper is funded by Natural Science Basic Research Plan in Shaanxi Province of China (2016JQ6078) and the Basic Research Funds of Chang'an University (0009-2014G6114024).

References

1. Bertino, E., Fovino, I.N., Provenza, L.P.: A framework for evaluating privacy preserving data mining algorithms. Data Min. Knowl. Discov. **11**(2), 121–154 (2005)
2. Fung, B.C.M., et al.: Privacy-preserving data publishing: a survey of recent developments. ACM Comput. Surv. **42**(4), 14:1–14:53 (2010)
3. Kisilevich, S., et al.: Efficient multidimensional suppression for K-anonymity. IEEE Trans. Knowl. Data Eng. **22**(3), 334–347 (2010)
4. Last, M., et al.: Improving accuracy of classification models induced from anonymized datasets. Inf. Sci. **256**, 138–161 (2014)
5. Thakkar, A., Bhatti, A.A., Vasa, J.: Correlation based anonymization using generalization and suppression for disclosure problems. Adv. Intell. Sys. Comput. **320**, 581–592 (2015)
6. Hoogh, S., et al.: Practical secure decision tree learning in a teletreatment application. In: Lecture Notes in Computer Science vol. 8437, pp. 179–194 (2014)
7. Emekci, F., et al.: Privacy preserving decision tree learning over multiple parties. Data Knowl. Eng. **63**(2), 348–361 (2007)
8. Duda, R., Hart, P., Stork, D.: Pattern Classification, 2nd edn. Wiley, New York (2000)

9. Wang, J., et al.: A novel data distortion approach via selective SSVD for privacy protection. Int. J. Inf. Comput. Secur. **2**(1), 48–70 (2008)
10. Xu, S., et al.: Singular value decomposition based data distortion strategy for privacy protection. Knowl. Inf. Syst. **10**(3), 383–397 (2006)
11. Witten, I.H., Frank, E., Hall, A.M.: Data Mining: Practical Machine Learning Tools and Techniques. Morgan Kaufmann, Burlington (2016)

Gesture Recognition Algorithm Based on Fingerprint Detection

Ge-Yan Ru[1(✉)], Zhang-Guo Wei[1,2], and Lu-Qiu Hong[2]

[1] Shanghai University of Electric Power, Shanghai, China
1032485466@qq.com
[2] Shanghai HRSTEK Co., Ltd, Shanghai, China

Abstract. In order to overcome the influence of traditional gesture recognition on the surrounding environment, illumination change and background, a gesture recognition method based on fingertip detection is proposed. Firstly, the depth image of Kinect is collected, and the threshold segmentation and color space are combined to complete the hand segmentation. And then calculate the curvature value of the palm of your hand, and obtain the fingertip and depression points by defining the value of curvature. Finally, the fingertip detection algorithm with the center of gravity distance is set, the distance from the palm to the fingertip and the depression is set, Fingertip mark points, complete gesture recognition. The experimental results were verified by 100 experiments, and the recognition rate was 97%. Experimental results show that this method can accurately carry out hand segmentation and fingertip recognition.

Keywords: Color space · Hand segmentation · Curvature · Center of gravity distance · Fingertip detection

1 Introduction

Gesture As the simplest and most intuitive way of communication in everyday life, it is widely used in human -computer interaction, and gesture recognition technology is applied to the motion control of mobile robot. Xu et al. [1] designed a wearable finger controller, composed of sensors, communication modules. The controller will receive the finger data sent to the computer through wifi, get the hand of the characteristic components, combined gesture attitude to complete gesture recognition. Weng et al. proposed a monocular vision [2] gesture recognition algorithm, the use of YCgCr color space for hand segmentation, through the mark convex block to complete the detection of the fingertips, and according to the size of the contour to further identify. Hasan and Abdul-Kareem [3] proposed a human gesture recognition technique based on shape analysis to explore gesture recognition based on neural networks. Qin et al. [4] used the depth information obtained by the depth sensor to detect and locate the gesture. The convex decomposition method of Radius Morse function was used to decompose the hand in real time. Such as the use of convolution neural network [5] for gesture recognition, directly to the hand of the data information input to the network, do not need to pre-morphological processing of the image, the output part has been sorted, eliminating the complex. The handling, simple and convenient operation. Dominio

F. Xhafa et al. (eds.), *Advances in Intelligent Systems and Interactive Applications*, Advances in Intelligent Systems and Computing 686,
https://doi.org/10.1007/978-3-319-69096-4_25

et al. [6] proposed a gesture recognition scheme for depth information, extracted hands from the acquired data, and divided the hand into palm and finger areas. Then the four different sets of feature descriptors are extracted, the distance between the fingertip and the center of the palm, the distance of the palm of the hand, the curvature of the hand contour, the geometry of the palm region, and the multi-class SVM classifier to identify the gesture.

There are two stages in the paper: the depth of gesture segmentation, the first use of Kinect depth information, the threshold segmentation and HSV color space combined to quickly split the gesture; fingertip detection stage, the center of gravity and the focus of the combination of fingertips, Complete the detection of fingertips, and then achieve gesture recognition.

2 Gesture Extraction

2.1 Threshold Segmentation

Gesture segmentation occupies a very important position throughout the gesture recognition process, and its segmentation results directly affect subsequent gesture recognition. Before the gesture segmentation, first use Kinect sensor to obtain the depth image, as shown in Fig. 1.

Fig. 1. Depth image

Kinect sensor directly to the depth of the image, the hand in the depth of the image and the body and the depth of the background map is intermittent, but also to the next recognition caused a lot of interference. In order to remove the background information in the depth map, only to retain the depth of the hand image, the paper first on the depth of the image OSTU maximum interclass variance threshold segmentation processing.

OTSU maximum interclass variance algorithm will hand area and background area as two parts, automatically select the appropriate threshold, both to make the maximum variance of hand and background, but also to ensure that the variance between the hand area minimum, the maximum possible. To remove the background interference information at the same time, to ensure the integrity of the hand segmentation, according to formula 1 and formula 2 variance formula, automatically calculate the appropriate threshold, the binarization of the image processing.

$$U = W_0 * U_0 + W_1 * U_1 \tag{1}$$

$$G = W_0 * (U_0 - U) * (U_0 - U) + W_1 * (U_1 - U) * (U_1 - U) \tag{2}$$

W_0 represents the ratio of the foreground pixels to the image, U_0 is the average gray level of the foreground pixel, W_1 is the ratio of the number of background pixels to the image, U_1 is the average gray level of the background pixel, U is the total gray level of the image, variance. By calculating the appropriate threshold to ensure that the hand and the background area of the largest variance, but also make the difference within the hand area of the smallest, and then the maximum part of the palm part of the split out, the threshold segmentation results shown in Fig. 2.

Fig. 2. Threshold segmentation results

2.2 Hand Extraction Based on Hsv Color Space

Figure 2 split the results mapped to the RGB color space, to further improve the gesture segmentation. 3 and 4 are the coordinates of the point at the depth of the camera, and the coordinates of the point on the image plane are the internal matrix of the depth camera. Set the coordinates of the same point in the RGB camera coordinates, for the point in the RGB image plane projection coordinates, RGB camera for the internal matrix. Although the depth of the camera coordinates and RGB camera coordinates are different, but they can be linked by the formula 5. And finally on the projection, get the corresponding RGB coordinates.

$$p\_ir = H\_irP\_ir \tag{3}$$

$$P\_ir = H\_ir^{-1}p\_ir \tag{4}$$

$$P\_rgb = RP\_ir + T \tag{5}$$

$$p\_rgb = H\_rgbP\_rgb \tag{6}$$

After the above formula, the threshold segmented image is reflected to the RGB color space. Since the RGB color space is particularly susceptible to external conditions, the RGB color space is transformed into the HSV color space according to the conversion formula of the color space. HSV color space H, S, V three components, H value by the background and lighting conditions of the smaller, compared to distinguish, you can only take the H value. Figure 3 shows the threshold segmentation result is reflected in the color space, and then get the HSV color space H value, Fig. 4 is a single background of the hand H value.

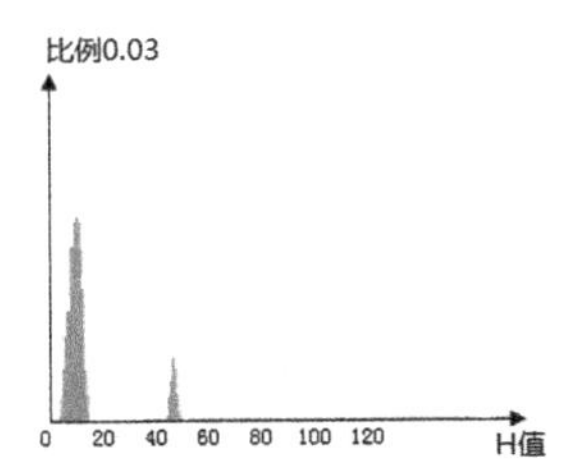

Fig. 3. Image H value

In Fig. 3, the H value of the image is distributed in (2, 18), (35, 40), (45, 57) and (77, 82); the H value of the hand in Fig. 4 is (3, 18). H (3, 18) as the standard value, the default in Fig. 3 within the scope of the hand, the standard as the background information, to be deleted, the resulting segmentation results shown in Fig. 5.

3 Fingertip Detection

3.1 Fingertips Based on Curvature Detection

After the gesture segmentation is complete, the Canny operator is used to obtain the contour of the hand. At the fingertip recognition section, the paper first uses the curvature to carry on the preliminary fingertip detection, the curvature in the setting threshold range thinks that the current point is the finger point, and then marks the finger condition coincides with the screening point collection. Curvature of the fingertip detection of the basic idea: any take the gesture on the outline of the point, remember that the coordinates of the set, set a constant N = 8, and points before the N points, recorded as coordinates, after the N Point, remember, the coordinates are shown in Fig. 6. Then the points and sum are connected together, the composition and the vector, so that the previous eight components of the eight directional vector, and the back also formed eight directional vector, to get a total of 16 before and after such a vector, As a group. The entire palm of the hand there are numerous such groups, there are numerous, follow-up to its screening.

The curvature of the fingertip is the largest, corresponding to the smallest, each group has a minimum value, but not each is a fingertip. Therefore, we set the value must be acute angle, and cannot be greater than 30°. Will not meet the conditions of exclusion, and will meet the conditions marked. The test results are shown in Fig. 7.

3.2 Excluding the Depression Point

Based on the curvature of the fingertip detection, roughly the fingertip point filter out, but which also contains the palm finger dents. In order to rule out the depression, the

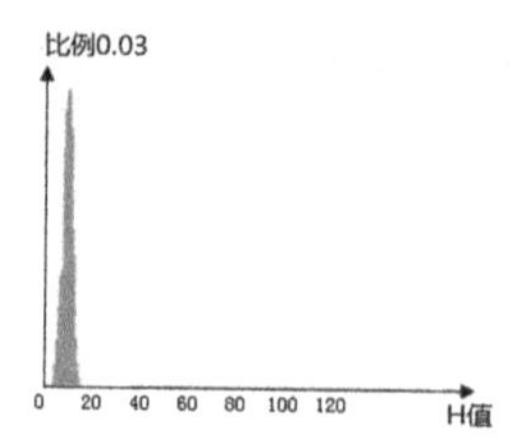

Fig. 4. Hand H value

Fig. 5. Gesture segmentation final result

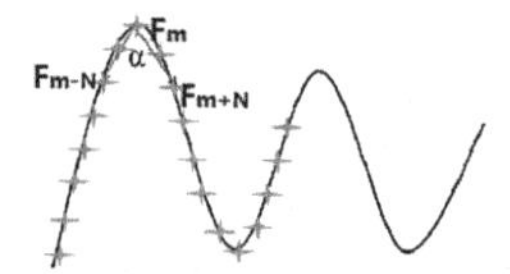

Fig. 6. Curve fingertip detection

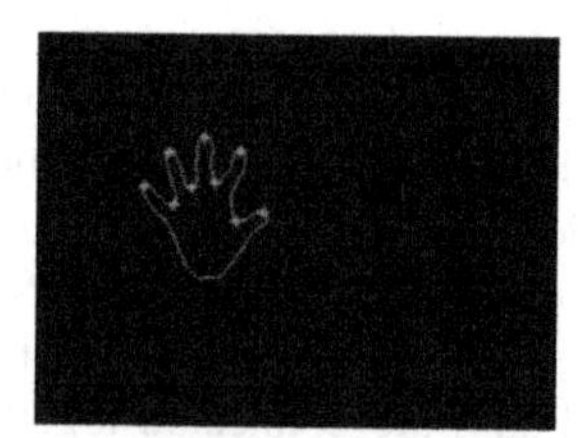

Fig. 7. Curve fingertip test results

paper then uses the fingertip detection based on the center of gravity to improve the detection process. Using the center of gravity formulas to calculate the palm center of gravity coordinates, set the palm coordinates, as shown in Fig. 7 any mark point coordinates. Connect the center of gravity with the fingertip, as shown in Fig. 8, to calculate the distance between the two.

$$D1 = \sqrt{(x_0 - x_m)^2 + (y_0 - y_m)^2} \tag{7}$$

Fig. 8. Center of gravity and distance

Assuming that the length of the center of gravity is the distance from the true fingertip, the distance from the depressed point is, by comparison, it is found that the length is much longer than 1.5 times. After measuring the distance between the center of gravity and all the fingertip points, the paper chooses the longest distance, uses the representation, and multiplies the distance between the other fingertips and the center of gravity by the scale factor 1.5, expressed by the formula 8 show.

$$D1_G = 1.5D1_F \tag{8}$$

When the value is greater than the value, that the current fingertips point for the real fingertips, if less than the value, does not meet the set criteria, that the current fingertips for the depression point, and give up. After screening, the depression is discarded, leaving the real fingertips.

4 Experimental Results

In order to verify the feasibility and accuracy of the algorithm, the experiment is based on 100, the accuracy of the gesture statistics, the experimental results shown in Table 1, the accuracy of the identification is relatively possible, but there will be a failure Recognition, which is the place to be improved later.

Table 1. Gesture recognition success rate statistics

Gesture	Total number	Success	Failure	Success rate
'5'	100	97	3	97%

5 Conclusion

In this paper, the use of Kinect to obtain the current depth of the background information, and the depth of the image threshold segmentation, the segmentation results reflected to the color space, and a single background by hand H value comparison, not in the range of the default background Hand split. Fingertip detection phase, by comparing the value of curvature to get the fingertip position, combined with the center of gravity distance, excluding the interference point, only the remaining fingertips.

Through the gesture experiment results, we can find that the algorithm can accurately identify the gesture under the condition that the ambient light is relatively stable. The results show that the recognition rate of gestures is high and can be applied to human—computer interaction and gesture control for robot. Since this experiment is carried out under the condition that the light is relatively stable and the gesture is constant from the Kinect distance, the experiment will be carried out under different light and distance conditions to determine the threshold under different conditions.

References

1. Xu, J., Liu, C., Meng, Y., et al.: Wearable gesture recognition controller [J]. Electron. Techn. App. **42**(7), 68–71 (2016)
2. Wen, H.-L., Y-W, Zhan: Improved multi-feature gesture recognition. Comput. Eng. Sci. **34** (2), 123–127 (2012)
3. Hasan, H., Abdul-Kareem, S.: Static hand gesture recognition using the neural networks. Artif. Intell. Rev. **41**(2), 147–181 (2014)
4. Qin, S., Zhu, X., Yang, Y., et al.: Real-time hand gesture recognition from depth images using convex shape decomposition method. J. Signal Process. Sys. **74**(1), 47–58 (2014)
5. Lu, Z., Chen, X., Li, Q., et al.: A hand gesture recognition framework and wearable gesture-based interaction prototype for mobile devices. IEEE Trans. Human-Machine Sys. **44** (2), 293–299 (2014)
6. Dominio, F., Donadeo, M., Zanuttigh, P.: Combining multiple depth-based descriptors for hand gesture recognition. Pattern Recogn. Letters **50**, 101–111 (2014)

Parka: A Parallel Implementation of BLAST with MapReduce

Li Zhang(✉) and Bing Tang

School of Computer Science and Engineering, Hunan University of Science and Technology, Xiangtan 411201, Hunan, China
zlhncdsy@163.com

Abstract. Bioinformatics applications have become more data-intensive and compute-intensive, which requires an effective method to implement parallel computing and get a high-throughput. Although there exists some tools to realize parallelization of BLAST, but most of them depend on complex platforms or software. A parallel BLAST is implemented using Spark, which is called Parka. The parallel execution time and speedup of Parka are evaluated in a cluster environment. Then, it is compared with Hadoop-based parallelization method. Results show that it is a scalable and effective parallelization approach for sequence alignment.

Keywords: BLAST · MapReduce · Data-intensive computing · Bioinformatics applications

1 Introduction

In the current era of big data, in the field of scientific research and engineering applications, such as the climate science, energy, high-energy physics, chemistry and astronomy, people are demanding more and more high-performance computing resources for complex engineering calculations and problem solving, as well as system simulations. In the field of bioinformatics, we need large-scale computing and storage capacity to deal with DNA structure and sequence analysis. As an important and fundamental tool in bioinformatics research, Basic Local Alignment Search Tool (BLAST) is widely used for sequence alignment analysis [1].

BLAST is a typical data-intensive and compute-intensive application, and BLAST computing time relies on the length of query sequences and database size. In order to speed up sequence searches, parallel or distributed sequence alignment solutions have been studied by computer science and bioinformatics scientists. These solutions utilized new parallel software platforms or new hardware environment and infrastructures, such as cluster, grid, and cloud environment.

mpiBLAST [2] is a MPI version of parallelized BLAST that uses the message passing library and achieves super-linear speedup through splitting the database into chunks and then each computing node searches a unique chunk. TurboBLAST [3] is another parallel version of BLAST which is fit for execution on interconnected heterogeneous computers in network environment. GPU-BLAST [4], is another multi-core-based improvement for BLAST application, which fully uses the

F. Xhafa et al. (eds.), *Advances in Intelligent Systems and Interactive Applications*, Advances in Intelligent Systems and Computing 686,
https://doi.org/10.1007/978-3-319-69096-4_26

computation capability of graphics processing unit (GPU), However, it is also confronted with the bottleneck of disk I/O rate and memory speed.

In grid environments, parallel execution of BLAST jobs could be enabled through deploying all kinds of grid middleware. Several implementations were reported, including ABCGrid [5], G-BLAST [6], BioGAT [7], etc. Our previous work [8] also studied a data-driven Master/Worker execution model of BLAST in Desktop Grids environment with BitDew middleware [9]. However, parallelized solutions of BLAST in grid environments cannot meet the requirement of current era of biological gene big data.

MapReduce is a popular parallel programming framework for data processing which attracts a lot of attentions in recent years. It is quite suitable for data-intensive bioinformatics applications, such as BLAST. Some BLAST implementations over Hadoop MapReduce have been proposed recently. However, some existing MapReduce-based solutions have very poor scalability, such as CloudBLAST [10], which does not consider database separation. Another work CloudBurst [11], only considers short reads mapping, while it is not fit for long DNA sequence alignment.

Spark is an effective distributed big data processing framework, which is developed by AMPLab of the University of California at Berkeley [12]. Spark realizes effective and reliable in-memory computing in cluster environment. It has been proved that it runs programs faster than Hadoop MapReduce [13]. Therefore, this paper studies on using Spark platform to improve the efficiency of parallel BLAST.

The main contributions of this paper are summarized as follows: firstly, it proposes a parallel BLAST approach combining sequence segmentation and database segmentation; secondly, the popular big data processing engine Spark is used to implement the proposed approach, which is proved to be more effective than Hadoop MapReduce-based implementation through performance comparison.

2 Parallelization of BLAST

There are two main approaches for the execution of parallel BLAST application: (1) *Query segmentation*. In this approach, the query sequences are segmented and each node only searches part of the total query sequences. Therefore, BLAST can be executed in parallel in a cluster environment. (2) *Database segmentation*. In this approach, the gene database is segmented into several chunks and each node only stores and searches independent chunks of the total database.

Unlike previous work, in this paper, query segmentation and database segmentation are combined together to realize large-scale sequence search. The idea of parallel BLAST is presented as follows. The input query sequences are segmented into i subsets and the database is segmented into j subsets, each pair of subsets is processed by a worker, results from each database subsets are sorted and merged. It can be achieved by following the MapReduce paradigm. Algorithm 1 demonstrates the algorithm of combining query segmentation and database segmentation in MapReduce paradigm-based BLAST.

Algorithm 1: MapReduce paradigm-based BLAST

Input: Query Sequences S
Input: Genebase D
Output: Final result R

1. write S to distributed file system;
2. database *segment* to generate D_j;
3. assign query sequence subset to each Worker;
4. ***Mapper*:**
5. **for all** query sequence S_i in S **do**
6.   **for all** database chunk D_j in D **do**
7.   execute BLAST with (S_i , D_j);
8.   obtain result R_{ij};
9.   **end for**
10. **end for**
11. ***Reducer*:**
12. collect R_{ij} to obtain the result R_i of S_i;
13. aggregate the result R_i of each sequence S_i to form final result R.
14. **return** R

3 Parallel BLAST with Spark MapReduce

Because of the low efficiency of network traffic and disk I/O in traditional Hadoop MapReduce implementation, Spark uses memory for data storage to realize quick processing, effectively ensuring the real-time processing and enabling interactive and iterative analysis. Fault-tolerance is achieved by the elastic resilient distributed datasets (RDD). It is entirely possible to re-implement MapReduce-like computation in Spark, thanks to Spark's primary abstract RDD, which exposes map(), reduce(), reduceByKey (), groupByKey() operations, etc. It can read from HDFS, HBase, Cassandra, and any Hadoop data source.

The same as "blastx" program from the NCBI BLAST suite, Parka also compares nucleotides sequences against protein sequences by translating the nucleotides into amino acids. In this paper, we define map and reduce operation to realize Parka with Spark MapReduce implementation. Parka uses a seed-and-extend alignment algorithm to efficiently find alignments. The alignment algorithm is implemented following MapReduce paradigm, and can be executed in parallel to many workers. Figure 1 demonstrates the overview of Parka. At the initiation steps, query sequences are input to the HDFS first, the database is also segmented and stored. Input query sequences have been converted to the format of "one-sequence-per-line" using a conversion tool written by Java, to eliminate newlines. Each line is divided into two parts by "tab" character. The first part (information about the DNA sequence) is the *key*, and the second part is the *value*. Then, build the words list and construct the scanner. There are three Map/Reduce phases: (1) Map: scan for hits; (2) Shuffle: sort results; (3) Reduce: extend hits. It should be noted that during the processing of the data input and output, Mapper reads one sequence at a time from the HDFS, and output the results which is stored in memory.

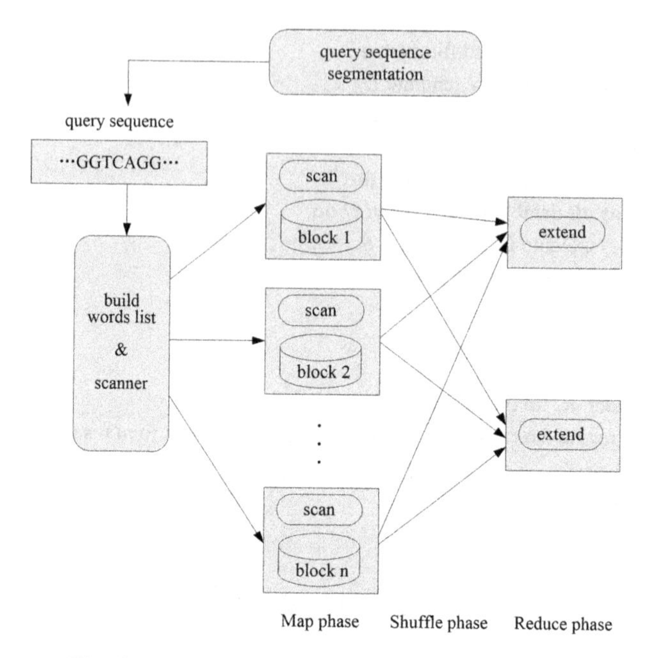

Fig. 1. The overview of map/reduce phase in Parka.

4 Results

4.1 Experiment Configuration

All the experiments were conducted in a cluster located at Hunan University of Science and technology. For each computing node, the hardware is configured as follow: Intel Xeon E5-1650 CPU at 3.5 GHz, with 32G DDR3 Memory, and 1TB SATA hard driver, and 1 Gb Ethernet. In our experiments, 11 nodes are used for performance evaluation, and 64 bits Ubuntu 14.04 as their operating systems. We use 64 bits Java JDK 1.7 and NCBI BLAST 2.2.14 for evaluation. The Spark platform we installed and configured is spark-1.0.0 version.

The input query sequences we used were taken from the standard GeneBank database, which was downloaded from the NCBI FTP site. In our performance evaluations, the query sequences are compared against the non-redundant (NR) protein sequence database (the total size is around 4.7 GB before formatting). In our BLAST performance evaluations, two sets of nucleotide sequences are used as input. One is called Short Query Sequence (SQS) which means short sequences take part in relative large proportion, and the other is Long Query Sequence (LQS) which means most of them are long sequences. Both SQS and LQS consist of 500 different sequences, and their characteristics are shown in Table 1.

Table 1. Description of query sequence sets

Query sequence set type	SQS	LQS
Number of sequences	500	500
Length of the longest sequence	850	4713
Length of the shortest sequence	128	230
Total number of nucleotides	215,743	502,294
Sequential execution time	8 h28 m	21 h10 m

4.2 Parallel Execution Over Hadoop

Hadoop Streaming utility is adopted in our experiment. We use Python to wrap the standard NCBI BLAST to obtain an executable BLAST mapper. Input query sequences have also been converted to the format of “one-sequence-per-line”. We run a streaming job by distributing large data through the “-file” option, so the BLAST application files and the Genebase files are packaged in streaming job first, and then shipped to worker nodes. The query sequences are input to the Hadoop Distributed File System first.

4.3 Parallel Execution Over Spark and Comparison of Speedup

In order to evaluate the performance of Parka, first we measured the sequential execution time. Therefore, we ran standard NCBI BLAST on one single worker, and ran for three times, and then calculate the average value as the final sequential execution time. The results of sequential execution time of all the 500 sequences in SQS and LQS have been shown in Table 1.

Then, we configured different worker number to measure the scalability of Parka, varying the number of nodes to perform BLAST computing. We use 1 nodes as Master, and the number of workers is varied from 1 to 10, increased 1 workers each step. We measured the total execution time of SQS and LQS. The relationship between the worker number and the total execution time are shown in Fig. 2. We observed that when the number of workers is smaller than 6, as the increase of worker number, the total execution time decreases slowly, while when the number of workers is larger than 6, it decreases rapidly.

Then, the speedup is calculated when SQS and LQS compared against NR Genebase. As you can see in Fig. 3, we also compared the proposed Spark approach with Hadoop approach. From this figure, it is observed that the value of speedup reaches 3.18 when using 6 workers, while the value is 5.5 when using 8 workers and we get 8.91 when we use 10 workers. With the Hadoop MapReduce approach, we cannot get a high speedup due to the weakness of shipping large size database chunks to worker nodes, and a longer time consumed for data reads and data writes. Parka outperforms Hadoop-based approach, and the reason is that with Spark MapReduce framework, intermediated data are stored in memory instead of local disk.

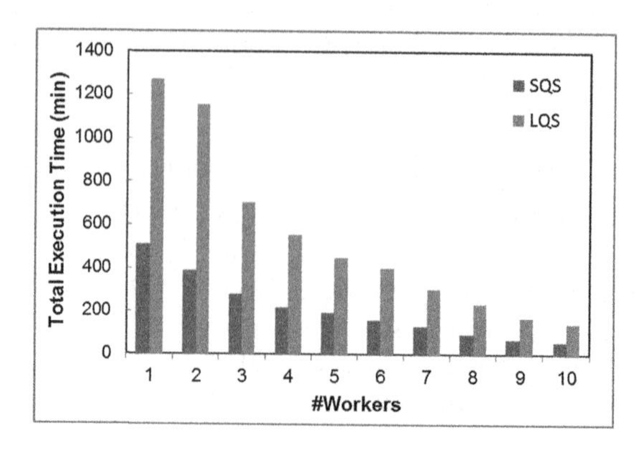

Fig. 2. Total execution time as the increase of worker number.

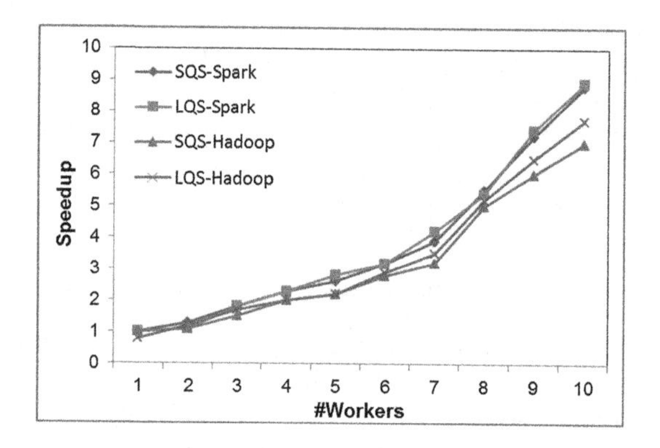

Fig. 3. The result of speedup comparison.

5 Conclusions

As an important programming model, MapReduce has attracted more and more attentions. We have implemented a new parallel BLAST based on MapReduce model, called Parka, with the help of Spark big data processing platform. The parallelization strategy is that query sequences are split into small pieces and database is also partitioned into blocks, and sequence alignments are performed on multi-workers, and then the output files are collected. In a cluster environment, the total execution time and speedup for the proposed Parka are evaluated. Experimental results demonstrate using Spark MapReduce implementation can shorten the execution time obviously, and achieves high speedup in a variety of processor configurations.

Acknowledgments. This work is partly supported by the National Natural Science Foundation of China (No. 61602169), the Natural Science Foundation of Hunan Province (No. 2015JJ3071), and the Scientific Research Fund of Hunan Provincial Education Department (No. 16C0643).

References

1. Altschul, S.F., Gish, W., Miller, W., Myers, E.W., Lipman, D.J.: Basic local alignment search tool. J. Mol. Biol. **215**(3), 403–410 (1990)
2. Darling, A.E., Carey, L., Feng, W.: The design, implementation, and evaluation of mpiBLAST. In: ClusterWorld Conference & Expo and the 4th International Conference on Linux Clusters: The HPC Revolution (2003)
3. Bjornson, R.D., Sherman, A.H., Weston, S.B., Willard, N., Wing, J.: TurboBLAST: a parallel implementation of BLAST build on the TurboHub. In: Proceedings of the International Parallel and Distributed Processing Symposium (IPDPS'02) (2002)
4. Vouzis, P.D., Sahinidis, N.V.: GPU-BLAST: using graphics processors to accelerate protein sequence alignment. Bioinformatics **27**(2), 182–188 (2011)
5. Sun, Y., Zhao, S., Yu, H., Gao, G., Luo, J.: ABCGrid: application for bioinformatics computing grid. Bioinformatics **23**(9), 1175–1177 (2007)
6. Yang, C.T., Han, T.F., Kan, H.C.: G-BLAST: a grid-based solution for mpiBLAST on computational Grids. Concurrency Comput. Pract. Exper. **21**(2), 225–255 (2009)
7. Mirto, M., Fiore, S., Epicoco, I., Cafaro, M., Mocavero, S., Blasi, E., Aloisio, G.: A bioinfomatics grid alignment toolkit. Future Gener. Comput. Syst. **24**(7), 752–762 (2008)
8. He, H., Fedak, G., Tang, B., Cappello, F.: BLAST application with data-aware desktop grid middleware. In: Proceedings of the 9th IEEE International Symposium on Cluster Computing and the Grid (CCGrid'09), pp. 284–291 (2009)
9. Fedak, G., He, H., Cappello, F.: BitDew: A data management and distribution service with multi-protocol file transfer and metadata abstraction. J. Netw. Comput. Appl. **32**(5), 961–975 (2009)
10. Matsunaga, A., Tsugawa, M., Fortes, J.: CloudBLAST: combining MapReduce and virtualization on distributed resources for bioinformatics applications. In: Proceeding of the Fourth IEEE International Conference on e-Science, pp. 222–229 (2008)
11. Schatz, M.C.: CloudBurst: highly sensitive read mapping with MapReduce. Bioinformatics **25**(11), 1363–1369 (2009)
12. Zaharia, M., Chowdhury, M., Franklin, M.J., Shenker, S., Stoica, I.: Spark: cluster computing with working sets. In: HotCloud 2010, USENIX Association, pp. 1–7 (2010)
13. Zaharia, M., Chowdhury, M., Das, T., Dave, A., Ma, J., McCauley, M., Franklin, M.J., Shenker, S., Stoica, I.: Resilient distributed datasets: a fault-tolerant abstraction for in-memory cluster computing. In: NSDI 2012, USENIX Association, pp. 15–28 (2012)

Research of Moving Target Tracking Algorithms in Video Surveillance System

Xu Lei[1(✉)], Peng Yueping[1,2], and Liu Man[1]

[1] Graduate Management Unit, Engineering University of PAP, Shanxi, China
734052869@qq.com

[2] Department of Information Engineering, Engineering University of PAP, Weiyang District, Xi'an 710086, Shanxi, China

Abstract. Although the target tracking is a difficult problem in the field of computer vision, it has a bright development prospect with the continuous development of video surveillance technology, and it adopts many kinds of algorithms whose effects are various. Based on the bibliography, the thesis introduces four kinds of tracking algorithms and analyzes their advantages and limitations. Adopting two or more algorithms, it can take the tracking research and also can improve the accuracy and efficiency of target tracking. Finally, it summarizes the problems and research trends of target tracking which need to be improved and also raise the prospective future of tracking algorithms.

Keywords: Moving target tracking · Tracking algorithm · Mean shift · Particle filter

1 Introduction

Moving target tracking refers to the method by using the location, color and other features in video image sequence to get trajectory information with the spatial and temporal variation by matching relationship between image frames [1]. It is widely used in military reconnaissance, traffic system and etc, but the background interference, target appearance changes and other factors in practical application will directly affect tracking result. Tracking algorithm takes the image processing as the core in its realization, it always locates and tracks through the three part of feature extraction, matching and location prediction. This paper mainly compares some target tracking algorithms which are practical now and analysis their advantages and disadvantages.

2 Moving Target Tracking Algorithm

At present, there are three technical sides of difficulty in target tracking algorithm:

1. The shadow of target is easy to be extracted as foreground information.
2. In the video images with high density crowd, targets are relatively crowded, and there is a mutual occlusion between them, which may leads to the difficulty of target extraction and accurate tracking.

F. Xhafa et al. (eds.), *Advances in Intelligent Systems and Interactive Applications*, Advances in Intelligent Systems and Computing 686,
https://doi.org/10.1007/978-3-319-69096-4_27

The amount of information in images is large, so the improvement of reliability and accuracy leads to higher requirement about the used algorithm in the process of tracking moving targets, especially in the real-time problem.

2.1 Particle Filter Tracking Algorithm Based on Compressed Sensing

At present, there are many tracking algorithms, but it still have challenges. Target tracking in compressed sensing is to extract the foreground and background as the samples are projected to low dimensional space, then classify information by using the Bias classifier to achieve the tracking purpose. Particle filter algorithm is based on the Bias estimation algorithm which can effectively solve the state estimation problem of nonlinear system. For partial occlusion of moving target, the robustness is better. The particle filter tracking algorithm based on compressed sensing is the theory of compressed sensing which is about compressed sampling of Haar-like characteristics, then using particle filter tracking algorithm to predict [1]. The timeliness of target tracking can be improved by compressed sensing technology, and it applies linear projection which is less than traditional Nyquist sampling to achieve approximate reconstruction of signal. The algorithm model is shown in Fig. 1:

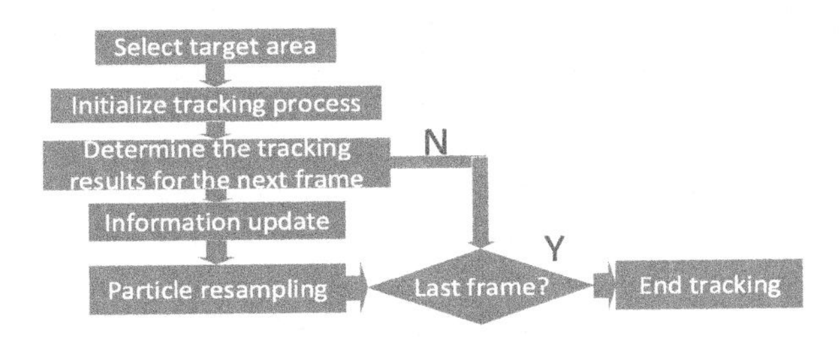

Fig. 1. Algorithm model

Using measurement matrix to project compressed or transformed signal to the low dimensional space, then recover original signal through the optimization to preserve the information of the original signal with high probability. The essence of feature extraction is to obtain the weight of image blocks extracted from the sample images.

The method uses the independent of measurement matrix between datum to reduce the complexity of algorithm, it can meet the tracking requirements of timeliness and its accuracy and robustness are higher than the traditional algorithm. But when the most of target is blocked, it will appear separation errors. In addition, the stability needs to be improved, and it has the fault that need to select target manually.

2.2 Mean Shift Tracking Algorithm Based On Rotation Invariant LBP Features

It's one of the tracking algorithm which based on target region, that can process iterative algorithm by pattern matching based on Mean-shift, then calculate the search target's location. In view of the color histogram has weak sensitivity on the information about the fuzzy degree of image, scale change and others, which leads to the judgment of defects while analyzing on color distribution of target, so we can see the drawbacks of traditional Mean Shift tracking algorithm: high computational complexity and poor timeliness. It still couldn't work well in distinguishing the pixel that have same gradient but different textures, even when the COV operator and HOG operator have been introduced, however the rotational correlation of LBP features could distinguish and solve problems well [2].

The Mean-Shift algorithm is a density estimation method with no parameter and complete the iterative algorithm through searching for maximum matching degree of target position's range in the previous frame. It can change the gradient direction of probability density by moving constantly until reaches the maximum value of local density [3]. Its overall idea: select the target area in the first frame of video image manually, and calculate the rotation invariant LBP eigenvalue of that area and candidate regions which is achieved by shifting circularly with LBP data expansion.

Then search for the minimum distance between moving and candidate target in LBP feature space, is to find out the maximum similarity [4]. And read the next video frame, form a circular track model through iterating the position of previous frame. That algorithm can overcome the drawback of traditional algorithm which may lost target easily and have poor robustness. It also can track video target under the situation with light mutation and similar background.

This algorithm is different from traditional algorithm which is matching point to point, it uses adaptive iteration to finish pattern matching and calculation fast, can carry on accurate search for high probability density and rough search for low probability density. Without prior parameters it can achieve fast convergence, but when the target size has serious changes, there will come the circumstance of low accuracy in positioning as well as poor tracking effect.

The key of such method is to take first frame as reference template and compute the correlation between that and subsequent image sequence to confirm the target position. It can track on the basis of a single or multiple features of moving target in the region. Although it has good accuracy and stability in the unobstructed conditions, there are still some problems in the processing of shadow and occlusion. And that algorithm ignores the coherence of target's motion, only search for target matched under the limited condition, so the matching accuracy should be enhanced [5].

2.3 Moving Object Tracking Algorithm Based on Camshift and EKF

Camshift algorithm, namely continuous adaptive Mean shift algorithm, it can avoid the influence of target scaling and other situations on the tracking effect by extend Mean shift to continuous image sequences. It uses color information as tracking features and reduces the impact of environmental factors on the process of tracking. It is robustness, but the tracking effect will be offset when the color of target and background is similar or target is severely blocked. Its block diagram is as follows (Fig. 2):

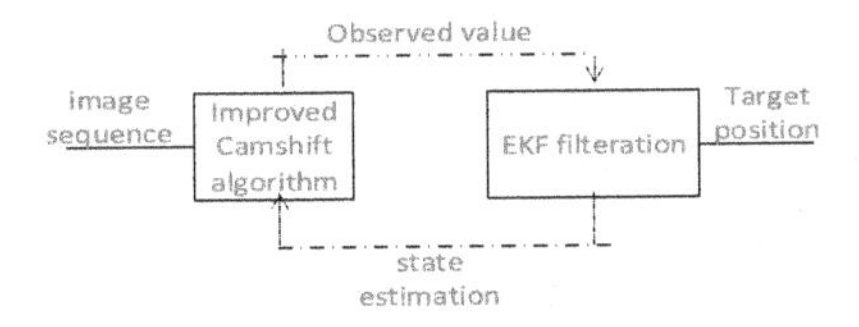

Fig. 2. The block diagram of improved Camshift algorithm

Extended Kalman filter (EKF) can accurately predict the target position situation in the case of similar background or object occlusion, and thus put forward the moving target tracking algorithm based on the combination of Camshift and EKF. The principle is: firstly extracting the moving target by using a Gauss mixture model for background modeling and use it to initialize the search window of Camshift algorithm; then initialize the EKF filter and obtain the position of search window by using its predictive function, then we can judge whether have interference or occlusion in target background. If not, we can take the target position that is calculated by Camshift algorithm as observation value to update the EKF filter; otherwise should take the position predicted by EKF filter as observation value to update the EKF filter, and then use it to track the target [6].

This method can improve tracking performance if background have color interference or target occlusion is severely or other circumstances, it also suitable for multiple target tracking. However it has a drawback of too much prior parameters and its research direction is how to solve the parameter adaptive problem. In addition, EKF filtering algorithm may appear error leading to tracking failure when dealing with the problem of long-time occlusion.

2.4 Target Tracking Algorithm Based on Multiple Feature Fusion

It belongs to the target tracking method based on feature. The feature information of gray level image is only 1/3 of color images. Taking the gray value as characteristic value simply will lead to the loss of target and unable to locate and identify. So it puts forward the method which can form multiple feature histogram by fusing gradient feature with gray feature images of X and Y direction for tracking stability on the target. The flow chart is as follows (Fig. 3):

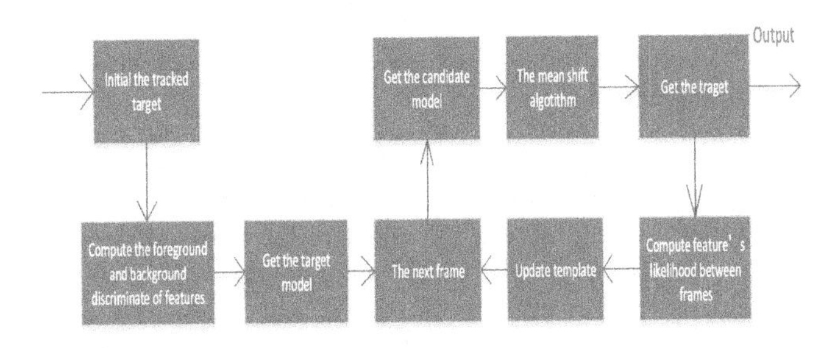

Fig. 3. The flow chart of purposed approach

Firstly building a feature model contains two parts of gray and gradient features, to find out texture features of X and Y direction by using the Sobel operator from the image of matching template, then using background modeling method to extract main foreground prospects [7], that can help to reduce interference information which is produced by background characteristic value. When the pixel position of the image to be matched belongs to the foreground image, it will estimate current probability density; if that belongs to background image, it will fetch the small probability number of Gauss probability distributional function for probability density image. Finally, in order to reduce the influence of illumination, deformation and other external factors act on matching degree, it will do histogram statistics after the scale transformation of matching image pixels. Then calculate characteristic value with the Mean shift algorithm and determine final position of the target by iteration.

The fusion of gradient and gray features in X and Y directions can make the contrast between target and background information become more obvious, so as to optimize the tracking effect. The method based on feature always focus on detection and tracking of moving target's elements with invariability, then to match and track the following video images [8]. In order to make the results more accurate, we usually combine with multiple factors to track, but complementary features may also have a negative impact, so feature selection is particularly important.

3 Summary Outlook

At present, the research of moving target tracking algorithm based on scene mode always make the ideal environment as the premise, not fully represent the factors of influence and interference in practical application. It can quickly identify and understand the behavior of simple semantic information, but there is a lack of understanding of deep motion patterns in complex scenes, so it can be combined with the mixed Gauss model, Markov model and more complex mathematical models to analyze, then we can improve the ability of analysis and understand.

As the key technology of intelligent video surveillance system, moving target tracking can realize the real-time tracking of military targets in the military field, can keep track of illegal vehicles in the field of intelligent transportation, can track the microscopic cells in medical field and so on, its research value is more and more high, also means that the algorithm need further optimize. Good robustness and high accuracy of target tracking algorithm is a measure of whether the algorithm is applicable, therefore we need to find efficient, fast and stable algorithm to deal with the target tracking in the complex environment. The following problems need for further solutions in the future research:

1. For the target occlusion or overlapping problem, it has been a difficult problem in video surveillance and target tracking, especially embodied in the detection and tracking of multiple moving targets, the effect of the above algorithms have some defects, so using multiple cameras to achieve all-round monitoring is still the most effective means.

2. For the small density crowd we can achieve a simple and efficient tracking algorithm with small calculation, but when appearing a large number of targets in the scene at the same time, it will have target separation error, large amount of calculation, and the limitation of the tracking speed and real-time.
3. The algorithm with a variety of methods combined can effectively solve the problem of target trajectory change in the background of complex circumstances, but when the target stops moving or move slowly, it will be difficult to obtain the stable background model [9]. In order to satisfied the tracking of target mutation and finish the acquisition of background model when target change slowly, we should consider the various factors to design algorithm.

References

1. Xu, H.: Research of moving target tracking algorithms in video surveillance system. TV Technol. **38**(19), 202–206 (2014)
2. Gao, Y.: Research on key technologies of high efficient video object tracking in complex scenes. Doctoral thesis, Yunnan University, December 2014
3. Xiao, J., Chen, X., Ding, L.: An improved mean shift moving target tracking algorithm. Inf. Technol. 127–130, January 2017
4. Li, J.: Detection and tracking of moving objects in complex background. Doctoral thesis, Hefei University of Technology, Hefei, April 2015
5. Zhang, Y., Zeng, Y., Zhang, Y.: Multiple target tracking based on mean shift and filtering, 10–13, December 2016
6. Shi, Z.: Research on moving object detection and tracking in video sequences. Master thesis, Xi'an Electronic and Science University, March 2014
7. Jiang, S., Zhang, R., Han, G., et al.: Multiple feature fusion moving object tracking in complex background gray image. Chin. Opt. **3**, 320–328
8. Gao, R.: Research on moving object detection and tracking algorithm in intelligent video surveillance. Master thesis, Nanjing University of Aeronautics and Astronautics, Nanjing, March 2014
9. Wang, H.: Detection and tracking of moving objects in complex scenes. Doctoral thesis, Beijing Institute of Technology, June 2015

Study on Tumble Behavior Recognition Based on Mining Algorithm for Potential Behavior Association

Zhang Qiusheng(✉), Lin Mingyu, and Ju Jianping

School of Electrical and Information Engineering, Hubei Business College, Wuhan 430079, Hubei, China
m13297979869@163.com

Abstract. Under the background of global ageing and empty nest families, the senior tumble behavior has becomes a focus problem in today's society. In order to provide timely help for seniors, and relieve the injury of tumble to them, a judgment method of senior tumble behavior based on potential behavior association mining is proposed in this paper. Using clustering algorithm and the mining algorithm for potential behavior association rule, it calculates the similarity of the senior behavior feature, then extracts the senior behavior features, calculates correlations between behavior features in seniors, which can complete data mining in senior tumble behavior. The experimental results show that the proposed algorithm could greatly improve the accuracy of the senior tumble identification, so as to provide security for senior trips.

Keywords: Tumble behavior · Clustering algorithm · Feature extraction · Data mining · Feature correlation

1 Introduction

Tumble is a common traumatic event for seniors, also is one of the leading causes of disability or death for them. That will bring not only mental stress and health threats, but also financial burden for their families and society. Therefore, how to ensure senior safety has become a hot issue to research on monitoring seniors [1, 2]. In the field of monitoring seniors, the main mining methods of senior tumble behavior include method based on association rules algorithm [3], based on genetic algorithm and based on K-means clustering algorithm [4].

The behavior pattern recognition algorithm which is more popular at present for mining senior tumble behavior, assumes that there are many differences between the ages of the monitoring target group in monitoring environment, or strong similarities between the tumble behavior features, would lead to chaos in correlation of senior tumble features, which is hard to identify accurately the senior tumble behaviors according to the senior tumble behaviors, and reduces the accuracy of mining senior tumble behavior.

In order to avoid the defects of the traditional algorithms above, it proposes a mining algorithm for senior tumble behavior based on potential behavior association.

F. Xhafa et al. (eds.), *Advances in Intelligent Systems and Interactive Applications*, Advances in Intelligent Systems and Computing 686,
https://doi.org/10.1007/978-3-319-69096-4_28

It calculates the similarity of senior behavior features, then extracts senior behavior features, calculates correlations between behavior features in seniors, which can complete data mining of senior tumble behavior.

2 The Recognition Principle of Behavior Feature

Behavior recognition is a hotpot in computer vision, the goal of human behavior recognition is automatically analysis the ongoing behaviors from an unknown video or image sequences. Simple behavior recognition is motion classification from a given video or image sequences, only classify correctly the behaviors into the several known categories, complex recognition is not just containing only one category, but multiple, it needs system to identify automatically the categories of motions and the start time and end time of them, and mark them.

This article analyzes senior tumble behaviors, selects the appropriate images about senior tumble as sample set, extracts behavior features to do clustering processing, and combines the mark results of senior tumble behaviors obtained with optimization operation [5, 6], and then do iterative processing operations. According to the results of the iterative processing, it could obtain the population of senior tumble behavior, so as to complete mining senior tumble behavior. But in the process of mining, there is very big difference in senior tumble behaviors due to the different ages, different environments, causing that correlation between the features of senior tumble behaviors is in the state of chaos, thereby it is hard to mining accurately of senior tumble behavior, and reduce the accuracy of senior tumble recognition.

3 Optimizing Identification of Mining Tumble Behavior

3.1 Behavior Feature Extraction

It could calculate the similarity of the features of senior behaviors by the clustering algorithm, and select one feature in each cluster according to the calculated results, which is used to describe the feature of this cluster, and delete the other elements of this cluster, reduce the redundancy at senior behavior features. Then the expected crossover operation is proposed for the rest of features of senior behaviors, so as to realize feature selection.

The parameters of senior behavior feature calculate as the following formula:

$$x_{lk} = \frac{ug_{lk} \times \log(P/p_k + 0.01) \times JH_k}{\sqrt{\sum_{k=1}^{q} (g_{lk} \times \log(P/p_k + 0.01) \times JH_k)^2}} \tag{1}$$

ug_{lk} is described as the frequency of senior behavior feature, p is the number of the monitoring target group in the monitoring area, q is the quantity of senior behavior features.

Set every feature of senior behavior as a sample to do clustering process, the weight coefficient corresponding to the feature is one-dimensional. Assuming that the senior behavior features are respectively u_1, u_2, the similarity coefficient between them can be calculated by the following formula:

$$sim(u_1, u_2) = \frac{\sum_{l=1}^{p} x_{1l} \times x_{2l}}{\sqrt{\left(\sum_{l=1}^{p} x_{1l}\right)^2 \left(\sum_{l=1}^{p} x_{2l}\right)^2}} \tag{2}$$

Sorting senior behavior features taking advantage of the excepted crossover operation, select the sample with the highest classification coefficient of the feature as senior behavior feature, which constitutes the subset of senior behavior features.

3.2 Calculation Correlation of Behavior Feature

Setting the number of senior behavior features is q, the l behavior feature is described as f_l. Assuming that the senior behavior features above are classified, whose subclass feature is described as $(x_1, x_2, \ldots, x_p)$. p is the number of senior behavior feature in subclass. Among them, each feature can is described as $f_l(x_{l1}, x_{l2}, \ldots, x_{lp})$. So, the set consists of behavior features of suspected tumble of all seniors could be described as $f_l = (x_{l1}, x_{l2}, \ldots, x_{lp})$, the mining algorithm based on potential association rule could classify multiple subsets such as f_l, and put them into the two different categories of senior normal behavior feature and tumble behavior feature [7–9]. The probability of feature correlation that senior behavior feature d_k belongs to g_l could calculate by the following formula:

$$r(g_l \cdot d_k) = (r(g_l) + r(g_l \cdot \delta_w))/r(d_k) \tag{3}$$

$l = 0, 1, \ldots,$q. $r(d_k)$ is the prior probability of mining senior tumble behavior, $r(g_l \cdot d_k)$ is the transcendental conditional probability of the behavior feature of senior tumble. Assuming that there is no any relation between features in the subcategory of senior behavior features, it could get the following results:

$$\begin{aligned} r(g_l \cdot d_k) &= r(x_1 \cdot d_k) + \cdots + r(x_n \cdot d_k) = \sum_{l=2} r(x_l \cdot d_k) \\ r(d_k) &= \sum_{l=2}^{q} r(g_l)/r(d_k \cdot g_l) \end{aligned} \tag{4}$$

According to some related parameters of senior tumble behavior, it could obtain the probabilistic value of the feature correlation of the different tumble behaviors, whose formula as described below:

$$r(d_k \cdot g_l) = U(q(d_k) + g_l) \Big/ U_{fl}^2 \tag{5}$$

Setting that the correlation probability threshold values of senior tumble behavior can be used to describe by γ, if $r(d_k \cdot g_l) > \gamma$, the result is that the feature belongs to the senior tumble behavior, otherwise, it belongs to the normal one, so as to complete mining the behavior of senior tumble.

4 Analysis of the Experimental Results

It makes Matlab as the experiment platform to simulate system, to verify the feasibility and validity of the mining algorithm based on the potential behavior association which is proposed in this paper.

There are a total of 100 pieces images as sample set acquired in this paper, including the images of senior tumble postures such as forward, backward, sideways and so on, the resolution of every image is not lower than 1024 by 756. It selects the top-50 samples for training, the rest samples for the tumble behavior recognition. Respectively using traditional mining algorithm and the algorithm proposed in this paper for senior tumble behavior, it is obtained the mining results described in Figs. 1 and 2. The experimental errors about mining of the senior tumble behaviors using different algorithms is described in Fig. 3.

Fig. 1. The result of the traditional algorithm

Fig. 2. The result of the algorithm in the paper

According to the experimental results, the algorithm proposed in this paper mines senior tumble behavior, can avoid the defects of the feature association of senior tumble behaviors caused by ages and difference monitoring environments, so as to improve the accuracy of the mining behavior of senior tumble.

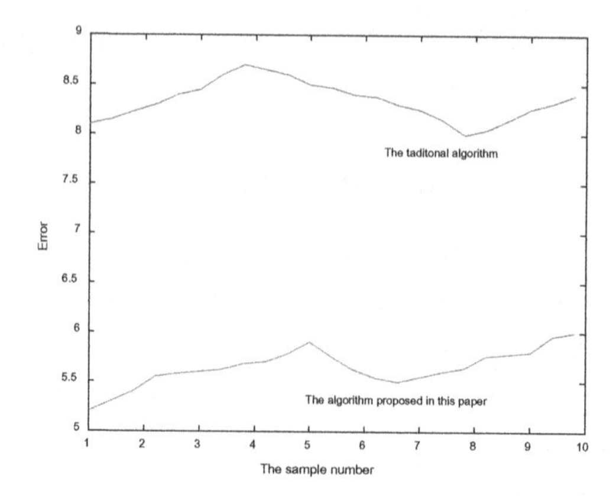

Fig. 3. Error trend chart between the mining algorithms

5 Conclusion

Aiming at the defects of traditional algorithms in mining behavior of senior tumble, a mining algorithm based on potential behavior association is proposed in this paper. It calculates the similarity of the features of senior behaviors using clustering algorithm, and extracts the behavior features of senior tumble. And then calculates the correlations between behavior features in seniors using the mining algorithm based on potential behavior association rule. The experimental results show that the proposed algorithm could greatly improve the accuracy of the senior tumble identification, so as to provide security for senior trips in public places.

Acknowledgments. Supported by Science and Technology Research Project of Hubei Provincial Educational Commission (No. B2016482).

References

1. Yuqing, P., Qingqing, G., Nanna, L., et al.: Fall behavior recognition based on multi-feature fusion. J. Data Acquisition Process. **31**(5), 890–902 (2016)
2. Bingqian, S., Zhiyong, W., Qianhua, H.E., et.al.: Detection method of human body posture judgment falls. J. Comput. Appl. **34**(51), 223–227 (2014)
3. Li, C., Licheng, J.: Association rule mining algorithm based on relation Algebra theory. J. Northwest Univerisity (Natural Science Edition) **35**(6), 692–694 (2005)
4. Weijun, Li: Summary of K-means algorithm for Clustering. Mod. Comput. **8**, 31–32 (2014)
5. Guangchao, Q., Ruiyu, J., Ran, Z., et al.: One optimized method of apriori algorithm. Comput. Eng. **34**(23), 196–198 (2008)
6. Zhangyan, X., Shichao, Z.: An optimized Apriori algorithm for mining association rules. Comput. Eng. **29**(19), 83–84 (2013)
7. Shucai, S., Aihua, Q., Jianxiong, W.: Facing the web site of the data mining technology in the optimization of personalized recommendation methods of research and application. Bull. Sci. Technol. **2**(28), 118–119 (2012)

8. Jianwen, T.: Distributed web log mining system based on mobile agent technology. Comput. Simul. **23**(10), 109–113 (2006)
9. Baowen, X., Weifeng, Z.: Applying data mining to web pro-fetching. Chin. J. Comput. **24**(4), 430–436 (2001)

Image Geometric Correction Based on Android Phone Sensors

Zhengyi Xiao(✉), Zhikai Fang, Liansheng Gao, and Baotao Xu

Beijing Engineering Research Center of the Intelligent Assembly Technology and Equipment for Aerospace Product, Beijing Institute of Spacecraft Environment Engineering, Beijing 10094, China
xzy020822@163.com

Abstract. For the purpose of image distortion caused by the oblique photography of Android CMOS camera, a auto-rectification approach was proposed in this thesis. According to the camera stance information took into geometric correction model and provided by the phone built-in acceleration sensor and geomagnetism sensor, the relationship of pixel coordinates between distortion image and standard image was obtained. The pixel coordinates was resampled to obtain orthorectified images without the squint distortion by OpenCV4Android perspective projection transformation. The experiment results show that the approach is efficient, real-time and has great significance for implementing precise position of the target.

Keywords: Geometric correction · Sensor · OpenCV4Android

1 Introduction

As the Android phone integrated camera, sensor and other hardware and has a good development environment, suitable for image recognition and positioning of the target. In the process of shooting the phone gesture inevitably changes, the camera will be a certain tilt angle imaging, resulting in the shooting of the target rotation, scaling and deformation, and the actual target between the obvious shape difference [1]. Therefore, it is necessary to correct the distorted image of the squint image as a standard image and to eliminate various geometric deformations in the image before object recognition and precise positioning using the image.

2 Strabismus Image Geometric Correction

2.1 Parameter Calibration Within the Camera

Zhang et al. [2] proposed a calibration method based on a planar template, which is between the traditional calibration method and the self-calibration method, and does not need to know the basic information of the camera's motion. The template image obtained by at least three different camera positions can be calculated Camera all inside and outside parameters. Compared with the traditional calibration method, the method

F. Xhafa et al. (eds.), *Advances in Intelligent Systems and Interactive Applications*, Advances in Intelligent Systems and Computing 686,
https://doi.org/10.1007/978-3-319-69096-4_29

is more flexible than the self-calibration method, so this method is used to calibrate the parameters in the camera.

Use the Panasonic LUMIX DMC-CM1 to start the camera program, from 10 different angles on the calibration plate shown in Fig. 1 to shoot, save 10 pictures, enter the calibration program.

Fig. 1. Calibration plate for the camera

Through the calibration procedure developed based on Zhang Zhengyou calibration method, the internal parameters obtained are shown in Table 1.

Table 1. Camera calibration results

Internal reference	f_u	f_v	u_0	v_0
Value	1058.82	1058.06	646.25	364.71

2.2 Camera Attitude Information to Obtain

2.2.1 Introduction to Android Phone Sensor

Android phone sensor using the natural coordinate system shown in Fig. 2. When the device moves or rotates, these axes are not changed, i.e. they are following the phone.

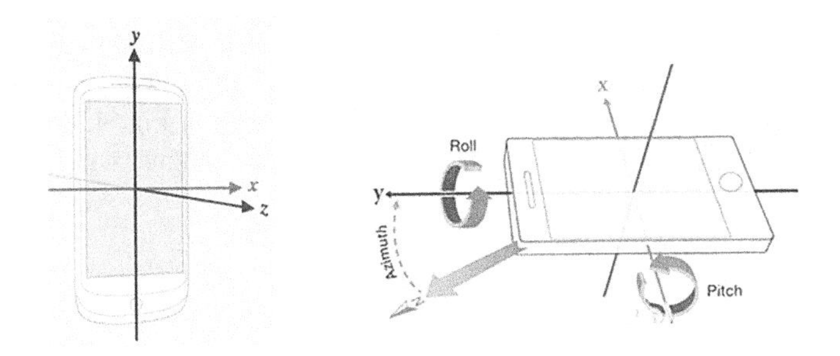

Fig. 2. Mobile phone and direction sensor natural coordinate system

The acceleration sensor and geomagnetic sensor can be used to derive a software-based direction sensor, which is represented by azimuth, pitch angle pitch and roll angle roll as shown in Fig. 2.

Azimuth is the rotation angle around the z axis, the actual is the y-axis positive axis and the geographical north of the angle, the range is −180 to 180, clockwise rotation is positive. Pitch angle is the rotation angle around the x-axis, the actual y-axis and horizontal elevation angle, the range is −90 to 90, clockwise rotation is positive. Rolling angle roll is the rotation angle around the y axis, the actual is the x-axis and horizontal angle, the range is −180 to 180, counterclockwise rotation is positive.

2.2.2 Camera Attitude Angle

In practical applications, the camera attitude angle is shown in Fig. 3. $O_w - X_wY_wZ_w$ is the coordinate system of the image pixel, which is the coordinate system of the image, and $O_c - X_SY_SZ_S$ is the coordinate system of the camera, which is the coordinate system of the image.

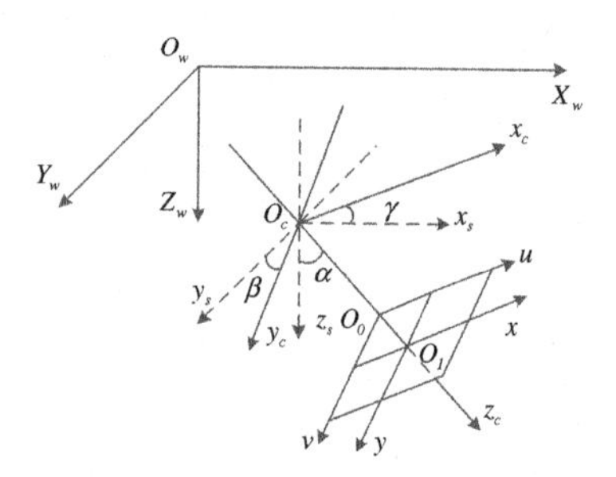

Fig. 3. Camera attitude angle

The angle between the camera coordinate system and the auxiliary coordinate system is the angle of the yaw angle, the counterclockwise is positive; the camera coordinate system axis and the auxiliary coordinate system. The angle of the pitch angle, counterclockwise for the positive.

2.2.3 Relationship between Sensor Value and Camera Attitude Angle

Since the camera is on the back of the Android phone, the camera's optical axis is facing down when the phone screen is facing up. As shown in Fig. 4, $O_c - x_cy_cz_c$ for the camera coordinate system, $O_c - xyz$ for the mobile phone natural coordinate system.

From the Fig. 4 you can get the following relationship:

$$\begin{aligned} azimuth &= \gamma \\ pitch &= \beta \\ roll &= -\alpha \end{aligned} \tag{1}$$

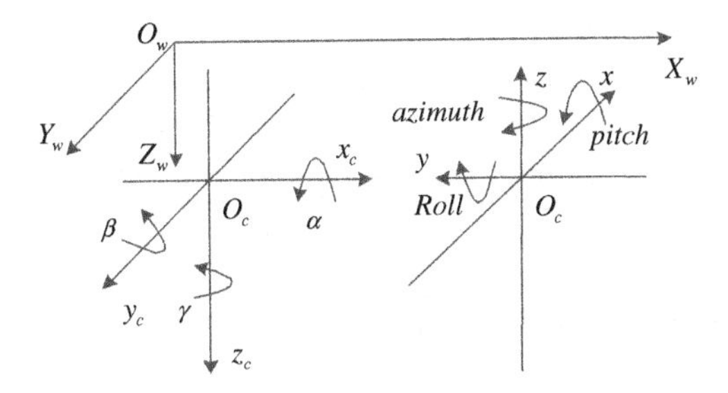

Fig. 4. Mobile phone natural coordinate system

2.2.4 Calculate the Rotation Matrix

The transformation of the camera coordinate system to the world coordinate system is done by rotating the matrix and the translation matrix. The translation matrix is a three-dimensional column vector, and the rotation matrix is the total product of the three conversion matrices formed by the axis of the camera, which is the axis of the auxiliary coordinate system [3, 4].

The conversion matrix is:

$$\mathbf{R}_X(\gamma) = \begin{bmatrix} \cos\gamma & \sin\gamma & 0 \\ -\sin\gamma & \cos\gamma & 0 \\ 0 & 0 & 1 \end{bmatrix}; \quad \mathbf{R}_Y(\beta) = \begin{bmatrix} \cos\beta & 0 & -\sin\beta \\ 0 & 1 & 0 \\ \sin\beta & 0 & \cos\beta \end{bmatrix};$$
$$\mathbf{R}_Z(\alpha) = \begin{bmatrix} 1 & 0 & 0 \\ 0 & \cos\alpha & \sin\alpha \\ 0 & -\sin\alpha & \cos\alpha \end{bmatrix} \tag{2–4}$$

When the three posture angles change at the same time, in accordance with the pitch angle, roll angle, yaw angle in the order of conversion [5]. The rotation matrix between the camera coordinate system and the world coordinate system can be constructed by the transformation matrix of three attitude angles. The rotation matrix between the camera coordinate system and the world coordinate system is as follows:

$$\mathbf{R} = \begin{bmatrix} \cos\gamma & \sin\gamma & 0 \\ -\sin\gamma & \cos\gamma & 0 \\ 0 & 0 & 1 \end{bmatrix} \begin{bmatrix} \cos\beta & 0 & -\sin\beta \\ 0 & 1 & 0 \\ \sin\beta & 0 & \cos\beta \end{bmatrix} \begin{bmatrix} 1 & 0 & 0 \\ 0 & \cos\alpha & \sin\alpha \\ 0 & -\sin\alpha & \cos\alpha \end{bmatrix} \tag{5}$$

(1) into the formula (5) available: (for the convenience of writing, where, with the first letter instead)

$$\mathbf{R} = \begin{bmatrix} \cos p\cos a & \cos r\sin a - \sin r\sin p\cos a & -\sin r\sin a - \cos r\sin p\cos a \\ -\cos p\sin a & \cos r\cos a + \sin r\sin p\sin a & -\sin r\cos a + \cos r\sin p\sin a \\ \sin p & \sin r\cos p & \cos r\cos p \end{bmatrix} \tag{6}$$

2.3 Image Correction Using Perspective Projection Transformation

2.3.1 Solve the Mapping Matrix

After the completion of the internal and external parameters matrix, if you want to use the perspective projection transformation for image correction need to meet the following conditions:

- the *camera* lens nonlinear distortion obtained by the above-mentioned *Zhang Zhengyou* calibration method is smaller than the image distortion caused by squint shooting,
- by the coordinate system translation, the camera lens heart and heart of the phone can overlap, and the camera coordinate system and the mobile phone coordinate system of the axis can also coincide;
- camera shooting area without ups and downs and ignore the camera strabismus shooting relative to the shooting of the optical center position changes.

The coordinates of a point p is $[u \quad v \quad 1]^T$, the coordinates of the corresponding coordinates of the world coordinate system is $[x_w \quad y_w \quad z_w \quad 1]^T$, we can see:

$$z_c p = \mathbf{K}[\mathbf{R}|\mathbf{t}]P \tag{7}$$

Using the perspective projection transformation to transform the distorted image into the standard image, the coordinates of the point corresponding to the point in the world coordinate system are invariant in the process of transformation. The change is only the pixel coordinates of the image of the distorted image and the standard image [6]. Therefore, for the same point, set it in the strabismus of the distortion of the image and the face of the standard image under the coordinates of the image, from the formula (7) available:

$$\begin{aligned} z_c p &= \mathrm{K}[\mathbf{R}|\mathbf{t}]P \\ z_{c0} p_0 &= \mathbf{K}[\mathbf{R}_0|\mathbf{t}_0]P \end{aligned} \tag{8}$$

Which, respectively, for the object in the distortion of the image and the standard image where the camera coordinate system under the axis coordinates (i.e., camera shooting height). For the distortion of the image and the standard image where the camera coordinates to the world coordinate system rotation matrix and translation vector. For the camera's internal matrix. (8) is satisfied by the assumption that the above-mentioned perspective projection transformation correction is satisfied:

$$p_0 = \mathbf{K}\mathbf{R}_0\mathbf{R}^{-1}\mathbf{K}^{-1}p \tag{9}$$

This gives the mapping matrix:

$$\mathbf{M}_0 = \mathbf{K}\mathbf{R}_0\mathbf{R}^{-1}\mathbf{K}^{-1} \tag{10}$$

The internal matrix is determined by the internal parameters obtained in Table 1, and the rotation matrix is determined by Eq. (6). The perspective transformation of the distorted image using the mapping matrix can be transformed into a standard image.

2.3.2 Pixel Gray-Scale Interpolation

In the field of digital image processing, it is meaningless for non-integers if the coordinates of a point in the pixel coordinate system are integers [7]. The projection pixels in the image space after the projection transformation are usually mapped to the non-integer position in the distorted image, and can't be directly assigned to the gray value. The gray-scale interpolation operation must be performed.

Pixel grayscale interpolation produces a grayscale value of an unknown pixel with a gray value that is known to be a neighboring pixel. At present, the gray image interpolation algorithm of digital image mainly includes nearest neighbor method, bilinear interpolation method and cubic convolution method. The results of the nearest neighbor method are the smallest, the interpolation accuracy is low, the interpolation effect is usually not ideal; the third convolution method has higher interpolation precision, but the computation is large; the accuracy and computation of the bilinear interpolation method are moderate and There is a low-pass filter effect, the edge of a certain degree of smooth effect. Under the premise of ensuring the interpolation effect, taking into account the practical application of real-time requirements, this paper uses bilinear interpolation method.

3 Image Recognition in Android Phone Environment

3.1 Use the Service Component to Obtain the Sensor Value in the Background

Service components are Android components of the four components with the most similar components, are representative of the executable program. Service has been running in the background, there is no user interface, have their own life cycle. Considering that this application requires the background to acquire the sensor value in real time and display the corrected image in the interface, it creates and configures a Service component that captures the mobile acceleration sensor and the geomagnetic sensor service and monitors its numerical changes, based on both values in real time Get the azimuth, pitch angle and roll angle that describe the gesture information of the phone.

Because the Service needs to exchange data with MainActivity, use bindService () and unbindService () to start and shut down the Service. MainActivity real-time communication with Service via the Ibinder object.

3.2 Perspective Transformation Function Warpperspective ()

OpenCV4Android performs a perspective transformation on the image as *warpPerspective*(). The official document in the function statement is as follows:

Where *src* is the input image, *dst* is the output image, *M* is the mapping matrix, *dsize* is the output image size, *flags* is the interpolation method, and this article defines *flags* = *INTER_LINEAR* + *WARP_FILL_OUTLIERS* that bilinear interpolation is used and the pixels outside the boundary are filled. If the partial pixels of the output image fall outside the boundaries of the input image, their values are set to zero.

As the standard image of the three posture angle (pitch angle, roll angle, yaw angle) are 0°, so that $R_0 = I_{3\times3}$, the standard image rotation matrix is a unit matrix, so by (10)

$$\mathbf{M} = \mathbf{K}\mathbf{R}^{-1}\mathbf{K}^{-1} \tag{11}$$

4 Oblique Distortion Error Analysis and Experimental Verification

4.1 Error and Experimental Verification of 3 Strabismus Distortion

When the six external parameters $(x_{c0}, y_{c0}, z_{c0}, \alpha_0, \beta_0, \gamma_0)$ deviate from the standard position when the spatial position (x_c, y_c, z_c) and the attitude angle (α, β, γ) are imaged, the image is distorted and the distortion error can be described by the coordinate error of the target point.

According to the principle of photo grammetry, the imaging geometrical relationship of the projected image of the center of the pinhole model can be expressed by the following collinear equation:

$$\begin{aligned} x = g_x = -f\frac{a_1(x_w - x_c) + b_1(y_w - y_c) + c_1(z_w - z_c)}{a_3(x_w - x_c) + b_3(y_w - y_c) + c_3(z_w - z_c)} \\ y = g_y = -f\frac{a_2(x_w - x_c) + b_2(y_w - y_c) + c_2(z_w - z_c)}{a_3(x_w - x_c) + b_3(y_w - y_c) + c_3(z_w - z_c)} \end{aligned} \tag{12}$$

$a_1, a_2, a_3, b_1, b_2, b_3, c_1, c_2, c_3$ which is composed of nine parameters composed of attitude angle α, β, γ, as shown in Eq. (5). Equation (12) can be derived from the external parameters of the displacement caused by the displacement of the formula:

$$\begin{aligned} dx = \frac{\partial g_x}{\partial x_c}dx_c + \frac{\partial g_x}{\partial y_c}dy_c + \frac{\partial g_x}{\partial z_c}dz_c + \frac{\partial g_x}{\partial \alpha}d\alpha + \frac{\partial g_x}{\partial \beta}d\beta + \frac{\partial g_x}{\partial \gamma}d\gamma \\ dy = \frac{\partial g_y}{\partial x_c}dx_c + \frac{\partial g_y}{\partial y_c}dy_c + \frac{\partial g_y}{\partial z_c}dz_c + \frac{\partial g_y}{\partial \alpha}d\alpha + \frac{\partial g_y}{\partial \beta}d\beta + \frac{\partial g_y}{\partial \gamma}d\gamma \end{aligned} \tag{13}$$

In the formula,

$$\frac{\partial g_x}{\partial x_c} = -\frac{f}{\overline{Z}^2}\left(\frac{\partial \overline{X}}{\partial x_c}\overline{Z} - \frac{\partial \overline{Z}}{\partial x_c}\overline{X}\right) = -\frac{f}{\overline{Z}^2}(-a_1\overline{Z} + a_3\overline{X}) = \frac{1}{\overline{Z}}(a_1 f + a_3 x)$$

among them,

$$\overline{X} = a_1(x_w - x_c) + b_1(y_w - y_c) + c_1(z_w - z_c)$$

$$\overline{Z} = a_3(x_w - x_c) + b_3(y_w - y_c) + c_3(z_w - z_c)$$

Similarly: $\frac{\partial g_x}{\partial y_c} = \frac{1}{\overline{Z}}(b_1 f + b_3 x)$, $\frac{\partial g_x}{\partial z_c} = \frac{1}{\overline{Z}}(c_1 f + c_3 x)$, $\frac{\partial g_x}{\partial \gamma} = y$,,

$$\frac{\partial g_x}{\partial \alpha} = -f \sin\gamma - \frac{x}{f}(x\sin\gamma + y\cos\gamma),\quad \frac{\partial g_x}{\partial \beta} = y\sin\alpha - [\frac{x}{f}(x\cos\gamma - y\sin\gamma) + f\cos\gamma]\cos\alpha;$$
$$\frac{\partial g_y}{\partial x_c} = \frac{1}{\overline{Z}}(a_2 f + a_3 y),\quad \frac{\partial g_y}{\partial y_c} = \frac{1}{\overline{Z}}(b_2 f + b_3 y),\quad \frac{\partial g_y}{\partial z_c} = \frac{1}{\overline{Z}}(c_2 f + c_3 y),\quad \frac{\partial g_y}{\partial \gamma} = -x;$$
$$\frac{\partial g_y}{\partial \alpha} = -f\cos\gamma - \frac{y}{f}(x\sin\gamma + y\cos\gamma),\quad \frac{\partial g_y}{\partial \beta} = -x\sin\alpha - [\frac{y}{f}(x\cos\gamma - y\sin\gamma) - f\sin\gamma]\cos\alpha;$$

Under vertical photography conditions, $\alpha = \beta = \gamma \approx 0$, the above formula can be simplified as:

$$\begin{aligned} dx &= (\frac{f}{\overline{Z}})dx_c + (\frac{x}{\overline{Z}})dz_c - (f + \frac{x^2}{f})d\beta - (\frac{xy}{f})d\alpha + yd\gamma \\ dy &= (\frac{f}{\overline{Z}})dy_c + (\frac{y}{\overline{Z}})dz_c - (f + \frac{y^2}{f})d\alpha - (\frac{xy}{f})d\beta - xd\gamma \end{aligned} \tag{14}$$

As can be seen from the above equation, the displacement of the image point is related to the position of the image point in the image coordinate system. $d\alpha$, $d\beta$ (the pitch angle and the roll angle change) make the image produce a nonlinear distortion error.

4.2 Geometric Distortion Analysis of Inclination

The pitch angle α and the roll angle β of the aforementioned non-linear distortion error of the image are collectively referred to as inclination angles θ. The inclination θ reduces the resolution of the image along the dip direction, as shown in Fig. 5.

In the Fig. 5, the image height is displayed, and the image resolution at the time of vertical shooting indicates the image resolution when the image is tilted, and the offset angle corresponding to the unit pixel. Then the triangular relationship can be obtained vertical resolution and tilt resolution of the formula are:

$$\begin{aligned} res_0^{-1} &= 2h\tan\frac{\varphi}{2} \\ res_c^{-1} &= h[\tan(\theta + \frac{\varphi}{2}) - \tan(\theta - \frac{\varphi}{2})] \end{aligned} \tag{15}$$

From the above formula available $\frac{res_c^{-1}}{res_0^{-1}} = \frac{\frac{1}{\cos^2\theta}}{1 - \tan^2\theta\tan^2\frac{\varphi}{2}} \underset{\theta \gg \varphi \to 0}{\approx} \frac{1}{\cos^2\theta}$

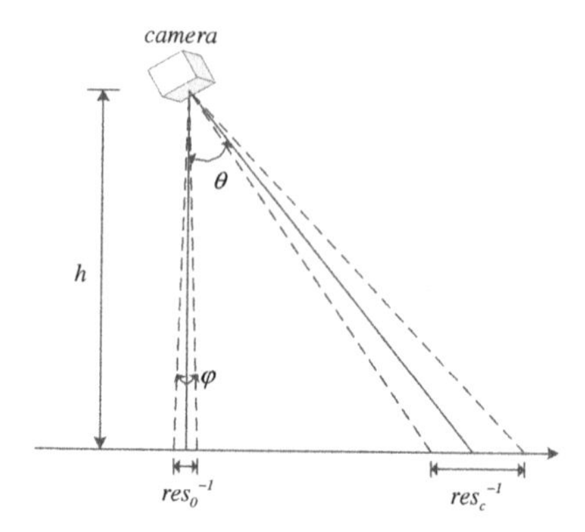

Fig. 5. Inclination causes a change in resolution

That is, when the inclination is much larger than the unit pixel offset angle, the relationship between the vertical resolution and the tilt resolution is:

$$\frac{res_0}{res_c} = \frac{1}{\cos^2 \theta} \tag{16}$$

The following conclusions can be drawn from the graph: the image information of the strabismus imaging loss is nonlinear. With the increase of the inclination angle, the distortion rate of the distortion error is increasing.

4.3 Target Recognition and Positioning Experiment Validation

In the practical application of image target recognition and precise positioning using the camera, it is required to detect the circle quickly and accurately. A very effective method is Hough transform, which is one of the basic methods for recognizing geometrical shapes from images in image processing, and is widely used. Hough Transform is a feature extraction technique in image processing. The process obtains a set of specific shapes that satisfy the specific shape in a parameter space by calculating the local maximum of the cumulative result as the Hough transform result.

In this paper, the Hough circle method is implemented by the Hough gradient method provided by OpenCV4Android, which can recognize the circle and display the coordinates of the center pixel.

Table 2 records the 10 sets of different posture information under the distortion of the image detected in the center pixel coordinates, using the algorithm to calculate the standard image in the center of the pixel coordinates. After the image correction is performed using the perspective transformation under the same attitude information, the center pixel coordinates detected after the correction are recorded.

Table 2. The central coordinates of the distorted image processing

$(pitch, roll)$	(x_0, y_0)	(x_i', y_i')	(x_i, y_i)
(−3.55,−5.31)	(706.5,420.5)	(642.24,322.98)	(640.5,321.5)
(−2.87,−2.58)	(672.5,390.5)	(619.50,342.87)	(621.5,340.5)
(5.35,−4.55)	(620.5,405.5)	(717.93,320.82)	(720.5,317.5)
(−2.80,3.91)	(718.5,319.5)	(668.45,392.90)	(668.5,391.5)
(4.43,1.02)	(534.5,394.5)	(619.53,415.07)	(618.5,413.5)
(−7.8,3.21)	(785.5,335.5)	(646.04,398.42)	(642.5,396.5)
(7.05,2.35)	(539.5,348.5)	(673.23,393.88)	(675.5,393.5)
(0.72,6.02)	(651.5,342.5)	(662.57,452.30)	(665.5,454.5)
(−6.91,1.10)	(712.5,351.5)	(584.84,371.94)	(581.5,372.5)
(−3.10,5.20)	(709.5,364.5)	(650.64,459.71)	(650.5,460.5)

Use the Root Mean Square Error when evaluating the image correction effect. The axial direction and the axial direction and the total root mean square error are expressed by Eq. (17), respectively.

$$
\begin{aligned}
RMSE_x &= \sqrt{\frac{1}{n}\sum_{i=1}^{n}(x_i - x_i')^2}; \quad RMSE_y = \sqrt{\frac{1}{n}\sum_{i=1}^{n}(y_i - y_i')^2}; \\
RMSE_{all} &= \sqrt{\frac{1}{n}\sum_{i=1}^{n}\left[(x_i - x_i')^2 + (y_i - y_i')^2\right]}
\end{aligned}
\tag{17}
$$

According to the experimental data calculated by the above formula: $RMSE_x = 2.283\,pixels$, $RMSE_y = 1.813\,pixels$, $RMSE_{all} = 2.916\,pixels$.

Set up the experiment condition as shown in Fig. 6.

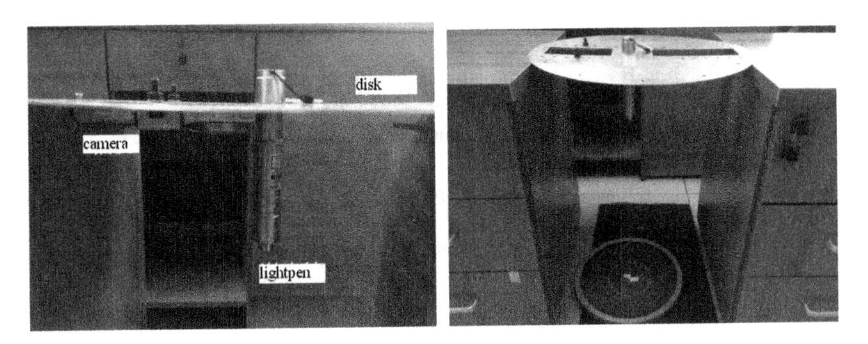

Fig. 6. Target location experiment scenario

Adjust the disc tilt angle, repeat the above experimental process, record the experimental data shown in Table 3.

According to the experimental data by the formula (17) calculated: $RMSE_x = 0.968\,\text{mm}$, $RMSE_y = 1.041\,\text{mm}$, $RMSE_{all} = 1.421\,\text{mm}$. Since the

Table 3. The deviation of the center of the circle after the distortion of the image

$(pitch, roll)$	(x_0, y_0)	(x_i', y_i')	(x_i, y_i)
(−1.26,0.68)	(523.5,363.5)	(61.745,1.855)	(60.1,0)
(−2.51,0.82)	(499.5,358.5)	(74.465,0.795)	(73.9,1.4)
(−3.92,0.91)	(474.5,359.5)	(87.715,0.265)	(88.4,0)
(0.71,0.80)	(558.5,337.5)	(43.195,11.925)	(42.3,12.4)
(1.81,1.36)	(573.5,378.5)	(35.245,9.805)	(34.1,9.3)
(−2.47,0.34)	(519.5,352.5)	(63.865,3.975)	(64.1,2.1)
(−5.01,0.33)	(472.5,353.5)	(88.775,3.445)	(89.1,2.1)
(−4.19,0.39)	(485.5,355.5)	(81.885,2.385)	(80.5,1.6)
(0.71,0.57)	(564.5,373.5)	(40.015,7.155)	(40.6,6.5)
(0.29,2.15)	(709.5,364.5)	(38.955,10.865)	(37.8,11.2)

experimental height is 560 mm and the scale is 0.529 mm/pixel, the target positioning experiment error is expressed as pixels. $RMSE_x = 1.830\,pixels$, $RMSE_y = 1.968\,pixels$, $RMSE_{all} = 2.686\,pixels$.

Acknowledgments. An image correction method based on sensor attitude information is proposed in this paper. By using the OpenCV4Android open source computer vision library on the Android platform, the real-time automatic correction of the strabismus image is realized and the error of the strabismus distortion is analyzed by using the pixel transformation of the same target in the distorted image and the standard image. The experimental results show that the root mean square error of the method in the axial direction and the axial direction is not more than 2 pixels. The geometric correction method does not need to detect the feature points and collect the reference image, the correction efficiency is high, and can be applied to the real-time accurate positioning of the image target based on the mobile intelligent device.

References

1. Zhou, Q., Liu, J., Gou, B., et al.: Geometric correction of squares image of aerial camera for aerial array CCD. Liq. Cryst. Disp. **30**(3), 506–513 (2015)
2. Zhang, Z.A.: Flexible new technique for camera calibration. IEEE Trans. Pattern Anal. Mach. Intell. **22**(11), 1330–1334 (2000)
3. Wang, C.: Geometric correction and target recognition technology of remote sensing image [D]. Harbin Institute of Technology, Harbin (2014)
4. Nixon, M.S., Aguado, A.S.: Feature extraction and image processing [M]. Li S., Yang G., translation .2 version of Beijing, Electronic Industry Press, 295–298 (2010)
5. Changchun: Changchun Institute of Optics, Fine Mechanics and Physics, Chinese Academy of Sciences, Department of Computer Science and Technology, Tsinghua University, Beijing 100084, China (2015)
6. Fu, S., Chen, S., Zhang, J., et al.: Image projection based on camera attitude information. Comput. Digit. Eng. **43**(10), 1858–1860 (2015)
7. Gonzalez, R.C., Woods, R.E.: Digital image processing [M]. Ruan, Q., Ruan, Y., translation. 3 version of Beijing, Electronic Industry Press, 51–53 (2011)

Fish-Eye Camera Model and Calibration Method

Cuilin Li(✉), Guihua Han, and Jianping Ju

School of Electrical and Information Engineering, Hubei Business College, Wuhan, China
licl_2004@126.com

Abstract. Fish-eye camera has been widely used in computer vision, mobile robots, photogrammetry and other fields for its wide-angle imaging. Fish-eye camera should be calibrated before being used, and in this paper, we first introduce several projection models for wide angle cameras, and choose the most commonly used equidistant projection, combined with the lens distortion model, to establish a calibration model for fish-eye camera. Then, by the linearization of the model, the least square solution of both the calibration parameters and the exterior orientation elements is given. Finally, through the experiment, it is concluded that the fish-eye camera calibration model can reach the ideal calibration precision with 0.3 pixels.

Keywords: Fish-eye camera · Lens distortion · Calibration · Equidistant projection

1 Introduction

In addition to traditional plane camera, fish-eye camera has also been widely used in computer vision,mobile robots [1, 2], and other fields for its wide-angle imaging. As the fish-eye camera has special projection and high distortion, it needs strict geometric calibration before use. Geometric calibration includes two key steps: first, to establish a sensor calibration model; second, to solve the calibration parameters with sufficient control points. The first step includes two aspects, a geometric projection model and a lens distortion model. The projection model needs to consider the light refraction of the spherical surface, rather than the simple collinearity [3–6]. At the second step, after obtaining the mapping between control points in the world coordinate system and their corresponding point locations in the image plane, iterative non-linear least squares methods are applied to solve the calibration parameters and the exterior orientation elements. In this paper, we will establish a calibration model suitable for fish-eye camera, and use sufficient control points to calculate interior parameters and distortion parameters, and finally compare the calibration results of collinear projection with that of equidistant projection.

F. Xhafa et al. (eds.), *Advances in Intelligent Systems and Interactive Applications*, Advances in Intelligent Systems and Computing 686,
https://doi.org/10.1007/978-3-319-69096-4_30

2 Fish-Eye Camera Calibration Method

2.1 Fish-Eye Camera Model

Perspective transformation is one of the most common projection transformations, reporting the collinear relationship among object points, corresponding image point and projection center, i.e., a pin-hole model. Geometric relation of a traditional plane camera is shown in Fig. 1a, where incident angle α equals refraction angle β. For the fish-eye camera, the projection no longer meets collinearity. Commonly, incident angle is larger than refraction angle, for showing larger scene. This effect is equivariant to a ray entering high-density material, as water, from low-density material, as air. As in Fig. 1b, the geometric relationship between incident angle α and refraction angle β reflects the physical properties of a fish-eye lens itself, that is, the lens refracts how much of the incident light.

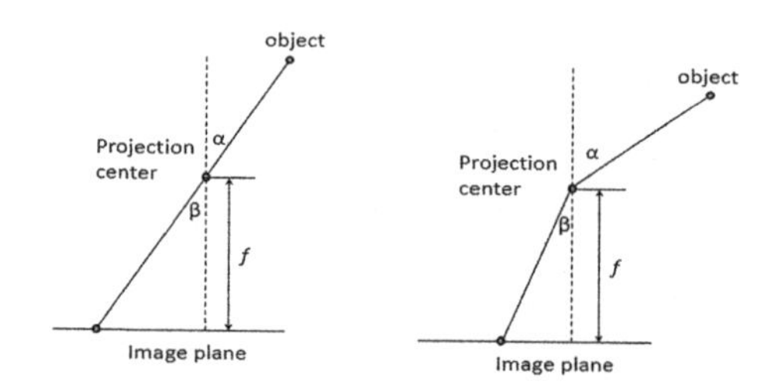

Fig. 1. (a) Collinear projection (b) fish-eye projection

For different types of physical lenses, a specific projection model may be required for suitable mathematical fitting. Up to now, there are several typical projection models have been explored.

(1) Perspective projection.
a pin-hole imaging is used for ideal plane camera. Incident angle is equal to reflection angle. In Eq. (1), r is the radius corresponding to the image coordinates $x = [x, y]$ of a certain point, and f is the focal length.

$$r = f \cdot \tan \alpha \tag{1}$$

(2) Equidistant projection.
The incident angle is linearly correlated to the radius. If an image point coordinate x is measured, and the corresponding radius r calculated, angle α of the incident light can be calculated linearly.

$$r = f \cdot \alpha \tag{2}$$

(3) Equal-area projection.
The projection ensures that the ratio of a solid incidence angle and the projection area in the image is constant, independent of the orientation of the incident angle.

$$r = 2f \cdot \sin 0.5\,\alpha \tag{3}$$

(4) Orthogonal projection.
When an angle is projected on the imaging surface, the angle value remains the same. Gauss projection is a common conformal projection.

$$r = f \cdot \sin\alpha \tag{4}$$

(5) Conformal projection.
When an angle is projected on the imaging surface, the angle value remains the same. Gauss projection is a common conformal projection.

$$r = 2f \tan 0.5\,\alpha \tag{5}$$

Among these models, equidistant projection is the most widely used in fish-eye calibration [7, 8].

3 Lens Distortion Model

Regardless of plane camera or fish-eye camera, lens distortion is a systematic error should be considered. Additional parameters, usually directly added to image point coordinate x, introduced to certain camera model to compensate for the distortion. Combination parameters of tangential and radial deformation are usually applied [9].

$$\begin{aligned} dx &= (x - x0) * (K1 * r^2 + K2 * r^4 + K3 * r^6) + P1 * (r^2 + 2 * (x - x0)^2) + 2 * P2 * (x - x0)(y - y0) \\ dy &= (y - y0) * (K1 * r^2 + K2 * r^4 + K3 * r^6) + P2 * (r^2 + 2 * (y - y0)^2) + 2 * P1 * (x - x0)(y - y0) \end{aligned} \tag{6}$$

In Eq. (6), $x0 = [x0, y0]$ is the principal point of camera. $K1$, $K2$ and $K3$ are radial distortion parameters, and $P1$ and $P2$ are tangential distortion parameters.

Equation (6) is proposed for the traditional plane camera, however, it has also been shown effective for fish-eye cameras [3]. In this paper, we also use this lens distortion model for geometric compensation.

4 Camera Calibration Model

According to the above fish-eye camera model and the lens distortion model, the calibration model of the fish-eye lens used in this paper is established. The 3D world coordinates of the object point are denoted by M, and the rotation matrix and translation vector between world and camera coordinates are R, T, respectively, and camera coordinate system $X = [X, Y, Z]$. We have,

$$X = R^{-1}(M - T) \tag{7}$$

For the ideal plane camera, the final calibration equations, combined with the additional distortion parameters, is,

$$\begin{aligned} x &= f\frac{X}{Z} + x_0 + dx \\ y &= f\frac{Y}{Z} + y_0 + dy \end{aligned} \tag{8}$$

For fish-eye lens, taken Eq. (2) equidistant projection as example we obtain,

$$\sqrt{x^2 + y^2} = f\alpha = f\ \arctan\frac{\sqrt{X^2 + Y^2}}{Z} \tag{9}$$

The corresponding points in two coordinate systems have the same projection angles on the XY plane (the projection angle is irrelevant to refraction), that is $\frac{Y}{X} = \frac{y}{x}$. Using this geometric relationship, it can be further deducted,

$$\begin{aligned} x &= fX\frac{\arctan\frac{\sqrt{X^2+Y^2}}{Z}}{\sqrt{X^2+Y^2}} + x_0 + dx \\ y &= fY\frac{\arctan\frac{\sqrt{X^2+Y^2}}{Z}}{\sqrt{X^2+Y^2}} + y_0 + dy \end{aligned} \tag{10}$$

For the other projection models, similar mathematical derivation can be used to obtain their corresponding calibration equation.

5 Solving the Camera Calibration Equation

Equations (8) and (10) both contain three types of unknowns, interior orientation elements, distortion parameters and exterior orientation elements. A total of 14 unknowns: which are denoted as l = *[f, x0, y0, K1, K2, K3, P1, P2, Xs, Ys, Zs,* φ, ω, κ*]*. Taken equidistant projection as example, firstly, substitute (7) into Eq. (10), calculate first derivatives of the 14 unknowns to obtain $k_1 \sim k_{28}$. Then, Eq. (10) is linearized as,

$$\begin{bmatrix} \Delta x \\ \Delta y \end{bmatrix} = \begin{bmatrix} k_1 & k_2 & k_3 & k_4 & k_5 & k_6 & k_7 & k_8 & k_9 & k_{10} & k_{11} & k_{12} & k_{13} & k_{14} \\ k_{15} & k_{16} & k_{17} & k_{18} & k_{19} & k_{20} & k_{21} & k_{22} & k_{23} & k_{24} & k_{25} & k_{26} & k_{27} & k_{28} \end{bmatrix} \cdot \Delta \mathbf{l}^{\mathrm{T}} \tag{11}$$

In Eq. (11), Δl is the incremental of the unknown parameter l, Δx, Δy is the rectified observations. After setting the initial value for l, the optimal solution can be achieved by using the least square iterations directly from Eq. (11). The solution requires at least seven control points. In fact, in order to obtain accurate calibration results, we need more evenly distributed control points, dozens to hundreds.

6 Experiment

6.1 Test Data

To verify the calibration method of the paper, the calibration fields were captured by a Nikon fish-eye lens camera with 10.5 mm focal length and 180° FOV. Control points in three images (two side views from left and right, one front view) were measured manually. The average number of control points is 70, and measurement accuracy is about 0.3 ~ 0.5 pixel. Image size is 6016 × 4000 in pixels, and 23.2 × 15.4 in millimeters. The initial value of focal length f is set to 2727.3 pixels, and $x0$ = 3088 pixels, $y0$ = 2000 pixels. For distortion parameters, Because the main distortion of fish-eye lens (relative to a plane camera) is in the direction of radiation, set the tangential distortion $P1$ and $P2$ to 0, The higher order term of radial distortion parameter $K2$, $K3$ is also set to 0. According to available FOV, the maximum radius r_{max} of image (set to 2000 pixels) corresponds to arc length $0.5\pi r_{max}$, that is, the maximum deformation of the edge dx_{max} is $(1\text{–}0.5\pi)\ r_{max}$. By substituting $P1$, $P2$, $K2$, $K3$, r_{max} and dx_{max} into Eq. (6), $K1$ can be calculated. The initial value of exterior orientation elements is provided by observations from a total station. With the initial value, the optimal solution can be obtained by stepwise iteration according to Eq. (11). The solution of this paper is completed on Matlab11; the average number of iterations is 5.

6.2 Result Analysis

Table 1 shows the calibration results of the interior orientation elements and the distortion parameters, and Table 2 is the result of the corresponding exterior orientation elements. If pin-hole model is used to calibrate the fish-eye lens, it produces a large systematic error. The residual mean square error (RMSE) of adjustment is 0.93 pixels, and residuals of many control points, especially those in the edge of images, are bigger than 3 pixels. On the contrary, under equidistant model, RMSE reaches 0.43 pixels, and the residuals of all control points are also smaller and the maximum residual does not exceed 1.5 pixels. Further, we utilized weight decaying method to eliminate a small amount of gross errors caused by manual measurement. Now RMSE dropped to 0.33 pixels, which is consistent with the best accuracy of artificial stereo measurement.

Table 1. The interior orientation and the distortion parameters

Methods	x_0	y_0	f	K_1	K_2	K_3	P_1	P_2
Initial value	3088	2000	2727.30	1.4e-9	0	0	0	0
Collinear equation	2388.36	1615.23	2704.86	−6.3e-9	−1.5e-016	−9.3e-24	−2.8e-8	−8.6e-8
Equidistant projection	2738.57	2041.26	3013.59	−5.9e-9	−2.0e-16	1.1e-23	−1.4e-7	1.9e-8
Equidistant projection (eliminating gross error)	2387.04	1614.55	2704.43	−6.78e-9	−1.47e-16	−7.30e-23	−1.81e-8	−5.89e-8

Table 2. The exterior orientation parameters and RMSE

Methods	σ_0	Xs	Y_s	Z_s	φ	ω	κ
Initial value	/	131.645	10.255	−497.052	2.567	0.003	0.019
Collinear equation	0.93	131.644	10.256	−497.051	2.565	0.003	0.019
Equidistant projection	0.43	131.645	10.256	−497.051	2.567	0.003	0.019
Equidistant projection (eliminating gross error)	0.33	131.645	10.256	−497.051	2.567	0.003	0.019

Table 2 also expresses that, the exterior orientation elements obtained by equidistant projection are consistent with the exterior orientation elements of the actual measurement by the total station. The line element and the angle element are almost the same as the actual measurement. Furthermore, a high accuracy of exterior orientation elements, along with intensive control condition, ensures that, the adjustment system can well distinguish interior orientation elements (including distortion parameters) from the exterior elements. In other words, the two types of parameters are weakly related, which in turn verifies the reliability of the calibration parameters obtained in Table 1. It is not surprising that, pin-hole model also reflects the distinguishability between these parameters, due to the distribution of reasonable control conditions rather than geometric model itself (Table 3).

Table 3. Comparison of the equidistant projection results from three sets of data

Data	x_0	y_0	f	K_1	K_2	K_3	P_1	P_2
Front image	2387.04	1614.55	2704.43	−6.78e-9	1.47e-16	−7.30e-8	−1.81e-8	−5.89e-8
Left image	2387.50	1613.61	2705.12	−5.55e-9	−6.53e-016	1.08e-22	9.71e-9	−6.35e-8
Right image	2385.65	1613.18	2702.71	−5.60e-9	−5.39e-16	8.68e-23	1.45e-8	−4.12e-8

7 Conclusions

This paper analyzes several geometric projection models suitable for the wide-angle lens, and establishes the calibration model from fish-eye camera model and distortion model. Using three sets of data for experiment, it is verified that the method can fulfill the calibration task of fish-eye camera, and it reaches good calibration precision of 0.3 pixels which is nearly the best calibration precision can be achieved by manual control point measurement.

References

1. Han, S.B., Kim, J.H., Myung, H.: Landmark-based particle localization algorithm for mobile robots with a fish-eye vision system. IEEE/ASME Transactions on Mechatronics, vol. 18, Jun. pp. 1745–1756, (2013)
2. Xiong, X., Choi, B.J.: Position estimation algorithm based on natural landmark and fish-eyes' lens for indoor mobile robot. In: IEEE International Conference on Communication Software & Networks, 596–600, 2011
3. Schneider, D., Schwalbe, E., Maas, H.: Validation of geometric models for fisheye lenses. ISPRS J. Photogrammetry and Remote Sensing **64**, 259 (2009)
4. Ying, X., Hu, Z. Zha, H.: Fisheye lenses calibration using straight-line pp. 61–70, 2006
5. Kannala, J., Brandt, S.: A generic camera model and calibration method for conventional, wide-angle, and fish-eye lenses. IEEE Trans. Pattern. Anal. Machine Intell. J. **28**, 1335–1340, 2006
6. Ricolfe-Viala, Carlos, Sanchez-Salmeron, Antonio-Jose: Lens distortion models evaluation. Appl. Opt. **49**, 5914–5928 (2010)
7. Hughes, C., Denny, P. Glavin, M.: Jones, E.: Equidistant fish-eye calibration and rectification by vanishing point extraction. IEEE Trans. Pattern Anal. Machine Intell. **32**(12), 2289–2296, (2010)
8. Lou, J., Li, Y., Liu, Y., Xu, W.: Hemi-cylinder unwrapping algorithm of fish-eye image based on equidistant projection model. In: IEEE International Conference on Multimedia & Expo Workshops, vol. 370, pp. 1–4 (2013)
9. Brown, D.: Close-range camera calibration. Photogrammetric Eng. **37**(8), 855–866 (1971)

Algorithm for Digital Location and Recognition of Digital Instrument in Complex Scenes

Hao Zhou[1], Juntao Lv[2], Guoqing Yang[1](✉), Zhimin Wang[2], Mingyang Liu[3], and Junliang Li[4]

[1] Shandong Luneng Intelligence Technology Co., Ltd., Jinan 250101, China
sdygq2004@163.com
[2] State Grid Shandong Electric Power Company, Jinan 250001, China
[3] School of Information Science and Engineering, Lanzhou University, Lanzhou 730000, China
[4] State Grid Jilin Electric Power Co., Ltd., Changchun 130000, China

Abstract. Algorithm for digital location and recognition of digital instruments in complex scenes is proposed. First of all, the digital instrument images collected during the inspection process of the substation inspection robot are preprocessed by gray scale, filtering and morphology. Secondly, looking for extreme area, and using extreme values for coarse positioning, and then extract the ROI area of the number. In the ROI region, the preprocessing operation such as filtering and morphology are performed again, and the extreme region is searched, the error window is eliminated which can be achieve accurate positioning and segmentation of digital. Finally, the digital feature vectors are extracted and sent into the SVM classifier in sequence. The determination of significant figure of result based on the first nonzero value, and the correct reading of the meter is exported. The experimental results show that the algorithm can realize the digital location and recognition of digital instrument in complex scene. The algorithm has high accuracy and strong robustness, which can be met the requirement of digital instrument for substation inspection robot.

Keywords: Complex scenes · Digital instrument · Digital location · Digital recognition · SVM

1 Introduction

With the advent of the information age, the images as the most visual and direct way to describe objects, have gradually become an important source of human access to information and an important means of using information. In recent years, more and more scholars have transferred their eyes to such fields as image processing, computer vision, etc. As a kind of image classification, scene image classification has become an indispensable tool in scientific research. It has important theoretical value and application value in image retrieval, intelligent robot scene recognition and other fields [1].

Substation is the core of power grid at all levels, and the routine inspection of station equipment is the key technical means to ensure the safe operation of power grid.

F. Xhafa et al. (eds.), *Advances in Intelligent Systems and Interactive Applications*, Advances in Intelligent Systems and Computing 686,
https://doi.org/10.1007/978-3-319-69096-4_31

The existing manual inspection mode has high labor intensity, poor quality of detection and large interference from bad weather, which cannot meet the current high-speed development of power system [2, 3]. With the rapid development of robot technology and combining robot technology with power application, it is possible to carry on the inspection of equipment instead of manual inspection based on mobile robot platform.

The digital instrument in the substation is a kind of important equipment in the station, and the reading of the instrument reflects the working condition of the current equipment. Changes in instrument values often represent changes in the operating state of the device. Because of the special environment of substation and the large number of equipments in the substation, it is difficult to locate and identify the target equipment. Therefore, the identification of digital instrument readings is an important task.

2 Digital Positioning

The positioning of the digital area is the basis of digital identification, its positioning accuracy directly affect the digital segmentation and recognition results. An algorithm based on maximally stable extremal regions (MSER) for license plate location is proposed in the reference [4]. However, the paper does not deal with very fine on the license plate positioning and segmentation after extracting the MSER, but only uses the appearance of the standard license plate, which positioning is easy to fail. When the background is complex, or the MSER extraction is not accurate enough, it is difficult to locate and segment.

2.1 Preprocessing

In order to improve the image quality and facilitate the later processing to get better results, it is necessary to preprocess the image. Grayscale is the first operation of the preprocessing, and gray image can improve the efficiency of the algorithm. Morphological processing refers to the use of digital morphology as a tool to extract image components useful for expressing and plotting area shapes from images, as well as morphological filtering, thinning and pruning for pretreatment or post-treatment. In this paper, morphological closed operations are used to manipulate binarization images. Figures 1 and 2 represent the source and grayscale images of the digital instrument I and the digital instrument II respectively.

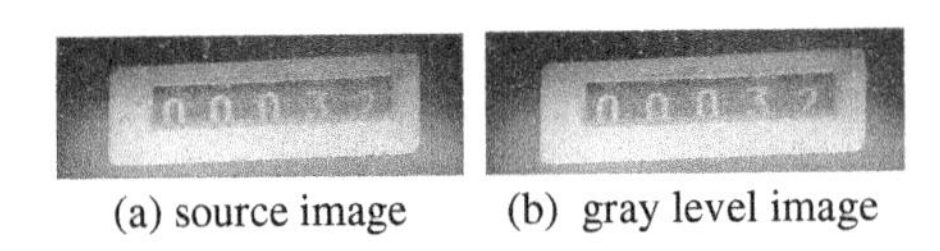

(a) source image (b) gray level image

Fig. 1. Digital instrumentation I

Bilateral filter is a non-linear filtering method, which is a compromise between the spatial proximity and the similarity of pixel values, taking into account the spatial information and gray similarity to achieve the purpose of denoising and edge

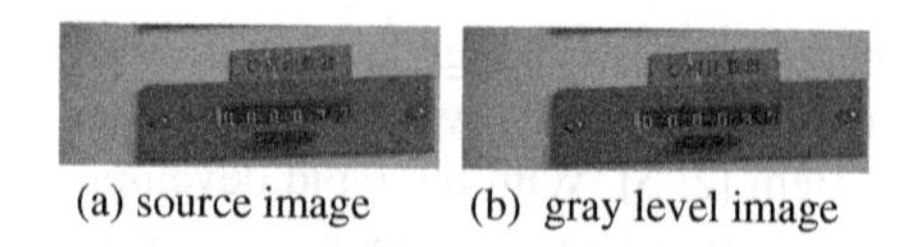

(a) source image (b) gray level image

Fig. 2. Digital instrumentation II

preserving. The filtering has a simple, non-iterative, local feature [5, 6]. In bilateral filtering, the value of the output pixel depends on the weighted sum of the neighborhood pixel values:

$$g(i,j) = \frac{\sum_{k,l} f(k,l)w(i,j,k,l)}{\sum_{k,l} w(i,j,k,l)} \tag{1}$$

where the weighting coefficient $w(i,j,k,l)$ is determined by the product of the definition domain kernel and the value kernel:

$$d(\mathrm{i,j,k,l}) = \exp(-\frac{(i-k)^2 + (j-l)^2}{2\sigma_d^2}) \tag{2}$$

$$r(\mathrm{i,j,k,l}) = \exp(-\frac{||f(i,j) - f(k,l)||^2}{2\sigma_d^2}) \tag{3}$$

Multiplied by the above two equations, the bilateral weight function can be obtained as follows:

$$w(\mathrm{i,j,k,l}) = \exp(-\frac{(i-k)^2 + (j-l)^2}{2\sigma_d^2} - \frac{||f(i,j) - f(k,l)||^2}{2\sigma_d^2}) \tag{4}$$

2.2 MSER Candidate Region Selection

Before the candidate regions are extracted, the image is processed first so as not to interfere with noise extraction results. MSER algorithm proposed by Matas is one of the better target region detection algorithms in the field of computer vision. By defining the sorting relationships of the appropriate pixel features, the method can highlight key features [7, 8]. Its mathematical definition is as follows:

$$q(i) = \frac{|Q_{i+\Delta} - Q_{i-\Delta}|}{Q_i} \tag{5}$$

where Q_i is a certain connected region at a threshold i, Δ is a small amount of change in the gray scale threshold, and $q(i)$ is the change rate of the region Q_i when the threshold is i. When $q(i)$ is a local minimum, then Q_i is the maximally stable extremal regions.

3 Digital Recognition

After the digital region is precisely located, the numbers are segmented and the location information of each digit is recorded so that the sequence of subsequent identification results is out of order. The digital image processing is performed for each number, and the extracted digital features are sent to the support vector machine (SVM) classifier to obtain the recognition result. In the field of machine learning, SVM is a supervised learning model that is commonly used for pattern recognition, classification, and regression analysis [9, 10].

SVM develops from the optimal classification surface in the linear separable case. The optimal classification surface requires that the classification line not only divide the two classes correctly (the training error rate is 0), but also make the classification interval maximum. SVM considers looking for a hyperplane that meets the requirements of the classification, and makes the point distance classification surface of the training set as far as possible, that is, looking for a classification surface to maximize the margin on both sides. In the two class of samples, the training samples of H1 and H2 on the hyperplane closest to the classification plane and parallel to the optimal classification plane are called support vectors.

4 Experimental Results and Analysis

The experimental data are obtained from the wheeled digital instrument image which collected by the patrol robot during the patrol process. As shown in Figs. 1 and 2. The process diagram of the algorithm is shown in Fig. 3, where the process diagrams of process 1 and process 2 are shown in Figs. 4 and 5, respectively. Figure 6 is a digital recognition process chart.

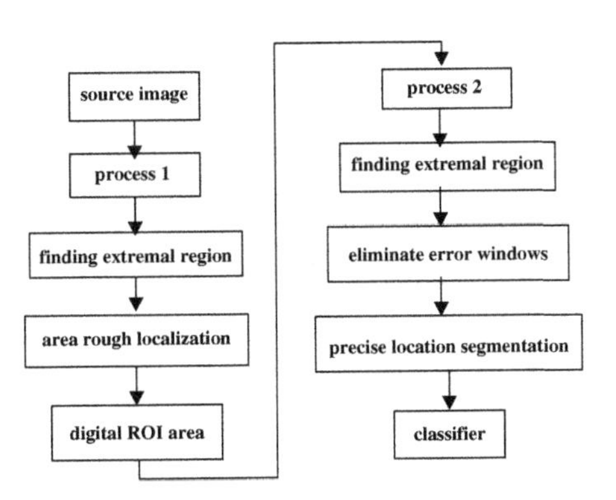

Fig. 3. Algorithm process diagram

In the experiment, the adaptive multi-scale threshold segmentation is used to segment the digital instrument image, as shown in Fig. 7. The binary image is subjected to morphological closing operation is shown in Fig. 8. The result of detecting the

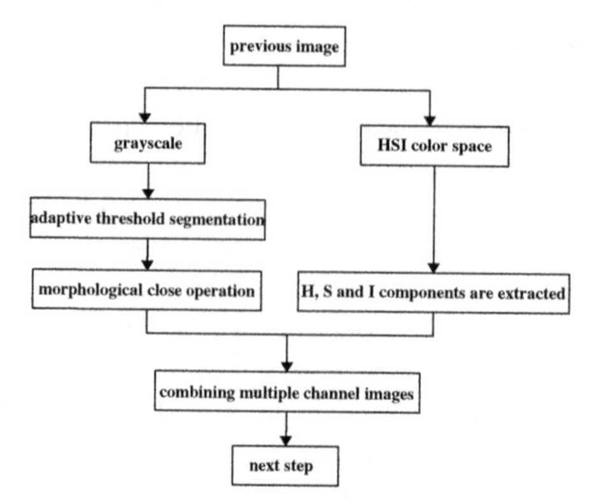

Fig. 4. Process chart of process 1

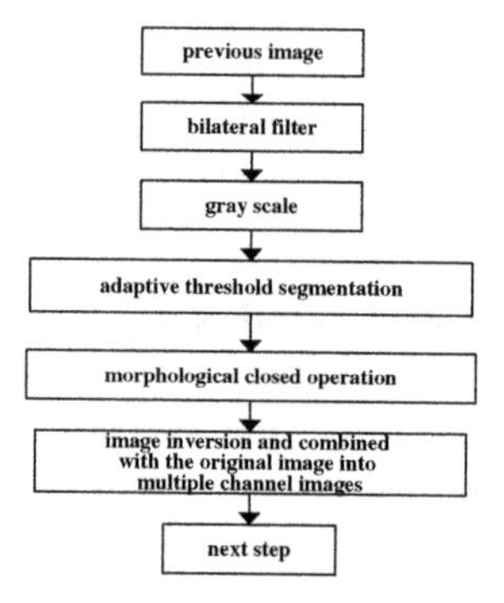

Fig. 5. Process chart of process 2

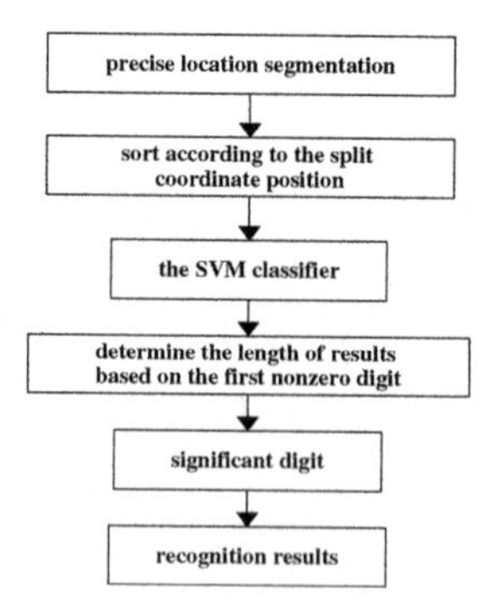

Fig. 6. Process chart of digital recognition

extreme value region is shown in Fig. 9. Locate an extremal region group that may be a character position and make a coarse location. Efficient pruning search method is used to exhaustive search all character sequences in real time. The digital region is shown in Fig. 10.

(a) digital instrument I (b) digital instrument II

Fig. 7. Adaptive threshold segmentation

(a) digital instrument I (b) digital instrument II

Fig. 8. Morphological closed operation

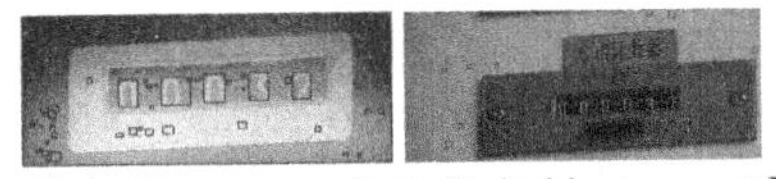

(a) digital instrument I (b) digital instrument II

Fig. 9. Detection results of extreme region

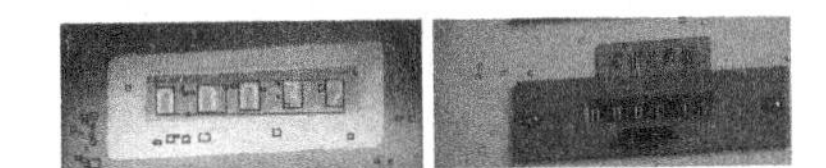

(a) digital instrument (b) digital instrument II

Fig. 10. Digital coarse positioning

After the coarse positioning of the digital region, the digital ROI region can be extracted from the original image. In the experiment, the ROI region has undergone bilateral filtering, adaptive multi-scale threshold segmentation and morphological closed operation, as shown in Figs. 11, 12, 13 and 14.

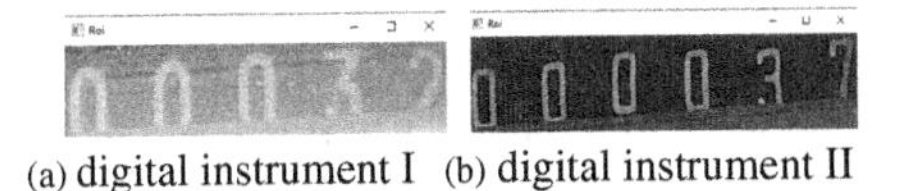

(a) digital instrument I (b) digital instrument II

Fig. 11. The digital ROI region

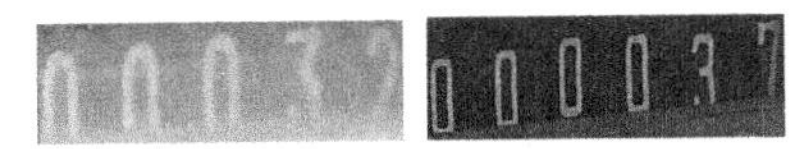

(a) digital instrument I (b) digital instrument II

Fig. 12. Bilateral filter

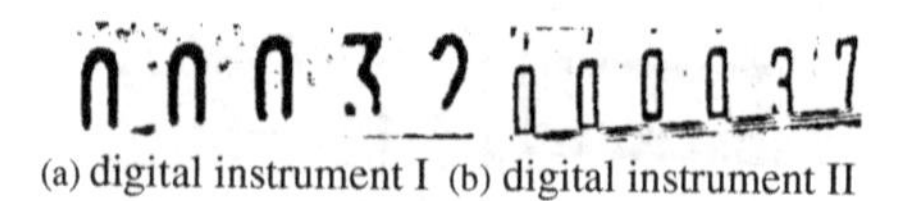

(a) digital instrument I (b) digital instrument II

Fig. 13. Adaptive threshold segmentation

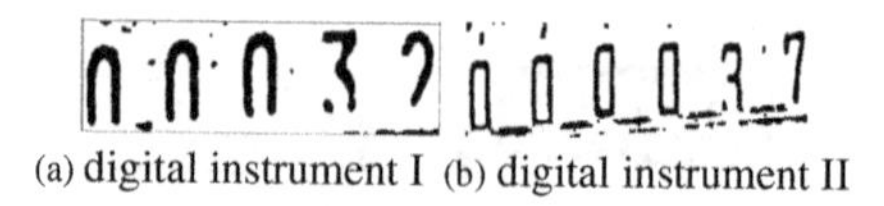

(a) digital instrument I (b) digital instrument II

Fig. 14. Morphological closed operation

According to the extremal region detection algorithm, and the result of extremal region is shown in Fig. 15. The result of the extreme area detection still has an error area, as shown in Fig. 15. Therefore, in order to get the correct position of the number, we need to eliminate the error window and get the precise positioning of the number, as shown in Fig. 16.

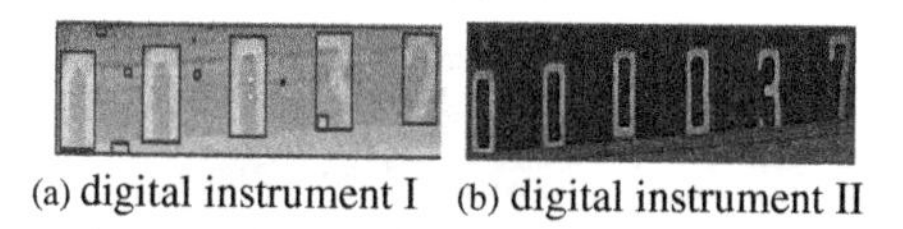

(a) digital instrument I (b) digital instrument II

Fig. 15. The result of extremal region

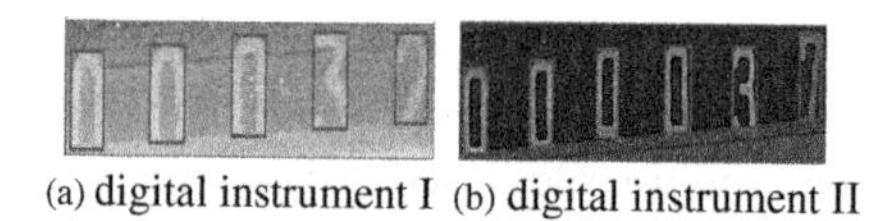

(a) digital instrument I (b) digital instrument II

Fig. 16. Accurate positioning

The feature vector of each digital small block after segmentation is extracted and sent into the trained SVM classifier to identify. The result of the recognition is shown in Fig. 17. As shown in Fig. 17, the numbers in the instrument are more, but not all of them are valid numbers. In order to make the recognition result more accurate, the natural numbers are synthesized according to the number of digits of the first nonzero number to determine the meter reading, as shown in Fig. 18.

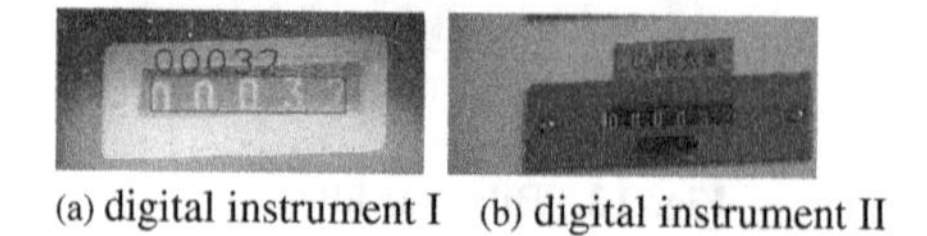

(a) digital instrument I (b) digital instrument II

Fig. 17. Digital recognition results

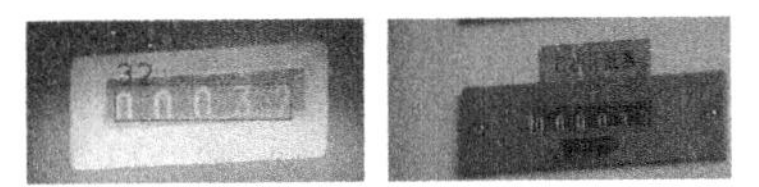

(a) digital instrument I (b) digital instrument II

Fig. 18. Correct reading of instrument

5 Conclusion

In this paper, the positioning and identification of the digital data area in the complex environment are realized under the platform of the substation inspection robot. The image processing technique is used to denoise the image, and the ROI region is extracted from the digital region of the extreme region and the digital eigenvector is sent into the SVM classifier to obtain the digital result. According to the effective number of bits to get the final correct instrument readings. Experimental results show that the algorithm is efficient and the algorithm is highly accurate. And the implementation of the algorithm to expand the scope of inspection of substation inspection robot, promote the unattended process, which lays a theoretical foundation and technical guarantee for automatic and intelligent analysis and identification of substation equipment.

Acknowledgements. This work was supported by research on pattern recognition function based on indoor and outdoor integrated patrol (item serial number: ZY-2017-11, management and control number: B306HQ17000L). Thank you for your support of the paper.

References

1. Wang, Y.B.: Registration of images with affine distortion based on MSER and phase congruency. Xidian University, pp. 1–10, (2014)
2. Lu, S.Y., Qian, Q.L., Zhang, B., et al.: Development of a mobile robot for substation equipment inspection. Autom. Electric Power Syst. **30**(13), 94–95 (2006)
3. Yang, D.X., Huang, Y.Z, Li, J.G., et al.: Research status review of robots applied in substations for equipment inspection. Shandong Electr. Power, **42**(1), 30–31 (2015)
4. Li, B., Tian, B., Yao, Q.M., et al.: A vehicle license plate recognition system based on analysis of maximally stable extremal regions. In: Proceedings of 2012 9th IEEE International Conference on Networking, Sensing and Control, pp. 399–404, (2012)
5. Liu, G.J., Ma, Y.M.: Textural image filtering method by hybridizing wave atoms shrinkage with bilateral filtering. Appl. Res. Comput. **30**(3), 942–943 (2013)
6. Yan, S.H., Geng, G.H., Li, K., Hao, N.: Feature line extraction method for heritage image. Appl. Res. Comput. **30**(11), 3500–3501 (2013)
7. Tang, Y.B., Bu, W., Wu, X.Q.: Natural scene text detection based on multi-level MSER. J. Zhejiang Univ. (Eng. Sci.), **50**(6), 1134–1137 (2016)

8. Xiao, Y., Jiang, J.: License plate location and character segmentation based on maximally stable extremal regions. Comput. Digit. Eng. **43**(12), 2271–2274 (2015)
9. Wang, H.Y., Li, J.H., Yang, F.L.: Overview of support vector machine analysis and algorithm. Comput. Digital Eng. **31**(5), 1281–1286 (2014)
10. Ding, S.F., Qi, B.J., Tan, H.Y.: An overview on theory and algorithm of support vector machines. J. Univ. Electron. Sci. Technol. China **40**(1), 1–4 (2011)

The Visualization Analysis on Present Research Situation and Trend of Tackling Overcapacity of Energy

Na Zheng, Dangguo Shao(✉), Lei Ma, Yan Xiang, Ying Xu, Wei Chen, and Zhengtao Yu

Key Lab of Computer Technologies Application of Yunnan Province, Kunming University of Science and Technology, Kunming 650500, China
23014260@qq.com

Abstract. As the pillar industry in the national economy, energy is facing a serious problem of overcapacity. Therefore the visualization study is demonstrated to more clearly understand the situation of tackling overcapacity of energy. The data source is based on the 4527 articles included in China National Knowledge Infrastructure database from 2000 to 2016, whose topics are all tackling overcapacity of energy. The tool SATI for statistical analysis of articles information is used to have frequency statistics of keyword and literature source, and to generate keywords co-occurrence matrix. Then the softwares of Ucinet and Netdraw are utilized to further draw the keyword co-occurrence of knowledge mapping for articles visualization, and to analyze the hotspots and tendencies in the area of tackling overcapacity of energy. From the research trend, tackling overcapacity of energy will be the hotspot for a period in the future. The paper is expected to provide the corresponding theoretical reference for the researchers in the field of tackling overcapacity of energy.

Keywords: Tackling overcapacity of energy · Knowledge mapping · SATI · Articles visualization

1 Introduction

Energy is the necessary material basis for social development [1]. In recent years, China's economic development has slowed down, the traditional energy industry (such as coal, steel, etc.) there is a more serious situation of excessive capacity [2]. Considering the importance of the energy industry, the problem of overcapacity of energy has been given more and more attention [3].

Based on the knowledge source of Chinese literature, this paper uses the statistical data analysis method SATI to carry on the literature statistics of quantity and source, the keyword frequency statistics, and to obtain the keyword co-occurrence knowledge map. Through the research situation interpreted from the above in the field of tackling overcapacity of energy, it can be predicted the development trend, aiming at providing reference for the relevant scholars in the field and learn from.

F. Xhafa et al. (eds.), *Advances in Intelligent Systems and Interactive Applications*, Advances in Intelligent Systems and Computing 686,
https://doi.org/10.1007/978-3-319-69096-4_32

2 Methods and Data

The basic method of visualization of literature is analysis of the development history, current situation and the further forecast of the development trend in a research field through citation analysis, keyword frequency analysis, cluster analysis [4]. In this paper, the data from China Knowledge Network (CNKI) is used to obtain the visual mapping through the data analysis by software SATI and drawing by visual software Ucinet and Netdraw. Developed on the NET platform using C# programming. SATI is a more mature analysis tool for domestic literature information statistics [5].

The data used in this paper is derived from CNKI, and the data format is EndNote. The search time is on January 03, 2017. The search term is tackling overcapacity of energy, the source of the literature is from 2000 to 2016 where a total of 4527 documents were retrieved. In order to avoid effect of the synonyms of the keywords in the literature on the analysis, these sensitive data have adequate preprocessing in the early stage.

3 Characteristics of Tackling Overcapacity of Energy

3.1 Time Distribution of Tackling Overcapacity of Energy Research

Changes in the number of academic literature is an important indicator of the development of quantitative disciplines. By using the number of documents per year, we can get the line graph of the literature distribution in the calendar year, which facilitate us to understand the research hotspot of the subject in a relatively clear way, to analyze the reason of the change of the poly line and to forecast the next developmental trend [6]. It can be seen from Fig. 1 where concerns of academic journals and papers have been grown since 2000, and the results have been shown that there are 4527 published papers at present. Still showing a sustained upward trend. Can be found the concept that "tackling overcapacity of energy" has just put forward in 2000–2009, and the relevant research in the initial stage was demonstrated in a small number of documents. The number of documents has a substantial increase since 2009. Tackling overcapacity of energy arouses the attention of more scholars and media. The decline in the number of documents included in 2015 was resulted from to a series of production capacity adjustments prior to the country, but did not affect the heat of research in capacity, and the heat gave a rise in 2016. It is believed that with the further implementation of energy production strategy for China, tackling overcapacity of energy is still a social research hotspot in the future.

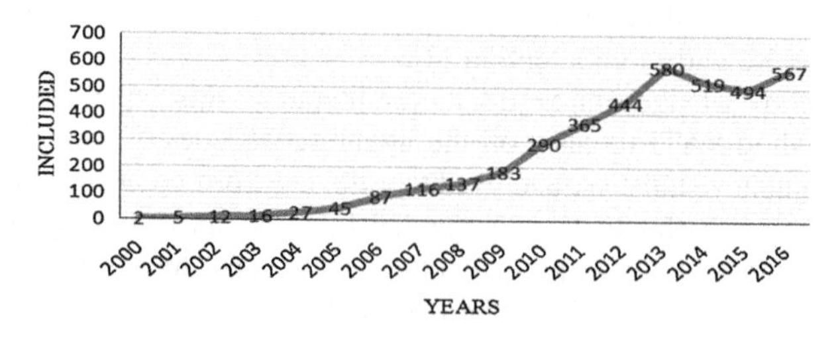

Fig. 1. The number of indexed articles about "Tackling overcapacity of energy"

3.2 Analysis of Source of Tackling Overcapacity of Energy

The research situation in the area can be measured using the source of the literature. According to the distribution of literature sources, it is helpful to understand the research status of the subject field, and then to predict the further developmental trend. Using SATI software to analyze the literature data, as shown in Fig. 2, the literature currently retrieved mainly from journals, papers, newspapers and conferences. The proportion has reached 63 and 28% respectively for journals and papers, which accounted for the largest proportion. It is mainly due to the wide range of research of tackling overcapacity of energy in the institutions of social research and higher education. With the nation's promotion to continuous production strategy, It is believed that the study of tackling overcapacity of energy in the conference and the newspaper will continue to heat up.

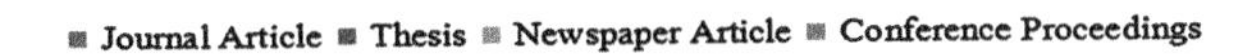

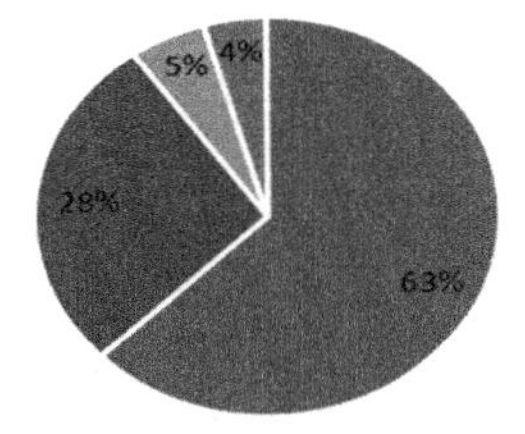

Fig. 2. The sources of articles about "Tackling overcapacity of energy"

3.3 Energy to the Production of Hot Spots Analysis

Keywords can be a good reflection of the article's subject, which is the highly summarized content of the article by the author. By analyzing the keywords, we can grasp the core theme of the article and the whole field for a specific period of time in the literature of high frequency keywords for analysis, and can dig out the field in this period. Figure 3 consists of main key words obtained from research literature analysis in the field of tackling overcapacity of energy. The top thirty key words related to "tackling overcapacity of energy" are: Excessive Capacity, Energy Conservation and Emission Reduction, Renewable Energy, Energy Consumption, National Development and Reform Commission, Shale-gas, Ministry of Industry and Information Technology, Energy Structure, Economic Growth, Development Strategy, Chemical Processing of Coal, Industrial Structure, Energy Use, The Economic Development, Photovoltaic Enterprises, Cyclic Economy, Photovoltaic, Emerging Industry, Plate Glass, Energy Price, Energy Management, Low-carbon Economy, Numerical Simulation, Feed-in Tariffs, Production Base, Energy Production, Chinese Light, Restructuring of Energy Industry, Adjustment of Industrial Structures, Coal Resources. It can be found that the study of overcapacity is mainly concentrated in several industries: the coal industry, flat glass industry and the photovoltaic industry, which are all excessive capacity industries. With the impact of excessive energy capacity on the economy, the concerns of the tackling overcapacity of energy is also grow.

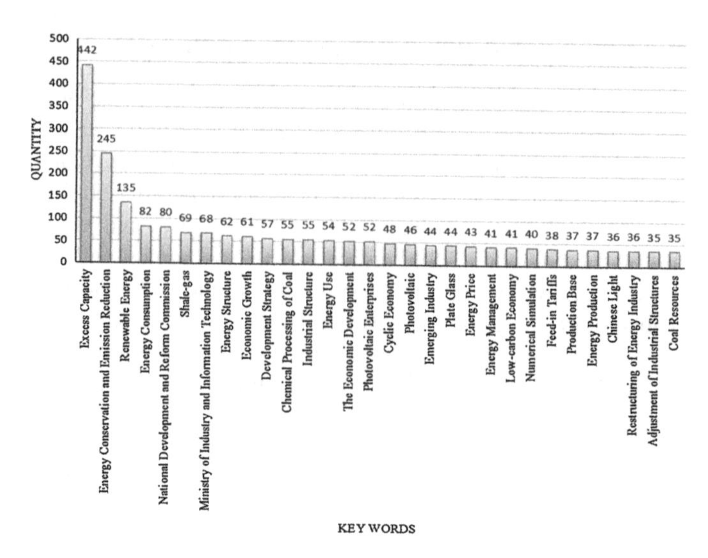

Fig. 3. Top 30 keywords of academic attention about "Tackling overcapacity of energy"

Through the clustering of multiple keywords with strong correlation in the field of tackling overcapacity of energy, we can get the co-occurrence relation knowledge map of the keywords, and we can get the research direction formed by several related keywords. To get the keyword co-occurrence knowledge map, we first get 30 * 30 high-frequency keyword co-occurrence matrix, which is needed to set the SATI row/column option unit number is 30, and then the matrix is obtained to import into Ucinet, which result into **h file. And then Netdraw software is adopted to calculate centrality of the key words, which demonstrates the size of node in the network. The node is in circle shape, is used for clustering algorithm (k-cores) clustering, and ultimately visualization knowledge map for the top 30 keyword is obtained in the field of tackling overcapacity of energy (Fig. 4) [7]. From Fig. 4, it can be seen that the Excessive Capacity of the node surplus, followed by the larger nodes also include Energy Conservation and Emission Reduction, Renewable Energy, National Development and Reform Commission, etc., but also more center, indicating that these keywords in tackling overcapacity of energy has been there, and occupy an important position, and has been widely concerned. And in Fig. 4 can also be found in the excessive capacity of the node and the nodes above have a close connection, indicating that overcapacity and these nodes have a tight link. The nodes in the central part of Fig. 4, are mostly in red, mainly because these nodes are a lot of overcapacity industry. The edge of the blue node is mainly linked with the production capacity of the in-depth implementation, which have been a number of regulatory excessive capacity of emerging industries and economic developmental methods. The coal resources, energy prices, energy structure adjustment and other different colors of the nodes. They were from the industrial resources, prices, structure and other aspects of the energy to show the impact of tackling overcapacity of energy process.

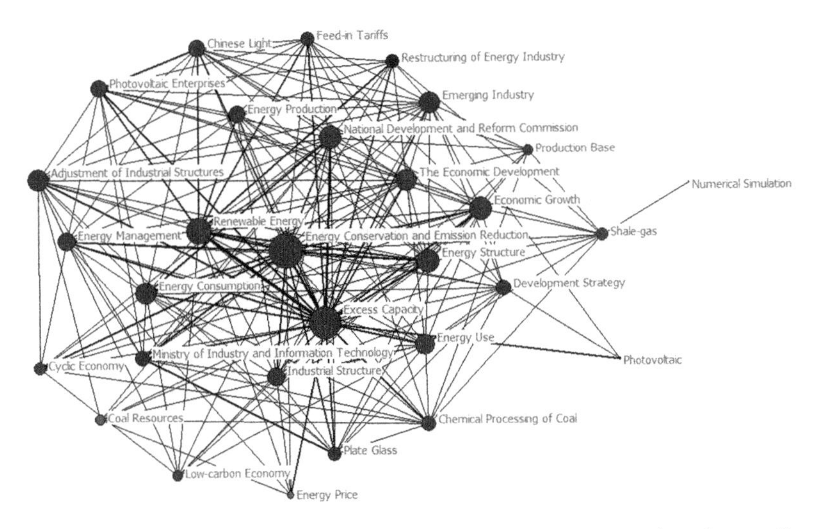

Fig. 4. Knowledge map of keywords about "Tackling overcapacity of energy"

Figures 3 and 4 show that tackling overcapacity of energy is mainly related to overcapacity industries such as coal enterprises, photovoltaic industry. At the same time as the emergence of excessive capacity, the state has developed strategy for the production capacity, so from the figure we see the National Development and Reform Commission, Ministry of Industry and Information Technology and some other national institutions have also issued the related policy of industrial restructuring and energy conservation and emission reduction. As a result of the country's high degree of concern and the introduction of a number of new industries, such as Renewable Energy, Low-carbon Economy, the energy for the country's development has a pivotal position, affecting the development of the national economy, so Economic Growth, and The Economic Development also appear in Fig. 4. At the same time energy is also related to the daily life of the people, so the Feed-in Tariffs and some elements related to the people's livelihood keywords also appear in Fig. 4. As the tackling overcapacity of energy also need statistics and other mathematical knowledge, so the analysis has shown the Numerical Simulation in Fig. 4 and other related words. Table 1 has more intuitively described the above situation.

Table 1. The influence of "Tackling overcapacity of energy" in the related fields

S/N	Related field	Related subject words
1	Overcapacity industry	Coal enterprises, photovoltaic
2	Relevant state departments	National Development and Reform Commission Ministry of Industry and Information Technology
3	Related policy	Industrial structure adjustment Energy conservation and emission reduction
4	Emerging industry	Regenerative resources Low-Carbon economy
5	Economy	Economic growth Economic development
6	Areas of livelihood	Electricity price

4 Conclusion

Through the statistics of the literature over the years, we can find that the number of related literatures on tackling overcapacity of energy has increased year by year, which shows that society's attention to tackling overcapacity of energy continues has risen. The source of tackling overcapacity of energy to the relevant literature shows that tackling overcapacity of energy is still a hotspot in social research institutions and higher educational institution. According to the keyword frequency statistics and the co-occurrence of the knowledge map, we can do more in-depth research in the field of tackling overcapacity of energy and dig out more valuable information.

The prediction of energy to production capacity will continue to be a part of the future hot spots in research. Aiming at developing the economy, digestion of the excessive capacity is necessary [8]. 2017 will witness real sense of energy to build energy. At the same time as the leading international energy development trend of the new energy industry, China has a huge space for development. So we should also pay attention to the development of new energy industry, optimize the industrial structure, to create a good economic developmental model.

Acknowledgements. This work was supported by China Postdoctoral Science Foundation (2016M592894XB) and the Science and Technology Office of Yunnan Province(KKS020170 3015). We also appreciated the valuable comments from the other members of our department.

References

1. Liu, H.J., Liu, C.M., Sun, Y.N.: Spatial correlation network structure of consumption and its effect in China. China Industrial. Economics **5**, 83–95 (2015)
2. Feng, M., Chen, P.: Quantitative analysis and warning on the excessive capacity of Chinese iron and steel industry. China Soft Sci. **5**, 110–116 (2013)
3. Wei, Z., Yu, B.Q.: China's photovoltaic industry today and recommended countermeasures to problems faced. Sino-Global Energy **18**(6), 15–25 (2013)
4. Liu Q.Y, Ye, Y.: A study on mining bibliographic records by designed software SATI: case study of on library and information science. J Inf. Res. Manage. (1), 50–58, (2012)
5. Zhong, Y.: Bibliometrics analysis of overseas research on institutional repositories from 2004 to 2013. Inf. Sci.**V34**(6), 71–76 (2016)
6. Ren, J.L., Li, H.: Visualization of research papers on eco-friendly water development in China —literature analysis tools based on CiteSpace. China. Water Resour. **5**, 55–58 (2016)
7. Zhao, R.Y., Li, F.: A comparative study of informetrics in China and in foreign countries based on social network analysis. Information. Science **2**, 7–12 (2013)
8. Han, G.G.: Analysis of the situation of China's industrial overcapacity and capacity reduction. Sci. Technol. Dev. **5**, 625–630 (2015)

Hybrid Algorithm for Prediction of Battlefield Rescue Capability of Brigade Medical Aid Station

Wen-Ming Zhou[1(✉)], San-Wei Shen[2], Wen-Xiang Xia[1], Chun-Rong Zhang[1,2], and Hai-Long Deng[1]

[1] Simulation Training Center, The Logistics Academy, Taiping Rd, No. 23, Beijing 100858, China
zwmedu@163.com

[2] Graduate Department of Academy of Military Science, The PLA, Xianghongqi Rd, No. 1, Beijing 100091, China

Abstract. Key links and main influencing factors of battlefield rescue are analyzed and studied. The index system of prediction of battlefield rescue capability of brigade medical aid station is established. Through integrating queuing theory, fuzzy comprehensive evaluation, analytic hierarchy process and mathematical definition with exponential method, a hybrid algorithm for prediction of battlefield rescue capability is proposed. A simulation example of evaluation is presented. The validity and practicability of the comprehensive method is verified, and the logistics command decision and system optimum design of the battlefield rescue health service support are strongly supported by the prediction algorithm.

Keywords: Battlefield rescue · Capability prediction · Exponential method · Queuing theory · Fuzzy comprehensive evaluation

1 Introduction

Battlefield rescue of brigade medical aid station is an important part of the logistics support in wartime. Analyzing and studying the characteristic and laws, and evaluating its support capability, can help to optimize the support method and health server force deployment. It can strongly promote resume battle effectiveness quickly [1, 2], and affect the process of the war, even the result.

Currently, study reports about battlefield rescue health service support simulation and evaluation is very less, joint theater-level simulation (JTLS) [3, 4] and joint warfare system (JWARS) [5] of American only simply described and simulated health service support, it couldn't embody the complex and law of it in modern information war. Through analyzing key links, main influencing factors and main proceeding of battlefield rescue health service support, this paper propose a hybrid prediction algorithm for the capability of battlefield rescue health service support.

F. Xhafa et al. (eds.), *Advances in Intelligent Systems and Interactive Applications*, Advances in Intelligent Systems and Computing 686,
https://doi.org/10.1007/978-3-319-69096-4_33

2 The Index System of Battlefield Rescue Capability Prediction

Analyzing the process of battlefield rescue, we find that the key links of it include medical aid station maneuver, take in classification, curing classification, the wounded curing, medical evacuation and so on. Furthermore, in information war, information level is a multiplicator of logistics support capability. The support capability of Leechdom and iatrical equipment is expressed by the rate of supplied amount to need. The full rate and the intact rate of medical equipments, maneuver capability, the wounded medical evacuation, the ability and level of medical personnel are all considered factors for evaluation. Considering the principle of systematicness, objectivity, completeness, measurability and independence during establishing index system, adopt similarity compress and singularity compress of information theory to optimize the index. The optimized index system is as Fig. 1.

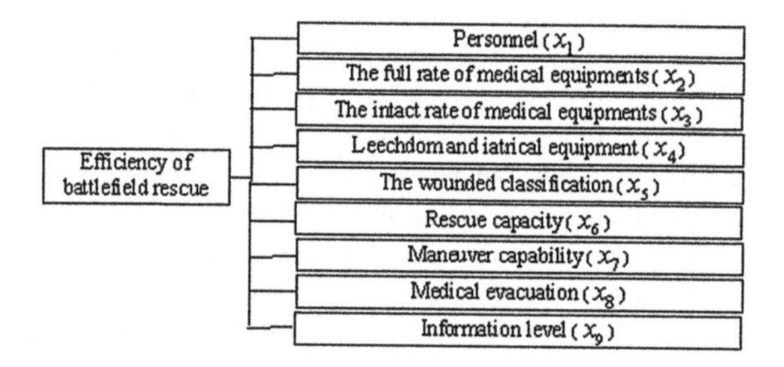

Fig. 1. Efficiency index of battlefield rescue

3 The Method for Computing Index Value

According to designed index system, adopt three kind of method to compute index value as follows: mathematics definition, operations research method and examination.

3.1 Mathematics Definition

Definition 1 The index of leechdom and iatrical equipment (x_4) is defined as the rate of practical reserves (n_m) to need for battlefield rescue (n_d). It is as follows:

$$x_4 = \begin{cases} \frac{n_m}{n_d}, & n_m \leq n_d \\ 1, & n_m > n_d \end{cases} \tag{1}$$

Definition 2 The full rate of medical equipments (x_2) is defined as the rate of the number of medical apparatus and instruments (n_v) to the allocated number (n_w). It is as follows:

$$x_2 = \begin{cases} \frac{n_v}{n_w}, & n_v \le n_w \\ 1, & n_v > n_w \end{cases} \quad (2)$$

The value $x_2$of can be calculated by analytic hierarchy process.

Definition 3 The intact rate of medical equipments (x_3) is defined as the rate of the number of intact medical equipment (n_1) to the allocated number (n_2). It's as follows:

$$x_3 = \begin{cases} \frac{n_1}{n_2}, & n_1 \le n_2 \\ 1, & n_1 > n_2 \end{cases} \quad (3)$$

3.2 Operations Research Method

Maneuver capability of medical aid station is a fuzzy index (x_7). It includes personnel maneuver (m_1) maneuver mode (m_2), maneuver environment (m_3), maneuver distance (m_4), carrying load (m_5) etc. $m_i(i = 1, 2, \ldots, 5)$ is confirmed by specialists.

$$x_7 = \sum_{i=1}^{5} a_i m_i \quad (4)$$

where a_i is weight coefficient [6], and $a_1 + a_2 + \cdots + a_5 = 1$.

Medical evacuation is a fuzzy index (x_8). It is defined as a synthesis by transportation facilities (m_1), transportation manner (m_2), transportation environment (m_3), transportation distance (m_4), transportation object (m_5) etc. m_i $(i = 1, 2, \ldots, 5)$ is confirmed by specialists.

$$x_8 = \sum_{i=1}^{5} a_i m_i \quad (5)$$

where a_i is weight coefficient, and $a_1 + a_2 + \cdots + a_5 = 1$.

The factor of medical personnel always play key and decisive role in all support. Staff with higher degree and stronger capability will be a multiplicator to efficiency of health service support. Personnel expressed as x_1. Its index system [7] is expressed in Table 1, adopt fuzzy comprehensive evaluation to calculate the value of index x_1 [8].

Information level expressed as x_9, include level of command, control and communication information system, level of specialist knowledge database, information management level of medical warehouse etc. Its value is also calculated through fuzzy comprehensive evaluation.

The index of Rescue capacity (x_6) is defined as the rate of number rescued to waiting number in limited time. *It is as follows*:

Table 1. Index classification pattern of health service support personnel

The ability and level of medical personnel		
Knowledge structure	Capacity and skill structure	Quality structure
Educational background	Political discrimination	Political quality
Training situation	Self-discipline	Military quality
Cultural science knowledge	Overall planning capability	Physical and mental qualities
Military science knowledge	Organizational training ability	Tenure
	Learning innovation ability	
	Psychological regulation ability	
	Ability to manage troops	
	Quick-reaction capability	
	Planning control capability	
	Planning decision making ability	
	Medical treatment ability	

$$x_6 = \frac{n_c}{N} \tag{6}$$

where n_c is the number of the rescued, and N is the total waiting number in limited time. n_c is calculated by queuing theory as the latter.

3.3 Examination

According to examination and inspection method of different unit, The value of wounded classification (x_5) is calculated by examination and inspection. Where it include take in classiffication and rescue classiffication.

4 The Method for Calculating Parameter N_c

The key to calculate the index x_6 lie in confirming the value of $n_c$in limited time. We will confirm its value on two case as follows [9]:

4.1 The Wounded Rescued Only Through One Kind of Rescue Desk

Hypothesis model:

(1) The input of queuing system subject to *Poisson* distribution with parameter λ.

$$P\{X(t)=n\}=\frac{(\lambda t)^n}{n!}e^{-\lambda t}\quad(n=0,1,2,\ldots) \tag{7}$$

(2) Random arrival of the wounded at every rescue point.
(3) The wounded only can be classified into m rescue group. That is to say that all arrival wounded belong to each group of m.
(4) Having n wounded arrival at rescue site at the same time at the moment t.
(5) The wounded arrivals n at most queues in m line. The number of being waiting for rescue in every line is n_1, n_2, …, n_m, and $\sum_{i=1}^{m} n_i = n$.
(6) First-come first-served for the wounded in every queue. Service time is independent of arrival interval, and subject to negative exponent distribution with parameter $\mu_i(\mu_i > 0)$ $(i = 1, 2, \ldots, m)$.
(7) Having one or more rescue site in every queue, that is service desk. The number of service desk is supposed as $s_i(i = 1, 2, \ldots, m)$, and every service desk is independent of other. Supposed $\rho_i = \lambda/s_i\mu_i$, it is service factor of every kind of service. It expresses the service ability ($s\mu$) of every service system is utilized averagely by the arrival custom (λ), and supposed $\rho_i < 1$.

The brigade medical aid station is an opening queue system by hypothesis model. Each queue is an independent queue. Because the wounded only get across one service desk, the steady state probability of queue system is as follows:

$$\begin{cases} \mu_i P_{i(1)} = \lambda P_{i(0)} \\ (n_i+1)\mu_i P_{i(n_i+1)} + \lambda P_{i(n_i-1)} = (\lambda + n_i\mu_i)P_{i(n_i)} \\ \quad (1 \le n_i \le s_i) \\ s_i\mu_i P_{i(n_i+1)} + \lambda P_{i(n_i-1)} = (\lambda + s_i\mu_i)P_{i(n_i)} \\ \quad (n_i > s_i) \end{cases} \tag{8}$$

where $\sum_{j=1}^{\infty} P_{i(j)} = 1(i = 1, 2, \ldots, m)$ and $\rho_i \le 1$

Through solving the difference equation by recurrence method, we can get the state probability as follow:

$$P_{i(0)} = \left[\sum_{k=0}^{s_i-1}\frac{1}{k!}\left(\frac{\lambda}{\mu_i}\right)^k + \frac{1}{s_i!}\frac{1}{\left(1-\frac{\lambda}{s_i\mu_i}\right)}\left(\frac{\lambda}{\mu_i}\right)^{s_i}\right]^{-1} \tag{9}$$

$$P_{i(n_i)} = \begin{cases} \frac{1}{n_i!}\left(\frac{\lambda}{\mu_i}\right)^{n_i} P_{i(0)}, 0 \le n_i \le s_i \\ \frac{1}{s_i! s_i^{n_i-s_i}}\left(\frac{\lambda}{\mu_i}\right)^{n_i} P_{i(0)}, n_i > s_i \end{cases} \tag{10}$$

Operating indexes of the system include: average queue length (L_i), expected length of the waiting line (except the custom being rescued) (L_{iq}).

$$L_{iq} = \sum_{n=s_i}^{\infty} (n - s_i) P_{i(n)} = \sum_{j=0}^{\infty} j P_{i(s_i+j)} \\ = \frac{P_{i(0)} (\lambda/\mu_i)^{s_i} \rho_i}{s_i!(1 - \rho_i)^2} \tag{11}$$

$$L_i = L_{iq} + \frac{\lambda}{\mu_i} \tag{12}$$

The expected waiting time (W_{iq}) and expected staying time (W_i) are as follows:

$$\left. \begin{aligned} W_{iq} &= \frac{L_{iq}}{\lambda} \\ W_i &= W_{iq} + \frac{1}{\mu_i} \end{aligned} \right\} \tag{13}$$

The number of average rescued wounded n_c is as follows:

$$n_c = \sum_{i=1}^{m} INT \left[\frac{T}{W_i} \right] \tag{14}$$

where $INT[\frac{T}{W_i}]$ express rounding, then the battlefield rescue capability of brigade medical aid station is as follows:

$$x_6 = \frac{\sum_{i=1}^{m} INT \left[\frac{T}{W_i} \right]}{N} \tag{15}$$

4.2 The Wounded Rescued Through One or More Kind of Rescue Desk

In the course of rescue, the wounded could be displaced from slight injury rescue site to serious injury rescue site or from critical injury rescue site to medical evacuation. That is the wounded would pass one or more service systems. It is a complex opening queuing system.

Hypothesis model:

(1) The n wounded would arrive at the moment t, and every wounded could pass K ($K \geq 2$) kind of rescue site.
(2) The probability of one wounded rescued pass from rescue desk i to rescue desk j is $P_{i,j}$.
(3) Every queuing space is unlimited.

The other condition is like the case 1.

One wounded would pass from rescue desk i to rescue desk j with probability $P_{i,j}$ or leave with probability q_i.

$$q_i = 1 - \sum_{j=1}^{m} p_{i,j} \tag{16}$$

Supposed $\delta_j (j = 1, 2, \ldots, m)$ It is the total arrival rate of queue j. It is the non-negative solution for equation set (17).

$$\delta_j = \lambda + \sum_{i=1}^{m} \delta_i p_{i,j} \quad (i = 1, 2, \ldots, m) \tag{17}$$

Supposed $\rho_i = \delta_i / s_i \mu_i (i = 1, 2, \ldots, m)$, $\rho_i (i = 1, 2, \ldots, m)$ is service factor of rescue site i. Supposed $m_i (i = 1, 2, \ldots, m)$ is the length in front of rescue site i. Then the distribution probability of steady length of the queue network is as follows:

$$p(m) = \prod_{i=1}^{m} p_i(m_i) \, m = (m_1, m_2, \ldots, m_m) \tag{18}$$

where

$$p_i(m_i) = \begin{cases} \dfrac{(s_i \rho_i)^{m_i}}{n_i} b_i & (m_i \leq s_i) \\ \dfrac{(s_i \rho_i)^{m_i}}{s_i! s_i^{m_i - s_i}} b_i & (m_i > s_i) \end{cases} \qquad b_i = [\sum_{j=0}^{s_i - 1} \tfrac{(s_i \rho_i)^j}{j!} + \tfrac{(s_i \rho_i)^{s_i}}{s_i!} \tfrac{1}{1 - \rho_i}]^{-1} \quad (i = 1, 2, \ldots, m)$$

It is independent queue system *M*/*M*/*s* Get the solution of the expected waiting time (W_{iq}) and expected staying time (W_i) of the wounded in every queue respectively.

$$L_{iq} = \frac{P_{i(0)} (\delta_i / \mu_i)^{s_i} \rho_i}{s_i! (1 - \rho_i)^2} \tag{19}$$

$$L_i = L_{iq} + \frac{\delta_i}{\mu_i} \tag{20}$$

Suppose the arrival rate of the queue network ω and $\omega = m\lambda$. It is the total arrival rate out of the system. Thus,

$$W = \sum_{i}^{m} \frac{L_i}{\omega} \tag{21}$$

The battlefield rescue capability of brigade medical aid station is as follows:

$$\begin{cases} n_c = \dfrac{T}{W} \\ x_6 = \dfrac{n_c}{N} \end{cases} \tag{22}$$

5 Exponential Method Modeling

The relationship among factors is embodied better by the exponential method [10, 11]. It considers the battlefield rescue efficiency of brigade medical aid station possesses the relationship with each indexes. It is express as follow function:

$$Y = F(X) = Kx_1^{\alpha_1} x_2^{\alpha 2} \ldots x_n^{\alpha_n} \tag{23}$$

where Y the efficiency of battlefield rescue; $X = (x_1, x_2, \ldots, x_n)$ is n efficiency index; $\alpha_i (i = 1, 2, \ldots, n)$ is power exponent, $\sum_{i=1}^{n} \alpha_i = 1$, expressed as weight; K adjustment coefficient.

Among the nine indexes of battlefield rescue, there is a strong coupling between x_4 and x_6. x_1, x_2, x_3, x_5, x_9 would not seriously effect on the overall efficiency, x_7 and x_8 are independent. Thus, $x_1, x_2, x_3, x_4, x_5, x_6$ and x_9 are chosen as exponential terms. x_7, x_8 are chosen as adjustment coefficient terms. The exponential method is as follows:

$$Y = x_1^{\alpha_1} \bullet x_2^{\alpha_2} \bullet x_3^{\alpha_3} \bullet x_4^{\alpha_4} \bullet x_5^{\alpha_5} \bullet x_6^{\alpha_6} \bullet x_7 \bullet x_8 \bullet x_9^{\alpha_7} \tag{24}$$

where $\alpha_i (i = 1, 2, \ldots, n)$ is confirmed through analytic hierarchy process.

6 Prediction Example

Supposed a brigade medical aid station has sixty medical staff after intensified. It is organized into slight injury rescue group, serious injury rescue group, critical injury rescue group and medical evacuation group etc. That is four kind of service system. One or two medical staff forms a service desk in slight injury group. More than two

Table 2. The case of battlefield rescue service classification of brigade medical aid station

Kind of service	Number of service desk	Service rate	Number of the waiting wounded	Limited time
Slight injure	9	0.96	37	18
Serioud injury	3	0.89	19	
Critical injury	1	0.91	7	
Medical evacuation	2	0.97	16	

Note The hour is the unit of time. Service rate express the rescued number in an hour, that is μ_i

medical persons can form a service desk in other three groups. The number of service desk and service rate of four service system is seen in Table 2. Supposed the arrival of the wounded in battlefield subjects to *Poisson* distribution with parameter $\lambda = 0.42$. Through calculating we get that $x_1 = 0.87$, $x_2 = 0.95$, $x_3 = 0.97$, $x_4 = 0.93$, $x_5 = 0.96$, $x_7 = 0.86$, $x_8 = 0.97$ and $x_9 = 0.76$, confirm the value of x_6 by function (9)–(15), the result can be seen in Table 3.

Table 3. The basic case of battlefield rescue

Queue	L_{iq}	L_i	W_{iq}	W_i	Number rescued	x_6
1	4.445×10^{-11}	0.4271	1.084×10^{-10}	1.0417	19	0.7468
2	0.0022	0.4629	0.0054	1.1290	17	
3	0.3912	0.8418	0.9542	2.0531	7(9)	
4	0.0198	0.4425	0.0483	1.0792	16(18)	

Note X(Y), Y-can complete rescue number in the limited time, X-practical number of rescued wounded

Confirmed the value of weights by analytic hierarchy process: $\alpha_1 = 0.18$, $\alpha_2 = 0.12$, $\alpha_3 = 0.125$, $\alpha_4 = 0.12$, $\alpha_5 = 0.075$, $\alpha_6 = 0.3$, $\alpha_7 = 0.08$. The support capability of battlefield rescue is as follows:

$$\begin{aligned} Y &= x_1^{\alpha_1} \cdot x_2^{\alpha_2} \cdot x_3^{\alpha_3} \cdot x_4^{\alpha_4} \cdot x_5^{\alpha_5} \cdot x_6^{\alpha_6} \cdot x_7 \cdot x_8 \cdot x_9^{\alpha_7} \\ &= 0.87^{0.18} \cdot 0.95^{0.12} \cdot 0.97^{0.125} \cdot 0.93^{0.12} \cdot \\ &\quad 0.96^{0.075} \cdot 0.7468^{0.3} \cdot 0.86 \cdot 0.97 \cdot 0.76^{0.08} \\ &= 0.7130 \end{aligned}$$

From Table 3, queue 1 and queue 2 don't complete the rescue task. One reason could be that the service rate is low. The other reason is that the service desks are less. Thus, according to concrete rescue task, we can add medical treatment staff or improve their medical treatment ability to improve the service rate. For example, the value of x_1 is up to 0.95, the value of x_6 is up to 0.85. At the same time, x_7 is up to 0.90, x_9 is up to 0.85. The battlefield rescue capability is calculated as follows:

$$\begin{aligned} Y &= x_1^{\alpha_1} \cdot x_2^{\alpha_2} \cdot x_3^{\alpha_3} \cdot x_4^{\alpha_4} \cdot x_5^{\alpha_5} \cdot x_6^{\alpha_6} \cdot x_7 \cdot x_8 \cdot x_9^{\alpha_7} \\ &= 0.95^{0.18} \cdot 0.95^{0.12} \cdot 0.97^{0.125} \cdot 0.93^{0.12} \cdot \\ &\quad 0.96^{0.075} \cdot 0.85^{0.3} \cdot 0.90 \cdot 0.97 \cdot 0.85^{0.08} \\ &= 0.8032 \end{aligned}$$

It is quite clear that the result coincides with the fact.

7 Conclusion

According to the result of prediction example, the relationship among factors is embodied better by the index method. The proposed hybrid algorithm for prediction of battlefield rescue capability of brigade medical aid station is valid and applicable. The logistics command decision making and system optimum design of the health service support information system will be strongly supported by it.

References

1. Zhang, S.H., Guo, S.S., Sun, J.S.: Health Service Tutorial, pp. 20–30. PLA Publishing House, Beijing (2010)
2. Chen, W.L.: Modern Health Service Front Along Theories, pp. 1–20. Military Medicine Science Publishing House, Beijing (2006)
3. U.S. Joint Forces Command Joint Warfighting Center. JTLS(3.4.3.0) Version Description Document. March, 2010
4. U.S. Joint Forces Command Joint Warfighting Center. JTLS Executive Overview. April, 2008
5. Stone, G.F. III, McIntyre, G.A.: The Joint Warfare System (JWARS): A Modeling and Analysis Tool For the Defense Department. In: Proceedings of the 2001 Winter Simulation Conference. 2001
6. Liao, A.H., Hou, F.J.: City security evaluating based on approximate ideal point and AHP. China Public Security·Acad Ed. **14**(1), 35–39 (2009)
7. Haiquan, R.: Science of Military Command Decision, pp. 1–10. National Defence University Publishing House, Beijing (2007)
8. Heping, S.H.I., Tao, H., et al.: Evaluation in support ability of equipment maintenance personnel based on fuzzy comprehensive evaluation. Mod. Electron. Technol. **246**(1), 96–98 (2008)
9. Hiller, F.S., Lieberman, G.J.: Introduction to Operation Research, pp. 783–787. Tsinghua University Publishing House, HU Yunquan translated (2007)
10. Shoulin, T., Shuangka, Y., Xuesong, C.: Fighting efficiency evaluation model of cruise missile based on index method. Fire Control and Command Control **35**(5), 173–179 (2010)
11. Rui, W., An, Z., Zhaowei, S.: Effectiveness evaluation of advanced fighter plane based on power series and fuzzy AHP. Fire Control and Command Control **33**(11), 73–80 (2008)

The Design of Optimal Synthesis Filter Bank and Receiver for the FBMC System

Yan Yang[1(✉)] and Pingping Xu[2]

[1] Department of Electronic and Electrical Engineering, Bengbu University, Bengbu 233030, China
yangetyan@126.com
[2] National Mobile Communications Research Lab, Southeast University, Nanjing 210096, China

Abstract. A new structure of synthesis filter bank with optimal synthesis filter and a receiver based on carrier frequency offset (CFO) are designed for the multicarrier (FBMC) system. Firstly, A synthesis filter bank (SFB) of a FBMC system is designed with dimensions. but the complexity will increase. An optimal dimensionality reduction method is proposed to overcome it. Secondly a receiver of FBMC system is proposed based on frequency-domain compensation technology (RDCT). Finally, the simulations are given to prove the advantages and the effectiveness for designed FBMC system.

Keywords: Filter bank multicarrier · Analysis filter bank · Multirate filter bank frequency-domain compensation

1 Introduction

In recent years one of key technologies of 5G, the filter bank multicarrier (FBMC) systems with high data-rate transmission in wireless channels have been studied widely by many researchers [1]. FBMC systems have low sidelobes and high spectrum efficiency without no CP [2]. Some optimization methods and structures appeared that the multirate Kalman analysis filter was used to replace the conventional filter to get optimal signal reconstruction in [3]. The optimal design of prototype filter is proposed in [4]. The frequency-despreading (FD) structure and fast-convolution (FC) structure have been designed in [5] etc. In this paper, a multirate synthesis filter bank is designed for FBMC system with transmission delay based on the correlation theory. Mean square errors of different algorithms are compared with different noise situation. And the receiver of FBMC system based on the frequency spreading (FS) is formed and FD CFO compensation by simulation will be discussed.

2 Multirate Synthesis Filter Bank Design

The structure of an optimal mulitrate synthesis filter bank is shown in Fig. 1.

$\tilde{f}(k)$ is the input signal and $u_i(k)$, $(k = 0, 1, 2, \cdots)$ is the convolution output between the $\tilde{f}(k)$ and the $h_i(k)$, $h_i(k)$ is the impulse response of the ith FIR analysis filter.

F. Xhafa et al. (eds.), *Advances in Intelligent Systems and Interactive Applications*, Advances in Intelligent Systems and Computing 686,
https://doi.org/10.1007/978-3-319-69096-4_34

The extracted factor of the sub-band signal sequence is R, because of noise-affected and interpolated the receiver signal of ith sub-band channel signal is $\tilde{y}(k)$. It can $\tilde{y}(k)$ written as follows:

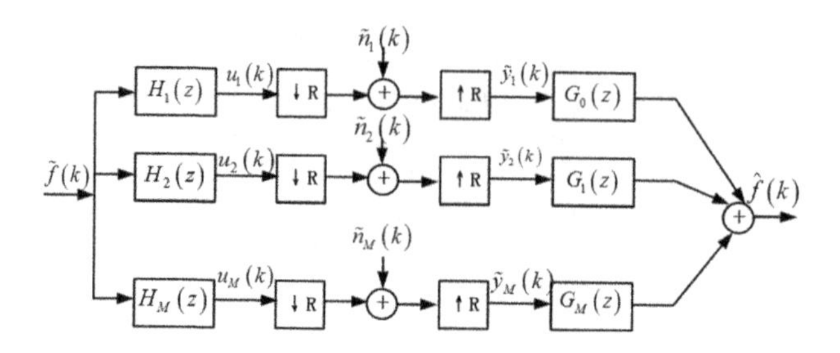

Fig. 1. The filter bank with optimal synthesis filter of M-channel

$$\tilde{y}(k) = u(k) + \tilde{n}(k)\, k = 0, R, 2R \tag{1}$$

An optimal synthesis filter is designed by a linear unbiased estimator in [6] the state vector $\bar{f}(k)$ is defined as follows:

$$\bar{f}(k) = [I_{2L}\, 0]\begin{bmatrix} f(k) \\ n(k) \end{bmatrix} = T\bar{f}(k) \tag{2}$$

where $f(k)$ and $n(k)$ are defined by the input signal $\tilde{f}(k)$, and the interfering noise $n(k)$ of the ith sub-band respectively. L is the length of the prototype filter.

$$f(k) = \begin{bmatrix} \tilde{f}(k) \\ \tilde{f}(k-R) \end{bmatrix}, n(k) = \begin{bmatrix} \tilde{n}(k) \\ \tilde{n}(k-R) \end{bmatrix} \tag{3}$$

Letting

$$e(k) = f(k) - f(k) \quad Q(k) = E\{e(k)e^T(k)\} \tag{4}$$

If the initial conditions $f(0) = 0$, $Q(0) = 0$ $Q_i = \sigma_i^2$, $i = 0, 1, \cdots, N-1$. σ_i^2 is mean square errors. The length of filter reduced is $2L$. To get the optimal estimation, $Q(k+1)$ need to be minimized. Define

$$A = \begin{bmatrix} \tilde{A}_R 0 \\ I\, 0 \end{bmatrix}, B = \begin{bmatrix} \tilde{B}_R \\ 0 \end{bmatrix}, C(k) = \left[\left(1 - \zeta(k)\tilde{C}\right) \zeta(k)\tilde{C}\right], \\ D(k) = [(1 - \zeta(k))\, \zeta(k)] \tag{5}$$

where $\tilde{A}, \tilde{B}, \tilde{C}$ are the same as defined in [6]. Bermoulli-distributed white nose sequence is $\zeta(k)$ satisfying Prob $\{\zeta(k) = 1\} = \alpha(k)$. If $M_2(k)$ is defined as the gain of the linear optimal filter and must satisfy the condition

$$M_2(k)D^0(K+1)\psi(k) = 0 \tag{6}$$

where $\psi(k) = [00I_0]$, once $M_2(k)$ is confirmed $M_1(k)$ and $Q(k+1)$ can be calculated as follows:

$$M_1(k) = A - M_2(k)C^0(k+R)A \tag{7}$$

$$\begin{aligned} Q(k+R) =& AQ(k)A^T + BQ_R(k)B^T \\ & - M_2(k)\Gamma^T(k) - \Gamma(k)M_2^T(k) \\ & + M_2(k)\Phi(k)M_2^T(k) \end{aligned} \tag{8}$$

If $f(k)$ is the output signal of the optimal synthesis filter. It can be derived by applying the following equation:

$$\begin{bmatrix} f(k-R+1) \\ f(k-R+2) \\ \vdots \\ f(k) \end{bmatrix} = [0I_R00]f(k), k = 0, R, 2R, \cdots \tag{9}$$

3 The Design of Receiver of Fmbc System in the Presence CFO Based on Frequency-Despreading Method

If $\tilde{f}(k)$ is the input signal in Sect. 2 the received signal of FBMC-OQAM system $r(k)$ can be expressed

$$r(k) = \lambda \tilde{f}(k)e^{j(2\pi/M)\theta_k} + \eta[k] \tag{10}$$

where $\lambda(0 < \lambda < 1)$ is the channel gain, normalized CFO to the subcarrier spacing is θ, $\eta(k)$ is additive noise which can be assumed a white Gaussian noise. Its mean and variance are zero and σ^2 respectively. If we assume that a normalized CFO value affects the signal of interest the decision variable $D_{n,i}^{(\mathrm{Re})}$ can be derived as follows:

$$D_{n,i}^{(\mathrm{Re})} = \frac{1}{A} j^{-i} e^{-j2\pi\theta n} \sum_{k=0}^{KM-1} r[nM+i] e^{-ji(2\pi/M)k} h_\theta^*(k) \tag{11}$$

where, the symbol "Re" denotes the real part of the complex data symbols which are transmitted on the kth subcarrier for the nth FBMC/OQAM symbol. Superscript $(\cdot)^*$

indicates the complex conjugation. $A = \lambda \varepsilon_g, \varepsilon_g \triangleq \sum_{k=0}^{KM-1} h(k)$, $h_\theta(k)$ is the modified prototype filter. So the FD CFO compensated version can be derived $h_\theta(k) = h(k)e^{j(2\pi/M)\theta k}$. Then (11) can be as follow:

$$D_{n,i,}^{(\mathrm{Re})} = \frac{1}{A} j^{-i} e^{-j2\pi\theta n} KM \sum_{l \in A_1} \tilde{H}^*_{l-iK-K\theta} R_{n,l} \tag{12}$$

where $A_1 \subseteq \{0, 1, \ldots KM - 1\}$ indicates the set of indices of $R_{n,i}$ as the DFT componets. It is obvious that the coefficients $\tilde{H}^*_{l-iK-K\theta}$ may be modified to get small errors estimation of θ. The coefficients of DFS of pth prototype filter is $H_p^{\mu M} \triangleq \frac{1}{KM} \sum_{k=0}^{KM-1} h[k + \mu M] e^{-j(2\pi/KM)kp}$ by uM samples and truncated in the interval $k = \{0, 1, \ldots KM - 1\}$. Then

$$\rho_i \triangleq \angle \left[\sum_{l \in A_1} \tilde{H}^*_{l-iK-K\theta} H_{l-iK-K\theta} \right] \tag{13}$$

$\angle[\cdot]$ indicates the argument of complex number in $[-\pi, \pi)$. δ_i the Kronecker delta.

If the receiver coefficients of FBMC system are updated to account with $\varepsilon \neq 0$ the block diagram of receiver structure in Fig. 2 is designed by proposed CFO compensation. The decision variable $D_{n,i,}^{(\mathrm{Re})}$ of (12) can be modified as follows:

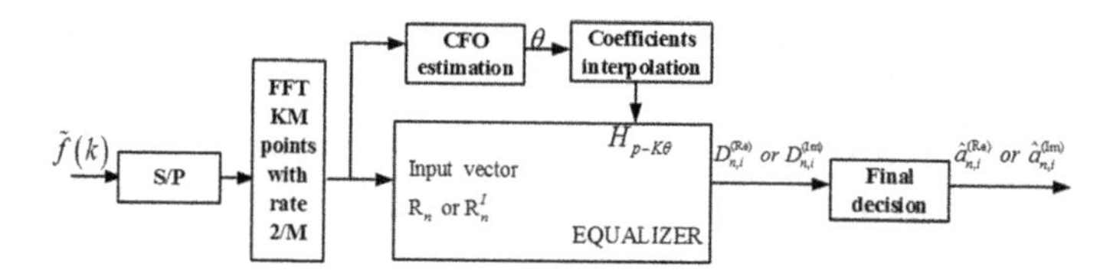

Fig. 2. CFO compensation block diagram of receiver for FBMC system

$$\begin{aligned}\frac{A}{KM}e^{-j\rho_i}D_{n,i,}^{(\mathrm{Re})} =&\lambda a_{n,i}^{\mathrm{Re}}\left|\sum_{l\in A_1}\tilde{H}_{l-iK-K\theta}^{*}H_{l-iK-K\theta}\right| \\ &+\lambda\sum_{n_1}\sum_{i_1\in A_1}(1-\delta_{n-n_1}\delta_{i-i_1})a_{n_1,i_1}^{\mathrm{Re}}e^{-j\rho_i}j^{(i_1-i)} \\ &\times\sum_{l\in A_1}\tilde{H}_{l-iK-K\theta}^{*}H_{l-iK-K\theta}^{(n-n_1)M} \\ &+j\lambda\sum_{n_1}\sum_{i_1\in A}a_{n_1,i_1}^{I}(-1)^{i_1}e^{-j\rho_i}j^{(i_1-i)} \\ &\times\sum_{l\in A_1}\tilde{H}_{l-iK-K\theta}^{*}H_{l-iK-K\theta}^{(n-n_1)M-2/M}\end{aligned} \tag{14}$$

Once the ptototype filter is designed very well then $\hat{a}_{n,i}^{\mathrm{Re}} = \mathrm{Re}\left[D_{n,i}^{(\mathrm{Re})}\right] \simeq a_{n,i}^{\mathrm{Re}}$ we can get follow

$$\sum_{p=0}^{KM-1}|H_P - K\theta|^2 = \frac{1}{KM}\varepsilon_g \tag{15}$$

On the ith subcarrier of FBMC system the Signal to Interference Ratio (SIR) is defined to be

$$SIR_I^{FBMC} = \frac{\left|\sum_{l\in A_1}\tilde{H}_{l-iK-K\theta}^{*}H_{l-iK-K\theta}\right|^2}{P_{self,\mathrm{Re}}^{FBMC}[i] + P_{self,\mathrm{Im}}^{FBMC}[i]} \tag{16}$$

4 Simulation Results

4.1 The Synthesis Filter Bank of FBMC System

To test the validity of the multitrate synthesis filter bank proposed in part 2 a simple two-channel filter bank system is considered and let $KM = R = L = 2$, and $\tilde{f}(k) = [f(k-1)f(k)]^T, \tilde{y}(k) = [y_1(k)\,y_2(k)]^T$, $\tilde{f}(k) + a_1\tilde{f}(k-1) + \cdots + a_p\tilde{f}(k-p) = v(k)$. The paraments of system are set to be as follows:

$$\bar{A} = \begin{bmatrix} 01 \\ 0.60.8 \end{bmatrix}, b = \begin{bmatrix} 0 \\ 1 \end{bmatrix}, \tilde{C} = \begin{bmatrix} 12 \\ 34 \end{bmatrix}$$

$\tilde{n}(k) = [n_1(k)\,n_2(k)]^T, v(k) \sim N[0, 0.25], n(k) \sim N[0, \sigma_1^2], n(k) \sim N[0, \sigma_2^2]$ $v(k)$, $n(k)$ are all the a zero-mean Gaussian white nose. The initial condition as follows:

$\tilde{f}(0) = [00]^T, P(0) = diag[\sigma_v^2, \sigma_v^2, \sigma_v^2, \sigma_v^2], \sigma_v^2 = 0.2$. Based on the above paraments the absolute errors are computed for the state estimation of the different methods in Table 1.

Table 1. The mean square errors (MSE) comparison under different noise situations

Results	Different noise	MSE of the original filter	MSE of the modified filter	Accuracy increase (%)
Situation 1	$\sigma_1^2 = 0.15$, $\sigma_2^2 = 0.25$	0.3800	0.2574	32.26
Situation 2	$\sigma_1^2 = 0.25$, $\sigma_2^2 = 0.35$	0.5138	0.3522	31.45
Situation 3	$\sigma_1^2 = 0.45$, $\sigma_2^2 = 0.35$	0.6835	0.4623	32.36

4.2 SIR of Receiver for FBMC System

According to designed receiver of FBMC system in part 3, the SIR expression is shown as in (18) and the receiver structure with FD CFO compensation is designed as the Fig. 2. The number of subcarrier is 512, a bandwidth is $1/T_s = 20\,\text{Mhz}$ and the value of the overlap parameter is $K = 4$. Figure 3 shows the different values of SIR for FBMC system with and without FD CFO compensation in downlink.

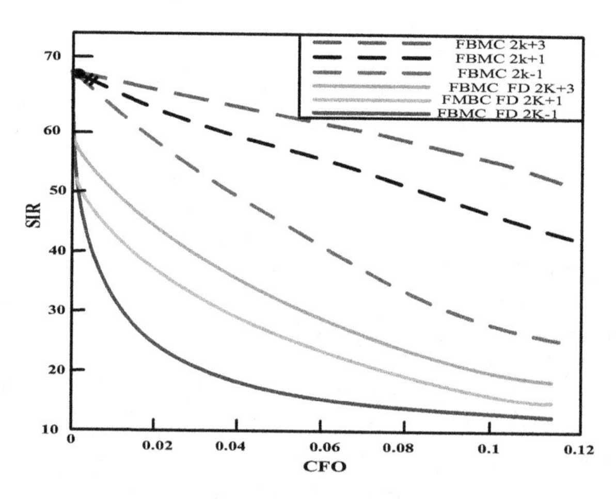

Fig. 3. SIR of FBMC receivers with and without FD CFO compensation in downlink

5 Conclusion

A synthesis filter bank of FBMC system is designed by dimension reducing method. The computed results show that high complexity of the prototype filter is not only reduced but the estimated accuracy of the modified filter is better than others.

The receiver of FBMC system is considered based on RDCT. The simulation results prove that the new designed structure of FBMC system is more effective and competitive than the original system without FD CFO compensation in downlink of FBMC system. In the future designed system will be apply into actual systems such as software defined radio system, multicarrier wireless communication and portable computing systems etc.

Acknowledgements. The above work is supported not only by Nature research fund for key project of Anhui higher education (No: KJ2016A455) but fund of Engineering Center of Bengbu University (No: BBXYGC2016B04).

References

1. Kang, X.A.S., Vigo, R.: Simulation analysis of prototype filter bank multicarrier cognitive radio under different performance parameters. J. Indonesian J. Electr. Eng Informatics **3**, 157–166 (2015)
2. Farhang-Boroujeny, V.B.: OFDM versus filter bank multicarrier. IEEE Signal Process. Mag. **28**(3), 92–112 (2011)
3. Wen, T., Q-b, Ge: Research on real-time block Kalman filtering estimation methods for the un-modeled system based on output measurement. Chin J Electron **40**(10), 1958–1964 (2012)
4. Li, P., Yan, Y., Guo, Y.: Modified IFIR-NPR design of prototype filter. J. of Chongqing Univ. Posts and Telecommunications (Natural Science Edition) **27**(2), 197–203 (2015)
5. Renfors, E.M., Harris, F.: Highly adjustable multirate digital filters based on fast convolution, in 20th European Conference on Circuit, Theory and Design (ECCTD), pp. 9–12, 2011
6. Wang, W., Li, D., Jiang, L.: Best linear unbiased estimation algorithm with Doppler measurements in spherical coordinates. J. Syst. Eng. Electron **27**(1), 128–139 (2016)

A Tool for IMA System Configuration Verification and Case Study

Lisong Wang(✉), Ying Zhou, Mingming Wang, and Jun Hu

Department of Computer Science and Technology University, Nanjing University of Aeronautics and Astronautics, Nanjing 211106, China
wangls@nuaa.edu.cn

Abstract. It is of a great importance for ensuring the correctness of system reconfiguration information and the satisfiability of partition time requirement in safety and reliability of critical systems such as integrated modular avionics (IMA). This paper considers a configuration information model transformation and verification approach and scheduling validation of IMA systems in the model-driven architecture with ARINC653 specification. Considering the features of IMA systems such as time or space multi-partition, this paper firstly defines a semantic mapping from the core elements of reconfiguration information (e.g. modules, partitions, memory, process and correspondence, etc.) to the MARTE model elements, and proposes a transformation approach between system configuration information and MARTE models. Then, design a scheduling validation framework of IMA partition system and then use MAST tool to make simulation for the MARTE model to verify the schedulability. Finally, a case study is illustrated to show the effectiveness of above proposed approach.

Keywords: Verification of system configuration information · MARTE · Model Driven Engineering · ARINC653 · Integrated Modular Avionics (IMA)

1 Introduction

Integrated Modular Avionics (IMA) [1] is a kind of typical complex embedded systems in the field of safety-critical applications, and can reconfigure the system according to the different operation requirements. Some other typical safety critical applications (such as: automotive electronics, air traffic control system command system etc.) similar features in complex control systems with IMA. Design and analysis for the IMA system has become one of the most important challenges of complex embedded system engineering research in recent years.

The configuration information in IMA system is usually up to ARINC653 standard [2]. It includes all the data information of the abstract level in system architecture, which can be used to configure the parameters of the IMA system such as hardware resources, the operating system interface and the environment of application; when the hardware and software on computing platform change, we can adjust the system configuration information so that existing applications can effectively correct running on the new platform. Therefore, how to ensure the correctness of the system

F. Xhafa et al. (eds.), *Advances in Intelligent Systems and Interactive Applications*, Advances in Intelligent Systems and Computing 686,
https://doi.org/10.1007/978-3-319-69096-4_35

reconfiguration information is an important problem in the current complex embedded system design field. The scheduling information of IMA is included in the system configuration information, schedulability verification analysis also has the important research significance.

Model Driven Engineering (MDE) [3] is the mainstream method in the field of system engineering and software engineering for the past ten years, its basic idea is centered on model design, model transformation, model analysis and verification, improving the ability and efficiency of development and maintenance of the complex engineering system. In the aviation software airworthiness standards for the latest version of the DO-178C has been officially proposed the requirement of model-based system development and formal methods [4–6]. MARTE [7, 8] (Modeling and Analysis of Real-Time and Embedded Systems) is a real-time system modeling and analysis of language published by Object Management Group (OMG), to support the modeling and analysis of functional modeling for design of complex embedded systems and widely exist in the time constraints, resource allocation and other non-functional properties.

In this paper, we design a tool for verifying resource allocation information for IMA system and perform a specific example to verify the correctness of this tool design. The remainder of the paper is organized as follows: In Sect. 2 we present a summary of the IMA system architecture, ARINC653 configuration information and MARTE modeling elements. In Sect. 3 we design and implements a prototype tool, CC653 (Configuration Checker for ARINC653), for modeling and verification of the ARINC653 configuration file, and uses it to analyze corresponding examples in Sect. 4. Finally, we draw conclusion and future work.

2 ARINC653 and MARTE

The IMA architecture of ARINC653 consists of several levels of structure: the application software layer, ARINC653 system interface layer, real-time operating system layer, hardware interface layer and hardware layer. The system functionality including partition management, process management, memory management, time management, interpartition communication, intrapartition communication and health monitoring.

The system configuration information is a very important part of ARINC653 system, including information and parameter configuration related to all of the above level. Configuration information usually includes the module level and partition level, respectively, to describe the interpartition and intraparition resource allocation [9] introduced the configuration information table in detail, including 13 modules and 16 partitions configuration table.

MARTE (Modeling and Analysis of Real-Time and Embedded Systems) is an industry-standard language the complements standard UML, refines standard UML concepts to provide their real-time interpretations. MARTE is a domain-specific modeling language for specifying and analyzing real-time and embedded software applications and systems. It is intended to replace the existing UML Profile for schedulability, performance and time. MARTE is structured around 3 main concerns: MARTE foundations, MARTE design model and MARTE analysis model.

3 Tool Design of IMA Configuration Information

The following will introduce the reassignment of prototype tool in the papyrus platform. The basic function is that user can edit the MARTE model which generated from the analysis of the configuration information, and to verify the correctness and schedulability of the configuration information.

3.1 Design Framework of the Tool

According to correctness and schedulability verification method of the ARINC653 system configuration information [10], we design a prototype verification system CC653 (Configuration Checker for ARINC653). Including the configuration information model extraction, configuration information correctness verification and partition schedulability verification, etc., the design framework as shown in Fig. 1:

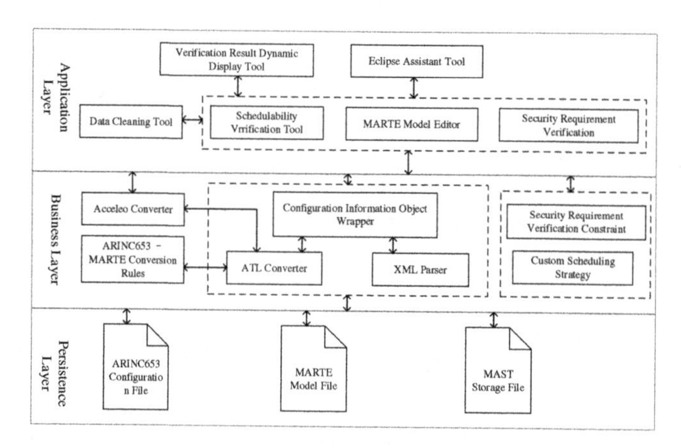

Fig. 1. The design framework of prototype tool

The CC653 tool is proceed from the software requirement design, the main design framework is divided into three levels: application layer, business layer and persistence layer.

(1) The application layer provides the functions like the MARTE model viewing and editing, scheduling verification, requirement correctness verification and verification results displaying. Data cleaning tool will judge whether the ARINC653 configuration file and the model file are available, and the data is complete, and will give the relevant error tips. The validation results dynamic displaying tool will show users the multi-dimensional validation results, such as scheduling simulation results, scheduling simulation Gantt chart.
(2) The business layer is processing the operation instruction come from the application layer, such as the conversion of the configuration information to the MARTE model, the data processing of the persistent layer, etc. And the business

layer also needs to provide scheduling strategies and verification documents to the application layer.

(3) The persistence layer manages the user data, such as reading the configuration file, accessing the MARTE file, generating the MAST-1 file, etc.

3.2 Execution Process of the Tool

Section 3.1 introduces the design framework of the prototype tool from the application layer, business layer and persistent layer; combined with the verification method, we will give the execution process of the prototype tool, as shown in Fig. 2.

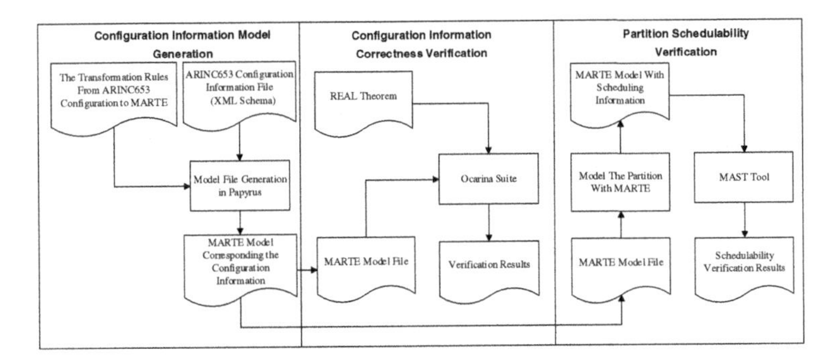

Fig. 2. The execution flow of prototype tool

(1) The module of configuration information model generation provides the functions like ARINC653 configuration information (XML format) editing and parsing, MARTE model generation, configuration information conversion rule management, verification function selection and other functions. The user can view the ARINC653 configuration information according to the integrated XML tool, and display and analyze the converted MARTE model.

(2) The module of configuration information correctness verification provides the functions like viewing REAL theorem, calling Ocarina model analysis suite to analyze and verify the MARTE model generated from the module of configuration information model generation.

(3) The module of partition schedulability verification provides the functions like modeling the scheduling information, the conversion of MARTE model to the text recognized by MAST, calling mast tool to judge the schedulability of the converted model.

To verify the correctness of ARINC653 system configuration information, first need to parse the configuration file, and save the converted MARTE model according to the transition rules; then construct REAL theorem through the configuration information requirement constraints, and import to Ocarina tools to verify with the MARTE model; then build model converted to text format identified by MAST tool for the scheduling information of task sets; Finally, output the verification results and the corresponding error message through the visual interface.

4 Example of Partition System Schedulability Verification

In this section we will give an example of modeling the partition system scheduling information with MARTE and the judgment of system schedulability, and introduce how to use MARTE to model the scheduling information of ARINC653 system with the help of custom scheduling strategy of MAST, and use MAST for the simulation of scheduling, according to the simulation results we know that the ARINC653 partition system is schedulable or not.

Table 1 describes the interpartition and intrapartition scheduling information of ARINC653 system in a time frame (10 ms). The system contains two partitions: Pr1 and Pr2, the scheduling strategies are DMS and RMS and the allocation time slice size is 6 and 4 respectively, each partition contains the task set parameters of tasks (T), period (P), execution time (ET) and deadline (DT).

Table 1. The scheduling information of ARINC653 system

Partition	Time slice	Clock offset	Period	Scheduling strategy	Parameters			
Pr1	6	0	10	DMS	T	P	ET	DT
					T1	10	3	10
					T2	5	1	5
Pr2	4	0	10	RMS	T	T	Ti	Td
					T3	20	2	20
					T4	10	2	10

According to the description of model transition described above, the MARTE model corresponding to the partition scheduling information in Table 1 is shown in Fig. 3, take this model as the input to the MAST tool, meanwhile, edit the custom scheduling strategies, the scheduling results is shown in Fig. 4.

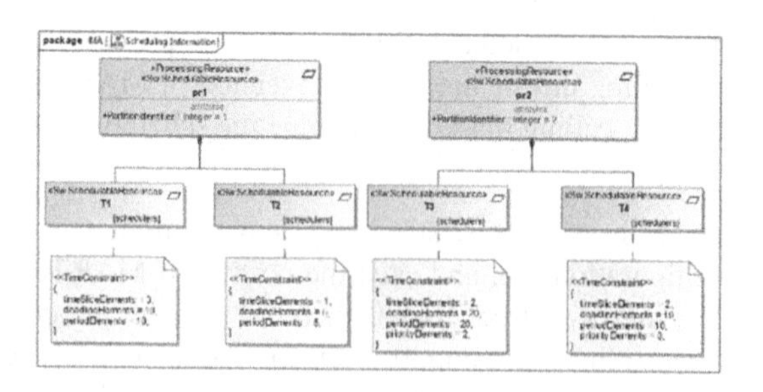

Fig. 3. Partition process scheduling of MARTE

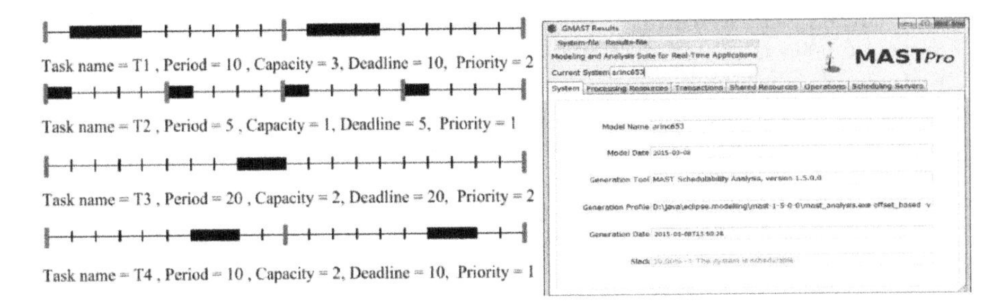

Fig. 4. The gantt chart and the result of partition process scheduling

The judgment results of schedulability divided into two parts, the first part is the scheduling Gantt of the task in the time of simulation as shown in the left side of Fig. 4, the second part is the determination result as shown in the right side of Fig. 4. A small lattice in scheduling Gantt represents a time slice, in this case the time slice size was 1 ms, the period of each partition obtained was 10 ms.

The right side of Fig. 4 shows the judgment results of the system schedulability, the results indicates that the system is schedulable and the system scheduling idle time accounted for 20% of the total scheduling time, namely in the scheduling process, 4 ms system time was in idle state.

5 Conclusion

According to correctness and schedulability verification method of the ARINC653 system configuration information, we design a prototype verification system. Its functions including the configuration information model extraction, configuration information correctness verification and partition schedulability verification, etc. Then we give an example of using MARTE to model the partition system scheduling information and system schedulability. We also describe how to use MARTE to model the scheduling information of ARINC653 system with the help of MAST's custom scheduling strategy, and use MAST simulation scheduling, according to the simulation results we know whether the ARINC653 partition system can be scheduled.

Acknowledgment. Supported by: The National Basic Research Program of China (973 Program) (2014CB744904 and No. 2014CB744901); Funding of Jiangsu Innovation Program for Graduate Education (SJZZ16_0062), the Fundamental Research Funds for the Central Universities.

References

1. Watkins, C.B., Walter, R.: Transitioning from federated avionics architectures to Integrated Modular Avionics. In: Digital Avionics Systems Conference, DASC'07. IEEE/AIAA 26th (2007)

2. Prisaznuk, P.J.: ARINC 653 role in Integrated Modular Avionics (IMA). In: Digital Avionics Systems Conference, DASC 2008. IEEE/AIAA 27th. 1. E. 5-1-1. E. 5-10 (2008)
3. Rutle, A., et al.: A formal approach to the specification and transformation of constraints in MDE. J. Logic Algebraic Program. **81**(4), 422–457 (2012)
4. Rierson, L.: Developing Safety-Critical Software: A Practical Guide for Aviation Software and DO-178c Compliance. CRC Press, Boca Raton (2013)
5. Rushby, J.: New challenges in certification for aircraft software. In: Proceedings of the ninth ACM international conference on Embedded software. ACM, New York (2011)
6. Moy, Y., et al.: Testing or formal verification: Do-178c alternatives and industrial experience. IEEE Softw **30**(3), 50–57 (2013)
7. Graf, S., et al.: Modeling and analysis of real-time and embedded systems. In: Satellite Events at the MoDELS 2005 Conference. Springer, Heidelberg (2006)
8. Object Management Group (OMG): Modeling and Analysis of Real-time and Embedded Systems Specification version 1.1 formal-11-06-02 (2011)
9. Ott, A.: System Testing in the Avionics Domain (2007)
10. Hu, J., Ma, J., et al.: Model-driven reconfiguration information verification for safety-critical systems. J. Frontiers Comput. Sci. Technol. **9**(4), 385–402 (2015)

Using Deep ConvNet for Robust 1D Barcode Detection

Jianjun Li[1(✉)], Qiang Zhao[1], Xu Tan[2], Zhenxing Luo[3], and Zhuo Tang[3]

[1] School of Computer Science and Technology, Hangzhou Dianzi University, Hangzhou, China
jianjun.li@hdu.edu.cn
[2] Institute of Mechanical Engineering, Zhejiang University, Hangzhou, China
[3] The 36th Research Institute of China Electronics Technology Group Corporation, Jiaxing, China

Abstract. Barcode has been widely adopted in many aspects, it is the unique identification and contains important information of goods. Regular barcode scanning device usually requires human being ¡¯s aids and is not suitable for multiple barcode scanning, especially in a complex background. In this paper, a cascaded strategy is proposed for accurate detection of 1D barcode with deep convolutional neural network. The work contains three parts: Firstly, a faster Region based Convolutional Neural Net (Faster R-CNN) framework is used to train a barcode detection model. Secondly, a powerful lo-level detector called Maximally Stable Extremal Regions (MSERs) is developed to eliminate the background noisy and detect the direction of the barcode. Thirdly, a postprocessing with like bilateral filter, called Adaptive Manifold (AM) filter, is applied when the image is blurred. We have carried out experiments on both Muenster Barcode Database and ArTe-Lab Barcode Database and compared with the previous barcode detection methods, the result shows that our method not only can get a higher barcode detection rate but also more robustness.

Keywords: Barcode detection · MSER · AM Filter · CNN · RCNN · Faster R-CNN

1 Introduction

Over the past few years, with the sharply increase in the number of smart phones, more and more people began to use the phone to scan the barcode on goods or other items. When we use the smart phone to scan the barcode, we need to close to the barcode with camera lens and trying to keep the camera not shaking. In the above scenes, all need to manually look for the location of the barcode and manually adjust the distance to get a clear picture of the barcode. However, this method is only "read" the barcode, and not really "Detection-Identification Barcode".

There have many methods proposed to solve the problem so far. Katona and Nyl [1] uses the method of combining mathematical morphology operation to detect barcode. This method can detect various types of 1D barcode and 2D barcode. Zamberletti

F. Xhafa et al. (eds.), *Advances in Intelligent Systems and Interactive Applications*, Advances in Intelligent Systems and Computing 686,
https://doi.org/10.1007/978-3-319-69096-4_36

et al. [2] proposed a method of using Hough transform and MLP combined to detect barcode. Wang et al. [3] uses the LBP algorithm to obtain the feature of barcode and background, then use SVM machine learning technology to establish the classification model and predict the image patches. Creusot and Munawar [4] uses the MSER algorithm can detect the relatively stable and invariant region of the image to detect the barcode area. However, the robustness of these methods are not strong enough when the image quality is not good.

In this paper, we propose a method combined the deep ConvNet with the traditional image processing method to detect barcode. We have carried out experiments on both Muenster and ArTe-Lab Barcode Datasets and compared with the previous methods, we find our method has better robustness and better performance of barcode detection compared to previous methods.

2 Related Works

2.1 Faster R-CNN Object Detection

In the field of traditional object detection, artificial design features such as HOG, SIFT, etc. are common methods. But over the last two years, deep learning method based on R-CNN [5] has become one of the main development direction of object detection. Faster R-CNN [6] is the newest version of R-CNN. The main steps of it can divided into 3 steps: 1. feature extraction; 2. RPN candidate region generation; 3. object classification and regression. It highly reduced the running time and enhanced the mAP (mean Average Precision) compared to R-CNN and Fast R-CNN.

2.2 AM Filtering

High-dimensional filtering has received significant attention in image processing, denoising and so on. The most popular one is the bilateral filter. In image denoising, the traditional Gauss filter only consider the distance between pixels, so the filtering result in some fuzzy edge details of the image. High-dimensional filtering can overcome the disadvantages of the Gauss filtering. It can hold the details information while smoothing the image. Gastal and Oliveira [7] gives a more time efficient real-time high-dimensional filter: Adaptive Manifold Filter which is a kind of bilateral filter. It has better filtering effect than the bilateral filter and higher efficiency.

When using MSER to detect the barcode in the image. The detection result is not satisfactory when the image quality is poor or slight fuzzy. Therefore, we consider the image enhancement with the AM filter before using the MSER algorithm. The enhancement effect comparison of original images and MSER detection result are shown in Figs. 1 and 2. It can obviously see the effect of our enhance method.

Fig. 1. AM filtering enhancement comparison

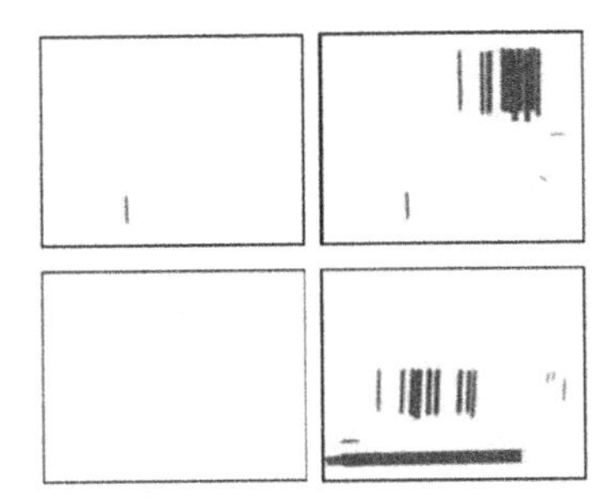

Fig. 2. MSER detection results comparison

3 Proposed Method

The detailed description is illustrated as the following Fig. 3. At first, an image is input to the MSER processor for barcode orientation detection as mentioned in Sect. 2.2. For those blurred images, the AM filter is used for enhancement. Furthermore, the image with an angle correction is fed into the deep ConvNet for barcode detection.

3.1 MSER Orientation Correction

We proposed a method to detect the barcode angle, according to the number of detected MSERs to judge whether the angle of the barcode is detected. If the number of detected MSERs are too small, such as the detection result of original image in Fig. 2, the image enhancement will be used. This can achieved for the adaptive enhancement of fuzzy image processing.

In order to accurately verify the effectiveness of our method, we processed the contrast experiment on the Muenster and the ArTe-Lab Barcode Dataset (extended version). They are 1D Barcode Datasets and widely used in testing the barcode detection accuracy, the details about the two datasets are shown in Table 1.

Using the MSER detection to the original and enhanced image in the two datasets. We statistics the angle detection accuracy in Table 2. It shows our method is very robust to the barcode images with low quality in angle detection.

The angle detection accuracy in different threshold compared with the previous method was shown in Fig. 4.

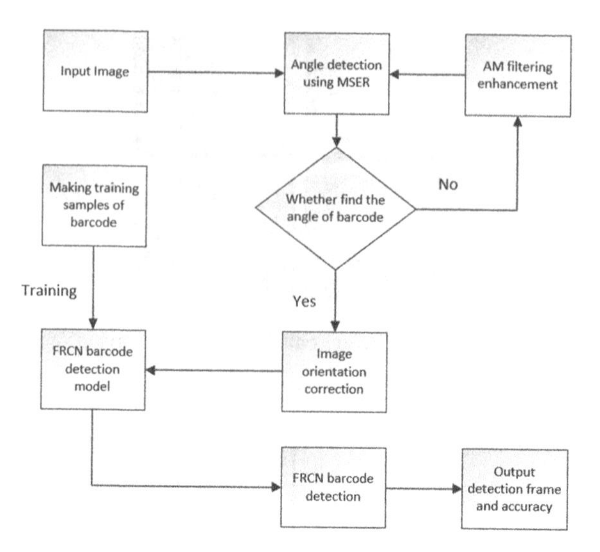

Fig. 3. Barcode detection procedure

Table 1. 1D barcode datasets

Datasets parameter	Muenster [8]	ArTe-Lab [9]
Numbers	1055	365
Dimensions	640 * 480	640 * 480
Device	Nokia N95	Nokia 5800 + others
Quality	Good	A little worse

Table 2. Barcode angle detection accuracy contrast

Compares datasets	Before enhance	After enhance	Increase rate (%)
Muenster	1014	1047	3.3
ArTe-Lab	330	360	9.1

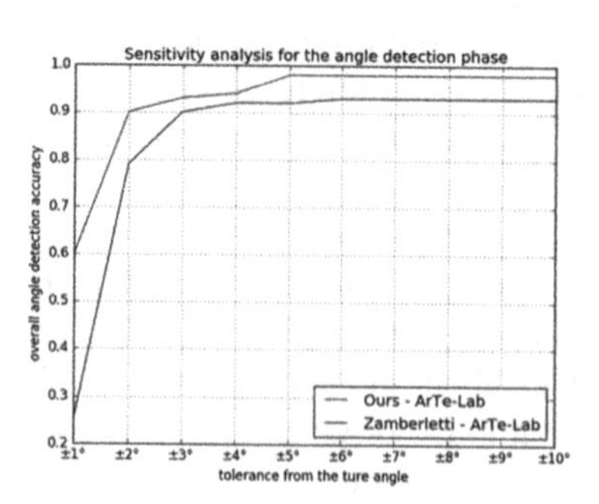

Fig. 4. The angle detection accuracy

Therefore, according to the angle of barcode to achieve the orientation correction which making the direction of the barcode to maintain the horizontal or vertical. It is very helpful to the later barcode detection.

3.2 Train Model and Barcode Detection Test

For the barcode detection, we proposed a method using the deep ConvNet framework to accomplish this task. As mentioned in the previous section, we use the Faster R-CNN detection framework for barcode detection, As shown in the Fig. 3, We train a detection model with the barcode datasets, after train finished, we use it to complete the experiment test.

4 Experiments and Results Analysis

4.1 Experiments Details

In experiments, we take 1000 images from the Muenster Barcode Database for training, then making the annotation according to format of VOC2007.

Due to the constraints of GPU memory size, the network used in our experiments is Z-F network. For the same reason, the batch size we set to 128, and momentum to 0.9. For the learning rate, the initial learning rate is set to 0.0001 due to the less training samples.

In order to make a better comparison with the previous methods, we use the same evaluation method as [2, 4].

4.2 Results and Analysis

As we can see, the results of three of all four methods: Clement, Faster R-CNN and our method, are very good on the Muenster dataset. We find the reasons as: Firstly, the image of Muenster Dataset are clear and have good quality. The deep convolutional network can learn better features of barcode. Secondly, the orientation of barcode in the Muenster Dataset are almost either horizontal or vertical, they do not need the angle correction. An additional test is also implemented on the ArTe-Lab Barcode Dataset. The test result is shown as Table 3 too.

Table 3. Detection result on Muenster and Arte-Lab datasets

Methods datasets	Zamberletti (%)	Clement (%)	FRCN (%)	Ours (%)
Muenster	82.9	96.3	98.1	99.4
ArTe-Lab	80.5	89.3	94.5	98.9

The complete detection profile can be seen for these methods on ArTe-Lab dataset in Fig. 5.

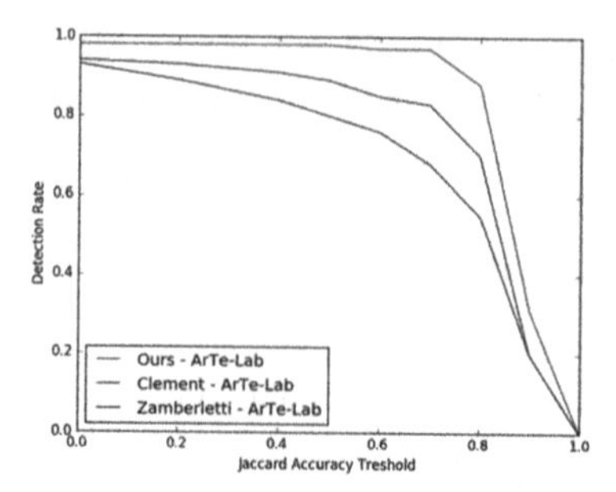

Fig. 5. Comparison of detection rates

As can be seen from Table 3, only the FRCN method is able to improve the detection accuracy up to 5.2% compared with the MSER method. It fully shows that the depth of convolutional network can learn image features much better than the traditional method. However, there are multiple directions of barcodes included in the ArTe-Lab dataset. Therefore, the ground truth can not cover the bounding box of barcodes very well as shown in Fig. 6. According to the evaluation criteria mentioned earlier, it will lead to a decline in the final detection accuracy.

Fig. 6. Detection without angle correction

Compared only using the FRCN method, our method improves the detection accuracy about 4.4%. From the previous analysis, we know that the bounding box is not able to cover the real area of barcode very well when only using the FRCN method. Therefore, it is better to correct image orientation according to the barcode orientation and then the detection bounding box can surround the ground truth very well. As shown in Fig. 7, compared to only using the FRCN method, the detection bounding boxes of our method fits the ground truth very well and the detection effect has a significant improvement.

Fig. 7. Detection result compared on ArTe-Lab

5 Conclusion

In this paper, we present an effective method to detect barcode location by combining the traditional image processing methods and the faster R-CNN deep learning framework. We prove the effectiveness and robustness of the proposed approach by using the Muenster and ArTe-Lab Barcode Dataset. The proposed method is able to precisely detect the barcode location even though there rotation or blurred happen in the image. The experimental results show that our method can achieve around 4% higher detection accuracy compared with the previous barcode detection methods. The proposed method also has a better robustness in a complex background and suitable for state of art using.

References

1. Katona, M., Nyl, L.: Efficient 1d and 2d barcode detection using mathematical morphology. In: Mathematical Morphology and Its Applications to Signal and Image Processing, LNCS, vol. 7883, pp. 464–475. Springer, Heidelberg (2013)
2. Zamberletti, A., Gallo, I., Albertini, S., et al.: Neural 1D barcode detection using the Hough transform. IMT **10**, 157–165 (2015)
3. Wang, Z., Chen, A., Li, J., et al.: 1D barcode region detection based on the hough transform and support vector machine. In: International Conference on Multimedia Modeling, pp. 79–90. Springer International Publishing, Heidelberg (2016)
4. Creusot, C., Munawar, A.: Real-time barcode detection in the wild. In: IEEE Winter Conference on Applications of Computer Vision, pp. 239–245. IEEE (2015)
5. Girshick, R.B., Donahue, J., Darrell, T., Malik, J.: Rich feature hierarchies for accurate object detection and semantic segmentation. In: IEEE Conference on Computer Vision and Pattern Recognition, CVPR 2014 (2014)
6. Ren, S., He, K., Girshick, R., Sun, J.: Faster R-CNN: towards real-time object detection with region proposal networks. In: NIPS (2015)
7. Gastal, E.S.L., Oliveira, M.M.: Adaptive Manifolds for Real-Time High-Dimensional Filtering, vol. 31, p. 33. ACM, New York (2012)
8. Wachenfeld, S., Terlunen, S., Jiang, X.: Robust recognition of 1-d barcodes using camera phones. In: Proceedings of 19th International Conference on Pattern Recognition (ICPR 2008), pp. 1–4 (2008)
9. Zamberletti, A., Gallo, I., Albertini, S.: Robust angle invariant 1d barcode detection. In: 2nd IAPR Asian Conference on Pattern Recognition (ACPR), pp. 160–164, November 2013

Study of Agricultural Machinery Operating System Based on Beidou Satellite Navigation System

Fanwen Meng[1,2,3,4](✉), Dongkai Yang[1,3,4], Youquan Wang[1,2,3,4], and Yuxiang Zhang[2]

[1] Shandong Orientation Electronic Technology Co., Ltd, Shandong, China
fanwenmeng@163.com
[2] Mechanical and Electrical Engineering Department, Ji'ning Polytechnic, Ji'ning, China
[3] Electronic and Information Engineering College, Beihang University, Beijing, China
[4] Post-doctoral Work Station of Innovation Center of Ji'ning Hi-Tech District, Ji'ning, China

Abstract. Real-time positioning system, circuit system, data transferring and remote monitor and control used for farmland operation are studied with Beidou satellite navigation system signal, and Locomotive operation management system used for crop harvesting is designed. Precise Beidou satellite positioning system, accurate area calculation method, Intelligent mobile phone APP software measurement are integrated in the system. Farm area, soil temperature and humidity and locomotive speed can be measured in real time. It also has the function of locomotive management.

Keywords: Beidou navigation system · Agricultural machinery operation · Upper software · Mobile phone APP software

1 Introduction

Beidou satellite navigation system is developed by our country, and it is the regional navigation satellite system with self-owned intellectual property and autonomous control [1, 2]. Beidou satellite navigation system has the functions of quick positioning, short message communication and precision timing, and it can supply full-time, high-accuracy and quick positioning service. User terminal has the communication capacity of Bidirectional digital message. Short message including 120 Chinese characters can be transferred once. The system also supplies 20 ns time synchronization precision. Beidou satellite navigation system works steadily, and it is widely used in vehicle positioning, rescue detection, data collecting, topographical reconnaissance, fire prevention, goods transportation, water level detection, etc [3–7].

With the gradual construction of Beidou satellite navigation system and the implementation of major national science and technology projects. The application of Beidou satellite navigation system enters a rapid development period. For Agricultural

F. Xhafa et al. (eds.), *Advances in Intelligent Systems and Interactive Applications*, Advances in Intelligent Systems and Computing 686,
https://doi.org/10.1007/978-3-319-69096-4_37

machinery operation, the area of machinery operation is a key factor, especially in cross-regional operation of grain combine harvester. In fine agriculture, operation area is the basic parameter, which decides the input amount of seeds, pesticide, chemical fertilizer, and it is also the basis of work quantity and charge. The project is from engineering practice and practical needs. It can adjust the computational method of locomotive working area according to locomotive trace, which forms locomotive working system that can position precisely, calculate real-time and get a high precision. It can supply theoretical basis and implementation method for calculating wording area, monitoring locomotive speed and soil tillage information.

2 System Design

Locomotive working system is composed of front monitoring point, application service platform and client terminals. Front monitoring point is composed of sensors and Beidou navigation chips. It can be used as portable set, and it can be connected with control center with wireless transferring. The platform receives the data and processes, then stores the data into the local special database. The data is used by the GIS to provide visible data management and application analysis. It combines with farm operation expert system to calculate farm area. Mobile phone APP software is developed to realize data collecting and farm area display (Fig. 1).

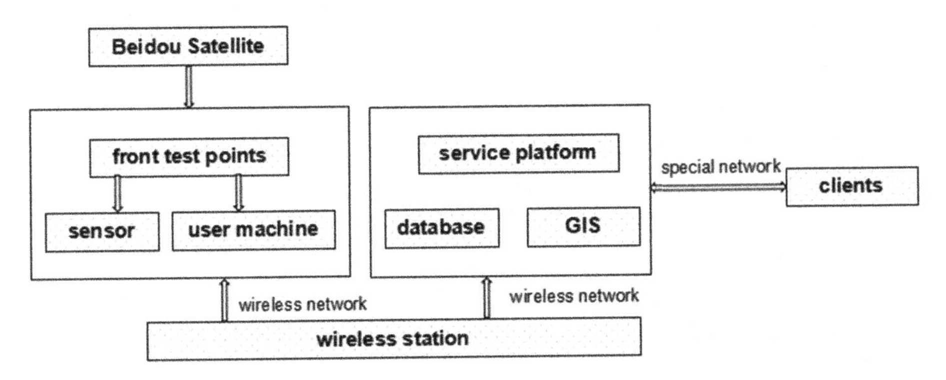

Fig. 1. General design frame of the system

3 Hardware Design

3.1 Beidou Chip VK2525M3G5

Beidou chip VK2525M3G5 produced by Shenzhen V.KEL Communication Equipment C0., LTD is used in the system. The chip has 25 * 25 * 4 mm bd2 antenna and UART/TTL interface that meets the requirement of the occupation standard. High precision TCXO with KDS 0.5 PPM is used, RTC crystal and hot start quicker than picofarad capacitance are embedded. LNA and low noise amplifier are embedded in the chip. Pins function and parameters description of VK2525M3G5 chip are displayed in the Tables 1 and 2.

Table 1. VK2525M3G5 chip pins function

Pin name function	PPS	VCC	TXDA	RXDA	GND
	Time standard pulse output	System power, voltage: 3.3–5 V	UARR interface	UART interface	Ground

Table 2. VK2525M3G5 parameters description

Parameters	Description
Receiving band	GPS, GALILEO, QZSS: L1 1575.42 MHz, BEIDOU: B1 1561.098 MHz
Receiving channel	Support 99 channels (33 Tracking, 99 Acquisition)
Positioning property	<10 m (Independent positioning; <5 m (WAAS)
Velocity	<0.1 m/S
Direction	<0.5°
Timing precision	30 uS
Reference coordinate system	WGS-84
Maximum altitude	50,000 m
Maximum travel speed	500 m/s
Accelerated speed	<4 g
Tracking sensitivity	−165 dBm
Acquisition sensitivity	−148 dBm
Start time	30 S
Operation temperature	−40°–80°
Standard clock pulse	0.25 Hz–1 KHz
Location update rate	1–10 Hz

3.2 Beidou Chip VK2525M3G5

SCM AT98C52 is utilized to realize data collecting and data communication of locomotive operation [8]. TTL level is used in SCM, but Beidou chip VK2525M3G5 uses RS-232 level.Two kinds of levels are not compatible, so MAX232 chip is used to realize the transformation of different levels. Two serial ports, COM1 and COM2 are used in the system. COM1 is used to connect Beidou chip VK2525M3G5 to receive satellite information. It is connected with SCM. COM2 is utilized to realize the connection of SCM and computer, and it is used for debugging and the collecting of operation area, temperature and soil humidity. Level transformation and interface circuit with MAX232 chip is shown as Fig. 2.

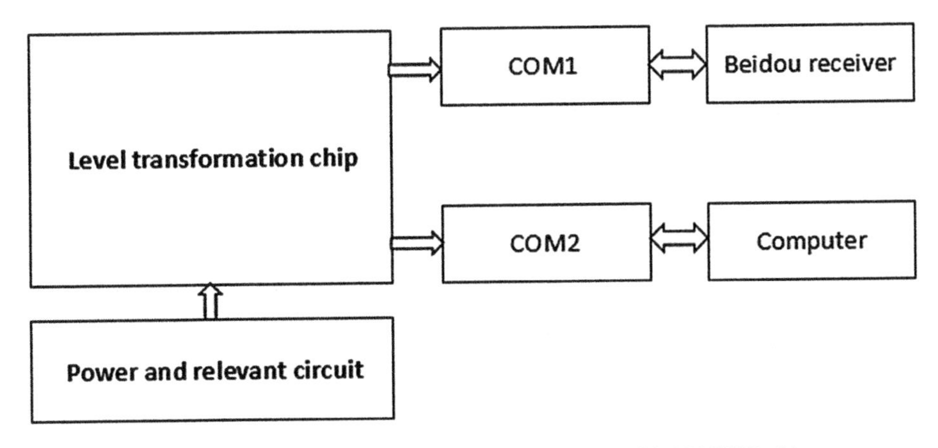

Fig. 2. Level transformation and interface circuit with MAX232 chip

4 Upper Computer Software Design and Mobile Phone APP Design

4.1 Upper Computer Software Design

Upper computer software is designed by high-level language to realize the communication with lower SCM. And the locomotive operation area, locomotive speed, temperature, soil humidity and illumination can be displayed real-time. The system also has the function of add/remove locomotives to implement the management and monitoring. Upper computer monitoring is shown as Fig. 3.

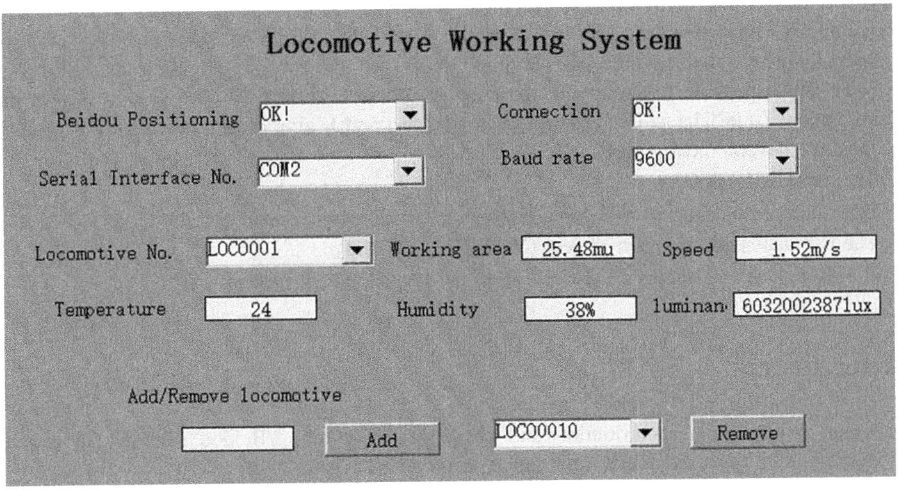

Fig. 3. Upper computer monitoring interface of the system

4.2 Mobile Phone APP Software

Operation area is usually calculated by Green formula, and double surface integral calculation is replaced by integral calculation on boundary curve. For the area calculation with regular shape, locomotive width can be input, and the locomotive working length can be measured by the Beidou satellite. Finally the filed area can be worked out by the formula. Intelligent mobile phone APP software is designed, and the locomotive operation area is calculated and displayed by satellite positioning system and mobile wireless network. The software has the function of searching blue teeth, starting measurement and displaying measuring results. By inputting the locomotive width, the working area can be calculated out by the software automatically.

5 Conclusions

Farm area, locomotive speed can be obtained by Beidou positioning system, GIS, remote sensing technique, and locomotive operation management system are designed used for crops harvest. Locomotive working system based on Beidou system can improve working efficiency and quality, reduce operating cost. It has the function of recording locomotive speed and displaying working area, which is convenient for operation and information management. If the system is used for agriculture harvest in the future, it can improve the intelligence of the operation, reduce human resource investment, and make better economic results to provide theory guidance of the farm operation system. Following-up work includes database development of locomotive operation and upper software optimization.

Acknowledgments. My deepest gratitude goes first and foremost to Professor Yang Dongkai, my supervisor, for his constant encouragement and guidance. He has walked me through all the stages of the writing of this paper. Second, I would like to express my thanks to Dr. Wang Youquan, who has helped me a lot in information knowledge. Third, my thanks would go to my beloved family for their loving considerations. I also owe my sincere gratitude to the following projects which provide fund support, they are Postdoctoral fund project of Ji'ning High-tech Development Zone, The national spark program project and Ji'ning Industry-Academy-Research cooperative development project.

The paper is funded by:

Postdoctoral fund project of Ji'ning High-tech Development Zone;

The national spark program project (No:2012GA740103);

Ji'ning Industry-Academy-Research cooperative development project.

References

1. Yang, Y.: Progress, contribution and challenges of compass/Beidou satellite navigation system. Acta Geodaetica et Cartographica Sinica **39**(1), 1–6 (2010)
2. Zhang, L., Yuan, B.: Algorithm research and implementation of GNSS multi-constellation navigation positioning. Acta Geodaetica et Cartographica Sinica **35**(5), 195–197 (2012)

3. Qiang, L., Yongkui, L.: The GPS navigation technology of farm machine in China. J. Agric. Mechanization Res. **31**(8), 242–244 (2009)
4. Weidong, Z., Chao, Y., Chun, W.: Agriculture machinery operation guidance system based on GPS and GIS. J. Heilongjiang August First Land Reclam. Univ. (5), 32–34
5. Dekui, X., Caojun, H., Tingting, D.: Farm operations locomotive condition monitoring system design. J. Agric. Mechanization Res. (9), 122–125 (2015)
6. Wei, X., Dan, Z., Sun, H.: Development of vehicular embedded information processing system for map-based precision farming. Trans Chin. Soc. Agric. Eng. **29**(6), 142–148 (2013)
7. Xia, Z., Gu, C., Li, B.: Research and implementation of wireless sensor network in precision agriculture. Mech. Electr. Eng. Mag. **32**(3), 439–442 (2015)
8. Zhai, X., Wang, B., Fan, M.: Programming of serial communication based on visual C#. Electron. Sci. Technol. **24**(2), 24–26 (2011)

Detection Algorithm of Slow Radial Velocity Ship based on Non-Negative Matrix Factorization by Over-The-Horizon Radar

Hui Xiao[1(✉)], Jun Yuan[2], Shaoying Shi[1], and Runhua Liu[1]

[1] Air Force Early Warning Academy, Wuhan 430019, China
xiaohui_nwpu@163.com
[2] Naval University of Engineering, Wuhan 430033, China

Abstract. In order to detect radial velocity ship in complex sea state, a novel detection algorithm is proposed. Considering that the frequencies of the ship and first-order sea clutters meet the requirements of sparse in the Doppler domain, certain sparseness constraints are imposed on Non-negative Matrix Factorization (NMF) to improve the efficiency of extracting the frequency spectrum of ship from that of the sea clutter. The following two topics are discussed: (1) the detection algorithm based on NMF (NMFSCC); (2) the detection results of NMFSCC, compared with Singular Value Decomposition (SVD) algorithm. Simulation results demonstrate the usefulness of the proposed approach.

Keywords: Detection algorithm · Non-negative matrix factorization · Sparseness constraints · Radial velocity ship

1 Introduction

Over-the-horizon Radar (OTHR) is based on sky wave transmitting and surface wave receiving. Therefore, OTHR has a wide coverage of detection and no range blind zone. However, due to the complex sea state, the two strongest Bragg peaks always have Doppler shift. And a short Fourier transform cannot provide enough resolution to detect ship with slow radial velocity from the Bragg peak nearby.

Clutter suppression method is useful and widely used to solve this problem. And these methods can be divided into three types. The first type is the characteristic decomposition methods such as EVD [1] which needs to judge the clutter subspace dimension accurately. The second type is the sea clutter cancellation methods such as block matrix suppression method [2]. These algorithms need to estimate noise parameters (frequency, amplitude and initial equivalent) and to determine the cancellation number or cancellation in area to avoid the loss of the ship target energy. The third type is the singular value decomposition (SVD) [3, 4]. This algorithm needs to determine the clutter subspace dimension or the number of singular values. Unfortunately, the determination is always not accurately.

F. Xhafa et al. (eds.), *Advances in Intelligent Systems and Interactive Applications*, Advances in Intelligent Systems and Computing 686,
https://doi.org/10.1007/978-3-319-69096-4_38

Due to the frequencies of the ship and first-order sea clutters meet the requirement of sparse in the Doppler domain, we present the NMFSCC algorithm to extract frequency spectrum of ship from the echo spectrum. In Sect. 3, we present NMFSCC algorithm. In Sect. 4, we present the simulation results, compared with SVD algorithm. Concluding remarks are given in Sect. 5.

2 The Model of Sea Clutter [2] and NMF [5]

Based on sea clutter reflectivity mechanism, the frequencies of the two strongest Bragg peaks can be expressed as

$$f_{\mathrm{B}} = \pm 0.102\sqrt{f_0 \cos\alpha} \tag{1}$$

where f_0 is frequency of radar signal, α is radar grazing angle. Consider the affection of sea breeze and tide, the Doppler frequency of Bragg peaks has frequency shift f_c. So the measured value can be written as $f_{\mathrm{B}} + f_c$.

In HF sea echo, the two strongest Bragg peaks can be described with sine signals. And the high order sea clutter is a set of sine signal. Consider the observed data sequence of CIT can be expressed as $\boldsymbol{X} = \left[x(t-(m-1)\Delta_t) \quad \cdots \quad x(t-\Delta_t) \quad x(t)\right]^{\mathrm{T}}$, where m is the number of coherent integration, Δ_t is sampling interval, and $[\bullet]^{\mathrm{T}}$ denotes matrix transpose. If different order of sea clutter is described with $f_1, f_2, \cdots, f_r$ and the Doppler frequency of signal is f_T, then sea echo by OTHR can be expressed as

$$\boldsymbol{X}_{m\times n} = \begin{bmatrix} 1 & \cdots & 1 & 1 \\ e^{j2\pi f_1} & \cdots & e^{j2\pi f_r} & e^{j2\pi f_T} \\ \vdots & \ddots & \vdots & \vdots \\ e^{j(m-1)2\pi f_1} & \cdots & e^{j(m-1)2\pi f_r} & e^{j(m-1)2\pi f_T} \end{bmatrix} \begin{bmatrix} c_1 \\ \vdots \\ c_r \\ s_T \end{bmatrix} + \boldsymbol{N}_{m\times n} \tag{2}$$

where $c_1, \cdots, c_r$ is amplitude of sea clutter, s_T is amplitude of signal, $\boldsymbol{N}$ denotes white noise and n is sampling number.

NMF is always used in processing high-dimensional data. It aims to find two nonnegative matrixes $\boldsymbol{U} \in \boldsymbol{R}^{m\times k}$ and $\boldsymbol{V} \in \boldsymbol{R}^{n\times k}$, where $k \le mn/(m+n)$. Then we can obtain the objective function

$$O_F = \left\|\boldsymbol{X}' - \boldsymbol{U}\boldsymbol{V}^{\mathrm{T}}\right\|^2 \tag{3}$$

where $\|\bullet\|_{\mathrm{F}}$ is Frobenius norm. And $\boldsymbol{X}' \ge 0$, $\boldsymbol{U} \ge 0$, $\boldsymbol{V} \ge 0$.

When $\boldsymbol{U}\boldsymbol{V}^{\mathrm{T}} = \boldsymbol{X}'$, the objective function gets the local optimal solution. And the iterative formula can be expressed as

$$u_{ik} \leftarrow u_{ik}\frac{(\boldsymbol{X}\boldsymbol{V})_{ik}}{\left(\boldsymbol{U}\boldsymbol{V}^{\mathrm{T}}\boldsymbol{V}\right)_{ik}} \tag{4}$$

$$v_{ik} \leftarrow v_{ik} \frac{\left(X^{\mathrm{T}} V\right)_{ik}}{\left(V^{\mathrm{T}} V U^{\mathrm{T}}\right)_{ik}} \tag{5}$$

where $\boldsymbol{U} = [u_{ik}]$, $\boldsymbol{V} = [v_{ik}]$.

3 NMFSCC Algorithm

To meet the requirement $\boldsymbol{X}' \geq 0$, we make FFT on echo data matrix $\boldsymbol{X}_{m \times n}$. Then $\boldsymbol{X}'$ can be factorized into non-negative basis matrix $\boldsymbol{U}$ and non-negative coefficient matrix $\boldsymbol{V}$. However, due to complex sea state, the dimension of frequency spectrum is high, as shown in Fig. 1.

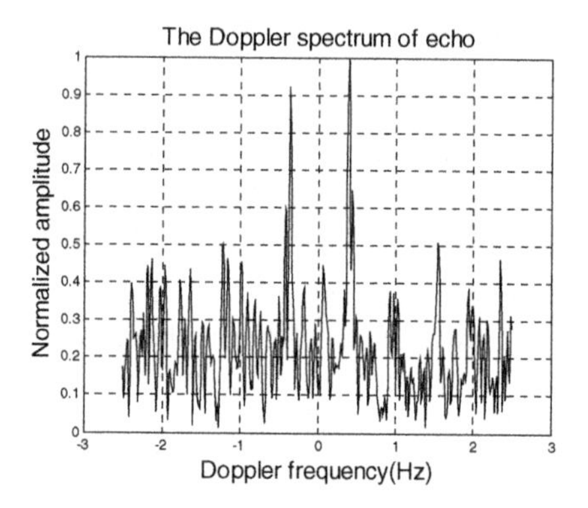

Fig. 1. The Doppler spectrum of echo

In Fig. 1, the first-order Bragg peaks appear at −0.39 and 0.35 Hz respectively. The target Doppler frequency is 0.4 Hz. And the Fourier transform cannot provide enough resolution to detect ship with slow radial velocity from the Bragg peak nearby. At the same time, the spectrum peak of noise affects the detection of target. So the computation speed and spectrum sensing accuracy of NMF is unstable.

Considering that the frequencies of the ship and first-order sea clutters meet the requirements of sparse in the Doppler domain, we impose the certain sparseness constraints on both of the basis matrix and coefficient matrix. And the sparse factor can be written as

$$sparseness(\mathbf{y}) = \frac{\sqrt{p} - \left(\sum |y_i|\right) / \sqrt{\sum y_i^2}}{\sqrt{p} - 1} \tag{6}$$

where p is the dimension of vector $\boldsymbol{y}$ and $0 \leq sparseness(\boldsymbol{y}) \leq 1$.

Then the iterative formula can be rewritten as

$$U = U - \mu_U (UV - X) V^{\mathrm{T}} \tag{7}$$

$$V = V - \mu_V U^{\mathrm{T}} (UV - X) \tag{8}$$

4 Simulation

To evaluate the detection performance of NMFSCC algorithm, the following scenario is generated: the target Doppler frequency is 0.4 Hz; the first-order Bragg peak appears at −0.39 and 0.35 Hz respectively due to complex situation; noise is white noise; the sampling number is 128. And the scenario is the same with that of Fig. 1.

Figure 2 shows the detection result of NMFSCC algorithm, compared with the performance of SVD algorithm. From Fig. 2(a), due to complex sea state, the Doppler frequency of target using SVD is almost submerged by the noise. The peak appears at 0.46 Hz and the error is 0.06Hz. While, the frequency detected by NMFSCC is 0.362 Hz. The accuracy of NMFSCC is higher.

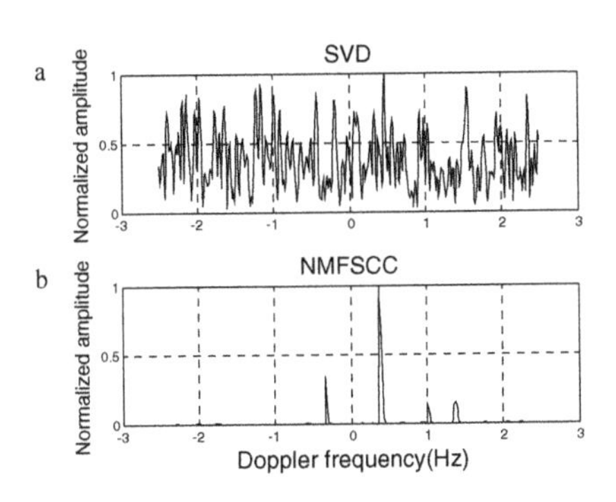

Fig. 2. a Normalized spectrum using SVD algorithm; b Normalized spectrum using NMFSCC algorithm

Figure 3 shows the spectrum of target and two first-order Bragg peak using NMFSCC algorithm. And Fig. 3(c) is exactly the same as Fig. 2(b). Figure 4 is the detail spectrums corresponding to Fig. 3.

In Fig. 3, it shows that NMFSCC algorithm can extract three peaks from the noise with a small residue. Figure 4 shows the Doppler frequencies of two first-order Bragg peaks detected by NMFSCC are −0.38 and 0.352 Hz. So NMFSCC algorithm can provide not only the efficient detection of target but also useful prior information of sea clutter to the process of cell under test.

Figure 5 shows the detection result of NMFSCC algorithm, with sampling number 512. Figure 6 is the detail spectrums corresponding to Fig. 5.

In Fig. 5, the residue in the spectrum of three peaks reduces quickly compared with that of in Fig. 3. From the detail of Fig. 6(a'), (b') and (c'), it shows the Doppler frequencies of two first-order Bragg peaks detected by NMFSCC are −0.39 and

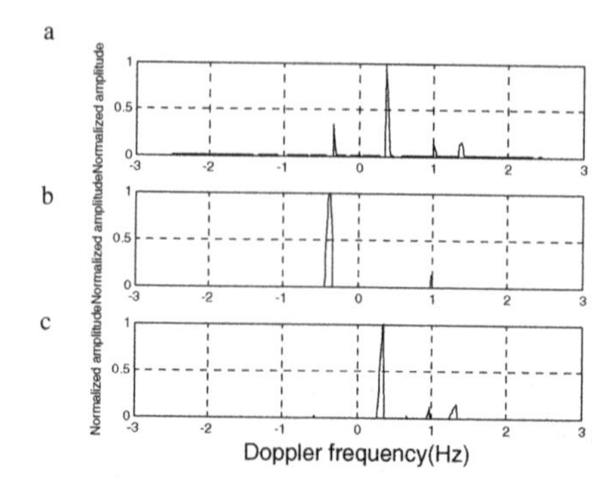

Fig. 3. a Spectrum of target; b Spectrum of first-order Bragg peak (negative frequency); c Spectrum of first-order Bragg peak (positive frequency)

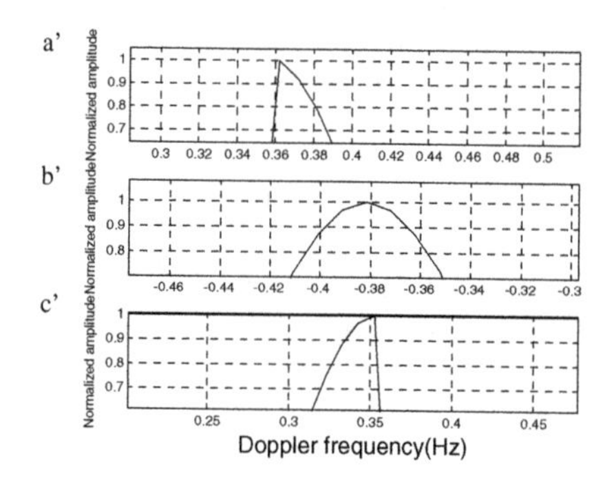

Fig. 4. Detail spectrums corresponding to Fig. 3(a), (b), (c) respectively

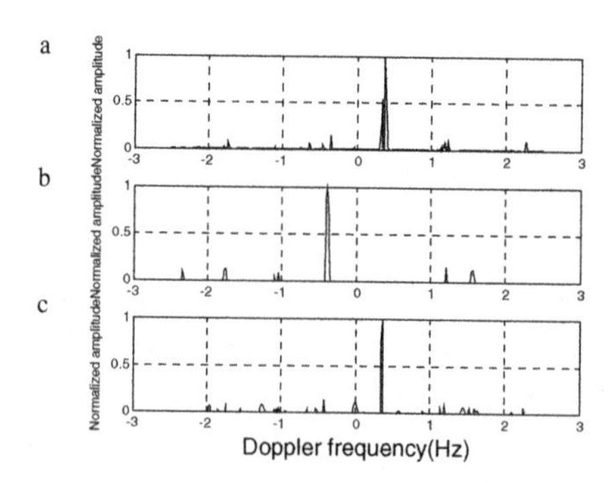

Fig. 5. a Spectrum of target; b Spectrum of first-order Bragg peak (negative frequency); c Spectrum of first-order Bragg peak (positive frequency)

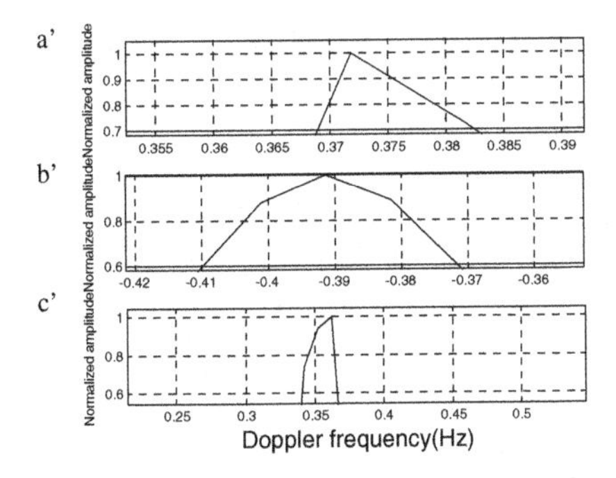

Fig. 6. Detail spectrums corresponding to Fig. 5(a), (b), (c) respectively

0.353 Hz, and the Doppler frequency of target is 0.372 Hz. The error is only 0.028 Hz. So NMFSCC algorithm is an efficient algorithm when detecting target of Slow Radial Velocity in complex sea state.

5 Summary

In this paper, an efficient implementation of NMFSCC algorithm for detecting ship of Slow Radial Velocity in complex sea state is proposed. The detection performance compared with SVD is then evaluated in a simulation study. It is concluded as following: (1) the accuracy using NMFSCC is higher than that of SVD when detecting target of Slow Radial Velocity; (2) NMFSCC algorithm can provide useful prior information of sea clutter to the process of cell under test; (3) the detection performance of NMFSCC in spread Doppler clutter is looking forward to analyze.

Acknowledgments. The authors would like to thank the National Natural Science Foundation of China for financial support (Grant No. 51309232).

References

1. Anderson, S.J., Abramovich, Y.I.: A unified approach to detection, classification, and correction of ionospheric distortion in HF sky wave radar systems. Radio Sci. **33**(4), 1055–1067 (2016)
2. Luo, H., Chen, J., Zhao, Z.: Ship target detection in Over-the-Horizon radar based on signal block matrix. Radar Sci. Technol. **9**(2), 161–165 (2011)
3. Poon, M.W.Y., Khan, R.H., Le-Ngoc, S.: A singular value decomposition (SVD) based method for suppressing ocean clutter in high frequency radar. IEEE Trans Sign. Process. **41** (3), 1421–1425 (1993)
4. Lu, K., Liu, X., Liu, Y.: Ionospheric decontamination and sea clutter suppression for HF skywave radars. IEEE J. Oceanic Eng. **30**(2), 455–462 (2005)
5. Lee, D.D., Seung, H.S.: Learning the parts of objects by non-negative matrix factorization. Nature **401**(6), 755–791 (1999)

A ν-Twin Bounded Support Tensor Machine for Image Classification

Biyan Dai, Huiru Wang, and Zhijian Zhou(✉)

College of Science, China Agricultural University, Beijing 100083, China
zzjmath@163.com

Abstract. Support Vector Machine (SVM) is an effective tool for classification problems. With the advent of the information age, tensor data problems are common in pattern recognition field. However, SVMs may lead to structural information loss and the curse of dimensionality when encounter into tensor data. In this paper, we propose a novel tensor-based classifier called the ν-Twin Bounded Tensor Machine (ν-TBSTM). It is an extension of ν-TBSVM. ν-TBSVM solves two smaller Quadratic Programming Problems (QPPs) instead of a larger one, meanwhile, it adopts the structural risk minimization principle. Compared with existing SVMs, ν-TBSVM has certain advantages. ν-TBSTM inherits all the advantages of ν-TBSVM, moreover, it utilizes the structural information of tensor data more directly and effectively, thus it gains better performance. The experimental results indicate the effectiveness and superiority of the new algorithm.

Keywords: Classification problems · Tensor data · Structural information

1 Introduction

Pattern recognition problems (classification problems) are a vital branch of machine learning. SVM invented by Vapnik [1] in 1990s, is a common and effective tool for classification problems [2]. The main idea of SVM is to seek an optimal hyper-plane by solving a Quadratic Programming Problem (QPP). Solving a big QPP may make the learning speed of SVM be very slow. Considering the large time consuming of traditional SVM algorithms particularly on high dimensional datasets, Jayadeva et al. proposed Twin SVM (TSVM). TSVM determines two nonparallel planes by solving two related SVM-type problems, each of which is smaller than in a conventional SVM [3]. Later many scholars turn to research on TSVM, and some extensions are proposed including LSTSVM [4], TBSVM [5], ν-TSVM [6] and so on.

SVM algorithm requires input data for vector data, but in many practical application fields, we often encounter multidimensional arrays namely tensor data. For example, gray face image is expressed as a second-order tensor, grayscale video sequence are usually represented as third-order tensor [7]. When this situation occurs, SVMs can also settle by transforming the tensor into vector data. But problems follow as well, one is it may destroy the structural information of tensor data, another is it may cause the curse of the dimension, because the converted vector very likely to be high dimensional. Considering the aforementioned drawbacks, many algorithms are

F. Xhafa et al. (eds.), *Advances in Intelligent Systems and Interactive Applications*, Advances in Intelligent Systems and Computing 686,
https://doi.org/10.1007/978-3-319-69096-4_39

proposed based on the tensor space. Tao et al. proposed Supervised Tensor Learning (STL) framework which took tensor as input data for the first time [8]. Based on it Cai et al. proposed Support Tensor Machine (STM) [9]. In recent years, there have been more and more researches evolving on SVMs to STMs, and all have obtained good experimental results and wide applications. For instance, Reshma et al. extended the Proximal SVM [10] to Proximal STM [11]. Chen et al. extended one-class SVM [12] to STM [13]. Shi et al. extended the algorithm of TBSVM to TBSTM [14].

ν-TBSVM [15] proposed by Xu et al. has many advantages compared with traditional SVM. It adopts the structural risk minimization principle and solves two smaller-sized QPPs, which gains a better performance and consumes less time. However, when encountered to tensor space, ν-TBSVM ignores the underlying tensor data structural information. Motivated by above studies, in this paper, we propose a new tensor-based classifier, called the ν-Twin Bounded Support Tensor Machine (ν-TBSTM), which is an extension of ν-TBSVM. Similar to ν-TBSVM, ν-TBSTM minimizes of structural risk and constructs two non-parallel hyper-planes. Meanwhile, ν-TBSTM extends the vector-based algorithm into the tensor space, which can overcome the limitations of the ν-TBSVM in tensor space. Firstly, ν-TBSTM avoids the missing of data structural information caused by converting the tensor into vector data. Moreover ν-TBSTM avoids the over-fitting problem and the curse of dimensionality. We verify our proposed method ν-TBSTM by experiments on thirteen benchmark vector and seven benchmark tensor datasets.

The rest of this paper is arranged as follows. In Sect. 2 we give a brief overview of ν-TBSVM. The model and algorithm of ν-TBSTM are introduced in Sect. 3. Section 4 shows the experimental results on thirteen datasets. Conclusion and future work are given in Sect. 5.

2 Brief Overview of ν-Twin Bounded Support Vector Machine

Xu et al. proposed ν-TBSVM based on ν-TSVM by adding the regularization term. Considering a binary classification problem with a training set $T = \{(\boldsymbol{x}_1, +1), \cdots, (\boldsymbol{x}_p, +1), (\boldsymbol{x}_{p+1}, -1), \cdots, (\boldsymbol{x}_{p+q}, -1)\}$, where $\boldsymbol{x}_i \in R^n, i = 1, 2, \cdots, p$ belongs to class +1, $\boldsymbol{x}_j \in R^n, j = p+1, \cdots, p+q$ belongs to class -1. ν-TBSVM searches for two non-parallel hyper-planes: $<\boldsymbol{w}_1 \cdot \boldsymbol{x}> + b_1 = 0, <\boldsymbol{w}_2 \cdot \boldsymbol{x}> + b_2 = 0$. where $\boldsymbol{w}_i \in R^n,\ b_i \in R, i = 1, 2$. The optimization problems of ν-TBSVM is deduced as:

$$
\begin{aligned}
\nu\text{ - TBSVM 1: } \min_{\mathbf{w}_1, b_1, \xi_-} \ & {}^1\!/_2 c_1(\|\mathbf{w}_1\|^2 + b_1^2) + {}^1\!/_2\|\mathbf{A}\mathbf{w}_1 + \mathbf{e}_1 b_1\|^2 - \nu_1\rho_1 + {}^1\!/_q \mathbf{e}_2^T \xi_- \\
\text{s.t. } & -(\mathbf{B}\mathbf{w}_2 + \mathbf{e}_2 b_1) \geq \rho_1 \mathbf{e}_2 - \xi_- \\
& \rho_1 \geq 0, \xi_- \geq \mathbf{0}
\end{aligned}
\tag{1}
$$

$$\nu\text{ - TBSVM 2: } \min_{\mathbf{w}_2, b_2, \xi_+} \ {}^1\!/_2 c_2(\|\mathbf{w}_2\|^2 + b_2^2) + {}^1\!/_2\|\mathbf{B}\mathbf{w}_2 + \mathbf{e}_2 b_2\|^2 - \nu_2\rho_2 + {}^1\!/_p \mathbf{e}_1^T \xi_+$$
$$\text{s.t. } \mathbf{A}\mathbf{w}_2 + \mathbf{e}_1 b_2 \geq \rho_2 \mathbf{e}_1 - \xi_+$$
$$\rho_2 \geq 0, \xi_+ \geq \mathbf{0} \tag{2}$$

where $A = [x_1, \cdots, x_p]^T, B = [x_{p+1}, \cdots, x_{p+q}]^T$. $\xi_+ = (\xi_1, \cdots, \xi_p)^T, \xi_- = (\xi_{p+1}, \cdots, \xi_{p+q})^T$ express slack variables, and $e_1 \in R^p$, $e_2 \in R^q$ are column vector with all one element, ν_1, ν_2, c_1, c_2 are parameters, and ν_1, ν_2 are used to control the bounds of the fractions of support vectors and the error margins.

3 ν-Twin Bounded Support Tensor Machine

In this section we first build the model of the second-order rank-one ν-TBSTM, and then give the detailed algorithm. The input data of ν-TBSTM are tensor data, suppose the dataset is: $T = \{(X_1, +1), \cdots, (X_p, +1), (X_{p+1}, -1), \cdots, (X_{p+q}, -1)\}$, where $X_i \in R^{n_1 \times n_2}$, $i = 1, 2, \cdots, p$ denotes a second-order tensor (matrix) belonging to class +1, $X_j \in R^{n_1 \times n_2}, j = p+1, \cdots, p+q$ belonging to class -1. Similarly, ν-TBSTM seeks two nonparallel hyper-planes: $< W_1, X > + b_1 = 0$, $< W_2, X > + b_2 = 0$, where different from vector space, W_1 and $W_2 \in R^{n_1 \times n_2}$ are second-order rank-one tensors, they can be decomposed as: $\mathbf{W}_1 = \mathbf{u}\mathbf{v}^T$, $\mathbf{W}_2 = \tilde{u}\tilde{v}^T$, where $u, \tilde{u} \in R^{n_1}$, $v, \tilde{v} \in R^{n_2}$. Then the decision functions can be rewritten as: $u^T X v + b_1 = 0$, $\tilde{u}^T X \tilde{v} + b_2 = 0$.

Now we give the optimization problems of second-order rank-one ν-TBSTM:

$$\nu\text{ - TBSTM1 : } \min_{\mathbf{W}_1, b_1, \xi_-} \ {}^1\!/_2 c_1(\|\mathbf{u}\mathbf{v}^T\|^2 + b_1^2) + {}^1\!/_2\|\mathcal{A} \times_1 \mathbf{u} \times_2 \mathbf{v} + \mathbf{e}_1 b_1\|^2 - \nu_1\rho_1 + {}^1\!/_q \mathbf{e}_2^T \xi_-$$
$$\text{s.t. } -(\mathcal{B} \times_1 \mathbf{u} \times_2 \mathbf{v} + \mathbf{e}_2 b_1) \geq \rho_1 \mathbf{e}_2 - \xi_-$$
$$\rho_1 \geq 0, \xi_- \geq \mathbf{0} \tag{3}$$

$$\nu\text{ - TBSTM2: } \min_{\mathbf{W}_2, b_2, \xi_+} \ {}^1\!/_2 c_2(\|\tilde{u}\tilde{v}^T\|^2 + b_2^2) + {}^1\!/_2\|\mathcal{B} \times_1 \tilde{u} \times_2 \tilde{v} + \mathbf{e}_2 b_2\|^2 - \nu_2\rho_2 + {}^1\!/_p \mathbf{e}_1^T \xi_+$$
$$\text{s.t. } \mathcal{A} \times_1 \tilde{u} \times_2 \tilde{v} + \mathbf{e}_1 b_2 \geq \rho_2 \mathbf{e}_1 - \xi_+$$
$$\rho_2 \geq 0, \xi_+ \geq \mathbf{0} \tag{4}$$

Different from ν-TBSVM, $\mathcal{A}, \mathcal{B}$ is third-order tensor which represents the positive and negative points respectively. ν_1, ν_2, c_1, c_2 are parameters.

Owning to ν-TBSTM1 and ν-TBSTM2 have the same status, we only discuss the model of ν-TBSTM1. Like to ν-TBSVM, since: $\|uv^T\|^2 = trace(uvv^T u^T) = (v^T v)(u^T u)$, we obtain the Lagrange function:

$$L(\boldsymbol{v}, \xi_-, \boldsymbol{\alpha}, \boldsymbol{\beta}, \gamma) = {}^1\!/_2 c_1((\boldsymbol{v}^T\boldsymbol{v})(\boldsymbol{u}\boldsymbol{u}^T) + b_1^2) + {}^1\!/_2||\mathcal{A} \times_1 \boldsymbol{u} \times_2 \boldsymbol{v} + \boldsymbol{e}_1 b_1||^2 - \nu_1\rho_1 + {}^1\!/_q \boldsymbol{e}_2^T\xi_- - \boldsymbol{\alpha}^T(-(\mathcal{B} \times_1 \boldsymbol{u} \times_2 \boldsymbol{v} + \boldsymbol{e}_2 b_1) - \rho_1\boldsymbol{e}_2 + \xi_-) - \boldsymbol{\beta}^T\xi_- - \gamma\rho_1 \tag{5}$$

Normally, we use Alternate Iterating Algorithm [16] to solve tensor problems.

Firstly, we initialize $\boldsymbol{u}$, set $K_1 = ||\boldsymbol{u}||^2$, $\boldsymbol{A}_1 = \mathcal{A} \times_1 \boldsymbol{u}$, $\boldsymbol{B}_1 = \mathcal{B} \times_1 \boldsymbol{u}$, then we can obtain:

$$L(\boldsymbol{v}, \xi_-, \boldsymbol{\alpha}, \boldsymbol{\beta}, \gamma) = {}^1\!/_2 c_1(\boldsymbol{K}_1(\boldsymbol{v}^T\boldsymbol{v}) + b_1^2) + {}^1\!/_2||\boldsymbol{A}_1\boldsymbol{v} + \boldsymbol{e}_1 b_1|||^2 - \nu_1\rho_1 + {}^1\!/_q \boldsymbol{e}_2^T\xi_- - \boldsymbol{\alpha}^T(-(\boldsymbol{B}_1\boldsymbol{v} + \boldsymbol{e}_2 b_1) - \rho_1\boldsymbol{e}_2 + \xi_-)) - \boldsymbol{\beta}^T\xi_- - \gamma\rho_1 \tag{6}$$

The Karush-Kuhn-Tucker (K.K.T) conditions are obtained as follows:

$$\partial L/\partial \boldsymbol{v} = c_1 K_1 \boldsymbol{v} + \boldsymbol{A}_1^T(\boldsymbol{A}_1\boldsymbol{v} + \boldsymbol{e}_1 b_1) + \boldsymbol{B}_1^T\boldsymbol{\alpha} = 0 \tag{7}$$

$$\partial L/\partial b_1 = c_1 b_1 + \boldsymbol{e}_1^T(\boldsymbol{A}_1\boldsymbol{v} + \boldsymbol{e}_1 b_1) + \boldsymbol{e}_2^T\boldsymbol{\alpha} = 0 \tag{8}$$

$$\partial L/\partial \xi_- = {}^1\!/_q \boldsymbol{e}_2 - \boldsymbol{\alpha} - \boldsymbol{\beta} = 0 \tag{9}$$

$$\partial L/\partial \rho_1 = -\nu_1 + \boldsymbol{e}_2^T\boldsymbol{\alpha} - \gamma = 0 \tag{10}$$

$$\boldsymbol{\alpha} \geq \boldsymbol{0}, \boldsymbol{\beta} \geq \boldsymbol{0}, \gamma \geq 0 \tag{11}$$

From Eqs. (9) and (11) we can obtain:

$$\boldsymbol{0} \leq \boldsymbol{\alpha} \leq {}^1\!/_q \boldsymbol{e}_2 \tag{12}$$

Form Eqs. (10) and (11) we can obtain: $\boldsymbol{e}_2^T\alpha \geq \nu_1$. From the Eqs. (7) and (8) we can obtain:

$$\left(\begin{bmatrix} \boldsymbol{A}_1^T \\ \boldsymbol{e}_1^T \end{bmatrix}[\boldsymbol{A}_1 \quad \boldsymbol{e}_1] + c_1\begin{bmatrix} K_1\boldsymbol{I} & \boldsymbol{0} \\ \boldsymbol{0} & 1 \end{bmatrix}\right)\begin{bmatrix} \boldsymbol{v} \\ b_1 \end{bmatrix} + \begin{bmatrix} \boldsymbol{B}_1^T \\ \boldsymbol{e}_2^T \end{bmatrix}\boldsymbol{\alpha} = \boldsymbol{0} \tag{13}$$

Let $\boldsymbol{H}_1 = [\boldsymbol{A}_1 \quad \boldsymbol{e}_1]$, $\boldsymbol{G}_1 = [\boldsymbol{B}_1 \quad \boldsymbol{e}_2]$, $\boldsymbol{P}_1 = \begin{bmatrix} K_1\boldsymbol{I} & \boldsymbol{0} \\ \boldsymbol{0} & 1 \end{bmatrix}$. We can get the Eq. (14) from (13):

$$[\boldsymbol{v}^{\mathrm{T}} \quad b_1]^{\mathrm{T}} = -(\boldsymbol{H}_1^T\boldsymbol{H}_1 + c_1\boldsymbol{P}_1)^{-1}\boldsymbol{G}_1^T\boldsymbol{\alpha} \tag{14}$$

Simultaneously, we can get the Wolfe dual problem of (3), then we can acquire $\boldsymbol{v}$, b_1:

$$\min_{\alpha} \ ^1/_2\alpha^T G_1 \left(H_1^T H_1 + c_1 P_1\right)^{-1} G_1^T \alpha$$
$$\text{s.t.} \quad 0 \le \alpha \le {}^1/_q e_2 \qquad (15)$$
$$e_2^T \alpha \ge v_1$$

Then we update u, let $K_2 = ||v||^2, A_2 = \mathcal{A} \times_2 v, B_2 = \mathcal{B} \times_2 v, H_2 = [A_2 \quad e_1], G_2 = [B_2 \quad e_2], P_2 = \begin{bmatrix} K_2 I & 0 \\ 0 & 1 \end{bmatrix}$. In the same way we can obtain u and b_1 through:

$$[u^T \quad b_1]^T = -\left(H_2^T H_2 + c_1 P_2\right)^{-1} G_2^T \alpha \qquad (16)$$

$$\min_{\alpha} \ ^1/_2\alpha^T G_2 \left(H_2^T H_2 + c_1 P_2\right)^{-1} G_2^T \alpha$$
$$\text{s.t.} \quad 0 \le \alpha \le {}^1/_q e_2 \qquad (17)$$
$$e_2^T \alpha \ge v_1$$

Next, the algorithm of ν-TBSTM1 is presented. ν-TBSTM2 is the same.

Algorithm 3.1 Linearν-TBSTM1.

Input: the dataset: $T = \{(X_1, +1), \cdots, (X_p, +1), (X_{p+1}, -1), \cdots, (X_{p+q}, -1)\}$, $X_i \in R^{n_1 \times n_2}$

Output: u_1, v_1 and b_1, $W_1 = u_1 v_1^T$, output the positive hyper-plane $<W_1, X> + b_1 = 0$.

Step1. Initialization. Let $k = 0$, $\varepsilon > 0$ be small enough and take $u^k = (1, \cdots, 1)^T \in R^{n_1}$.

Step2. Calculate v. Put $u = u^k$ and let $K_1 = || u ||^2$, $A_1 = \mathcal{A} \times_1 u$, $B_1 = \mathcal{B} \times_2 u$, then v^k, b_1^k can be obtained by solving (15) and (14).

Step3. Update u. Put $v = v^k$, solve the Wolfe dual problem (17), and then calculate u^{k+1}, b_1^{k+1} by formulas (16).

Step4. Update v. Put $u = u^{k+1}$ and let $K_2 = || v ||^2$, $A_2 = \mathcal{A} \times_2 v$, $B_2 = \mathcal{B} \times_2 v$, then we can get the value of v^{k+1}, b_1^{k+1} by solving (15) and (14).

Step5. If the termination criterion that $|| u^{k+1} - u^k || \le \varepsilon$, $|| v^{k+1} - v^k || \le \varepsilon$ and $|| b^{k+1} - b^k || \le \varepsilon$ are reached or the maximum number of iteration is achieved, put $u_1 = u^{k+1}$, $v_1 = v^{k+1}$, and $b_1 = b^{k+1}$; otherwise, set $k = k + 1$ and return to step 2.

---

4 Experimental Results

In this section the experimental results are presented. All the algorithms have been implemented in MATLAB R2012a on Windows 7 running on a CPU with Intel(R) Core(TM) 2Duo CPU E7500 @ 2.93 GHz 2.94 GHz. The results of ν-TBSTM are compared with ν-TSVM, ν-TBSVM and ν-TSTM. There are four running parameters in the ν-TBSTM model which are ν_1, ν_2, c_1, c_2. The optimal parameters c_1 and c_2 were searched from $[2^{-5}, 2^{-4}, \cdots, 2^5]$ and the parameters ν_1 and ν_2 were searched from $[0.1, 0.2, \cdots, 0.9]$. We set $\nu_1 = \nu_2$, $c_1 = c_2$, and all the parameters in this paper are Ten-fold cross-validation on the training set to find the best value. We analyze the performance of ν-TBSTM by the averaged testing accuracy and running time.

In our experiments, thirteen different datasets in total are used to ensure the reliability of the experimental results including six vector datasets and seven tensor datasets. Six vector datasets are Lung cancer, Wine, Heart disease, Balance Scale, Bupa and German dataset. Seven tensor datasets are Libras movement, Eyes, Orl1 face, Robot, Orl2 face, Yale face and Hand writing dataset respectively. All the datasets are downloaded from UCI datasets.

The experimental results on vector datasets indicate that our improved algorithm performs better than ν-TSTM and almost nearly to ν-TBSVM on most vector datasets. The results on tensor datasets indicate that our improved algorithm is the best classifier among ν-TBSVM, ν-TSVM and ν-TSTM. ν-TBSTM takes the structured risk into account so it performs better than ν-TSTM. Simultaneously, it is better than ν-TSVM and ν-TBSVM because it utilizes the information of the tensor data adequately. It is worth noting that when encounter to high dimension datasets the running time of ν-TBSTM classifier is reduced to a large extent compared to vector-based algorithms. In general, the experiments illustrate that the proposed algorithm not only has obvious advantages in dealing with tensor-based problems especially for high dimensional problems but can solve vector-based problems. Table 1 shows the detailed information of datasets. The averaged testing accuracy and computing time on datasets are presented in Table 2.

Table 1. Description of thirteen datasets

Datasets	Number of samples P + N	Training-number $p+q$	Attribute	Matrix-size
Lung	10 + 13	2 + 2	56	(7,8)
Wine	59 + 71	6 + 7	13	(3,5)
Heart	48 + 18	5 + 2	13	(3,5)
Balance	288 + 288	30 + 30	4	(2,2)
Bupa	145 + 200	15 + 15	6	(2,3)
German	700 + 300	7 + 3	24	(4,6)
Libras	24 + 24	2 + 2	(45,2)	(45,2)
Eyes	40 + 40	4 + 4	(84,56)	(84,56)
Orl1	10 + 10	2 + 2	(32,32)	(32,32)
Robot	20 + 27	2 + 3	(15,6)	(15,6)
Orl2	10 + 10	1 + 1	(64,64)	(64,64)
Yale	11 + 11	2 + 2	(100,100)	(100,100)
Hand	20 + 20	2 + 2	(16,16)	(16,16)

Table 2. Averaged testing accuracy and running time on thirteen datasets

Datasets	ν-TSVM	ν-TBSVM	ν-TSTM	ν-TBSTM
	Accuracy(%) ± std	Accuracy(%) ± std	Accuracy(%) ± std	Accuracy(%) ± std
	Time(s)	Time(s)	Time(s)	Time(s)
Lung	75.79 ± 16.86	75.79 ± 16.86	78.95 ± 5.54	**82.63 ± 6.10**
	0.066	0.057	0.484	0.316
Wine	92.28 ± 8.24	**96.13 ± 2.69**	93.10 ± 2.44	95.13 ± 2.82
	0.069	0.056	0.266	0.309
Heart	76.27 ± 11.96	80.51 ± 13.38	81.69 ± 1.56	**82.04 ± 1.99**
	0.082	0.064	0.282	0.280
Balance	97.35 ± 1.09	**97.54 ± 1.04**	94.57 ± 1.35	95.37 ± 1.05
	0.264	0.230	1.258	0.486
Bupa	**79.65 ± 8.17**	78.51 ± 7.08	66.89 ± 6.11	67.90 ± 9.08
	0.128	0.112	0.389	0.494
German	71.17 ± 13.38	**73.81 ± 12.90**	70.85 ± 1.20	71.10 ± 0.06
	0.104	0.093	0.470	0.434
Libras	69.09 ± 23.43	70.00 ± 21.37	72.50 ± 1.68	**75.00 ± 2.14**
	0.078	0.068	0.230	0.282
Eyes	**100.00 ± 0.00**	**100.00 ± 0.00**	93.75 ± 1.18	**100.00 ± 0.00**
	115.541	58.786	0.877	0.889
Orl1	97.50 ± 3.23	97.50 ± 3.23	90.00 ± 3.23	**100.00 ± 0.00**
	81.590	43.400	0.457	0.425
Robot	67.14 ± 27.14	67.14 ± 27.14	**78.57 ± 2.97**	67.62 ± 8.99
	0.075	0.067	0.268	0.291
Orl2	78.33 ± 17.85	82.78 ± 17.85	**95.00 ± 1.76**	**95.00 ± 1.76**
	81.590	43.400	0.305	0.301
Yale	**100.00 ± 0.00**	**100.00 ± 0.00**	**100.00 ± 0.00**	**100.00 ± 0.00**
	982.395	456.095	0.693	0.683
Hand	93.61 ± 15.50	93.61 ± 15.50	**97.35 ± 0.93**	96.94 ± 2.05
	0.116	0.107	0.272	0.282

5 Conclusion and Future Work

In this paper we extend ν-TBSVM to ν-TBSTM which not only inherit the advantages of ν-TBSVM but has better performance and wider applications especially in tensor space. The benefits of the classifier we proposed are wide-ranging. Firstly, it improves the accuracy owing to retaining the structural information of tensor data and adding regularization term. Secondly, it is less-time consuming than vector-based algorithms especially when gets into high dimensional problem. Thirdly, it helps overcome the curse of dimension which is encountered mostly in vector-based algorithms. However, only the performance of ν-TBSTM in linear second-order rank-one case is certified in this work. The performance in nonlinear or higher order situation is not tested. Therefore our future work is to extend the proposed algorithm into nonlinear case and second-order problems to high-order.

Acknowledgements. This work is supported in part by National Natural Science Foundation of China (No. 11671010).

References

1. Vapnik, V.N.: The Nature of Statistical Learning Theory. New York (1995)
2. Han, J., Kamber, M., Pei, J.: Data Mining Concept and Techniques. China Machine Press, Beijing (2012)
3. Jayadeva, Khemchandani, R., Chandra, S.: Twin support vector machines for pattern classification. IEEE Trans. Pattern Anal. Mach. Intell. **29**(5), 905–910 (2007)
4. Kumar, M.A., Gopal, M.: Least squares twin support vector machines for pattern classification. Expert Syst. Appl. **36**(4), 7535–7543 (2009)
5. Shao, Y.H., Zhang, C.H., Wang, X.B., Deng, N.Y.: Improvements on twin support vector machines. IEEE Trans. Neural Netw. **2**(6), 962–968 (2011)
6. Peng, X.: A ν-Twin Support Vector Machine (v-TSVM) classifier and its geometric algorithms. Inf. Sci. **180**(20), 3863–3875 (2010)
7. Etemad, K., Chellappa, R.: Discriminant Analysis for Recognition of Human Face Images, vol. 8, no. 14, pp. 1724–1733. Springer, Heidelberg (1997)
8. Tao, D., Li, X., Hu, E., Maybank, S.J.: Supervised tensor learning. Knowl. Inf. Syst. **13**(1), 450–457 (2005)
9. Cai, D., He, X., Han, J.: Learning with tensor representation. Pattern Recogn. (2006)
10. Oraon, D.: Study on proximal support vector machine as a classifier. Mach. Learn. **59**(1), 77–97 (2015)
11. Khemchandani, R., Karpatne, A., Chandra, S.: Proximal support tensor machines. Int. J. Mach. Learn. Cybern. **4**(6), 703–712 (2013)
12. Schölkopf, B., Platt, J.C., Taylor, J.S., Smola, A.J., Williamson, R.C.: Estimating the support of a high-dimensional distribution. Neural Comput. **13**(7), 1443–1471 (2001)
13. Chen, Y., Wang, K., Zhong, P.: One-class support tensor machine. Knowl.-Based Syst. **96**, 14–28 (2016)
14. Shi, H., Zhao, X., Zhen, L., Jing, L.: Twin bounded support tensor machine for classification. Int. J. Pattern Recogn. Artif. Intell. **30**(1) (2015)
15. Xu, Y., Guo, R.: An improved v-twin support vector machine. Appl. Intell. **41**(1), 42–54 (2014)
16. Cai, D., He, X., Wen, J.R., Han, J., Ma, W.Y.: Support tensor machines for text categorization. Int. J. Acad. Res. Bus. Soc. Sci. **12**(2), 2222–6990 (2006)

Super-resolution Reconstruction of Face Image Based on Convolution Network

Wenqing Huang(✉), Yinglong Chen, Li Mei, and Hui You

Zhejiang Sci-Tech University, Hangzhou, China
patternrecog@163.com

Abstract. A Face Image Super-Resolution (SR) reconstruction based on Convolution Neural Networks is constructed. Firstly, extract two-level feature map by multiple convolution kernel images. Secondly, after each feature map is extracted, mapping the extracted features to another plane by means of a non-linear mapping method. Lastly, rebuilding the final SR images through adding all the second-level feature maps and plus a constant. The experimental results show that our method can get a better result in single face image SR reconstruction.

Keywords: Single image super-resolution (SR) · Convolution neural network · Deep learning

1 Introduction

Single image super-resolution is designed to recover high-resolution (HR) images from a single low-resolution (LR) image, which is a classic problem in computer vision. This problem is inherently incorrect due to the existence of a variety of solutions for one low-resolution image. That is to say, this is an uncertain inverse problem, and its solution is not the only. The solution space is usually constrained by powerful prior information, which usually alleviates the problem.

Image resolution is an important indicator of evaluating an image quality. In general, the higher the image resolution, the better the image quality. The SR reconstruction of face image in the field of security is particularly important, the police extract the suspect's clear appearance and other features quickly by the HR face images and it helps solve the case.

In this paper, we apply the method based on SRCNN [1]; we learn a mapping from LR image to HR image directly, to change the network structure and training data of SRCNN for face SR image reconstruction. Our method and SRCNN have several differences: (a) Network structure parameters: Both the two methods have three layers of CNN structure; in SRCNN, each layer's filter size are 9, 1, 5 respectively, while our method are 11, 3, 1. (b) The Processing method of training data: SRCNN gets the LR image set by down-sampling the SR image with Bicubic interpolation method directly, and our method get the LR image set with the same method, besides, we add Guass noise to a small part of the LR image set with a small probability to prevent overfitting in data training. (c) Optimization method: Our method use different optimization

F. Xhafa et al. (eds.), *Advances in Intelligent Systems and Interactive Applications*, Advances in Intelligent Systems and Computing 686,
https://doi.org/10.1007/978-3-319-69096-4_40

methods in data training, SRCNN only use Stochastic gradient descent (SGD). Our method uses Nesterov's accelerated gradient (NAG) [2] optimization method to make the loss of rapid decline firstly, and then use SGD to stabilize Optimization.

2 Related Work

2.1 Image Super-Resolution

Many methods of image reconstruction have been proposed, such as SRCNN [1], MC-SRCNN [3], Sparse Coding based Network (SCN) [4], Anchored Neighborhood Regression (ANR) [5] and Adjusted Anchored Neighborhood Regression (A+) [6] etc. These methods have achieved good results. SRCNN through the Bicubic interpolation method to down-sampling the HR image to obtain LR image, through the construction of a three-layer CNN network, learn a mapping between LR image and HR image directly to achieve LR image of ultra-high-definition reconstruction; The key difference between SRCNN and MC-SRCNN is that the MC-SRCNN uses a multi-channel input image. ANR learns exemplar neighborhoods offline using ridge regression and to precompute projections to map LR patches onto the HR domain using these neighborhoods. A+ is an improved variant of ANR, which is based on the characteristics from ANR and anchored regression, rather than learning the dictionary's regression device, which uses the complete training material [6].

2.2 Convolution Neural Network

Convolutional Neural Networks (CNN) [7] is one of the research hotspots in many fields of science, especially in the field of pattern classification. Since the CNN avoids the complex image preprocessing, it can input the original image directly, so it has been widely used in many fields.

In brief, CNN consists of two important steps. The first step is feature extraction, extracting the feature maps through convoluting the previous layer's output with one or more convolution kernels. The second step is feature mapping. Each layer's output is composed of multiple feature maps. Each feature map is a feature plane.

2.3 Deep Learning for Image Restoration

There are many methods of image reconstruction based on deep learning, SRCNN and MC-SRCNN have already been introduced in Sect. 2.1, DRCN (Deeply-Recursive Convolutional Network for Image Super-Resolution) proposed to use more convolution layers to increase network receptive fields (41×41) and use recurrent neural network (RNN) to avoid excessive network parameters [8]. In SRCNN and DRCN, the LR images are first interpolated to obtain the same size as the HR image, and then as a network input, meaning that the convolution operation is performed at a higher resolution. ESPCN (Real-Time Single Image and Video Super-Resolution Using an Efficient Sub-Pixel Convolutional Neural Network) proposed a high efficiency method for calculating convolution directly to obtain HR images on LR images [9].

3 Convolutional Neural Networks for Face Image Super-Resolution Restoration

3.1 Network Structure

Our method structure is similar to SRCNN. SRCNN has three main steps: feature extraction, feature mapping and image reconstruction [1]. As shown in Fig. 1, our method is also a three-layer structure: two feature extraction layers and a feature mapping layer.

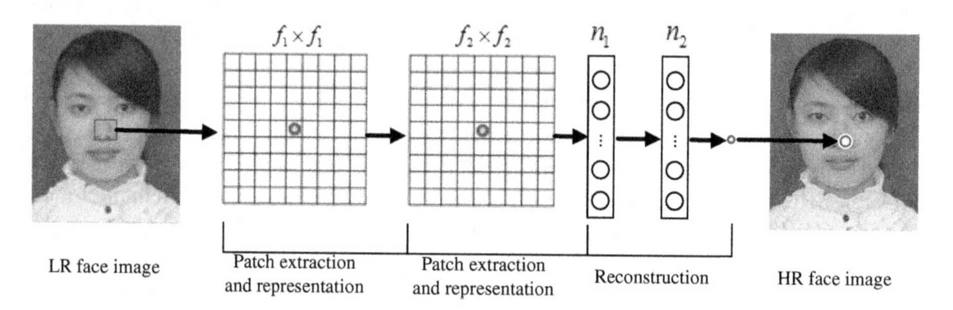

Fig. 1. The network structure of our method

1. Feature extraction

Feature extraction is a convolution process to the image. The network structure of our method contains two feature extraction layers, as shown in the Formula 1, Input the LR face image X, obtained the first-order feature map Y_1 through convolution operation and Rectified Linear Units(ReLU) operation.

$$Y_1 = \max(X \otimes F_1 + b_1, 0) \tag{1}$$

After obtaining the first-level feature map Y_1, acquired the second-level feature maps by the same way.

$$Y_2 = \max(Y_1 \otimes F_2 + b_2, 0) \tag{2}$$

Where F_i contains n_i convolution kernels, the kernel size is $f_i \times f_i$, $n_1 = 64$, $n_2 = 32, f_1 = 11, f_2 = 3$.

2. Image reconstruction

The reconstruction method is similar to the feature mapping step of SRCNN, through a convolutional layer with a convolution kernel size of 1, and the second-lever feature map mapped to another space. The image reconstruction process can be simplified as shown in Formula 3.

$$Y = f_3 \sum_{y_i \subset Y_2} y_i + b_3 \tag{3}$$

Where y_i is one feature map in Y_2, f_3 and b_3 are real numbers. In fact, the reconstruction step is the sum of all second-level feature maps, and plus the offset term b_3.

3.2 Network Layer Parameters

Our method consists of three convolution layers, the first two layers followed by a Rectified Linear Units (ReLU) layer to ensure that the results does not appear negative after the convolution layers.

1. Number of map

   Our method and SRCNN are composed of three layers of convoluted layers, SRCNN the number of feature maps of the three layers are 64, 32 and 1 respectively, also tried a larger feature map size 128, 64 and 1, but the number of the larger feature map will lead to a sharp rise in reconstruction time [1], so the number of feature map used in this paper is 64, 32 and 1.

2. Filter size

   In SRCNN, the size of each convolution kernel is 9 × 9, 1 × 1 and 5 × 5 respectively, In our method, the size of each convolution kernel is 11 × 11, 3 × 3 and 1 × 1, the size of receptive field and SRCNN are the same size, and achieved a better reconstruction effect.

3.3 Solver Optimization Method

Solver is an optimization method used to minimize training errors. In our experiment, for a data set D, we divided D into n batches, the objective function that needs to be optimized is the average of a batch's error in the entire data set:

$$L(w) = \frac{1}{|D_j|} \sum_{i}^{|D_j|} f_w(X_i) + \lambda r(w), D_j \in D, j \leq n \tag{4}$$

Where D_j is the first j batch in D, X_i is the first i sample in D_j, $f_w(X_i)$ is the data X_i error, $r(w)$ is the regular term, used to suppress over-fitting, λ is the coefficient of the regular term. Using a batch's error as the D's error improve the training speed.

In Caffe [10], there are many optimization options, such as Stochastic gradient descent (SGD) and Nesterov's accelerated gradient (NAG), etc. SGD can make the loss of stability declined, but the speed is relatively slow. NAG is the ideal method for convex optimization, and its convergence rate is very fast, and its convergence rate can be achieved $O(1/t^2)$.

In our experiment, firstly, we use NAG to reduce the loss to a steady state quickly, and then use SGD to a stable decline. When the equipment configuration is not very good, use NAG can determine whether a network is feasible quickly.

3.4 Quantitative Evaluation

The training process of the network is the process of minimizing the loss function. Improve the similarity between the reconstructed image and the label image gradually by minimizing the loss function. The SR reconstruction of face images is essentially a regression problem, and the mean square error is suitable for the regression task. Therefore, the mean square error (MSE) is used as the loss function in this experiment. The formula is as follows:

$$E = \frac{1}{2n}\sum_{i=1}^{n} \|Y_i - X_i\|_2^2 \tag{5}$$

Where n is number of samples, X_i is reconstructed image, Y_i is original image.

4 Experiment and Results Analysis

4.1 Experiment Platform

The system platform is Windows Server 2008 R2 Enterprise, processor is Inter(R) Xeon(R) CPU E5-2620 v2 @ 2.10Ghz, installation memory is 64 GB, graphics card model is NVIDIA Tesla K20m. We use Caffe as the deep learning framework in our experiment.

4.2 Data Set

Our method used a data source of 500 different face images set S, S only contains the luminance channel, this experiment randomly divided S into three parts, T_1, T_2 and faceSet10, including 460, 30 and 10 face images respectively. We regarded T_1 as training set, T_2 as training test set, faceSet10 as test set.

4.3 Results and Conclusions

This experiment was conducted on the platform described in Sect. 4.1, we trained the network and stopped train after 80 million iteration. Table 1 shows a comparison of faceset10 dataset with various image reconstruction algorithms. It can be seen that our method is superior to SRCNN in each image quality evaluation index.

In the figures below, we just reconstructed the Y channel, we get the Cb and Cr channel through up-sampling LR image by the Bicubic interpolation method directly. Figures 2, 3, 4 is the comparison of the effect graph after reconstruction.

The experimental results show that our method has better effect on face SR image reconstruction.

Table 1. The average results of PSNR (dB), SSIM, IFC, WPSNR (dB) and MSSIM on faceSet10

Eval. Med	Scale	Bicubic	SCN	ANR	A+	SRCNN	Ours
PSNR	2	39.5137	41.8270	41.6728	42.0689	42.0803	**42.1037**
	3	36.2854	38.1538	37.9448	38.3006	38.3181	**38.3383**
	4	34.2130	35.0904	35.7048	36.1937	36.2028	**36.2126**
SSIM	2	0.9747	0.9831	0.9825	0.9838	0.9837	**0.9840**
	3	0.9500	0.9528	0.9583	0.9616	0.9590	**0.9691**
	4	0.9277	0.9350	0.9362	0.9419	0.9417	**0.9420**
IFC	2	5.2650	5.8170	6.6981	7.1090	7.0120	**7.1163**
	3	3.3797	3.5383	4.2805	**4.3257**	4.3052	4.3086
	4	2.2964	2.3989	3.0492	**3.0982**	2.8619	3.0868
WPSNR	2	55.1887	59.4364	62.5361	63.1935	62.2572	**63.5598**
	3	47.2934	51.2981	51.8966	52.4634	51.6789	**52.5261**
	4	42.9005	46.4585	46.0341	46.5486	46.5237	**46.5303**
MSSSIM	2	0.9962	0.9978	0.9982	0.9983	0.9982	**0.9984**
	3	0.9873	0.9912	0.9912	0.9913	0.9887	**0.9918**
	4	0.9794	0.9824	0.9851	0.9867	0.9848	**0.9863**

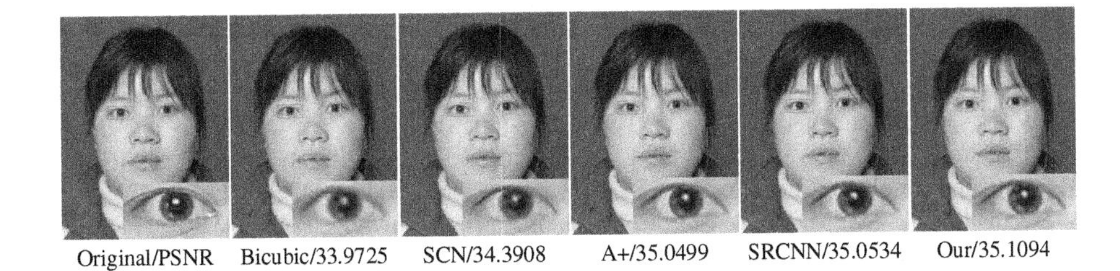

Fig. 2. The face image from faceset10 with an upscaling factor 2.

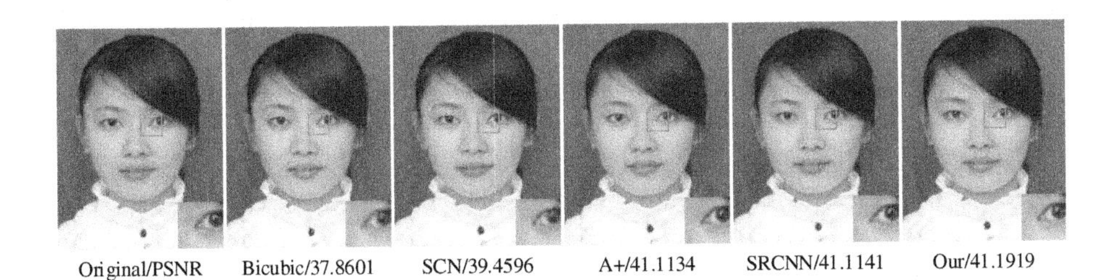

Fig. 3. The face image from faceset10 with an upscaling factor 3.

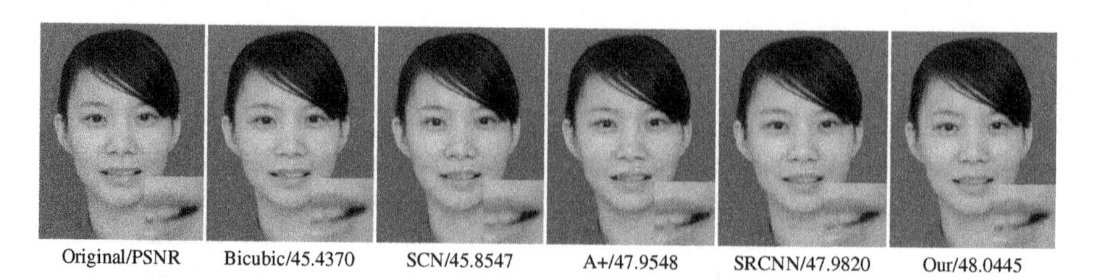

Fig. 4. The face image from faceset10 with an upscaling factor 6.

References

1. Dong, C., Loy, C.C., He, K., Tang, X.: Learning a Deep Convolutional Network for Image Super-Resolution. Springer International Publishing, Berlin (2014)
2. Botev, A., Lever, G., Barber, D.: Nesterov's Accelerated Gradient and Momentum as approximations to Regularised Update Descent (2016)
3. Youm, G.Y., Bae, S.H., Kim, M.: Image super-resolution based on convolution neural networks using multi-channel input. In: Image, Video, and Multidimensional Signal Processing Workshop IEEE, pp. 1–5 (2016)
4. Wang, Z., Liu, D., Han, W., Yang, J., Huang, T.S.: Deep networks for image super-resolution with sparse Prior. In: International Conference on Computer Vision (ICCV) (accepted) (2015)
5. Timofte, R., De Smet, V., Van Gool, L.: Anchored neighborhood regression for fast example-based super-resolution. In: IEEE International Conference on Computer Vision IEEE Computer Society, pp. 1920–1927 (2013)
6. Timofte, R., De Smet, V., Van Gool, L.: A+: adjusted anchored neighborhood regression for fast super-resolution. Lecture Notes in Computer Science **9006**, 111–126 (2015)
7. Sainath, T.N., San Martin, M.C.P.: Convolutional neural networks. US20160283841 (2016)
8. Kim, J., Lee, J.K., Lee, K.M.: Deeply-Recursive Convolutional Network for Image Super-Resolution, pp. 1637–1645 (2015)
9. Shi, W., et al.: Real-Time Single Image and Video Super-Resolution Using an Efficient Sub-Pixel Convolutional Neural Network (2016)
10. Jia et al.: Caffe: Convolutional Architecture for Fast Feature Embedding. Eprint Arxiv, pp. 675–678 (2014)

An Image Retrieval Algorithm Based on Semantic Self-Feedback Mechanism

Lang Pei[1(✉)], Jia Xu[1], and Jinhua Cai[2]

[1] College of Computer Science, Wuhan Qinchuan University, Wuhan, China
11286978@qq.com
[2] Department of Information Engineering, Wuhan International TkadeL University, Wuhan, China

Abstract. According to the image retrieval process the user need to submit many times and feedback the query, based on semantic correlation self-feedback algorithm, a user submits the demand, active feedback query, finally get to meet the needs of the results, simplify the query process, Enhanced the "fool" feature of the image retrieval system. The test results show that the algorithm has higher accuracy and higher flexibility.

Keywords: Semantic Relevance · Self-Feedback · Image Retrieval

1 Introduction

Content-based Image Search system (Content Based Image Search) and content-based Image retrieval system (Content Based Image Retrieval) is the fundamental difference between the former is facing WEB environment [1], while the latter is for a particular application in the field of one type of Image to build Image library. This difference is caused by: facing the network search system on the distribution of a large number of species complex image data, from the system point of view, its response speed, the accuracy of feature extraction and matching algorithm and so on must be more accurate, stronger anti-interference ability, etc.; From the user's point of view, the most important performance indicator of its concern is the response time, which should be interested in the accuracy of the search. To search feature extraction module in the system that put forward the serious challenge, must be designed to adapt to the large amount of data environment accurately capture the image of essential content extraction method; In addition, for a specific application which is different from traditional image retrieval system, in the field of image search system based on WEB is any network user oriented, rather than a specific or experts in the field of technical personnel, as a result, the human-computer interaction system is simple, intuitive, convenient and efficient "fool" feature is its ideal goal.

In this paper, the WEB image search system under the environment of architecture made simple introduction, and mainly introduced the detailed structure of user interaction and function of each part of the final gives a detailed description of user interaction of query process and test results.

F. Xhafa et al. (eds.), *Advances in Intelligent Systems and Interactive Applications*, Advances in Intelligent Systems and Computing 686,
https://doi.org/10.1007/978-3-319-69096-4_41

2 Module Introduction

In an image search system, the interactive module is the platform for the user's entire search system, and is the only part of the system that is directly aimed at the user. On the one hand, it queries the service for the user, and the user tells the system his query request and the expected description through the interaction module. On the other hand, it receives user awareness information, which is the image clustering service in the index database; After the system is returned to the user search results, the user can submit further feedback information, to collect the information system, study hard and get the user's psychological needs, and according to the requirements in terms of query adjustment, makes them get more ideal results [2].

The entire interaction system consists of interaction interfaces, query analysis and processing, syntax transformation, log management, result processing, and so on, as shown in Fig. 1.

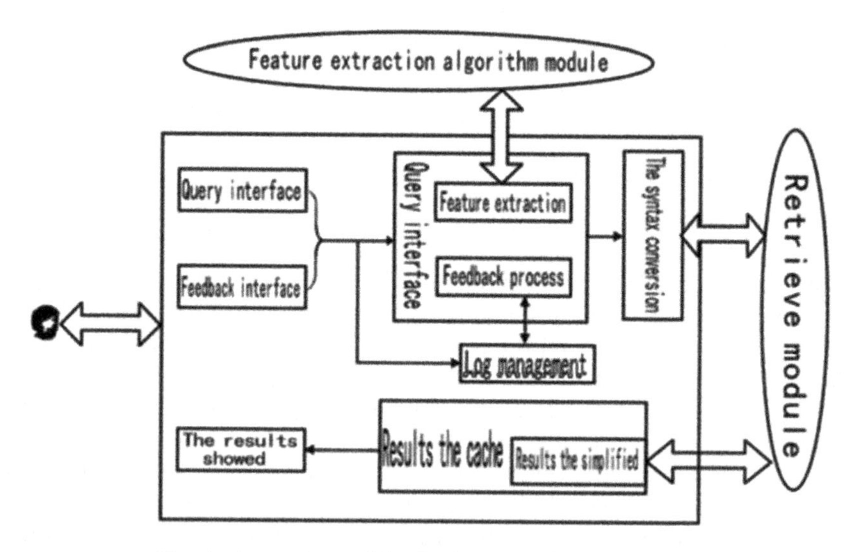

Fig. 1. Interactive and feedback system function module

2.1 Interaction Interface

The user interaction interface includes the query interface, the feedback interface, and the interface display interface.

Query interface, the user submits a query interface conditions, according to the sample images provided by the user's choice of different query mode selection, upload, provide sketch rendering tools in the sketch query mode, provide the key words in the keyword query mode function such as input, provide more than several functions in the integrated query mode combination. Mainly provide sample query interface to the user to submit a query image, when a user to find and a pair of existing images are similar to other images, this is the defining characteristic content-based image search system;

Sketches mainly provide a sketch of the user query interface mapping function, when the user needs according to their own psychological need to describe a picture to query, rather than using some image which thread of query, map query images using this interface, this mainly give some professional users such as some industry experts use; The keyword query interface is used for a single keyword query or a comprehensive query when the user input text markup.

Information about the choice of the standard template. When the user submits the feedback information, the system will adjust according to the feedback information of query demand, then use the new query requirements in terms of the next query, query or feedback. Now using more feedback methods are: the query point movement, a similarity measure criteria update, based on the traditional statistical learning theory, based on the theory of machine learning, etc. We will present a new feedback model based on user semantic understanding.

Results show that: the thumbnail image and text, image size, the degree of similarity with the query image, image and function of the web page address information such as the paging display. Image thumbnails are shown to the user in a fixed small size for the image of the query. The text markup is the relevant text description associated with the image, which may be the image name or the relevant text information about the page of the image. Said that the similarity of images and image, the similar degree between user query examples are normalized to a real number between [0, 1], the value is calculated through the system selected similarity measure function; The image and the address of the page where it is located are connected, and if the user CARES about the other information about the page that appears in the image, you can open the link and browse. Other relevant information such as the number of images per page, related links, etc.

2.2 Query Analysis and Processing

Query condition analysis and treatment including query analysis, feature extraction and feedback of three parts.

Query analysis: the analysis is the original query or feedback query, query condition if it is the original query, call the original query database retrieval matching, for the first time any user query belong to this type, it is often only depends on the user first submit search query condition; The feedback query adjusts the query vector based on the feedback information and converts it directly to the retrieval module.

The feedback to the user interaction part here is different, the feedback is the system according to the different feature extraction algorithm of automatic image features description of fine-tuning, the characteristic components such as weight adjustment, etc. The other purpose of query analysis is to parse the image information and text information, to make the text parsing of the key parts submitted by the user to facilitate subsequent processing. Image feature extraction interface will be parsed to feature extraction, feature extraction module and returns the corresponding eigenvectors, the feature vector is image content abstract Numbers, is the main index in the content-based image retrieval field.

2.3 The Syntax Conversion

Syntax conversion module is the interface between query and retrieval systems, it is converted to the query conditions retrieval system can identify the query, the query information of the system is just the key word or image feature vector, and how to transform the original query data into a retrieval system can meet the query expression is the main function of the syntax conversion.

2.4 Log Management

Log management include two kinds of historical records and analysis. Image query historical information: including image characteristics over a period of time the query hotspots and keyword query hotspot. Boring information feedback logs: boring information record and analysis, This contributes to the analysis of semantic relations between images, image classification and improve the accuracy of the query. Log management and effective utilization of intelligent system into an integral part of a large number of users using the system, and user requirements, querying, important information such as size distribution could be attained only by tracking log statistics, and then develop the system to provide information support for us.

2.5 The Query Results

Query results cache and simplify the processing including two parts.

Cache the result: a user submits a query request, may need to get the results of the query again in order to improve the accuracy, such result cache avoids the query image database again. In addition, when a large number of users are using the system at the same time, it is inevitable that different users will make similar queries at almost the same time, If several different users of several similar queries occurred in smaller time intervals, so that later can't original database retrieval query demand, and will cache the first a similar to the results of a query returned directly to the user, the advantage of this is to ensure the response time and greatly reduce the burden of the system and the system hardware, avoid the repeated the happening of the query.

Results simplified: any image in the image library, similar to the query image or not, there will be a query image similarity measure value, and truly valuable data may only be a large part of the similarity, and therefore the query results is not necessary to also can't completely one-time show to the user, the results of simplified can extract the key information, and according to the similarity between groups, the user is most likely to interest, according to the image information of the return if the user needs to check the detailed information or more results can be further submitted to display.

2.6 The Query Process

In content-based image search system, any a submit query process can be attributed to the following steps, as shown in Fig. 2. A detailed description is as follows.

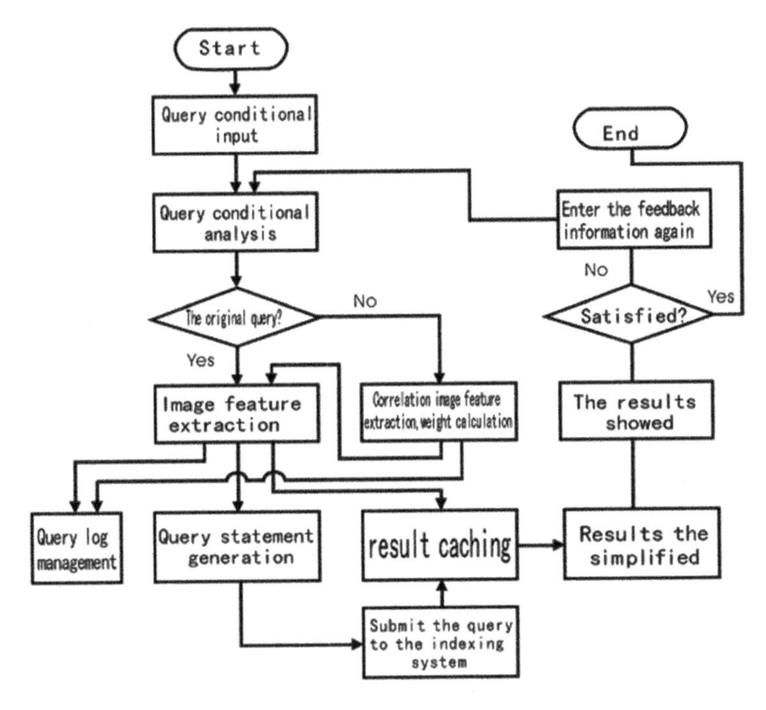

Fig. 2. System query flowchart

The user submits a query conditions (example, sketches and keywords, etc.); Query system analysis of query conditions, then the sample images or sketch for feature extraction; User information such as the feature vector and the key word to syntax conversion module, at the same time to the query log analysis of storage management module; To generate the query syntax conversion module, submit a query to retrieve system task; The result of retrieval system cache; Simplify and ultimately displayed to the user; If the user is satisfied with the results the end; Users are not satisfied, again depending on the input from the user's feedback information query, feedback information including interest and not interested in the image (or area), the standard template information, etc.; Through the query analysis submitted to feedback processing module after restructuring the query conditions, including regional feature weights, recalculate, etc. According to the different feedback model will have a different concrete implementation method; According to the feedback after the adjustment the query requirements; The new query results and display; Go to step 7);

3 The Experimental Analysis

3.1 Feedback Requirements

In the system, the user feedback information contains only is image, image and counterexample and system according to this feedback the strong relevant feedback model set, weak relevant set, negative set three kinds of feedback information, in the database to build three related respectively, users complete semantic understanding way of feedback.

3.2 Feedback Processing

Users to submit feedback information, the system according to the query point to set up the new query vector, and feedback to query, assuming queryimagefeature for the original query vector.

```
MyNewFeature=(double[])session.getValue("queryimageFeature")//To obtain the original query vector;
If(queryrequest.getParameter("RadioGroup1"+Integer.toString((currentpage-1)*20
+4*i+j))!="null"//To determine the current page shows the first results;
{double[]}tempFeature=myQueryResultSet0.getQueryResultitem((currentPage-1)
*20+4*i+j).getFeature()//Get the results of each feature vector;
Doublesimilarity=myQueryResultSet0.getQueryResultitem((currentPage-1)*20
+4*i+j).getSimilarity();//By looking at the object to obtain the P/N mark;
For(int k=0;k<N;k++){// Obtain the feedback after the adjustment the new feature vector;
MyNewFeature[k]=myNewFeature[k]+tempFeature[k];//out.println("the relevant vector is :"+tempfeature[1]);}// Call the query point movement algorithm calculating the feature component;
NewSimilarity=newSimilarity+similarity;}// Get a new similarity;
Submit new feedback information and make a new query;
```

3.3 Retrieval Algorithm Performance Testing

In the query point movement feedback strategy, we experimented on a number of different feedback system, in the result due to the query point movement is still according to the characteristics of the underlying adjustment and to update the query vector, and cannot take into account the user's psychological needs, thus showing increased performance instability, namely: for some more suitable for feature extraction algorithms that query examples performance is high, and for some is not suitable for image performance is low, and this is the reason why we brought forward a new model of feedback, the specific data as shown in Table 1.

Table 1. Feedback times and precision rate tables

Feedback times	1	2	3	4	5	6
Precision rate (%)	13.1	24.7	28.6	31.4	33.7	37.1
Feedback times	1	2	3	4	5	6
Precision rate (%)	19.3	24.7	28.6	21.4	13.7	7.1

4 The Conclusion

Overall architecture of image search based on WEB environment system are described, and emphatically introduced the function of user interaction system of the connection between the component and module, interactive interface for users to produce visual sensory stimulation, the user first experience of the whole system, so the interaction system of beauty and simple, it is important for personality, meet different users; In addition, the feedback module is established on the basis of improve query readiness of function module, beautiful and easy, of course, is very important, but it is definitely not only, more is not the main concern of the user, the user use the search system of the ultimate goal is to find what they want pictures, so the accuracy is the essence of the system and the improvement of accuracy depends on two aspects: one is to make full use of the knowledge of computer vision, can make the most fully the underlying characteristics, the nature of the most accurate description and expression of image content, is the feature extraction algorithm is more accurate; system according to their previous query "learning experience" to query and prayed to conduct a preliminary adjustment is also important. Therefore, we designed the feedback module to fine-tune the query description. Log management provides query log management and tracking, its role is to the query demand of users often do cache handling, to ensure that the small interval period again several times repeated or similar queries don't have to query the database every time, and will cache the data returned directly to the user, so that to ensure the utilization of hardware resources and query speed, but also better guarantee the accuracy of the query.

References

1. Zhu, L.: Geoblock: a LVQ-Based Framework for Geographic Image Retrieval, Information Technology: Coding and Computing, Proceedings. ITCC 2004, pp. 156–163 (2004)
2. Praks, P., Snasel, V., Dvorsky, J., Cernohorsky, J.: On SVD-free latent semantic indexing for image retrieval for application in a hard industrial environment industrial technology. In: IEEE International Conference on Volume 1, 10–12 Doc, vol. l, pp. 466–471 (2003)
3. Jin, D., Hu, Z.: A study on image retrieval based on semantics. J. Wuhan Univ., 1255–1259 (2009)
4. Wang, C., Yang, Y., Chen, S.F.: Algorithms of high-level semantic-based image retrieval. J. Softw., 1461–1469 (2004)

Using Supervised Machine Learning Algorithms to Screen Down Syndrome and Identify the Critical Protein Factors

Bing Feng[1,2], William Hoskins[2], Jun Zhou[1,2], Xinying Xu[3], and Jijun Tang[2(✉)]

[1] School of Computer Science and Technology, Tianjin University, Tianjin, People's Republic China
[2] Department of Computer Science and Engineering, University of South Carolina, Columbia, USA
jtang@cse.sc.edu
[3] College of Information Engineering, Taiyuan University of Technology, Taiyuan, People's Republic China

Abstract. Down syndrome (DS) is a genetic disorder caused by trisomy of all or part of the human chromosome 21. Since currently there is no cure for Down syndrome, the screening tests became the most efficient ways for DS prevention. Here, we used various supervised learning algorithms to build DS classification/screening models based on the protein/protein modification expression level of mice DS model Ts65Dn. Furthermore, we applied an adaptive boosted Decision Tree method to identify the most correlated and informative proteins factors that were associated with DS biological processes and pathways. Moreover, we improved the DS classification/screening models by using these selected DS related critical proteins. Finally, we used unsupervised learning algorithms to confirm the results we obtained above. These selected DS related proteins could be further used for protein-related and coding gene-related drugs developments.

Keywords: Down syndrome · Supervised learning · Unsupervised learning · Feature selection

1 Introduction

Down syndrome (DS) is a genetic abnormality caused by trisomy of all or part of the human chromosome 21 [1]. It is usually associated with various diseases and disorders including intellectual disability, heart diseases, early-onset Alzheimer's disease, childhood leukaemia, early aging, hypotonia, motor disorders, and other physical anomalies [2]. Even though DS happens at a very high rate (1 per 1000 live births in the world) [3], and it has been well studied, researchers still face many difficulties in implementing studies and experiments on human bodies4. Furthermore, there is no cure for DS [4]. As a consequence, DS screening is the most efficient way in human DS prevention. Currently, DS have been studied in different ways for therapies on human being and mouse models [5].

F. Xhafa et al. (eds.), *Advances in Intelligent Systems and Interactive Applications*, Advances in Intelligent Systems and Computing 686,
https://doi.org/10.1007/978-3-319-69096-4_42

Mice models can be used to dissect the Hsa21 gene interactions and study the DS phenotype. Experimental pathway perturbations studies on the mouse model with segmental trisomy for Hsa21 provide unprecedented opportunity to decipher human DS pathway studies [6]. Ts65Dn mice model has three copies of over half of the Hsa21 orthologs genes on chromosome 16 [7], and it is a widely used DS model of human [8]. Some achievements have been accomplished in the DS therapies in mouse models. Laura et al found that prenatal treatments with NAPVSIPQ and SALLRSIPA could prevent developmental delay and the glial deficit [9]. Alexander et al discovered that Kcnj6 gene was a critical factor in synaptic and cognitive dysfunction. Its reduction could restore the Kir3.2 hippocampal level to normal [10].

In this study, we used various supervised learning algorithms to build DS screening models based on the protein/protein modification expression level in mouse DS model Ts65Dn cells, which showed very high accuracies and low false positive rates. Then we built an adaptive boosted (AdaBoost) feature selection method to identify the most correlated and informative proteins factors that was involved in the complex perturbations of DS biological processes and pathways. Furthermore, we improved the DS classification models by using these selected critical protein factors, which showed even better performances than the previous models build in this study. Moreover, we used the unsupervised learning algorithms to prove that the selected proteins were the critical factors that cause disorders of various DS phenotypes. These selected DS related proteins could be further used for protein-related and coding gene-related drugs developments.

2 Result

DS Screening Model Built from the Original Protein Expression Dataset

In this section, we used three supervised learning algorithms to construct the DS screening models based on the original proteins expression datasets of mouse DS model Ts65Dn, which contained 77 proteins (features) and 1080 instances. We first applied Random Forest, Support Vector Machine (SVM), and Decision Tree three algorithms to construct the DS screening models respectively. We randomly selected 880 instances as the training set, and used the rest 200 instances as testing sets. In order to build robust and reliable classification models, we also ran a 10-folds cross-validation for each model to provide a systematic evaluation. As Table 1

Table 1. Performances of DS classification models built from the original protein dataset

Classifier	Accuracy (%)	Precision (%)	Recall (%)	F-score (%)	Cross-validation (%)
Random Forest	97.0	97.1	96.8	96.9	96.8
SVM	93.5	93.5	93.3	93.4	95.6
Decision Tree	86.5	86.4	86.6	86.4	91.4

shown, all these three screening models could achieve good performances in the DS classification and in the 10-folds cross-validation, which all achieved over 80% accuracies. The screening model built by Random Forest had the best performance over all of the three models, which could achieve an accuracy of 97% in DS prediction and an accuracy of 96.8% in the 10-folds cross-validation. The Decision Tree screening model had the worst performance, which only had an accuracy of 86.5% in the DS prediction and an accuracy of 91.4% in the 10-folds cross-validation. SVM performed in the middle of the other two models. Figure 1 showed the confusion matrix for the DS predictions from these three models that constructed from discrete algorithms. All of these models not only achieved high accuracies, but also had very low false positive rates. We provided a novel and independent way for the DS screening based on the DS related protein expression datasets of mice DS model Ts65Dn.

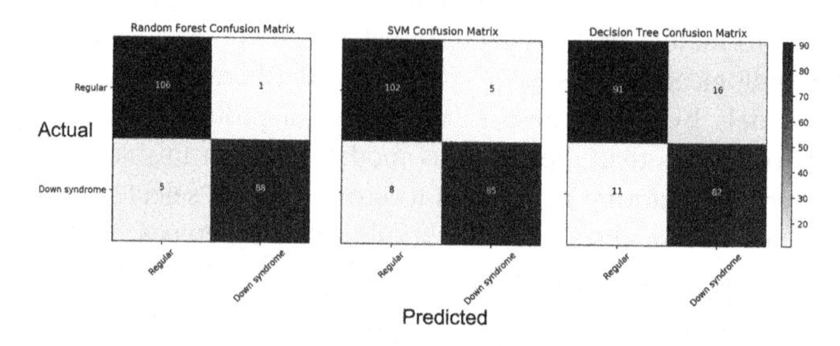

Fig. 1. Confusion matrix of three different models built from the original protein dataset

Ds Screening Models Built from the Selected Protein Expression Dataset

In this part, we built a feature selection method to search for the most correlated and informative proteins subset from the original protein expression dataset. Our algorithm was based on the combination of AdaBoost (Adaptive boosting) and Decision Tree algorithms. We used a set of Decision Tree classifiers with depth one to represent the original raw feature, and then used them as the weak classifiers in the Adaboost algorithm. The weights of the training instances would be updated on each weak learning cycle adaptively by decreasing the weight of correctly classified samples and increasing the weight of incorrectly classified samples. We used this feature selection method to fit the original protein expression datasets and output the importance of each feature by order, and select the most important n feature as a subset. Our method identified subset with 30 proteins (features), which removed more than half of the original features. We further constructed the improved screening models based on the datasets with selected proteins. As Table 2 shown, the new screening models built from selected proteins could achieve better performances than the previous models built from the original protein datasets, which indicated the validities and effectiveness of the selected proteins. The model built from SVM had the same performance with the previous model. We provided a novel and independent machine learning based, a non-invasive method to detect DS at a very high accuracy (98%) and low false positive rate (1.5%) (Fig. 2).

Table 2. Performances of DS classification models built from selected protein datasets

Classifier	Accuracy (%)	Precision (%)	Recall (%)	F-score (%)	Cross-validation (%)
Random Forest	98.0	98.0	97.9	97.9	97.7
SVM	93.5	93.3	93.4	93.4	95.6
Decision Tree	90.0	89.9	89.9	89.9	91.4

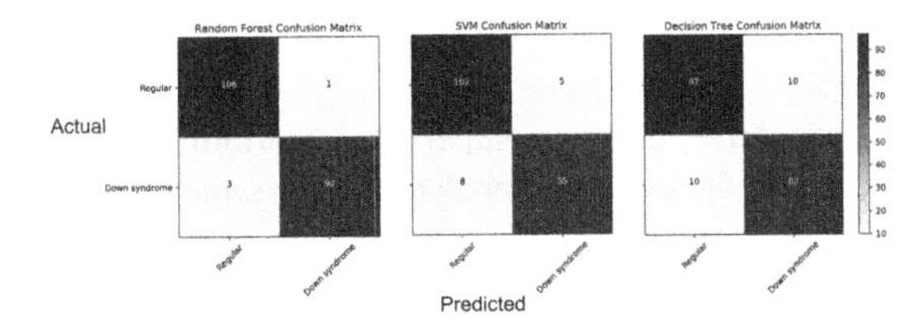

Fig. 2. Confusion matrix of three different models built from the selected protein dataset

Validation of Selected Proteins Subsets

Here, we used unsupervised learning algorithms to further validate the selected protein subset obtained in last section. We demonstrated that the selected proteins were critical biomarkers and pathway components that related with the phenotypes of DS. We first did the Principal component analysis (PCA) for both of selected DS related protein expression datasets and original protein expression datasets, and projected these datasets into 3D spaces. As Fig. 3 shown, the PCA of selected protein datasets (Fig. 3b) could better discriminate the non-DS and DS mice instances when compared with the PCA of the original protein dataset (Fig. 3a).

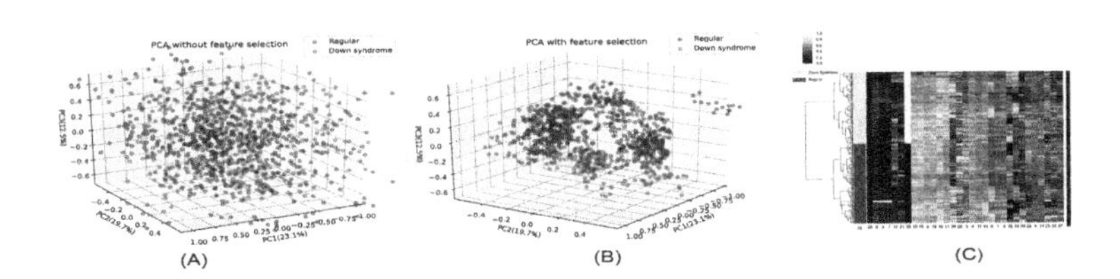

Fig. 3. PCA and hierarchical/agglomerative clustering heat map analyses of selected protein subsets based on unsupervised learning algorithms

Next, we used hierarchical/agglomerative clustering to cluster the original datasets and construct a hierarchical tree based on the selected features, which employed the Ward variance minimization algorithm to calculate the distances. We also provided a visualization of these clustering results in a cluster heat map, which could show the correlations and differences between any two instances. As Fig. 3c shown, the selected proteins could successfully cluster 510 DS instances and 570 non-DS instances into two sub-trees of the hierarchical tree topology. These two sub-trees are also matched with their original class labels on the left side of Fig. 3c. The above results demonstrated that our selected proteins were the most important and informative proteins that correlated with the DS phenotype and biological pathways, which could discriminate classes of DS mice and non-DS mice effectively. That's also why we could construct better DS classification/screening models by using the selected DS related protein subset in the last section.

3 Discussion

In this study, we used three different supervised learning algorithms to build DS screening models based on the protein expression data of mice model Ts65Dn. We also built a combined Adaboost and Decision Tree feature selection methods to find the most correlated and informative proteins that were related to the DS phenotype and biological pathways, which could discriminate classes of DS mice and non-DS mice effectively. All proteins identified in this study were associated with known functions in the biological pathways of DS [12]. Unsupervised PCA analysis and hierarchical/agglomerative clustering analysis also confirmed that these proteins were the critical factors and pathway components that were correlated with various symptoms of DS. DS relevant proteins subset revealed the molecular mechanisms and critical factor of DS biological pathways. Their abnormalities and over expressions of these selected proteins result in various symptoms of DS. A number of drugs were developed based on the functions and response mechanisms of DS relevant proteins, which suggested feasible prevention and amelioration ways to the cognitive deficits for human being [11]. Therefore, the selected DS related protein could be utilized on protein and their coding gene related and drug design.

Our DS screening models built from these selected proteins achieved very high accuracies and low false positive rates. Therefore, in practice, we didn't need to measure the protein expression levels for all DS related proteins since our methods only need a subset of them to construct accurate DS screening models. Only selected proteins needed to be measured in the clinical data collection process, and they could be further used in our machine learning DS screening method, which significantly reduced the cost and improved the precision of diagnosis. When applied to high-risk pregnancies, our methods could detect the DS based on the protein expressions levels from isolated fetal cells, which could substantially reduce the requirements of invasive diagnostic procedures and other high risks diagnostic procedures.

4 Methods

4.1 Data

The dataset used in this study contained the expression levels of 77 proteins/protein modifications in DS mice model Ts65Dn. All proteins used in this study were associated with known functions in the biological pathways of DS [12]. There were total 72 mice including 38 non-DS mice and 34 DS mice in this dataset. Each mouse was measured for 15 times. Each measurement was considered as an instance with known class labels. As a consequence, there were total 1080 instances in this datasets. 570 instances were non-DS mice, and 510 instances were the DS mice. This dataset was available in UCI Machine Learning Repository [13].

4.2 Protein/Feature Selection

This method was implemented by Python and Scikit Learn package [14]. We utilized the AdaBoost classifier from Scikit Learn package to build a strong classifier based on a set of weak classifiers. We used a set of Decision Trees with depth one as the weak classifiers to represent the raw features. The goal was to create an AdaBoosted model by learning simple decision rules inferred from raw data features. We used the "SAMME" algorithm as the boosting algorithm and set the learning rate to 1.5. The method would output the importance for each feature. We also kept the most important n raw features as the termination condition.

4.3 Learning Algorithms

All learning algorithms (Random Forest, SVM, Decision Tree and Principal component analysis (PCA)) that used in this study were implemented by Python and Scikit Learn package [14].The clustering was performed by hierarchical/agglomerative clustering of SciPy package, which used the Ward variance minimization algorithm to calculate the distances. The hierarchically clustered heat map was implemented by Python and Seaborn package.

References

1. Patterson, D.: Molecular genetic analysis of down syndrome. Hum. Genet. **126**, 195–214 (2009)
2. Antonarakis, S.E.: Down syndrome and the complexity of genome dosage imbalance. Nat. Rev. Genet. **18**, 147–163 (2017)
3. Weijerman, M.E., De Winter, J.P.: Clinical practice. Eur. J. Pediatr. **169**, 1445–1452 (2010)
4. Wuang, Y.-P., Chiang, C.-S., Su, C.-Y., Wang, C.-C.: Effectiveness of virtual reality using wii gaming technology in children with down syndrome. Res. Dev. Disabil. **32**, 312–321 (2011)
5. Smith-Hicks, C.L., Cai, P., Savonenko, A.V., Reeves, R.H., Worley, P.F.: Increased sparsity of hippocampal ca1 neuronal ensembles in a mouse model of down syndrome assayed by arc expression. Front. Neural Circuits 11 (2017)
6. Gardiner, K., et al.: Down syndrome: from understanding the neurobiology to therapy. J. Neurosci. **30**, 14943–14945 (2010)
7. Reinholdt, L.G., et al.: Molecular characterization of the translocation breakpoints in the down syndrome mouse model ts65dn. Mamm. Genome **22**, 685–691 (2011)
8. Kuehn, B.M.: Treating Trisomies: Prenatal Down's Syndrome Therapies Explored in Mice (2016)

9. Toso, L., et al.: Prevention of developmental delays in a down syndrome mouse model. Obstet. Gynecol. **112**, 1242 (2008)
10. Kleschevnikov, A.M., et al.: Evidence that increased kcnj6 gene dose is necessary for deficits in behavior and dentate gyrus synaptic plasticity in the ts65dn mouse model of down syndrome. Neurobiol. Dis. **103**, 1–10 (2017)
11. Gardiner, K.J.: Pharmacological approaches to improving cognitive function in down syndrome: current status and considerations. Drug. Des. Devel. Ther. **9**, 103–125 (2015)
12. Higuera, C., Gardiner, K.J., Cios, K.J.: Self-organizing feature maps identify proteins critical to learning in a mouse model of down syndrome. PloS one **10**, e0129126 (2015)
13. Lichman, M.: UCI Machine Learning Repository (2013) URL http://archive.ics.uci.edu/ml
14. Pedregosa, F., et al.: Scikit-learn: machine learning in python. J. Mach. Learn. Res. **12**, 2825–2830 (2011)

A Multi-view Approach for Visual Exploration of Temporal Multi-dimensional Vehicle Experiment Data

Xianglei Zhu[1(✉)], Haining Tong[2], Shuai Zhao[1], Quan Wen[1], and Jie Li[2]

[1] China Automotive Technology and Research Center, Tianjin, China
zhuxianglei@catarc.ac.cn
[2] School of Computer Software, Tianjin University, Tianjin, China

Abstract. We introduce a new visual analytics approach to the vehicle experiment data with multi-dimensional and time-series characteristics. Our approach integrates four visualization techniques to explore potential patterns hidden in huge amounts of actual experiment data. Overview view is the main view, which is used to show high level data characteristics of the experiment process and provide an intuitive and compact view for analyzing temporal and multi-dimensional patterns. Other three visualization techniques can be viewed as the complementary views, specialized in analyzing temporal features, relationships between different attributes and vehicle status variation, which are also important domain tasks. A case study has been conducted using a real-world dataset to verify the effectiveness and scalability of our approach.

Keywords: Vehicle experiment data · Temporal multi-dimensional visualization · Multi-view visualization · Visual analysis system · Visual analytics

1 Introduction

Vehicle running experiment is one of the main means to obtain vehicle performance and driving experience. Furthermore, the research of self-driving vehicle [1] also depends on such kinds of experiment data which simultaneously contain driving operations, vehicle status and road condition.

Although the experiment can effectively improve the maturity of the vehicles, also imposes great challenges for analyzing and understanding the potential patterns hidden in them. To fully utilize the data, effective visualization techniques are needed [2]. Although there exist many types of visual analytics techniques work on analyzing vehicle running data, such as Trajectory wall [3] and GeoTime [4], they mainly focus on trajectories, few of them can be directly used to represent and analyze vehicle running experiment data which explicitly has temporal multi-dimensional characteristic. The complex data structure makes it difficult to find a general approach that can cater for visualizing the overview, as well as find the potential patterns in terms of the vehicle and driver.

F. Xhafa et al. (eds.), *Advances in Intelligent Systems and Interactive Applications*, Advances in Intelligent Systems and Computing 686,
https://doi.org/10.1007/978-3-319-69096-4_43

This paper introduces a visualization framework that can intuitively and interactively help the analyst restore the experiment process. The main objective is to find various potential patterns of vehicles and drivers, and the relationships between experiment results and different vehicle parameters. The major contributions include:

- A multi-view visual analytics framework for discovering the patterns in vehicle experiment data patterns.
- Three visual analytic views specialized in visualizing overview, temporal, multi-dimensional features of vehicle experiment data.
- A case study with actual vehicle running records, which proves the effectiveness and the scalability of the approach.

2 Approach

2.1 Objectives

We conducted several interviews with four researchers from China Automobile Technology & Research Center, through which we have identified different analytical tasks. We identify the most important tasks into three major categories:

- **Experiment Process Overview**: understanding the general experiment process by comprehensively visualizing the distributions of different attributes.
- **Time Series/Multi-dimensional Characteristic**: finding the period having specific attribute features, highlighting key time points, showing the distribution of multiple attributes and identifying the relationships between different attributes.
- **Abnormal Case**: discovering the abnormal driving actions and vehicle statuses.

2.2 Visual and Interactive Design

Our approach consists of four views, i.e. overview view, multi-dimensional view, temporal view, and status view. The main interface of our approach is illustrated in Fig. 1. We will introduce the four views in the next four sections.

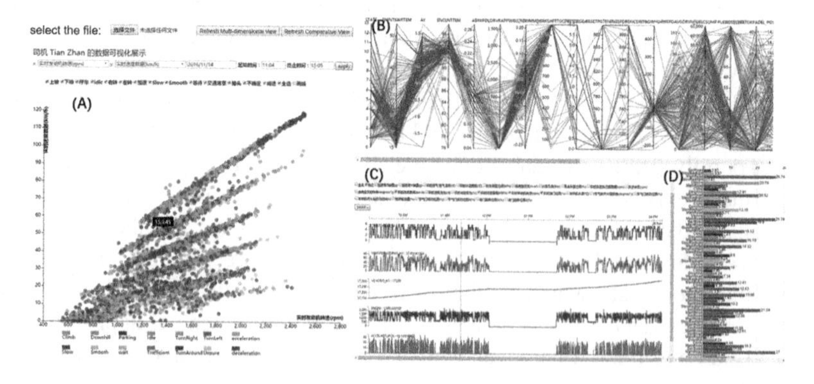

Fig. 1. The main interface of our approach. **a** Overview view. **b** Multi-dimensional view. **c** Temporal view. **d** Status view.

Overview View
Overview view, as in Fig. 1a, is the main view of our approach, which shows the high-level spatiotemporal features of the analyzed dataset. In general, the overview view is a scatter plot. The basic idea is to create a series of snapshots of subsequent points, each snapshot contains an instance of the original data.

Multi-dimensional View
The vehicle status has always been changing, reflected as the variations of the values of different sensor parameters. We thus introduce a multi-dimensional view specialized in analyzing the relationships between different attributes and the vehicle status. The Correlation Detection View is constructed based on the parallel coordinates, as shown in Fig. 1b.

Temporal View
Identifying the temporal features of different attributes is the prerequisite for finding and understanding patterns hidden in the data. We therefore design the temporal view, as in Fig. 1c. The temporal view is constructed based on line chart. The simple and intuitive visual design enables the easy understanding of the variation of different attribute values.

Status View
To show the status variation, we design the status view, as in Fig. 1d. We divided the vehicle status into 14 categories, which can represent the general moving feature of the vehicle. Vertical direction represents time axis. Each bar represents a continuous vehicle status, and the width represents the duration. We assign a color to each status. By comprehensively viewing the color and the bar length, the user can intuitively understand the experiment process and determine the possible road condition.

3 Case Study

3.1 Verifying the Verified Conclusions

We first use the overview view to review the experiment process. Because the experiment data of different drivers have similar structure, we use the data of two drivers to illustrate the effectiveness of our approach, as in Fig. 2. The vertical and horizontal axes are respectively set to represent the remaining fuel and time. We found that the fuel amounts of the both the vehicles has been continuously decreasing during the experiment processes, while the first vehicle was refueled at 3:15 pm (see the steep line in Fig. 2a). At noon, both two experiments were suspended for one hour, because the fuel amount did not decrease during that period. Furthermore, we find that the slope of the temporal line of the second driver is much bigger, meaning that the driving style or vehicle of the second driver is more gas-guzzling.

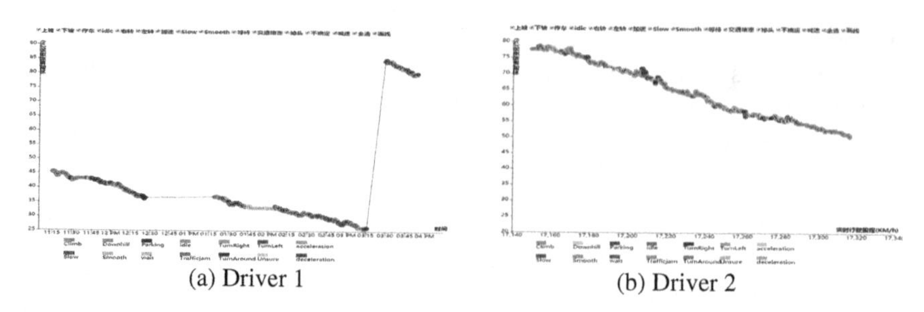

(a) Driver 1 (b) Driver 2

Fig. 2. Experiment processes of two drivers.

We continue to set different attributes to the axes of the overview view. Figure 3a shows the point distribution of steering wheel corner and real-time speed. In general, the visualization is in the form of normal distribution, meaning that the steering wheel corner gradually reduces with the real-time speed increases. This is consistent with the real situation that only when the vehicle is at a low speed can we greatly turn the steering wheel. We also set the two axes to represent real-time speed and engine speed to view the corresponding point distribution, as in Fig. 3b. The data points form 6 linear distribution areas, which represent different gear patterns. Obviously, the top area represents the high-speed gear, because the vehicle speed increase much faster than other areas, while the bottom area represents the low-speed gear.

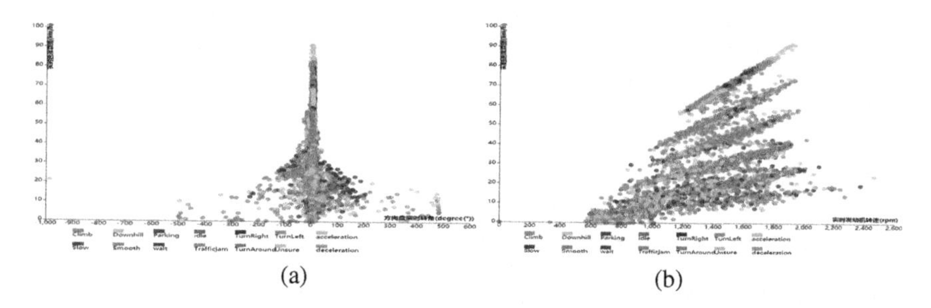

(a) (b)

Fig. 3. Point distributions of two different axis settings. **a** X-axis and y-axis represent steering wheel corner and real-time speed. **b** X-axis and y-axis represent real-time speed and engine speed.

We continue to use the status view to analyze the possible road conditions of two experiments. Two representative periods are selected, as in Fig. 4. In the First visualization (Fig. 4a), the acceleration and deceleration statuses are frequently switched and the duration of a status is relatively short, which implies a congested traffic. The second visualization (Fig. 4b) contains many downhill and climb statuses, indicating that the terrain variation is more complex. The domain experts confirm our findings. In fact, the two experiments were conducted in downtown and mountain area, resulting in two different status variation patterns.

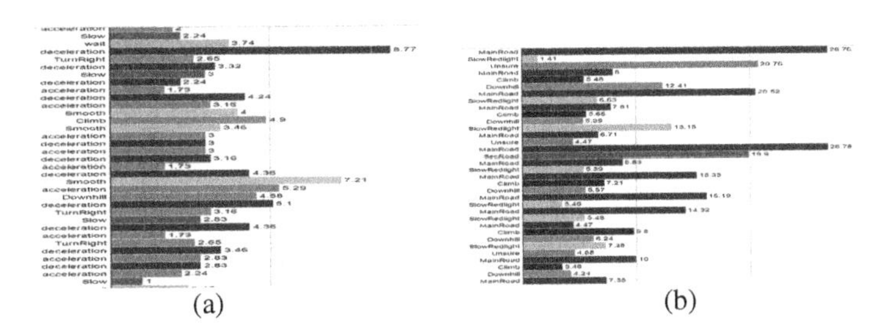

(a) (b)

Fig. 4. Status variations of two drivers.

3.2 Interesting Findings

We attempt to find several abnormal distribution in overall view. First, the vertical and horizontal axes are set to represent accelerator position and brake pedal power. As we know, it is impossible for the driver to simultaneously step on brake and accelerator and all the data points should be distributed on two axes. However, we find there are a number of points that are not on the two axes. Figure 5 shows such distributions of two driver. Obviously, the second driver have more abnormal points, indicating that his driving habit is not as normative as the first driver.

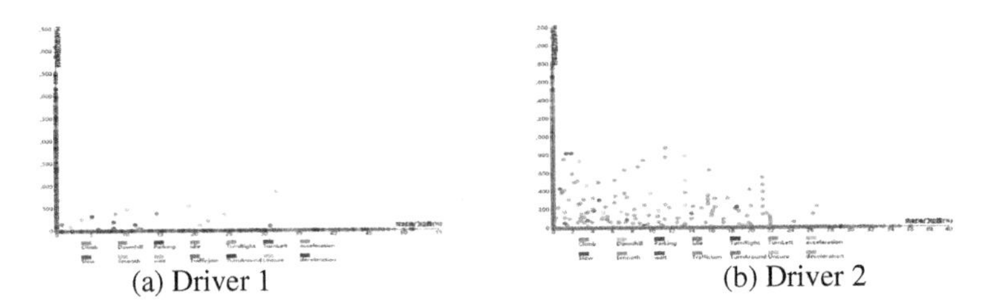

(a) Driver 1 (b) Driver 2

Fig. 5. Abnormal data distribution. Vertical and horizontal axes represent accelerator position and brake pedal power.

To further comparative the driving styles of the two drivers, we use the multi-dimensional view to visualize the gear distribution of the two drivers when turning right, as in Fig. 6. We interactively select the fifth status, which represents turning right (see the first axis in Fig. 6), and view the gear distribution on the second axis. The visualization results also confirm that the first driver has better driving habit, because he always uses low gear when turning right, while the second driver frequently uses high gear.

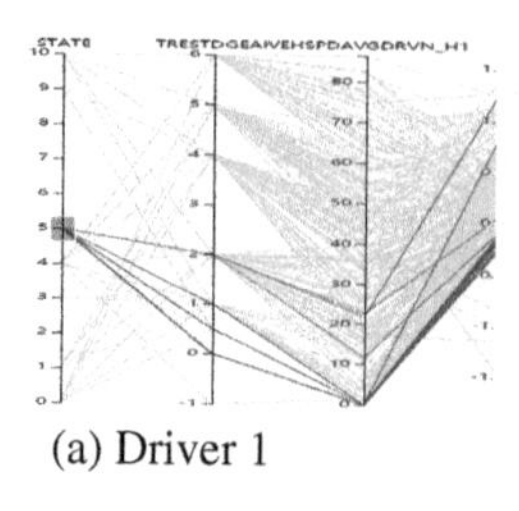

(a) Driver 1

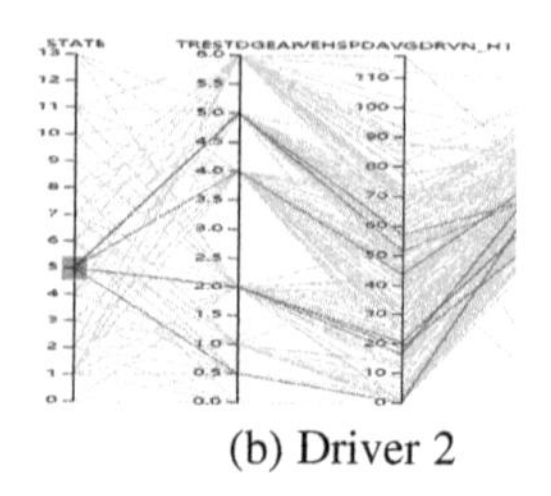

(b) Driver 2

Fig. 6. The gear distribution when turning right.

We choose several attributes to analyze their variation patterns. Figure 7 shows the temporal variation of three attributes: gear (blue), engine speed (orange), and accelerator position (green). Having drawn the temporal lines of the three attributes, their variation patterns are clearly shown. We find that when the gear is switched from low to high, the engine speed will significantly reduce, and the engine speed will increase if the gear gets down. When the gear remains unchanged, the engine speed is proportional to the accelerator position.

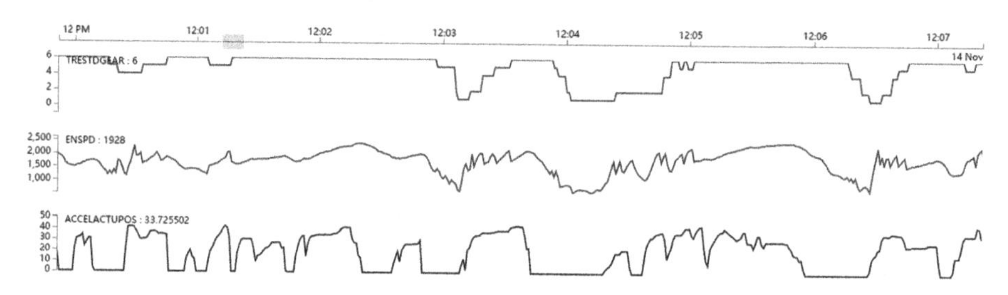

Fig. 7. Variation patterns of gear, engine speed, and accelerator position.

4 Conclusion

This paper has shown a multi-view visualization framework to the analysis of vehicle experiment data. Three visualization views, specialized in showing different aspects of the experiment data, have been designed and used to identify the patterns hidden in huge amounts of experiment data. Having analyzed the various temporal and multi-dimensional patterns, we consider our approach to be effective and useful in real-world scenarios. In the future, our approach could be improved in two aspects. First, we plan to introduce more interactive and rendering techniques to improve the exploration experience. Second, we aim to conduct a thorough empirical user study to evaluate our approach in both laboratory and practical settings.

Acknowledgements. The work is partially supported by National NSFC project (Grant number 61602340), National NSFC project (Grant number 61572348) and National High-tech R&D Program (863 Grant number 215AA020506).

References

1. Kim, J., Kim, H., Lakshmanan, K., Rajkumar, R.R.: Parallel scheduling for cyber-physical systems: analysis and case study on a self-driving car. In: Proceedings of the ACM/IEEE 4th International Conference on Cyber-Physical Systems, pp. 31–40 (2013)
2. Andrienko, G., Andrienko, N., Wrobel, S.: Visual analytics tools for analysis of movement data. In: Proceedings of ACM SIGKDD Explorations Newsletter, vol. 9, no. 2, pp. 38–46 (2007)
3. Tominski, C., Schumann, H., Andrienko, G., Andrienko, N.: Stacking-based visualization of trajectory attribute data. IEEE Trans. Vis. Comput. Graph. **18**(12), 2565–2574 (2012)
4. Kapler, T., Wright, W.: GeoTime information visualization. Inf. Vis. **4**(2), 136–146 (2005)

Impact Analysis of Geometry Parameters of Buoy on the Pitching Motion Mechanism and Power Response for Multi-section Wave Energy Converter

Biao Li and Hongtao Gao(✉)

The College of Marine Engineering, Dalian Maritime University, Dalian, China
gaohongtao@dlmu.edu.cn

Abstract. This paper presents the motion characteristic and power response of a multi-section wave energy converter (WEC) consisting of two hinged cylindrical buoys and two power take-off (PTO) damping units at the joint according to a heaving and pitching motion model. The influences of geometry parameters of buoy (length, radius and draft) on the pitching motion and energy conversion ability are analyzed and summarized. The analysis results show that the relative pitching angle of two buoys depends on the pitching angles and the phase difference. The resonance frequency of pitching motion of buoy is mainly related to the radius and draft, but the phase difference generally depends on the length of buoy. The resonance of buoy and phase difference of odd multiples of 180° are necessary conditions for attaining maximum relative capture width.

Keywords: Wave energy · WEC · Geometry parameter · Relative capture width · Pitching motion · Power response

1 Introduction

Ocean wave energy is being increasingly regarded as one of the most promising renewable and alternative resources around the world [1–3]. In major coastal countries, wave energy is integrated in the government's policies, although it is still in an initial stage of development. In the white paper on China's energy policy (2012) published by the Information Office of the State Council in October 2012, the renewables must account for the overall target of a 15% share of total energy by 2020, in which the wave energy provides an important source [4]. For the advantages of widely distribution and large energy flux density [5], various wave energy converter (WEC) designs and techniques have been proposed for energy-harvesting since the late 18th century [6]. So far, there have been more than 1000 WEC patents granted in Europe, North America and East Asia [7]. However, only a minority of them have been actually used. Among these devices, the multi-section WEC have been proven to have high wave energy conversion efficiency and good survivability in rough sea conditions. They mainly use the relative rotation around the hinge(s) to drive electricity generation system. On the principle of kinetic energy conversion, the multi-section WEC is generally classified as

F. Xhafa et al. (eds.), *Advances in Intelligent Systems and Interactive Applications*, Advances in Intelligent Systems and Computing 686,
https://doi.org/10.1007/978-3-319-69096-4_44

line absorber with multi-degree of freedom (DOF) motions of buoys. Compared with the point absorber, there is little wave energy losses in the line absorber [8].

So far, to the authors' knowledge, for the multi-section WEC, no work has been reported about the correlation studies between the motion mechanism and energy conversion efficiency [9]. In this work, a two hinged buoys WEC is taken as the research subject. A multi-DOF motion (heaving and pitching) model with motion constraint condition is introduced based on hydrodynamic theory. Then the influences of geometric parameters of buoy on the pitching motion and energy conversion ability are analyzed and summarized.

2 Motion Model

As shown in Fig. 1, the WEC is constituted of two cylindrical buoys (*Buoy (1)* and *Buoy (2)*) connected by two PTO damping units. The springs are installed to provide part of restoring force for pitching motion of buoy. The device partially submerged in the seawater. The two buoys are equal of length (L) and weight (m) that evenly distributed. The length of the hinge is $2l$.

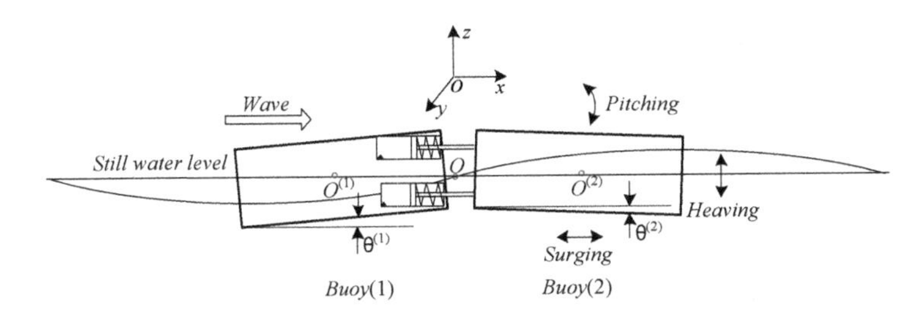

Fig. 1. Configuration diagram of the wave energy converter

Only the heaving and pitching motions are considered in the stable running with an incidence angle of 0. It is assumed that the fluid flow is inviscid, incompressible, and irrotational. The heaving-pitching motion of each buoy can be expressed as follows:

$$\begin{cases} \left(m^{(n)} + m_A^{(n)}\right) \cdot \ddot{z}^{(n)}(t) = F^{(n)}(t) \\ \left(J_{yy}^{(n)} + J_A^{(n)}\right) \cdot \ddot{\theta}^{(n)}(t) = M^{(n)}(t) \end{cases} \tag{1}$$

Where $n = 1$ and $n = 2$ represent the physical variables of the *Buoy* (1) and *Buoy* (2) respectively. $m^{(n)}$ is the mass of buoy, $m_A^{(n)}$ is the added mass in the z-axis. $J_{yy}^{(n)}$ and $J_A^{(n)}$ are the moment of inertia and added mass moment of inertia to y-axis. $\ddot{z}^{(n)}(t)$ is acceleration of heaving motion, $\ddot{\theta}^{(n)}(t)$ is acceleration of pitching motion. $F^{(n)}(t)$ is resultant force acting on buoy, $M^{(n)}(t)$ is resultant moment acting on the geometric center of the WEC.

In addition, a motion constraint condition must also be satisfied.

$$z^{(1)} - z^{(2)} = -(L/2 + l) \cdot \left(\sin\theta^{(1)} + \sin\theta^{(2)}\right) \tag{2}$$

Together, the heaving-pitching coupling motion model of the WEC is described by Eqs. (1) and (2).

In this paper, RAO (Response Amplitude Operator) is a dimensionless parameter which is defined to describe the motion amplitude relative to the wave.

$$RAO = \frac{\eta_i}{A} \tag{3}$$

Where η_i is the motion amplitude of buoy in the i DOF (Degree of Freedom), A is the wave amplitude.

3 Pitching Motion Characteristic and Power Analysis

In this paper, a fixed constant damping coefficient of PTO damping units and stiffness coefficient of restoring springs are adopted that are set as $C = 5.0 \times 10^4$ N s/m and $K = 2.5 \times 10^4$ N/m respectively. Figure 2 is the frequency-domain curves of pitching motion and phase difference. The buoy length L is 8.0 m, radius R is 1.0 m, draft D_r is 1.0 m and $2l$ is 1.0 m. It is easy to understand that the relative pitching angle depends on the pitching angle and phase difference.

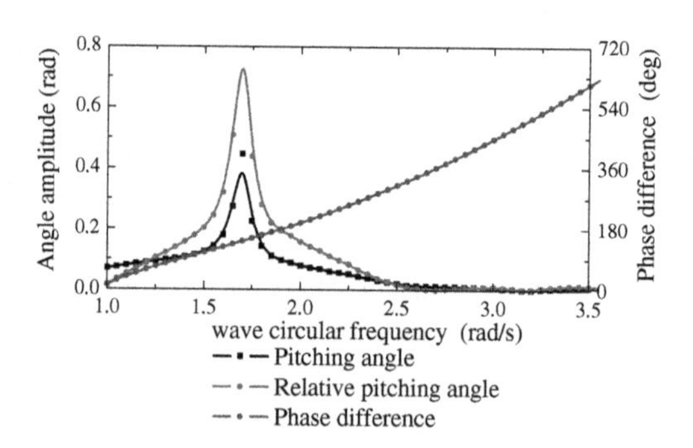

Fig. 2. Frequency-domain curves of pitching motion and phase difference

When the peak of pitching angle reaches a maximum, the resonance can be considered to occur. In the combined actions of pitching motion and phase difference, when the phase differences are 180° or the odd multiples of 180°, the two pitching motions display in symmetrical forms. When the phase differences are 0° or the even multiples of 180°, the relative pitching angles are 0.

3.1 Influence of Length of Buoy

Figure 3 shows the RAO and phase difference with wave circular frequency. For the buoys at radius R = 1.0 m and draft D_r = 1.0 m, the RAO increase with wave circular frequency, then decrease after reaching maximum values at the same wave circular frequency. It can be considered that the length does not affect the natural vibration frequency of buoy under setting conditions. And as the length increases, the phase difference becomes greater.

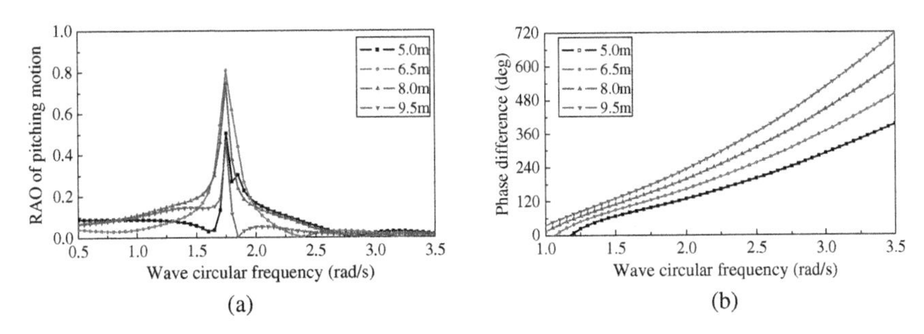

Fig. 3. RAO and phase difference of pitching motion with wave circular frequency in several lengths of buoy

Figure 4 shows the relative capture width with length of buoy. It is indicated that the highest relative capture width is founded at the lengths of 7.9 m and wave circular frequency of 1.9 rad/s. In this case, the phase difference is about 180°.

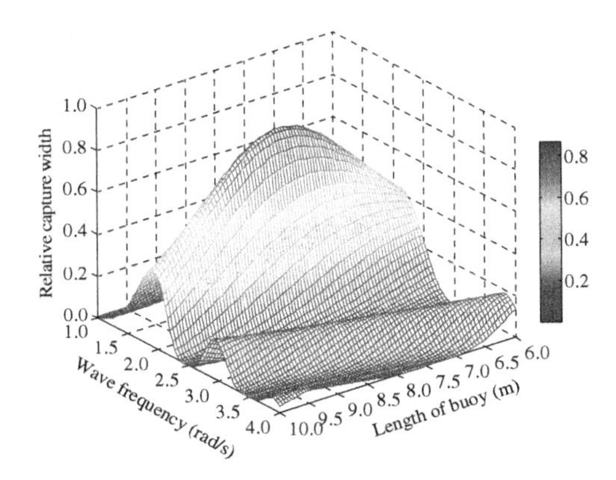

Fig. 4. Frequency-domain curves of relative capture width with length of buoy

3.2 Influence of Radius of Buoy

The RAO and phase difference with wave circular frequency in several radiuses of buoy is given in Fig. 5. With the increase of the radius, the wave circular frequency that corresponds to the peak RAO decreases. When the wave circular frequency exceeds 1.3 rad/s, the radius has little impact on the phase difference.

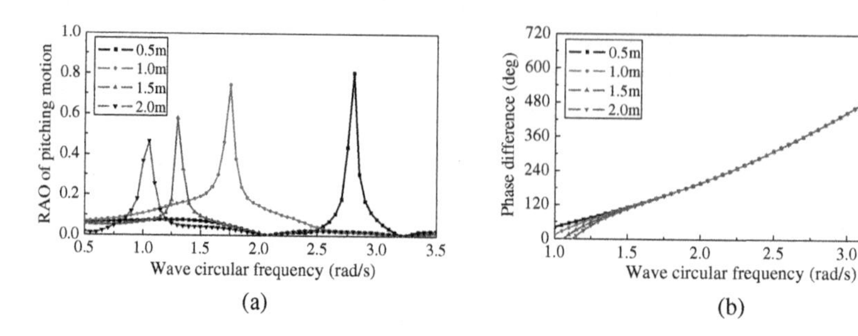

Fig. 5. RAO and phase difference of pitching motion with wave circular frequency in several radiuses of buoy

The relative capture width with radius of buoy is shown in Fig. 6, the peak relative capture width is about 0.98 which is discovered at the wave circular frequency of 1.9 rad/s and radius of about 1.2 m. The phase difference is about 180°.

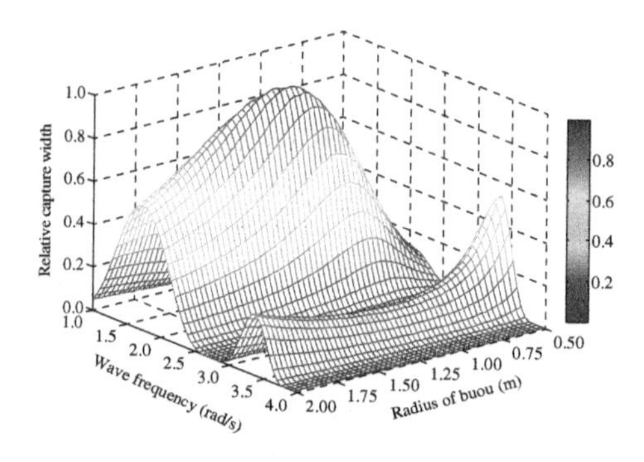

Fig. 6. Frequency-domain curves of relative capture width with radius of buoy

3.3 Influence of Draft of Buoy

Figure 7 shows the RAO and phase difference with wave circular frequency. With increase of draft, the wave circular frequencies of the curve peaks decrease continuously. The draft also has almost no influence on the phase difference.

Figure 8 presents the relative capture width with draft of buoy, when the draft is 1.25 m, the maximum relative capture width is about 1.0 at the frequency of about 1.9 rad/s. In this case, the phase difference is also about 180°.

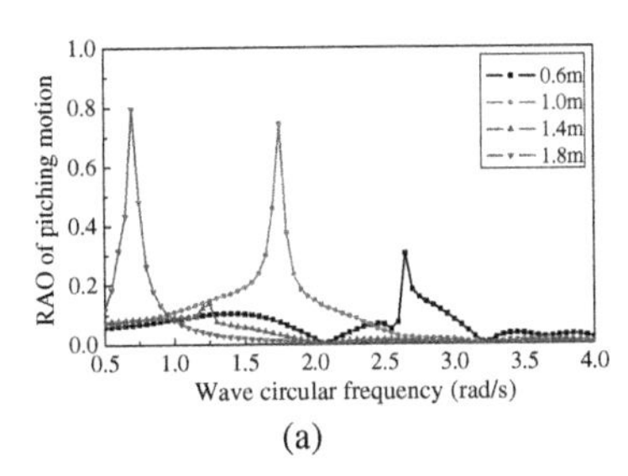

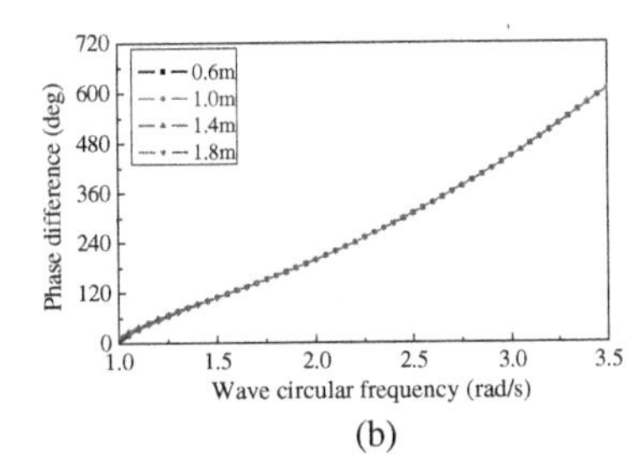

Fig. 7. RAO and phase difference of pitching motion with wave circular frequency in several radiuses of buoy

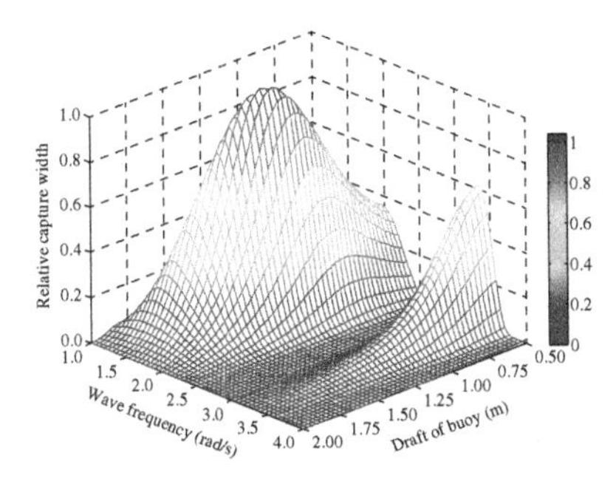

Fig. 8. Frequency-domain curves of relative capture width with draft of buoy

4 Conclusion

The conclusions of this study can be summarized as follows:

(1) The pitching motion amplitudes of the two buoys are the same with a phase difference by the theoretical analysis. The relative pitching angle depends on the pitching angle and phase difference of two buoys.
(2) The resonance frequency of pitching motion of buoy is mainly related to radius and draft. The phase difference generally depends on the length of buoy.
(3) The resonance of buoy and phase difference of odd multiples of 180° are necessary conditions for attaining maximum relative capture width.

Acknowledgments. The authors extend their thanks to the financial support by the Maritime Safety Administration of the People's Republic of China (No. 2012_26). Special thanks are given to Liaoning Maritime Safety Administration of the People's Republic of China for the help to make the project go through.

References

1. Falcão, A.F.O.: Wave energy utilization: a review of the technologies. Renew. Sust. Energ. Rev. **14**, 899–918 (2010)
2. Behrens, S., Hayward, J., Hemer, M., et al.: Assessing the wave energy converter potential for Australian coastal regions. Renew Energ **43**, 210–217 (2012)
3. Stansell, P., Pizer, D.J.: Maximum wave-power absorption by attenuating line absorbers under volume constraints. Appl. Ocean Res. **40**, 83–93 (2013)
4. State Council of the People's Republic of China: China's Energy Policy. Beijing, China (2012)
5. Lund, H.: Renewable energy strategies for sustainable development. Energy **32**, 912–919 (2007)
6. Wan, N.W., Sulaiman, O.O., Rosliza, R., et al.: Wave energy resource assessment and review of the technologies. Int. J. Energ. Environ. Econ. **2**, 1101–1112 (2011)
7. Drew, B., Plummer, A.R., Sahinkaya, M.N.: A review of wave energy converter technology. Proc. Inst. Mech. Eng. Part A J. Power Energ. **223**, 887–902 (2009)
8. Yemm, R., Henderson, R.: Pelamis: experience from concept to connection. Philos. Trans. R. Soc. Math. Phys. Eng. Sci. **370**, 365–380 (2012)
9. Lee, C.H., Newman, J.N.: An assessment of hydroelasticity for very large hinged vessels. J. Fluids Struct. **14**(7), 957–970 (2000)

Binary Tree Construction of Multiclass Pinball SVM Via Farthest Centroid Selection

Qiangkui Leng[1(✉)], Fude Liu[2], and Yuping Qin[3]

[1] College of Information Science and Technology, Bohai University, Jinzhou 121000, China
qkleng@gmail.com

[2] Research and Teaching Institute of College Basics, Bohai University, Jinzhou 121000, China

[3] College of Engineering, Bohai University, Jinzhou 121000, China

Abstract. The paper generalizes PinSVM to multiclass version by using binary tree structure. At each internal node, all inherited classes are first divided into two groups via farthest centroid selection. Then, PinSVM is constructed between two groups. When each group contains only one class, the leaf node can be identified. The experimental results show that binary-tree multiclass PinSVM is very competitive with one-versus-one PinSVM and one-versus-one SVM. Especially in terms of computational time, it has clear superiority than them.

Keywords: Multiclass PinSVM · Binary tree structure · Centroid selection · SVM

1 Introduction

As a successful classification method, support vector machine (SVM) has been widely concerned and has gained great progress [1]. It is built on the principle of structural risk minimization and VC dimension theory. SVM employed the hinge loss in general, which may be sensitive to noise, and exhibits instability during resampling [2]. Lately, Huang et al. [3] proposed SVM with pinball loss (PinSVM). It is related to the quantile distance and enjoys noise insensitivity. PinSVM was initially designed for binary classification. However, most real-world scenarios correspond to multiclass problems. Therefore, extending PinSVM to multiclass version is an issue worthy of concern.

The well-known methods for multi-class SVMs include one-versus-one [4], one-versus-all [5], directed acyclic graph [6], and binary tree structure [7]. For one-versus-one and one-versus-all, they require a lot of binary tests for making a final decision. Although directed acyclic graph may be able to reduce the testing number, it is still undesirable when the number of classes is large. As for binary tree structure, it only needs to train K-1 binary classifiers and test log_2^K times, where a fast decision is achieved.

In the paper, we will generalize PinSVM to multiclass version by using binary tree structure. At each internal node that contains more than two classes, all inherited classes are first divides into two groups via farthest centroid selection. Then, PinSVM is constructed between two groups. When each group contains only one class, the leaf node can

F. Xhafa et al. (eds.), *Advances in Intelligent Systems and Interactive Applications*, Advances in Intelligent Systems and Computing 686,
https://doi.org/10.1007/978-3-319-69096-4_45

be identified. Eventually, a binary tree is created which represents a hierarchical partition. The numerical experiments will demonstrate the validity of the proposed method.

2 Preliminaries

The hard-margin SVM is the simplest form in SVM community. For two finite sets X, $Y \subseteq \mathbf{R}^n$, if they are linearly separable, a hyperplane $w{\cdot}x + b = 0$ can be obtained, and the classification margin is denoted as $2/\|w\|$. The optimization goal can be denoted as:

$$\begin{aligned} &\min \frac{1}{2}\|w\|^2 \\ &\text{s.t.}\ \ y_i(w \cdot x_i + b) \geq 1,\ \ i = 1, 2, \ldots, l. \end{aligned} \tag{1}$$

l is the number of training samples. The dual problem of (1) can be expressed as:

$$\begin{aligned} \min\ \ & \frac{1}{2}\sum_{i=1}^{l}\sum_{j=1}^{l}\alpha_i\alpha_j y_i y_j (x_i \cdot x_j) - \sum_{j=1}^{l}\alpha_j \\ \text{s.t.}\ \ & \sum_{i=1}^{l}\alpha_i y_i = 0,\quad \alpha_i \geq 0,\ \ i = 1, 2, \ldots, l. \end{aligned} \tag{2}$$

In order to deal with nonseparable problems, soft-margin SVM is further presented:

$$\min \frac{1}{2}\|w\|^2 + C\sum_{i=1}^{l} L(x_i, y, f(x_i)). \tag{3}$$

where C is a trade-off parameter and $L(x, y, f(x))$ denotes the loss function. When we introduce hinge loss into (3), the following optimization goal is generated:

$$\begin{aligned} &\min \frac{1}{2}\|w\|^2 + C\sum_{i=1}^{l}\xi_i \\ &\text{s.t.}\ \ y_i(w \cdot x_i + b) \geq 1 - \xi_i,\ \ \xi_i \geq 0,\ \ i = 1, 2, \ldots, l. \end{aligned} \tag{4}$$

The hinge loss is related to the shortest distance, denoted as

$$L_{hinge}(x, y, f(x)) = \max\{1 - yf(x), 0\} \tag{5}$$

However, it is sensitive to noise and exhibits instability during resampling. By the relationship of hinge loss and shortest distance, the pinball loss [3] is proposed as follows:

$$L_\tau(x, y, f(x)) = \begin{cases} 1 - yf(x), & 1 - yf(x) \geq 0 \\ -\tau(1 - yf(x)), & 1 - yf(x) < 0. \end{cases} \tag{6}$$

Thus, pinball loss can be used in (3), and the new optimization goal is obtained:

$$\begin{aligned} &\min \frac{1}{2}\|w\|^2 + C\sum_{i=1}^{l}\xi_i \\ &\text{s.t.}\ \ y_i(w\cdot\phi(x_i)+b)\geq 1-\xi_i,\ i=1,2,\ldots,l \\ &\qquad y_i(w\cdot\phi(x_i)+b)\geq 1+\frac{1}{\tau}\xi_i,\ i=1,2,\ldots,l. \end{aligned} \tag{7}$$

The dual form of (7) is expressed as follows,

$$\begin{aligned} \max\quad & -\frac{1}{2}\sum_{i=1}^{m}\sum_{j=1}^{m}\lambda_i y_i K(x_i,y_j)y_j\lambda_j + \sum_{i=1}^{m}\lambda_i \\ \text{s.t.}\quad & \sum_{i=1}^{m}\lambda_i y_i = 0,\ \ \lambda_i + (1+\frac{1}{\tau})\beta_i = C, i=1,2,\ldots,m, \\ & \lambda_i+\beta_i\geq 0, \beta_i\geq 0, i=1,2,\ldots,m. \end{aligned} \tag{8}$$

The key superiority of PinSVM over SVM with hinge loss is its strong insensitivity to the effects of noise. Moreover, it has the similar time consumption with hinge loss SVM.

3 Binary Tree Structure for Multiclass PinSVM

To intuitively illustrate the proposed method, we introduce a 4-class example in Fig. 1a. Suppose $K(\geq 2)$ stands for the number of classes. We can denote a class $Class_i = \{X_l^{\ i} \mid l = 1, 2, \ldots, N_i\}$, which is composed of samples in n-dimensional Euclidean space,

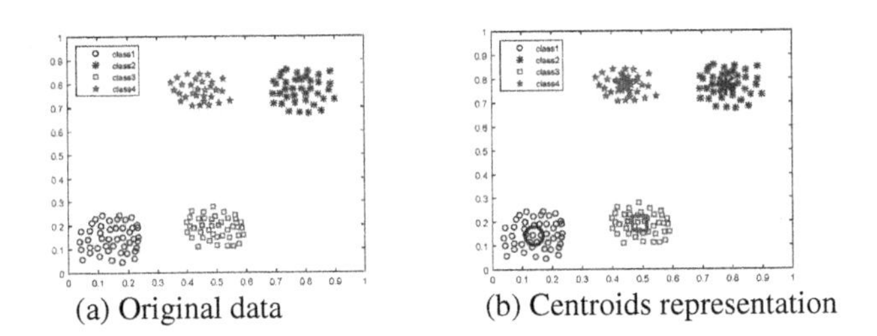

(a) Original data (b) Centroids representation

Fig. 1. An example of four classes

$$X_l^i = \left(x_{1l}^i, x_{2l}^i, \ldots, x_{nl}^i\right), i = 1, 2, \ldots, K. \tag{9}$$

First, we want to divide all the classes into two groups. Inspired by the trick in [8], we compute the centroids of all classes as their representations,

$$C^i = \left(c_1^i, c_2^i, \ldots, c_j^i, \ldots, c_n^i\right) \tag{10}$$

where

$$c_j^i = \frac{1}{N_i}\sum_{l=1}^{N_i} x_{jl}^i, \quad i = 1, 2, \ldots, K. \tag{11}$$

Thus, we can establish the centroids concerning all classes, denoted as

$$C = \left\{C^i \middle| i = 1, 2, \ldots, K\right\}. \tag{12}$$

In Fig. 1b, we generate four centroids, namely $C = \{C^1, C^2, C^3, C^4\}$. Then, we will determinate two farthest centroids, and mark them as C^p and C^q. Each can identity a group, G_C^p or G_C^q. For the remaining centroids, we can assign them to G_C^p or G_C^q by their distance to C^p and C^q. For $C^i(i \neq p\&i \neq q)$, if $d(C^i, C^p) \leq d(C^i, C^q)$, then $C^i \in$ G_C^p; $C^i \in$ G_C^q otherwise. Fig. 1b shows two marked groups, G_C^1 = $\{C^1, C^3\}$ and G_C^2 = $\{C^2, C^4\}$.

Next, we start to generate each node of binary partition tree. If each of groups has more than one class, we can build an internal node, and then employ PinSVM to execute classification. In Fig. 2a, the $PinSVM_1$ is generated, and it divides two groups G_C^1 and G_C^2.

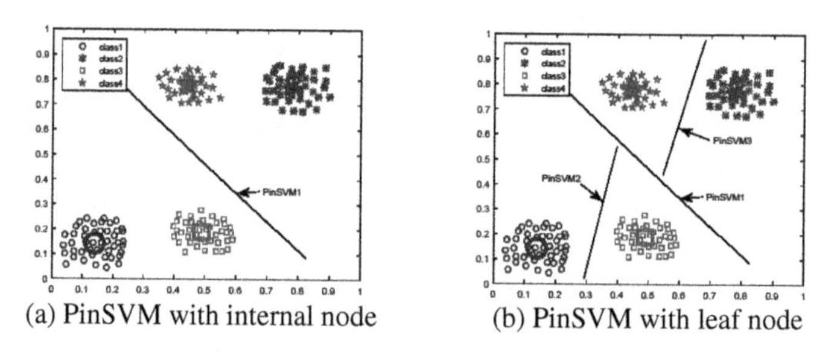

(a) PinSVM with internal node (b) PinSVM with leaf node

Fig. 2. Generation of a binary partition tree

When each group contains one class, the leaf node can be identified as shown in Fig. 2b. Now we give an informal description of binary-tree multiclass PinSVM in Algorithm-1.

Algorithm-1 Binary partition tree for multiclass PinSVM (BPT-PinSVM)
Input: Training set with K classes, parameters C, σ, ε, τ
1: Generate the K-centroid set, namely $C=\{C^1, C^2, \ldots, C^K\}$;
2: Divide C into two groups, marked G_C^p and G_C^q according to the Euclidean distance to two farthest centroids C^p and C^q;
3: Compute a PinSVM by using all samples concerning G_C^p and G_C^q. Create leaf node(s) when a group contains only one class;
- **If** the number of G_C^p containing centroids is greater than one, then C^{left}= G_C^p and construct left subtree;
- **If** the number of G_C^q containing centroids is greater than one, then C^{right}= G_C^q and construct right subtree;

4: **Return** until all the K classes reaching the leaf nodes.
Output: A binary partition tree.

In testing process, a testing sample first go through the root node, and then it is classified into the left or right subtree. A final label is obtained when it reaches leaf node. For a K-class problem, K-1 binary classifiers are generated. If let l be the number of training samples, the time complexity of BPT-PinSVM can be roughly estimated as $O((K-1)l^2)$.

4 Experimental Results

To evaluate BPT-PinSVM, we perform the experiments on eight datasets from UCI [9] listed in Table 1. We compare BPT-PinSVM with one-versus-one PinSVM (1-Vs.-1 PinSVM) and one-versus-one SVM (1-Vs.-1 SVM). The parameters are set as follows: penalty parameter $C = 1$, kernel parameter σ = 1/dimension, precision parameter $\varepsilon = 10^{-3}$, loss parameter $\tau = 0.5$. PC environment: I5–3230 M CPU, 4 GB memory, and Windows 8.1 OS.

Table 1. Datasets used in the experiments

Datasets	#class	#training	#testing	#dim
Iris	3	75	75	4
Wine	3	88	90	13
Glass	6	105	109	10
Vowel	11	528	462	10
Sensorless	11	29249	29260	48
Vehicle	4	422	424	18
Segment	7	1155	1155	19
Letter	26	15000	5000	16

Table 2 shows that BPT-PinSVM obtains higher accuracies than 1-Vs.-1 PinSVM on two datasets (glass and vowel), but a bit worse on four datasets (sensorless, vehicle, segment, letter). The possible reason is that 1-Vs.-1 PinSVM needs to go through *K*(*K*-1)/2 sub-classifiers, and the result is a statistical average, resulting in more stable performance. Compared with 1-Vs.-1 SVM, BPT-PinSVM obtains higher acccucies on four datasets (iris, glass, vowel, vehicle).

Table 2. Classification accuracies of BPT-PinSVM with 1-Vs.-1 PinSVM, 1-Vs.-1 SVM

Datasets	BPT-PinSVM	1-Vs.-1 PinSVM	1-Vs.-1 SVM
Iris	**96.00**	**96.00**	94.37
Wine	**100.00**	**100.00**	**100.00**
Glass	**68.81**	65.14	59.63
Vowel	**83.33**	82.25	80.87
Sensorless	67.93	71.77	**74.71**
Vehicle	71.46	**72.41**	70.28
Segment	90.30	**94.29**	90.82
Letter	77.58	80.46	**82.28**

The computational time is reported in Table 3. BPT-PinSVM generally takes less training time than 1-Vs.-1 PinSVM and 1-Vs.-1 SVM. As for a *K*-class classification problem, BPT-PinSVM needs to train *K*-1 binary classifiers. However, 1-Vs.-1 methods have to train *K*(*K*-1)/2 binary classifiers. For instance, on dataset "sensorless", it costs 1-Vs.-1 PinSVM and 1-Vs.-1 SVM 4.27×10^5 and 1.34×10^5 ms, respectively. But it only costs BPT-PinSVM about 4.30×10^4 ms. In terms of testing time, BPT-PinSVM has an obvious advantage over 1-Vs.-1 PinSVM and 1-Vs.-1 SVM. This is important for integrating it into intelligent devices, enabling real-time response for fast classification.

Table 3. Computational time of BPT-PinSVM with 1-Vs.-1 PinSVM, 1-Vs.-1 SVM

Datasets	Training time (ms)			Testing time (ms)		
	BPT-PinSVM	1-Vs.-1 PinSVM	1-Vs.-1 SVM	BPT-PinSVM	1-Vs.-1 PinSVM	1-Vs.-1 SVM
Iris	1.82	1.29	**0.85**	**0.26**	0.87	0.52
Wine	2.98	1.76	**0.87**	**0.53**	2.03	1.50
Glass	**7.04**	15.64	12.68	**0.81**	2.54	1.77
Vowel	**23.81**	99.32	57.42	**0.46**	27.40	25.94
Sensorless	$\mathbf{4.30 \times 10^4}$	4.27×10^5	1.34×10^5	**3105.77**	2.19×10^5	1.56×10^5
Vehicle	**22.13**	53.88	23.85	**9.19**	23.66	23.47
Segment	**109.37**	201.75	157.55	**21.04**	107.89	88.83
Letter	**1737.62**	3.98×10^4	1.11×10^4	**359.18**	1.70×10^4	1.18×10^4

5 Conclusion

In this paper, we use a binary tree structure to extend PinSVM to its multiclass version, which represents a hierarchical partition of given classes. The experimental results have confirmed that the proposed BPT-PinSVM has clear superiority in terms of computational time. As future work, we plan to introduce some heuristic processes to improve the classification accuracy of the method, so that it is given priority in some practical applications.

Acknowledgements. This work was supported in part by the National Natural Science Foundation of China under grant 61602056, the Doctoral Scientific Research Foundation of Liaoning Province under grant 201601348, and the Scientific Research Project of Liaoning Provincial Committee of Education under grant LZ2016005.

References

1. Vapnik, V.: The Nature of Statistical Learning Theory, pp. 123–179. Springer, New York (1995)
2. Bi, J., Zhang, T.: Support vector classification with input data uncertainty. In: International Conference on Neural Information Processing Systems, pp. 161–168. MIT Press, Cambridge, MA (2004)
3. Huang, X., Shi, L., Suykens, J.A.: Support vector machine classifier with pinball Loss. IEEE Trans. Pattern Anal. Mach. Intell. **36**(5), 984–997 (2014)
4. Hsu, C.W., Lin, C.J.: A comparison of methods for multiclass support vector machines. IEEE Trans. Neural Networks **13**(2), 415–425 (2002)
5. Lorena, A.C., de Carvalho, A.C.: A review on the combination of binary classifiers in multiclass problems. Artif. Intell. Rev. **30**(1), 19–37 (2008)
6. Kijsirikul, B., Ussivakul, N.: Multiclass support vector machines using adaptive directed acyclic graph. In: Proceedings of the 2002 International Joint Conference on Neural Networks, vol. 1, pp. 980–985. IEEE, New York (2002)
7. Fei, B., Liu, J.: Binary tree of SVM: a new fast multiclass training and classification algorithm. IEEE Trans. Neural Networks **17**(3), 696–704 (2006)
8. Kostin, A.: A simple and fast multi-class piecewise linear pattern classifier. Pattern Recogn. **39**(11), 1949–1962 (2006)
9. Frank, A.: A. Asuncion. UCI machine learning repository, 2010. URL http://archive.ics.uci.edu/ml

Research on Fractal Feature Extraction of Radar Signal Based on Wavelet Transform

Shen Lei(✉), Han Yu-sheng, and Wang Shuo

Army Officer Academy of PLA, Anhui Hefei 230031, China
2692348248@qq.com

Abstract. Fractal feature can measure the complexity of radar signal. Radar signal is a kind of time series, which can be represented by fractal dimension. Wavelet transform is the signal processing microscope, which can observe the general situation and details of the signal and reduce the influence of noise. In this paper, the fractal characteristics of radar signals are extracted by combining the wavelet transform and fractal theory with the advantage of the two. Firstly, the wavelet transform is used to decompose the radar signals, and then the fractal dimension of radar signals under different decomposition levels is calculated. The radar signals with different complexity are obtained according to the difference of the fractal dimension. Finally, the validity of this method is verified by Matlab platform.

Keywords: Wavelet transform · Complexity · Feature extraction · Fractal dimension

1 Introduction

Radar interception system is widely used in various weapon platforms because of its advantages of good concealment and long detection range. With the rapid development of radar technology, the signal environment faced by radar acquisition system is becoming more and more dense and complex, which makes it difficult to separate and identify radar signals. Radar signals are functions of time, space, relative position and various complex modulations. The modulation of the frequency, amplitude and phase of the signal, together with the signal specific polarization and arrival angle, constitutes the characteristics of the radar signal.

The purpose of electronic reconnaissance is to hunt these features and analyze these characteristics to determine the nature of the target and the threat level. Signal intra pulse characteristics include radar transmitter modulation of frequency and phase, and modulation of envelope shape. Especially the emergence of advanced radar systems. Intra pulse sorting is a very important method for obtaining the fine features of signals and distinguishing the signal types.

The intra pulse characteristics of radar signals are mainly represented by the changes and distributions of frequency, phase and amplitude. The fractal dimension can be used to characterize it. The fractal dimension is an effective feature of radar signals [1].

F. Xhafa et al. (eds.), *Advances in Intelligent Systems and Interactive Applications*, Advances in Intelligent Systems and Computing 686,
https://doi.org/10.1007/978-3-319-69096-4_46

The key of fractal theory is to calculate the fractal dimension. It can recognize the general situation and details of complex phenomena through the resolution of different scales. Therefore, the multi-scale characteristics of wavelet transform are necessarily related to the statistical self similarity of signals. Based on this, wavelet transform and fractal theory can be combined to extract the fractal characteristics of radar signals [2].

2 Wavelet Transform

Wavelet analysis is a new method developed rapidly in recent years, its biggest advantage is the advantage of multi-resolution, can provide a variety of basis functions, can be changed according to the actual needs of any scale, so as to improve the resolution. In particular, orthogonal wavelets like Daubechies classes have compactly supported properties.

Daubechies wavelet is a series of orthonormal wavelets with compact support constructed by Daubechies in 1988. These wavelets have good local characteristics and have been widely used in signal analysis. Therefore, Daubechies wavelet is used to perform wavelet transform of radar signals [3].

Daubechies puts forward $I(x)$ as trigonometric polynomial. This is $I(x) = \sum_{l} i(l)e^{-jxl}$, the coefficient $i(l)$ is the real number. Then $I(x)$ can be represented as:

$$I(x) = [\frac{1}{2}(1 + e^{jx})]^{N} R(e^{-jw}) \quad N \in Z_{+} \tag{1}$$

Where R is the real coefficient, so $R^{*}(e^{-jx}) = R(e^{-jx}) = R(e^{jx})$ So $|R(e^{-jx})|^{2} = R(e^{jx})R(e^{-jx})$ is an even function of x, and because $\cos x$ is also even function. So you can represent $|R(e^{-jx})|^{2}$ as a polynomial of $\cos x$. Because $(1 - \cos x)/2 = \sin^{2}(x/2)$, polynomial that can be expressed as $\sin^{2}(x/2)$. We mark as $Q(z)$, $z = \sin^{2}(x/2)$. Note that $|(1 + e^{jx})/2|^{2} = |\cos^{2} x/2|$, then $I(x)$ may be represented as:

$$|I(x)|^{2} = |\cos^{2}(x/2)|^{N} |R(e^{-jx})|^{2} = |1 - \sin^{2}(x/2)|^{N} |R(e^{-jx})|^{2} = (1 - z)^{N} Q(z) \tag{2}$$

And

$$\begin{aligned} |I(x + \pi)|^{2} &= |\sin^{2}(x/2)|^{N} |R(e^{-j(x+\pi)})|^{2} = |\sin^{2}(x/2)|^{N} Q[\sin^{2}(x + \pi)/2] \\ &= z^{N} Q[\cos^{2}(x/2)] = z^{N} Q[1 - \sin^{2}(x/2)] = z^{N} Q(1 - z) \end{aligned} \tag{3}$$

Also because $|I(x)|^{2} + |I(x + \pi)|^{2} = 1$.

$$(1 - z)^{N} Q(z) = z^{N} Q(1 - z) = 1 \quad Q(z) \geq 0, \; z \in [0, 1] \tag{4}$$

It can be obtained by (4) formula and Reisz theorem:

$$\left|R(e^{jx})\right|^2 = \sum_{l=0}^{N-1} C_{N+l-1}^{l} \left|\sin^2\frac{x}{2}\right|^l + \left|\sin^2\frac{x}{2}\right|^l S\left(\frac{1}{2} - \sin^2\frac{x}{2}\right) \tag{5}$$

Daubechies wavelet filter to find the algorithm can be expressed as:

(1) The selection of natural numbers $N \geq 0$, obtained $\left|R(e^{-jx})\right|^2$.
(2) The $\left|R(e^{-jx})\right|^2$ is transformed into $\cos x$, and x is used instead of $\cos x$ to obtain a unary higher equation $W(x)$ about x.
(3) Solve all roots of $W(x)$.
(4) The corresponding z is obtained by $x = (a + 1/a)$.
(5) Take a real root in each pair, every two of a pair of complex roots to form $R(a) = B\prod(a - a_j)$.
(6) The $R(j) = \sqrt{W(j)}$ determines the B.
(7) Seeking $l(o)$ from $R(a)$.

3 Fractal Characteristics

If an object is a part of it in a way similar to the whole, it is called fractals. It has the following typical geometric properties: (1) The fractal set has proportional details. (2) The fractal cannot be described by the traditional geometric language, that is, the locus of the points that do not satisfy certain conditions, nor the set of solutions of some simple equations. (3) A fractal set has some form of self similarity, which may be approximately self similar or statistically self similar. There are many definitions and calculation methods of fractal dimension, and the Moorhouse dimension, box dimension, information dimension, similarity dimension, correlation dimension and generalized dimension are often used [4]. The box dimension describes the geometric properties of the fractal set (the density of the midpoint of the geometry or the degree of material distribution), so the box dimension is used to characterize the intra pulse characteristics of the radar signal [5].

Let (Y, e) be a metric space. I is a compact set of Y. α is a real number. The $C(y, \alpha)$ represents closed ball with a center at y and a radius of α. Let's say B is a nonempty compact set in Y. For each positive number α, make $N(B, \alpha)$ mean the minimum number of closed balls that cover B, that is:

$$N(B, \alpha) = \{M : B \subset \bigcup_{j=1}^{M} N(y_j, \alpha)\} \tag{6}$$

Among them, $y_1, y_2, \ldots, y_M$ is a different point in Y.

Suppose B is a compact set, and a nonnegative real number, if it exists

$$E_c = \lim_{\alpha \to 0} \frac{\ln N(B, \alpha)}{\ln(1/\alpha)} \tag{7}$$

E_c is called the fractal dimension of set B. It is called $E_c = E_c(B)$, and B is called fractal dimension E_c. This dimension is called box dimension.

If the signal sequence is $\{h(j), j = 1, 2, \ldots, N\}$, where N is the length of the signal, the following method is used to compute the box dimension:

Place the signal sequence $\{h(j)\}$ in a unit square, and the minimum interval of abscissa is $r(1/N)$

$$N(r) = N + \{\sum_{j=1}^{N-1} \max\{h(j), h(j+1)\}/r - \sum_{i=1}^{N-1} \min\{h(j), h(j+1)\}/r\}/r^2 \tag{8}$$

The box dimension of the signal is:

$$E_c = -\frac{\ln N(r)}{\ln r} \tag{9}$$

4 Calculation of Fractal Dimension of Radar Signal

With the characteristics of wavelet transform, the radar signals are decomposed at different scales, and the signal details of different intra pulse modulation types are different at the same level. In order to represent the difference of each kind of detail signal, the complexity of the detail signal can be used as their characteristic. The fractal dimension can describe signal complexity, therefore, as the feature signal recognition is feasible.

The radar signal sequence with finite length L is $s_0 = \{s_{0,n} | n = 0, 1, \ldots, L-1\}$. The calculation of fractal dimension is carried out in two steps.

(1) According to the construction method of the Daubechies orthogonal wavelet bases in the second part, the low-pass filter coefficients of Daubechies are obtained, and the coefficients are $\{h_0, h_1, \ldots, h_N\}$, N = 5. According to the Mallat algorithm, the s_0 wavelet is decomposed into

$$\begin{cases} s_{m+1,n} = \sum_{K=2n}^{2n+N} h_{k-2n} s_{m,k} & \text{Even, } n = \frac{1-N}{2}, \frac{1-N}{2}+1, \cdots, \frac{L}{2}-1 \\ & \text{Odd, } n = \frac{1-N}{2}, \frac{1-N}{2}+1, \cdots, \frac{L}{2}-1 \\ d_{m+1,n} = \sum_{K=2n+1-N}^{2n+1} (-1)^{k+1} h_{2n-k+1} s_{m,k} & \text{Even, } n = 0, 1, \cdots, \frac{L+N-3}{2} \\ & \text{Odd, } n = 0, 1, \cdots, \frac{L+N-2}{2} \end{cases} \tag{10}$$

Among them, s_m and d_m respectively represent discrete approximation and discrete detail under decomposition level M.

(2) Under different decomposition levels, the discrete approximation is $\{s_m(i), i = 1, 2, \ldots, N\}$, and the box dimension of each stage is calculated according to formulas (8) and (9).

Because of the discrete approximation, s_m preserves the main characteristics of the signal, and d_m is the signal detail information. After the wavelet transform, the energy is mainly concentrated on the discrete approximation, while the additive random noise in the transmission and processing of the wavelet transform, the energy is dispersed in discrete detail. Therefore, the fractal dimension of the discrete approximation is not sensitive to the noise at different decomposition levels [6]. In addition, the amount of discrete approximation is less than that of the original signal after wavelet decomposition, and the computation of the fractal dimension will be greatly reduced.

The complexity of the signal can be effectively measured by the box dimension, and the amplitude, frequency and phase information of the signal can be effectively changed. Therefore, box dimension can be used as an effective feature for signal sorting and identification.

5 Simulation Experiment

This paper chooses three kinds of radar signal is widely used for simulation experiments: fixed carrier frequency rectangular pulse modulation signal, LFM signal and phase encoding pulse compression phase encoding signal in radar pulse compression radar. Firstly, the three radar signals are decomposed into 6 orders by wavelet decomposition, and the discrete approximation of s_m under the decomposition and decomposition levels is obtained. Next, the box dimensions of each s_m are computed, and the results are shown in Fig. 1.

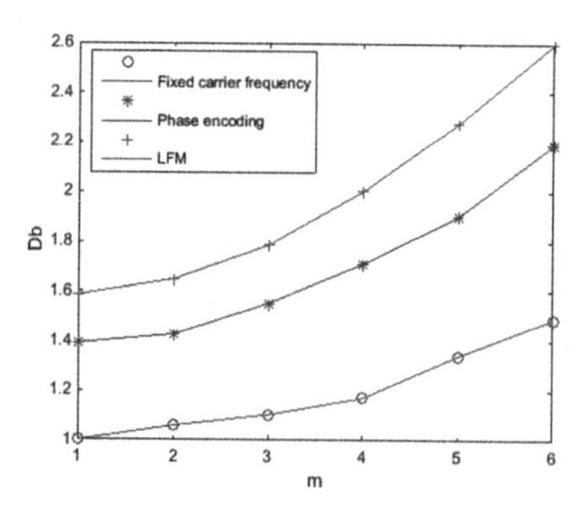

Fig. 1. Three signals box dimension in different decomposed level

In order to consider the influence of noise, the phase coded signal is taken as an example, and the random noise is added to the signal, and the signal-to-noise ratio is 0 and 10 dB respectively. The box dimension is calculated by the same method and compared with the results without noise interference, as shown in Fig. 2.

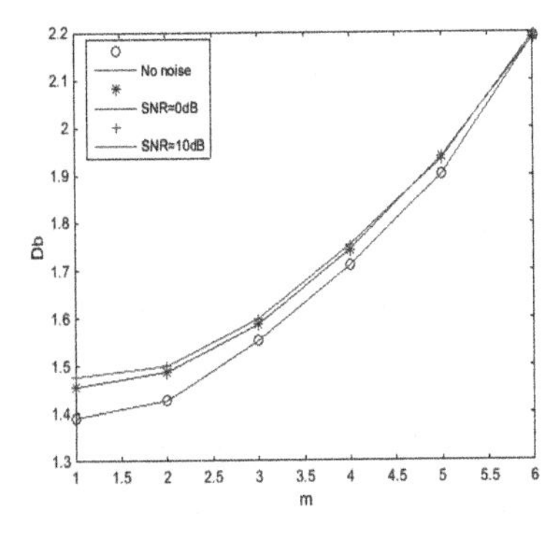

Fig. 2. LMF signals box dimension in different SNR

The simulation results can be seen from Figs. 2 and 3:

(1) The fractal dimension can quantitatively describe the complexity of the signal. The more complex the signal, the greater the fractal dimension; on the contrary, the simpler the signal, the smaller the fractal dimension.
(2) The fractal dimension of signal is related to the series of wavelet decomposition. The higher the series of wavelet decomposition is, the bigger the fractal dimension is. The reason is that the higher the decomposition order, the more detailed the signal is decomposed, the more complex the signal is formed, and the more the fractal dimension is;
(3) The fractal dimension is not sensitive to noise. That is, the noise has little influence on the fractal dimension, the main reason is that the computation of the fractal dimension is carried out by discrete approximation after wavelet decomposition, and the noise is concentrated on the discrete detail. Therefore, wavelet decomposition also has the function of noise elimination. As can be seen from Fig. 3, the larger the series of wavelet decomposition, the closer the fractal dimension under different SNR, the smaller the influence of noise.

6 Conclusion

In this paper, wavelet transform and fractal theory are used to analyze the fractal characteristics of radar signals. Wavelet transform cannot only observe the details of the signal, but also reduce the impact of noise. On this basis, the box dimensions under different decomposition levels can be used to characterize the complexity of radar signals and provide important data support for sorting and identification of radar signals.

References

1. Han, J., Ming-Hao, H.E., Mao, Y, et al.: Extracting of radar signal in-pulse characteristics in low SNR. Radar Sci. Technol. (2007)
2. Guo, S.: A method of feature extraction for modulated signals based wavelet and fractal theory. Sign. Process. (2005)
3. Chen, D.J., Fan, J.W., Luo, T., et al.: Application of Daubechies wavelet on feature extraction and identification of dynamic error of machine tool. J. Beijing Univ. Technol. **38**(10), 1467–1473 (2012)
4. Kang, S., Gang, J., Wang, C.Y., et al.: Modeling sea clutter in radar scanning mode by multifractional brownian motion. J. Electron. Inf. Technol. **37**(4), 982–988 (2015)
5. Yang, S., Tian, B., Zhou, R.: A jamming identification method against radar deception based on bispectrum analysis and fractal dimension. J. Xian Jiaotong Univ. (2016)
6. Fan, Y., Luo, F., Ming, L.I., et al.: The multifractal properties of AR spectrum and weak target detection in sea clutter background. J. Electron. Inf. Technol. (2016)

A Method of Moving Target Detection Based on Scaling Background

Yu-chen Tang(✉) and Xu-dong Yang

College of Computer Science, Beijing University of Posts and Telecommunications, Beijing, China
tangyuchenkk@126.com

Abstract. Moving Object Detection Based on Global Motion Video Sequence, Background's movement mainly behave in the translation and scaling. Due to camera movement caused the background produces a zoom. This paper present an improved gray-scale projection algorithm. This method can quickly estimate the scaling parameters between the two images before and after, and then refer to the background of the reference frame. The background is mapped to the current frame image according to the scaling parameters. Using this algorithm to improve the three-frame difference method can accurately detect the moving target. Experimental results show, this method can effectively extract the target in the video sequence.

Keywords: Object detection · Gray-scale projection algorithm · Scale parameters

1 Introduction

Moving target detection is a hotspot in the field of machine vision. In the intelligent transportation, safety monitoring, medical diagnosis, military navigation, human–computer interaction and other fields have a broad application prospects [10]. Its purpose is to automatically separate the moving pixels in the video image, make the change area extracted from the background image. Now there are three main method: optical flow method [6], background subtraction method [1] and interframe difference method [1, 2]. The first method has a good adaptability to the background changes, but the algorithm is too complex to ensure real-time. The second one is simple but poor applicability. The third method can get good effect in static background, and is not sensitive to changes in light, the algorithm is simple. But in the actual moving target detection system, the camera itself will produce movement. It is impossible to detect the moving target directly by the frame difference method, the camera moves to cause the background to produce both scaling and translation. For translation, gray-scale projection algorithm [7–9] because of its simple principle, good real-time has been widely used in the calculation of translational motion parameters. This paper improves the gray scale projection algorithm so that it can estimate the scaling parameters between two images. By analyzing the gray scale projection curve, can get the scaled

F. Xhafa et al. (eds.), *Advances in Intelligent Systems and Interactive Applications*, Advances in Intelligent Systems and Computing 686,
https://doi.org/10.1007/978-3-319-69096-4_47

gray projection vector feature. Using the method of mapping to the logarithmic coordinate space [3–5], Reduce the dimension of the zoom vector as a translation vector. And then through the traditional one-dimensional vector operation to calculate the scaling parameters. Using this algorithm to improve the interframe difference method, which is able to eliminate the interference information.

2 Grayscale Projection Algorithm

The gray scale projection algorithm uses the variation rule of the gray distribution of the image to obtain the motion vector of the current frame with respect to the reference frame. The principle is to project the rows and columns of the image into x, y directions, and a projection vector is formed in the respective direction, and then match the projection vector to get the image offset.

2.1 Image Rows and Columns Gray Scale Projection

Filtered pretreated for each frame image in the input sequence, and map its gray value to two independent one-dimensional waveforms. The method is expressed as follows:

$$G_k(j) = \sum_i G_k(i,j) \tag{1}$$

$$G_k(i) = \sum_j G_k(i,j) \tag{2}$$

$$G_k(j) = \sum_i G_k(i,j) - E \tag{3}$$

$$G_k(i) = \sum_j G_k(i,j) - E \tag{4}$$

$G_k(j)$ is the gray value of the jth column of the k-th frame. ; $G_k(i,j)$ is the gray value of the k-th frame at coordinates (i, j); $G_k(i)$ is the gray value of the i-th row of the k-th frame; Adds each pixel gray value of the jth frame of the k-th frame to a value $\sum_j G_k(i,j)$; E is the mean of the projection vector. Projection vector do mean normalization process to eliminate the influence of light changes. The formula (3) (4) is the modification of the formula (1) (2).

2.2 Calculate the Offset

L the gray scale projection curves of adjacent two frames are calculated by cross correlation. Compare the valleys of the two curves is able to determined the displacement offset of the current frame relative to the reference frame, Calculated as follows:

$$C(w) = \sum_{j=1}^{cl} [G_k(j + w - l) - G_r(m + j)]^2 (1 \le w \le 2m + 1) \tag{5}$$

The formula $G_k(j)$ $G_r(j)$ is the gray scale projection value of the k-th frame and the reference frame j column; cl is the length of the selected matching vector; m is the search width of the displacement vector on one side with respect to the reference frame.

3 Scale Parameter Estimation

After the image has been scaled, the projection vector will be shortened or stretched, the scale of the vector has changed, so cannot use the traditional vector-related technology to solve.

3.1 The Feature of Scale the Projection Vector

Assume that the scaled projection vector is G_k, reference frame projection vector is G_r, Scaling factor is a. So then:

$$G_k(j') = \left(1 - \frac{a-1}{a}\right) \cdot G_r(j) + n(a, G_r(j)) = a^{-1} G_r(j) + n(a, G_r(j)) \tag{6}$$

Where $n(a, G_r(j))$ represents the noise caused by the zoom, the difference between scaled projection vector G_k and reference frame projection vector G_r as follows:

$$\Delta = G_r(j) - \left[a^{-1} G_r(j) + n(a, G_r(j))\right] = \left(1 - a^{-1}\right) G_r(j) - n(a, G_r(j)) \tag{7}$$

When doing the motion estimation, the difference between adjacent two frames is very small, in the above tolerable range can be considered: If $\forall |i' - j'| = a|i - j|$, and $G_k(i') = G_r(i)$ so then $G_k(j') = G_r(j)$.

3.2 Projection Vector Coordinate Space Transformation

According to the multiplication, that the logarithmic transformation can be converted into the addition operation. In this paper, the logarithmic transformation method is used to map the projection vector to the new logarithmic space, the conversion formula is as follows:

$$\begin{aligned} &\text{If } \; G_1(x) = G_2(ax) \; \text{ then} \\ &G_1(logx) = G_2(\log(a)x) = G_2(logx) + G_2(loga) \Rightarrow G_1(y) = G_2(y + b) \end{aligned} \tag{8}$$

$y = logx$, $b = loga$, So that the expansion of the vector is converted into a translation. And then use the above-mentioned method to carry on the correlation computation to the transformed vector. Calculate the offset b, and $\log a = b \Rightarrow a = (base)^b$.

4 Improved Three—Frame Difference Method and Moving Target Detection

4.1 Ordinary Interframe Difference Method

Select three consecutive images in the video. Calculates a difference between two adjacent image. The calculation formula is below:

$$d_{(i,i-1)}(x,y) = |I_i(x,y) - I_{i-1}(x,y)| \tag{9}$$

$$d_{(i+1,i)}(x,y) = |I_{i+1}(x,y) - I_i(x,y)| \tag{10}$$

The resulting difference image is binarized by the appropriate threshold value T. The formula is as follows:

$$b_{(i,i-1)}(x,y) = \begin{cases} 1 \; d_{(i,i-1)}(x,y) \geq T \\ 0 \; d_{(i,i-1)}(x,y) < T \end{cases} \tag{11}$$

$$b_{(i+1,i)}(x,y) = \begin{cases} 1 \; d_{(i+1,i)}(x,y) \geq T \\ 0 \; d_{(i+1,i)}(x,y) < T \end{cases} \tag{12}$$

Traverse each pixel (x, y), let the resulting binary image do the ‘and’ operation. The formula is as follows:

$$B_i(x,y) = \begin{cases} 1 \; b_{(i,i-1)}(x,y) \cap b_{(i+1,i)}(x,y) = 1 \\ 0 \; b_{(i,i-1)}(x,y) \cap b_{(i+1,i)}(x,y) \neq 1 \end{cases} \tag{13}$$

4.2 Improved Algorithm Process

The method steps are as follows:

1. Using the improved gray scale projection algorithm to calculate the kth frame relative to the k−1 frame background scaling parameters. Use this parameter to correct the background. The same way to deal with the background in k + 1;
2. Using the frame difference method to compensate the image after the difference, the difference images d1, d2 are obtained;
3. D1 and d2 do ‘and’ operator ‘to get d;
4. The noise reduction and the binarization processing for the difference image;

The process is as follows (Fig. 1).

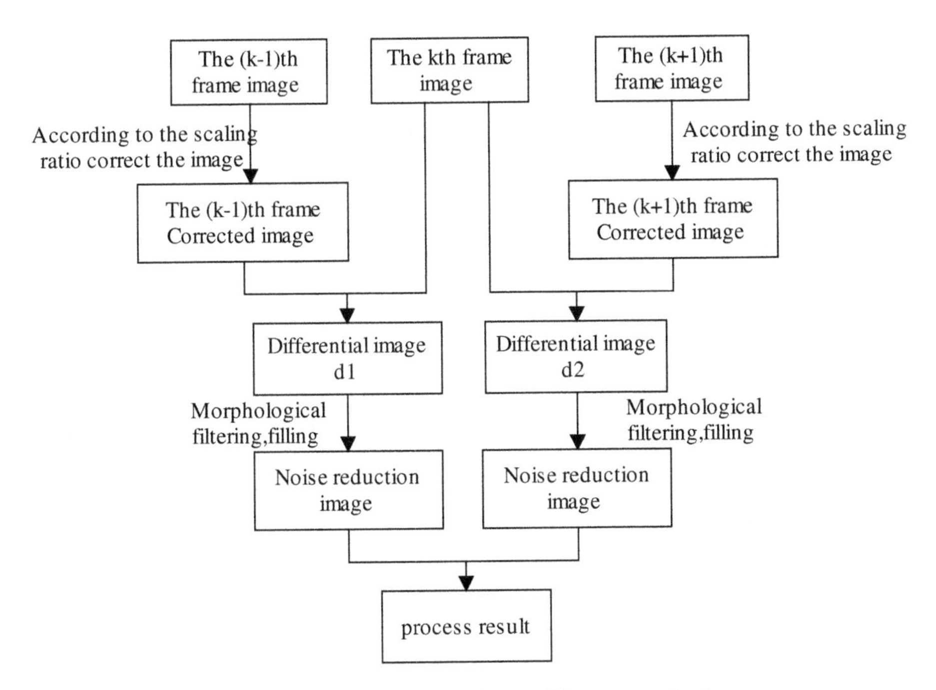

Fig. 1. Improved three-frame difference method

5 Analysis of Result

According to the new algorithm proposed in this paper, experiments were carried out in an outdoor scene. The experimental environment is the Windows operating system, the algorithm language is Matlab.

The vertical lines of the blue lines in Fig. 2a, b are the vertical projection vector characteristic curves of Fig. 3a, b. The red lines in Fig. 2a, b, are the vertical projection characteristic curves after the dashed lines in Fig. 2a, b. The scaling method can be solved by using the technique of translation vector calculation. Figure 3a, b and c is a continuous three-frame resolution of 256×256 RGB image conversion after the grayscale image; Fig. 3d is the result of the improved front frame difference method. Using the improved gray scale projection algorithm can calculated the background scale ratio of Fig. 3a, b is 1.1085; the background scale ratio of Fig. 3c relative to Fig. 3b is 0.9065; Fig. 3e is the result of the k-th frame and previous frame differential processing; Fig. 3f is the result of the k-th frame and behind frame differential processing; Fig. 3g and h are the results of Fig. 3e and f do noise reduction processing images; Fig. 3i is the result of doing the ‘and’ operation in Fig. 3g and h.

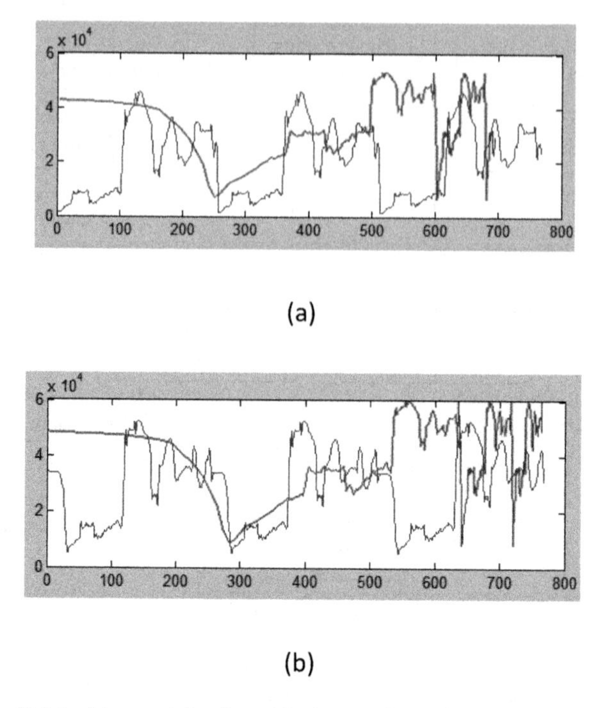

Fig. 2. Original image (after logarithmic transformation) Vertical projection

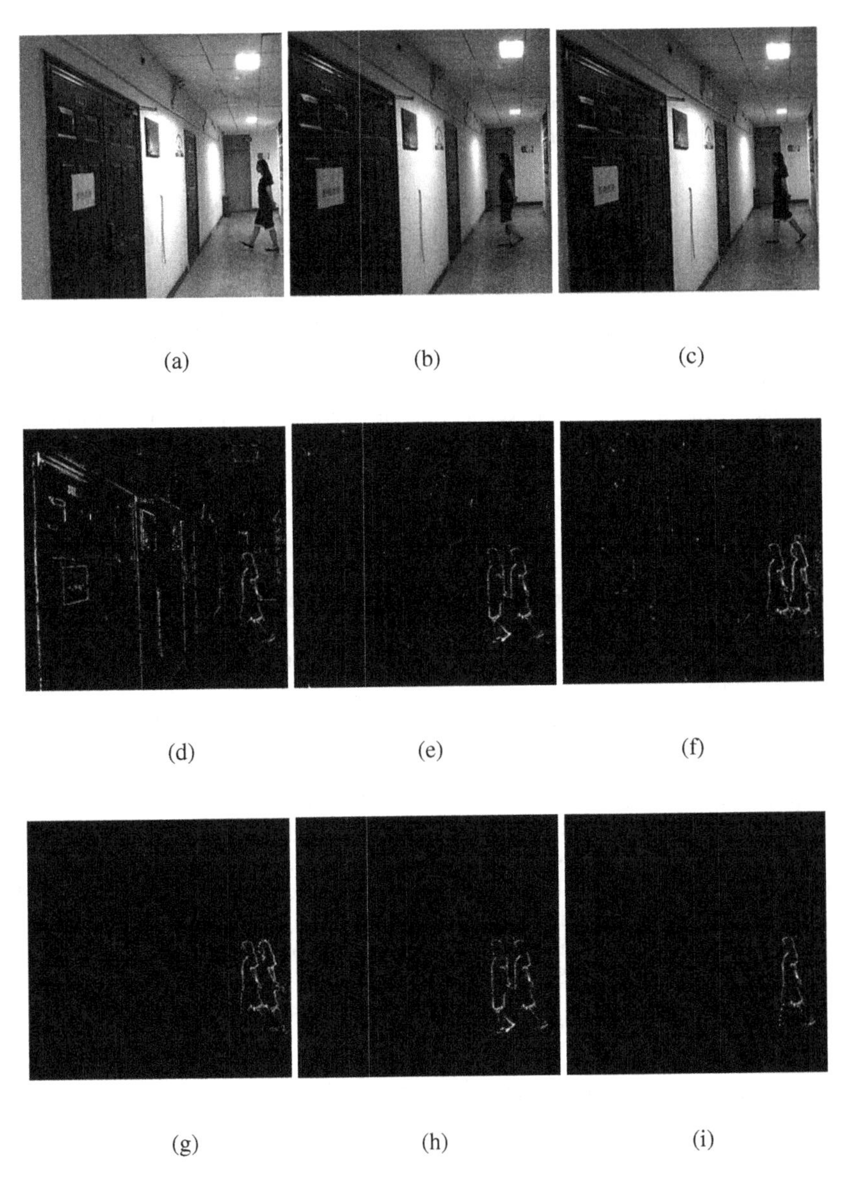

Fig. 3. Experimental results

Experiments show that the improved frame difference method can complete the moving target detection in the case of background scaling.

6 Conclusion

In this paper, the improved gray scale projection algorithm is used to calculate the motion background scaling parameters, and compensate the global motion vector, so that the background of several consecutive frames is stabilized on an image. Effectively suppressing the interference of false motion information due to background movement. The experimental results show that the method can effectively accomplish the target detection.

Acknowledgments. This article is completed by Professor Xudong Yang's careful guidance and strong support. Mr. Yang gave me a lot of help in both topic and experimentation, so that I had a good experimental environment. In addition, thanks to my classmate Jun Zhe Zhang's selfless help Mr Zhang help me carry out research work and complete the experiment. So I can successfully complete the paper.

References

1. Zhang, X., Sun, W.: Moving object extraction in complex background. Photoelectric Eng. **33**(4), 10–13 (2006)
2. Qing, T., Zhou, Z.: A New method for detecting moving targets of sequential images. Comput. Softw.
3. Chang, H.-C., Lai, S.-H., Lu, K.-R.: A robust real-time video stabilization algorithm. J. Vis. Commun. Image R **17**, 659–673 (2006)
4. Lai, Z., Wang, J.: A target detection algorithm based on robust background motion compensation. Comput. Appl. Res. **24**(3), 66–68 (2007)
5. Sun, Bin, Huang, Shen-yi: Research on moving target detection and tracking in mobile background. J. Electron. Meas. Instrum. **25**(3), 206–209 (2011)
6. Gao, Y., He, M.: Target detection based on frame difference method and template matching method in dynamic background. Electron. Des. Eng. **20**(5), 142–145 (2012)
7. Bugeau, A., Perez, P.: Detection and Segmentation of Moving Objects in Highly Dynamic Scenes. Technical report, IRISA (PI 1864) (2007)
8. Comaniciu, D.: An algorithm for data-driven handwith selection. IEEE Trans. Pattern Anal. Mach. Intel. **25**(2), 281–288 (2003)
9. Luan, Q., Chen, Z.: A method of detecting moving targets in sports background. Comput. Math. Eng. **36**(10), 165–168 (2008)
10. Tian, F., Yan, J.: Research on target location in machine vision. Comput. Meas. Control **18**(4), 900–901 (2010)

Facial Expression Recognition Based on Deep Learning: A Survey

Ting Zhang(✉)

School of Information Engineering, Minzu University of China, Beijing 100081, China
tozhangting@126.com

Abstract. Facial expression recognition (FER) enables computers to understand human emotions and is the basis and prerequisite for quantitative analysis of human emotions. As a challenging interdisciplinary in biometrics and emotional computing, FER has become a research hotspot in the field of pattern recognition, computer vision and artificial intelligence both at home and abroad. As a new machine learning theory, deep learning not only emphasizes the depth of learning model, but also highlights the importance of feature learning for network model, and has made some research achievements in facial expression recognition. In this paper, the current research states are analyzed mostly from the latest facial expression extraction algorithm and the FER algorithm based on deep learning a comparison is made of these methods. Finally, the research challenges are generally concluded, and the possible trends are outlined.

Keywords: Facial expression · Deep learning · Feature extraction · Computer vision

1 Introduction

As the psychologist Mehrabian said "the facial expression sends the 55% information during the process of human communications while language sends only 7%". Therefore, the expression in the daily communication of mankind occupies an important position in the daily communication of mankind, is an important way to exchange emotional information among people. In order to have the ability of human thinking, it is necessary for the computer to understand the human emotions, that is the reason why after several decades of development the researches on facial expression recognition have been becoming the focus.

Usually facial expression recognition is divided into four processes: image preprocessing, face detection, feature extraction and facial expression classification, as shown in Fig. 1. The basic task of feature extraction is to produce some basic features according to the object to be identified, and then find the most effective representation from these characteristics, so that the difference is big in the different categories of expressions, and is small in the same type of expression difference is small. Feature extraction is a key problem in pattern recognition and one of the most difficult tasks in facial expression recognition. How to extract these features effectively is the key to realize accurate and feasible expression recognition. Due to the different research

F. Xhafa et al. (eds.), *Advances in Intelligent Systems and Interactive Applications*, Advances in Intelligent Systems and Computing 686,
https://doi.org/10.1007/978-3-319-69096-4_48

background, the researchers have different definitions of basic emotions. Among them, Ekman's emotional theory has the greatest impact, where there are six basic emotions: anger, disgust, fear, pleasure, sadness and surprise. The expression classifier divides the extracted features into the corresponding categories according to the corresponding classification mechanism.

Fig. 1. Facial expression classification block diagram

2 Expression Feature Extraction Algorithm

At present, there are many kinds of feature extraction methods, which can be divided into geometric shapes, texture features and the combination of the two.

2.1 Methods Based Geometric Features Based

In the expression recognition algorithm based on geometric deformation, ASM (Active Shape Model) [1] and FAUs (Facial Action Units) [2] are the two most typical methods. ASM constructs geometric features by extracting facial feature points. The geometric features were extracted by the improving ASM and the use of the triangle features in the model, and achieved good results [3]. Combined with the texture and shape, AAM (Active Appearance Model) as a statistical model is used to model the face, to identify and analyze the expression. An expression recognition algorithm is proposed, combining AAM and ASM to improve the problem of feature point convergence in AAM [4].

According to FACS, as another important geometric feature, by analyzing the constant facial features and instantaneous characteristics. In the image sequence, a variety of features were tracked and modeled in order to recognize facial expression [5]. Combined with multi-scale feature extraction and multi-core learning algorithm, the contour of the face shape was outlined [6]. The face geometry and the activity appearance of the activity in the Candide grid frame are feature vector of expression [7]. Through the self-organizing mapping, the feature points that describes the shape of the eye, lips, and eyebrows in the face constitutes a low-dimensional eigenvector for expression recognition [8].

2.2 Method Based on Texture Feature

LBP (Local Binary Patterns, LBP) as a local operator can be used for image texture analysis, face recognition and facial expression recognition. Based on the idea of LBP, an LDP (Local Directional Patterns, LDP) operator is proposed, which is used to extract facial texture features [9]. The LDP feature is weighted by LDP variance, and then the most significant direction of LDP is calculated by PCA (Principle Component

Analysis) to reduce the feature dimension [11]. The proposed Gradient Directional Pattern calculates the texture information of the local area by quantifying the gradient direction to form the facial representation [12].

Lyons et al. applied the Gabor feature to calculate the expression characteristics, combining with elasticity graph matching and linear analysis [13]. But the high-dimensional Gabor feature is an obstacle to quickly identify facial expressions. The image is divided into local sub-regions to extract multi-scale Gabor features in order to describe the topology structure of human visual [14]. The algorithm combined Gauss-Laguerre wavelet and face reference to describe the expression feature, the former in which extracted the texture feature. The latter computed geometric features [15].

2.3 Multi-feature Fusion Method

Multi-feature fusion is an effective way of the facial expressions representation, and the different features in many algorithms are combined to describe facial expressions. The combination of Gabor and LBP is used in [16], and the minimized error rate and feature dimension is selected by genetic algorithm (GA). Li et al. Used SIFT and PHOG (Pyramid Histogram of Oriented Gradient) to calculate local texture and shape information [17]. TC (Topographic Context) is used as a feature operator to describe the static facial expression feature [18]. The haar-like feature was used to characterize facial expression changes over time, and was encoded to form a dynamic feature according to the binary pattern [19]. According to the main direction extracted in the video sequence and the average optical flow characteristics in the face block, MDMO (Main directional mean optical flow feature, MDMO) was proposed in [20]. Table 1 lists the main feature extraction methods in facial expression recognition.

Table 1. Main feature extraction methods for expression recognition

Literature	Feature	Category	Dataset	Rate of recognition (%)
[3]	ASM	7	JAFFE	86.96
[5]	FACS	7	Cohn-Kanade	96.7
[7]	Candide	4	Cohn-Kanade	85
[9]	LDP	7	Cohn-Kanade	93.4
		7	JAFFE	85.4
[10]	LDN	7	Cohn-Kanade	95.1
		7	JAFFE	91.1
[14]	Gabor	7	JAFFE	89.67
		7	Cohn-Kanade	91.51
[15]	Gauss-Laguerre	6	Cohn-Kanade	90.37
		7	MMI	85.97
[16]	LBP + Gabor	7	Cohn-Kanade	99.2
		6	MMI	94.1
[17]	PHOG + SIFT	6	Cohn-Kanade	96.33
		7	JAFFE	96.2

3 Facial Expression Recognition Algorithm Based on Deep Learning

By constructing a nonlinear neural network with multiple hidden layers, the deep learning simulates the human brain for analysis and learning. By transforming the input data by layer, the feature representation of the sample in the original space is transformed into a more abstract feature space, so as to study the essential characteristics of the data set and further to imitate the human brain to interpret the data such as image, sound and text. At present, the theory research for facial expression recognition based on the deep learning mainly focused on the two methods: the method based on DBN (deep belief network) and the method based on CNN (convolution neural network). Block diagram of FER based on deep learning as shown in Fig. 2.

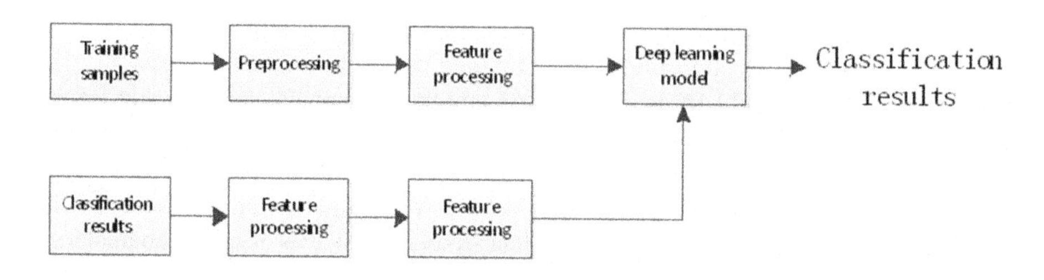

Fig. 2. Block diagram of FER based on deep learning

3.1 Method Based on DBN

As a classic method deep learning, DBN has aroused widespread concern, which can contain more hidden layer, and can better learn a variety of complex data structure and distribution. As shown in Fig. 3, the structure of DDBN (discriminative deep belief networks), includes an input layer h^0, N hidden layers h^1, h^2, …, h^N and a label layer with C units at the top, equivalent to the number of classes in the tag data y. Look for the mapping function $X \rightarrow Y$, where it can be converted into a deep architecture to find the parameter space. The problem of finding the mapping function $X \rightarrow Y$ can be transformed into a deep architecture to find the parameter space W.

Combined with other methods, DBN has achieved significant results for the facial expression recognition. In [23], first the face was cut at different scales, and then the HOG (Histogram of Oriented Gradient) feature was extracted from these local cut, finally DBN were detected through the DBN. The Gabor wavelet features were extracted from the face cuts detected by DBN, and the fused feature fusion is input to the automatic coding machine for expression recognition [24].The strong classifier combined DBN and the adaboost method got a higher expression recognition accuracy rate for FER [25, 28]. A set of deep confidence network model was proposed, which can improve the performance of expression classification. MLDP-GDA (Modified Local Directional Patterns, Generalized Discriminant Analysis) was proposed, which was a novel method to extract salient features from depth faces was applied to DBN for training different facial expressions [31].

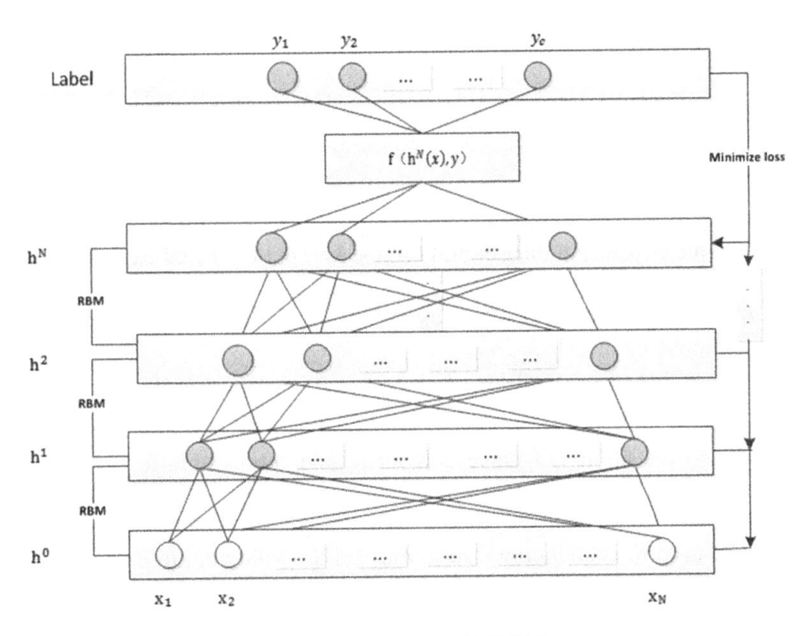

Fig. 3. Structure of DDBN

3.2 Method Based on CNN

As early as 2002, in order to solve the often encountered problems of the face gestures and the uneven illumination, the convolution neural network was applied to identify the facial expression. A set of FER system based on CNN was developed, which could predict the location of facial feature points recognize the facial expression by training a multi-task deep neural network. A novel approach based on based on the combination of optical flow and a deep neural network-stacked sparse autoencoder could analyze video image sequences effectively and reduce the influence of personal appearance difference on facial expression recognition [29]. Combined region of interest (ROI) and K-nearest neighbors (KNN), a fast and simple improved method called ROI-KNN for facial expression classification was proposed, which relieves the poor generalization of deep neural networks due to lacking of data and decreases the testing error rate apparently and generally [30]. A new architecture had the two hard-coded feature extractors: a Convolutional Autoencoder (CAE) and a standard CNN, which could can significantly boost accuracy and reduce the overall training time [32]. A 3D Convolutional Neural Network architecture consists of 3D Inception-ResNet layers followed by an LSTM unit that together extracts the spatial relations within facial images as well as the temporal relations between different frames in the video [33]. The attention model, which consisted of a deep architecture which implements convolutional neural networks to learn the location of emotional expressions in a cluttered scene, greatly improved the expression recognition [34].

4 Challenge and Future Work

Facial expression recognition is a very challenging subject, and is still in the elementary exploratory stage. With the non-rigid of the face and the complexity of FER, there are still many challenges:

(1) Improve the algorithm robustness of the expression recognition. At present, most research is the expression dataset build in the lab, but the facial expressions in real life are more complex and constantly changing.
(2) The difference differences in the race, the geography, and the culture.
(3) Gender differences in expression.
(4) Difficult to real-time recognition.

With the study of computer graphics, machine learning, artificial intelligence and other disciplines, the ability to make computers and robots able to understand and express emotions like humans will fundamentally change the relationship between people and computers, Facial expression recognition, as a long-term challenging subject, as a long-term challenging issues, may be the following trends:

(1) 3-d facial expression recognition.
(2) Robust facial expression representation algorithm and generalization recognition classification algorithm.
(3) Improve the deep learning model.

The deep learning model is widely used because of its strong feature learning ability. However, most of the models are time-consuming in training. It is necessary to train the model in a shorter time and improve the real-time ability.

Acknowledgments. This work was partially supported the support programs for leading personnel of State Ethnic Affairs Commission (p.n. 10301-017004020403).

References

1. Alemy, R., Shiri, M.E., Didehvar, F., Hajimohammadi, Z.: New facial feature localization algorithm using adaptive active shape model. Int. J. Pattern Recogn. Artif. Intell. **20**(1), 1–18 (2012)
2. Kotsia, I., Pitas, I.: Facial expression recognition in image sequences using geometric deformation features and support vector machines. IEEE Trans. Image Proc. **16**(1), 172–187 (2007)
3. Huang, K.C., Kuo, Y.H., Horng, M.F.: Emotion recognition by a novel triangular facial feature extraction method. Int. J. Innovative Comput. Inf. Control **8**(11), 7729–7746 (2012)
4. Sung, J.W., Kaneda, T., Kim, D.: A unified gradient-based approach for combining ASM into AAM. Int. J. Comput. Vis. **75**(2), 297–310 (2007)
5. Tian, Y.I., Kanade, T., Cohn, J.F.: Recognizing action units for facial expression analysis. IEEE Trans. Pattern Anal. Mach. Intell. **23**(2), 97–115 (2001)
6. Rapp, V., Bailly, K., Senechal, T., Prevost, L.: Multi-kernel appearance model. Image Vis. Comput. **31**(8), 542–554 (2013)

7. Patil, R.A., Sahula, V., Mandal, A.S.: Features classification using geometrical deformation feature vector of support vector machine and active appearance algorithm for automatic facial expression recognition. Mach. Vis. Appl. **25**(3), 747–761 (2014)
8. Majumder, A., Behera, L., Subramanian, V.K.: Emotion recognition from geometric facial features using self-organizing map. Pattern Recogn. **47**(3), 1282–1293 (2014)
9. Jabid, T., Kabir, M.H., Chae, O.: Robust facial expression recognition based on local directional pattern. ETRI J. **32**(5), 784–794 (2010)
10. Rivera, A.R., Castillo, J.R., Chae, O.: Local directional number pattern for face analysis: face and expression recognition. IEEE Trans. Image Process. **22**(5), 1740–1752 (2013)
11. Kabir, M.H., Jabid, T., Chae, O.: Local Directional Pattern Variance (LDPV): a robust feature descriptor for facial expression recognition. Int. Arab J. Inf. Technol. **9**(4), 382–391 (2012)
12. Ahmed, F.: Gradient directional pattern: a robust feature descriptor for facial expression recognition. Electron. Lett. **48**(19), 1203–1204 (2012)
13. Lyons, M.J., Budynek, J., Akamatsu, S.: Automatic classification of single facial images. IEEE Trans. Pattern Anal. Mach. Intell. **21**(12), 1357–1362 (1999)
14. Gu, W.F., Xiang, C., Venkatesh, Y.V., Huang, D., Lin, H.: Facial expression recognition using radial encoding of local Gabor features and classifier synthesis. Pattern Recogn. **45**(1), 80–91 (2012)
15. Poursaberi, A., Noubari, H.A., Gavrilova, M., Yanushkevich, S.N.: Gauss-Laguerre wavelet textural feature fusion with geometrical information for facial expression identification. EURASIP J. Image Video Process. **17**, 1–13 (2012)
16. Zavaschi, T.H.H., Britto, A.S., Oliveira, L.E.S., Koerich, A.L.: Fusion of feature sets and classifiers for facial expression recognition. Expert Syst. Appl. **40**(2), 646–655 (2013)
17. Li, Z., Imai, J., Kaneko, M.: Facial-component-based bag of words and PHOG descriptor for facial expression recognition. In: Proceedings of the 2009 IEEE International Conference on Systems, Man, and Cybernetics, vol. 64(2), pp. 1353–1358. San Antonio, TX, USA (2009)
18. Wang, J., Yin, L.J.: Static topographic modeling for facial expression recognition and analysis. Comput. Vis. Image Underst. **108**(1–2), 19–34 (2007)
19. Yang, P., Liu, Q.S., Metaxas, D.N.: Boosting encoded dynamic features for facial expression recognition. Pattern Recogn. Lett. **30**(2), 132–139 (2009)
20. Liu, Y.J., Zhang, J.K., Yan, W.J., Wang, S.J., Zhao, G.Y., Fu, X.L.: A main directional mean optical flow feature for spontaneous micro-expression recognition. IEEE Trans. Affect. Comput. **7**(4), 299–310 (2016)
21. Fasel, B.: Mutliscale facial expression recognition using convolutional neural networks. In: Indian Conference on Computer Vision, Graphics and Image Processing (ICVGIP 02), pp. 1–9. Ahmedabad, India (2002)
22. Liu, Y., Hou, X., Chen, J., et al.: Facial expression recognition and generation using sparse autoencoder. In: International Conference on Smart Computing, pp. 125–130. Hong Kong, China (2014)
23. Lv, Y., Feng, Z., Xu, C.: Facial expression recognition via deep learning. Int. Conf. Smart Comput. **32**(5), 347–355 (2015)
24. Lv, Y., Feng, Z., Xu, C.: Facial expression recognition via deep learning. In: International Conference on Smart Computing, pp. 303–308. Hong Kong, China (2014)
25. Jung, H., Lee, S., Park, S., et al.: Development of deep learning-based facial expression recognition system. In: 21st Korea-Japan Joint Workshop on Frontiers of Computer Vision, pp. 1–4. Mokpo, South Korea (2015)
26. Liu, P., Han, S., Meng, Z., et al.: Facial expression recognition via a boosted deep belief Network. In: IEEE Conference on Computer Vision and Pattern Recognition, pp. 1805–1812. Columbus, USA (2014)

27. Devries, T., Biswaranjan, K., Taylor, G.W.: Multi-task learning of facial landmarks and expression. In: Proceedings of IEEE International Conference on Computer and Robot Vision (CRV), pp. 98–103. Montréal, Canada (2014)
28. Kim, Y., Lee, H., Provost, E.M.: Deep learning for robust feature generation in audiovisual emotion recognition. In: Proceedings of IEEE International Conference on Speech and Signal Processing (ICASSP), pp. 3687–3691. Vancouver, Canada (2013)
29. Liu, Y., Hou, X., Chen, J., et al.: Facial expression recognition and generation using sparse autoencoder. In: Proceedings of IEEE International Conference on Smart Computing (SMARTCOMP), pp. 125–130. Hong Kong, China (2014)
30. Xiao, S., Ting, P., Fu-Ji, Rn: Facial expression recognition using ROI-KNN deep convolutional neural networks. Acta Automatica Sinica **42**(6), 883–891 (2016)
31. Uddin, M.Z. et al.: A facial expression recognition system using robust face features from depth videos and deep learning, Computers and Electrical Engineering. Available online 29 Apr 2017
32. Hamester, D., Barros, P., Wermter, S.: Face expression recognition with a 2-channel convolutional neural network. In: Proceedings of International Joint Conference on Neural Networks (IJCNN), pp. 1787–1794 (2015)
33. Hasani, B., Mahoor, M.H.: Facial expression recognition using enhanced deep 3D convolutional neural networks. In: IEEE Conference on Computer Vision and Pattern Recognition. https://arxiv.org/pdf/1705.07871.pdf (2017)
34. Barros, Pablo, et al.: Emotion-modulated attention improves expression recognition: a deep learning model. Neurocomputing **30**(253), 104–114 (2017)

A Fast Connected Components Analysis Algorithm for Object Extraction

Dai Dehui(✉) and Li Zhiyong

Army Officer Academy PLA, Hefei Anhui, China
ddh0318@163.com

Abstract. When the traditional connected components labeling algorithm is used to label the connected components, it is necessary to scan the image multiple times, resulting in unnecessary repetition. In addition, additional analysis is required to obtain the characteristic information of the connected components. Aiming at the above problems, this paper proposes a connected components labeling algorithm for one scan based on the idea of "linked list" and "Multi-tree". It only needs one scan to complete the merging process of the connected components, and can get the connected components feature information. The experimental results show that the algorithm has strong anti-noise interference performance and fast processing speed, and can be applied to the actual target extraction.

Keywords: Connected components labeling · One-scan · Linked list · Pointer · Object extraction

1 Introduction

Target extraction is a key link from target detection (treating the pixel as the content) to target classification (identifying the object with the target area) [1]. The connected components labeling [2] specifically implements the conversion from the pixel point to the area to be recognized, and its processing performance directly affects the subsequent operation, which is of great significance.

Connected components labeling refers to assigning a unique label to each connected component under the definition of connectivity, which is the basis of the field of computer vision, target extraction and image processing. At present, the connected components labeling algorithm can be divided into pixel-based method, run-based method and regional growth method. Among them, according to the number of scanning, pixel-based method can be divided into one-scan method [3], two-scan method [4, 5] and multi-scan method [6]. The run-based method [7] is mainly to use the straight line as the basic unit of detection. This method makes full use of the structural characteristics of the regional connectivity, which is more efficient than the pixel-based method. The method of regional growth [8] is not affected by the number of targets and the overall complexity of the image, so it has a strong robustness. However, since the connectivity judgment is required for each target pixel, the efficiency of the algorithm is reduced.

F. Xhafa et al. (eds.), *Advances in Intelligent Systems and Interactive Applications*, Advances in Intelligent Systems and Computing 686,
https://doi.org/10.1007/978-3-319-69096-4_49

In this paper, through the establishment of linked list pointer, only a scan of the image, you can achieve connected components labeling. The speed of the algorithm compared to the traditional two-pass and multi-pass has been greatly improved. In addition, it is possible to obtain the feature information such as the number of pixels, the area and the minimum external matrix of the connected components after the scanning, so as to realize the extraction of the target.

2 General Idea

The traditional method of connected components labeling generally requires two scans. The first time is to scan the images one by one row, to determine the connectivity between the pixels, to label the same connected component pixels and assign the corresponding connected tags. But this scan will cause some pixels to be marked with different markings. Therefore, it is necessary to scan the image for the second time to eliminate the duplicate connected tags, and merge the adjacent connected components, and finally get accurate and complete connected components.

Based on the idea of "Multi-tree" generation, this paper designs the linked list operation algorithm and classifies the connected pixels. The pixel coordinates, the number, the circumscribed rectangle, etc. of the connected components of a frame image can be realized by a progressive scan. Compared to other methods of connected components scans, there are the following characteristics: (1) It only searches for the connectivity of the pixels before the current pixel position, so it applies to the way in which the pixels are entered one by one by line. It requires less memory storage, especially suitable for hardware resources limited embedded processing systems; (2) It uses a pointer rather than an integer to mark each foreground pixel, so all the connected pixels share the label of the pointer, and no longer need to modify the label by pixel; (3) Through a scan, it implements the classification management of the connected pixels, which in general reduces the number of operations and improves the speed of the algorithm. The specific process of the algorithm is shown in Fig. 1.

Step 1. The images are scanned from top to bottom, from left to right. When the foreground pixel is found, go to Step 2, otherwise go to Step 5.
Step 2. Search of connected pixels. In order to search the current pixel of the upper left corner, just above, the top right and left direction if there is a adjacent foreground pixel. If there is, go to Step 2.1; if it does not exist, then go to Step 3.
Step 2.1. When the neighboring pixel is found for the first time, the current pixel is added to the secondary node where the adjacent pixel is located and the connected tags corresponding to the current pixel position is set to be the same as the adjacent pixel.
Step 2.2. When the adjacent pixels are found again, compare the same level of connected tags of the two pixels. If the same, continue searching for the next adjacent pixel. If different, you need to merge the connected segments, go to Step 4.
Step 3. Create a linked list pointer.
Step 3.1. Generate a new first-level node in a list, and include forward and backward pointers for each first-level node that is generated.

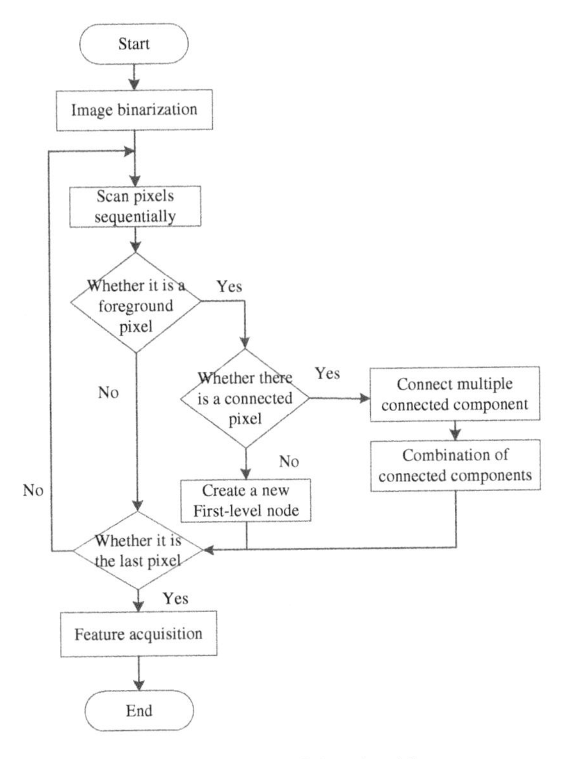

Fig. 1. Process of the algorithm

Step 3.2. Create a new second-level list in the first-level node data. The process includes: Firstly, create a new second-level node, the searched current pixel points added to the second-level node, and the connected tags of the second-level node plus 1. Then, the current connected tags is recorded in the first and second nodes respectively, and the pointer of the first-level node is recorded in the tag array corresponding to the label. Finally, the address of the tag array is recorded in the marker of the foreground pixel.
Step 4. Consolidation of connected components.
Step 4.1. The merging of the connected components is the most important link in the connected components labeling. The method is to add all the second-level node of the first-level node where the next neighboring pixel is to the second-level nodes of the first-level node where the preceding adjacent pixel is located and update the pixel pointer of the marker. The pointer points to the new first-level node address.
Step 4.2. Remove the merged first-level nodes from the first-level list. If the first node of the first-level list is deleted, the next node of the deleted node is used as the first node of the first-level list; If the middle node is deleted, the latter node and the previous node must be connected through the front and rear instructions; If the end node is deleted, the tail pointer of the first-level list must be set to the node before the node was deleted. Finally, the total number of nodes in the first-level list is reduced by one.

Step 4.3. Update the labeling array. The secondary connected tag in all the second-level nodes of the merged first-level node is used as the array subscript and the original first-level nodes address in the corresponding pointer array is changed to the new first-level nodes address.
Step 5. Determine whether the pixel is the last pixel of the image, and if so, then go to Step 6, otherwise return to Step 1.
Step 6. After the merging of the above-mentioned connecting segments, the number and area of the pixels in the connected components can be obtained by connecting the position information of the connected components, and the minimum circumscribed rectangle of the connected components is obtained according to the position information of the most marginal pixels in the four directions. You can extract the target and end the operation.

3 Implementation of the Method

3.1 Create a Linked List

In this paper, inspired by the idea of the "linked list" and "Multi-tree", each connected component is assigned a pointer address, and the connected components labeling for each pixel location is no longer an integer label used in other methods, but an address pointer. When merging the connected components, you do not have to modify the marker graph for each pixel position, and you only need to modify the address pointed to by the pointer in the marker.

This article has created a total of two levels of linked list. The first-level list is composed of connected components separated from each other in the image. Each first-level node represents a connected component and points to a second-level list. The second-level list contains the join segments to be merged in the connected component, and each second-level node points to a bifurcated segment.

This paper gives the control relationship shown in Fig. 2 to more intuitively express the relationship between the connected component and the linked list pointer. Figure 2a shows the connecting sections C [0], C [1] and C [2] separated from each other in the

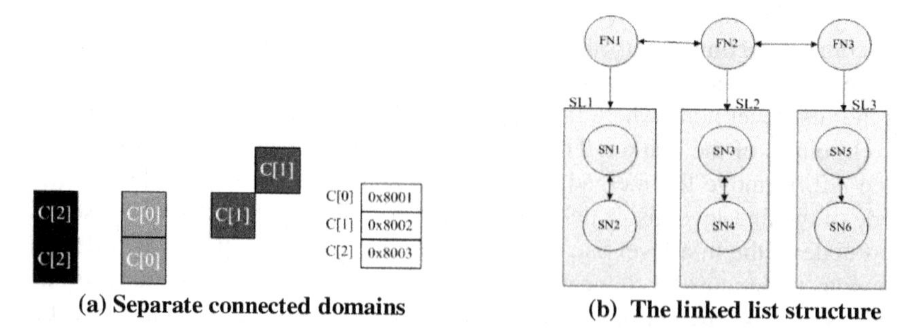

(a) Separate connected domains (b) The linked list structure

Fig. 2. The process of creating a linked list, **a** Separate connected domains. **b** The linked list structure

three segments of the image, which correspond to the first-level nodes FN1, FN2 and FN3 respectively in the linked list pointer, and the three first-level nodes together form a first-level list structure. SL1, SL2, and SL3 represent a second-level node, which represents the pointer address, pixel coordinates, and label of the specific pixel of the connected component. Each second-level node represents a connected segment in the image.

In the process of creating a linked list, you also need to create a label variable, and create an unsigned integer array SOURCE as a first-level node address storage space. When a new first-level node is generated, the label variable *first_lable* is incremented by one and the first-level node address is stored in the array with the subscript variable *first_lable* as the subscript.

In the corresponding image pixel position, it is necessary to establish a connectivity map, and each pixel location is located in the first-level node address of the connected component. The map is generated step by step based on the connected component search and guides the subsequent connected component search operation.

3.2 Search of Connected Components

The connected components are usually divided into 8-connected, 4-connected and M-connected, and the conventional method typically performs two or more times on the image, and usually use 8-connected and 4-connected. Because this article uses a one-pass method for connected component analysis, in the process from top to bottom, from left to right, the current pixel only in the upper left corner, just above, the upper right corner and left side direction may exist adjacent prospects pixel. Therefore, based on the rule of 8-connected rule, a new method of connected components search is improved. As shown in Fig. 3, only the adjacent pixel of the upper left corner, the top, the upper right and the left of the current pixel will be done connectivity detection. This method is in accordance with the principle of 8-connected identification, and reduces the number of redundant neighborhood searches and improves the search speed of the algorithm.

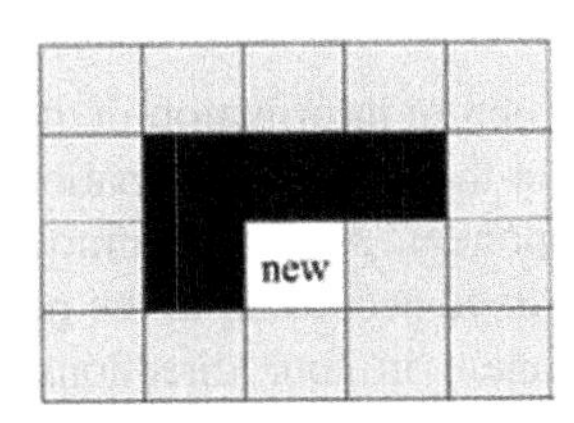

Fig. 3. The search area for adjacent pixels

3.3 Combination of Connected Components

During the scanning process, if the connected component is an upwardly bifurcated shape, it is necessary to merge when the bifurcations intersect. When merging the connected components, only the created linked list pointers need to be manipulated.

The second-level list of one of the connected segments is added to the second-level list of the other component to be merged, and the first-level node of the connected component is deleted from the first-level list, then the merge of the connected components is realized.

Figure 4 is a schematic diagram of the connected components merging in the linked list structure where is the original linked list state, and FN1 and FN2 need to be merged when scanning to a new foreground pixel that belongs to both FN1 and FN2. Firstly, the pointer of the second-level node SL2 is directed to SL1, so that the SL1 and SL2 information can be obtained by the SL1 pointer, and the FN2 node is deleted from the first-level list, thus completing the merging of the FN1 and FN2 join segments. Similarly, when a new foreground pixel belonging to FN1 and FN2 is scanned, the SN3 pointer address is directed to SN1 and the FN3 is deleted from the first-level node, thus completing the merging of the FN1 and FN3 connected components. Through the above process, a complete connection component can finally be achieved.

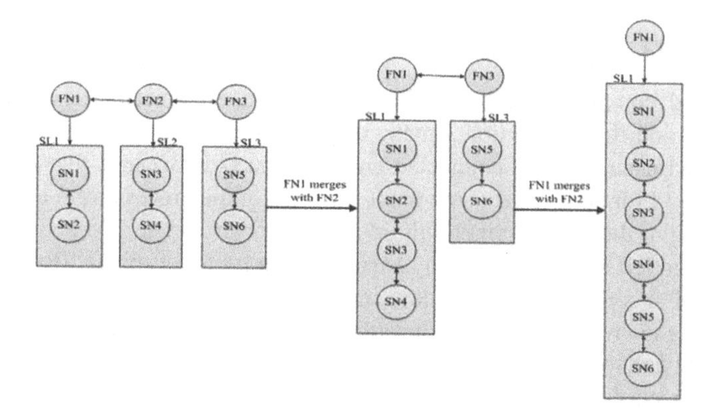

Fig. 4. The process of merging linked lists

3.4 Feature Extraction

After the end of the scan, the structure information in the two-level list pointer defined in Sect. 3.1 can be used to obtain the feature information such as the number and area of pixels in the connected component. And the minimum circumscribed rectangle of the connected component is obtained according to the position information of the edge pixels of the connected component in four directions, so as to achieve the target extraction.

4 Experimental Results and Analysis

The experimental environment is as follows: CPU is the Intel Core i5-6200U 2.30GHZ and the memory is 4.00 GB, and the operating system is Windows 7 (64-bit), and programming development environment is the VS 2010.

In order to quantitatively analyze the effect of this algorithm, the algorithm is integrated into the process of moving object detection. In this paper, two sets of test sequences were selected, and the motion of human and cartons existed in sequence 1 and sequence 2, respectively, and motion shadows were present in the scene. Finally, the algorithm is compared with two-scan algorithm and region filling method in computing speed. The specific process shown in Fig. 5. Figure 5a is the current frame of the test sequence, and Fig. 5b is the result of the foreground detection obtained by the codebook modeling method [9]. Due to the presence of motion shadows in the scene, the shadow part is usually detected when the moving object is detected. Thus, there is usually a real target area and a false area due to motion shadows in the results of foreground detection. Based on the analysis of the foreground detection results, the algorithm obtains the characteristic information of the connected domain, and combines the target's own characteristics to distinguish the real target and the false shadow area, thus eliminating the interference of the motion shadow. In addition, we can get the object's smallest circumscribed rectangle to extract the object, as shown in Fig. 5c.

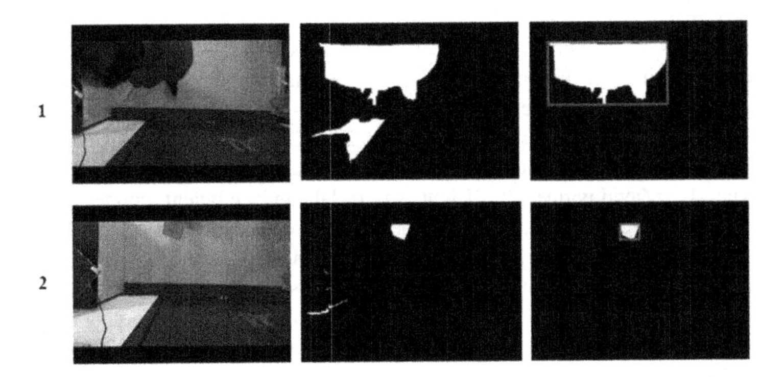

Fig. 5. Target extraction results. Where **a** is the current frame, and **b** is foreground detection results, and **c** is algorithmic processing.

According to the detection effect of Fig. 5c, we compare the time of the connected component labeling in the detection process, and obtain the average detection time of several algorithms in the two sequences, as shown in Table 1.

Table 1. Comparison of the running time of several algorithms

	Region filling method/ms	Two-scan method/ms	Our method/ms
Sequence 1	267	185	95
Sequence 2	98	65	30

5 Conclusions

Based on the linked list pointer, this paper only needs to scan the image once to achieve the connected component labeling. Compared with the traditional algorithm, our algorithm can not only improve the speed of the operation by reducing the number of times of scanning, but also can get the area of the connected component, the number of pixels and the minimum external matrix and other feature information. The experimental result shows that our algorithm can distinguish the real and shadow noise from the scene and extract the real object area, which can lay a good foundation for the realization of the follow-up object recognition and tracking.

References

1. Xiaofeng, L., Tao, Z., Zaiwen, L., et al.: An improved method for moving human detection. Comput. Simul. **28**(2), 308–311 (2011)
2. Zuo, Z.G., Xuyan, T., et al.: Image connection domain tagging algorithm based on equivalence pair. Comput. Simul. **28**(1), 14–16 (2011)
3. He, L.F., Chao, Y.Y., Suzuki, K., et al.: A run-based one-scan labeling algorithm. In: Proceedings of the 6th International Conference on Image Analysis and Recognition, pp. 93–102. Springer, Heidelberg (2009)
4. Bock, D.J., Philips, W.: Fast and memory efficient 2-D connected components using linked lists of line segments. IEEE Trans. Image Process. **19**(12), 3222–3231 (2010)
5. Lacassagne, L., Zavidovique, B.: Light speed labeling: efficient connected component labeling on RISC architectures. J. Real-Time Image Process. **6**(2), 117–135 (2011)
6. Suzuki, K., Horiba, I., Sugie, N.: Linear-time connected-component labeling based on sequential local operations. Compu. Vis. Image Underst. **89**(1), 1–23 (2003)
7. Lianqiang, N., Min, P., Zhongli, S., et al.: Fast connected components labeling by propagating labels of run sets. J. Comput. Aided Des. Comput. Graphic **1**, 128–135 (2015)
8. Xie, Z., Peng, Q., Bai, Y.: Algorithm research and SoPG implementation of connected component labeling of edge image. Electron. Technol. Appl. **37**(3), 35–37 (2011)
9. Liang, X., Weiming, L.: Foreground detection algorithm based on background codebook model. Sci. Technol. Eng. **10**(9), 2118–2121 (2010)

Bibliometrics and Visualization Analysis of Knowledge Map in Metallurgical Field

Ying Xu, Yan Xiang, Dangguo Shao(✉), Zhengtao Yu, Na Zheng, Wei Chen, and Lei Ma

Key Lab of Computer Technologies Application of Yunnan Province, Kunming University of Science and Technology, Kunming 650500, China
23014260@qq.com

Abstract. In this paper, we used the methods of bibliometrics and information visualization to quantitatively analyze papers in the field of metallurgy in the past 20 years. Methods of bibliometrics and information visualization can make scholars intuitively understand the research hot spots in the academic field and the trend of disciplines. This paper analyzed the overall study of the domestic and foreign technical development, we adopted keyword coexistence and the co-citation network map of literature. We revealed the distribution of metallurgical papers and the new hot spot on the field of metallurgy in recent years both here and abroad, and mastered the development trend of metallurgy field.

Keywords: Metallurgy · Bibliometrics · Knowledge map · Visualization

1 Introduction

The metallurgical industry provides the necessary materials for machinery, energy, chemicals, transportation etc. Domestic and foreign scholars explore the field of metallurgy from the qualitative or technical point of view as usual, they rarely apply visualization techniques in the field of literature to analyze the important hidden information, then explore the field of metallurgical focus points and the development of the forefront of exploration, with intuitive image method to deal with disorganized data, not only can save time, but also to complete the study efficiently.

Mapping is an expression of abstract data visualization which is convenient for analyzing data, discovering regularities and supporting decisions [1]. The method of bibliometrics is using mathematical and statistical methods to study the quantitative and statistical analysis of article publication and distribution. At the beginning of the twentieth century, the method was first proposed by the American scholar A.J. Lotka to describe the correlation between the paper, the author and institution, etc. [2, 3]. The visualization software is utilized to visualize the overall status of metallurgical research and to understand the development of international metallurgy.

In this paper, the Citespace can be utilized to show the characteristics and research contents of the metallurgical field from the perspective of bibliometrics [4]. Through the intuitive tables and maps, we can find the change of metallurgical mode and the development trend of metallurgy, we learn the experience of metallurgy better.

F. Xhafa et al. (eds.), *Advances in Intelligent Systems and Interactive Applications*, Advances in Intelligent Systems and Computing 686,
https://doi.org/10.1007/978-3-319-69096-4_50

2 Methods and Data

The data source is the web of Science core collection, it includes the notorious three major Citation Indexes (SCI, SSCI, A&HCI), and contains numerous Chinese literatures. Then the search range is set to "subject", the search term is set to "metallurgy", and the retrieval time is from 1997 to 2016, as a result 15,011 search results are obtained, while any uncorrelated literatures are excluded. In order to make sure that the research more accurate, the literature type is selected as "article". In the end, we obtained 11,060 articles. This paper uses Citespace and excel for qualitative analysis of data. The research hot spot in the field of metallurgy has been investigated in the past two decades. In addition, we analyze Co-Citation literatures to get important references of the year. We can get critical literatures by analyzing the co citation network map of literatures.

3 Research Focus of Metallurgical Discipline and the Frontier Analysis

3.1 High Frequency Word Statistics

Keywords were extracted in the field of metallurgy, which facilitates us understand the research hot spot. Word frequency statistics can predict trends in disciplines, industries, and technologies.

This paper is based on word frequency statistics. We extracted 196 keywords. Keywords with higher frequency can reflect hot spots. We do data cleaning, including the merging synonyms, removing meaningless words, and getting ten important keywords. The results are shown in Table 1. Centrality analysis recognizes the 'core' and the 'broker' in the local network. "Powder metallurgy" has the highest frequency and the biggest centrality which occupies a dominant position in the metallurgy field. Secondly, further comprising a commonly used reinforcement keyword "mechanical properties", "aluminum", "microstructure" and "metal matrix composite", the iron-based composite materials such as a metal matrix composite, material properties.

Table 1. High frequency keyword statistics

Keywords	Frequency	Year	Centrality
Powder metallurgy	3258	1997	0.16
Mechanical property	1776	1997	0.16
Aluminum	423	1998	0.15
Composite	788	1997	0.14
Microstructure	2131	2000	0.1
Alloy	1222	1997	0.1
Metal matrix composite	659	2001	0.09
Deformation	449	1997	0.09
Aluminum alloy	270	1997	0.09
Oxidation	91	1997	0.08

The present study was to explore the domestic and foreign applications. Therefore, the statistics of high-frequency keywords allow us to grasp focus points of powder metallurgy.

3.2 Keyword Concurrence Networks

Keywords concurrence network is used to present the keyword distribution of subjects or previous and current research topics [5]. Articles are imported into Citespace and keyword is set as a node to analyze, constructing keyword concurrence map. The result is shown in Fig. 1, the "ring" represents words, the size of ring represents the appearance frequency of keywords, the greater the ring, is the higher the frequency of occurrence of the keyword, and the thicker the connection between nodes, is the deeper the correlation between the keywords. The highest frequency one is "powder metallurgy". According to Table 1, we know the frequency is 3258, it is made into a composite material of metal materials and various types of products of the technology by forming and sintering. "Powder metallurgy" is related to keywords "microstructure", "mechanical property", and microstructure is studied for a period focusing on metal.

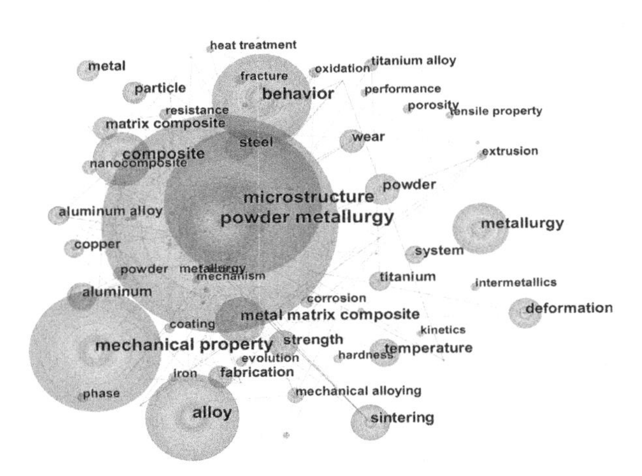

Fig. 1. Keyword co-occurrence network

4 Literature Cited Analysis

4.1 Document Default Clustering

Literatures are based on co-citation have more or less similarity in the subject. The total number of cited times can measure the relevance of literatures in the content. Great citation rates are indicating that the higher the correlation between literatures, the more closer the relationship is. In this paper, 11,060 articles were analyzed by Citespace from

1997 to 2017, and the time slices were generated for one year to generate the literature cluster. Each node in the figure is a citation. The size of node represents numbers of cited references, The more times is taken, the greater the node is, the year in the parentheses represents the time of the first quote of the paper, the color and thickness of the inner circle of the node indicate the frequency of the cited time in different periods. The connection between nodes indicates a co-occurrence relationship, and its thickness indicates the strength of the co-occurrence. The color corresponds to the time when the node first shares for the first time. According to statistics, citations are 4578 psalms, and the highest one was cited 66 times, the average number of citations per paper is 6, showing that metallurgy has been the subject of research both at home and abroad, and literatures in metallurgy field have a high reference value [6] (Fig. 2).

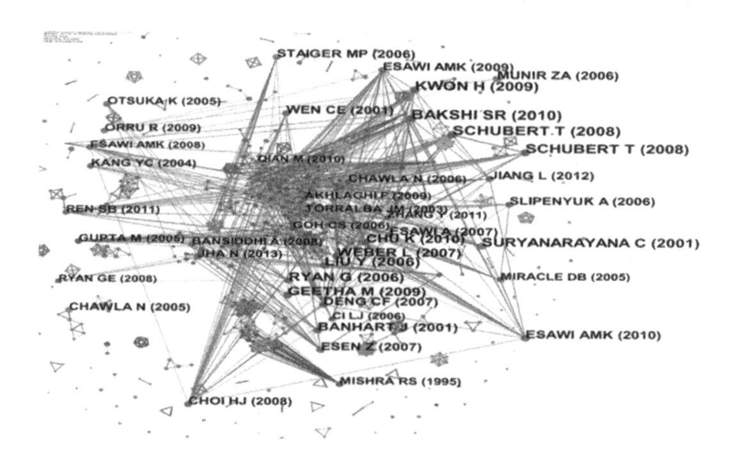

Fig. 2. Literature cited network

4.2 Document Automatic Tag Clustering

On the basis of the default clustering, automatic clustering of titles, and some of the critical literatures are obtained. The results are shown in Table 2. These documents represent the frontier of the research. For example, we create a literature "towards sustainability for recovery of critical metals from electronic waste", the use of metallurgical technology to recover key metals from spent electronic materials can save metal resources, and preserve the environment. The number of articles for the clustering of 3 can be seen in the latest research in the field of metallurgy in 2016. "Distribution control of particles in Mg-Al/AIN composites", the technology of particle-reinforced composites has been studied at home and abroad, and there are new developments [7].

Table 2. Title automatic cluster

Clustering number	Number of documents	Important literatures	First cited time
0	131	Toward sustainability for recovery of critical metals from electronic waste: the hydrochemistry processes	2012
1	100	Compression fatigue behavior and failure mechanism of porous titanium for biomedical applications	2006
2	89	Carbon fiber reinforced metal matrix composites: fabrication processes and properties	2012
3	24	Distribution control of aln particles in mg-al/aln composites	2016

On the basis of the default clustering, keywords were extracted for automatic clustering of tags. The citation network is divided into 616 common clusters, and the largest group consists of 70 keywords, which are tagged by the index terms from their own references. What LLR stands for is not the simple relationship between words and word's frequency, based on the LLR (Logarithmic Likelihood Ratio), but the distance between the words is relatively accurate [8]. LLR indicates the representativeness of the group, the bigger LLR is, the greater influence is. Through the LLR algorithm, we get the knowledge map shown in Fig. 3, the literature cluster number 0 is the representative of the word "interfacial reaction", which is similar to the text where the meaning of these articles have most relevant words. These documents are mainly in the study of crystal or solid interface in the phenomenon. These representative words can represent the forefront of the field of metallurgical terms.

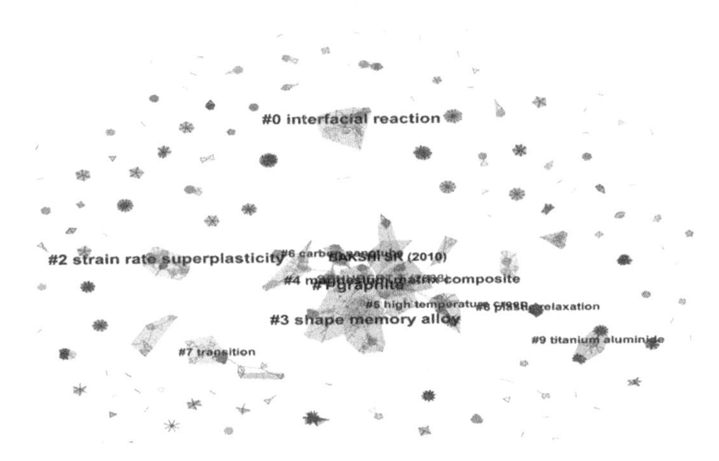

Fig. 3. The literature were cited–Key words automatic clustering

5 Conclusion

I would be grateful to teachers and students to complete this paper together. In this paper, metallurgical literatures in the web of science are analyzed with methods of bibliometrics and visualization. High frequency keywords and co-citation literatures are analyzed to explore the hot topics and research frontiers.

According to the statistical data of the paper, metallurgy has been extensively used in the field of manufacturing, aerospace industry and other fields. We calculated high frequency keywords, and drew a conclusion which is the golden period of powder metallurgy in recent years. It is the golden age of "powder metallurgy" research in recent years, which is for the automotive industry, equipment manufacturing, metal industry, aerospace, military industry, instrumentation and other fields of spare parts production. Key words and titles can show the research focus and the frontier of metallurgy field, we calculate co-citation and draw a conclusion that the researchers attach great importance to the recycling of metals, especially precious metals, as well as environmental protection. But the experimental data are different, the results of software analysis also vary, perhaps through further experiments, more exciting results will be found than this article. In order to obtain more valuable informations we will analyze foreign metallurgical articles and domestic metallurgical literatures in the next experiment.

Acknowledgements. This work was supported by China Postdoctoral Science Foundation (2016M592894XB) and the Science and Technology Office of Yunnan Province (KKS0201703015). We also appreciated the valuable comments from the other members of our department.

References

1. Zhao, K.: Data visualization. Data Anal. **7**(12), 35–54 (2012)
2. Kumar, S., Sharma, P., Sharma, P., Garg, K.C.: Lotka's law and institutional productivity. Inf. Process. Manage. **34**(6), 775–783 (1998)
3. Tang, J.: Analysis on the international situation and evolution of research: quantitative analysis on the SSCI theses from 2005 to 2014. Adv. J. Commun. **03**(4), 100–109 (2015)
4. Chen, C.: CiteSpace II: detecting and visualizing emerging trends and transient patterns in scientific literature. J. Assoc. Inf. Sci. Technol. **57**(3), 359–377 (2006)
5. Le, L., Yu, H.: A new method for evaluating node importance in complex networks based on data field theory. In: First International Conference on Networking and Distributed Computing, pp. 133–136. IEEE (2010)
6. Wallace, D.P., Fleet, C.V., Downs, L.J.: The research core of the knowledge management literature. Int. J. Inf. Manage. J. Inf. Prof. **31**(1), 14–20 (2011)
7. Chen, J., Bao, C., Ma, Y., et al.: Distribution control of AlN particles in Mg-Al/AlN composites. J. Alloy. Compd. **695**, 162–170 (2017)
8. Chen, C., Ibekwe-Sanjuan, F., Hou, J.: The structure and dynamics of cocitation clusters: a multiple-perspective cocitation analysis. J. Assoc. Inf. Sci. Technol. **61**(7), 1386–1409 (2010)

Hierarchical Decision Tree Model for Human Activity Recognition Using Wearable Sensors

Cheng Xu[1,2], Jie He[1,2(✉)], and Xiaotong Zhang[1,2]

[1] School of Computer and Communication Engineering, University of Science and Technology Beijing, Beijing, China
hejie@ustb.edu.cn

[2] Beijing Key Laboratory of Knowledge Engineering for Materials Science, Beijing, China

Abstract. Motion related human activity recognition using wearable sensors can potentially enable various useful daily applications. In this study, we start from a deep analysis on natural physical properties of human motions, and then extract the implied commonly prior knowledge. With the prior knowledge, a hierarchical decision tree (H-DT) model has been proposed to recognize human motions and activities. H-DT has a multi-layer heuristic structure that is easy to understand. Support Vector Machine (SVM) has been selected as sub-classifier of each layer in H-DT. The experiment results indicate that the proposed H-DT methods performs superior to those adopted in related works, such as decision tree, k-NN, SVM, neural networks, and the H-DT has achieved a general classification rate of 96.4% ± 0.025.

Keywords: Activity recognition · Prior knowledge · Decision tree

1 Introduction

Human activity recognition (HAR) is one of the most promising research topics for a variety UMAN motion related activity recognition (HAR) is of areas and has been drawing more and more researchers' attention. The maturity of Machine Learning and Ubiquitous Computing promotes the adhibition of various models and algorithms into human activity recognition. All these improvements enhance the advancement of HAR techniques with high precision, high real time capability, and has been widely used in a variety application areas, such as medical care [1], emergency rescue [2, 3], intelligent monitoring [4] and smart home surveillance [5, 6].

Due to the advantages of no need to deploy in advance, smaller data volume, lower cost and power consumption, sensors-based HAR stands out among various technologies and has been drawing tremendous attention and applied into many fields. The overall objective of HAR is, based on the given activity set and observation data, to pursue a certain classification model, aiming at a good generalization performance as far as possible. Thereinto, generalization performance refers to the capability of predicting unseen instances with the use of designed classifier given some training data under certain phenomenon (human activities in this study). Classifier's generalization performance are determined by two factors: data and prior knowledge [7–9].

F. Xhafa et al. (eds.), *Advances in Intelligent Systems and Interactive Applications*, Advances in Intelligent Systems and Computing 686,
https://doi.org/10.1007/978-3-319-69096-4_51

In this paper, we present a conceptual model of human motions with which a new approach is put forward to recognize human motion related activities. By deeply mining commonly understanding of motions, a conceptual hierarchical decision tree model (H-DT) is proposed which intuitively presents the knowledge contained in motions. With applying this prior knowledge into motion classification and the integration of Support Vector Machine (SVM) using RBF Kernel, H-DT improves the performance of traditional decision tree method and makes up for the inadequacy of data itself. In this way, key features are extracted and the classification result shows that our proposed H-DT method works better than traditional methods such as C4.5, SVM, BP and has achieved a general true classification rate of 96.4% $\pm$ 0.025.

2 Hierarchical Decision Tree Model

Features used in the classifier present different activities' similarities and differences, and play a big part on the classification performance. To solve aforementioned problem, we try to bring more expert knowledge into the classifier to achieve the goal of extracting and using key features to improve classification performance in the motion recognition process. In this section, we present a new approach based on exploring rich domain knowledge for activity classification.

2.1 Conceptual Motion

As for activity recognition problems, prior knowledge is reflected in our understanding of motions. Unless we have already had a clearly definition and description on a certain activity, there is no chance we can tell if from the others. It's commonly believed that a human motion can be described from several attributes, such as intensity, orientation, velocity, and so on. These attributes, in some aspects, embody characteristics of motions and can be related with a series of key features that most eminently reflect the physical difference among activities. These key features may be used to group different kinds of activities into several subclasses as they have various distribution overlap on the same attribute. We model a human motion with attributes of intensity, orientation, velocity, body position and duration. Each attribute represents human motions in a side view from a particular angle.

2.2 Hierarchical Decision Tree Model

The proposed conceptual model above establish links between activities and conceptual information through activity-based attributes and make it possible to understand and distinguish different motions in finer perspectives. At the same time, multi-class classification could be done in steps one of which adopt one attribute as a basis. In this way, hierarchical relationships are constructed that link conceptual information with sensor observations through activity attributes. Above mentioned considerations similarly make decision tree classifier a first choice with the advantage of easier to build multi-level heuristic structure as decision tree is a set of if-then rules which are successively applied to the input data. Based on the analysis of activity attributes, we propose a fusion method, Hierarchical Decision Tree (H-DT), to achieve the goal of classification in a hierarchical way which at the same time avoid over-fitting and dimension disaster problems of traditional decision tree method and pursue a better generalization performance. The decision tree method is simple, easy to understand and has a clear hierarchy. However, it also suffers the drawback that may lead to serious over-fitting problem and have low generalization performance as the model construct process depends on the training data too much. Considering these problems, we take the advantage of proposed conceptual motion, to improve traditional decision tree method's performance.

Meanwhile, the conceptual activity model mentioned above is knowledge-based, so it may be feasible to replace the internal node in a decision tree structure with a specific classifier according to activity attributes. Taken these into consideration, we propose a Hierarchical Decision Tree (H-DT) method against activity model with the combination of decision tree and internal classifiers, shown in Fig. 1.

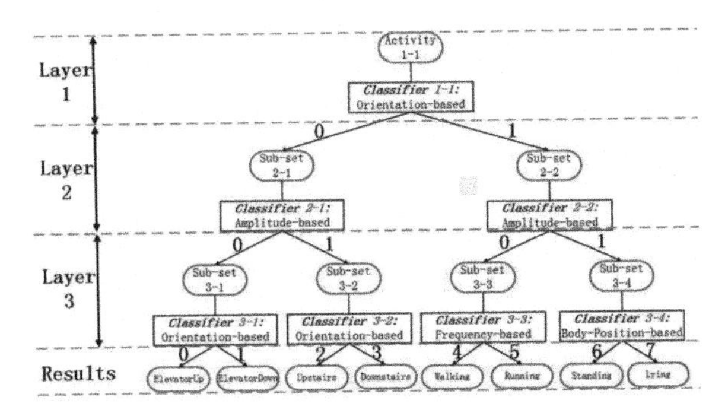

Fig. 1. Hierarchical Decision Tree. Global distinguished attributes (orientation and intensity) based sub-classifier are chosen as root node classifier, and local distinguished attributes (frequency and body-position) are distributed as leaf nodes.

2.3 Classifier

As for the internal classifier, SVM stands out because it should work better in solving binary classification problems. Support Vector Machine is originally designed to solve binary classification problems and when it comes to multi-classification, usually several binary classifiers are constructed in methods of one-against-one or one-against-all. However, one-against-all method has seriously drawbacks as it chooses one kind of samples as a class and the rest as the other class. Then k classes will construct k SVMs. Under this condition, the training data set will lead to data bias problem and don't have much practical value. So one-against-one classifier is a better choice. Besides, in H-DT, all internal classifier view the input as two sub-classes so all work is binary classification which can to the most extent reflects SVM's advantage.

LibSVM [10] is a practical tool realized by one-against-one method. The classifier constructs k binary-classification rules where the mth function $\omega_m^T \phi(x) + b$ separates training vectors of the class m from the other vectors, where ω is the weighted coefficient vector that is normal to the hyper plane and b is the bias term [10]. Hence we obtain k decision functions but they are all used to solve the same problem. Thus it can be considered that the SVM solves the following problem:

$$\min_{\omega,b,\xi} \frac{1}{2}\sum_{m=1}^{k} \omega_m^T \omega_m + C\sum_{i=1}^{l}\sum_{m \neq A_i} \xi_i^m \omega_{A_i}^T \phi(x_i)$$
$$subject\ to\ \omega_{A_i}^T \phi(x_i) + b_{A_i} \geq \omega_m^T \phi(x_i) + b_m + 2 - \xi_i^m$$
$$\xi_i^m \geq 0, i = 1, \ldots, n, m \in \{1, \ldots, k\} - A_i$$

where the training data xi are mapped to a higher dimensional space by the function $\varphi(\bullet)$ when data are not linear separable, C is the penalty parameter and $C\sum_{i=1}^{l}\sum_{m \neq A_i} \xi_i^m \omega_{A_i}^T \phi(x_i)$ is a penalty term which can reduce the number of training errors. Equation (11) also means there need to be a balance between the regularization term $\frac{1}{2}\sum_{m=1}^{k} \omega_m^T \omega_m$ and the training errors.

3 Recognition of Motion Related Human Activities Using H-DT

For the purpose of activity recognition listed in Activity case set, two most widely used sensors in related works are taken into consideration, namely a triaxial accelerometer, a triaxial gyroscope and in addition a barometer, which can be denoted by:

$$SensorUnit = \{Accelerometer, Gyroscope, Barometer\}$$

These sensors are mounted to several parts of human body and it can be defined as:

$$Location = \{Ankle, Knee, Waist, Shoulder, Wrist\}$$

3.1 Experimental Setting

On the basis of above considerations, our hierarchical decision tree is conducted through the process shown in Fig. 2(a) and (b) to classify an unknown activity by analyzing data collected by each sensor node. The entire process can be divided into two phases: modeling phase and prediction phase.

3.2 Comparison with Existing Approaches

Our proposed Hierarchical Decision Tree (H-DT) method links conceptual information, namely common knowledge with activities through activity-based attributes and construct a hierarchical decision tree to simplify the classification process with exploring common knowledge to extract key features. To verify the validity of H-DT on HAR problem, we take decision trees (C4.5), support vector machine and BP neural work algorithms which are the most widely used four algorithms in the study of HAR to

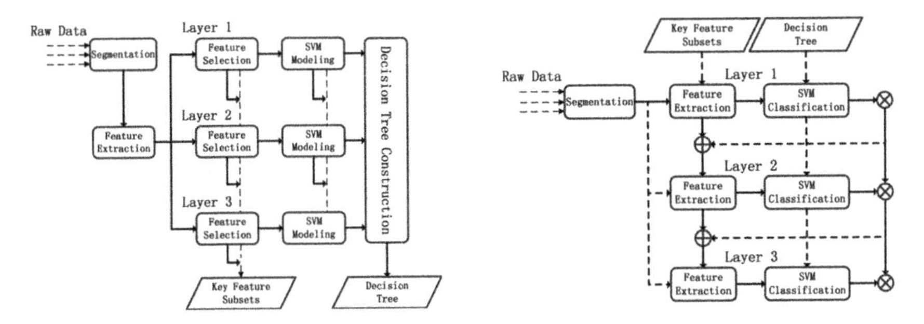

Fig. 2. System processing procedures. (a) Training phase. (b) Prediction phase.

make a brute-force comparison. To compare the classifiers and to identify a principal classifier, we used the experimenter environment in the WEKA toolkit. C4.5 is a most widely used decision tree algorithm providing good classification accuracy. We take a confidence factor of 0.25 to address the issue of tree pruning. Pruning of the decision tree is done by replacing a whole subtree by a leaf node as there may be anomalies due to noise or outliers in the training data. And a radial basis kernel (RBF) based SVM is adopted using LibSVM [10] with automatic parameter selection through grid searching techniques. For the BP neural work, we take the standard approach of recursively evaluating values for the learning rate and momentum using cross validation. A 10-folder cross validation is applied to each classifier independently and the experiment results are shown in the following table.

From Table 1 we can see, the four algorithms show different classification accuracy and variance with the 10-folder experiments and according to the performance, the four can be sorted in the following order: H-DT > BP > SVM > C4.5. Furthermore, H-DT shows the highest global average accuracy and the lowest variance reflecting a high stability during the classification. And in each independent activity, H-DT also presents a better performance in classification accuracy and stability.

Table 1. Classification accuracy and variance

	C4.5	SVM	BP	H-DT
Standing	0.639 ± 1.170	0.973 ± 0.032	0.952 ± 0.351	0.993 ± 0.017
Lying	0.923 ± 0.020	1 ± 0	0.967 ± 0.211	0.997 ± 0.008
Elevator up	0.894 ± 0.035	0.961 ± 0.069	0.974 ± 0.025	0.997 ± 0.002
Elevator down	0.91 ± 0.034	0.947 ± 0.080	0.937 ± 0.020	0.97 ± 0.078
Upstairs	0.907 ± 0.137	0.514 ± 0.285	0.917 ± 0.160	0.933 ± 0.043
Downstairs	0.765 ± 0.523	0.296 ± 0.208	0.889 ± 0.133	0.905 ± 0.080
Walking	0.842 ± 0.129	0.568 ± 0.330	0.886 ± 0.286	0.918 ± 0.173
Running	0.931 ± 0.089	1 ± 0	0.967 ± 0.075	0.998 ± 0.001
Global average	0.851 ± 0.267	0.7824 ± 0.126	0.936 ± 0.131	0.964 ± 0.025

4 Conclusion

The major contribution of this work is the proposal of a knowledge-driven method to recognize motion related human activities. In this study, we construct a hierarchical decision tree (H-DT) is constructed. H-DT can be viewed as a recognition method with knowledge applied into the dealing of data which at the same time covers the advantages of data-driven methods. With a set of hierarchical rules successively applied to the recognition process, H-DT shows a better recognition accuracy (96.4% on average) and lower time consumption (0.02 s on average) compared with most widely used methods such as decision tree, k-NN, SVM and neural networks.

Acknowledgments. This work is supported by National Natural Science Foundation of China (NSFC) project No. 61671056, No. 61302065, No. 61304257, No.61402033, The National Key R&D Program of China, No. 2016YFC0901303, Beijing Natural Science Foundation project No.4152036 and Tianjin Special Program for Science and Technology No. 16ZXCXSF00150.

References

1. Weng, S., Xiang, L., Tang, W., et al.: A low power and high accuracy MEMS sensor based activity recognition algorithm. In: IEEE International Conference on Bioinformatics and Biomedicine (BIB-M), pp. 33–38. IEEE Computer Society (2014)
2. Kau, L., Chen, C.: A smart phone-based pocket fall accident detection, positioning and rescue system (2014)
3. Fairchild, D.P., Narayanan, R.M.: Classification of human motions using empirical mode decomposition of human micro-Doppler signatures. IET Radar, Sonar Navig. **8**(5), 425–434 (2014)
4. Turaga, P., Chellappa, R., Udrea, O.: Machine recognition of human activities: a survey. IEEE Trans. Circuits Syst. Video Technol. **18**(11), 1473–1488 (2008)
5. Singla, G., Cook, D., Schmitter-Edgecombe, M.: Recognizing independent and joint activities among multiple residents in smart environments. J. Ambient Intell. Humanized Comput. **1**(1), 57–63 (2010)
6. Minor, B., Doppa, J.R., Cook, D.J.: Data-driven activity prediction: algorithms, evaluation methodology, and applications. In: Proceedings of the 21th ACM SIGKDD International Conference on Knowledge Discovery and Data Mining, pp. 805–814. ACM (2015)
7. Bousquet, O., Boucheron, S., Lugosi, G.: Introduction to statistical learning theory, volume. Lect. Notes Artif. Intell. **2004**, 169–207 (2004)
8. Schölkopf, B., Smola, A.J.: Learning with kernels: support vector machines, regularization, optimization, and beyond. J. Am. Stat. Assoc. **98**(3), 781 (2003)
9. Lauer, F., Bloch, G.: Incorporating prior knowledge in support vector machines for classification: a review. Neurocomputing **71**, 1578–1594 (2008)
10. Chang, C.C., Lin, C.J.: LIBSVM: A library for support vector machines. ACM Trans. Intell. Syst. Technol. (TIST) **2**(3), 1–27 (2011)

Human Motion Monitoring Platform Based on Positional Relationship and Inertial Features

Jie He[1,2(✉)], Cunda Wang[1,2], Cheng Xu[1,2], and Shihong Duan[1,2]

[1] School of Computer and Communication Engineering, University of Science and Technology Beijing, Beijing, China
hejie@ustb.edu.com
[2] Beijing Key Laboratory of Knowledge Engineering for Materials Science, Beijing, China

Abstract. Human motion monitoring widely used in medical rehabilitation, health surveillance, video effects, virtual reality and natural human–computer interaction. Existing human motion monitoring can be divided into two ways, non-wearable and wearable. Non-wearable human motion monitoring system can be used when it is not contact with user, so it has no influence on users' daily life, but its monitoring range is constrained. Wearable human motion monitoring system can solve this problem well. But as the dataset is not complete, it is inconvenient for the research of related algorithms, and many researchers choose to build their own data acquisition platform. On the other hand, the traditional motion recognition platform which is based on inertial features, is difficult to identify static pose. This limits the application of wearable motion recognition technology. In this paper, we design a human motion recognition platform based on inertial features and positional relationship, which can provide a platform for data acquisition and dataset for the researchers who study algorithm of wearable motion recognition.

Keywords: Positional relationship · Wearable sensor · Human motion monitoring · Hardware platform

1 Introduction

Due to human motion monitoring has a wide range of application in health care [1], film and television special effects, virtual reality, natural human–computer interaction, etc., it has been attracting more and more research workers and related businesses. At present, human motion monitoring technology could be divided into two kinds according to the scope of the monitoring system and the movable attributes: non-wearable monitoring technology and wearable monitoring technology.

Non-wearable monitoring technique means that the primary monitoring device is immovable or the monitoring device is non-contact with the monitored object. Such monitoring techniques are partly bases on optical image recognition [2], electromagnetic tracking, or acoustic positioning tracking.

F. Xhafa et al. (eds.), *Advances in Intelligent Systems and Interactive Applications*, Advances in Intelligent Systems and Computing 686,
https://doi.org/10.1007/978-3-319-69096-4_52

Wearable monitoring system means that the monitoring system is worn on the monitored person [3]. The monitoring device is in contact with the monitored person. Wearable monitoring technology according to the different combinations of sensors can be divided into different monitoring programs [4]. Typical wearable system monitoring techniques include mechanical tracking and MEMS sensor-based monitoring methods [5]. There are also wearable attitude detection systems using other technologies [6] such as bio-fibers [7].

The two monitoring methods have their own advantages and disadvantages, and we compare the two techniques at this stage, show in Table 1.

Table 1. Comparison of non-wearable and wearable human motion recognition techniques

Current technology	Non-wearable	Wearable
Accuracy	High	Low
Portability	Low	High
Identify the type of gesture	Much	Less
System complexity	High	High
Identify the difficulty of the algorithm	Higher	High
Anti-jamming capability	Medium	Very high
Computing capacity requirements	Very high	Medium

By contrast, wearable human motion recognition has a strong portability, anti-interference ability, etc., but the recognition rate is low and the identification type is less, which limits its own application. Therefore, in this paper, we design a motion recognition platform based on position relationship and inertial feature. The accuracy is increased from the viewpoint of the sensor type. On the other hands, it provides researchers with a convenient data acquisition platform and then facilitate the researchers to identify the algorithm.

The rest of the paper is organized as follows. Section 2 puts forward the classification model of simple human motion. Section 3 introduces the hardware design of the wearable motion monitoring system. We verify the platform from the two aspects of data acquisition and motion recognition in Sect. 4. Section 5 comes the conclusion.

2 Attributes Based Classification Model

2.1 Human Motion Attributes

In order to distinguish the physical characteristics of different motions, we firstly make a standard to describe these motions. This standard is called motion attribute, which is used to describe human motions. For a variety of motion attributes, we need to abstract these attributes to be applied to describe the characteristics of any motion. The key attributes that describe human motion include rate, intensity, relative positional relationship between nodes, node absolute position relationship, etc.

2.2 Absolute Positional Relationship

Some people's actions will lead to changes of body position in space, either horizontal or vertical. These changes can be reflected by the absolute position relationship (APR) of node. (APR) of the node refers to the position change, which is relative to the outside of the system. For some of the motion, it is difficult to use inertial characteristics to realize identification, such as up and down the stairs, because from the aspects of inertia characteristics, there is no difference between walking and up and down the stairs. When use the absolute position of the node changes can be distinguished up and down the stairs and walking.

2.3 Relative Positional Relationship

Relative positional relationship (RPR) refers to the positional relationship between multiple nodes. For some actions, we do not need to know the absolute position of the nodes, but only need to consider the relative position of some parts of the body. For instance, we can use the distance between the feet to identify the changes when walking needs to be classified. It is easier to identify this type of motion by using the relative positional relationship between nodes. Sensor data, which can represent the relative positional relationship of the nodes include the distance between nodes, the difference in air pressure between nodes and so on.

3 Platform Design

The goal of the platform is to conveniently and accurately collect the inertial sensor data, the distance between nodes, the pressure value and other data. Considering the cost of the platform and data redundancy and other issues, it should reduce the number of nodes, the location information of each node complement each other, to ensure that the characteristics of motion integrity. Hardware platform is to collect the human motion data information, so it is should consider the data storage methods and formats and mark the different motion conversion. Finally, it is also should consider the synchronization between nodes and minimize the power consumption and other issues.

3.1 Hardware Design

The motion recognition platform is divided into two parts, one part is a handheld device, and the other part is a wearable node. The main function of the handheld device is to control the wearable node to collect data or not, timing to send synchronization packets to the node, synchronize the clocks of each node. Handheld devices mainly include the main control module, button, RF module, power management module. The node includes a master module, a sensor module, a data storage module, a radio frequency module and a power management module. The system block diagram is shown in Fig. 1.

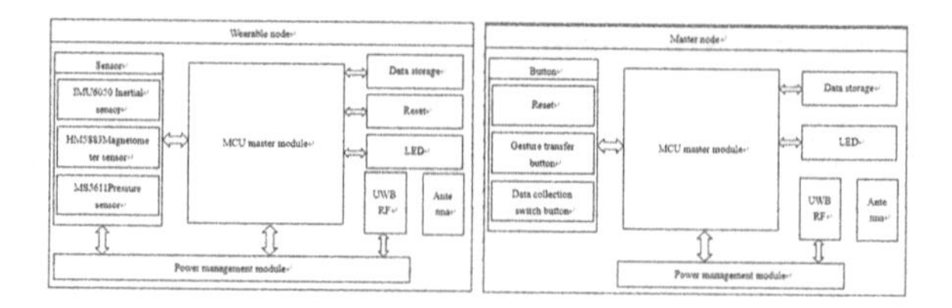

Fig. 1. System block diagram of the platform

For the sake of wearing, the node should be as small as possible. We choose a smaller device, and the layout of device is designed as dense as possible. The size of the node circuit board is 3.4 cm × 3.0 cm. Node physical map is shown in Fig. 2.

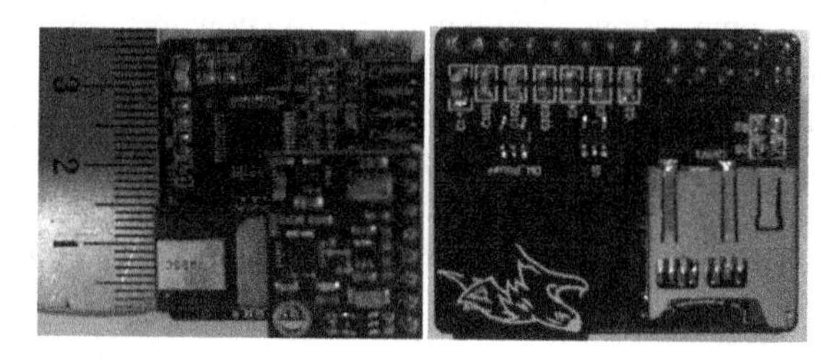

Fig. 2. Node physical map

Our designed platform uses a three-axis accelerometer, three-axis gyroscope, three-axis magnetometer. We choose InvenSense's MPU6050 six-axis sensor. MPU6050 combination of three-axis accelerometer and three-axis gyroscope. Compared to multi-component solutions, it eliminates the difference between the combined gyroscope and the accelerator timeline and reduced a lot of packaging space. Three-axis magnetometer select Honeywell's HMC5883L. In the sensor for positional relationship monitoring, MEAS's MS5611 barometer is selected.

4 Experiments and Result Analysis

4.1 Data Collection

This paper studies the simple human motion recognition. It needs to collect data of 10 motions, including standing, lying, squatting, sitting, up and down the stairs, up and down the elevator, walking, running, 6 sets of data were collected for each motion, and each group of action duration is 5–7 min or so. The data collection process is accompanied by another person to record the duration time and type of motion. The actual locations of nodes are shown in Fig. 3.

Fig. 3. Node actual location

In the ten-axis sensor data, accelerometer is used to measure the acceleration value in three directions. In the state of motion, the more intense the movement in which direction, the greater absolute value of acceleration in that direction. From Fig. 4 we can see how the acceleration changes in static and movement state. The gyroscope is a measure of angular velocity changes, the difference between the attitude of the movement and the static also clearly stands out. Pneumatic pressure reflects the height of the node changes, in the upper and lower elevators, up and down the stairs can clearly see the trend of changes in pressure. For a constant attitude, the pressure remains essentially constant. In the enlarged part, we can see that lying and standing state of the shoulder pressure changes significantly, and the foot pressure is basically the same as the actual body parts, which is also consistent with the height of the actual body parts.

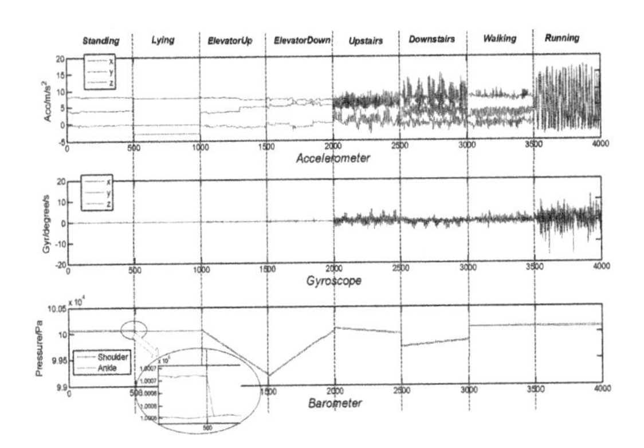

Fig. 4. Raw data

4.2 Algorithm Validation

Select the appropriate set of features, using the commonly used motion recognition algorithm, like neural network BP, decision tree C4.5, SVM classifier and other algorithms to verify the correctness of the platform. C4.5 confidence factor is set to 0.25, SVM kernel function select RBF, BP uses the default configuration of a single hidden layer. SVM uses the LibSVM toolbox, C4.5 and BP use Weka toolbox.

The experimental results show in Table 2 that BP is more accurate than C4.5, C4.5 is higher than SVM. In general, the recognition accuracy is relatively high. The description of the platform to meet the motion recognition inertia feature data acquisition work.

Table 2. BP, C4.5, SVM recognition accuracy comparison

Motion	Algorithm		
	SVM	C4.5	BP
Standing	0.973 ± 0.032	0.635 ± 1.170	0.952 ± 0.351
Lying	1 ± 0	0.923 ± 0.020	0.967 ± 0.211
Up elevator	0.961 ± 0.069	0.894 ± 0.035	0.974 ± 0.025
Down elevator	0.947 ± 0.080	0.91 ± 0.034	0.937 ± 0.020
Up stair	0.514 ± 0.285	0.907 ± 0.137	0.917 ± 0.160
Down stair	0.296 ± 0.208	0.765 ± 0.523	0.889 ± 0.133
Walking	0.568 ± 0.330	0.842 ± 0.129	0.886 ± 0.286
Running	1 ± 0	0.931 ± 0.089	0.967 ± 0.075
Average correct rate	0.7824 ± 0.126	0.851 ± 0.267	0.936 ± 0.131

5 Conclusion

In view of the lack of wearable gesture recognition data sources, this paper designs a more complete data acquisition platform. To use of handheld devices to control the collection of node data and motion calibration, making motion data collection more convenient. Through the analysis of human motion attributes, we proposed a more reasonable data collection program and node deployment program. According to this program, we can collect data, analyze data characteristics and build data sets. In the end we use different algorithms of the motion classification to verify the correctness of the platform.

Acknowledgements. This work is supported by National Natural Science Foundation of China (NSFC) project No. 61671056, No. 61302065, No. 61304257, Beijing Natural Science Foundation Project No. 4152036 and the Fundamental Research Funds for the Central Universities No. FRF-TP-15-026A2.

References

1. Pantelopoulos, A., Bourbakis, N.G.: A survey on wearable sensor-based systems for health monitoring and prognosis. IEEE Trans. Syst. Man Cybern. Part C Appl. Rev. **40**(1), 1–12 (2010)
2. Shian-Ru, K., Le Uyen, Thuc H., Yong-Jin, L., et al.: A review on video-based human activity recognition. Computers **2**(2), 88–131 (2013)

3. Moeslund, T.B., Hilton, A., Krüger, V.: A survey of advances in vision-based human motion capture and analysis. Comput. Vis. Image Underst. **104**(2), 90–126 (2006)
4. Lara, O.D., Labrador, M.A.: A survey on human activity recognition using wearable sensors. IEEE Commun. Surv. Tutorials **15**(3), 1192–1209 (2013)
5. Shaolin, W., Luping, X., Weiwei, T., et al.: A low power and high accuracy MEMS sensor based activity recognition algorithm pp. 33–38 (2014)
6. Tolkiehn, M., Atallah, L., Lo, B., et al.: Direction sensitive fall detection using a triaxial accelerometer and a barometric pressure sensor. In: Annual International Conference of the IEEE Engineering in Medicine and Biology Society, pp. 369–372. IEEE (2011)
7. Lorincz, K., Chen, B., Challen, G.W., et al.: Mercury: a wearable sensor network platform for high-fidelity motion analysis. SenSys **9**, 183–196 (2009)

Three-Order Computational Ghost Imaging and Its Inverse Algorithm

Wenbing Zeng(✉), Yi Xu, and Dong Zhou

University of Electronic Science and Technology, Chengdu, China
870470218@qq.com

Abstract. This paper mainly studies the characteristics of third-order computational GI whose imaging system is a single-arm system using a spatial light modulator (SLM) instead of a rotating ground glass in the classic GI system. To simplify the theoretical analysis of three-order computational GI, we assume that the two reference beams are equal and find that it is very similar to the theoretical results of the two-order GI. We analyze the impact of signal intensity fluctuation of three light beams on imaging. The numerical simulation shows that the intensity fluctuation of the light is the source of imaging information. We propose an inverse algorithm, which can improve the visibility of imaging effectively.

Keywords: Three-order · Computational GI · Single-arm · Algorithm · Visibility

1 Introduction

Ghost imaging, also called coincidence imaging, is one of the frontiers and hotspots of quantum optics in recent years [1, 2]. In 1995, using the entangled photon pairs generated by transforming their entangled behavior, Pittman et al. [3] completed the experiment of coincidence imaging based on quantum entanglement. Later in 2002, people achieve experimentally coincidence imaging based on classical thermal light source [4]. On this basis, people have completed experiments of GI based on pseudothermal source and thermal source, respectively [5, 6]. Compared with the quantum GI, the source of the thermal light GI can be got more easily, so it has a wider application prospect. In 2008, Baris and Jeffrey developed a Gaussian-state framework to unify the theory of GI between classical source and quantum source and proposed the single-arm GI theory [7]. Jeffrey H. Shapiro described a computational GI arrangement that uses only a single-pixel detector [8]. Later compressive techniques are used in order to gain a successful reconstruction with a number of observation. In 2014, Chi Zhang et al. discussed some of inverse algorithms for object reconstruction of GI [16], which had a problem that the studied matrix has a size of $M \times N$ and M is related to the number of iterators, so when M dose not equal to N, the system is ill-conditional and calculating the inverse of the matrix is not straightforward. In this paper, we transform the matrix of object into a one-dimension vector, then find it is an inverse problem to reconstruct the image where the size of the studied matrix is $N \times N$ so as to avoid the ill-condition.

F. Xhafa et al. (eds.), *Advances in Intelligent Systems and Interactive Applications*, Advances in Intelligent Systems and Computing 686,
https://doi.org/10.1007/978-3-319-69096-4_53

2 Observation Model Establishment

With the development of liquid crystal technology, it is entirely possible to replace the rotating rough glass with a spatial light modulator. Using Huygens-Fresnel's theory we can calculate the light field distribution after the source propagates L-distance. The single-arm system of computational GI is shown in Fig. 1 [8].

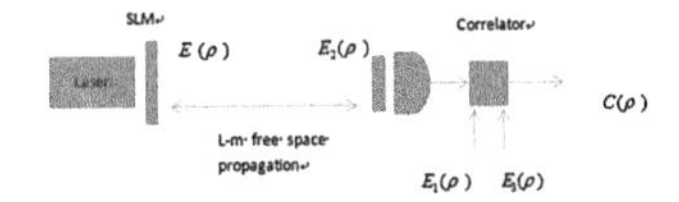

Fig. 1. The single-arm setup

In the single-arm system, the result of $E_2(\rho)$ interaction with object function $T(\rho)$ can be expressed as $P_2(\rho) = E_2(\rho)T(\rho)$. The output of bucket detector is $p_2(\rho) = \int_{A_2} d\rho |P_2(\rho)|^2$ and A_2 is the square of the bucket detector. According to analysis in literature [13, 14], the cross correlation of three-order computational GI is:

$$G(\rho_1, \rho_2, \rho_3) = \int_{A_2} d\rho_2 < E_1^*(\rho_1) P_2^*(\rho_2) E_3^*(\rho_3) E_1(\rho_1) P_2(\rho_2) E_3(\rho_3) > \tag{1}$$

To study the relationship between two-order GI and three-order GI, we suppose $E_2(\rho) = E_3(\rho)$ and the theoretical analysis of three-order computational GI can be simplified. Considering that $E(\rho)$ is a complex value Gaussian random process with zero-mean, therefore, the Gaussian-Schell model phase-insensitive correlation function is used to describe the characteristics of computation GI. The correlation of $E(\rho)$ can be defined as $< E^*(\rho_1) E(\rho_2) > = K(\rho_1, \rho_2)$, and we can simplify the third-order computational GI as follow:

$$\begin{aligned} G(\rho_1, \rho_2, \rho_3) = G(\rho_1, \rho_2, \rho_1) &= \int_{A_2} \begin{array}{l} d\rho_2 \{ 2 < I_1(\rho_1) >^2 < P_2^*(\rho_2) P_2(\rho_2) > \\ + 4 < I_1(\rho_1) > | < E_1^*(\rho_1) P_2(\rho_2) > |^2 \} \end{array} \\ &= 2\left(\frac{P}{\pi a_L^2}\right)^3 \int_{A_2} d\rho_2 T(\rho_2) + 4\left(\frac{P}{\pi a_L^2}\right)^3 \int_{A_2} d\rho_2 e^{-2|\rho_1 - \rho_2|^2 / 2\rho_L^2} T(\rho_2) \end{aligned} \tag{2}$$

The result of Eq. (2) is very similar to that of the second-order GI in Literature [8]. Measuring the correlation of intensity fluctuations in time-space-average can eliminates the background term and enhances the visibility of GI [10, 11]. Now, we use intensity fluctuation to express Eq. (1),we have:

$$G(\rho_1,\rho_2,\rho_3) = \int_{A_2} d\rho_2 <I_1 p_2 I_3> = G_0(\rho_1) + G_1(\rho_1) \tag{3}$$

$$G_0(\rho_1) = \int_{A_2} d\rho_2 \{ <p_2> <\Delta I_1 \Delta I_1> + <I_1> <I_1> <p_2> \} \tag{4}$$

$$G_1(\rho_1) = \int_{A_2} d\rho_2 \{ <\Delta I_1 \Delta I_1 \Delta p_2> + 2 <I_1> <\Delta I_1 \Delta p_2> \} \tag{5}$$

It is clear that signal term is the score of correlation between the intensity fluctuation of signal light and the intensity of reference light, but background term is the score of correlation between the intensities of the two light paths.

3 Discrete Diffraction Transform (DDT) and Inverse Algorithm

Discrete propagation transform (DDT) is the discretization of continuous light waves and then using the transformation theory to simplify the signal processing. We digitize the third-order computational GI theory in the previous section. The laser generated by the spatial light modulator (SLM) can be expressed as a $N_1 \times N_2$ matrix. For easily handling, we further splice it into a one-dimensional matrix $u[k]$ with a length of $n = N_1 \times N_2$. According to the DDT theory of light propagation proposed in Literature [12], $u[k]$ free propagates L-distance and can be expressed as $u_\ell = A_d u$, $A_d \in C^{n \times n}$ is Discrete Forward Propagation Operator, $u \cdot u_\ell$ are the light field column vector at the source and the L-distance with $\ell = [1,2,3]$. We use multiple samples to find the average with a total experimental number K. In the signal path, the score of bucket detector is

$$o_r = \sum_{k=1}^{n} |u_{2,r}[k]|^2 |T[k]|^2 = b_r^T c, r = 1, \ldots, K \tag{6}$$

$b_r = |u_{2,r}|^2$ is the light field intensity and $c = |T|^2$, T is the column vector of object function. Equation (1) can be re-expressed as

$$|g[k]|^2 = \frac{1}{K} \sum_{r=1}^{K} o_r |u_{1,r}[k]|^2 |u_{3,r}[k]|^2, r = 1, \ldots, K \tag{7}$$

3.1 Apply Inverse Algorithm to $<\Delta I_1 \Delta I_1 \Delta p_2>$ and $<I_1> <\Delta I_1 \Delta p_2>$

So we can add a differentiator converter (DC) behind the detector to obtain the intensity fluctuation information. The first term $<\Delta I_1 \Delta I_1 \Delta p_2>$ of Eq. (5) can be expressed as

$$|\Delta g[k]|^2 = \frac{1}{K}\sum_{r=1}^{K}(o_r - \overline{o_K})(b_{1,r} - \overline{b_{1,K}})(b_{3,r} - \overline{b_{3,K}}) \tag{8}$$

$\overline{o_K} = \frac{1}{K}\sum_{r=1}^{K} o_r$ is the average of the signal of bucket detector in multiple experiments, and $\overline{b_{\ell,K}} = \frac{1}{K}\sum_{r=1}^{K} b_{\ell,r}, \ell = [1,3]$ is the average of the signal of CCD in multiple experiments with $b_{\ell,r} = |u_{\ell,r}[k]|^2, \ell = [1,3]$. Bringing $\overline{o_K}, o_r$ into Eq. (8) and $b_r = |u_{2,r}|^2 = |u_{1,r}|^2 = |u_{3,r}|^2$, we have

$$|\Delta g[k]|^2 = \frac{1}{K}\sum_{r=1}^{K}(b_{2,r} - \overline{b_{2,K}})^T c(b_{1,r} - \overline{b_{1,K}})(b_{3,r} - \overline{b_{3,K}}) \tag{9}$$

Observing Eq. (9) we find that the right side of the equation contains the information of object c, and the Expectation of Eq. (9) is calculated as $E\{|\Delta g[k]|^2\} = D_K c$ and $D_K = E\{\frac{1}{K}\sum_{r=1}^{K}(b_{2,r} - \overline{b_{2,K}})(b_{1,r} - \overline{b_{1,K}})(b_{3,r} - \overline{b_{3,K}})^T\}$. Let $\hat{D}_K$ is an asymptotically unbiased estimate of D_K, and $\hat{D}_K^-$ is the pseudoinverse (or regularized inverse) of $\hat{D}_K$, We have $|\Delta\hat{g}[k]|^2 = \hat{D}_K^- \frac{1}{K}\sum_{r=1}^{K}(b_{2,r} - \overline{b_{2,K}})^T c(b_{1,r} - \overline{b_{1,K}})(b_{3,r} - \overline{b_{3,K}})$ and $\hat{D}_K = \frac{1}{K}\sum_{r=1}^{K}(b_{2,r} - \overline{b_{2,K}})^T(b_{1,r} - \overline{b_{1,K}})(b_{3,r} - \overline{b_{3,K}})$. According to random process theory

$$E[|\Delta\hat{g}[k]|^2] = E[\hat{D}_K^-]E[D_K^T]c \tag{10}$$

So when $K \to \infty$, $|\Delta\hat{g}[k]|^2 \to c$, and $|\Delta\hat{g}[k]|^2$ is the unbiased estimator of c. So the inverse algorithm makes the imaging result closer to the object function and we get a distincter image. Since $\hat{D}_K$ is $n \times n$ matrixs, the pseudoinverse of them must exist. So our algorithm can avoid the problem in Literature [15, 16].

3.2 Peak Signal-to-Noise Ratio and Visibility

It is easily to get the variance and root mean square of a single pixel in simulation and practice, so we define the peak signal-to-noise radio as $PSNR = 20\log_{10}(\frac{\max(|T[k]|^2)}{sqrt(|\Delta\hat{g}[k]|^2 - |T[k]|^2)})$, sqrt(*) means square root, the unit of PSNR is dB.

Next, let's study the visibility of image. The visibility is related to the background noise of the restored image,and according to previous work [13, 14], we define the visibility as $v = \frac{G_1(\rho_1)}{G(\rho_1)} = \frac{G_1(\rho_1)}{G_0(\rho_1)+G_1(\rho_1)}$.

4 Numerical Simulation

We use a 26×44 image TUT to represent the object to be tested, so the length of vector c is $n = 26 \times 44$. We set the single pixel of SLM to a $\Delta x \times \Delta x$ square with $\Delta x = 8\ \mu\text{m}$, the wavelength of source $\lambda = 0.532\ \mu\text{m}$. Since the imaging system is lensless, we use equation $L = Ny * \Delta_x^2/\lambda, Ny = 64$ to calculate propagation distance. The SLM simply modulates the phase of the light without changing the intensity, its modulation function is $M_r[x_1, x_2] = \exp(j2\pi\phi_r[x_1, x_2])$, r is the experiment numbers, $r = 1, \ldots, K$, K is the total number of experiments, $\phi_r[x_1, x_2]$ is a computer-generated pseudo-random matrix with a size of $N_1 \times N_2$, which obtains a uniform distribution between [0,1] and is independent between different $[x_1, x_2]$ and between different experiments. The light modulated by SLM and after propagating L-distance can be expressed as $ifft[A_d * fft(M_r[x_1, x_2])]$, A_d is the transfer operator of free space [12]. We can restore the image of object by correlating among signals of the bucket detector and the two CCDs.

Figure 2b, c are the images of the two parts in Eq. (4), we find that the two parts which all contains $<p_2>$ are the correlation among the light intensity of three paths. It means that the intensity fluctuation of the signal light is the source of imaging. Figure 3a, c are the images of the two parts in Eq. (5), these two parts are showing the image of the object, and the imaging effect of $<\Delta I_1> <\Delta I_3 <\Delta p_2>$ is more superior than $<I_1> <\Delta I_3 \Delta p_2>$. We find that these two all contain the intensity fluctuation Δp2, which proves the correctness of the previous conclusion and further indicates the intensity fluctuation of the two reference lights are not the source of imaging.

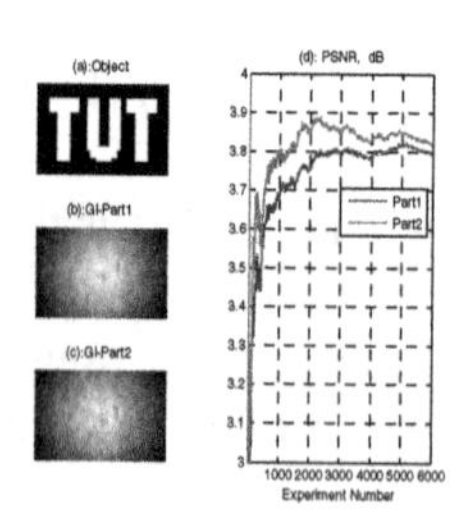

Fig. 2. The images of background terms for TUT, **a** object image, **b** for $<I_1> <I_3> <p_2>$, **c** for $<\Delta I_1 \Delta I_3> <p_2>$ and **d** for PSNRs of the two parts

We further apply the inverse algorithm to $<\Delta I_1> <\Delta I_3 <\Delta p_2>$ and $<I_1> <\Delta I_3 \Delta p_2>$, and Fig. 3b, d are the images. Figure 3b is more clear and has a higher PSNR compared with Fig. 3d. And the PSNR of the former shows an upward

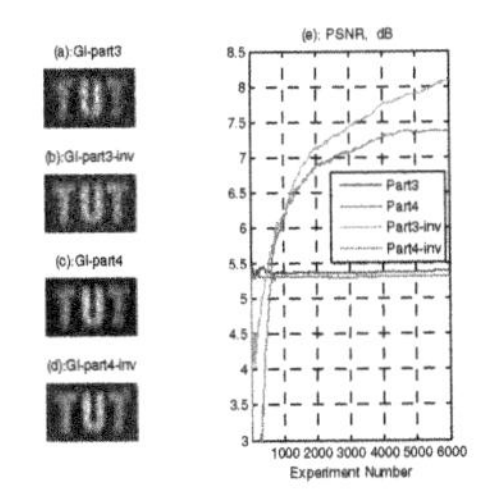

Fig. 3. The images of signal terms for TUT, **a** for $<\Delta I_1> <\Delta I_3 <\Delta p_2>$, **b** for **a** used inverse algorithm, **c** for $<I_1> <\Delta I_3 \Delta p_2>$, **d** for **c** used inverse algorithm, **e** the PSNRs of each part.

trend after 6000th experiment, but the latter's PSNR become flat after 2000th experiment, so the image quality of $<\Delta I_1> <\Delta I_3 <\Delta p_2>$ is superior to $<I_1> <\Delta I_3 \Delta p_2>$. Comparing Vertically Fig. 3a, c with Fig. 3b, d, it is found that the inverse algorithm improves the PNSR of imaging. By comparison, we find the inverse algorithm has a great improvement to computational GI in PSNR and visibility, and it is also proved that this algorithm has a great advantage in improving the imaging quality of computational GI (Fig. 4).

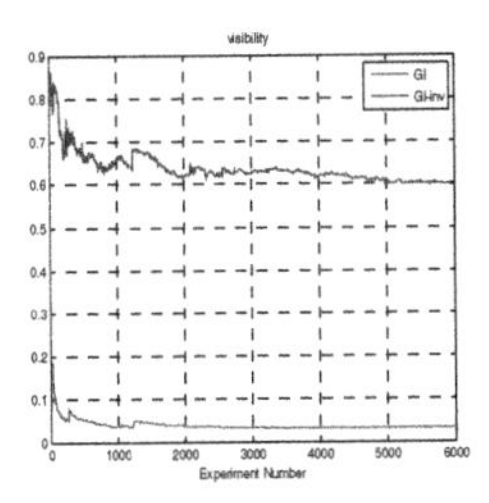

Fig. 4. The visibility of GI using inverse algorithm before and after for TUT.

5 Conclusion

We suppose the two reference arms are the same in the traditional third-order GI system in the single-arm system, which simplifies the theoretical analysis of the third-order computational GI. The three-order intensity correlation function indicates that the third-order computational GI is similar to the traditional second-order GI with background term and signal term. Besides, we express the three-order intensity correlation function as the intensity fluctuation correlation, finding that the intensity fluctuation of the signal light is the source of imaging but not the intensity fluctuation of the two reference lights combined with numerical simulation. At last, we propose a inverse algorithm to correct the signal part, and find this method has great potential to improve the visibility and PSNR of third-order computational GI.

References

1. Klyshko, D.N.: Combine EPR and two-slit experiment: interference of advanced waves. Phys. Lett. A **132**(6), 299–304 (1988)
2. Pittman, T.B., Shih, Y.H., Sergienko, A.V., et al.: Experimental tests of Bell's inequalities based on space-time and spin variables. Phys. Rev. A **51**(5), 3495–3498 (1995)
3. Bennik, R.S., Bentley, S.J., Boyd, R.W.: "Two-photon" coincidence imaging with a classical source. Phys. Rev. Lett. **89**(11), 113601 (2002)
4. Gatti, A., Brambilla, E., Bache, M., et al.: Ghost imaging with thermal light: comparing entanglement and classical correlation. Phys. Rev. Lett. **93**(9), 093602 (2004)
5. Valencia, A., Scarcelli, G., D' Angelo, M., et al.: Two-photon imaging with thermal light. Phys. Rev. Lett. **94**(6), 063601 (2005)
6. Xiong, J., Cao, D.Z., Huang, F., et al.: Experimental observation of classical subwavelength interference with a pseudothermal light source. Phys. Rev. Lett. **94**(17), 173601 (2005)
7. Erkmen, B.I., Shapiro, J.H.: Unified theory of ghost imaging with Gaussian-state light. Phys. Rev. A **77**, 043809 (2008)
8. Shapiro, J.H.: Computational ghost imaging. Phys. Rev. A **78**, 061802(R) (2008)
9. Bai, Ya., Han, S.: Ghost imaging with thermal light by third-order correlation. Phys. Rev. A **76**, 043828 (2007)
10. Yu, J., Chang, F., et al.: The visibility of the third-order ghost imaging. J. Changchun Univ. Sci. Technol. Nat. Sci. Ed. **39**(3), 1672–9870 (2016)
11. Gong, W., Han, S.: A method to improve the visibility of ghost images obtained by thermal light. Phys. Lett. A **374**(8), 1005–1008 (2010)
12. Katkovnik, V., Astola, J., Egiazarian, K.: Discrete diffraction transform for propagation, reconstruction, and design of wave field distributions. Appl. Opt. **47**, 3481–3493 (2008)
13. Bache, M., Magatti, D., Ferri, F., Gatti, A., Brambilla, E., Lugiato, L.A.: Coherent imaging of a pure phase object with classical incoherent light. Phys. Rev. A **73**, 053802 (2006)
14. Gatti, A., Bache, M., Magatti, D., Brambilla, E., Ferri, F., Lugiato, L.A.: Coherent imaging with pseudo-thermal incoherent light. J. Mod. Opt. **53**, 739 (2006)
15. Sun, B., Welsh, S.S., Edgar, M.P., Shapiro, J.H., Padgett, M.J.: Normalized ghost imaging. Opt. Soc. Am. **20**, 16892–16901 (2012)
16. Zhang, C., Guo, S., Cao, J., Guan, J., Gao, F.: Object reconstitution using pseudo-inverse for ghost imaging. Opt. Soc. Am. **22**, 30063–30073 (2014)

Face Recognition Using Deep Convolutional Neural Network in Cross-Database Study

Mei Guo[1], Min Xiao[1](✉), and Deliang Gong[2]

[1] College of Software and Communication Engineering, Xiangnan University, Chenzhou 423000, Hunan, China
xnxyxm@163.com
[2] Comprehensive Experimental Training Center, Xiangnan University, Chenzhou 423000, Hunan, China

Abstract. In this paper we study face recognition using convolutional neural network. First, we introduced the basic CNN neural network architecture. Second, we modify the traditional neural network and adapt it to another database by fine tuning its parameters. Third, the network architecture is extended to the cross database problem. The CNN is first trained on a large dataset and then tested on another. Experimental results show that the proposed algorithm is suitable for building various real world applications.

Keywords: Face recognition · Deep neural network · Image processing

1 Introduction

Face recognition has been a popular topic in computer science and artificial intelligence in the past decades [1–6]. There are many popular deep neural network models for image classification. Such as AlexNet [7], VggNet [8], GoogleNet [9], etc. These models are trained on the large image databases, and the fine-tuning on these models can be adapted to specific applications.

In many image classification problems, it is challenging to transfer one model to the other. The training process requires a large amount of data and the resulting model sometimes is lack of generalization ability. Therefore it is important to fine-tuning the existing deep models and adapt it to wider applications on cross-database situation.

We study the face recognition based on convolutional neural network. We modified the traditional network and adapt it to another database by fine tuning its parameters. The proposed algorithm is suitable for building various real world applications.

In this paper, we propose to use a modified convolutional neural network architecture for face recognition. A modified convolutional neural network architecture is demonstrated in Fig. 1. The rest of the paper is organized as follows: Sect. 2 gives the details on the basic network. Section 3 gives the modified face recognition algorithm. Experimental results are provided in Sect. 4. Finally, conclusions are drawn in Sec. 5.

F. Xhafa et al. (eds.), *Advances in Intelligent Systems and Interactive Applications*, Advances in Intelligent Systems and Computing 686,
https://doi.org/10.1007/978-3-319-69096-4_54

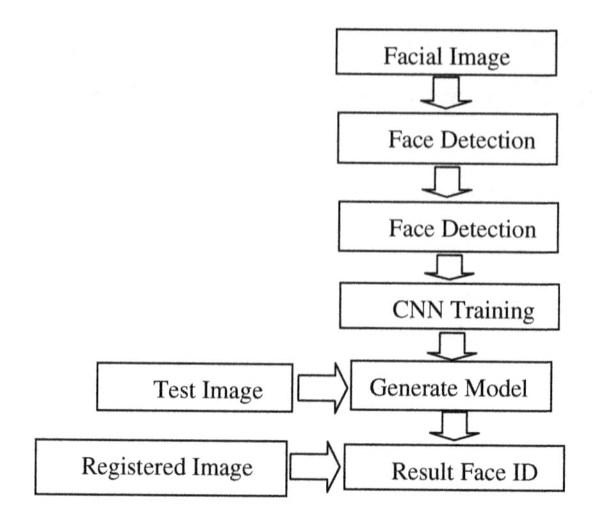

Fig. 1 A depiction of the face recognition system layout.

2 Deep Neural Network

The facial image features are constructed by the deep neural network. DNN has the ability to generate highly abstracted features that are advantageous compared to the traditional hand crafted features. The traditional face model is presented. Alex Krizhevsky in Toronto University proposed the deep convolutional neural network, also known as AlexNet [10]. It has a popular convolutional structure and it is widely used in the face recognition model. Many current models are inspired by AlexNet, such as Vgg-Face model. It extend the original structure to deeper levels. In the contest of Imagenet LSVRC-2012 contest, 1,200,000 images are classified into 1000 classes [10]. The convolutional neural network achieves satisfactory results. The structures are demonstrated in Fig. 2.

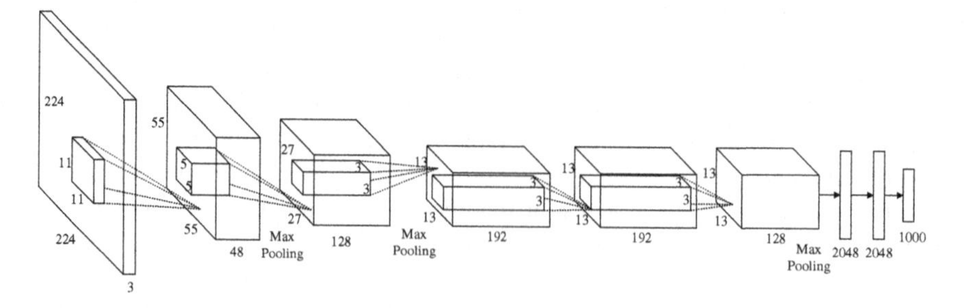

Fig. 2 Traditional convolutional neural network architecture [10].

The input image is first processed by five convolutional layers for feature abstraction. It is then send into the fully connected layers for modelling. The final

Softmax classifier is used for face classification. The convolutional layer involves convolution, activation, pooling and local response normalization.

3 Improved Neural network Architecture

The existing neural network model can be adapted to specific recognition tasks with modified network architecture and fine turning. The traditional network training method requires a very large number of training samples which is not practical in real application. In real world the training sample is limited to hundreds with reasonable human annotation. Therefore we propose to modify the network architecture to better suit this cross database problem. The CNN is first trained on a large dataset and then tested on another in the experimental section.

The basic neural network nodes are shown in Fig. 3.

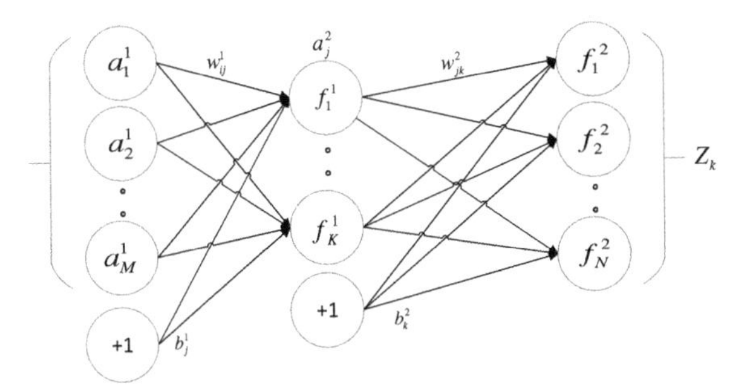

Fig. 3 Neural network structure

The input layer is denoted as a_i^1, the hidden layer is denoted as a_i^2. b_i^1 and b_i^2 is the bias parameter and f_i^1 and f_i^2 are the non-linear activation function of each nodes. The output results is denoted as Z_i.

The calculation of Z_i is:

$$\begin{cases} a^2 = f^1\left(w^1 * a^1 + b^1\right) \\ a^3 = f^2\left(w^2 * a^2 + b^2\right) \\ \qquad\qquad\qquad\vdots \\ Z = f^i\left(w^i * a^i + b^i\right) \end{cases} \tag{1}$$

The activation function is sigmoid:

$$y = f(x) = \frac{1}{1 + e^{-x}} \tag{2}$$

tanh:

$$y = \tanh(x) = \frac{\sinh(x)}{\cosh(x)} \tag{3}$$

The neural network approximates the relation between target and feature through the activation function [11, 12].

In order to improve the cross database performance and adapt to different real world data, the modified structure is presented in Table 1.

Table 1. The modified neural network structure

Input (224 × 224 RGB)
Conv3-64
Conv3-64
MaxPooling
Conv3-128
Conv3-128
MaxPooling
Conv3-256
Conv3-256
Conv3-256
Conv3-256
MaxPooling
Conv3-512
Conv3-512
Conv3-512
Conv3-512
MaxPooling
Conv3-512
Conv3-512
Conv3-512
Conv3-512
MaxPooling
FullContact-4096
FullContact-4096
FullContact-4096
FullContact-1000
SoftMax

4 Experimental Results

The fine tuning is based on the Vgg_face model and retraining each layer in the model is not practical. Therefore we make use of the convolutional parts in the model ant using the new training samples to adjust the parameters. The Vgg-Face model is trained on 2.6M images and the retraining is carried out on the CASIA database [13], examples of the database is shown in Fig. 4. It includes different subjects with different expression and poses. In each class there are around 400 images and the least coherent has 20 images. We preprocess the coherents for balanced training sets. The classes that have more then 50 samples are chosen and form the training set. There are 10,575 people in total in the original database.

Fig. 4 CASIA-WebFace Image Examples

The established model is tested on LFW database [14]. In this database the faces are captured in the wild. It is not some posed images used in the lab, but real data from real world. Therefore it is more accurate for the face recognition study compared to other simple face images taken in lab environment which is not suitable for real applications.

There are illumination changes in the image sample in LFW database. We first prepocess the images by equalization techniques. Then we use the proposed neural network to extract the face features. The feature vectors are then compared in distance and the final ID is achieved in view of similarity.

In our test, the hardware platform configuration is given in Table 2. The LFW database provides test dataset that contains 10 groups. In each group, there are 600 pairs of images from the same people and 300 pairs from different people. We calculate the image feature vectors from CNN and the distance measure. After normalize the feature vectors to [0, 1], we test the true positive and false negative rates, as well as the false positive and true negative rates. Results are shown in Table 3. We can see that the proposed algorithm is effective for the adaptation on LFW dataset.

Table 2. Hardware configuration of the face recognition experiment.

Item	Parameter
CPU	Intel i5 3450 3.1GHz 6M
Memory	Kinston DDR3 16G
GPU	NVIDIA GeForce GTX Titan X 12G

Table 3. Face recognition results

Normalized threshold	True positive/False negative	False positive/True negative
0.3700	2732/280	53/2946
0.4263	2833/141	154/2846
0.5200	2931/94	240/2760
0.6320	2932/13	1270/1730

5 Conclusion

In this paper, we propose to use a modified convolutional neural network architecture for face recognition. Gives the details on the basic network, and gives the modified face recognition algorithm. Experimental results show that the proposed algorithm is effective for the adaptation on LFW dataset.

References

1. Schroff, F., Kalenichenko, D., Philbin, J.: Facenet: A unified embedding for face recognition and clustering. In: Proceedings of the IEEE Conference on Computer Vision and Pattern Recognition, pp. 815–823 (2015)
2. Parkhi, O.M., Vedaldi, A., Zisserman, A.: Deep face recognition. BMVC **1**(3), 6 (2015)
3. Haig, N.D.: The effect of feature displacement on face recognition. Perception **42**(11), 1158–1165 (2013)
4. Sagar, G.V., Barker, S.Y., Raja, K.B., et al.: Convolution based face recognition using DWT and feature vector compression. In: Third International Conference on Image Information Processing (ICIIP), pp. 444–449. IEEE, New York (2015)
5. Lu, J., Tan, Y.P., Wang, G.: Discriminative multimanifold analysis for face recognition from a single training sample per person. IEEE Trans. Pattern Anal. Mach. intell. **35**(1), 39–51 (2013)
6. Sun, Y., Liang, D., Wang, X., et al.: Deepid3: Face recognition with very deep neural networks. arXiv preprint arXiv:1502.00873 (2015)
7. Ballester, P., Araujo, R.M.: On the performance of GoogLeNet and AlexNet applied to sketches. In: Thirtieth AAAI Conference on Artificial Intelligence (2016)
8. Wu, X., He, R., Sun, Z.: A lightened cnn for deep face representation. In: IEEE Conference on IEEE Computer Vision and Pattern Recognition (CVPR) (2015)
9. Zhong, Z., Jin, L., Xie, Z.: High performance offline handwritten Chinese character recognition using GoogLeNet and directional feature maps. In: 13th International Conference on Document Analysis and Recognition (ICDAR), pp. 846–850. IEEE, New York (2015)

10. Krizhevsky, A., Sutskever, I., Hinton, G.E.: Imagenet classification with deep convolutional neural networks. In: Advances in Neural information Processing Systems, pp. 1097–1105 (2012)
11. Deng, L., Hinton, G., Kingsbury, B.: New types of deep neural network learning for speech recognition and related applications: An overview. In: IEEE International Conference on Acoustics, Speech and Signal Processing (ICASSP), pp. 8599–8603. IEEE, New York (2013)
12. Kingsbury, B., Sainath, T.N., Soltau, H.: Scalable minimum Bayes risk training of deep neural network acoustic models using distributed Hessian-free optimization, pp. 10–13. In: Interspeech (2012)
13. Yi, D., Lei, Z., Liao, S., Li, S.Z.: Learning face representation from scratch. arXiv preprint arXiv:1411.7923 (2014). Available: http://www.cbsr.ia.ac.cn/english/CASIA-WebFace-Database.html
14. Huang, G.B., Ramesh, M., Berg, T., et al.: Labeled Faces in the Wild: A Database for Studying Face Recognition in Unconstrained Environments. Technical Report 07-49, University of Massachusetts, Amherst (2007)

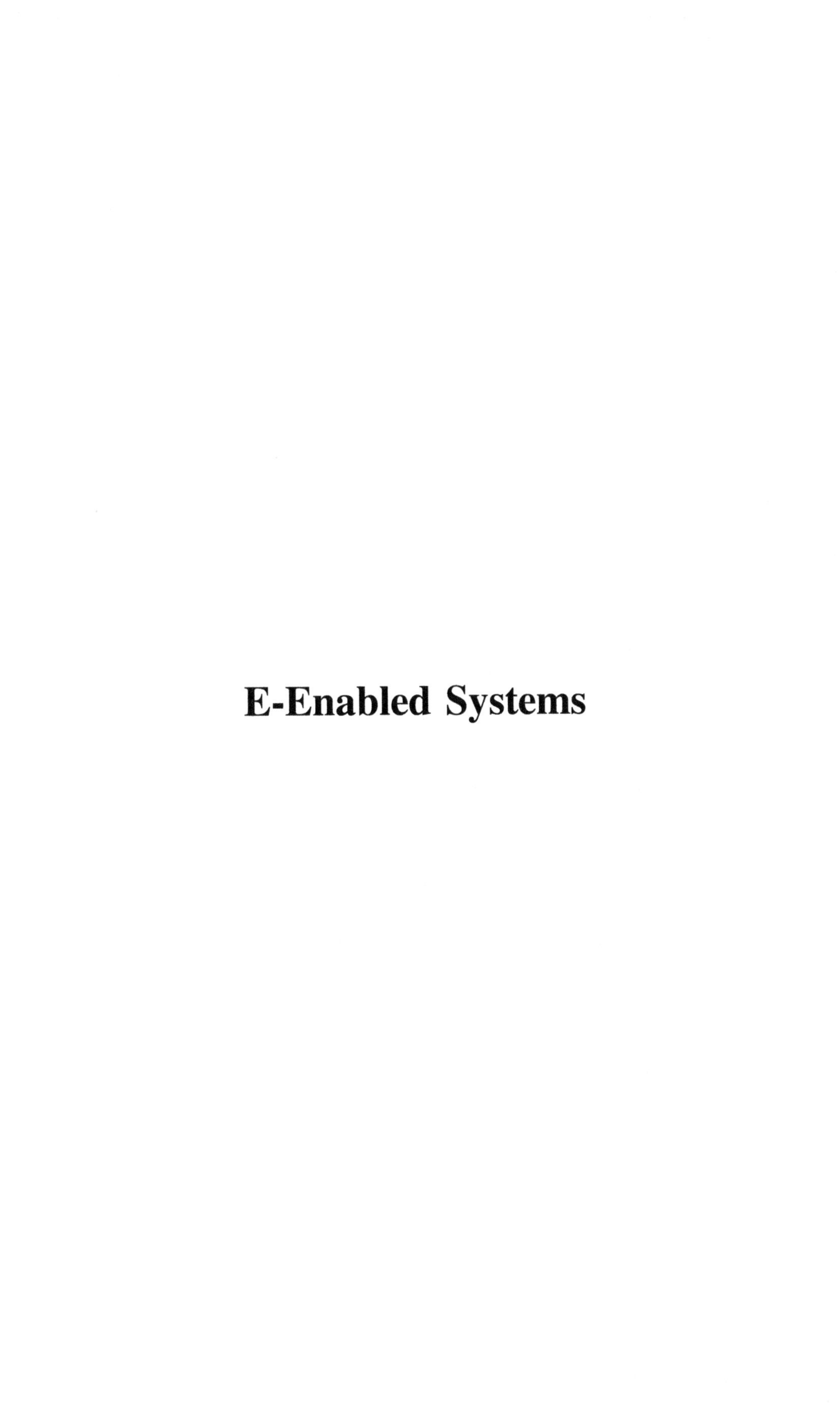

E-Enabled Systems

Stock Market Forecasting Using S-System Model

Wei Zhang and Bin Yang(✉)

School of Information Science and Engineering, Zaozhuang University,
Zaozhuang 277160, China
batsi@126.com

Abstract. In this paper, S-system model is first presented to forecast stock market index. An improved additive tree model named restricted additive tree (RAT) is proposed to represent S-system model. A hybrid evolutionary algorithm based on structure-based evolutionary method and cuckoo search (CS) is used to evolve the structure and parameters of RAT model. Shanghai stock exchange composite index is used as example to evaluate the performances of S-system model. Results show that S-system model outperforms other traditional models, including neural network, wavelet neural network, flexible neural tree and ordinary differential equation.

Keywords: Stock market · Restricted additive tree · Cuckoo search · Forecasting

1 Introduction

Stock market is the place where the stock could be issued, traded and exchanged. It could reflect the economic condition of an enterprise and makes an important role in the country's economy [1]. However it behaves like a random walk, time varying and complex process [2], which is influenced by economic and noneconomic factors, such as market rumors, the company's performance and national policy. Thus stock market prediction is very difficult [3].

In the past, some intelligent models have been presented to forecast stock market. Artificial neural networks (ANN) with self-learning, associative storage function and searching optimal solutions quickly, have been widely applied for stock market prediction [4], such as fuzzy neural networks (FNN) [5], back propagation neural network (BPNN) [6] and radial basis function (RBF) neural network [7].

Neural network (NN) is very powerful, but it has some disadvantages. When the data is insufficient, NN could not work efficiently. Before using NN, the features of all problems must be translated into numbers, so the results will lose some information. The most serious problem about NN is the lack of ability to explain their reasoning [8]. The system of differential equations could describe the mathematical expression among independent variable, dependent variable and external factors, which could explain reasoning of these variables in the complex systems, and forecast the future behavior. Ordinary differential equations (ODEs) were proposed to forecast the populations of USA and stock index, and the results reveal that the ODE model performed better than other classical methods [9, 10].

F. Xhafa et al. (eds.), *Advances in Intelligent Systems and Interactive Applications*, Advances in Intelligent Systems and Computing 686,
https://doi.org/10.1007/978-3-319-69096-4_55

S-system model is an effective and nonlinear ODE system and has been used for modeling the complex molecular biological systems. In this paper, S-system model is proposed to forecast stock market. The restricted additive tree model is proposed to optimize the structure of S-system model. Cuckoo search is used to evolve the parameters of model.

2 Method

2.1 S-system Model

The S-system model is a well-known nonlinear ODE set and well suited for simulating complex systems. It is a set of ODEs consisting of power law functions:

$$X_i'(t) = \alpha_i \prod_{j=1}^{N} X_j^{g_{ij}}(t) - \beta_i \prod_{j=1}^{N} X_j^{h_{ij}}(t). \tag{1}$$

Where N is the number of state or characteristic variables, and the number of ODEs in the system. X_i is the state variable, α_i and β_i are non-negative rate constants, and g_{ij} and h_{ij} are kinetic orders, which represent the impact of variable i on variable j.

2.2 Restricted Additive Tree Model

Additive tree models were proposed by Chen in 2005, which have been widely applied for system identification especially linear and nonlinear ordinary differential equation inference [11]. S-system model has two terms and each term only has power law operator. In order to search the optimal S-system model quickly, an improved additive tree model, namely restricted additive tree (RAT) model (Fig. 1), is proposed. RAT model has two improved points:

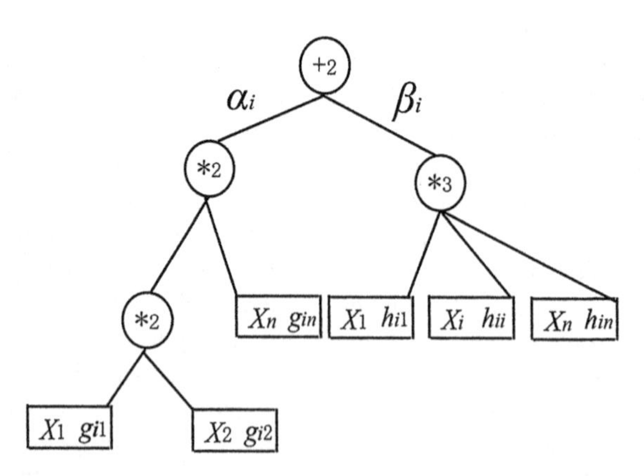

Fig. 1. An example of RAT model, its mathematical equation: $f_i = \alpha_i X_1^{g_{i1}} X_2^{g_{i2}} X_n^{g_{in}} + \beta_i X_1^{h_{i1}} X_i^{h_{ii}} X_n^{h_{in}}$.

(1) The node in the first layer of RAT model is set as +2, which means that the expression has only two terms.
(2) The nodes in other layers could be created randomly, which are selected from operator set I

$$I = \{*2, *3, \ldots, *n, x_1, x_2, \ldots, x_n, R\}. \tag{2}$$

Where $*n$ is the arithmetic node and presents the product of n state variables, $x_i$is state variable (leaf node), and R is constant [11].

2.3 Evolving Structure of S-system Model

A structure-based evolutionary method is proposed to find an optimal RAT model. In order to create new individuals for the next generation, three genetic operators are used.

(1) Selection: roulette Wheel Selection is used to select the individuals with high fitness to the next generation.
(2) Mutation: mutation operator is used to change partial nodes of individual. In an individual, one node is chosen randomly, which is called mutation point. Mutation point could be the arithmetic node in RAT model, or the leaf node (variable or constant). Then, delete mutation point and the subtree whose root node is mutation point. Finally a subtree generated randomly is inserted into the mutation point. In this part, three kinds of mutation operators (change one leaf node, grow and prune) are applied to generate offspring population, which are described detailedly in Ref [11].
(3) Crossover: exchange the information of two individuals from parent population. First two individuals from parent population are selected. Each individual selects randomly one subtree, and swap them.

2.4 Evolving Parameters of S-system Model

S-system model contains two kinds of parameters: rate constants (α_i and β_i) and kinetic orders (g_{ij} and h_{ij}), which need to be optimized. Because cuckoo search (CS) adds information exchange among groups, improve convergence speed, and has few parameters, so CS is proposed to optimize the parameters of S-system model.

The flowchart of cuckoo search is described in detail as follows [12].

(1) CS uses the host nest as solution. Suppose X_i^t represents the position of i-th host nest at the t-th generation. Firstly generate population randomly $X_i^1\,(i = 1, 2, \ldots n)$.
(2) Evaluate population and judge whether the optimal solution is sought out. If the optimal or near-optimal solution is found, algorithm is over.
(3) Lévy flight is used to search randomly path. The path and position of cuckoo in the process of searching nest is updated as follows.

$$X_i^{t+1} = X_i^t + \alpha \oplus Lévy(\lambda). \tag{3}$$

Where $\alpha$is the step controlled size. In general, $\alpha = 1$. $\oplus$ means entrywise multiplications.

(4) When nest positions are updated, create a random value r from [0, 1]. If $r < p_a$, keep the updated position X_i^{t+1}. Otherwise generate new position randomly instead of X_i^{t+1}. The random position could be calculated as follows.

$$X_i^{t+1} = X_i^t + r(X_m^t - X_n^t). \tag{4}$$

Where $X_m^t$and X_n^t are solutions selected randomly at t-th generation. Go to step (2).

3 Experiments

Shanghai composite index (Shanghai index) is used to test the performance of S-system model. The data are corrected from 01 January, 2011 to 31 December, 2014. In order to forecast stock price, five stock parameters (opening price, closing price, low price, high price and trading volume) are used to train S-system model. The root mean squared error (*RMSE*), maximum absolute percentage error (*MAP*), mean absolute percentage error (*MAPE*) are proposed to evaluate the forecasting performance, which are described as follows.

$$RMSE = \sqrt{\frac{1}{N}\sum_{i=1}^{N}\left(f_{target}^{i} - f_{forecast}^{i}\right)^2} \tag{5}$$

$$MAP = \max\left(\frac{\left|f_{target}^{i} - f_{forecast}^{i}\right|}{f_{forecast}^{i}} \times 100\right) \tag{6}$$

$$MAPE = \frac{1}{N}\sum_{i=1}^{N}\left(\frac{\left|f_{target}^{i} - f_{forecast}^{i}\right|}{f_{forecast}^{i}}\right) \times 100 \tag{7}$$

Where N is the number of sample points, f_{target}^{i} is the real stock index in the i-th day, and $f_{forecast}^{i}$ is the prediction stock index in the i-th day.

In this part, four classic methods including neural network (NN), wavelet neural network (WNN), flexible neural network (FNT) [13] and ODE [10] are also used to predict Shanghai index. The parameters in these methods are set as same as in the corresponding literatures. To optimize S-system model, the used instruction set $I_1 = F \cup T = \{*1, *2, *3, *4\} \cup \{x_1, x_2, x_3, x_4, x_5, x_6\}$ in RAP. The training and test sets are chosen uniformly. In CS, population size is set as 30, and probability$p_a$is set as 0.25.

The forecasting result of Shanghai index is illustrated in Figure 2, which depicts the one day future real stock index and forecasting stock one. Figure 3 shows the result of

Shanghai index plotted as target stock index versus forecasting one. From Figs. 2 and 3, we could clearly see that S-system model could predict the one day ahead stock index well and the error is relatively small.

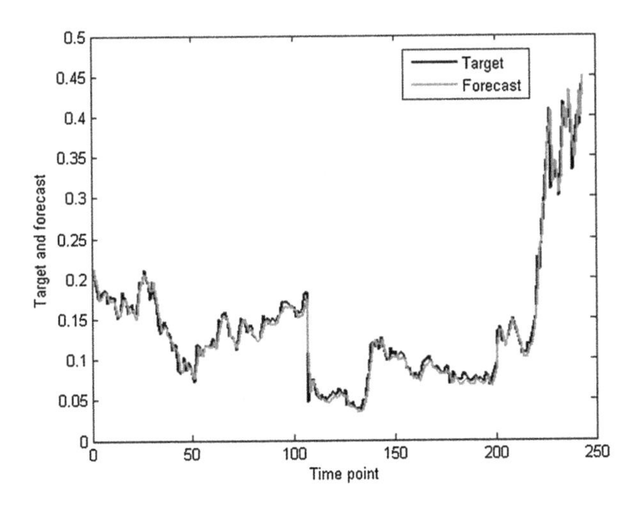

Fig. 2. The forecasting result for the Shanghai index with the target one

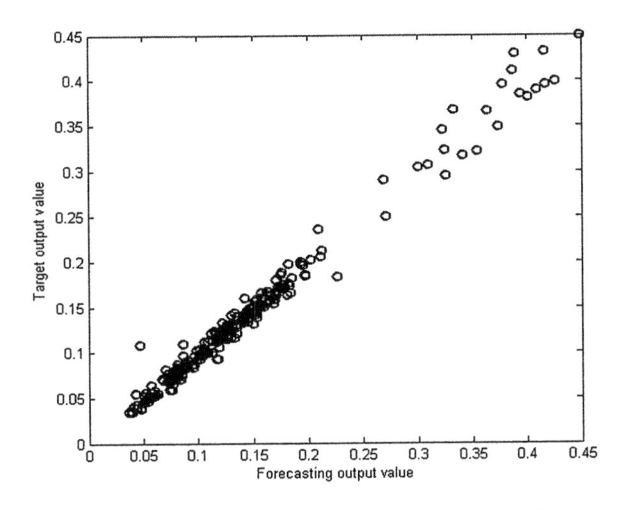

Fig. 3. Target vs forecasting stock index with test set of the Shanghai index

Four methods are also used to forecast Shanghai index data set. The results are listed in Table 1. Results show that S-system model performs better than NN, WNN, FNT and ODE and is an efficient model for forecasting stock prices.

Table 1. Forecasting results of five different methods for Shanghai index

	RMSE	MAP	MAPE
NN	0.01275	45.865	7.092
WNN	0.01191	35.778	6.4279
FNT	0.01182	27.688	6.2696
ODE	0.01102	22.171	5.8652
S-system	**0.00984**	**20.622**	**4.6012**

4 Conclusions

In order to forecast accurately, a novel time series prediction approach based on S-system model is propose. An improved additive tree model and cuckoo search are presented to optimize the structure and parameters of S-system model, respectively. Shanghai stock index is used to test prediction ability of S-system model. Results show that compared with neural network, wavelet neural network, flexible neural tree and ordinary differential equation, S-system model performs better.

Acknowledgments. This study was funded by the PhD research startup foundation of Zaozhuang University (No.2014BS13, No.2015YY02).

References

1. Roy, S.S., Mittal, D., Basu, A., Abraham, A.: Stock market forecasting using LASSO linear regression model. Adv. Intell. Syst. Comput. **334**, 371–381 (2015)
2. Majhi, B., Anish, C.M.: Multiobjective optimization based adaptive models with fuzzy decision making for stock market forecasting. Neurocomputing **167**, 502–511 (2015)
3. Jonathan, L., Ticknor, A.: Bayesian regularized artificial neural network for stock market forecasting. Expert Syst. Appl. **40**, 5501–5506 (2013)
4. Angeline, P.J., Saunders, G.M., Pollack, J.B.: An evolutionary algorithm that constructs recurrent neural networks. IEEE Trans. Neural Netw. **5**, 54–65 (1994)
5. Chang, P.C., Liu, C.H.: A TSK type fuzzy rule based system for stock price prediction. Expert Syst. Appl. **34**(1), 135–144 (2008)
6. Wang, J.Z., Wang, J.J., Zhang, Z.G., Guo, S.P.: Forecasting stock indices with back propagation neural network. Expert Syst. Appl. **38**(11), 14346–14355 (2011)
7. Zheng, W.L., Ma, J.W.: Diagonal Log-Normal Generalized RBF Neural Network for Stock Price Prediction. Lect. Notes Comput. Sci. **8866**, 576–583 (2014)
8. Yeh, C.Y., Huang, C.W., Lee, S.J.: A multiple-kernel support vector regression approach for stock market price forecasting. Expert Syst. Appl. **38**, 2177–2186 (2011)
9. Cao, H., Kang, L., Chen, Y., Yu, J.: Evolutionary Modeling of Systems of Ordinary Differential Equations with Genetic Programming. Genet. Program. Evolvable Mach. **1**, 309–337 (2000)
10. Liao, C.Z., Jiang, M.Y., Yang, B., Ben, X.Y.: Stock Index Modeling Using the System of Differential Equations. Appl. Mech. Mater. **543–547**, 4304–4307 (2014)
11. Chen, Y.H., Yang, J., Zhang, Y., Dong, J.: Evolving additive tree models for system identification. Int. J. Comput. Cognition **3**(2), 19–26 (2005)

12. Yang, X.S., Deb, S.: Cuckoo search: recent advances and applications. Neural Comput. Appl. **24**(1), 169–174 (2014)
13. Chen, Y.H., Peng, L.Z., Abraham, A.: Stock Index Modeling Using Hierarchical Radial Basis Function Networks. Lect. Notes Artif. Intell. **4253**, 398–405 (2006)

Online Detection Approach to Auto Parts Internal Defect

Lin Mingyu(✉) and Ju Jianping

School of Electrical and Information Engineering, Hubei Business College, Hubei, China
linmingyu2007@163.com

Abstract. Aiming at the detection requirements for a certain type of auto parts production and processing, an online detection approach to defect auto parts internal defect based on industrial endoscope, is studied in this paper. In this method, auto parts internal detection device is designed based on hard tube industrial endoscope and industrial camera, and the structure and working principle of the device are described detailedly, ultimately proceed in sliding defect detection making use of the texture block.

Keywords: Industrial endoscope · Defect detection · Homomorphic filtering · Texture block sliding detection

1 Introduction

Industrial endoscope used in nondestructive testing technology, can directly reflect the condition of the inside and outside surface of the detected object, without dismantling or destroy the detected objects.

Nondestructive testing based on the industrial endoscope is widely used in aerospace, auto industry and other fields. Industrial endoscope usually can be divided into hard tube endoscope, optical fiber endoscope and video endoscope. Hard tube endoscope does imaging by optical objective lens, and provides imaging light source through the optical fiber, not bend; Optical endoscope transmits images by the fiber bundles, enter into the interior softly, and change the sight direction; Video endoscope is integrating the miniature CCD sensors in the detected port, puts the outputs image signals to the video endoscope controller which detects, displays images [1].

2 Defect Detection Device Design and Image Capture

A certain type of auto parts is shaft hole parts, it needs to detect the painting conditions of the parts surface and internal daub by the machine vision technology in the process of production. For parts internal ring-shaped aperture, auto parts internal defect detection device is designed using hard tube endoscope, and it realizes to automatically capture and detect the parts internal images.

F. Xhafa et al. (eds.), *Advances in Intelligent Systems and Interactive Applications*, Advances in Intelligent Systems and Computing 686,
https://doi.org/10.1007/978-3-319-69096-4_56

2.1 Selection of the Endoscope and Industrial Camera

The hard tube endoscope selected in this paper is HSW which is resistant to water and oil, and 70°, the diameter is 4 mm, to match the rear LED light source.

For easy to process the images, we capture the endoscopic optical images using industrial camera through video imaging transform interface. This system chooses the color CCD camera with Basler Ethernet interface, whose highest frame rate is 90 FPS.

2.2 Defect Detection Device Design

Auto parts internal defects detection device based on the hard tube endoscope and industrial camera designed is shown in Fig. 1.

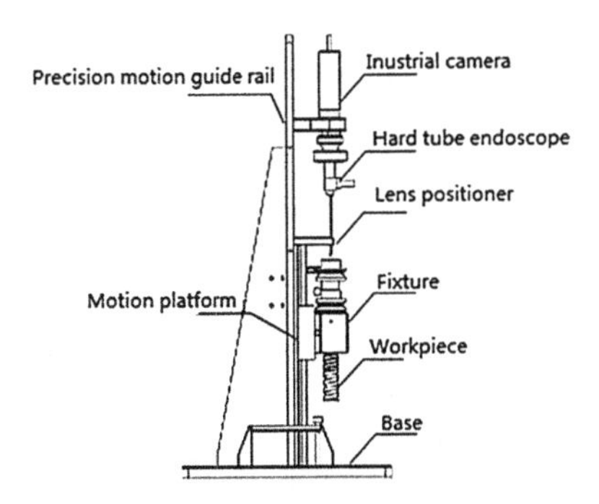

Fig. 1. The structure chart of auto parts internal defect detection device

This device is composed of base, workpiece, precision motion guide rail, industrial camera, hard tube endoscope, lens positioner, fixture and motion platform and so on, which is placed in an airtight environment, to avoid that the external light changes affect the detection results. Precision motion guide rail is installed on the base, hard tube endoscope is fixed in the end of the motion guide rail, the top of the hard tube endoscope fixes industrial camera, which connects IPC via gigabit Ethernet front-end port. The lens positioner is used to fix the endoscope lens, to ensure its concentricity with the workpiece. The characteristics of the workpiece fixture are radial soft and axial rigidity, which is in order to avoid obstruction phenomenon caused by the radial positioning deviation in the process of assembly. Motion platform is composed of the two parts such as the vertical motion control unit and rotary motion control unit: the vertical motion control unit composed of servo motor controls the workpiece fixture which clamps the workpiece moving from the bottom of the base to the detection target location along the precision motion guide rail; the rotary motion control unit controlled by stepper motor is mainly used for rotating the workpiece and taking pictures.

2.3 Image Acquisition and Detection Requirements

This detection device is used for automatically detecting the painting conditions of the auto parts surface and internal daub, when the detection of the workpiece surface is qualified, the mechanical arms of the production line place the workpiece on the fixture of the base, as the same time, the vertical motion control unit controlled by PLC moves to the detection target location. After the grating ruler confirms the workpiece position, PLC controls the rotary motion control unit beginning to rotate the workpiece and take pictures.

To ensure that it can get the workpiece internal complete images, thus each workpiece must rotate 6 times, and gets 6 pictures all which need to detect: when 6 images are all qualified, IPC sends qualified signals to PLC through the IO interface of the computer control card; when some image is unqualified, IPC sends unqualified signal to PLC and terminates the current detection; PLC controls the mechanical arms doing different processing by the qualified or unqualified signal received.

Actual acquired images are as shown in Fig. 2, (a) is workpiece internal image without daub, (b) is the qualified image with daub, taking the transverse groove in figure as a benchmark, above which the painting area is, if that area is not complete painted, then as unqualified product; the vertical groove area bans painting, once there is daub in this area, is considered unqualified; The area between the transverse and the vertical groove is transition zone, the area with or without painting both are OK.

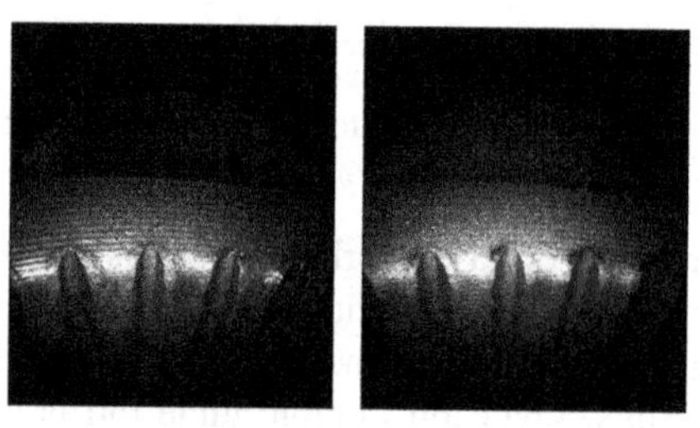

(a) No daub workpiece (b) Dualified workpiece

Fig. 2. Actual acquired images

3 Image Preprocessing

Because the distance between auto parts internal surface and endoscopic lens is very small, and its surface is metal strong reflective surface, which can produce strong reflected light, makes the luminance of the area which closes light source in the image is higher, the luminance of the area which slightly far for light source is lower, luminance distributes is nonuniform, this will affect the subsequent image processing, thus lead to appear false positives or false negatives during defect detection [2].

This article realizes the luminance correction of the original images using homomorphic filtering operation in frequency domain. In the illumination-reflectance model, the original image$f(x, y)$ is expressed with the incoming component $i(x, y)$and reflection component$r(x, y)$, the corresponding relation is [3].

$$f(x, y) = i(x, y)r(x, y) \tag{1}$$

Set$H(u, v)$is homomorphic filtering function of the homomorphic filter, and the results after homomorphic filtering are as follows:

$$s(x, y) = F^{-1}[S(u, v)] = F^{-1}[H(u, v)I(u, v)] + F^{-1}[H(u, v)R(u, v)] \tag{2}$$

In form (2), $s(x, y)$ is the output image after filtering, $I(u, v)$ and $R(u, v)$ are respectively the Fourier transform after the natural logarithm of the incoming component and reflection component. It comes true dynamic range compression and contrast enhancement of the image gray, by selecting $H(u, v)$ which decays the incoming component and enhances the reflection component. It usually adopts second-order Butterworth high-pass filtering functions as homomorphic filtering functions for endoscopic:

$$H(u, v) = (r_h - r_l)/(1 + (C \times D_0/D(u, v))^2) + r_1 \tag{3}$$

In form (3), r_h and $r_l$mean high-frequency gain and low-frequency gain respectively, $C$is sharpen coefficient, $D(u, v)$ is the distance between frequency (u, v) to the filter center (u_0, v_0), D_0 is the value of $D(u, v)$ when u_0 and v_0 both is 0, means the cut-off frequency. The luminance correction images after homomorphic filtering for the auto parts internal surface image capturing by endoscopic are shown in Fig. 3.

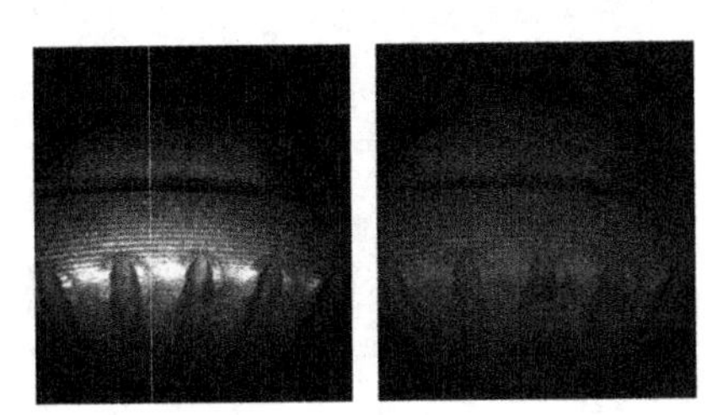

(a)Original image with illumination (b) Correction image heterogeneity

Fig. 3. Luminance correction images after homomorphic filtering

4 Defect Detection

The texture of the auto parts internal surface when not painting daub is consistent, once daub painted, it can make the texture in some area change, and the textures in different areas with painting daub are different, so extracting the texture features of the auto parts internal surface, can achieve to separate the some samples of different class and gather the similar samples, thereby proceed to defect detection effectively.

4.1 Texture Extraction

Texture extraction adopts Sobel operator to calculate the gradient value of the image, then does threshold processing to turn into binary image, high gradient is white, low gradient to black, so the texture can be extracted.

4.2 Selection of the Detection Area and Defect Identification

In Fig.4 (a), only detecting mask part which is the interested area. Mask area includes two parts: the upper part (corresponding grey value as $th1$) corresponds the top area of the transverse groove, mainly detects whether the daub in this area is painted completely; the bottom part (grey value as $th2$, $th2 < th1$) corresponds the area of the vertical groove, mainly detects whether there is the daub into this area.

(a) Mask gray image (b) Texture image of the qualified workpiece

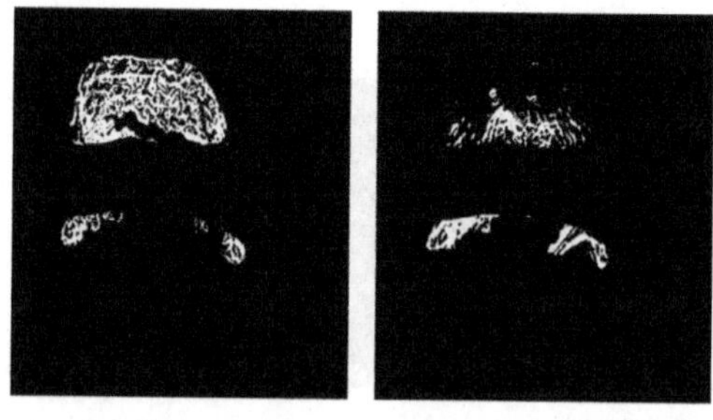

(c) Texture image of the unqualified workpiece in the upper area (d)Texture image of the unqualified workpiece in the bottomarea

Fig. 4. Mask image and workpiece texture image

Set $th1$is binary texture image of the image to be detected using fixed threshold value, as shown in Fig. 4. The normalization integral figures of the binary image to be detected and the mask image gain respectively on this basis. Then proceeds to sliding detection with using the texture block (the block size of 20 × 20), the grey value sum of the texture block of the binary image to be detected and the mask image denotes respectively $sum(X_g, Y_g)$and $sum(X_s, Y_s)$are defined as follows:

$$sum(X_g, Y_g) = \sum_{x<X_g, y<Y_g} I_g(x, y), sum(X_s, Y_s) = \sum_{x<X_s, y<Y_s} I_s(x, y) \quad (4)$$

Finally, the defect judgment is according to the following formula:

$$\begin{cases} \frac{2 \times sum(X_g, Y_g)}{sum(X_s, Y_s)} > \lambda_1 & \textit{the upper part not painted completely} \\ \frac{sum(X_g, Y_g)}{sum(X_s, Y_s)} < \lambda_2 & \textit{the bottom part with daub} \end{cases} \quad (5)$$

In form (5), λ_1, λ_2 are defect judgments coefficients of the two areas respectively, set according to the practical light and the texture distribution feature [4, 5].

5 Test and Result Analysis

In order to verify the approach designed in this paper, we test 200 choosed workpieces, and define the corresponding test indicators: miss rate refers to the rate that makes the defective workpieces as the qualified products; false rate refers to the rate that makes the qualified ones as defective. Test results are shown in Table 1.

Table 1. Test statistics of the online detection approach to defect the auto parts internal defect

Test records (Total 200)	Items		
	Qualified	Miss	False
Workpiece Number	172	7	21
Rate (%)	86	3.5	10.5

Known by the test results above, this detection method can effectively solve the online detection of the auto parts internal defect, the stability of the system and the test results can meet the needs of the practical production.

Acknowledgment. Supported by Science and Technology Research Project of Hubei Provincial Educational Commission (No. B2016482).

References

1. Hua, L., Guo, C., Xinbo, C., et al.: Study on automatic measurement method for aero-engine inner damage crac. Comput. Eng. Appl. **52**(11), 233–237 (2016)
2. Yang, Y., Fan, J., Zhao, J.: Preprocessing for highly reflective surface defect image. Opti. Precision Eng. **18**(10), 2288–2295 (2010)
3. Huixian, S., Luo, F., Zhang, Y.: Novel contrast enhancement method for industrial endoscope images. Opto-Electronic Eng. **35**(12), 107–111 (2008)
4. Di, J., Lu, M., Sun, J., et al.: Edge detection method of block distance combining with summed area table. J. Image Graph. **20**(10), 1322–1330 (2015)
5. Guan, S.Q., Shi, H.Y., Wang, Y.N.: Strip steel defect detection based on zero-mean method. J. Iron Steel Res. **4**, 014 (2013)

Globalized Translation Talent Training Model based on Artificial Intelligence and Big Data

Fang Yang(✉)

School of Foreign Language, Shenyang Ligong University, Shenyang, China
yang_fang1972@163.com

Abstract. With the continuous development of new technologies such as global data, the internet, machine translation and speech recognition, language service industry, language education industry and even relevant vertical areas have undergone a series of profound changes. How to integrate innovation and breakthrough in translation teaching, scientific research and practice has becomes an urgent problem to be solved. Big data and artificial intelligence has brought new opportunities for the development of education, and language services play the core values of language for the "going out" of Chinese culture.

Keywords: Artificial intelligence · Big data · Translation talents · Language services

1 Introduction

In the era of big data, the features of information, such as large quantity and various fields, present new challenges to the language service industry. Traditional translation by human individuals cannot meet the new demand of language service with high quality and efficiency in contemporary society. The research and teaching of traditional translation mainly focus on theoretical discussion and skill teaching, but neglect the tracking study and applications of information technology and fail to embed information technology, especially translation technical knowledge, into the curriculum system of translation, which is detrimental to cultivate technical leading translation talents. The form of translation is experiencing fission and reconstruction in the age of technical translation.

2 Big Data Age and Localization of Language Services

With the deepening of the Belt and Road initiative, along which for the foreign capital introduction and the Chinese enterprises' "going out" are gradually becoming the norm, and thus the domestic and foreign enterprises and governments demands for the language and big data service increase sharply. In the government work report in 2017, Li Keqiang, premier of the State Council, has pointed out the necessity of strengthening the emerging industries including artificial intelligence which is written in the government work report for the first time.

F. Xhafa et al. (eds.), *Advances in Intelligent Systems and Interactive Applications*, Advances in Intelligent Systems and Computing 686,
https://doi.org/10.1007/978-3-319-69096-4_57

Language processing is currently one of the most difficult problems in artificial intelligence. After the State Council published Action of promoting the development of big data in 2015, many industries have stirred waves to accelerate the development of big data industry. Under this circumstance, the localization of language service has become a powerful instrument of the economic and cultural globalization.

Language service means that "the service of language knowledge which provides customers with language service and helps communicate by different means" [1]. In a narrow sense, language service is translation service. The current language service blurs the line between literature and non-literature, which includes the translation of practical writings such as literature, culture, technology, government documents, business composition, network communication, social media, etc. It also includes the translation of immediate and fragmented content and information of complete translation text that does not constitute in the traditional sense as well as the so-called network text, cybertext or hypertext which does not use paper text and contains various symbols like words, audios, images, videos and so on [2].

3 Language and Translation

Language carries culture and thought; different historical background and social culture formed the difference between different ethnic groups, which is reflected not only in the languages, but also in thinking. Therefore, translation is a comprehensive process of analysis, and should take into account the setting, the identity, the cultural background and other factors.

The premise of translation work is to have a good command of the native language, and on this basis, cultivate multilingual thinking and ability. The advance in technology allows translators to access a wide range of multi-lingual, multi-regional, multi-domain materials, especially for interpreters, who can practice coping with different accents. Cross-linguistic data can help translators to increase awareness of different disciplines and fields and construct multilingual knowledge structures. Former Italian prime minister, former chairman of the European Commission, Romano Prodi, once said: "The language of Europe is translation." Globalization, immigration enhanced the multilingual situation in Europe. The future development trend is the human–machine coordination to meet the world's need for efficient and high-quality communication requirements.

For language industry, big data is an indispensable part of its internationalization process. Language workers in the new era have to accept training in language and learn how to use big data and related advanced technology to make their work more efficient and accurate.

4 Comparison of Translation Patterns

The study of the quality, prospects and assessment of "language and machine" has been lasted for long. Machine, online, crowdsourcing translation, the existence and complement of human translation model reflect the current changes of translation model under the background of Internet, big data, cloud computing and other techniques. Machine translation uses computer software to complete the automatic conversion between two or more natural language. Online translation relies on the Internet big data technology for online synchronization translation. Crowdsourcing translation, with the help of Web2.0, cloud computing, big data and other network technology innovation and the optimization of the network platform, achieves information exchange and sharing between different sites and the majority of users, which, in essence, is a social intensive virtual network translation factory. Human translation is basically a continuation of the traditional translation pattern in which individual translator or a small number of translators work manually, with support from reference on the internet.

Machine translation and crowdsourcing translation meet the massive translation needs created by the current economic, cultural and information dissemination; thus are widely accepted by various types of customers at the market. Machine translation, online translation, crowdsourcing translation have advantage in size, speed and cost, but are not satisfactory in the quality, style, customer satisfaction.

Adopting an integration of network technology platform, advanced technology, language talents, data resources, business resources, crowdsourcing reflects the function of real-time, open and community construction and sharing of information communication, realizes the technology and the socialized production of translation, therefore, is an important localized language service model.

Online translation, with its open, free, real-time and synchronized function meet the user's demand for translation on the most extensive coverage, the strongest service capacity and the greatest social benefits.

The advantages of quality, style and degree of satisfaction with human translation are obvious, but the scale, speed and cost cannot match machine translation, online translation and crowdsourcing translation. In language service industry, technology and tools of translation has the tendency of replacing traditional manual translation, but the output of machine and automatic online translation relies on human revision.

5 The Trend of Translation Talents Training

Language itself is a data, and data is a language of communication. Nowadays, the global language industry and language education undergo a profound change. Whether in education, translation or other fields, the achievement is made through constant communication and exploration.

Big data has a wide influence on each vertical field and brings so many opportunities, but the constraining factors should not be neglected. For example, in the application of technology, how to apply big data to the everyday teaching and scientific research; in the upgrading of technology, common universities and colleges could not afford the large sum of money needed in the development and application of new technology and big data. In the future exploration of big data, only by realizing the full union of the language data alliance, combining the excellent courses from different universities and achieving the sharing of the course resources, can teaching and scientific research be mutually promoted and truly realize the value of big data.

5.1 The Development of Big Data Shocking the Current Talents Training Mode in Colleges and Universities

The rapid development of big data, artificial intelligence and other technologies has a tremendous impact on the universities' traditional talents training mode. The principle from Guangdong University of Foreign Studies Zhong Weihe says the development of big data has brought much convenience to education. At the same time, the change of the technological means will necessarily demand the universities to change the past teaching style. As the educators, they should bravely welcome the coming of the era. The vice-principle of Sichuan Foreign Language University Zhu Chaowei holds that the combination of big data and education will become the necessary demand of the development of the times. The ultimate value of the big data in education should be shown in the deep integration into the educational mainstream business and further promotion of the intelligent change in the educational system [3].

5.2 The Development of Big Data Promoting the Change of Appeals of the Talents Training in Colleges and Universities

In the times of big data, the appeals of talents training in different colleges and universities also change a lot. The principle of Tianjin University Xiu Gang thinks the excellent foreign language teaching should not only cultivate students' language ability but should also teach students how to make use of the advanced methods and data. Only in this way, can the teaching reform develop more rapidly. The vice-president Yang Junfeng from Dalian Institute of Foreign Languages points out that the talent training model against the background of big data mainly focuses on two points: the change of teachers' idea and the spread of advanced technology. The impact brought by big data and artificial intelligence is becoming deeper and deeper. The traditional teaching mode of colleges and universities can no longer satisfy the needs of the times and is gradually replaced by big data and artificial intelligence. Only with the change of the concept can the development of education be promoted rapidly.

5.3 The Big Data Advocating the Exchanging and Sharing of the Educational Resources

Big data not only means the enormous, various and quick data processing, but also a means of subversive way of thinking and a technological reform of innovation, whose shock and impact on education is profound and ever-lasting. Currently, the big data and artificial intelligence have become the new impetus for the transformation of the talents training mold in colleges and universities, effectively allocating educational resources and forging smart and dynamic internet plus educational big data biological circle.

The mode of training in translation major can try a new model like "artificial intelligence translation plus human translation", which is first done by the artificial intelligence to finish the basic translation and then revised by human translation. In addition, the literariness and specialty of the translation script will be further revised for higher efficiency and quality. The computers will also follow up the results of artificial intelligence to help the translators avoid making low-level errors, thus generating the virtuous circle. Artificial intelligence can learn about the specialty of different translators as well. By collecting and analyzing of this type of data, artificial intelligence can allocate the translation tasks to different translators according to their specialty [4].

6 Conclusion

Artificial intelligence translation is the necessity of human exploration. However, the change in translation method due to new technological era is by no means the change of the essence of translation. The translators' subjectivity cannot be replaced.

In theoretical research, curriculum setting, compiling of textbooks, training of teachers, choice of teaching methods and other aspects in talents training, technological translation courses should be changed from subsidiary courses into major courses in order to cultivate translation learners to be professional translation talents familiar with translation knowledge and skills and project management to meet the new demands on translation talents posed by economic, cultural and social development in the information era.

References

1. Zhifang, T.: Internet plus language service reform of the times. Chin. Trans. J. **4**, 74 (2015)
2. Manyun, Liu: The fission and reconstruction of translation mode in the technological time of translation. Chin. Sci. Technol. Trans. J. **4**, 17–20 (2016)
3. http://www.gtcom.com.cn/gxlm/10/20170417/105939106938246.html
4. Liping, Z.h.: Artificial intelligence translation makes you understand the world more, Science News. **8**, 1–2 (2016)

A Study on the Relationship between Enterprise Education and Training and Operational Performance—A Cases Study of a Multinational Group

Yung Chang Wu, Lin Feng, and Shiann Ming Wu(✉)

College of Business Administration, National Huaqiao University, Quanzhou, Fujian, China
wumin.hqu@gmail.com

Abstract. Amid global competition, the environments of industry businesses are being tested. With the development of new technologies, which can be created anytime and anywhere, talents have become the most important resources for enterprises, and are the driving force behind the growth and profits of enterprises. The quality of human resources determines the operational performance of enterprises, and the success or failure of enterprise education and training will affect the quality of human resources. This study is designed to explore the impact of education and training, as implemented by enterprises, on performance evaluation, and develops a model for a quality education and training system in order to analyze the overall structural equation modeling (SEM), as based on the relationship between Kirkpatrick's four levels, organizational performance, and relevant literature. This study modifies Kirkpatrick's L1-L4 four-level training effectiveness evaluation model and Phillips' fifth level evaluation ROI model into a framework of "Education and Training Evaluation Scale" suitable for this case study. A questionnaire survey is conducted on the employees and directors of a multinational group, and verification and analysis are conducted based on AMOS SEM.

On the basis of the data analysis: (1) The model relationship of education and training effectiveness, organizational commitment, and employee productivity is verified, and the impacts of education and training effectiveness and organizational commitment on employee productivity and related models are offered; (2) The resulting level of education and training produces significantly positive impact on the improvement of organizational performance; (3) The more trainees are satisfied with training, the better the trainees perform in improving their personal abilities and practical work application effectiveness. The resulting levels also demonstrate that education and training can enhance the skills and productivity of employees, increase employee loyalty, and reduce turnover and absenteeism, thereby, making a practical contribution to organizational performance.

Keywords: Evaluation of education and training · Kirkpatrick · ROI · SEM

F. Xhafa et al. (eds.), *Advances in Intelligent Systems and Interactive Applications*, Advances in Intelligent Systems and Computing 686,
https://doi.org/10.1007/978-3-319-69096-4_58

1 Introduction

Most companies that understand the successful use of education, training, and operational performance have more outstanding financial performances, meaning employee training and capacity development could help companies cope with constant competitive pressures [1]. The study lends support to this argument, and argues that there is a clear correlation between investment in employee education and training and improved competitiveness. The reason why employee training is closely related to competitiveness is that employee training helps enterprises to respond to the challenges of business [2]. The importance of education and training is unquestionable, and the effectiveness of training transfer cannot be ignored. However, the study points out that trainees rarely apply the knowledge, skills, and work attitude, as acquired through training, to their work. After investigation and analysis, Newstrom [3] maintained that, after training, 40% of employees could immediately apply what they learned to work; 25% of employees could continue the application for 6 months; but only 15% of employees could continue the application one year after training.

Presently, there is extensive research into education and training both at home and abroad, and applying the Kirkpatrick model as an evaluation model facilitates empirical application. Nonetheless, only reaction, learning, and behavior levels are evaluated. The resulting levels cannot be clearly defined due to the effectiveness and performance of trainees, which are affected by many factors and cannot be attributed to training alone. Therefore, evaluating the resulting level is very difficult, and effectiveness is often undemonstrated in the resulting levels, which renders evaluation more difficult. Therefore, the reliability and validity of the results are questioned, contributing to fewer evaluations of the resulting level [4].

From the perspective of evaluation, the main motivation of this study is to examine whether enterprises' in-service training allows trainees to acquire the knowledge and attitudes related to work, and whether they can successfully apply the acquired knowledge and attitudes to work in order to help enterprises achieve organizational performance. Based on Kirkpatrick's evaluation model, and through literature review of the ROI evaluation model, this study constructs an association model, as well as its influencing variables, regarding the introduction of education and training by enterprises, in order to understand the strength of the relationship between enterprises' implementation of education and training and employee training demands.

2 Literature Review

This study primarily probes into the relationship between education and training and enterprise operational performance by introducing education and training, as based on the opinions of the ROI models proposed by Kirkpatrick [6] and Phillips [7], and through quantitative processes. Kirkpatrick published the 4-level model in the "Techniques for Evaluating Training Programs" in 1959 and 1960. This evaluation model is still frequently used by academia and industry. The various levels are presented, as follows:

1. Reaction level: The reaction level refers to the trainees' views regarding the overall training program, that is, their overall satisfaction with the implementation of training. The evaluation is generally conducted through questionnaire survey and observation.
2. Learning level: The learning level mainly measures the extent to which the trainees understand the training course and knowledge by the end of training, and whether such training can boost their self-confidence and improve their work attitude in order to understand the ways to inspect training effectiveness. Inspections are generally conducted through written tests, oral tests, and classroom performance.
3. Behavior level: The behavioral level mainly evaluates whether trainees can apply the learning outcomes to work after training. Moreover, training is intended to change behavior, and thus, improve work performance. This level can generally be measured through a behavior-oriented performance evaluation scale or observation after trainees return to work.
4. Result level: The evaluation of the result level is an attempt to explore the impact of training on organizational performance, and evaluate the specific contributions provided to organizations after trainees receive training. Evaluations can be measured by comparing the related information before and after training, such as increased productivity, improved quality of service, increased profits, increased return on investment, and reduced turnover rate. Chen [5] suggested that studies on result levels mostly employ cost-benefit analysis; however, cost-benefit analysis is particularly feasible in evaluating the effectiveness of training programs, due to difficulties in estimating costs, ownership of benefits, and other intangible benefits.

After adopting Kirkpatrick's evaluation for many years, Phillips [7] proposed a fifth level for the evaluation of "ROI" (return on investment): (1) Reaction and planned action; (2) Learning evaluation; (3) Job applications; (4) Business results; (5) Return on investment (ROI).

Performance is the criterion for measuring achievement. When an individual is evaluated, job performance is formed; when an organization is evaluated, organizational performance is formed. Kaplan and Norton [8] held that organizational performance indices include finance, customer, internal business processes, learning, and growth. The findings by Chang [4] are, as follows: (1). Education and training produces a positive impact on organizational performance. (2) Higher organizational commitment could lower employee turnover rate, enhance productivity and quality, shorten the production cycle, reduce the complaint rate, and increase the organizational performance of enterprises. Ngo, Turban, Lau, and Lui [10] also found significant positive correlations between structured training and corporate performance.

3 Research Structure and Method

This paper selects the research variables by reviewing domestic and foreign literature, and constructs the research framework (as shown in Fig. 1) based on the research background, purpose, literature review, and discussion.

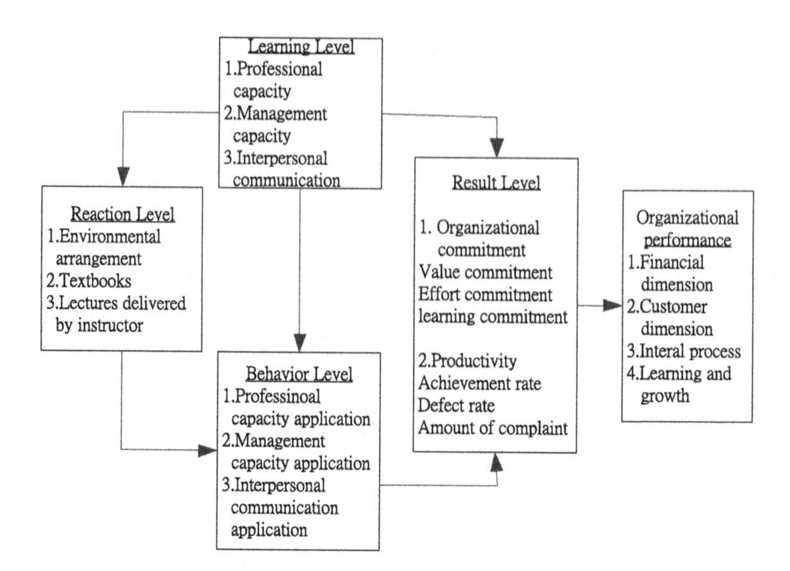

Fig. 1. Research Framework

The six hypotheses for the framework are described, as follows:

H1: The design of well-planned training courses and the arrangement of lecturers and environmental equipment will help trainees to improve their professional capacity, management capacity, and interpersonal communication.

H2: The design of well-planned training courses and arrangement of lecturers and environmental equipment will help trainees to become more proficient in their professional capacity, management capacity, and interpersonal communication.

H3: The abilities developed from learning levels can be applied to address work problems at the behavior level.

H4: The abilities developed from the learning levels can improve knowledge, skills, and attitudes, which can help to enhance trainee's loyalty to the organization, improve their productivity, and lower the defect rate.

H5: Proper applications of interpersonal communication, management capacity, and professional ability can enhance trainee's loyalty to the organization, improve productivity, and lower the defect rate

In addition, Warr and Birdi [11] pointed out that evaluation of a result level is exceedingly difficult, primarily because the changes in organizational performance, as caused by a single training activity, are not easily measured. However, after adopting Kirkpatrick's 4-level training evaluation model, Phillips proposed the evaluation concept of a fifth level for "ROI" [7]. Therefore, the influence of the result level on creating organizational performance can be strengthened.

H6: Reduced turnover rate, increased loyalty and productivity, and enhanced quality improvement capacity indicate that the impact of the result level on organizational performance is greater, and vice versa.

The questionnaire survey applies the balanced scorecard proposed by Kaplan and Norton [8], and refers to the opinions of experts and scholars, who revised and developed the final questionnaire. For instance, the Scale of Individual Performance for Education and Training is modified from the Effectiveness Scale of Education and Training, as based on the "Satisfaction Scale for Management Development Program" developed by Wu (1984) and the Effectiveness Scale of Education and Training revised by Chen (1997). The questionnaire design includes 5 constructs and 19 variables, and degree is measured according to seven scales, as based on a Likert Scale.

In this study, the valid questionnaire data, as obtained through preliminary screening of the recovered questionnaires, are input into Microsoft Excel, and SPSS 21.0 and AMOS 17.0 statistical software packages are used for statistical analysis of the related data.

4 Empirical Data Analysis

In this study, a total of 380 questionnaires are distributed in the form of e-mails, and 360 questionnaires are recovered, for a recovery rate of 94.74%. After invalid questionnaires are screened from the collected samples, the number of valid samples is 318, for a valid recovery rate of 88.33%. In the formal stage, the chi-square χ^2 test of the sample cluster regarding gender and age does not reach the significance level of $\alpha = 0.05$, indicating that the samples are representative, stable, and homogeneous.

Regarding the measurement of reliability, this study analyzes the items for total correlation, and deletes the variables with too few items for total correlation coefficient. Additionally, Cronbach's α value of internal consistency analysis is used to measure the internal consistency of the factors of each measured variable. A larger Cronbach's α suggests a greater correlation between each variable of this factor. In other words, the higher the internal consistency, the better the reliability. Scholars generally believe that the scale should be re-compiled if Cronbach's α value is less than 0.6; Cronbach's α value should be at least 0.7; 0.9–0.95 is ideal; the scale should be deleted if the item-total correlation is less than 0.4 (Hair et al. 1998).

Validity refers to the correctness of the measurement, that is, whether the questionnaire can measure the actual psychological responses of the subjects. Higher validity indicates the test results can better represent the actual responses of the subjects. To ensure rigorousness and completeness, the model's test for goodness of fit is conducted with the measurement of the structural equation model, and construction validity is tested in order to verify whether each dimension has sufficient convergent validity and discriminant validity.

This study determines whether there is a significant causal relationship between latent variables (η and ξ) through the significance testing of Beta (β) and Gamma (γ). In addition, whether the causal relationship between the latent variables reaches the significance level is judged based on the t value. In conclusion, the criteria used to evaluate the model's goodness of fit are organized in Table 1.

Table 1. Summary of Each Evaluation Item for Goodness of Fit of Overall Model

	Evaluation item	Ideal evaluation result	Result of this study	Consistency between study result and theoretical value
Basic criteria for goodness of fit of overall model	Error variance	Cannot be negative	All are positive	Yes
		Reach a significant level	All reach a significant level	Yes
	Standard error			Yes
Goodness of fit of internal structure	X^2		1055.641	—
	d.f.		586	—
	X^2/d.f.	<3	1.80	Yes
	GFI	>0.8	0.845	Yes
	AGFI	>0.8	0.824	Yes
	NFI	>0.9	0.894	Yes
	NNFI	>0.9	0.946	Yes
	IFI	>0.9	0.950	Yes
	CFI	>0.9	0.950	Yes
	RMSEA	<0.08	0.050	Yes
Goodness of fit of internal structure	Reliability of individual items	>0.5	36 variables > 0.5	Yes
	Estimated parameters	Reach a significant level	36 estimates reach a significant level	Yes

Direct effect: (1) The reaction level is an important factor that directly affects the learning level, with an impact effect of 0.89. The result is that, the greater the satisfaction with education and training in the reaction level, the greater the improvement in personal ability in the learning level. (2) The impact of the reaction level on the behavior level is not significant; the greater the improvement in personal ability in the learning level, the higher the degree of practical job application in the behavior level, and the impact effect is 0.75. (3) The impact of learning level on the result level is not significant. (4) The behavior level is an important factor that directly affects the result level, with an impact effect of 0.75. (5) The result level shows increased employee loyalty to companies and work efficiency, and decreased defect rate, suggesting better organizational performance, and the impact effect is 0.88.

Total effect: (1) The reaction level, meaning satisfaction with education and training, only produces a direct effect on organizational performance, thus, the indirect effect is the total effect on organizational performance (0.67). (2) The learning and behavior levels only have combined indirect effects on organizational performance. Specifically, the total effects on organizational performance are 0.63 and 0.66, respectively. (3)The reaction level only has significant direct relationship with the learning level, and its total effect is the direct effect (0.89). (4) The result level only has significant direct relationship with organizational performance, and its total effect is the direct effect (0.88).

It can be learned from the analysis of the research models that the variation explanatory power (R2) of each latent dependent variable to the overall model is learning level (0.80), behavior level (0.78), result level (0.78), and organizational performance (0.77), respectively. Moreover, the overall explanatory power of each latent dependent variable in the result level to the organizational performance is 0.77, indicating a sound explanation of the model for the latent variation degree.

This study establishes a preliminary research model through exploration into related literature theories, and then, through empirical analysis and research, obtains the results of the relation path diagram for the final overall framework, as shown in Fig. 2.

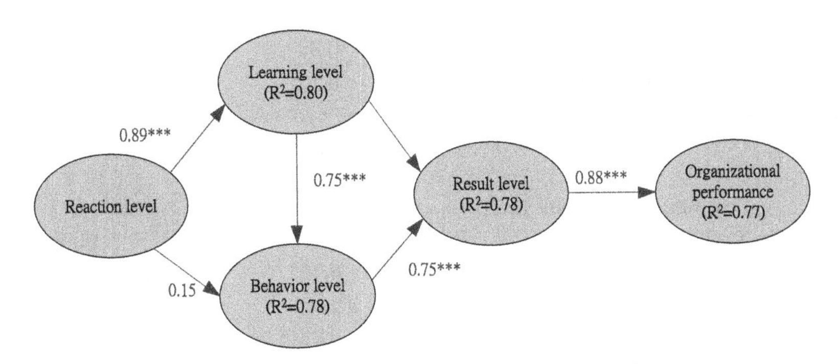

Fig. 2. Path diagram of path relationship results of structure model

To examine whether the research hypotheses of the association model are established, this study, as based on the empirical results, sorts the path relationship coefficients of the structural model of this study, as shown in Table 2, and the analysis results are, as follows:

1. H1 and H2: The estimated parameters of the positive direct impact of "reaction level" on "learning level" and "behavior level" are 0.89 and 0.15 (t = 12.107 *** and t = 1.37), respectively, which illustrates the hypothesis, meaning a significant positive correlation between the reaction level and the learning level is established. By contrast, as the t value does not reach the significance level, thus, the hypothesis, that there is significant positive correlation between the reaction level and the behavior level, cannot be established.
2. H3 and H4: The estimated parameters of the positive direct impact of "learning level" on "behavior level" and "result level" are 0.75 and 0.15 (t = 6.376 *** and t = 1.538), respectively, which illustrates the hypothesis, that there is significant positive correlation between the learning level and the behavior level, is established. By contrast, as the t value does not reach the significance level, the hypothesis, that

there is significant positive correlation between the learning level and the result level, cannot be established.

3. H5: The estimated parameters of the positive direct impact of "behavior level" on "behavior level" is 0.75 (t = 6.615 ***), indicating that the hypothesis, that there is a significant positive correlation between the behavior level and the result level, is established.
4. H6: The estimated parameters of the positive direct impact of "result level" on "organizational performance" is 0.88 (t = 12.475 ***), indicating that the hypothesis, that there is significant positive correlation between the result level and organizational performance, is established.

Table 2. Path relationship coefficients of the research model

Hypotheses	Latent variables	Latent dependent variable	Path relationship	t value	Results
H1	Reaction level	Learning level	0.89***	12.107	Supported
H2	Reaction level	Behavior level	0.15	1.370	Not supported
H3	Learning level	Behavior level	0.75***	6.376	Supported
H4	Learning level	Result level	0.15	1.538	Not supported
H5	Behavior level	Result level	0.75***	6.615	Supported
H6	Result level	Organizational performance	0.88***	12.475	Supported

Note: ** denotes $p<0.01$, *** denotes $p<0.001$

5 Conclusion

In order to gain an in-depth understanding of the comparison between the organizational performance evaluation of education and training in the studied multinational group and general enterprises, this study adopted the research results of the Performance Evaluation Model of Enterprises Applying ISO10015, as proposed by Chang [4], as a basis for comparison, and intended to identify whether enterprises introduced ISO 10015? Have employees received education and training regarding ISO 10015? The differences in the organizational performances of companies receiving education and training were compared, and the empirical results are shown in Table 3.

The comparative analysis of the studied multinational group and general enterprises whose employees received education and training regarding ISO10015 is, as follows:

1. The hypothesis is established, meaning there is significant positive correlation between the "reaction level" and the "learning level" in both the studied multinational group and general enterprises.

Table 3. A comparison table of path relationship result of whether the studied case received education and training about ISO 10015

Latent variables	Latent Dependent variables	Hypotheses		Path relationship		T Value		Results	
		Multinational group	General enterprises	Multinational group	General enterprises	Multinational group	General enterprises	Multinational group	General enterprises
Reaction level	Learning level	H1	H1	0.89***	0.35***	12.107	5.83	Supported	Supported
Reaction level	Behavior level	H2	H2	0.15	0.09	1.37	1.37	Not supported	Not supported
Reaction level	Result level		H3		0.05		1.09		Not supported
Learning level	Behavior level	H3	H4	0.75***	0.45***	6.376	7.39	Supported	Supported
Learning level	Result level	H4	H5	0.15	0.14**	1.538	2.95	Not supported	Supported
Behavior level	Result level	H5	H6	0.75***	0.68***	6.615	11.87	Supported	Supported
Result level	Organizational performance	H6	H7	0.889***	0.56***	12.475	8.27	Supported	Supported
Whether the employees received education and training about ISO10015		No	Yes	No	Yes	No	Yes	Yes	Yes

Source Compiled by this study

2. As the t value does reach the significance level, the hypothesis, that there is significant positive correlation between the "reaction level" and the "behavior level" in both the studied multinational group and general enterprises, cannot be established.
3. This study did not explore the path relationship between the "reaction level" and the "result level", thus, no comparison can be made.
4. The hypothesis is established, meaning there is significant positive correlation between the "learning level" and the "behavior level" in both the studied multinational group and general enterprises.
5. Regarding the hypothesis of significant positive correlation between the "learning level" and the "result level", the hypothesis is established in general enterprises, as the t value reaches the significance level. On the contrary, the hypothesis cannot be established in the studied case, as the t value does not reach the significance level. The studied case and general enterprises show opposite results.
6. The hypothesis of significant positive correlation between the "behavior level" and the "result level" is established in both the studied case and general enterprises.
7. The hypothesis of significant positive correlation between the "result level" and the "organizational performance" is established in both the studied case and general enterprises.

While the studied case did not introduce the ISO10015 education and training system, it has a sound corporate training system. As can be observed from analysis of the comparison results, the majority of empirical results of the studied case are the same as the results of general enterprises whose employees received education and training regarding ISO 10015.

Most of the study samples are aged 20–30 years, with an education background of senior high school (vocational high school); 79.56% have seniority of less than 5 years; the majority are unmarried employees in the manufacturing department. Therefore, personal factors, such as seniority and age, affect organization members' recognition, as well as their input levels of attitudes and behaviors. As a result, although the members have received the complete education and training offered by the companies, their specific contributions to the organizations do not produce significant impact.

References

1. McGraw-Hill.: Employee Training & Development, 2nd ed., Chien, C.Y. (trans), 2007, Taipei (2004)
2. Noe, R.A.: Employee Training and Development. McGraw-Hill, New York (2004)
3. Newstrom, J.W.: A role-taker time differentiation integration of transfer strategies. Paper presented at the meeting of the American Psychological Association, Toronto, Ontario (1984)
4. Chang, H.L.: Performance evaluation model of enterprise applying ISO10015. Master's thesis, Graduate Institute of Industrial Engineering and Management, National Chin-Yi University of Technology (2008)
5. Chen, S.C.: Research into the effectiveness evaluation of training for local civil servants. Master's thesis, Graduate Institute of Public Affairs, Tunghai University (2001)

6. Kirkpatrick, D.L.: Techniques for evaluating training programs. J. Am. Soc. Training Direct. **13**(3–9), 21–26 (1959)
7. Phillips, J.J.: Handbook of Training Evaluation and Measurement Methods, 3rd edn. Gulf Publishing Company, Houston, TX (1997)
8. Kaplan, R.S, Norton, D.: The balanced scorecard-measures that drive performance. Harvard Business Review, January/February, pp. 71–79 (1992)
9. Hsu, C.P.: A research of the effectiveness of employee training programs effect on the organizational commitment and employee productivity—an empirical study of domestic internet bank. Master's thesis, Graduate Institute of Business Management, Chang Jung Christian University (2002)
10. Ngo, H., Turban, D., Lau, C., Lui, S.: Human resource management practices an firm performance of multinational corporations: Influences of country of origin. Int. J. Hum. Resour. Manag. **9**, 632–653 (1998)
11. Warr, P., Allan, C., Birdi, K.: Predicting three levels of training outcome. J. Occup. Organ. Psychol. **72**, 351–375 (1999)

Online Handwritten Character Recognition of New Tai Lue Based on Online Random Forests

Yong Yu(✉), Pengfei Yu, Haiyan Li, and Hao Zhou

School of Information, Yunnan University, Kunming, Yunnan, China
yu1183688986@163.com

Abstract. The character recognition technology has been widely developed during these years, but the character recognition study for the new Tai Lue has lagged behind. As a result, the digital processes of new Tai Lue have been in troubles. To solve this problem, this paper proposed an online handwritten character recognition method of new Tai Lue based on online-random forests in on-line training model. Firstly, the handwritten new Tai Lue characters are preprocessed, and then the eigenvectors of these characters are extracted. Finally, the appropriate training samples and test samples are selected, online-random forests algorithm is used to train and test them. When it iterates ten times, the recognition rate reaches 87.86%. The experimental results show that online-random forests algorithm is effective in online handwritten character recognition for new Tai Lue.

Keywords: Online handwritten character recognition · New Tai Lue character · Online random forests

1 Introduction

At present, the new Tai Lue [1] handwritten recognition is not yet mature. In order to promote the information process of the new Tai Lue, the study on online handwritten recognition of new Tai Lue should overcome the complexity of handwritten character and the extraction of effective features.

Online handwritten character recognition of new Tai Lue is different from the printed character recognition of new Tai Lue [2] and offline character recognition [3]. The main process of online handwritten character recognition include: data collection, sample preprocessing, feature extraction, training the classifier model, and identify unknown samples with trained model. Nowadays, many online handwritten character recognition methods are developed. For example, Ren et al. [4] use BP Neural Network to identify online handwritten Uyghur character; Parui et al. [5] use HMM to identify online handwritten Bangla character; Ahmad et al. [6] use support vector machine to study online handwritten character recognition. These methods mentioned above for online handwritten character recognition have a common characteristic is that they are all trained in off-line mode.

F. Xhafa et al. (eds.), *Advances in Intelligent Systems and Interactive Applications*, Advances in Intelligent Systems and Computing 686,
https://doi.org/10.1007/978-3-319-69096-4_59

This paper use online random forests (ORF) [7] which works in an on-line mode to accomplish the online handwritten character recognition of new Tai Lue. Compared to off-line methods, on-line learning has a lot of advantages: for example, large amounts of data can be exploited by the online method which makes the memory requirements be much lower, because there is no samples need to be stored.

The rest of the paper is organized as follows. Section 2 describes the sample data preprocessing. Section 3 presents the character feature extraction. Section 4 illustrates the online random forests method. In Sect. 5, the experimental results are reported. Conclusions are drawn in Sect. 6.

2 Sample Data Preprocessing

After collecting numerous online handwritten sample data of new Tai Lue characters with different handwritten styles, it is necessary to preprocess for every sample, because when writing a new Tai Lue character, its size is not easy to control, and there are a lot of noise points. In this paper, the preprocess steps are described in the following subsections.

2.1 Removing the Noise Points

When the pen or the finger stays on a point of the screen in a long time, it will be sample many nearby points repeatedly in the location. There is no doubt that the duplicate points should be removed to retain only a point as an effective track point. Linear smoothing technique is a simple and suitable character stroke smoothing method for removing noise points, it has a very good effect on solving the distortion of the characters and the track which is not smooth and other issues. In this paper, three-point smoothing algorithm is adopted to perform this task. The new coordinate $(x_i^{'}, y_i^{'})$ of the point (x_i, y_i) after smoothing are calculated by Eq. (1).

$$\begin{cases} x_i' = 0.25x_{i-1} + 0.5x_i + 0.25x_{i+1} \\ y_i' = 0.25y_{i-1} + 0.5y_i + 0.25y_{i+1} \end{cases} \tag{1}$$

2.2 Normalization

In order to decrease the impact of the original character length and width ratio changes, different scale coefficients for linear normalization should be used to scale the handwritten characters to a same size. In this paper, the online input characters are scaled in the size of 50 × 50 by the linear normalization operation. Assume that the extremes of the original trajectory points of the new Tai Lue character as x_{max}, x_{min}, y_{max}, y_{min}. The coordinate (x^t, y^t) normalization formula is shown in Eq. (2).

$$\begin{cases} x_t' = [50/(x_{max} - x_{min})][x_t - (x_{max} + x_{min})/2] \\ y_t' = [50/(x_{max} - x_{min})][y_t - (y_{max} + y_{min})/2] \end{cases} \tag{2}$$

2.3 Interpolation Operation

During the stage of the new Tai Lue characters collection, the tablet is too sensitive and handwriting's speed is unstable. This will lead to the stroke points to be sampled uneven. To solve this problem, it is necessary to insert some points, as well as interpolation operation when the spacing of two adjacent points is relatively large. The interpolation method used in this paper is the midpoint interpolation algorithm based on average distance. Its steps are as follows:

(1) Calculate the average distance L between two consecutive points in each stroke of the collected online handwritten new Tai Lue characters.
(2) Calculate the distance l between two points in each stroke, if $l > L$, insert a point at the midpoint of these two points, and then continue to determine whether these two points with their midpoint distance is greater than L, if so, continue to insert or return to determine the next two points until the last point of each stroke is processed.

2.4 Re-sampling

Even after the preprocessing by the two methods mentioned above, the distance between two adjacent points in the stroke is not equal, and the number of points in a stroke is too larger to process. In order to make the interval between the sampling points in the stroke as much as possible equal and remove the non-critical points which have little contribution to the recognition, it is necessary to re-sample the track points of the new Tai Lue characters.

The principle of the re-sampling method based on distance summing is as follows: assume that the starting point of the new Tai Lue character stroke is p_s. Along the stroke direction of the track point, the distance between adjacent track points are accumulated in turn until the sum of the distance L_{sum} is bigger than or equal to the sampling interval ΔL. After that, the corresponding coordinate point p_e is selected as the new starting point. At the same time, all the track points between p_s and p_e are deleted. The above steps are repeated until all points in a stroke are processed.

The effect of online handwritten new Tai Lue character "ᦢ" after the preprocess is shown in Fig. 1.

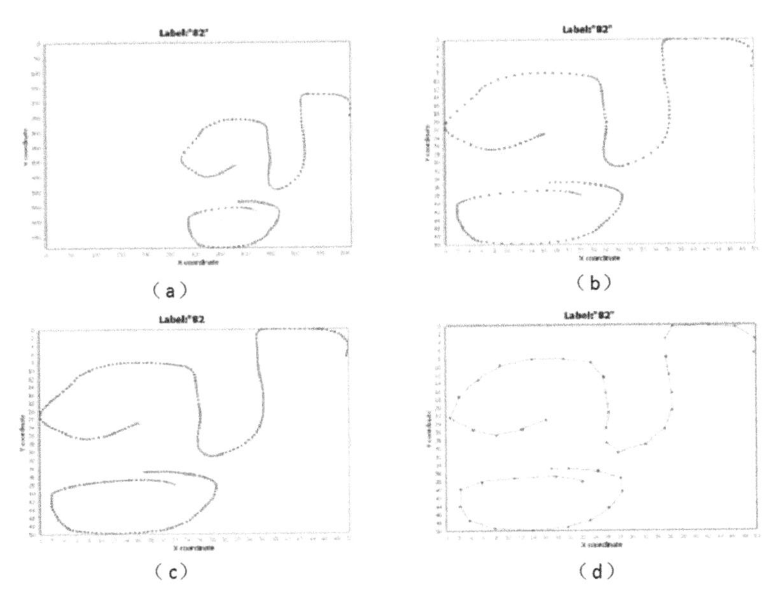

Fig. 1. The effect of character "ᦵ" after preprocessing. **a** Original character stroke, **b** Normalization result, **c** Interpolation result, **d** Re-sampling result.

3 Character Feature Extraction

3.1 The Number of Strokes

For all of the new Tai Lue characters, their stroke number are in the range of 1 to 3, For example, "cɔ" have one stroke, "ɔʒ" have two strokes, "ɔ̂ʒ" have three strokes. So the character stroke character can be divided into three subsets according their stroke numbers, and the number of characters in these subsets are: 52, 26, and 5.

3.2 The Position of Additional Strokes

The so-called additional stroke refers to the smallest stroke in a new Tai Lue character while the rest of the strokes are the main strokes. For a new Tai Lue character with an additional stroke, the position relationship between its main stroke and additional stroke can be selected as a character feature which is determined by the centroid of the additional stroke and the upper or lower boundary of the main stroke. New Tai Lue characters additional stroke and main strokes have three positions: the centroids of the additional stroke is located above, below and inside the main frame.

3.3 The Quadrant of the Start and End Points

To distinguish the quadrant feature clearly, the center (25, 25) of the character is shifted to the origin of the coordinates when a new Tai Lue character is normalized to

50×50. Assuming the normalized character stroke coordinate (x, y), the corresponding coordinate (x', y') of the stroke after translation are given by (3):

$$\begin{cases} x' = x - 25 \\ y' = y - 25 \end{cases} \tag{3}$$

at this time, the quadrant of the start point and the end point can be determined in the Cartesian coordinate system.

3.4 Steering Feature of the Beginning and the End of the Track

The cross product of two vectors could be used to determine the steering characteristics of a track. Assume the starting point $p_0(x_0, y_0)$ the next two track points $p_1(x_1, y_1)$ and $p_2(x_2, y_2)$, for the steering characteristics of the pen it only need to calculate whether the vector $p_0\,p_2$ is in the clockwise or counterclockwise direction of the vector $p_0\,p_1$. The cross product can be calculated by Eq. (4)

$$(p_2 - p_0) \times (p_1 - p_0) = (x_2 - x_0)(y_1 - y_0) - (x_1 - x_0)(y_2 - y_0) \tag{4}$$

If the result is positive, the vector $p_0\,p_2$ is in the clockwise direction of the vector p_0 p_1, then the corresponding steering characteristics of the pen is clockwise; If the result is negative, the vector $p_0\,p_2$ is in the counterclockwise direction of the vector $p_0\,p_1$, then the corresponding steering characteristics of the pen is counterclockwise; If the result is 0, then they are collinear, and no steering characteristics here.

3.5 Straight Line Intersecting Feature

The straight line intersecting feature of the online handwritten character of the new Tai Lue is defined as the times of all strokes that intersect with prescribed lines. Assume that the extremes of the original trajectory points of the new Tai Lue character as x_{max}, x_{min}, y_{max}, $y_{\min}$, the expression of the six straight lines specified in this paper are as follows:

$$L_1 : x = x_{min} + (x_{max} - x_{min})/4; L_2 : x = x_{min} + 2(x_{max} - x_{min})/4;$$
$$L_3 : x = x_{min} + 3(x_{max} - x_{min})/4; L_4 : y = y_{min} + (y_{max} - y_{min})/4;$$
$$L_5 : y = y_{min} + 2(y_{max} - y_{min})/4; L_6 : y = y_{min} + 3(y_{max} - y_{min})/4;$$

The straight line intersecting feature of character "[illegible]" is shown in Fig. 2. It can give a 6-dimensional feature [4,4,1,3,4,3] respectively, on behalf of the times of six straight lines intersect with character strokes.

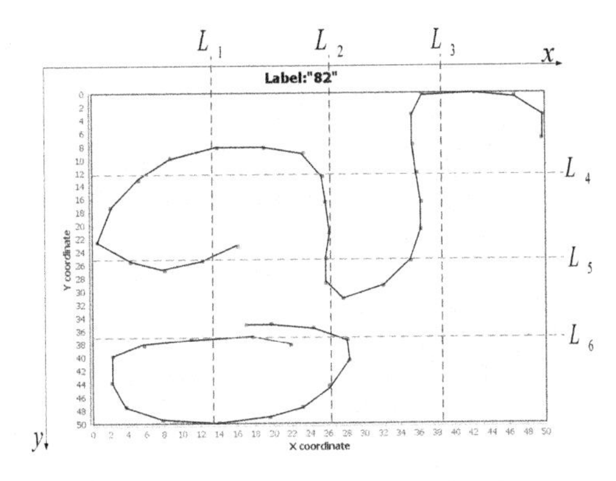

Fig. 2. The straight line intersecting feature

3.6 Rough Grid Feature

The rough grid feature [8] works by calculating the number of 1 pixel in a grid. So the size of the eigenvector is similar to the number of the grid. Online handwritten new Tai Lue character rough grid feature extraction is as follows:

(1) The new Tai Lue handwritten character should be converted to a 0-1 binary image. The size of the character after preprocessing is 50×50, so the number of pixels corresponding to the binary image is 50×50 too, then set the character track point pixel value to 1 and the rest to 0.
(2) Dividing the image into 25 regions by 10×10 grid, then the sum of pixels in each area are chosen as the value of the rough grid feature which are a total of 25 dimensions.

Taking character "[illegible]" for example, the rough grid feature is shown in Fig. 3.

	1-10	11	12	13	14	15	16	17	18	19	20	21-50
1-30	...	...	...	...	...	...	...	...	...	...	...	...
31	...	0	0	0	0	0	0	0	0	0	0	...
32	...	0	0	0	0	0	0	0	0	0	0	...
33	...	0	0	0	0	0	0	0	0	0	0	...
34	...	0	0	0	0	0	0	0	0	0	0	...
35	...	0	0	0	0	0	0	0	0	0	0	...
36	...	0	0	0	0	0	0	0	1	0	0	...
37	...	0	0	0	0	0	0	0	0	0	0	...
38	...	0	1	0	0	0	0	0	0	1	0	...
39	...	0	0	0	0	0	0	0	0	0	0	...
40	...	0	0	0	0	0	0	0	0	0	0	...
41-50	...	...	...	...	...	...	...	...	...	...	...	...

(a)

	1	2	3	4	5
1	0	2	1	4	4
2	3	0	2	2	0
3	3	2	3	3	0
4	1	3	5	0	0
5	4	1	4	0	0

(b)

Fig. 3. The rough grid feature of character "[illegible]". **a** part of 0-1 binarization result, **b** the sum of pixels in each region

In this paper the number of strokes, the position of additional strokes are 1-dimensional feature; The quadrant of the start and end points and steering characteristics of the beginning and the end of the track are 2-dimensional feature; The straight line intersecting feature is a 6-dimensional feature; The rough grid feature is a 25-dimensional feature. So a 37-dimensional feature vector is constructed for each new Tai Lue character.

4 Online Random Forests

Random forests (RFs) [9] are mainly for classification and regression, usually RFs are trained in off-line mode, and the entire training data is given in advance, the training and testing phases are separated. In practice, however, the training data may not be given beforehand, but arrive sequentially in the application phase, for example, in tracking applications where predictions need to be given on-the-fly. In this case, learning algorithms must work in the on-line mode. Online random forests can be used to solve the above problems. Each tree in an online random forest is created and tested independently from other trees, so that the training and testing procedures can be carried out at the same time. At the time of the training, a new boostrapped training set is generated by sub-sampling with the original training set for each tree.

Compared to RFs, ORF are made up of many on-line random decision trees. Specifically, there is a test in form of $g(x) > \theta$ for each decision node in a tree where $g(x)$ is a function which can generate test randomly and usually returns a scalar value and θ is a threshold based on the random feature which decides the left/right propagation of samples. Usually, for each node of the tree, its tests can be chosen with the following steps: (1) creating a set of random tests, (2) through the quality measurement to pick out the best of them. Generally, the selection for quality measures are the entropy

$$L(R_j) = -\sum_{i=1}^{K} p_i^j \log_2(p_i^j) \tag{5}$$

or the Gini coefficient

$$L(R_j) = \sum_{i=1}^{K} p_i^j (1 - p_i^j) \tag{6}$$

where K is the number of classes and p_i^j is the label density of class i in node j.

A node creates a set of N random tests $S = \{(g_1(x), \theta), \ldots, (g_N(x), \theta_N)\}$ when it is created. Then this node begins to collect the statistics of the samples that falling in it and the statistics of the splits made with each test in S. Specifically, for a random test in S, there are two sets of statistics: $p_{\mathrm{jls}} = \left[p_1^{\mathrm{jls}}, \ldots, p_K^{\mathrm{jls}}\right]$ and $p_{\mathrm{jrs}} = \left[p_1^{\mathrm{jrs}}, \ldots, p_K^{\mathrm{jrs}}\right]$, they represent the sample statistics that falling into the left(l) and right(r) partitions according to test s

The gain to a test s can be calculated as:

$$\Delta L(R_j, s) = L(R_j) - \frac{|R_{jls}|}{|R_j|} L(R_{jls}) - \frac{|R_{jrs}|}{|R_j|} L(R_{jrs}) \tag{7}$$

Where R_{jls} and R_{jrs} are the left and right partitions and $|\cdot|$ denotes the number of samples in a partition made by the test s. So, nodes when to split depends on 1) $|R_j| > \alpha$, α is the minimum number of samples that a node must be seen before splitting, 2) $\exists s \in S : \Delta L(R_j, s) > \beta$, β is the minimum gain a split has to achieve.

For ORF, a forest with T trees can be denoted as $F = \{f_1, \ldots, f_T\}$. And the estimated probability for predicting class k for a sample can be expressed as

$$p(k|x) = \frac{1}{T} \sum_{t=1}^{T} p_t(k|x) \tag{8}$$

where $p_t(k|x)$ is the estimated density of class labels of the leaf of the tth tree where x falls. So, the final multi-calss decision function of the forest can be defined as

$$C(x) = \arg\max_{k \in K} p(k|x) \tag{9}$$

5 Experimental Results and Discussions

This paper sets 200 trees in the forest, 5 random features and thresholds are selected for decision tests, uses a maximum tree-depth of 45, and sets the $\alpha = 50$ and $\beta = 0.1$ for the online random forests. In addition, this paper continue to carry out online random forest training process through shuffling the data until reaching the end of the dataset, and repeating the entire process for 10 times. The training and test of ORF are carried out with 18000 samples as the training set which comes from 15 people, and 2100 samples are chosen randomly to form the test set which is not coming from the training set. The experimental result is shown in Fig. 4.

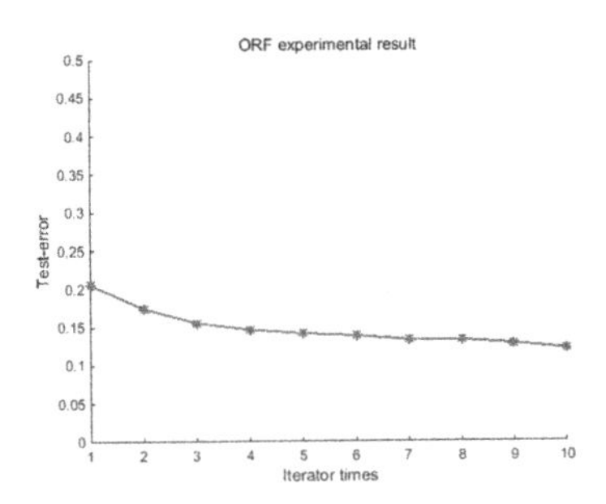

Fig. 4. The experimental result of ORF

In the experiment, the recognition rate of ORF reaches 87.86% when it iterates 10 times and it takes 325 seconds. In sum, the test-error of online handwritten character recognition of new Tai Lue can be classified for two reasons: (1) the reason of the new Tai Lue characters themselves, it is note that there are a lot of similar characters and such as "ᦓ" and "ᦒ", "ᦀ" and "ᦰ". These characters can not be distinguished easily. (2) The reason for handwriting. Some of the test sample handwriting deformation is particularly serious which is difficult to identify even by the human eyes.

6 Conclusion and Future Works

This paper mainly studies the online handwritten character recognition of new Tai Lue. First of all, the character samples are preprocessed by normalization, resampling and other methods, and then the six categories of typical features are extracted. In the classification stage, an on-line learning mode ORF is used, and experimental results show that the recognition rate of ORF is 87.86% at the completion of training after 10 times. The future work is focused on finding some valid features that can distinguish between similar characters. In addition, some other recognition algorithms could be applied to improve the recognition rate.

Acknowledgment. This work was supported by National Natural Science Foundation of China (Grant No. 61462094,61561050) and Applied Basic Research Project of Yunnan Province (Grant No. 2015FB116).

References

1. Tan, Y.T.: A study on the international standard of New Tai Lue characters code. In: The National Minority Youth Conference on Natural Language Processing (2008)
2. Li, D.F., Yu, P.F., Li, H.Y., Peng, G.: Printed New Tai Lue character recognition based on BP neural network. In: IEEE International Conference on Signal and Image Processing, pp. 339–342. (2016)
3. Zhu, X.Y., Shi, Y.F.: Research on handwritten character recognition. Pattern Recognit. Artif. Intell. **13**(2), 174–180 (2000)
4. Ren, H.Y., Yuan, B.S., Tian, Y.: On-line handwritten Uyghur character recognition based on BP neural network. Microelectron. Comput. **27**(8), 238–241 (2010)
5. Parui, S.K., Guin, K., Bhattacharya, U., Chaudhuri, B.: Online handwritten Bangla character recognition using HMM. In: International Conference on Pattern Recognition, pp. 1–4. (2008)
6. Ahmad, A.R., Khalia, M., Viard-Gaudin, C., Poisson, E.: Online handwriting recognition using support vector machine. In: Tencon IEEE Region 10 Conference, pp. 311–314. (2004)
7. Saffari, A., Leistner, C., Santner, J.: On-line random forests. In: Proceedings of 2009 IEEE 12th International Conference on Computer Vision Workshops, pp. 1393–1400. (2009)
8. Zhang, Y., Xie, S., Wei, S.: Industrial character recognition based on grid feature and wavelet moment. In: IEEE International Conference on Imaging Systems and Techniques, pp. 56–59. (2013)
9. Breiman, L.: Random forests. Mach. Learn. **45**(1), 5–32 (2001)

Web Information Transfer Between Android Client and Server

Baoqin Liu(✉)

Software and Information Engineering Institute, Beijing Information Technology College, Beijing, China
liubq@bitc.edu.cn

Abstract. To improve the transfer efficiencies when information is sent from Android to web server or vice versa, the information transmit method between them is studied. And to save information access time in Android, how to store the information in Android is proposed. How to send and extract the information between Android and web server is presented. JSON technology is put forward to achieve transfer information, the information storage way is analyzed and SQLite is used to store information in Android in this study.

Keywords: Extract · Android · Communication

1 Introduction

It is known to all that Android is a popular mobile devices operating system. It is first introduced by google corporation in 2007. It is based on Linux operating system and is open sources. As with all other operating system it is a bridge that connects the device hardware and the top level application. It can control the underlying hardware after the system start up. And it also manages top level software installed in the system. It is now used by more and more mobile device manufactures.

Because of the popularity of Android system, the application demand based on it is continues to increase. There are two difficulties when developing the applications, one is to transfer information from Android client to web server or vice versa, the other is information storage in Android. The solution to these two questions is presented in the research.

In this paper the menu application as a carrier is designed to explain how to solve the problems raised above. It is based on Android client and web server. The menu application can replace the original paper menu which demands more people and need more time to deal with. It can greatly increase working efficiency and decrease the manpower and material cost in the long run. So the menu application has very important practical application value. In the research the follow problems are solved: how the request is sent from Android to server, and how the server get the request information from the client and then how the menu data details are sent to Android, next how the menu information be extracted in Android, then followed how the information is displayed and stored in Android client.

F. Xhafa et al. (eds.), *Advances in Intelligent Systems and Interactive Applications*, Advances in Intelligent Systems and Computing 686,
https://doi.org/10.1007/978-3-319-69096-4_60

2 General Design of System

The C/S(Client/Server) architectures is chosen in the menu application. C/S architecture is one of the two software architectures, the other one is B/S(Browser/Server) architecture. Usually C/S architecture is built on special and small local area network, while B/S architecture is often used on wide area network. For the menu application is mainly used in local area network environment, so the C/S architecture is selected.

Obviously, there are two parts in C/S architecture: client and server. Software on android client need to be developed and will be installed on android client machine. Humanized interface should be presented, by which the customers can conveniently and rapidly look through the menu and choose foods they want, then finish their ordering. The waiter could send login and ordering information to the server, and accept and analyze information from the server.

On the other hand, the server application needs to be developed and installed in the server machine. In the research the server application is developed to support the client application. It can accept the login information sent by the android client, and then provides the authentication. It also deals with the other request from client such as opening a new table, renewing menu, ordering, and so on.

3 Server Part

The server application is developed and run on PC, Apache Tomcat is chosen as the web container. It is based on java, free and open sources, servlet and JSP can run on it. Java and JSP is used as developing language. MySQL is used as the database manage system. Local area network environment is needed.

Major modules are listed as below in Fig. 1:

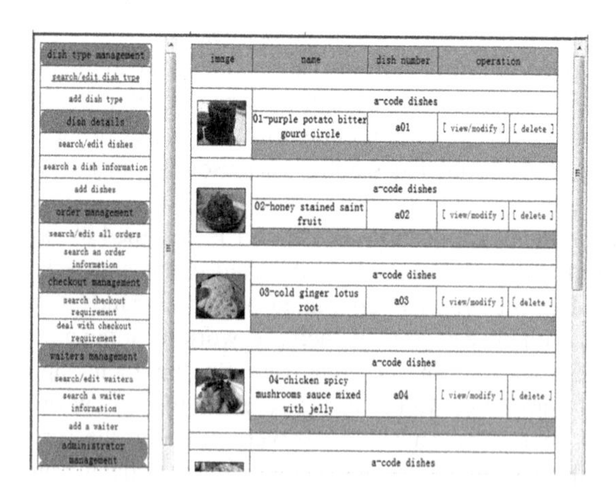

Fig. 1. Major modules at server

3.1 User Information Module

In the module user information can be managed. It can be searched, added, updated and deleted. The user information is stored in the corresponding tables of the database. There are two tables about user, one is about administrator and the other is about waiter. Administrators have the privilege to operate all the modules in server. The waiter information can also be operated.

When login request from client is received, the information will be compared with the waiter information table in the database. If authentication succeeds the customer at client can continue to order.

Java Bean is used to encapsulate the database access handling. The handling work includes connecting to the database, querying information, modifying information, deleting information and closing database.

Whether authentication is success or fail, the feedback message would be sent to the client.

3.2 Menu Management Module

Menu information is managed in the module, which includes menu information adding, modifying, querying and deleting. Menu information consists of foods types and foods details, which are stored in different tables in database.

Java Bean is used to encapsulate the database access handling. The handling work includes connecting to the database, querying information, modifying information, deleting information and closing database.

When the update request from the client is received it can send the menu information to the client.

3.3 Orders and Checking Out Module

Orders and checking out information are handled in this section. The order information can be viewed and printed. The checking request can be received and dealt with in real time.

4 Client Part

4.1 About Android

Android7.0 (i.e. Nougat) edition is used as the target developing environment. The programming language includes java and xml. Android SDK (software development kit) and Android Studio are used as development platform and tools.

Android SDK includes a variety of custom tools that help to develop mobile applications on the Android platform [1, 2]. It can be downloaded at the android official website. After downloading the software can be installed. But make sure that the JDK has been installed before because Android SDK adopted the java language. The installing method is just uncompressing the compression pack.

There are two common integrated development environment (IDE) in android development [3]. One is android studio, and the other is Eclipse. Because Google corporation no longer supports Eclipse, the android studio is used in the research.

Android Studio is released by Google corporation. The first edition is released in May 2013 [4]; the latest stable release edition is version 7.0. The android applications develop and debug can be implemented on it. It should be download firstly, then installed on the computer.

4.2 Login to the Server

When new customers come the waiter needs to login to the server and open a new table for the customers. The application provides an interface to input the waiter's login information and send it to the server. Then the feedback information will be received. If login success, the latest menu information will be received from the server and customer can start to order. otherwise the error prompt will be received.

As Fig. 2 shown, the waiter opens a new table and choose the table number, then the customers can begin to choose their foods.

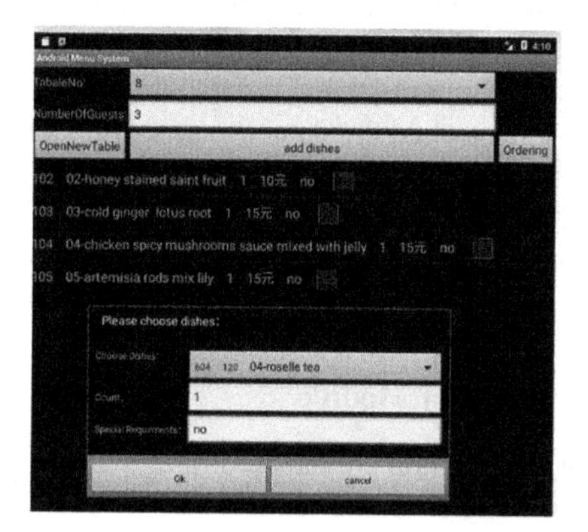

Fig. 2. Ordering interface at client

4.3 Ordering Section

When the 'add dishes' button is clicked, the dishes list is displayed according to different dish types such as cold dish, hot dish, staple food, snack and dessert. The dish can be chosen by clicking on it. Then the dish quantities can also be chosen, and special requirement can be written in the respondent text field. Finally, customer can click the 'ok' button then a dish is selected. Through repeating the above steps all the foods will be chosen.

When all the foods have been chosen, it needs to be confirmed by the customer again, then by clicking the 'ordering' button the order information would be sent to the server, and a feedback information can be received.

4.4 Pay the Bill

When customers want to pay the bill, the 'pay the bill' button can be clicked and the request will be sent to the server. The invoice information can also be written and sent to the server.

5 Problems to be Solved

As it is said at the beginning of this article, there are two problems will be solved:

The first one is when the information is sent between the two terminal, how to send and how to receive and extract on each terminal. During the client authenticates to the Web server, the android client must send the data in some way, and the data is accepted and analyzed in some other way in server, if authentication is successful, then the success information and the latest menu data will be sent to server.

The second one is how to raise the information access speed at Android terminal. If all the menu data is sent each time the authentication is successful, the access speed must be decreased. To solve the problem, the menu information can be stored in android client after first login. If the menu is updated, then use the update function at client to get the latest menu information. So how to store data in android client is also to be solved.

5.1 Data Transferred from Android to Web

There are two common ways to connect and interact data with server. One is using Socket, the other is using HttpURLConnection.

Socket is built at TCP/IP layer. It is one endpoint of a two-way communication link between two programs running on the network. It is bound to a port number so that the TCP layer can identify the application that data is destined to be sent to [5]. Once the connection is built, it is a forever connection. The server could send an initiative information to the client without request by the client.

HttpURLConnection is built at Application layer and based on HTTP. If a HTTP connection is set up, it is just once connection. That means the client sends a request, and the server accept the request and give a response, then the connection is released. If the client wants to send request to the server once more, the connection must be built once again.

In fact, HttpURLConnection is a relatively high-level connection to the Web and use sockets as part of the underlying implementation. Socket is low level than HttpURLConnection. HttpURLConnection can use a socket to connect with HTTP.

In this research the HttpURLConnection class is chosen because it is more proper and does not have to care the connection state and thread management.

It needs five steps to interact with the server, below are step details:

Step 1. Obtain a new HttpURLConnection object;
Step 2. set connection parameters;
Step 3. connect to server;
Step 4. send login information to server;
Step 5. read data from server.

5.2 Authentication

When the login information is received at server, the 'getInputStream' function of the 'HttpServletRequest' class is used to accept the authentication data, then the java bean object is used to access MySQL to authenticate. If the authentication proved to be correct, the latest menu information could be sent to the client.

As the menu data is relatively complex, the transfer way is to be considered. The relatively popular data exchange format JSON (i.e. JavaScript Object Notation) is chosen. It can easily encapsulate the data to be a regular form such as array, vector, list or sequence [6], and convenient to read and write.

To use JSON to send menu information is as below steps:

Firstly, create an instance of class JSONObject, for example named as 'obj1';

Secondly, use obj1 to call 'put' method to store one attribute's value of one kind of foods into the object, for example the food id, at the same time the name for the attribute need to be given. And so on, all attributes (such as name, type, image, price, description) of one food can be stored into the JSONObject instance obj1;

Thirdly, put obj1 into an array instance of 'JsonArray', for example named as 'ja', in which all the foods information will be stored.

Fourthly, repeat step two and three until all foods information been stored into 'ja'.

At last, the 'ja' object is sent to android terminal by the 'write' method in 'Printwriter' class.

5.3 Information Deal with on Android Client

The 'StreamTool' class is used on Android client to convert InputStream data type to byte array, then convert to String type, below is the core code:

byte[] info = StreamTool.read(in); String dishInfo = new String(info);
Then turns it to JSONArray type by construct JSONArray object.
JSONArray dishInfoArray = new JSONArray(dishInfo);
Then use getJSONObject method to achieve every json object which stores the menu details.
JSONObject dObj=array.getJSONObject(i); String na = dObj.getString('name');

Then all the foods information is extracted in the client.

But the above method cannot get the food picture, only the picture address is gotten. The food picture can be gotten by using the 'decodeFile' method of the class 'BitmapFactory', picture address gotten before is needed as the parameter of the method 'decodeFile'.

5.4 Improve Access Efficiency at Android Terminal

The access speed will be greatly decreased if the menu information is updated with each login at client, and it is also not necessary. The menu should be updated just as the server updates it. So the menu information should be stored at the android client.

There are two methods at Android client to store the menu information:

One method is to store it temporarily in memory and will disappear after the application quit, the data can be stored in an object array;

The other method is persistent data storage. Android provides five options to realize persistent data storage, includes Shared Preferences, Internal Storage, External Storage, SQLite Databases and Network Connection.

SQLite is chosen to store data in the Android client because it is small and convenient. There are many SQLite operation application interface in android system. By creating subclass of 'SQLiteOpenHelper' the databases can be created, the constructor must be overridden in which the super class constructor is called, and by overriding the onCreate and onUpgrade methods to specify the actions to create and/or update the database. The core steps are shown below:

Firstly, create a subclass of SQLiteOpenHelper; Secondly, get the database object;

Thirdly, create table and index by calling execSQL method; Finally, insert and query dish type data.

6 Conclusions

Information transfer between Android client and web server has been studied in the research, and the expected result has been achieved. The menu system is used as an example, a method of information transfer between web server and Android client is achieved. The data storage method on android client is presented. The application has passed the test in real machine, and runs well. The difficulties have been solved and the core codes have been given. The problems in the research are also problems lie in many other similar items, so the solution could also be used to those items. The research achievement is also helpful to other mobile applications. The follow up studies include optimizing parameters and algorithms, improve system efficiency.

References

1. 'Tools Overview'. Android developers. 10 Jan 2017
2. Tanapro GmbH, Tom Arn. 'JavaIDEdroid—Android Apps on Google Play'. google.com
3. 'Android software development'. https://en.wikipedia.org/wiki/Android_software_development
4. D.- Ducrohet, X., Norbye, T., Chou, K.: (May 15, 2013). 'Android studio: An IDE built for Android'. Android Developers Blog. Google. Retrieved 16 May 2013
5. 'What Is a Socket?' https://docs.oracle.com/javase/tutorial/networking/sockets
6. JsonOrg, 'Introducing JSON'. http://www.json.org/

Real-time Dynamic Data Analysis Model Based on Wearable Smartband

Xiangyu Li, Chixiang Wang, Haodi Wang, and Junqi Guo(✉)

Beijing Normal University, Beijing, China
guojunqi@bnu.edu.cn

Abstract. Since the traditional annual physical fitness test for adolescents in schools lacks of real-time dynamic data collection and deep analysis, we propose a data analysis model mapping from raw dynamic data to physical fitness evaluation, based on a self-developed wearable smartband by which we collect dynamic data from middle school students in Beijing. Firstly, the model presents a preprocessing algorithm which consists of a smoothness priors approach (SPA) and a median filter (MF), aiming for preprocessing of both photo plethysmo graphy (PPG) signals and three-axis acceleration data collected from the wearable smartband. Secondly, the model implements physiological and physical index estimation to acquire heart rate (HR), blood oxygen saturation ($\boldsymbol{SpO_2}$) and exercise amount estimates from the preprocessed data. Thirdly, the model extracts several key features closely related to physical fitness evaluation from the estimated HR, SpO_2 and acceleration data. Finally, a support vector machine (SVM) algorithm is employed for classification of physical fitness level of different smartband users. An application testing of our self-developed wearable smartband has been implemented in Tongzhou No. 6 middle school of Beijing. Experimental evaluation results demonstrate feasibility and effectiveness of both the smartband hardware/software and the proposed data analysis model.

Keywords: Dynamic data · Physical fitness evaluation · PPG signal · Smoothness prior approach · Median filter · Support vector machine

1 Introduction

Nowadays studies generally think that health-related physical fitness evaluation is an important part of physical education [1]. Traditional physical fitness test for adolescents, carrying out annually, cannot reflect young people's daily physical health status. Otherwise, its ability for data analysis and evaluation is also poor, lacking effective feedback and tracking mechanism. According to C.J.Caspersen's research, physical fitness is a set of attributes that are either health-related or skill-related and the degree of these attributes can be measured with specific test [2]. To improve the traditional physical fitness test, we consider wearable devices, which can monitor physical fitness condition, have set physiological information collection, storage, display and other functions in one. However, it is still basically stay in the stage of simple visualization of physiological data, but lack a model of analysis from raw data to physical health evaluation.

F. Xhafa et al. (eds.), *Advances in Intelligent Systems and Interactive Applications*, Advances in Intelligent Systems and Computing 686,
https://doi.org/10.1007/978-3-319-69096-4_61

After collecting real-time dynamic physiological data: real-time heart rate (HR), blood oxygen saturation (SpO_2)and three-axis acceleration through self-developed wearable smartband. The model proposed in this paper does preprocessing and analyzing, then extracts eigenvector and synthesized feature matrix with the static features obtained by traditional physical test. Finally, the result of the physical fitness evaluation is attained by Support Vector Machine (SVM) algorithm.

2 Existing Solutions

The dynamic data is a series of data ordered by time. It is available to extract feature index efficiently from the dynamic data to evaluate teenagers' health condition. Yang Z, Wang G. and others have done researches basing on the annual "dynamic" data of adolescents' health status by manual testing [3, 4]. However, this method can not reflect the real-time conditions reasonablely. So reseachers consider using wearable smartband to obtain real-time data. Photo Plethysmo Graphy (PPG), the most widely used method to extract dynamic data in wearable smartband, uses basic principle of photoelectric signal conversion, utilizing the character that the arterial blood has a different level of absorbing light from the muscles, bones, veins and other connective tissue, and extracting the AC signal from the light signal to reflect the trait of blood fluxion [5, 6].

The electro cardio signal (ECG) preprocess is divided into denoising and removing baseline drift. The former includes one-dimensional digital filtering, wavelet denoising, median filter denoising, and so on. Meanwhile, the most common way to deal with baseline drift is filter. H.S.Niranjana and others did researches on ECG denoising and proposed an optimized wavelet function [7]. B. Sharma and J. Suji analyzed the effects of different window sizes in ECG denoising [8]. As for evaluating the physical fitness, A.R. Lipu and others proposed a clustering algorithm using FPGA [9]. Y.Omard applied the double-hashing to sort [10]. Y. Dai, X. Yang utilized the fuzzy mathematics and established a sorting model applied to the evaluation of students' physical fitness [11].

3 A New Dynamic Data Analysis Model

Our model's main goal is establishing a mapping relation from raw physical data to physical fitness evaluation. This process contains five steps: raw data acquisition, data preprocessing, index estimation, extracting eigenvector and training SVM classifier (Fig. 1).

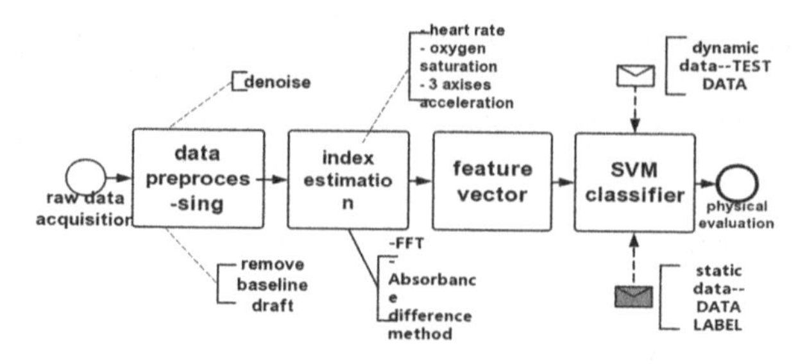

Fig. 1. Model framework

3.1 Data Acquisition

Basing on previous researches, a series of typical raw data are selected. This series includes static data from conventional physical fitness test: BMI, vital capacity, physical quality (strength, endurance, flexibility, speed, sensitivity) and dynamic data obtained from smartband: realtime PPG signal (HR, SpO2) and three-axis acceleration.

3.2 Data Preprocessing

3.2.1 SPA (Smoothness Priors Approach) Denoising

SPA [12] is effective as a non-linear approach to detrend. Raw PPG signal data contain periodic term and aperiodic term (trend term). SPA filters out the aperiodic term which has high frequency, and remains the periodic term as result.

$$z = z_{stat} + z_{trend} \tag{1}$$

where: z_{stat} is periodic term, z_{trend} is aperiodic term (trendterm).

We use single parameter to filter the target term. Calculate parameters by the optimal estimation method. In this way,we solve the equation below:

$$\lambda^2 \times \ddot{g} + g = f \tag{2}$$

If $f = \cos \omega t$, then $g = \frac{1}{(1+\lambda\omega^4)} \cos \omega t$. It is a typical lowpass filter for discrete time series. Tune the parameters λ to isolate terms and take out the high-frequency noise.

3.2.2 Median Filter

Intercept a data window from the raw data, reorder and replace the previous median with new median. In this way, we can filter the frequency component of baseline.

- Take k as midpoint, and define the data window which possesses N points. If N is even, the window is: $\left[k - \frac{N}{2}, k + \frac{N}{2} - 1\right]$, else the window is: $\left[k - \frac{N-1}{2}, k + \frac{N-1}{2}\right]$
- For every k, reorder the data in the window and replace x with the new midpoint.
- Tune the size N of the window till the frequency component of baseline is elected.

Get rid of it and then the PPG signal data become horizontal.

3.3 Index Estimation

3.3.1 Heart Rate Estimation

Discrete Fourier Transform (DFT) converts time-domain to frequency-domain so as to acquire the signal period. Heart rate can be attained from the signal period. Discrete Fast Fourier transform (FFT) is a fast algorithm of DFT, which uses low point' DFT to count high point' DFT repeatedly. In other words, FFT decomposes Fourier transform to a smaller one iteratively in order to reduce computing complexity. $F = \frac{\max|\mathrm{fft}(x)| * fs * \frac{60s}{min}}{N}$, N is the length of the data, fs is sampling frequency.

3.3.2 SpO_2

In human blood deoxyhemoglobin and oxyhemoglobin have different unique absorption spectrum in red and near infrared regions. Arteriopalmus can cause change of blood flow, so that the absorption spectrum will change meanwhile. On the other hand, non-blood tissue's absorption spectrum is constant, such as skin, muscle, skeleton. So detect the change of absorption spectrum caused by fluctuation in blood volume, eliminate effects of non-blood tissue, and then calculate SpO_2 [13]:

$$SpO_2 = A\frac{I_{AC}^{\lambda_1}/I_{DC}^{\lambda_1}}{I_{AC}^{\lambda_2}/I_{DC}^{\lambda_2}} - B,\ \mathrm{A} = \frac{\varepsilon_{Hb}^{\lambda 2}}{\varepsilon_{HbO_2}^{\lambda 1} - \varepsilon_{Hb}^{\lambda 1}},\ \mathrm{B} = \frac{\varepsilon_{Hb}^{\lambda 1}}{\varepsilon_{HbO_2}^{\lambda 2} - \varepsilon_{Hb}^{\lambda 2}} \tag{3}$$

light λ_i's absorptivity: $\varepsilon_{HbO_2}^{\lambda i}, \varepsilon_{Hb}^{\lambda i}$, intensity's DC and AC component is $I_{AC}^{\lambda_i}, I_{DC}^{\lambda_i}$

3.3.3 Three-axis Acceleration Estimation

- Speed estimate

i. Known three-axis acceleration x, y, z, $A = \sqrt{x^2 + y^2 + z^2}$, Compute A every 0.19 s, recognize pace by $A_{max} - A_{min}$. When the difference $\leq 0.13g$ and $\geq 0.07g$ ($g = 9.8\mathrm{ms}^{-2}$), record it as a single step.

ii. The distance is the sum of stride, which is computed by Based Stride Length (BSL)

$$BSL = Height \times GenderFactor \times 1.1 \tag{4}$$

$$Stride = BSL \times StepRateFactor \tag{5}$$

where: $GenderFactor$ is 0.415(male)/0.413(female). $StepRateFactor$ is based on speed.

iii. Define time window: $Speed = \frac{distance}{time}$, time is the length of the window.

- Energy expended per step

$$\text{Calories} = \frac{\mathit{MetabolicFactor} \times 0.00029}{\mathit{StepRate}} \times \mathit{Weight} \tag{6}$$

where: *weight* is measured in kilograms; MetabolicFactor depends on speed.

3.4 The Feature Extraction

After data preprocessing and estimation, attain HR, SpO_2 and three-axis acceleration ordered by time, which are still original and their features are not obvious. So extract features from them and build eigenvector to make evaluations on the physical conditions.

3.4.1 Define Eigenvector

A: Resting Heart Rate (**RHR**), Heart Rate Reserve (**HRR**) [14], Recovery Heart Rate (**RHR**) [15], Immediate Heart Rate after Exercise (**IHRE**)
B: Resting Blood Oxygen Saturation (**R-SpO_2**) [16], Falling Time of SpO_2<high-intensity exercise>(**FT- SpO_2**) [16]
C: Average Speed (**AS**), Maximum Speed (**MS**), Energy Expended Per Step (**EEPS**)

The eigenvector is consist of three parts: $S = \{A, B, C\}$. A is related to HR, B is related to SpO_2, C is related to three-axis acceleration.

3.4.2 Experimental Paradigm

Wearing the activated smartband, fitstly subjects sit for 3 minutes on the playground. Then they are requested to run a 2000-meter dash. After running, all subjects sit and rest for at least 2 minutes and then trun off the smartband. During the test, subjects are not allowed to drink and stop.

3.5 Classification

3.5.1 Classification Model

We regard the core from static data as data label for every sample, and use it to train our classifier. That is, input the eigenvector defined above, adjust the parameter till the output closes to its data label. Repeat this step to train the classifier. There will be 4 categories after classifying, standing for excellent, above-average, medium-low and bad.

3.5.2 SVM

It is a typical multi-classification problem, as we should give each sample a evaluation result. Because conventional SVM can only apply to binary classification problems, we choose one-against-one method to train the classifier. That is, build $k(k-1)/2$ binary classifier to solve k classification problem. Furthermore, we use C-SVC algorithm and radial basis function (RBF) to realize Multilinear map [17].

Further discussion—*Adding static data to the model*

Above, we mainly discuss dynamic data attained from smartband. Static data from conventional physical fitness test only provides data label and does not take part in the model. More valid index will improve the accuracy of out classifier. So consider to extract features indexes from static data, add the dimension of eigenvector and train the new SVM classifier. The new eigenvector is $S' = \{S, S_1\}^T$ S_1 is extracted from static data.

4 Experimental Evaluation

We conducted extensive experimental studies to verify the validity of this model, using MATLAB and MySQL. 302 students took part in the test who had done traditional physical fitness test recently. They weared smartband and tested in accordance with the experimental paradigm determined in advance. The sampling frequency of PPG signal and three-axis acceleration was 25 times/sec.

4.1 Data Preprocessing and Index Estimation

In the data preprocessing, frequently use Root-Mean-Square Error (RMSE), Signal-Noise Ratio (SNR) to assess preprocessing's effect. Larger RMSE means more effective preprocessing. Smaller SNR means more effective preprocessing. Meanwhile, in order to attain real-time physical fitness evaluation, we require superior time requirement.

- RMSE is the square roots of the variance between raw data and the preprocessing result. $\mathrm{RMSE} = \left\{\frac{[f(n)-f_1(n)]^2}{n}\right\}^{\frac{1}{2}}$, $f(n)$ is raw signal data, $f_1(n)$ is preprocessing result.
- SNR is also a traditional approach: $\mathrm{SNR} = 10\log_{10}\left(\frac{p_s}{p_z}\right)$, $p_s = \frac{[\sum_n f^2(n)]}{n}$ is the power of raw data, $p_z = RMSE^2$ is the power of noise.
- Compare our solution with wavelet transform, median filter and butterworth low pass filter. Then record the average RMSE, SNR and TIME of each algorithm in Table 1. Shown in which, our solution is better in RMSE and SNR, especially in TIME.

Table.1 Comparation of preprocessing

Denoising	SPA+ MF	Wavelet theory	Median filter	Butterworth
RMSE	4.79e + 05	1.17e + 06	8.81e + 05	7.99e + 05
SNR	112.0021	94.2014	99.8440	103.2775
TIME	0.04603	0.34622	0.12426	0.24355

To test accuracy of the model's index estimation, we compared HR and SpO_2 measured by professional medical equipment between our model based on the test. The accuracy reached 90%, which was enough for our evaluation.

4.2 Classification Evaluation

All samples were divided into two parts, 60% of samples were set as training set and 40% were set as test set. There were 302 subjects in the sample data set and sample's dimension was 9. All samples were divided into 4 categories. Compared the data label and classification result, accuracy in training set reached 93%, and in test set accuracy reached 90%.

Table 2 indicates the result of classification, all points are divided into 4 parts, which matches the 4 categories. Points in grey means the sample's data label, and points in blue means the sample's result of classification. Based on the Fig. 2, the clear majority of points is coincident (in grey), which means the result is up to standard. Some dominant blue points stand for the sample whose result is not accord with the data label.

Table.2 Result of experiment

Category	Category 1 (%)	Category 2 (%)	Category 3 (%)	Category 4 (%)
Training set-data label	34	39	20	7
Training set-result	34	42	17	7
Test set-data label	34	37	19	10
Test set-result	34	40	18	8

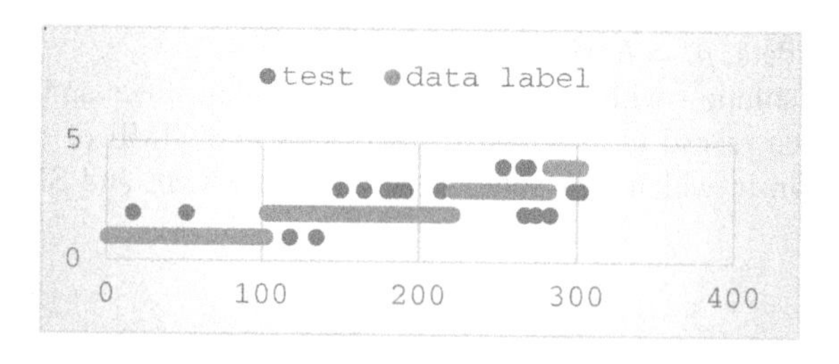

Fig. 2. Accuracy of experiment

5 Conclusion

A model is proposed to evaluate physical fitness in this paper. Based on the designed experimental paradigm, we obtain real-time dynamic data from smartband. Preprocess data using SPA to de-noises and MF to remove baseline drift. Then FFT is adopted to obtain the real-time sequence of HR. Make use of the difference of DC component between AC component of PPG signal to obtain the real-time sequence of SpO_2. Through three-axis acceleration we can obtain subjects' speed, etc. Based on the preprocessed dynamic data, we extract eigenvector and train SVM with known data label from traditional physical fitness test. Finally, we classify the eigenvector by trained SVM. Our work established a complete model from raw dynamic phyciological data to physical fitness evaluation. This model remedys the traditional physical fitness test's weakness and takes advantage of wearable smartband whose data is real-time. The actual test we conducted validate the usability of the model.

Acknowledgments. This research is sponsored by National Natural Science Foundation of China (No.61401029), and Beijing Advanced Innovation Center for Future Education (BJAICFE2016IR-004)

References

1. Corbin, C.B., Welk, G.J., Richardson, C., Vowell, C.: Youth physical fitness: ten key concepts. J. Phys. Educ., Recreat. Dance, 24–31 (30 Jan 2014)
2. Caspersen, C.J., Powell, K.E., Christenson, G.M.: Physical activity, exercise, and physical fitness: definitions and distinctions for health-related research. Public Health Rep. **100**(2), 126–131 (Mar–Apr 1985)
3. Zhang, Y., He, L.: Dynamic analysis of the physical health of Chinese adolescents—Four national health monitoring data in 2014. Beijing University of Aeronautics and Astronautics
4. Wang, G.: Students Constitution of He Nan Province in 1995-2010 to study the dynamic characteristics analysis and trends of research data. Zhengzhou University
5. Heerlein, J., Rueheimer, T.: LED Based Sensors for Fitness Tracking Wearables (2014)
6. Application document "SFH 7050—Photoplethysmography Sensor". Osram Opto Semiconductors (2014)
7. Niranjana Murthy, H.S., Meenakshi, M.: Optimum choice of wavelet function and thresholding rule for ECG signal denoising. In: 2015 International Conference on Smart Sensors and Systems (IC-SSS)
8. Sharma, B., Suji, J.: Analysis of various window techniques used for denoising ECG signal. In: 2016 Symposium on Colossal Data Analysis and Network (CDAN)
9. Lipu, A.R., Amin, R., et al.: Exploiting parallelism for faster implementation of Bubble sort algorithm using FPGA. Telecommunication Engineering (ICECTE)
10. Omar, Y.M.K., Osama, H., Badr, A.: Double hashing sort algorithm. Comput. Sci. Eng. (2017)
11. Dai, Y., Xu, Y.: The Fuzzy Clustering Analyzes in the Forecasting of College. College of Science, Liaoning Technical University
12. Decomposition forecasting of displacement of landslide based on SPA (Smoothness Priors Approach)
13. Lv, C.: Study on Non-invasive Pulse Monitor of Oxygen Saturation of Blood. Liaoning Technical University
14. http://www.shapesense.com/fitness-exercise/calculators/heart-rate-reserve-calculator.aspx
15. http://www.med-health.net/Recovery-Heart-Rate.html
16. https://en.wikipedia.org/wiki/Oxygen_saturation
17. Ma, Y., Guo, G.: Support Vector Machine Application. ISBN: 978-3-319-02299-4. Springer International Publishing

Investigation and Analysis on the Influencing Factors of Consumers' Trust to Fresh Agricultural Products in E-commerce

Yipeng Li and Yong Zhang(✉)

Zhongnan University of Economics and Law, Wuhan 430000, China
xian1996@163.com

Abstract. In the "Internet +" times, the E-commerce has rapidly developed. But it still remains a problem whether consumers who are accustomed to the traditional face-to-face shopping way are willing to cooperate with this rhythm to accept the brand new E-commerce consumption pattern. So it is critical to improve consumers' trust towards the E-commerce patterns and let consumers have the courage to try. This paper will take fresh produce as the breakthrough point and from two aspects of product characteristics and quality of service to analyze the influence of the quality of fresh agricultural products, the value of products, the quality of logistics services, the quality of interface design and the quality of after-sales service on customers' trust. And we carry out practical investigation and then use statistical software to analyze the results of the valid survey and reach a conclusion. These conclusions show that these five factors all have an impact on consumers' trust. The research results of this paper have guide meaning to the troubles how enterprises of E-commerce of fresh agricultural products improve trust in the minds of consumers and develop better.

Keywords: Internet + agriculture · Consumers' trust · Data analysis · Electronic commerce

1 Introduction

In the E-commerce environment, the transaction between consumers and businesses rely mainly on the trust of both sides to maintain because there is no formal contract. Although the E-commerce brings many conveniences to consumers, the network environment still exists uncertainty and risks, which makes the problem of consumers' trust more prominent. For the E-commerce enterprises of selling fresh agricultural products, it is particularly important for the entire enterprises' market expansion to build consumers' trust so that consumers and enterprises can maintain a stable relationship between the transaction.

In this paper, we predict a situation that will lead to the change of consumers' trust to E-commerce's agricultural products. Besides, this situation may be multi-factor. Firstly, we determine the product's characteristics and quality of service will affect the consumers' trust and then select five possible influencing factors. We did a practical survey (use the form of a questionnaire) to test the differences of impact of the five factors on consumers' trust. These five factors represent respectively the product's

F. Xhafa et al. (eds.), *Advances in Intelligent Systems and Interactive Applications*, Advances in Intelligent Systems and Computing 686,
https://doi.org/10.1007/978-3-319-69096-4_62

characteristics and quality of service. Finally, we use statistical software to analyze the results of the practical survey and reach a conclusion. These conclusions show that these five factors all have an impact on consumers' trust, but some factors have significant impact and some factors have little impact. It is obvious that more research is needed to determine the extent of these factors' enhancements and how best to combine them.

2 Theory

2.1 Overseas and Domestic Research Status

The author He Dehua published "The study on the consumer's willingness to buy fresh agricultural products through E-commerce", this paper proposed to build influencing factor model of China's consumer's purchasing intention to fresh agricultural products in E-commerce. In the foreign countries, Peter Kerkhof[1] published a paper, drawing the conclusion by the comparing experiment and based on the general principles, the conclusion of the study has certain theoretical guidance and application value to increase consumers' trust about the E-commerce enterprises of selling fresh agricultural products, enhance the rapid and healthy development of E-commerce.

The results of research at home and abroad are very efficient. However, the theories of domestic and foreign scholars still need to be further revised and perfected. For example, the existing researches focus mainly on the consumers' trust of pure E-commerce enterprises and ignore the research on consumers' trust of E-commerce related to entity. Different industries determine the different focus of their consumers' trust in E-commerce. E-commerce's consumers' trust research should pay more attention to the specific industries.

Based on these theories, we find that product characteristics and quality of service affect consumers' trust indeed. Therefore, we concentrate on investigating several influencing factors based on the two aspects.

2.2 Generate Hypotheses

Consumers' trust towards E-commerce of selling fresh agricultural products refers to the consumers' recognition towards enterprises' sincerity, ability and kindness. In E-commerce's environment, consumers rely mainly on the trust towards businesses to determine the purchasing intention. Therefore, the study of what factors will affect consumers of fresh electricity businessmen's trust, we can be targeted to develop the specific strategies to promote the development of E-commerce's enterprises.

After completing the preliminary investigation and consulting previous researches, this paper proposes the following hypotheses:

[1] UTZ S, KERKHOF P, VAN DEN BOS J. Consumers Rule: How Consumer Reviews Influence Perceived Trustworthiness of Online Stores.

H1: The high quality of fresh agricultural products will have a positive impact on the consumers' trust in E-commerce.
H2: The high value of goods will have a positive impact on the consumers' trust in E-commerce.
H3: The high quality of logistics services will have a positive impact on the consumers' trust in E-commerce.
H4: The high quality of interface design will have a positive impact on the consumers' trust in E-commerce.
H5: The high quality of after-sales service will have a positive impact on the consumers' trust in E-commerce.

Based on the five basic hypotheses, this paper constructs the model of influencing factors of consumers' trust in E-commerce of fresh agricultural products, as shown in Fig. 1.

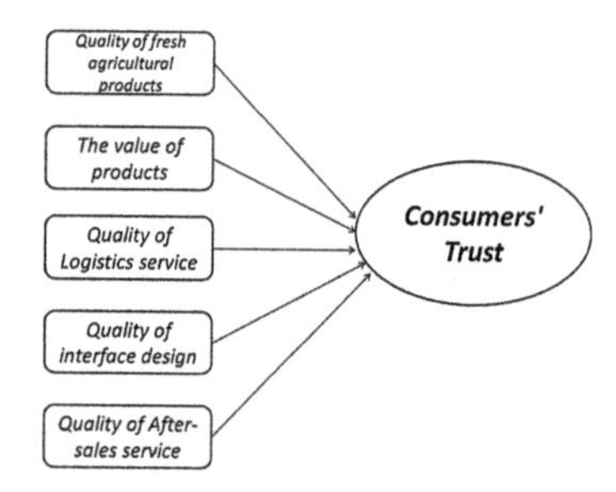

Fig. 1. The model of influencing factors of consumers' trust

3 Method

3.1 Survey Methodology

We conducted a controlled survey experiment with the five impact factors as the within-subject factors. The dependent variables representing customer's trust were whether you would buy it again on the Internet. Research uses the following method:

(1) *Exploratory factor analysis*

Exploratory factor analysis (EFA) is a statistical method used to find the essential structure of multiple observed variables. Each variable has multiple measurement indicators. And there is a certain correlation between the indexes of different variables, there are some differences between the indexes of different variables. So EFA is used to analyze the difference validity and convergence validity of multiple indexes of variables.

(2) *Cross-analysis*

Cross-analysis method, also known as the three-dimensional analysis, is in the vertical analysis and horizontal analysis on the basis of the cross. It is a method of analysis from three-dimensional point of view, shallow to deep, low to advance. We regard measure content as a line variable, consumers' trust (presented as whether to re-purchase the products in the questionnaire) as a column variable.

3.2 Experimental Procedure

This paper uses a questionnaire survey to validate the research hypotheses. In this paper, the quality of fresh agricultural products, the value of goods, logistics service quality, interface design quality and after-sales service quality, these five variables are used to be the independent variables of influencing the consumers' trust. Each variable is measured by two to three measure contents. Then, 30 consumers were selected to conduct the pre-test. According to their feedback, we finally formed a formal scale. The specific contents of the scale are shown in Table 1.

Table 1. The list of variables

Variables	Number	Measure content
Quality of fresh agricultural products	1.1	bright color without borers
	1.2	right size without germination
	1.3	Good quality and good taste
Value of products	2.1	Favorable price
	2.2	Value for the price
Quality of logistics service	3.1	Fast logistics speed
	3.2	Safe logistics service
	3.3	No loss situation
Quality of interface design	4.1	Beautiful interface
	4.2	Neat interface
	4.3	Good interaction between human and interface
Quality of after-sales service	5.1	Accurate and timely after-sales service
	5.2	Good atttitude
	5.3	24 hours online

The investigation object of this paper is the customers that had the experience of purchasing fresh agricultural products online. The survey received a total of 300 valid questionnaires, including 120 males, 40% and 180 females, accounting for 60% of the total. The proportion of people aged 18–24 is the highest in terms of age structure. Followed by people aged 24–30 years and over, and below 18 years of age is the smallest proportion of people, indicating that the survey mostly focuses on 18–24 years old between the crowd.

4 Results

Table 2 reports the number and percentage of subjects selected as "very important" among the measurement items. Compared with Table 1, we can see from Table 2 that the factors that e-commerce consumers consider "very important" are "quality of fresh agricultural products" (57.3%, average ratio = number of selections divided by total number), followed by "the value of products" (38.3%) and "quality of the logistics service" (37.3%), followed by "quality of the after-sales service" (18.6%), compared with "quality of interface design" (only 7.3 %).

Table 2. The initial result of the questionnaire

Measure content	Total	Rate(%)
Bright color without borers	150	50
Right size without germination	156	52
Good quality and good taste	210	70
Favorable price	72	24
Value for the price	156	52
Fast logistics speed	102	34
Safe logistics service	168	56
No loss situation	66	22
Beautiful interface	24	8
Neat interface	36	12
Good interaction between human and interface	6	2
Accurate and timely after-sales service	96	32
Good atttitude	66	22
24 hours online	6	2
Population of valid entries	300	

The data from Fig. 2 were analyzed by the method of cross analysis, which was based on the content of variables, and the degree of trust (whether or not it would appear again) as a column variable It displays the effect of this option on the consumer's purchase behavior, whether they will trust the E-commerce enterprises to buy again when the previous item is selected as the "Very Important" option. First of all, it can be seen from Table 3 that consumers who think "interaction between human and interface is good" is very important will choose to buy again (reach 100%). What is worthy of our attention is that the ratio of consumers who think "the value of products" is very important choosing to buy again (reach 63.5%) is relatively high. 50% of consumers who think "quality of fresh agricultural products" is very important will choose to buy again and similarly, 49.6% of consumers who think "quality of logistic service" is very important will choose to buy again.

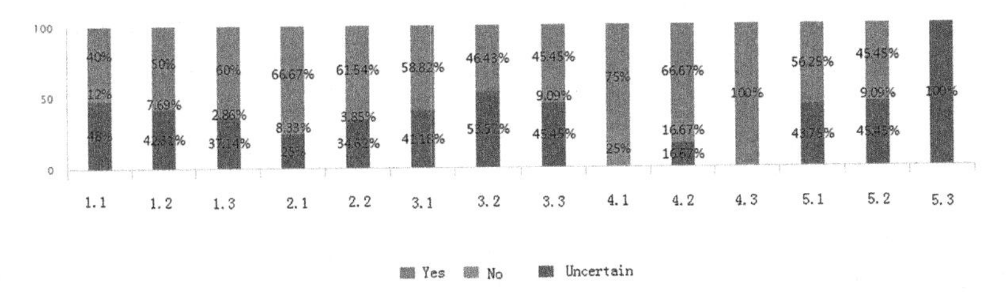

Fig. 2. The data analyzed by the method of cross analysis

5 Discussion

Through the questionnaire survey and statistical analysis, we can see that consumers are mostly concerned about "quality of fresh agricultural products", followed by "the value of products" and "quality of logistics services", and then the "quality of after-sales service", the least concern is "quality of interface design". In addition, the results showed that these five variables all positively influenced consumers' trust. Among them, the value of products is the most prominent (H1), followed by quality of fresh agricultural products (H2) and quality of logistics service (H3). Combining the theoretical analysis and practical test results, we can draw the following suggestions:

Firstly, it is important to increase the value of fresh agricultural products so that the usefulness of the commodity can be also enhanced. Secondly, it ensures that customers can buy fresh and clean fresh agricultural products by strict checks. Thirdly, the further improvement of logistics system is to ensure that goods quickly and efficiently reach the destination. Fourth, the e-commerce seller's website/APP page's design should be sophisticated, as friendly and convenient as possible. Fifth, after-sales service itself is also a promotional tool, and it plays a very important role in the consumer shopping process. We can improve customer's service of delivery procedures, initiatively offer logistics' details and improve the work efficiency of the e-shop to enhance the quality of after-sales service.

Acknowledgments. This research was financially supported by the National Natural Science Foundation of China [No.71401180].

References

1. Li, Y.M., Yeh, Y.S.: Increasing trust in mobile commerce through design aesthetics. Comput. Hum. Behav. **26**(4), 673–684 (2010)
2. Utz, S., Kerkhof, P., Van den Bos, J.: Consumers rule: how consumer reviews influence perceived trustworthiness of online stores. Electron. Commer. Res. Appl. **11**(1), 49–58 (2011)
3. Ogonowski, A., Montandon, A., Botha, E., et al.: Should new online stores invest in social presence elements? The effect of social presence on initial trust formation. J. Retail. Consum. Serv. **21**(4), 482–491 (2014)
4. Hong, I.B., Cha, H.S.: The mediating role of consumer trust in an online merchant in predicting purchase intention. Int. J. Inform. Manag. **33**(6), 927–939 (2013)

5. Li, H., Jiang, J., Wu, M.: The effects of trust assurances on consumers' initial online trust, a two stage decision-making process perspective. Int. J. Inform. Manag. **34**(3), 395–405 (2014)
6. Karimov, F., Brengman, M.: An examination of trust assurances adopted by top internet retailers, unveiling some critical determinants. Electron. Commer. Res. **14**(1), 1–38 (2014)
7. Kareem Abdul, W., Gaur, S.S., Penaloza, L.N.: The determinants of customer trust in buyer seller relationships, an empirical investigation in rural India. Aust. Mark. J. **20**(4), 303–313 (2012)
8. Harris, L.C., Goode, M.M.H.: The four levels of loyalty and the pivotal role of trust, a study of online service dynamics. J. Retail. **80**(2), 139–158 (2004)

A PROUD Methodology for TOGAF Business Architecture Modeling

Feng Ni(✉), Fang Dai, Michael J. Ryoba, Shaojian Qu, and Ying Ji

Business School, University of Shanghai for Science and Technology, Shanghai, China
nifeng@usst.edu.cn

Abstract. Concerning the business architecture as the center of the enterprise architecture development, the TOGAF business architecture deserve to be modeled with a simple and definite model suit for the subsequent modeling phases to follow a clear requirement. We propose a modeling method named PROUD based on tailoring & mapping of the content meta-model of TOGAF business architecture. Furthermore, a 2-dimensional iterative modeling matrix is defined to ensure the derivation of the PROUD model instances from meta-models in gradient granularity. A simplified use case of gift exchange in Disneyland with MagicBand is illustrated for PROUD modeling example.

Keywords: TOGAF · Meta-model · Business architecture · BPMN · MagicBand

1 Introduction

John A. Zachman, an IBM architect, put forward the Zackman framework as the first theoretical framework for enterprise architecture in his article titled *A Framework for Information Systems Architecture* (1987) [1]. Since then, the academic and industry community has taken effort and made progress to find practical methodologies for design and evolution of complex interoperated systems in the past three decades [2]. By applying Model Driven System Engineering (MDSE) theory to series of system design methodologies, organizations from military to industry all over the world have released dozens of versions of different enterprise architecture frameworks and standards. Most of them limit their focus to information system architecture level or divide the business and system layers separately without the consideration of the entire system development and evolution strategy.

Enterprises need to be able to rapidly adapt themselves to changes in their environment to seize opportunities and improve competitiveness. It is important to align the business process and information system architecture closely for the realization of enterprise strategy. The Open Group Architecture Framework (TOGAF) is proposed by The Open Group and revised in 2011 [3], which is primarily concerned with bridging the gap between strategy and implementation. TOGAF became one of the leading architecture frameworks worldwide where over 73% of international corporations over the world are currently using (or have used) TOGAF [2]. As the most important part of

F. Xhafa et al. (eds.), *Advances in Intelligent Systems and Interactive Applications*, Advances in Intelligent Systems and Computing 686,
https://doi.org/10.1007/978-3-319-69096-4_63

TOGAF, the Architecture Development Method (ADM) defines the enterprise architecture design as an iterative development circulation from business architecture to information architecture and specifies the inputs/outputs for every phase [3]. With respect to ADM, the capability of information system is designed to support the core business process requirements. Reversely, business architecture as the most important phase of enterprise architecture development circle should be the emphasis to improve productivity in the marketplace.

Although ADM provides 9 suggested steps and 18 alternative deliverables (inputs/outputs) for the business architecture phase [3], it fails to describe the development and utilization of these deliverables. Besides, there is a lack of effective methods, tools and typical cases to ensure the operability, practicability and feasibility of development of business architecture in current research [4].

Therefore, our research focus on practical modeling methodology of TOGAF business architecture. We propose a set of meta-model-based descriptive model combinations named PROUD (Process, Rule, Organization Unit and Data) to deliver a tailored enterprise architecture business viewpoint description framework and the corresponding modeling tools. Consequently, the TOGAF business architecture can be expressed in a concise form based on a combination of these four models, with adequate information at the business level stakeholder's concern.

2 Business Architecture of TOGAF

The TOGAF is divided into four architectural viewpoints in the ADM which depicts business processes, information services, application integration and technical standards in different phase. (1) Business Architecture: describes the business strategy, organizational structure, business activities, business processes, business rules, and information flows related. (2) Data Architecture: describes the details of an organization's logical and physical data assets and data resources management. (3) Application Architecture: provides a blueprint for the individual application systems to be deployed and their relationships to the core business processes. (4) Technology Architecture: describes the logical software and hardware capabilities that are required to support the deployment of business, data, and application services. Our research focus on the Business Architecture, which serve as the base of the subsequent three architecture modeling phases.

The corresponding entities and deliverables of each domains of architecture are described by the TOGAF architecture content framework (ACF) [3], which provides a conceptual meta-model for describing architectural artefacts. A meta-model is generally defined as a "model of models". Meta-models can help a modeler to ensure that a set of models are consistent. ACF meta-model is the soul of TOGAF that serves as a "glue" not only between each domain of architectures but also within them.

Given the importance and adoption of TOGAF content meta-model for enterprise architecture, the ACF particularly plays a crucial role in the business architecture development. A typical content meta-model of business architecture is illustrated in Fig. 1, which is a tailored version of the business involved part of ACF. The business architecture content meta-model contains six core content entities (Organization Unit,

Actor, Role, Business Service, and Process) and two extension entities (Event, Contract) of business architecture, and an entity (Data Entity) from the data architecture domain.

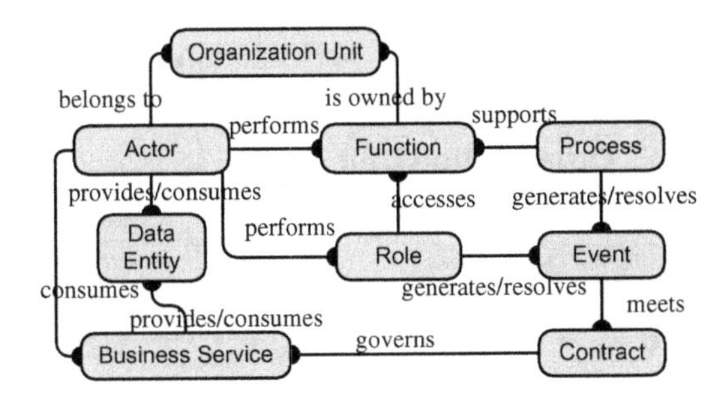

Fig. 1. Business architecture content meta-model derived from ACF

The business architecture content meta-model indicate what are supposed to be modeled in the business architecture model suit. Then how?

3 PROUD Methodology

A model is said to conform to its meta-model when all semantic elements in the model maps to corresponding entities in the meta-model [5]. So there should be a many-to-many mapping relationship between entities (with relationships) in a meta-model to corresponding elements (with relationships) in the model suit. Given the definition that model suit conform the meta-model when all of the meta-model entities be covered by the mapping, we propose a business architecture modeling methodology named PROUD (Process, Rule, Organization Unit and Data). PROUD divide the business architecture into a suit of four models separated and related to each other by tailoring the content meta-model into four fragments partially overlapped and non-contained to each other, as illustrated in Fig. 2.

(1) **Process Model**: The semantic domain of business process model covers six entities (with relationships) in the business architecture content meta-model, as shown in Fig. 2a. The business process model describes the tasks sequence performed by business participants to meet specific business functions, including the message/data interaction between and within each participant. Business Process Modeling Notation (BPMN) [6] is adopted to model the business process, which has four basic categories of graphical elements to build diagrams: flow objects, connecting objects, pools/lanes and artifacts.

(2) **Rule Model**: The semantic domain of business rule model covers three entities (with relationships) in the business architecture content meta-model, as shown in Fig. 2b. Rule model describe the infrastructure of the business process in details. Given the context of enabled tasks, stimulated events and received messages, the

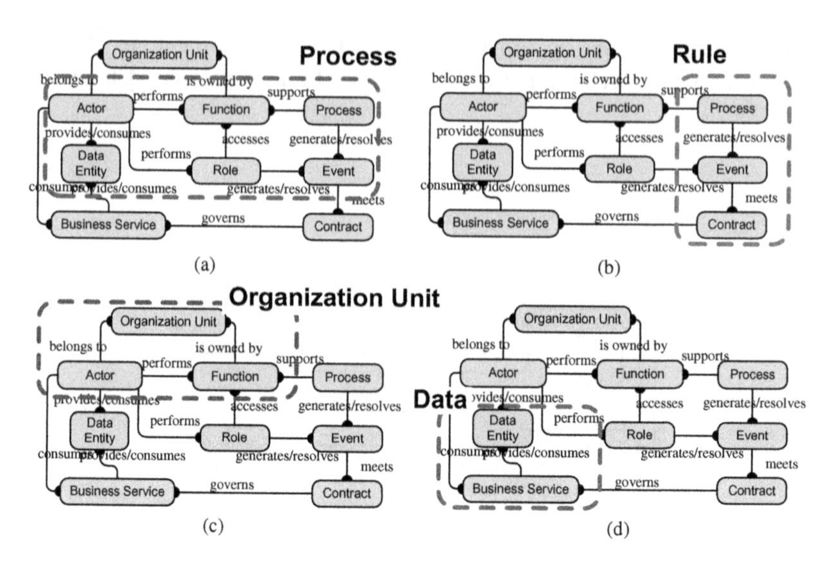

Fig. 2. PROUD methodology with a 4-models suit

corresponding business rule define the triggering condition and results. We model business rules with rule logic control statements in pseudo-code such as IF-THEN-ELSE and CASE.

(3) **Organization Unit Model**: The semantic domain of business organization unit model covers three entities (with relationships) in the business architecture content meta-model, as shown in Fig. 2c. Organization unit model describes the interoperation between participants of business process and performers of business functions. Including both information and logical association relationships. We adopt UML collaboration diagram to model business organization units related to each other.

(4) **Data Model:** The semantic domain of business data model covers two entities (with relationships) in the business architecture content meta-model, as shown in Fig. 2d. In business interoperation, information produced/consumed by business service participants is carried by both messages between participants and data objects within participants. The business data model defines the logical data structure of messages and data objects, including the entity class, attributes and relationships. IDEF1x is adopted to model the business data in entity relationship diagram.

The PROUD methodology adopts four interrelated, partially-overlapped and non-redundant descriptive models in TOGAF business architecture modelling on the basis of the corresponding tailored parts of the content meta-model, with adequate information in business domain covered. The development of PROUD model suit follows an iterative modeling matrix in 2 dimensions: the modeling domain dimension (horizontal) and modeling granularity dimension (vertical), as shown in Fig. 3. Typical PROUD modeling process follow two directions: (1) Horizontal modeling separately from process, rule, organization, and data domains of the business architecture. (2) Vertical generalization from the meta-model to the model instance and granularity refinement from level 1 to level m.

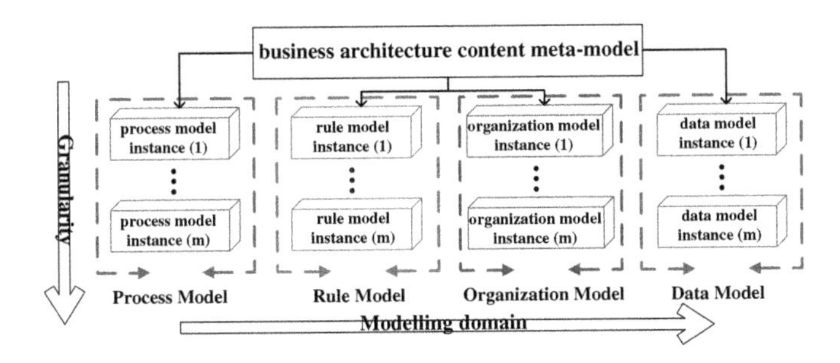

Fig. 3. PROUD 2-dimensional modeling matrix

4 Example Application

This section presents an example application of the PROUD method described in Sect. 3. The proposed use case deals with a typical business process for credit & gift exchange in Disneyland with its popular MagicBand. For briefness, the modeling granularity is frozen and we follow the horizontal modeling dimension from model to model. The process model for gift exchange case in BPMN is shown in Fig. 4. The organizational model and data model (in IDEF1x) is shown in Fig. 5.

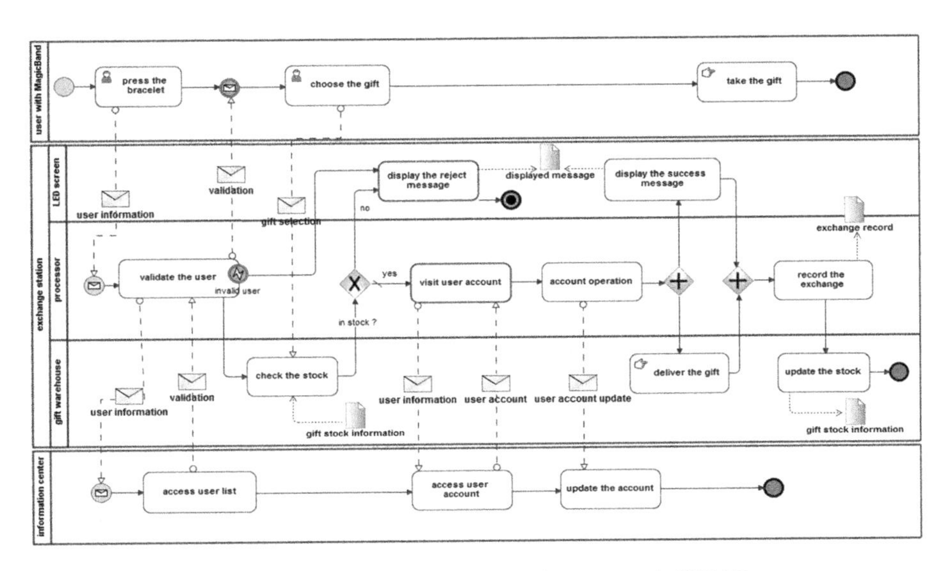

Fig. 4. Process model for gift exchange case in BPMN

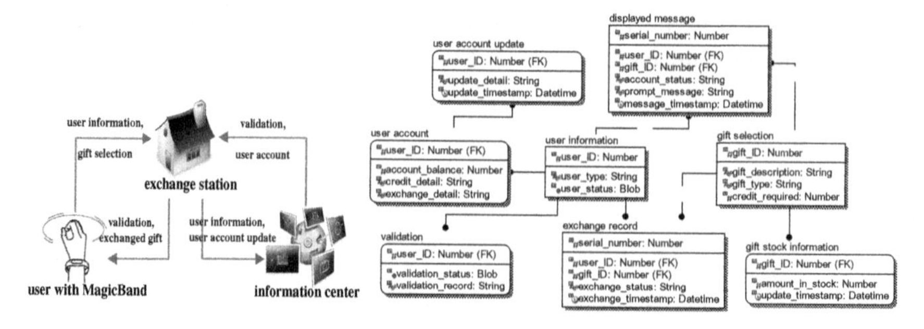

Fig. 5. Organization model and data model for gift exchange case

Finally, we establish a business rule model to specify further details of gift exchange process, which is associated with the process model, data model, and organization model in figures. Table 1 shows the rule "validate the user" partially in IF-THEN-ELSE pseudo-code.

Table 1. Rule model of the task "validate the user" (partial)

IF	user_information.user_status=1 AND user_information.user_ID is within the valid user list in database
THEN	validation.validation_status=1 AND validation.validation_record is updated ElSE validation_status=0 AND dispalyed_message. prompt_message="invalid user"

5 Conclusion

This paper presents an approach named PROUD for development of TOGAF business architecture. The PROUD model suit is based on the TOGAF business architecture content meta-model tailored from ACF, which ensure the consistency and completeness of the architecture model at the meta-model level. A two-dimension modeling matrix is also defined to support the derivation of the PROUD model instances from meta-models in gradient granularity to achieve granularity control, flexibility and traceability. We hope our work provide a useful reference of enterprise architecture development for both academic research and industry application.

Acknowledgments. The work is funded from the National Natural Science Foundation of China [Grant No. 61403255, and No. 71571055]. The work is also supported by The Program for Professor of Special Appointment (Eastern Scholar) at Shanghai Institutions of Higher Learning [grant number TP2014043].

References

1. Lapalme, J., Gerber, A., Van der Merwe, A., et al.: Exploring the future of enterprise architecture: a Zachman perspective. Comput. Ind. **79**(SI), 103–113 (2016)

2. Clarke, M., Hall, J.G., Rapanotti, L.: Enterprise architecture: a snapshot from practice. Int. J. IT/Bus. Alignment Gov. **4**(1), 1–10 (2013)
3. TOGAF Version 9.1.: The Open Group (2011)
4. Dietz, J.L.G., Hoogervorst, J.A.P.: A critical investigation of TOGAF—based on the enterprise engineering theory and practice. Springer Verlag (2011)
5. Gonzalez-Perez, C., Henderson-Sellers, B.: Modelling software development methodologies: a conceptual foundation. J. Syst. Softw. **80**(11), 1778–1796 (2007)
6. Chinosi, M., Trombetta, A.: BPMN: an introduction to the standard. Comput. Stand. Interfaces **34**(1), 124–134 (2012)

Application of SVM Based on Information Entropy in Intrusion Detection

Nuo Jia(✉) and Dan Liu

University of Electronic Science and Technology, Chengdu 610000, China
jianuo0123@qq.com

Abstract. Information entropy and SVM can be applied to the intrusion detection system, the combination of the two, the user measured the inherent nature of the audit data or the implementation of appropriate data deformation, so that it can be applied to the model when the training data, experimental results prove that the two combination can be more efficient detection of abnormal intrusion.

Keywords: Information entropy · SVM · Intrusion detection

1 Introduction

With the popularity of the Internet, the arrival of large data age, in general, the firewall as the first line of network security line of defense, but with the increasingly mature attack technology, attack tools and techniques are increasingly complex and diverse, simple firewall technology can not meet the information security is highly sensitive to the needs of departments, therefore, the network must be used to prevent a deep and diverse means. Intrusion detection, as one of the important technologies in the field of computer network security, has become a hot topic in the current oretical study of computer network.

In the process of network transmission the statistical model of data packets and their characteristics are studied. The essence of information is the uncertainty of information, and the information entropy is equivalent to its uncertainty. In order to quantify the data packets and features uniformly, We choose the information entropy as a measure of its importance. The combination of information entropy theory and SVM algorithm can make up for the shortcomings of SVM algorithm, on the other hand can make full use of information entropy to determine the advantages of high accuracy of intrusion.

Entropy is a very important concept in information theory, it is a measure of the value of a random variable uncertainty. In terms of data set, entropy can be used as a measure of the degree of irregularity of the data set. The so-called irregularity refers to the strength of the temporal dependencies between data elements before and after the set.

SVM detection technology is an anomaly detection technology, the goal is to find an optimal classification of the super-plane, it can not only be given a given input samples are correctly divided into two categories of normal and abnormal, but also

F. Xhafa et al. (eds.), *Advances in Intelligent Systems and Interactive Applications*, Advances in Intelligent Systems and Computing 686,
https://doi.org/10.1007/978-3-319-69096-4_64

divided into two categories of data between the classification The interval is as large as possible. Compared with the traditional anomaly detection technology, SVM pay more attention to the information from the sample itself rather than the law of the sample, and can automatically find the support vector with better ability to distinguish the classification.

2 Abnormal Intrusion Detection of SVM Based on Information Entropy

There are always difference between the open network and the monitored computer system by information entropy. The message often flow from high entropy to low entropy. In contrast, the direction of low entropy is infiltrated in the direction of high entropy Information, not only pure and the amount of data is small, easy to study its state profile. Thanks to this theory, Based on information entropy's SVM intrusion detection model can include the following aspects: SVM algorithm and information entropy theory, through reduce the eigenvector and find SVM kernel function's optimal parameters.

2.1 SVM Kernel Function Optimization

Internally detected computer system or network and external open network. Each network connection can be considered as a random sampling of the intrusion event, and the overall understanding of the incident can be achieved by statistical analysis of observations and experimental results. When an intrusion event occurs, it causes an instruction or data exception to the system or network. Assuming that the computer system or network is being monitored, the user and the process are in a series of rules that do not contain attacks, and do not contain a sequence of commands that compromise the system security policy. Generally in line with the statistical forecasting model. If internally detected computer system or network users and processes meet the statistical forecasting model, you can use external open network. Some of the characteristics of the parameters and closed value to define the normal user behavior and the system normal contour, and then the normal contour and the system compared to the transient profile, if there is a tolerance threshold difference, then identified as abnormal.

Step1 Extracts n features from each connection to stand for the connection, $x_i = (c_1, c_2, \cdots, c_n)$.

Step2 Process the raw data. Since there are 41 numeric attributes in the raw data, the range of values for each attribute is different, and the data needs to be normalized to normalize the values of each attribute to the set $\{-1, +1\}$. This paper replaces the characteristics with the information of the feature, get the information of $c'_{\mathrm{i}} = -lbpc_i$. The information quantity is used to solve the problem of unified measurement of feature information, and the factors of uncertainty in feature information are eliminated., $x'_{\mathrm{i}} = (c'_1, c'_2, \cdots, c'_{\mathrm{m}})$ indicates the connection of xi, Use the training sample set to get $En_{\mathrm{i}} = (EC'_1, EC'_2, \cdots, EC'_{\mathrm{m}})$ (N stands for external open network while N_{i} stands for internally detected computer system or network) and variance

$DN_i = (DC'1, DC'2, \cdots, DC'_m)$, through EN_i and DN_i can describe the contour values of the N_i normal mode.

Step 3 The confidence space is set by variance of the amount of information and the entropy. If $c'i$ is in this interval normal +1,otherwise normal -1. Enter $x'_i = (c'_1, c'_2, \cdots, c'_m)$, output $y'_i = (Ic'_1, Ic'_2, \cdots, Ic'_m)$.

Step 4 The optimal classification superplanar is found by SVM, and the given input samples are correctly divided into normal and anomalous categories, and the classification between the two types of data is made as big as possible. This problem has been translated into looking for a reflect:r: $Y' \to Y$, $y'_i \in Y'$, $Y = \{-1, +1\}$; at the same time make the least risk of misclassification. The transformation must be very complex, generally speaking the idea is hard to get. But pay attention to the optimal function involves only the inner volume between $(y'_i \bullet y'_j)$, In this way, in the high-dimensional space in fact only the inner volume operation, and this inner volume operation can be used to get the original space function. In the light of the functional theory. Only a kernel function $K(y'_i \bullet y'_j)$ meet Mercer conditions. It corresponds to the inner product in a certain transformation space. Relative classification function is that: $g(y') = \text{sgn}\left\{\sum_{i=1}^{n} a_i^* y'_i K(y'_i \bullet y') + b^*\right\}$. In general, the information entropy is introduced into the SVM detection model. Based on the statistical law of the feature, the feature information is used to characterize the feature itself, and the original data feature metric is unified.

2.2 SVM Intrusion Detection Process based on Information Entropy

The general process is shown in Fig.1.

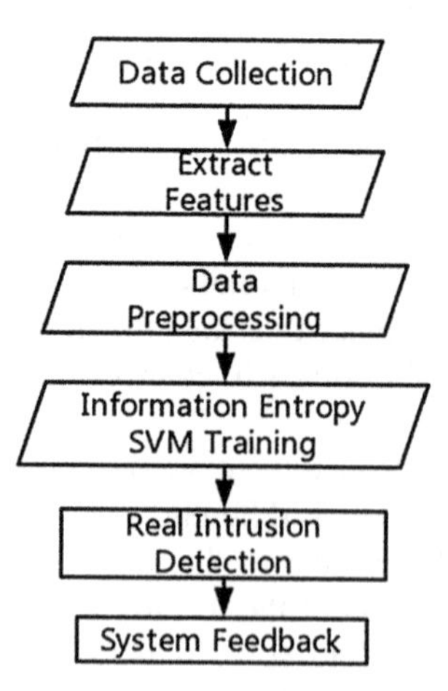

Fig. 1. SVM-based intrusion detection process on information entropy

The specific process are as follows

1. Data collection, with wirehark crawl network data and data processing into KDD99 data set format.
2. Feature extraction, extraction of connection status characteristics and uplink data characteristics.
3. Data preprocessing, the network connection records for vector processing.
4. Based on the SVM of information entropy, SVM based on information entropy is trained by training sample.
5. Based on the information entropy SVM detection, the trained SVM is used to classify the network connection samples.
6. System response, after detecting the abnormal connection, call the response module, take the corresponding response measures.

3 Verification Experiment

The data set used in the experiment is from KDD99, which is a simulated intrusion data in the military network environment and is a vector of some useful features extracted from the connection record. These features include the basic characteristics of a single TCP connection, the content characteristics of the domain-level connection, and the traffic characteristics calculated in the two-second time window.

Taking into account the data set the amount of intrusion data is a large proportion. This experiment uses three samples were 100000, 300000, 500000, select the large number of samples is to improve the accuracy of the experiment. There are two main aspects of the performance evaluation of an intrusion detection system: false rate and detection rate, The false positive rate refers to the proportion of normal data that is erroneously divided into human invasions to all normal data, the detection rate is the proportion of the detected intrusion data to all intrusion data. The results of the experiment are as follows (Table 1):

Table 1. SVM algorithm analysis

Samples	SVM		SVM and Entropy	
	Detection /%	False /%	Detection /%	False /%
100000	95.55	0.45	97.82	0.87
300000	96.57	0.66	96.93	0.97
500000	95.35	0.87	96.61	1.17

4 Summary

This paper discusses the application of information entropy theory in intrusion detection technology based on SVM. Experiments show that the application of information entropy theory in intrusion detection technology based on SVM not only guarantees the training sample set and the optimal classification surface compared with the traditional SVM-based intrusion detection algorithm And reduce the size of the sample, improve the SVM intrusion detection technology training speed and detection efficiency, which is an effective improvement attempt.

References

1. Alshammari, R., Sonamthiang, S., Teimourim, M., et al. Using neuro-fuzzy approach to reduce false positive alerts. In: Proceeding of the Fifth Annual Conference on Communication Networks and Services Research, pp. 345–349. Los Alamitos, CA, USA (2007)
2. Zeng, Q., Wu, S., Wang, M.: Application of information entropy in intrusion detection. J. Nanchang Univ. **31**(6), p619–p621 (2007)
3. Sommer, R., Paxson, V.: Outside the closed world; On using machine learning for network intrusion detection. In: Proceedings of 2010 IEEE Symposium on Security and Privacy, California, pp. 305–316 (2010)
4. Xiao, Y., Han C., Zheng, Q., et al.: Network intrusion detection method based on multi-class support vector machine. J. Xi'AN Jiaotong Univ. **39**(6), 562–565 (2005)
5. Li, K.L., Huang, H.K., Tian, S.F., Liu, Z.P., Liu, Z.Q.: Fuzzy multi-class support vector machine and application in intrusion detection. Chin. J. Comput. **28**(2), 274–280 (2005)
6. Xia, Q., Wang, Z., Lu, K.: The method of intrusion detection system using information entropy to detect network attack. J. Xi'an Jiaotong Univ. **47**(2), 14–19 (2013)
7. Spathoulas, G.P., Katsikas, S.K.: Reducing false positives in intrusion detection systems. Comput. Secur. **29**(1), 35–44 (2010)
8. Niu, G., Guan, X., Long, Y., et al.: Analysis method of multi-source flow characteristics and its application in anomaly detection. J. PLA Univ. Sci. Technol.:Nat. Sci. Ed. **10**(4), 350–355 (2009)
9. Mabonsy, V.M.A.: Machine learning approach to detecting attacks by identifying anomalies in network traffic. Ph.D. Thesis. Florida Institute of Technology, Melbourne, pp. 129–134 (2003)

Research of Music Recommendation System Based on User Behavior Analysis and Word2vec User Emotion Extraction

Qiuxia Li(✉) and Dan Liu

School of University of Electronic Science and Technology of China, Chengdu, China
1019597908@qq.com

Abstract. Aiming at the recommendation accuracy, diversity and timeliness of music recommendation system, this paper puts forward the music recommendation based on user behavior analysis and user emotion extraction. User behavior analysis can analyze the music preferences, and establish user's interest model. Collaborative filtering algorithm and user similarity calculation can be a good way to explore the user's new interests. In addition, by analyzing user's real-time text information in the social network. Using word2vec and clustering can help achieve the user's real-time feeling. Combining user's interests with user's emotional needs, filter out the user's current emotional music recommendation-list from the recommended music list. By this way, users get a better experience.

Keywords: Music recommendation · User behavior · Similarity calculation · Collaborative filtering · Word2vec · Emotion analysis

1 Introduction

With the rapid development of the network and the growing number of information, users can hardly get the information they really want, which is known as information overload [1]. The Music human listened to is always related to user's preferences, user behavior characteristics, and user's mood. The personalized recommendation system required to be based on the listener's own flavors and their real-time need, which makes user behavior analysis an effective method. Music personalized recommendation system is usually composed of three parts: user interest model, music resource description and recommendation algorithm [2].

Music recommendations generally can be based on the following ways: content filtering recommendations, collaborative filtering recommendations, context-based recommendations, Recommendation based on graph model, mixed model recommendations. Based on the album, genre, artist name, lyrics, audio [3], can use the content filter recommended algorithm, according to user's historical information to select the appropriate song for recommendation. Collaborative filtering is based on memory or model recommendations. The memory-based recommendation algorithm searches for similar users based on past user ratings data and searches for similar users

F. Xhafa et al. (eds.), *Advances in Intelligent Systems and Interactive Applications*, Advances in Intelligent Systems and Computing 686,
https://doi.org/10.1007/978-3-319-69096-4_65

based on similar user ratings and user similarity. This article mainly uses the recommended method of collaborative filtering. Early music personalized recommendation system mostly based on acoustic metadata, such as: Bogdanov [4], extract acoustic characteristics from the audio samples, then based on content filtering recommendation algorithm to achieve song recommendations. Gradually the music personalized recommendation system based on editing metadata and cultural metadata appeared. Schedl et al [5], Have created a MusiClef dataset that includes editing metadata information for songs, albums, artists, and MusicBrainz identifiers link to other datasets. These means can improve the recommended accuracy and credibility, but lack of personalization and scalability. Kim [6] classifies the tags according to the emotions in the cultural metadata, and proposes a music recommendation method based on semantic labels. This method have situational awareness and timely availability, but not very good at dealing with the long tail of music, and recommendation experience is not good for users.

To get user's music preferences, this article builds user's interest model based on user behavior analysis. On the other hand, through crawling user real-time message information from the social network or music platform, and using Word2vec representation and clustering method, this article achieve user's current emotional state and get the user's real-time changes in mood. Then combine collaborative filtering recommended list with emotional analysis results to get the integration of matching, generate music recommendations in line with the user's mood.

1.1 Emotional Extraction Based on Word2vec Clustering

The music recommendation method proposed in this article is based on the information posted on user's social network. According to user's blog text information, Word2vec is used to convert the word into vector form and clustering is used to get the keyword, and then the similarity of the vector space is calculated by calculating the similarity of the vector space. Combine user's blogs with user's behavior information like music play records. We can figure out the emotional state under blog, and use user's preference for music to establish user emotional music interest description matrix.

In the field of information retrieval and data mining, TF-IDF algorithm are commonly used as a feature extraction technology [7]. TF is the word frequency, indicates the number of a word appeared in the text, in general, the number of occurrences is proportional to the importance of its representation; and IDF is the reverse file frequency, it is the number of occurrences of this word in the corpus of all text, the more related text appears in the corpus, the more important this word will be (Fig. 1).

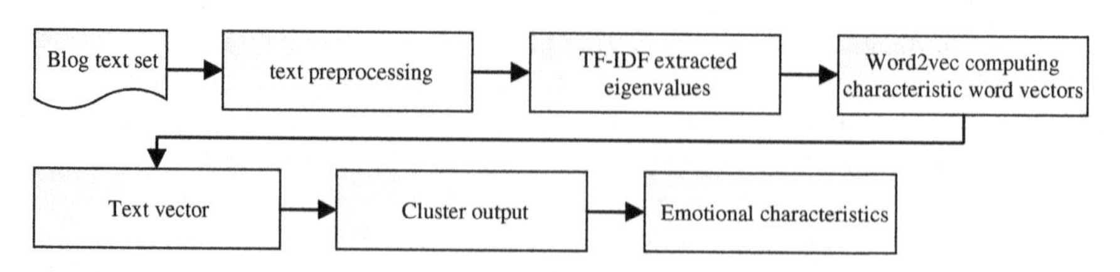

Fig. 1. Establish user emotion—music interest description model

TF-IDF algorithm can effectively find the keyword in the blog, and help Word2vec to express the emotional characteristic words of the text into the form of word vector to construct the vector space. Through the conversion of words to vectors, text content can be simplified as a vector in vector space.

The similarity of the blog texts information can be measured by calculating the cosine between the vectors: the smaller the cosine value is, the higher the similarity of the texts will be. After achieving the emotional feature vector, k-means algorithm is used to cluster the blog text. The K-means algorithm requires a custom classification quantity K, and the similarity of the sample can be calculated by the distance between the sample and the cluster center in the class. The algorithm steps are as follows:

(1) K samples are selected as the initial cluster center from the samples with the total number of n;
(2) For the remaining K-n samples, these samples are divided into the category of the nearest cluster center according to the distance (or similarity) of these samples from the K cluster centers;
(3) According to the class and the sample obtained in step 2, the clustering center of the category is recalculated using the sample mean method;
(4) Cycle steps 2 and 3 until the new cluster center is equal to or smaller than a previous iteration until the algorithm ends.

2 Music Recommendation Process based on User Behavior Analysis

User behavior data contains the user's behavior records, in the music recommended system, user behavior analysis including music play, search, purchase, collection, comment, share and other user activities, through the processing of user data, in line with these specific time to generate the user Interest characteristic data for music. The user profile analysis flow chart is shown below (Fig. 2).

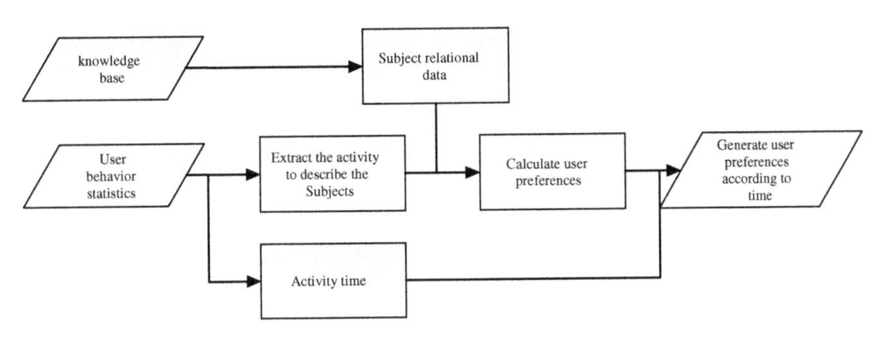

Fig. 2. Flow chart of user profile analysis

This paper will deeply analyze the user behavior data and modeling it, and mining user behavior data behind patterns and laws, through the user log and display feedback information to analyze the user listening habits. Through the calculation of user similarity, and integrated user real-time emotional analysis, we use the emotional analysis of the collaborative filtering recommendation algorithm to get the recommended results. Use the accuracy and recall rate to verify the pros and cons of recommended quality.

2.1 User Behavior Analysis and Modeling

User behavior can be divided into explicit and implicit behavior [8], explicit user behavior mainly refers to the user's evaluation of the song or collection, sharing, the behavior of like or do not like. User log behavior can be analyzed to obtain user information.

We can obtain user preferences for different types of music, WE also can get the user's preference for different types of music through the user's songs playlist. User preferences of different types of music formula are shown as below:

$$\mathrm{TP(j)}{=}\frac{\mathrm{SongType(j)}}{\mathrm{sumtypeCount}} \tag{1}$$

User implicit behavior data can be counted as follows (Table 1):

Table 1. Table of user implicit behavior data

user Id	Song Id	Behavior	Time
1	5146	share	10:26 am
1	5147	play	10:33 am
2	5146	comment	5:10 pm
3	6036	loop	8:20 pm
…	…	…	…

You can construct the user model from user preferences types and emotional labels of songs. By calculating the proportion of the number of times of songs with the emotion feature j, we can easily obtain the preference degree of the user i to the music feature j.

2.2 User Similarity Calculation

The collaborative filtering algorithm based on user needs to calculate the similarity between users. There are many methods for similarity calculation, such as Pearson correlation coefficient method, cosine similarity method and Euclidean distance method [9–11]. Assume that in the music list {A, B, C, D, E, F}, user U pay attention to songs of {A, B, C}, user V pay attention to songs of {A, F}. We use cosine similarity to calculate the interest similarity between these two users:

$$w_{uv} = \frac{|N(u) \cap N(v)|}{\sqrt{|N(u)|}\sqrt{|N(v)|}} = \frac{|\{A,B,C\}\{A,F\}|}{\sqrt{|\{A,B,C\}|}\sqrt{|\{A,F\}|}} = \frac{1}{\sqrt{6}} \tag{2}$$

2.3 Music Recommendation Algorithm Based on User Behavior Analysis and Word2vec Emotional Extraction

The traditional user-based collaborative filtering algorithm considers similar users share similar hobbies, we can get recommend music list through finding similar users and extract similar users songs which not contain in target users music list [12].

This article follow the user behavior to obtain similar users and similar users' songs list to get recommendation list, after that we need to classify the emotions extract from text, and matching them to get final recommend list. The specific calculation of the recommended list is described as follows:

(1) According to the user behavior analysis, establish emotional tag model to express the user - song relationship;
(2) Based on the user interest model to find similar users, generate recommended songs;
(3) Monitoring the user's real-time mood, using word2vec to calculate the user's emotional word vector in social network, combine the word vectors as the song feature vector D_i;
(4) For the candidate songs comes from similar users, use the tf-idf to extract the emotional signature words, use word2vec to compute their word vectors, and combine the word vectors as the recommended song vector d_k;
(5) Use the cosine similarity to compute the matching value of each recommended song with the real-time emotional feature:

$$match_{ki} = \frac{d_k D_i}{|d_k||D_i|}. \tag{3}$$

(6) Compare the size between each of the recommended music and real-time feature matching values, and add the matching values as the main recommendation to the recommended list.

3 Experimental Analysis

The experimental data comes from Netease cloud music list. We can get song-lists of friends and get play behaviors of them. Then, we count the data and classify the statistics data to establishment user models and emotional interest table. 20 associated users are extracted to this experiment, received 37021 songs in total. After that we found people to make artificial calibration emotional classification, counted user preferences and other information preferences. Last, devided emotional characteristics into sadness, quiet, sweet, inspirational and restless dimensions.

The recommended results are evaluated by commonly used evaluation criteria which for information retrieval: accuracy, recall rate [13]. The music recommendation based on the user behavior and the emotion characteristics extracted by word2vec of the classification matched the song emotion classification and the text emotion classification in the recommended song list. The experimental results are as follows (Table 2):

Table 2. Experiment comparison results

	Accuracy	Recall rate
Based on user behavior	82.06	73.65
Combine emotinal extract	87.52	78.33

4 Concluding Remarks

This paper is based on the user behavior analysis and word2vec emotional extraction of the recommendation, research the user behavior information and modeling, in the recommended effect has some advantages. The validity and stability of the method are verified by the experimental results.

References

1. Wang, G., Liu, H.: A summary of personalized recommendation system. Comput. Eng. Appl. **48**(7), 66–76 (2012)
2. Campbel, M.E., Clune Iii, J.E., Hicken, W.T. et al.: Music Recommendation System and Method. USA, WO 2005038666 A1[P]. 13 Aug 2004
3. Fang, M., Lin, H.: Social tag-based mobile music search. In: Proceedings of the 5th China Information Retrieval Conference, pp. 269–271, Chinese Information Processing Society of China, Shanghai (2009)
4. Bogdanov, D., Haro, M., Fuhrmann, F., et al.: Semantic audio content-based music recommendation and visualization based on user preference examples. Inform. Process. Manag. **49**(1), 13–33 (2013)
5. Schedl, M., Liem, C.C.S., Peeters, G., et al.: A professionally annotated and enriched multimodal data set on popular music. In: Proceedings of the 4th ACM Multimedia Systems Conference, pp. 78–83. ACM (2013)
6. Kim, H.H.A.: Semantically enhanced tag-based music recommendation using emotion ontology. In: Intelligent Information and Database Systems, pp. 119–128. Springer, Berlin, Heidelberg (2013)
7. Wang, Z., Jiang, W., Li, Y.: Research on the method of fixed feature collocation in emotional analysis. J. Manag. Eng. **28**(4), 180–186 (2014)
8. Qinjun, X., Zhenyang, W.: Research on behavior recognition in video sequences. J. Electron. Measur. Instrum. **28**(4), 343–351 (2014)
9. Wang, Y., Wang, Y., Zhao, X.: The similarity method of complex matrix structure for objective evaluation of image quality. J. Instrum. Instrum. **35**(5), 1118–1129 (2014)

10. Collaboration, C.: Description and performance of track and primary-vertex reconstruction with the CMS tracker. J. Instrum **9**(10), P10009 -1–P10009 -80 (2014)
11. Liu, S., Zhang, H., Mao, Z.: Target detection method based on HRM feature extraction and SVM. Foreign Electron. Measur. Technol. **10**, 38–41 (2014)
12. Meng, X., Hu, X., Wang, L.: Mobile recommendation system and its application. J. Softw. **24**(1), 91–108 (2013)
13. Li, Y., Gao, T., Li, X.: Design of video recommendation system based on cloud computing. In: China Education and Research Computer Network Academic Conference (2013)

Exploration of Information Security Education of University Students

Shi Wang(✉), Yongxin Qu, Likun Zheng, Yawen Xiao, and Huiying Shi

Harbin University of Commerce, Harbin, China
wangshi5331@163.com

Abstract. It is one of the basic and essential ability of university students to use computers and Internet to solve the problems of learning, living and working. University students are becoming the main victims of all kinds of security incidents due to the long time online, the large absolute number of Internet users, less social experience and the weak information security awareness. Some of them lose their money and some even lose their lives. University students should be taught the knowledge of information security so that they can deal with the increasingly serious forms of security to protect individual, social and national security. This paper discusses practical experience and exploration in the process of education from educational targets, forms of education, educational contents and assessment methods in information security.

Keywords: University students · Information security · Education

1 Introduction

As of June 2016, Internet users in China reached 710 million, mobile phone users reached 656 million, of which 10–29 age group, accounting for 50.5% overall, about 359 million people. The number of college students should be 100% of Internet users. It is one of the basic and essential ability for university students to use computers and the Internet to solve the problems of learning, living and working. With the improvement of the performance of mobile phones and the rapid development of mobile Internet technology, students can surf the Internet without time and space constraints. At the same time, online fraud, phishing sites, information leakage and other network security problems become increasingly severe with the deepening of network applications. Pornography offenses, gambling and drug dealing online and some other bad behaviours online, such as spreading violence and terror information and anti-China forces penetration and so on, are even more shocking. During these network security incidents, college students have become high-risk groups because they are simple, lack of social experience and the ability of self-protection, and easy to accept new things. University students should accept information security education so as to avoid harm to network security incidents and contribute to maintenance campus harmony, social stability and national security.

F. Xhafa et al. (eds.), *Advances in Intelligent Systems and Interactive Applications*, Advances in Intelligent Systems and Computing 686,
https://doi.org/10.1007/978-3-319-69096-4_66

2 The Necessity of Information Security Education

2.1 To Cope with Increasingly Serious Security Forms

In crimes in a variety of purposes, network has become the main criminal ways to get information, communicate, give orders and make profits. And in recent years cyber-crimes have become the high-risk areas of various kinds of security incidents because of its strong concealment, anonymity, difficult investigation, light penalties, low investment, low risk and high return. Among various kinds of network security events, University students are becoming the main victims of all kinds of security incidents due to the long time online, the large absolute number of Internet users, less social experience and the weak information security awareness. Some of them lose their money and some even lose their lives. Based on net platform statistics, in the distribution of the deceived people in 2015, post-90s accounted for 51.7%. The proportion of all kinds of security incidents is shown in Fig. 1.

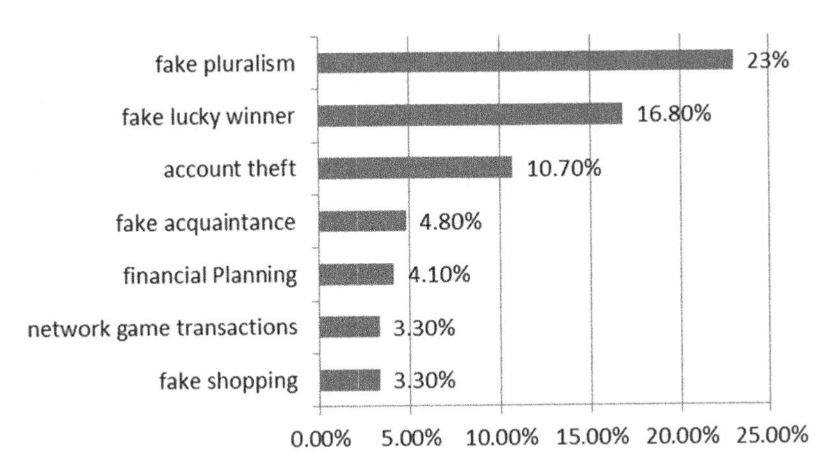

Fig. 1. The proportion of incidents

2.2 To Carry Out the National Policy for the Network Security Education

Beginning in 2014, our country paid great attention to network and information security, and it has become strategic issues. Our country has taken many policies and measures to achieve national network security maintenance and the goal of power construction. In 2014 and 2015, China has organized two publicity week activities in network security. In 2016, our country issued a notice on national publicity week activity in network security and determined the publicity week activities would be held in the third week in September each year.

In May 2016, the ministry of education issued a notice on strengthening campus lending risk prevention and education guidance work. It requests "strengthen the sense of students in financial and network security." In June 2016, the ministry of education issued on strengthening the network security guidance on subject construction and talent cultivation. It clearly requests "encourage colleges to offer public basic courses in network security, promote all the students acquire the network security knowledge and skills." "Talent is the first resource of network security. The personnel training in network security should be scheduled and sped up. Colleges should provide powerful talent guarantee to implement the strategy of network power and safeguard national network security.

3 Set Up a System of Information Security Education

3.1 Education Goals Must Be Clear

Goal is the starting point and end-result of information security education. And it plays important roles in guidance, motivation, regulation and evaluation. Overall goal is that students can fully realize the importance of information security by learning the knowledge of information security. They will have good sharpness and sense of responsibility and they will master the basic knowledge, basic skills and basic methods of information security. So they will have good ethics of information security and avoid the risk effectively.

3.2 Education Methods Must Be Various

According to the characteristics and targets, the knowledge of information security should be taught in a variety of ways, such as flexible theoretical explanation, case analysis, demonstration, operating experiment, role playing and Publicity etc. Make full use of the advantages of information technology teaching to make the difficulties easy to improve teaching effect. For example, the awareness in information security can be taught by holding information security warning education, by showing information security spy films, by discussing leakage cases and by posting posters etc. In these ways students can realize the importance of information security and strengthen the consciousness of information security crisis. Information security skills can be taught by setting up laboratories. Students can do experiments to simulate processes of attacks and defenses in information security. In this way students can master the necessary prevention and handling skills.

3.3 Education Content Should be Practical

Education content is the centre of the whole plan and it is an important foundation to check if the education goal can be realized. Practical education content is the key to ensure the students' learning interests. First of all, students' background knowledge must be considered when designing contents. Students are the main body and network security incidents data is the basis. Students should learn more cases about threats they may meet in their study, life and work. Secondly, compared with preventing from fire, robbery, earthquake and other national security awareness, people are short of awareness of preventing from information security. Information security is a lifelong course because there are more and more threats happening about information security. It is difficult to stop them because they need advanced professional techniques and the techniques progress rapidly. So its teaching contents must be designed comply with the principle of learning. The model of the content is shown in Fig. 2.

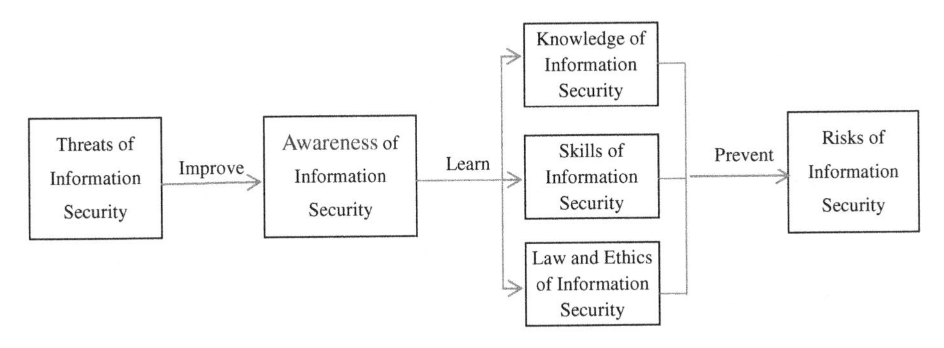

Fig. 2. The model of information security education

3.4 Test Ways Must Be Flexible

Information security is a course requiring stronger operation and offenses and defenses. So its test ways must be flexible and meet its characteristics. It must be tested in theory, operation, and practice ability. It also needs a higher level of selection, such as participating in related training camps and organizing all kinds of competitions to choose the information security personnel, to supplement the shortage of the information security professionals.

4 Conclusion

Information security has become the cornerstone of national overall security in the information age. "Without the network security, there is no national security". College students are regarded as the future of society, so their ability and level in information security will directly affect the goal of our country's network power. It also concerns China's economic security and even national security. Information security education is a complicated system, involving computer, communications, law, management, and other multi-disciplinary knowledge. At the same time, it is related to politics, economy, culture and etc. So the processes of information security education must be carried out continually, rather than loosely and optionally, even marginally, to make the students have the ability to keep pace with The Times to take appropriate security measures.

Acknowledgements. This thesis is a stage achievement of the project (No. 16GLD05) of the Heilongjiang Province Society Scientific Fund and the project (No. HSDJY03(Z) and

No. 201710240101) of Harbin University of Commerce. Many people have contributed to it. This thesis will never be accomplished without all the invaluable contributions selflessly from my group members. It is my greatest pleasure and honour working with my group. I'm deeply grateful to them.

References

1. Luo, Li.: The cultivation of national information security literacy. Library and information service, pp. 25–28. March 2012
2. Gao, D., Cai, H.: Design of assessment scale of students' network information security literacy. China medical education technology, pp. 173–177. April 2013
3. Cai, H.: Evaluation research on information security literacy for cadets based on AHP. Information technology, pp. 188–192. January 2013
4. Liu X., Song J.: Development of domestic information security training. Information security and technology, pp. 74–77, December 2010
5. Information on http://baike.baidu.com/view/65955.htm

Mobile and Wireless Communication

Research on the Path Monitoring Capability of Wireless Multimedia Sensor Network

Zhao Jing[1,2](✉), Liu Zhuohua[1], and Xue Songdong[2]

[1] College of Information Engineering, Guangdong Mechanical and Electrical Polytechnic, Guangzhou 510515, People's Republic of China
zhaojing_740609@163.com

[2] Division of Industrial and System Engineering, Taiyuan University of Science of Technology, Taiyuan 030024, People's Republic of China

Abstract. The Wireless Multimedia Sensor network (~WMSNs~) differs from the traditional Wireless sensor network (~WSNs~) because of the direction and rotation of the multimedia sensor sensing area. The perceptual model was established, and the path monitoring capability of wireless multimedia sensor network was researched, and the path monitoring conclusions adapted to the multimedia network were obtained, and the simulation experiments were carried out, and the simulation results verified the correctness of the conclusion.

Keywords: Wireless multimedia sensor network · Path monitoring capability · Sense area · Multimedia sensor · Coverage

1 Introduction

The development of wireless communication, microelectronics and embedded technology has spawned the wireless sensor network technology (Wireless Sensor network, WSNs). Wireless sensor networks consist of a large number of low-cost micro sensor nodes [1], complete specific monitoring tasks, Wireless sensor network (WSN) includes the traditional omni-sensing network, Multimedia sensor network (WMSNS) two categories. The traditional whole-to-sensing network realizes simple environmental data (such as temperature, humidity, intensity, etc.) collecting, transmitting and processing [2]. However, with the complication of monitoring tasks, more comprehensive and abundant environmental data (such as images, videos, etc.) are needed to accomplish more precise monitoring tasks, resulting in a network of direction perception. Wireless multimedia sensor networks, network capabilities (such as acquisition, storage, transmission, etc.) enhanced, processing data (such as images, videos, etc.) more complex, thus increasing the ability of data processing, such as complex image, video data compression, recognition, fusion processing, in short, multimedia sensor networks face more new challenges, urgently need new solutions to guide the network system effectively [3].

In WSNs, assuming that the node is deployed uniformly randomly, the probability that a sub area in the target area is monitored by K nodes conforms to the Poisson's point process (Poisson pointprocess) [1, 4, 5]:

F. Xhafa et al. (eds.), *Advances in Intelligent Systems and Interactive Applications*, Advances in Intelligent Systems and Computing 686,
https://doi.org/10.1007/978-3-319-69096-4_67

$$P[N(A) = k] = e^{-\lambda\mu(A)} \times \frac{(\lambda\mu(A))^k}{k!} \tag{1}$$

A represents the target area, λ indicating the strength of the sensor, which is the number of sensors per unit area, $\mu(A)$ indicating the area that a node can perceive, and the number k of sensors that are monitored to any point in the region. The theory can be applied to multiple coverage.

From the above theory, the perceived area of the node is $\mu(A)$, and the probability that the perception of at least one node at any point in the region A is:

$$f_{area} = P[N(A) \geq 1] = 1 - P[N(A) = 0] = 1 - e^{-\lambda\mu(A)} \tag{2}$$

In WSNs, the sensor's ability to monitor the path of the intruder relies on the intruder's knowledge of the network. When the intruder does not know the exact location of the sensor, and randomly selected point a specific path when at least one sensor exists in a tubular area of the distance path less than equal to r, the intruder is bound to be monitored from the point to the, as shown in Fig. 1. When taking a straight line from the point to the, the size of the tube is mini-mal and the area is $\mu(A) = 2r\|q_0 - q_1\| + \pi r^2$, with at least one sensor probability in this area:

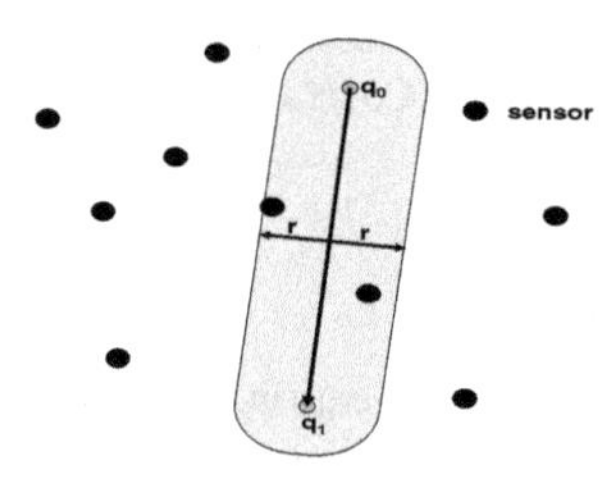

Fig. 1. Path monitoring of unknown networks

$$P_d = P[N(A) \geq 1] = 1 - e^{-\lambda(2r\|q_0 - q_1\| + \pi r^2)} \tag{3}$$

There are many researches on the path monitoring capability of WSNs network. such as [4–6], but these mature theories cannot be applied directly to the multi-media network, and there is no wireless multimedia sensor network path monitoring capability research, the current WMSNs network application is very widespread, it is urgent need to carry on the research of its path monitoring ability.

The establishment of perceptual model is the precondition of the whole study, if the perceptual model does not reflect the real application, then the research cannot be applied to the actual monitoring task. In the literature about two-dimensional wireless video sensor networks covering research, the fan-sensing model is generally used: the non-rotating fan-sensing model is established in the literature [7–11]. Typically, a four-tuple is represented: said that represents the location coordinates of the video node, indicates the sensing radius, for the sensing direction of the video sensor node, represents the video node angle offset. The literature [12, 13] simplifies the sector-sensing model, establishes the directional adjustable isosceles triangle perception model, represents the node coordinates; is the angular deflection of the left side of the triangle relative to the horizontal axis, which is used to denote the perception direction, expresses the perception triangle's vertex, expresses the biggest sense radius. But these two perceptual models only consider the sensor's instantaneous perception range, the rotation of the multimedia sensor is not considered, even if the assumed direction is adjustable, but in the coverage calculation, only the transient perception area is considered.

Firstly, this paper establishes the sensing model, then researches the network path monitoring capability of wireless multimedia sensor, and obtains some correctness of the results verified by the multimedia network.

2 Sense Model

The establishment of perceptual model is as follows: A four-tuple representing: in which represents the location coordinates of the video node, represents the sensing radius; is the sensing direction of the video sensor node, which does not rotate around the node coordinates; represents a video node offset from the perspective, as shown in Fig. 2.

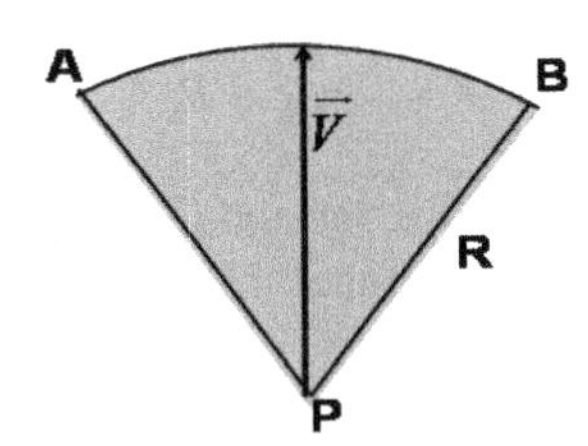

Fig. 2. Non-delayed covering (class A) perception model

3 Analysis of Path Monitoring Capability

When an intruder chooses to travel from q_0 point to q_1 Point, the probability of its path perception depends on the position and direction of the perceived node. When considering the perceived direction of the node, as shown in Fig. 3. From the graph, the sensing area of adjacent nodes O is the dark sector AOB, the path can be monitored by node o only when the sensing direction of the adjacent node O is rotated from the OM counterclockwise to on. Suppose the angle of AOB and POQ for α, Angle AOP is θ, and the angle of the node perception direction of the path is.

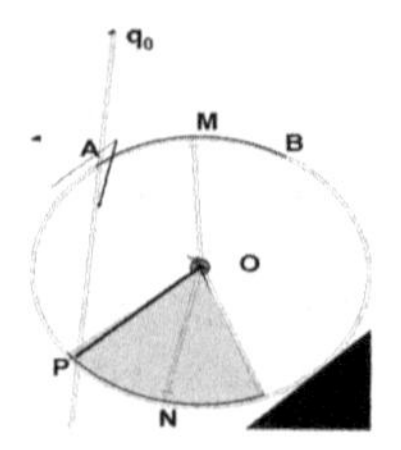

Fig. 3. Angle range of route covered by sensor

$$Ang_o = \alpha + \theta = \alpha + 2 \times \arccos(\frac{d_o}{r}) \tag{4}$$

The d_o represents the distance between the node o to the path, and r indicates the sense radius.

When the sensing direction of a node is within the range of OM and ON, the node can perceive the path, as the node perceives the randomness of the direction, the node may perceive the ratio of the path, that is, the ratio of the angle range to the entire angular range (2π) is:

$$P_{ao} = \frac{\alpha + \theta}{2\pi} \tag{5}$$

You cannot monitor the range of angles to the path:

$$PN_{ao} = 1 - P_{ao} \tag{6}$$

Assuming that the number of adjacent nodes of the path is n, all adjacent nodes cause the probability that the path cannot be monitored because of the perceived orientation is not in the perceptual area:

$$PNall = (PN_{ao})^n \tag{7}$$

Then, the probability that at least a nodes that can be obtained from the perception orientation problem can be monitored to the path:

$$P_{\geq 1} = 1 - PNall \tag{8}$$

Because the node can perceive that the path depends on the location and direction of the node, the probability that the intruder is monitored by a node in the course of the travel by two factors is:

$$P_{N(A) \geq 1} = P_{\geq 1} \times P_d \tag{9}$$

In the analysis of the probability that the appeal path is monitored at least one node, consider the actual number of adjacent nodes and the actual distance of each adjacent node to the path, which prevents the analysis of the network path monitoring performance unknown to a node coordinates. However, because nodes are evenly dispersed in the target area, it is possible to predict the number of nodes in a path-aware area of the pipe:

$$n \approx num = \lambda \times \mu(A) \tag{10}$$

Also, when the number of nodes is greater than a certain value, the proximity of the node to the path can be averaged, so we can estimate the distance:

That is $d_o = \frac{r}{2}$, then

$$\theta \approx 2 \times \arccos(\frac{1}{2}) \tag{11}$$

In this way, we can analyze the network path monitoring ability without knowing the specific location and angle of the node.

4 Simulation

Assumes the sensing radius of the sensor node is 60 m, sensing sector angle $\alpha = 45°$, the target area is 500 m × 500 m, the number of nodes is 50, the intruder randomly selects the starting point and the end point, Fig. 4 is the initial deployment of the

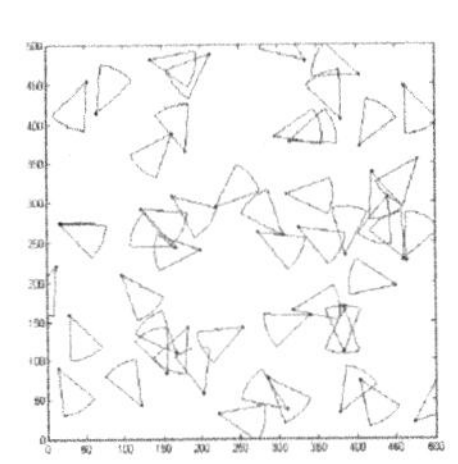

Fig. 4. Initial coverage

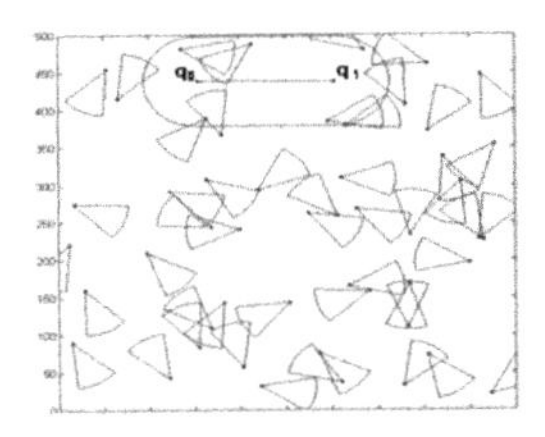

Fig. 5. Selection of start and final point

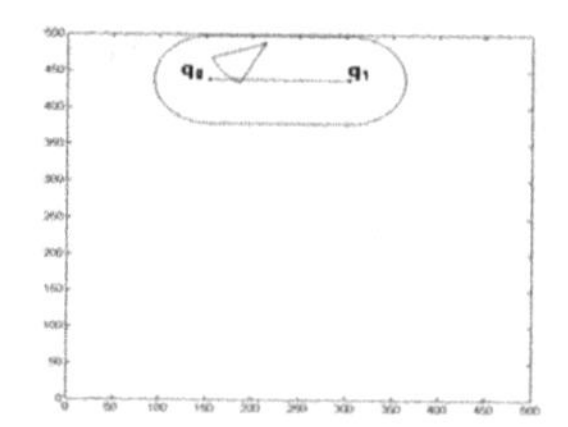

Fig. 6. Selection of sense node

network, Fig. 5 shows the beginning and end of the intruder choice. In Fig. 6, the path is perceived as a condition.

5 Conclusions

This paper establishes a perceptual model for the multimedia network, and researches the network path monitoring ability of wireless multimedia sensor based on perceptual model, obtains some conclusions of the path monitoring for multimedia network, and carries on the simulation experiment, the simulation result verifies the correctness of the conclusion.

Acknowledgments. This work is supported by the Shanxi Province Natural Science Foundation under Grant No. 2014011019-2; the school postdoctoral scientific research foundation of China under Grant No. 20142023; Shanxi Province Programs for Science and Technology Development under Grant No. 2015031004, Project from Shanxi Scholarship Council of China under Grant No. 2016-091.

About the Author: Zhao Jing (1974–), female, associate professor, Ph.D., Research Fields: Equipment Health Management and Wireless Sensor Network; LIU Zhuohua (1979–), male, lecturer, M.Sc. in Computer Science, Research Fields: Cloud Computting; XUE Songdong (1968–), male, associate professor, Ph.D., Research Fields: Robot and Wireless Sensor Network.

References

1. Karl, H., Willig, A., Carle, et al.: Protocol and Architecture of Wireless Sensor Networks. Electronic Industry Publishing House, China (2007)
2. Akyildiz, I.F., Su, W., Sankarasubramaniam, Y., et al.: Wireless sensor networks: a survey. Comput. Netw. Int. J. Comput. Telecommun. Networking **38**(4), 393–422 (2002)
3. Hua-Dong, M.A., Dan, T.A.O.: Multimedia sensor network and its research progresses. J. Softw. **17**(9), 2013–2028 (2006)
4. Henna, S.: Energy Efficient Fault Tolerant Coverage in Wireless Sensor Networks (2017)
5. Kilaru, S., Lakshmanachari, S., Kishore, P.K., et al.: An efficient probability of detection model for wireless sensor networks. In: Proceedings of the First International Conference on Computational Intelligence and Informatics, Springer, Singapore (2017)
6. Henna, S.: Energy Efficient Fault Tolerant Coverage in Wireless Sensor Networks (2017)

7. Ma, H., Liu, Y.: On Coverage Problems of Directional Sensor Networks. In: Mobile Ad-hoc and Sensor Networks, First International Conference, MSN 2005, December 13–15, Wuhan, China, Proceedings. DBLP:721–731 (2005)
8. Dan, T.A.O., Hua-Dong, M.A., Liang, L.I.U.: A virtual potential field based coverage-enhancing algorithm for directional sensor networks. J. Softw. **18**(5), 1152–1163 (2007)
9. Jing, A., Abouzeid, A.A.: Coverage by directional sensors in randomly deployed wireless sensor networks. J. Combinatorial Optim. **11**(1), 21–41 (2006)
10. Han, G., Jiang, J., Guizani, M., et al.: Green routing protocols for wireless multimedia sensor networks. IEEE Wirel. Commun. **23**(6), 140–146 (2017)
11. Chang, H.Y.: A connectivity-increasing mechanism of ZigBee-based IoT devices for wireless multimedia sensor networks. Multimedia Tools Appl. 1–18 (2017)
12. Tezcan, N., Wang, W.: Self-orienting wireless multimedia sensor networks for occlusion-free viewpoints. Elsevier North-Holland, Inc. (2008)
13. Adriaens, J., Megerian, S., Potkonjak, M.: Optimal Worst-Case Coverage of Directional Field-of-View Sensor Networks. 1:336–345 (2006)

Preopen Set and Preclosed Set on Intuitive Fuzzy Topological Spaces

Zhen-Guo Xu[1(✉)], Rui-Dan Wang[1], Zhi-Hui Chen[1], and Ying Zhao[2]

[1] National Science and Technology Infrastructure Center, Beijing 100862, China
zhenguoxu@126.com

[2] Department of Strategy and Development, Beijing North Vehicle Group Corporation, Beijing 100862, China

Abstract. In this paper, we shall give concepts of fuzzy preopen set, fuzzy preclosed set,fuzzy preinterior, fuzzy preclosure on intuitive fuzzy topological space and we shall introduce precontinuou, preirresolute, preopen (preclosed) mapping between two intuitive fuzzy topological spaces. Moreover, we shall give their equivalent characterizations.

Keywords: Intuitive preopen set · Intuitive preclosed set · Intuitive precontinuous · Intuitive preirresolute · Intuitive preopen (preclosed) mapping

1 Introduction

After the introduction of the concept of fuzzy sets by Zadeh in [1] several researches were conducted on the generalization of the notion of fuzzy set. The idea of "intuitive fuzzy set" was published by Atanassov in [2]. Later, this concept was generalized to "intuitive *L*-fuzzy sets" by Atanassov and Stoeva in [3]. Many works by the same author and his colleagues appeared in the literature [4–6]. In [7], the author introduced the concepts of intuitive fuzzy topological spaces and intuitive fuzzy continuous mappings.

In this paper, on base [7] we shall introduce the concepts of fuzzy preopen set, fuzzy preclosed set and fuzzy preinterior and fuzzy preclosure on intuitive fuzzy topological spaces and fuzzy precontinuous, fuzzy preirresolute mapping, fuzzy pre-open (preclosed) mapping between two intuitive fuzzy topological spaces. Moreover, we shall discuss their some properties.

Definition 1.1 ([3]). Let Y be a nonempty set. An intuitive fuzzy set (**ZMJ** for short) A is an object having the form $\Delta = \{\langle y, \mu_\Delta(y), \nu_\Delta(y)\rangle | y \in Y\}$ where the mappings $\mu_\Delta(x): Y \to I$ and $\nu_\Delta(x) : Y \to I$ denote the degree of membership and the degree of nonmembership of each element $y \in Y$ to the set Δ, respectively and $0 \le \mu_\Delta(y) + \nu_\Delta(y) \le 1$ for each $y \in \Delta$.

Definition 1.2 ([3]). $\tilde{0} = \{\langle y, 0, 1\rangle | y \in Y\}$ and $\tilde{1} = \{\langle y, 1, 0\rangle | y \in Y\}$.

Definition 1.3 ([7]). An intuitive fuzzy topological space (**ZMTK** for short) is a pair (Y, τ), where τ is a subfamily of **ZMJs** in X which contains $\tilde{0}, \tilde{1}$ and is closed for any suprema and finite infima. (Y, τ) is called an intuitive fuzzy topology on X. Each

F. Xhafa et al. (eds.), *Advances in Intelligent Systems and Interactive Applications*, Advances in Intelligent Systems and Computing 686,
https://doi.org/10.1007/978-3-319-69096-4_68

member of τ is called an intuitive fuzzy open set (**ZMKJ** for short) and its quasi-complementation is called an intuitive fuzzy closed set (**ZMBJ** for short).

Definition 1.4 ([7]). Suppose (Y, τ) be an **ZMTK** and $\Delta = \{\langle y, \mu_\Delta(y), \nu_\Delta(y)\rangle | y \in Y\}$ be an **ZMJ** in Y. Then the fuzzy interior and fuzzy closure of Δ are defined by
(1) $bb(\Delta) = \cap\{\Gamma|\Gamma \geq \Delta, \Gamma\,\text{is an}\,\mathbf{ZMKJ}\}$; (2) $nb(\Delta) = \vee\{\Gamma|\Gamma \leq \Delta, \Gamma\,\text{is an}\,\mathbf{ZMBJ}\}$.

Definition 1.5 ([7]). Suppose (Y_1, τ_1) and (Y_2, τ_2) be two **ZMTKs** and let $h : Y_1 \to Y_2$ be a mapping. Then

(1) h is said to be fuzzy continuous if and only if the primage of each **ZMJ** in τ_2 is an **ZMK** in τ_1.
(2) h is said to be fuzzy open (closed) if and only if the image of each **ZMJ** in τ_2, τ_2' is an **ZMK** in τ_1, τ_1'.

2 Intuitive Fuzzy Preopen Set and Preclosed Set

Definition 2.1. Suppose (Y, τ) be an **ZMTK** and $\Delta = \{\langle y, \mu_\Delta(y), \nu_\Delta(y)\rangle | y \in Y\}$ be an **ZMJ** in Y. Then Δ is called:

(1) Intuitive fuzzy preopen set (**ZMNKJ** for short) iff $\Delta \leq nb(bb(\Delta))$;
(2) Intuitive fuzzy preclosed set (**ZMNBJ** for short) iff $nn(nb(\Delta)) \leq \Delta$;
(3) $pnb(\Delta) = \cup\{\Lambda|\Lambda \leq \Delta, \Lambda\,\text{is an}\,\mathbf{ZMNKJ}\}$;
(4) $pbb(A) = \cap\{\Lambda|\Pi \geq \Delta, \Lambda\,\text{is an}\,\mathbf{ZMNBJ}\}$.

It can be also shown that $pbb(\Delta)$ is an **ZMNBJ** and $pbb(\Delta)$ is an **ZMNKJ** in Y

Remark 2.1. From Definition 2.1, we can know that **ZMKJ** (**ZMBJ**) is **ZMNKJ** (**ZMNBJ**), but the inverses is false is shown by the following Example 2.1.

Example 2.1. Let $Y = \{a_1, a_2\}$ and $\Delta_1 = \{\langle y, (\frac{a_1}{0.3}, \frac{a_2}{0.4}), (\frac{a_1}{0.3}, \frac{a_2}{0.5})\rangle | y \in Y\}$, $\Delta_2 = \{\langle y, (\frac{a_1}{0.2}, \frac{a_2}{0.4}), (\frac{a_1}{0.2}, \frac{a_2}{0.5})\rangle | y \in Y\}$. Then the family $\tau = \{\underline{0}, \underline{1}, \Delta_1, \Delta_2\}$ is an **ZMT** on Y. Let $\Lambda = \{\langle y, (\frac{a_1}{0.1}, \frac{a_2}{0.2}), (\frac{a_1}{0.5}, \frac{a_2}{0.5})\rangle | y \in Y\}$. Then Λ is not an **ZMKJ**, but $bb(\Lambda) = \Delta_1'$, $nb(\Delta_1') = \Delta'$, $\Lambda \leq \Delta_1$, so $\Lambda \leq nb(bb(\Lambda))$, i.e, Λ is an **ZMNKJ**.

By Definition 2.1, we have the following theorem.

Theorem 2.1. Suppose (Y, τ) be an **ZMTK** and $\Delta = \{\langle y, \mu_\Delta(y), \nu_\Delta(y)\rangle | y \in Y\}$ be an **ZMJ** in Y. We have

(1) Δ is an **ZMNKJ** if and only if $\Delta = pnb(\Delta)$, Δ is an **ZMNBJ** if and only if $\Delta = pbb(\Delta)$;
(2) $\underline{0} = pbb(\underline{0}), \underline{1} = pnb(\underline{1})$;
(3) $pnb(A) = (pbb(A'))'$;
(4) $pbb(\Delta) = pbb(pbb\Delta), pnb(\Delta) = pnb(pnb(\Delta))$;
(5) $pbb(bb(\Delta)) = bb(\Delta), pnb(nb\Delta) = nb(\Delta)$;
(6) If Δ is a **ZMNKJ** (**ZMNBJ**), so $nb(pnb\Delta) = nb(\Delta), cl(pbb(\Delta)) = bb(\Delta)$.

3 Some Mappings on Intuitive Fuzzy Preopen (Preclosed) Set

Using **ZMNKJ** and **ZMNBJ** we can obtain the following definitions:

Definition 3.1. Suppose (Y_1, τ_1) and (Y_2, τ_2) be two **ZMTKs** and suppose $h : Y_1 \to Y_2$ be a mapping. We have

(1) h is called to be intuitive fuzzy precontinuous $\Leftrightarrow$ the preimage of every **ZMJ** in τ_2 is an **ZMNKJ** inY_1;
(2) h is called to be intuitive fuzzy preirresolute $\Leftrightarrow$ the preimage of every **ZMNKJ** in Y_2 is an **ZMNKJ** in Y_1;
(3) h is called to be intuitive fuzzy preopen (preclosed) $\Leftrightarrow$ the image of every **ZMJ** in τ_1 is an **ZMNKJ (ZMNIBJ)** in Y_2;
(4) h is called to be intuitive fuzzy preirresolute open (closed) $\Leftrightarrow$ the image of each **ZMNKJ (ZMNBJ)** in Y_1 is an **ZMNKJ (ZMNBJ)** in Y_2.

By Definition 1.6, Definition 3.1–3.4, we can obtain following relations:

Theorem 3.1. Suppose (Y_1, τ_1) and (Y_2, τ_2) be two **ZMTKs** and let $h : Y_1 \to Y_2$ be a mapping.

(1) Suppose h is intuitive fuzzy continuous, so h is intuitive fuzzy precon-tinuous;
(2) Suppose h is intuitive fuzzy preirresolute, so h is intuitive fuzzy pre-continuous;
(3) Suppose h is intuitive fuzzy open (closed), so h is intuitive fuzzy pre-open (preclosed).

Remark 3.1. The inverse of Theorem 3.1 is false. The following examples can explain.

Example 3.1. Suppose $Y = \{a_1, a_2\}$, $Z = \{b_1, b_2\}$ and $\Delta_1 = \{\langle y, (\frac{a_1}{0.3}, \frac{a_2}{0.4}), (\frac{a_1}{0.6}, \frac{a_2}{0.5})\rangle | y \in Y\}$, $\Delta_2 = \{\langle z, (\frac{b_1}{0.6}, \frac{b_2}{0.5}), (\frac{b_1}{0.3}, \frac{b_2}{0.3})\rangle | z \in Z\}$. Let (Y, τ) and (Z, σ) be two **ZMTKs**, where $\tau = \{\underline{0}, \tilde{1}, \Delta_2\}$ and $\sigma = \{\underline{0}, \tilde{1}, \Delta_2\}$. $h : (Y, \tau) \to (Z, \sigma)$ defined by $h(a_1) = b_1$, $h(a_2) = b_2$ is not continuous, because $h^{-1}(A_2) = \{\langle y, (\frac{a_2}{0.6}, \frac{a_2}{0.5}), (\frac{a_1}{0.3}, \frac{a_2}{0.3})\rangle | y \in Y\}$ is not an **ZMKJ** in Y, but $nb(bb(h^{-1}(\Delta_2))) = \tilde{1}$. Therefore $h^{-1}(\Delta_2) \le nb(bb(h^{-1}(\Delta_2)) = \tilde{1}$, i.e, $h^{-1}(\Delta_2)$ is an **ZMNKJ** in Y. Therefore h is precontinuous.

Example 3.2. Suppose $Y = \{a_1, a_2\}$, $Z = \{b_1, b_2\}$ and $\Delta_1 = \{\langle y, (\frac{a_1}{0.3}, \frac{a_2}{0.4}), (\frac{a_1}{0.5}, \frac{a_2}{0.5})\rangle | y \in Y\}$, $\Delta_2 = \{\langle z, (\frac{b_1}{0.2}, \frac{b_2}{0.7}), (\frac{b_1}{0.3}, \frac{b_2}{0.3})\rangle | z \in Z\}$. Let (Y, τ) and (Z, σ) be two **ZMTKs**, where $\tau = \{\tilde{0}, \tilde{1}, \Delta_1\}$ and $\sigma = \{\tilde{0}, \tilde{1}, \Delta_2\}$. $h : (Y, \tau) \to (Z, \sigma)$ defined by $h(a_1) = b_1, h(a_2) = b_2$ is precontinuous. In fact, $h^{-1}(\Delta_2) = \{\langle y, (\frac{a_1}{0.2}, \frac{a_2}{0.7}), (\frac{a_1}{0.3}, \frac{a_2}{0.3})\rangle | y \in Y\}$, $bb(h^{-1}(\Delta_2)) = \tilde{1}$, so that $h^{-1}(\Delta_2) \le nb(bb(h^{-1}(\Delta_2)) = \tilde{1}$, thus $h^{-1}(\Delta_2)$ is preopen L-set. But h is not preirresolute, in fact, let $\Gamma = \{\langle z, (\frac{b_1}{0.5}, \frac{b_2}{0.3}), (\frac{b_1}{0.4}, \frac{b_2}{0.4})\rangle | z \in Z\}$ in Z. Then Γ is an **ZMNKJ**. But, we know $h^{-1}(\Gamma) = \{\langle y, (\frac{a_1}{0.5}, \frac{a_2}{0.3}), (\frac{a_1}{0.4}, \frac{a_2}{0.4})\rangle | y \in Y\}$, hence $nb(bb(h^{-1}(\Gamma)) = \tilde{0}$, therefore $h^{-1}(\Gamma) \not\le nb(bb(h^{-1}(\Gamma)) = \tilde{0}$, therefore h is not preirresolute.

Example 3.3. $Y = \{a_1, a_2\}$, $Z = \{b_1, b_2\}$ and $\Delta = \{\langle y, (\frac{a_1}{0.2}, \frac{a_2}{0.4}), (\frac{a_1}{0.6}, \frac{a_2}{0.7})\rangle | y \in Y\}$, $\Lambda_1 = \{\langle z, (\frac{b_1}{0.6}, \frac{b_2}{0.5}), (\frac{b_1}{0.4}, \frac{b_2}{0.4})\rangle | z \in Z\}$, $\Delta_2 = \{\langle z, (\frac{b_1}{0.4}, \frac{b_2}{0.4}), (\frac{b_1}{0.6}, \frac{b_2}{0.6})\rangle | z \in Z\}$. Let (Y, τ_1) and (Z, τ_2) be two **ZMTKs**, where $\tau_1 = \{\tilde{0}, \tilde{1}, \Delta\}$ and $\tau_2 = \{\tilde{0}, \tilde{1}, \Lambda_1, \Lambda_2\}$. $h : (Y, \tau_1) \to (Z, \tau_2)$ defined by $h(a_1) = b_1, h(a_2) = b_2$. Thus h is preopen maping, in fact, $h(\Delta) = \{\langle z, (\frac{a_1}{0.2}, \frac{a_2}{0.4}), (\frac{a_1}{0.6}, \frac{a_2}{0.7})\rangle | z \in Z\}$, $bb(h(\Delta)) = \Lambda_2', nb(\Lambda_2') = \Lambda_2$, so that $h(\Delta) \le nb(bb(h(\Delta)) = \Lambda_2$, thus $h(\Delta)$ is preopen L-set. $h(\Delta) \ne \Lambda_1, \Lambda_2$, i.e, h is not a fuzzy open mapping.

4 The Equivalent Characterizations for Correlative Mappings

Theorem 4.1. Suppose (Y_1, τ_1) and (Y_2, τ_2) be two **ZMTKs** and let $h : Y_1 \to Y_2$ be a mapping. Then (1)–(4) are equivalent.

(1) h is intuitive fuzzy precontinuous;
(2) $h^{-1}(\Delta_2)$ is an **ZMNBJ** in Y_1 for every **ZMBJ** Δ_2 in Y_2;
(3) $pbb(h^{-1}(\Delta_2)) \le h^{-1}(bb(\Delta_2))$ for every **ZMJ** Δ_2 in Y_2;
(4) $h^{-1}(nb(\Delta_2)) \le pnb(h^{-1}(\Delta_2))$ for every **ZMJ** Δ_2 in Y_2;
(5) $h^{-1}(\Delta_2) \le nb(bb(h^{-1}(\Delta_2)))$ for every **ZMJ** Δ_2 in Y_2.

Proof. (1)⇒(2) is obvious. (2) ⇒ (3). Let Δ_2 be an **ZMJ** in Y_2. So $bb(\Delta_2)$ is an **ZMBJ**, so by (2), $h^{-1}(bb(\Delta_2))$ is an **ZMNBJ** in Y_1. Noting $\Delta_2 \le bb(\Delta_2)$, we obtain $h^{-1}(\Delta_2) \le h^{-1}(bb(\Delta_2))$, therefore $pbb(h^{-1}(\Delta_2)) \le h^{-1}(bb(\Delta_2))$ from Definition 2.1.

(3)⇒(4). This proof is easily and therefore omitted. (4) ⇒ (5). For any **ZMKJ** Δ_2 of Y_2. Then $\Delta_2 = nb(\Delta_2)$, thus $h^{-1}(\Delta_2) = h^{-1}(nb(\Delta_2)) \le pnb(h^{-1}(\Delta_2))$, hence $h^{-1}(\Delta_2) = pnb(h^{-1}(\Delta_2))$, therefore $h^{-1}(\Delta_2)$ is an **ZMNKJ** from Theorem 2.1. Thus $h^{-1}(\Delta_2) \le nb(bb(h^{-1}(\Delta_2)))$ from Definition 2.1. (5) ⇒ (1). It follows immediately from Definition 2.1 and therefore omitted.

Theorem 4.2. Suppose (Y_1, τ_1) and (Y_2, τ_2) be two **ZMTKs** and let $h : Y_1 \to Y_2$ be a mapping. Then (1)–(4) are equivalent.

(1) h is intuitive fuzzy preirresolute;
(2) $h^{-1}(\Delta_2)$ is an **ZMNBJ** in Y_l for every **ZMNBJ** Δ_2 in Y_2;
(3) $h(pbb(\Delta_1)) \le pbb(h(\Delta_1))$ for every **ZMJ** Δ_1 in Y_1;
(4) $pbb(h^{-1}(\Delta_2)) \le h^{-1}(pbb(\Delta_2))$ for ever Δ_1 y **ZMJ** Δ_2 in Y_2;
(5) $h^{-1}(pnb(\Delta_2)) \le pnb(h^{-1}(\Delta_2))$ for every **ZMJ** Δ_2 in Y_2.

Proof. (1)⇒(2) is easy.

(2) ⇒(3). For any **ZMJ** Δ_1 in Y_1, we have $\Delta_1 \le h^{-1}(h(\Delta_1)) \le h^{-1}(pbb(h(\Delta_1)))$, we know that $pcl(f(A_1))$ is an **ZMNBJ** in Y from Definition 2.1, hence by (2), $h^{-1}(pbb(h(\Delta_1)))$ is an **ZMNBJ** in Y_l, thus by Definition 2.1(4) we obtain $pbb(\Delta_1) \le h^{-1}(pbb(h(\Delta_1)))$, therefore

$$h(pbb(\Delta_1)) \leq h(h^{-1}(pbb(h(\Delta_1)))) \leq pbb(h(\Delta_1)).$$

(3)⇒(4). For any **ZMJ** Δ_2 in Y_2, let $h^{-1}(\Delta_2) = \Delta_1$, by (3), we obtain

$$h(pbb(h^{-1}(\Delta_2))) \leq pbb(h(h^{-1}(\Delta_2))) \leq pbb(\Delta_2)$$

This indicates

$$pbb(h^{-1}(\Delta_2)) \leq h^{-1}(h(pbb(h^{-1}(\Delta_2)))) \leq h^{-1}(pbb(\Delta_2)).$$

(4)⇒(5). For any **ZMJ** Δ_2 in Y_2, by $pnb(\Delta_2) = (pbb(\Delta_2'))'$and (4), we have

$$\begin{aligned} h^{-1}(pnb(\Delta_2)) &= h^{-1}((pbb(\Delta_2'))') = (h^{-1}(pbb(\Delta_2')))' \\ &\leq (pbb(h^{-1}(\Delta_2')))' = pnb(h^{-1}(\Delta_2)) \end{aligned}$$

(5)⇒(1). Let Δ_2 be an **ZMNKJ** in Y_2, then $\Delta_2 = pnb(\Delta_2)$ from Theorem 2.1, by (5) we obtain $h^{-1}(\Delta_2) \leq pnb(h^{-1}(\Delta_2))$. Moreover $h^{-1}(\Delta_2) \geq pnb(h^{-1}(\Delta_2))$ by Definition 2.1, thus $h^{-1}(\Delta_2) = pnb(h^{-1}(\Delta_2))$, therefore $pnb(h^{-1}(\Delta_2))$ is an **ZMNKJ** in Y_1 from Theorem 2.1.

Theorem 4.3. Suppose (Y_1, τ_1) and (Y_2, τ_2) be two **ZMTKs** and let $h : Y_1 \to Y_2$ be a mapping. Then (1)–(4) are equivalent.

(1) h is intuitive fuzzy preirresolute open;
(2) $h(pnb(\Delta_1)) \leq pnb(h(\Delta_1))$ for every **ZMJ** Δ_1 in Y_1;
(3) $pnb(h^{-1}(\Delta_2)) \leq h^{-1}(pnb(\Delta_2))$ for every **ZMJ** Δ_2 in Y_2;
(4) For every **ZMJ** Δ_1 in Y_1, **ZMJ** Δ_2 in Y_2 and let Δ_1 be the **ZMNBJ** such that $h^{-1}(\Delta_2) \leq \Delta_1$. Then there exists a **ZMNBJ** Γ in Y_2 and $\Delta_2 \leq \Gamma$ such that $h^{-1}(\Gamma) \leq \Delta_1$.

Proof. (1)⇒(2) By Definition 2.1(3), we have $pnb(\Delta_1) \leq \Delta_1$, hence $h(pnb(\Delta_1)) \leq h(\Delta_1)$ and by Definition 2.1, we know $pnb(\Delta_1)$ is an **ZMNKJ** in Y_1, thus $h(pnb(\Delta_1)) \leq pnb(h(\Delta_1))$.

(2)⇒ (3). Let $\Delta_1 = h^{-1}(\Delta_2)$. Form (2) we have

$$h(pnb(h^{-1}(\Delta_2))) \leq pnb(h(h^{-1}(\Delta_2))) \leq pnb(\Delta_2)$$

This implies

$$pnb(h^{-1}(\Delta_2)) \leq h^{-1}(h(pnb(h^{-1}(\Delta_2)))) \leq h^{-1}(pnb(\Delta_2)),$$

i.e.,

$$pnb(h^{-1}(\Delta_2)) \leq h^{-1}(pnb(\Delta_2)).$$

(3)⇒(4). Let Δ_1 be an **ZMNBJ** in Y_1 and Δ_2 be an **ZMJ** in Y_2 such that $h^{-1}(\Delta_2) \leq \Delta_1$ hence $\Delta_1' \leq h^{-1}(\Delta_2')$ from Proposition 1.1(2) and Theorem 1.1(3), we know that Δ_1' is an **ZMNKJ**, thus $pnb(\Delta_1') = \Delta_1' \leq pnb(h^{-1}(\Delta_2'))$, therefore $\Delta_1' \leq pnb(h^{-1}(\Delta_2')) \leq h^{-1}(pnb(\Delta_2'))$, this implies

$$\Delta_1 \geq (h^{-1}(pnb(\Delta_2')))' = h^{-1}(pbb(\Delta_2))$$

Let $\Gamma = pnb(\Delta_2)$,then Γ satisfies conditions of (4).

(4)⇒ (1). Let Σ be an **ZMNKJ** in Y_1, $\Delta_2 = (h(\Sigma))', \Delta_1 = \Sigma'$. Then Δ_1 is an **ZMNBJ**, hence $h^{-1}(\Delta_2) = h^{-1}((h(\Sigma))') = (h^{-1}(h(\Sigma)))' \leq \Sigma' = \Delta_1$, by (4), there exists an **ZMNBJ** Γ and $\Delta_2 \leq \Gamma$ such that $h^{-1}(\Gamma) \leq \Delta_1 = \Sigma'$, thus $\Sigma \leq (h^{-1}(\Gamma))'$, so that $h(\Sigma) \leq h(h^{-1}(\Gamma')) \leq \Gamma'$. On the other hand by $\Delta_2 \leq \Gamma$, $h(\Sigma) = \Delta_2' \geq \Gamma'$, hence $h(\Sigma) = \Gamma'$. Since Γ' is an **ZMNKJ**, we have $h(\Sigma)$ is an **ZMNKJ**.

Analogously, we can prove following theorems:

Theorem 4.4. Suppose (Y_1, τ_1) and (Y_2, τ_2) be two **ZMTKs** and let $h : Y_1 \to Y_2$ be a mapping. Then (1)–(4) are equivalent.

(1) h is intuitive fuzzy preirresolute closed;
(2) $h(pbb(\Delta_1)) \geq pbb(h(\Delta_1))$ for every **ZMJ** Δ_1 in Y_1;
(3) $pbb(h^{-1}(\Delta_2)) \geq h^{-1}(pbb(\Delta_2))$ for every **ZMJ** Δ_2 in Y_2;
(4) For each **ZMJ** Δ_1 in Y_1, **ZMJ** Λ in Y_2 and let Δ_1 be the **ZMNKJ** such that $h^{-1}(\Delta_2) \leq \Delta_1$. Then there exists an **ZMNKJ** Γ in Y_2 and $\Delta_2 \leq \Gamma$ such that $h^{-1}(\Gamma) \leq \Delta_1$.

Theorem 4.5. Suppose (Y_1, τ_1) and (Y_2, τ_2) be two **ZMTKs** and let $h : Y_1 \to Y_2$ be a mapping. Then (1)-(3) are equivalent.

(1) h is intuitive fuzzy preopen;
(2) $h(nb(\Delta_1)) \leq pnb(h(\Delta_1))$ for every **ZMJ** Δ_1 in Y_1;
(3) $nb(h^{-1}(\Delta_2)) \leq h^{-1}(pnb(\Delta_2))$ for every **ZMJ** Δ_2 in Y_2.

Theorem 4.6. Suppose (Y_1, τ_1) and (Y_2, τ_2) be two **ZMTKs** and let $h : Y_1 \to Y_2$ be a mapping. Then (1)–(3) are equivalent.

(1) h is intuitive fuzzy preclosed;
(2) $pbb(h(\Delta_1)) \leq h(bb(\Delta_1))$ for every **ZMJ** Δ_1 in Y_1;
(3) $bb(h^{-1}(\Delta_2)) \leq h^{-1}(pbb(\Delta_2))$ for every **ZMJ** Δ_2 in Y_2.

Acknowledgments. The authors are very grateful to the referees for their valuable comments.

References

1. Zadeh, L.A.: Fuzzy sets. Inform. and Control **8**, 338–353 (1965)
2. Atanassov, K.: Intuitive fuzzy sets. Fuzzy Sets Sys. **20**, 87–96 (1986)
3. Atanassov, K.: Review and new results on intuitive fuzzy sets. Preprint IM-MFAIS-1-88, Sofia (1988)
4. Rajarajeswari, P., Moorthy, R.K.: On intuitive fuzzy weakly generalized closed set and its applications. Int. J. Comput. App. **27**, 9–13 (2011)
5. Rajarajeswari, P., Moorthy, R.K.: Intuitive fuzzy weakly generalized irres- olute mappings. Ultrascientist Phys. Sci. **24**, 204–212 (2012)
6. Santhi, R., Sakthivel, K.: Intuitive fuzzy alpha generalized connectedness in fuzzy topological spaces. Int. J. App. Math. Phy. **3**, 1–5 (2011)
7. Coker, D.: An introduction to inuitionistic fuzzy topological spaces. Fuzzy Sets Sys. **88**, 81–89 (1997)

Neural Network Method for Compressed Sensing

Zixin Liu[1,2(✉)], Yuanan Liu[2], Nengfa Wang[2], and Lianglin Xiong[3]

[1] School of Electronic Engineering, Beijing University of Posts and Telecommunications, Beijing 102209, China
xinxin905@163.com

[2] Department of Mathematics and Statistics, Guizhou University of Finance and Economics, Guiyang 550025, China

[3] School of Mathematics and Computer Science, Yunnan Minzu University, Kunming 650500, China

Abstract. By using approximate smooth method of l_1 norm, a new smoothed model to approximate traditional compressed sensing problem with small dense noise is established. Utilizing projection technique, two discrete-time projection neural networks are proposed for the optimization solution of concerned compressed sensing problem. Applying the definition of exponential stability, two exponentially stable criteria are also given to guarantee the state vector convergent to optimization solution.

Keywords: Compressed sensing · Neural network · Smooth method · Exponential stability

1 Introduction

Since compressed sensing (CS) theory [1, 2] can be widely applied to many different research fields, such as various compression imaging, MRI and CT, variable selection, biological computing, face recognition, LASSO, remote sensing and so on [3–5], it has aroused many researchers' interest. Compared with the usual measurements, CS theory can overcome the bottleneck problem of sampling method by using Shannon theory, and can simultaneously collect sample data and complete their compressing problem. This means that CS theory can reduce data collecting period, and has lower level requirement for the implementing hardware.

For compressed sensing theory, many reconstruction algorithms are established. These algorithms mainly use convex optimization theory, neural network method, greedy algorithms, machine learning and so on. As all know that, a typical sparsest solution with linear constraint can be expressed by "l_0-norm" minimization optimization problem as follows.

$$\min_{X \in \Omega} ||X||_0 \quad s.t. \quad AX = b \, , \tag{1}$$

F. Xhafa et al. (eds.), *Advances in Intelligent Systems and Interactive Applications*, Advances in Intelligent Systems and Computing 686,
https://doi.org/10.1007/978-3-319-69096-4_69

where $\Omega = \{(x_1, x_2, \cdots, x_n)|a_i \leq x_i \leq d_i, i = 1, 2, \cdots, n\}$ is a closed convex set, $A \in R^{k \times n}(k < < n), b \in R^k$, $||X||_0$ represents the number of nonzero elements in vector X. Obviously, expression (1) has been widely used in many different scientific fields, however, it is NP-hard, this makes the theoretical research for problem (1) becomes very difficult. To overcomes this flaw, an approximation model named basis pursuit (BP) becomes the best alternative one, since $||X||_1$ is the minimum convex hull of $||X||_0$, here $||X||_1 = \sum_i |x_i|$ denotes the l_1 norm of X. Among all of approximate substitution methods, BP model based on l_1 norm constrained is the most popular one. To further solve this popular approximate model, many various algorithms have been established. Among these various algorithms, an effective one is the split Bregman algorithm derived in [6]. "Basic Pursuit problem" [7] was proposed as follows.

$$\min_{X \in \Omega} ||X||_1 \quad s.t. \quad AX = b \quad . \tag{2}$$

For problem (2), if term b is contaminated by small dense noise, the following modified model for problem (2) can be expressed as

$$\min_{X \in \Omega} ||X||_1 \quad s.t. \quad ||AX - b||_2 < \varepsilon \quad , \tag{3}$$

where ε is a non-negative constant; $||X||_2$ denotes l_2 norm of X. Applying convex analysis tool, one can transform problem (3) into the following $l_1 - l_2$ optimization problem equivalently

$$\min_{X \in \Omega} \frac{1}{2} ||b - AX||_2^2 + \tau ||X||_1, \tag{4}$$

where $\tau > 0$ is a positive constant.

2 Approximate Smooth Method

From the expression of (4), one can see that $\frac{1}{2}||b - AX||_2^2 + \tau||X||_1$ is a convex function, but this function is non-differentiable. It is a non-smooth convex optimization problem. In order to solve problem (4), Liu and Hu [8] established a projection neural network to simulate the optimization solution by introducing a special transform. Notice that, in problem (4), $||b - AX||_2^2$ is a smooth term, and there is only one term $||X||_1$ is non-smooth. And term $||X||_1$ can be smoothed by function $\sum_{i=1}^{n} \left(1 - e^{-x_i^2 \sigma^2 / 2}\right)$ approximately. Thus, this paper introduces a smooth form for model (4) as follows.

$$\min_{X \in \Omega} \frac{1}{2} ||b - AX||_2^2 + \tau \sum_{i=1}^{n} \left(1 - e^{-x_i^2 \sigma^2 / 2}\right), \tag{5}$$

where $\sigma, \tau > 0$ are positive constants.

Set $H(X) = \frac{1}{2}||b - AX||_2^2 + \tau \sum_{i=1}^{n} \left(1 - e^{-x_i^2\sigma^2/2}\right)$, it follows that $H(X)$ is a smooth and convex function on closed convex set Ω. Before proceed, the following Lemmas are needed.

Lemma 1. Let Ω be a closed convex set of R^n. Then

$$(\upsilon - P_\Omega(\upsilon))^T(P_\Omega(\upsilon) - X) \geq 0, \upsilon \in R^n, X \in \Omega,$$

$$||P_\Omega(u) - P_\Omega(\upsilon)||_2 \leq ||u - \upsilon||_2, u, \upsilon \in R^n.$$

where $P_\Omega(u)$ is projection operator defined by $P_\Omega(u) = \arg\min_{\upsilon \in \Omega} ||u - \upsilon||$.

Lemma 2. Let $H : R^n \rightarrow R^n$ denote a continuous function, Ω denotes a subset of R^n, then vector X^* satisfying inequality $(X - X^*)^T H(X^*) \geq 0$ equivalents that, for all $X^* \in R^n$, equation $P_\Omega(X^* - \alpha H(X^*)) = X^*$ holds, where α is an arbitrary positive constant.

From Lemma 1 and Lemma 2, X^* is a solution of (5) if and only if it satisfied equation $X^* = P_\Omega(X^* - \alpha \nabla H(X^*))$, where projection vector $P_\Omega(X)$ is defined as $P_\Omega(X) = (P_\Omega(x_1), P_\Omega(x_2) \cdots P_\Omega(x_n))^T$, and for $i = 1, \cdots, n$,

$$P_\Omega(x_i) = \begin{cases} d_i & x_i > d_i \\ x_i & c_i \leq x_i \leq d_i \\ c_i & x_i < c_i \end{cases}.$$

Based on equality $X^* = P_\Omega(X^* - \alpha \nabla H(X^*))$, the following discrete-time neural network can be established to solve optimization problem (5)

$$X(i+1) = X(i) + \rho[P_\Omega(X(i) - \alpha \nabla H(X(i))) - X(i)], \tag{6}$$

where $i = 0, 1, 2, 3 \cdots$, and $\rho > 0$ is a sufficient small arbitrary positive constant. Similar to the proof in [9], it follows that

Theorem 1. X^* is an optimization solution of problem (5) iff it is an equilibrium point of neural network (6).

Proof. Similar to literature [9], one can easily obtain this result, thus it is omitted here.

Remark 1. In classical compressed sensing problem, the constraint set Ω is R^n. Obviously, this is a general case of this paper. When constraint set Ω becomes R^n, the expression of projection operator $P_\Omega(u)$ becomes $P_{R^n}(u) = u$. In this case, model (6) degenerates into a more simple form

$$X(i+1) = X(i) - \rho\alpha \nabla H(X(i)), \tag{7}$$

where $i = 0, 1, 2, 3 \cdots$, and $\rho > 0$ is a sufficient small arbitrary positive constant.

3 Stability Analysis

Applying stability theory, we will give out two stability results for the established neural network (6) and (7). Before proceed, the following definition is need.

Definition 1. If sequence $\{X(i)\}$ satisfies $||X(i) - X^*|| \leq \mu||X(0) - X^*||e^{-\eta i}$ $i = 0, 1, 2, 3 \cdots$, then neural network (7) is called to be convergent to equilibrium X^* exponentially with convergent rate η, where μ, η are any positive constants.

Theorem 2. Assume that solution X^* of system (5) is unique, then neural network (6) converges to X^* globally and exponentially.

Proof. From expression of $H(X)$, it follows that every elements of Hessian matrix $\nabla^2 H(X)$ are continuous on closed convex set Ω, thus, there is a positive definite diagonal matrix M and a positive constant $\alpha > 0$ such that

$$||I - \alpha M|| \geq \sup_{X \in \Omega} ||I - \alpha \nabla^2 H(X)||.$$

Notice that $X^* = P_\Omega(X^* - \alpha \nabla H(X^*))$, from Lemma 1 and Lemma 2, it obtains

$$\begin{aligned}
||X(i+1) - X^*|| &= ||X(i) + \rho(P_\Omega(X(i) - \alpha \nabla H(X(i))) - X(i)) - X^*|| \\
&= ||X(i) - X^* + \rho(P_\Omega(X(i) - \alpha \nabla H(X(i))) \\
&\quad - X^* + X^* - X(i))|| \\
&= ||(1-\rho)(X(i) - X^*) + \rho[P_\Omega(X(i) - \alpha \nabla H(X(i)) \\
&\quad - P_\Omega(X^* - \alpha \nabla H((X^*))]|| \\
&\leq (1-\rho) \cdot ||X(i) - X^*|| \\
&\quad + \rho||X(i) - X^* - \alpha(\nabla H(X(i)) - \nabla H(X^*))|| \\
&= (1-\rho) \cdot ||X(i) - X^*|| \\
&\quad + \rho||X(i) - X^* - \alpha \nabla^2 H(\widehat{X})(X(i) - X^*)|| \\
&\leq (1-\rho) \cdot ||X(i) - X^*|| \\
&\quad + \rho||I - \alpha M|| \cdot ||(X(i) - X^*)||.
\end{aligned}$$

Denote $\lambda_i > 0$ to be eigenvalues of positive definite diagonal matrix M, $\varsigma(\alpha) = \max_{1 \leq i \leq n} |1 - \alpha \lambda_i|$, since α is an arbitrary positive constant, thus when α is sufficient small, it can guarantee $1 - \alpha \lambda_i$ to be non-negative, in this case, it follows that

$$||X(i+1) - X^*|| \leq (1 - \alpha\rho\lambda_{\min})||X(i) - X^*||.$$

Set $\gamma = 1 - \alpha\rho\lambda_{\min}$, then $0<\gamma<1$ when $\rho<1/\alpha\lambda_{\min}$, where $\lambda_{\min}$ is the minimum eigenvalue of positive definite matrix M, and $\lambda_{\min} > 0$. This means that $\mu = -\ln\gamma > 0$, namely,

$$||X(i) - X^*|| \leq e^{-\mu i}||X(0) - X^*||.$$

This implies that neural network (6) converges to X^* globally and exponentially, which completes the proof.

Theorem 3. Assume that solution X^* of system (5) is unique, then neural network (7) converges to X^* globally and exponentially.

Proof. Set $F(X) = \tau\sum_{i=1}^{n}\left(1 - e^{-x_i^2\sigma^2/2}\right)$, from the expression of $H(X)$, it follows that $H(X) = \frac{1}{2}||b - AX||_2^2 + F(X)$, and $\nabla H(X) = A^TAX - A^Tb + \nabla F(X)$,

where $\nabla F(X) = (\tau x_1\sigma^2 e^{-x_1^2\sigma^2/2}, \tau x_2\sigma^2 e^{-x_2^2\sigma^2/2}, \cdots, \tau x_n\sigma^2 e^{-x_n^2\sigma^2/2})^T$. Similar to the proof of theorem 2, and notice that $\rho\alpha\nabla H(X^*)) = 0$, one can obtain that

$$||X(i+1) - X^*|| \leq (1 - \rho)||X(i) - X^*|| +$$

$$\rho||(I - \alpha(A^TA + \nabla^2 F(\widehat{X})))(X(i) - X^*)||,$$

where $\nabla^2 F(\widehat{X}) = (\tau\sigma^2(1 - x_1^2\sigma^2)e^{-x_1^2\sigma^2/2}, \cdots, \tau\sigma^2(1 - x_n^2\sigma^2)e^{-x_n^2\sigma^2/2})^T$. Since every component of vector $\nabla^2 F(\widehat{X})$ is continuous and bounded on set R^n, the upper bound is $\tau\sigma^2$, the lower bound is $-2\tau\sigma^2 e^{-1.5}$. It follows that every elements of Hessian matrix $\nabla^2 F(X)$ are bounded and continuous on set R^n, thus there must exist a positive definite matrix N and positive constant α such that

$$||I - \alpha(A^TA + N)|| \geq \sup_{X \in R^n} ||I - \alpha(A^TA + \nabla^2 F(X)||.$$

Denote $\mu_i > 0$ to be eigenvalues of positive definite matrix $A^TA + N$, $\zeta(\alpha) = \max_{1 \leq i \leq n} |1 - \alpha\mu_i|$, since α is an arbitrary positive constant, thus when α is sufficient small, it can guarantee $1 - \alpha\mu_i$ to be non-negative, in this case, it follows that

$$||X(i+1) - X^*|| \leq (1 - \rho\alpha\mu_{\min}) \cdot ||(X(i) - X^*)||$$

Set $\gamma = 1 - \alpha\rho\mu_{\min}$, then $0<\gamma<1$ when $\rho<1/\alpha\mu_{\min}$, where $\mu_{\min}$ is the minimum eigenvalue of positive definite matrix $A^TA + N$, and $\mu_{\min} > 0$, which means that $\eta = -\ln\gamma > 0$. This implies that the state vector of neural network (7) globally and exponentially converges to X^*, which completes the proof.

4 Conclusions

By using smooth technique, a new smoothed appropriate compressed sensing model with small dense noise is given. And two discrete-time projection neural network to solve this appropriate model are also established. Finally, two exponentially stable criteria are also derived to guarantee the state vector's convergence.

Acknowledgments. This project is supported partially by National Natural Science Foundation of China (Grant No. 61472093, 11461082), Scientific Research Fund Project in Guizhou Provincial Department of Education under Grants 2014[243], and Guizhou Provincial Soft Science Foundation ([2011]LKC2004).

References

1. Donoho, D.L.: Compressed sensing. IEEE Trans. Inform. Theor. **52**(4), 1289–1306 (2006)
2. Candés, E.J., Wakin, M.B.: An introduction to compressive sampling. IEEE Sig. Process. Mag. **25**, 21–30 (2008)
3. Jung, H., Ye, J., Kim, E.: Improved k-t blask and k-t sense using focuss. Phys. Med. Biol. **52** (11), 3201–3226 (2007)
4. Andrés, A.M., Padovani, S., Tepper, M., et al.: Face recognition on partially occluded images using compressed sensing. Pattern Recognit. Lett. **36**, 235–242 (2014)
5. Tibshirani, R.: Regression shrinkage and selection via the LASSO. J. R. Stat. Soc.: Ser. B. 58, 267–288 (1996)
6. Goldstein, T., Osher, S.: The split Bregman method for l_1–regularized problems. SIAM J. Imaging Sci. **2**(2), 323–343 (2009)
7. Chen, S.S., Donoho, D.L., Saunders, M.A.: Atomic decomposition by basis pursuit. SIAM J. Sci. Comput. **20**, 33–61 (1998)
8. Liu, Yongwei, Jianfeng, Hu: A neural network for l1–l2 minimization based on scaled gradient projection: Application to compressed sensing. Neurocomputing **173**, 988–993 (2016)
9. Yashtini, M., Malek, A.: A discrete-time neural network for solving nonlinear convex problems with hybrid constraints. Appl. Math. Comput. **195**, 576–584 (2008)

A Survey on Microwave Surface Emissivity Retrieval Methods

De Xing[1,2(✉)], Qunbo Huang[1,2,3], Bainian Liu[3], and Weimin Zhang[2]

[1] Academy of Ocean Science and Engineering, National University of Defense Technology, Changsha 410073, China
xd_wony@icloud.com
[2] College of Computer, National University of Defense Technology, Changsha 410073, China
[3] Weather Center of PLA Air Force, Beijing 100843, China

Abstract. Owing to its wide coverage and high observation density, remote sensing satellite microwave observation data has become the most data used in numerical weather prediction system. However, due to the influence of the uncertainty of the surface emissivity, most of the data collected from remote sensing satellites are the observations at higher-peaking channels over land. A large number of these observations at lower-peaking channels are discarded, so the accurate surface emissivity is the key of assimilation of remote-sensing satellite microwave observation data over land. This paper introduces several commonly retrieval methods used for passive microwave surface emissivity and discusses their advantages and shortcomings. At last, an evaluation of each retrieval method and a conclusion is made.

Keywords: Microwave surface emissivity · Retrieval method · Remote sensing

1 Introduction

The surface emissivity is defined as the ratio of the thermal radiation emitted by the surface and the thermal radiation emitted by a black body at the same temperature [1]. It reflects the thermal radiation capability of the surface. It is an important parameter for understanding the geophysical processes that control the surface energy balance and surface radiation [2]. In recent years, the surface emissivity has drawn continuous concern. Microwave surface emissivity not only depends on the soil chemical composition, structure and texture, but also affected by the vegetation type and season. It can be used to detect the change of surface type, and can also be used as land surface information into assimilation system [3].

With the development of atmospheric microwave remote sensing technology, satellite microwave observation has become an important means to obtain atmospheric information. But for those data observed over land, in most cases only the observations of upper-level channels or channels not sensitive to land surface are assimilated, which led to the utilization of observation over land is less than that over sea. The reason for this phenomenon is that the land surface emissivity has a greater uncertainty than the

F. Xhafa et al. (eds.), *Advances in Intelligent Systems and Interactive Applications*, Advances in Intelligent Systems and Computing 686,
https://doi.org/10.1007/978-3-319-69096-4_70

sea surface emissivity [4]. Therefore, if the surface emissivity cannot be accurately described, the surface emissivity component and the atmospheric radiation component cannot be distinguished from the observations. The accurate calculation of the surface emissivity is necessary for accurately determining the long-wave energy radiation from the surface and is a prerequisite for assimilating remote sensing satellite observations over land. Therefore, the surface emissivity retrieval method is of utmost importance in the land surface data assimilation.

2 Surface Emissivity Retrieval Methods

2.1 Empirical Statistical Methods

Empirical statistical method is based on the strong correlation between the surface emissivity and the satellite observed brightness temperature. Through the statistical methods to establish the empirical equation between the satellite observed brightness temperature and the corresponding pixel surface emissivity, and apply the equation to the whole study area to calculate the surface emissivity. The simple empirical method is to use the satellite observed brightness temperature as the only factor that affects the surface emissivity, and establish the regression equation between the surface emissivity and the satellite observed brightness temperature directly [5].

According to the detection data on the related channels of MSU oxygen absorption band, Grody put forward a statistical retrieval algorithm which is applicable for the microwave surface emissivity of 50.30, 53.74, 54.96 and 57.97 GHz channels [6]. That is,

$$\varepsilon = a_0(\theta) + a_1(\theta)T_{B50} - a_2(\theta)T_{B53}, \tag{1}$$

where ε is the surface microwave emissivity of the window channel; a_i (i = 0, 1, 2) is the zenith angle coefficient of different places; θ is the satellite zenith angle of the observation point; T_{B50} and T_{B53} are the bright temperature of 50.30 and 53.74 GHz channels, respectively. This algorithm uses the combination of bright temperature of low frequency channel to reflect the microwave surface emissivity of the window channel, and the coefficients in the equation are related to the observation angle. The retrieval results of the microwave surface emissivity of the window channels are basically between 0.8 ~ 0.95, and the microwave emissivity values of lower channels are often affected by precipitation and clouds. The results show that the lower the frequency, the better the retrieval results.

The NOAA National Environmental Satellite, Data, and Information Service (NESDIS) released the AMSU-A surface microwave emissivity product. Through polynomial combination statistics on microwave brightness temperature of AMSU-A channel 1, 2 and 3 to obtain the surface microwave emissivity of channel 3 [7]. The statistical retrieval equation is:

$$\varepsilon = b_0 + b_1T_{B1} + b_2T_{B1}^2 + b_3T_{B2} + b_4T_{B2}^2 + b_5T_{B3} + b_6T_{B3}^2, \tag{2}$$

where ε is the surface microwave emissivity of AMSU-A channel 3; b_i is the regression coefficient, i = 0, 1, …, 6; T_{BM} is the brightness temperature of the M channel. M represents the channel number, M = 1, 2, 3.

2.2 Retrieving Surface Emissivity Using Satellite Observations

In order to calculate the surface emissivity using satellite observations, it is usually necessary to suppose that the surface is under the plane parallel condition. This assumption is only valid when the transcendental information of the surface is absent. Mazler pointed out that it is not applicable to low-altitude observations, the Lambert component is more significant in this case, thus resulting in greater error. Karbou and Prigent [8] demonstrated that the surface emissivity error obtained by this assumption is less than 1% on non-snow-covered surfaces. Mazler suggests that this error can be controlled less by adjusting specific parameters.

Under the assumption of plane parallel atmosphere and given the satellite zenith angle θ, spectral frequency v, the bright temperature observed by satellite sensors can be denoted as [9]:

$$T_b(v,\theta) = T_s\varepsilon(v,\theta)\Gamma + [1-\varepsilon(v,\theta)]\Gamma T_a^{\downarrow}(v,\theta) + T_a^{\uparrow}(v,\theta), \tag{3}$$

$$\Gamma = \exp\left[\frac{-\tau(0,H)}{\cos\theta}\right], \tag{4}$$

where $\varepsilon(v,\theta)$ is the surface emissivity under frequency v and satellite zenith angle θ. T_s is the skin temperature. $T_a^{\downarrow}(v,\theta)$ and $T_a^{\uparrow}(v,\theta)$ are the downward and upward radiation of the atmosphere, respectively. Γ is the transmittance from surface to the top of the atmosphere, which can be expressed as a function of satellite zenith angle θ and atmospheric opacity $\tau(0,H)$. H is the atmosphere top height.

From Eq. 4 we can get the expression of surface emissivity as:

$$\varepsilon(v,\theta) = \frac{T_b(v,\theta) - T_a^{\uparrow}(v,\theta) - T_a^{\downarrow}(v,\theta)\Gamma}{\left[T_s - T_a^{\downarrow}(v,\theta)\right]\Gamma} \tag{5}$$

For some sensors like AMSU, the bright temperature is observed through a rotating antennae system, which makes the calculated emissivity a mixture of horizontally and vertically polarized emissivities. The relationship could be denoted as:

$$\varepsilon(v,\theta) = \varepsilon_p(v,\theta)\cos^2\varphi + \varepsilon_q(v,\theta)\sin^2\varphi, \tag{6}$$

$$\varphi = \arcsin\left(\frac{R}{R+H_{sat}}\sin\theta\right), \tag{7}$$

where θ is the satellite zenith angle, φ is the satellite scan angle, which can be expressed as a function of θ, earth radius R and satellite height H_{sat}.

2.3 One Dimensional Variational Method

One dimensional variational data assimilation (1D-Var) method is actually a physical retrieval method [10]. It can retrieve multiple physical parameters simultaneously using the observations of passive sensors as well as other ancillary data. This method has the advantage of emphasizing the physical constraints of the parameters, and the physical quantities of the parameters can maintain a consistent physical relationship. The theoretical basis of the 1D-Var method is the Bayesian principle, assuming that both the observation error and the background error obey the Gaussian error distribution, then by minimizing the cost function we can get the analysis value with a minimum error [11]. The goal of 1D-Var method is to obtain the best estimation of the retrieved parameters by combing observations and background values from a variety of data sources. Its cost function can be denoted as:

$$J(X) = \frac{1}{2}(X - X_0)^T B^{-1}(X - X_0) + \frac{1}{2}\left(Y^0 - H(X)\right)^T E^{-1}\left(Y^0 - H(X)\right), \quad (8)$$

where X is the analysis value we want to get, X_0 is the background value, Y^0 is the observation value, H is the observation operator, E is the observation error covariance and B is the background error covariance. The premise of solving this equation is to have a radiative transfer model with relatively smaller error, and the simulated radiation value has a good statistical result in the observation error covariance matrix E [12].

The key to retrieve microwave surface emissivity using 1D-Var method is to establish the global microwave surface emissivity first-guess in accordance with the channel characteristics of the satellite sensors. The European Centre for Medium-range Weather Forecasts (ECMWF) developed A Tool to Estimate Land Surface Emissivities at Microwave Frequencies (TELSEM) based on a monthly averaged surface emissivity calculated from Special Sensor Microwave/Imager (SSM/I) satellite observations. TELSEM can provide the estimation of microwave surface emissivity at the frequency of 19–100 GHz and its error covariance matrix [13]. By using TELSEM we can establish the global first-guess of microwave surface emissivity.

2.4 Neural Network Method

Neural network method uses a large number of interconnected neurons to approximate complex nonlinear relationship. This method does not need to know the interaction mechanism between surface emissivity and surface temperature, bright temperature or surface characteristics. It uses the train data set to directly establish the relationship between surface emissivity and input variables. However, this method requires a high degree of accuracy for the first guess, otherwise the result would be not satisfactory.

Aires et al. [14] used first-guess to develop neural network method. They retrieved land surface emissivity between the frequency of 19–85 GHz as well as skin temperature using SSM/I observations. The results show that the retrieval accuracy of surface emissivity for each channel under clear sky (or cloudy) condition is less than 0.008 (or 0.010), which indicates this method can retrieve surface emissivity accurately. Aires et al. [14] also retrieved land surface emissivity of daytime, and the result had a rather small root mean square error.

2.5 Index Analysis Method

If the atmospheric temperature, humidity and surface temperature are known, the microwave surface emissivity can be estimated using the radiative transfer equation. However, in most cases, the atmospheric and surface condition parameters are unknown, so it is necessary to construct a parameter that is sensitive to surface microwave emissivity while insensitive or relatively less insensitive to surface temperature and atmospheric parameters, and then establish a statistical relations between such parameters and microwave surface emissivity. This method is called as index analysis.

A variety forms of feature index can be defined such as soil wetness index (SWI), polarization ratio (PR) and microwave vegetation index (MVI) for different purposes. Morland et al. [15] used the normalized difference vegetation index (NDVI) calculated from satellite observations of visible wavelengths and surface humidity index to estimate the microwave surface emissivity of the African Sahel semi-arid region, which can be denoted as:

$$\varepsilon = a + b\log_c cN, \tag{9}$$

where ε is the surface emissivity; a, b and c are the empirical parameters, N stands for NDVI. The result shows that, there is a good agreement between the retrieved results and the ground observations under the clean dry atmosphere condition. But when the atmosphere is humid or precipitation occurred previously, the retrieved results are somewhat less consistent with the ground observations.

3 Conclusion and Evaluation

Due to the microwave can penetrate through clouds, the spaceborne microwave radiometer can provide information on clouds and precipitations in the atmosphere. Therefore, the microwave surface emissivity is important for microwave precipitation algorithm, atmospheric parameter retrieval and satellite data assimilation. The combination of land surface model and surface emissivity retrieval method can be used to study the changes of surface emissivity on a short time scale.

In the retrieval of microwave surface emissivity, the existing empirical method, neural network method and one-dimensional variational method have been universally recognized and applied, but each of them has its strengths and limitations. The empirical statistical method is one of the simplest methods to estimate the microwave surface emissivity, but which cannot solve non-linear problem [6]. The radiative transfer equation method is based on the radiation transmission theory, whose physical meaning is clear, but the physical procedure is very complex [16]. The index analysis method does not need the atmospheric and land surface conditions, but the application scopes of different indexes are limited [17]. The neural network method's biggest advantage is that theoretically it can be used to approximate any complex non-linear relationships while does not need a complex retrieval algorithm, but the its accuracy depends heavily on the estimation of first-guess. The one-dimensional variation method can effectively reduce the root mean square error of the model analysis field. However, only when the forward model's linearity is strong enough and the background information is good enough, this method can get good results. Further research can be done in the study of dynamic retrieval method and the advanced Kalman filter method of the land surface emissivity.

Acknowledgements. The work was supported by the Natural Science Foundation of China (Grant No. 41375113).

References

1. Felde, G.W., Pickle, J.D.: Retrieval of 91 and 150 GHz earth surface emissivities. J. Geophys. Res. **100**(D10), 20855–20866 (1995)
2. Weng, F.Z, Yan B.H, Grody, N.C.: A microwave land emissivity model. J. Geophys. Res. (2001)
3. Jackson, T.J.: Measuring surface soil moisture using passive microwave remote sensing. Hydro. Process **7**(2), 139–152 (1993)
4. Kazumori, M. English, S.J.: Use of the ocean surface wind direction signal in micro wave radiance assimilation. Q. J. Royal Meterol. Soc. **141**(689), 1354–1375 (2015)
5. Author list, paper title, journal name, vol. no. pages-, year
6. Grody, N.C.: Severe storm observations using the microwave sounding unit. Clim. App. Meteorol. **22**(2), 609–625 (1983)
7. NOAA Satellite-earth Observing Laboratory, Digital Earth Emissivity Information System (DEEIS). http://www.eol.ucar.edu/projects/gapp/dm/satellite/noaa_list.html
8. Karbou, F., Prigent, C., Eymard, L., et al.: Microwave land emissivity calculations using AMSU measurements. IEEE Trans. Geosci. Remote Sens. **43**(5), 948–959 (2005)
9. Karbou, F., Élisabeth, G., Rabier, F.: Microwave land emissivity and skin temperature for AMSU-A and -B assimilation over land. Q. J. Royal Meteorol. Soc. **132**(132), 2333–2355 (2010)
10. Ying, W.U, Wang, Z.H.: Advances in the study of land surface emissivity retrieval from passive microwave remote sensing. Remote Sens. Land Resour. **24**(4), 1–7, (2012)
11. Lorenc, A.C.: Analysis methods for numerical weather prediction. Q. J. Royal Meteorol. Soc. **112**(474), 1177–1194 (1986)
12. Eyre, J.R., Kelly, G.A., Mcnally, A.P., et al.: Assimilation of TOVS radiance information through one-dimensional variational analysis. Q. J. Royal Meteorol. Soc. **119**, 1427–1463 (1993)
13. Filipe, A., Catherine, P., Frédéric, B., et al.: A tool to estimate land-surface emissivities at microwave frequencies (TELSEM) for use in numerical weather prediction. Q. J. Royal Meteorol. Soc. **137**(656), 690–699 (2011)
14. Aires, F., Prigent, C., Rossow, W.B., et al.: A new neural network approach including first guess for retrieval of atmospheric water vapor, cloud liquid water path, surface temperature, and emissivities over land from satellite microwave observations. J. Geophys. Res. **106** (D14), 14887–14907 (2001)

15. Morland, J.C., Grimes, D.I.F., Hewison, T.J.: Satellite observations of the microwave emissivity of a semi - arid land surface. Remote Sens. Environ. **77**(2), 149–164 (2001)
16. Yang, H., Weng, F.: Error sources in remote sensing of microwave land surface emissivity. IEEE Trans. Geosci. Remote Sens. **49**(9), 3437–3442 (2011)
17. Anne, C.W, David, P.K, Shashi, K.G.: Surface emissivity maps for use in satellite retrievals of longwave radiation. NASA Langley Technical Report Server (1999)

Design of Multi-mode Switching Damping Shock Absorber for Active Suspension and Ride Comfort Test

Zhao Jing-bo(✉), Han Bing-yuan, Bei Shao-yi, and Liu Hai-mei

School of Automotive and Traffic Engineering,
Jiangsu University of Technology, Zhenjiang, China
zhaojb@jsut.edu.cn

Abstract. In order to enhance the damping adjustable working mode and the adjustment range, a new multi-mode switching adjustable shock absorber was designed based on a traditional hydraulic damper shock absorber. The structure characteristics and working principle were analyzed. The adjustable damper was controlled in the soft compression and soft rebound mode, the hard compression and soft rebound mode, the soft compression and hard rebound mode and the hard compression and hard rebound mode. The strut assembly and electro-magnetic valve assembly was designed. The ride comfort test of full vehicle under random input and pulse input were conducted respectively. The results shown that the structure design scheme of multi-mode switching adjustable shock absorber was feasible, and the mean square root value of the weighted acceleration of the driver's seat had a significant increase trend with the increase of the speed. The root mean square value of weighted acceleration increased with the increase of damping value under random road input. The adjustment range and comprehensive properties of multi-mode switching adjustable shock absorber for active suspension system are improved. It has important theory value and engineering application prospect for active suspension system and its control strategy.

Keywords: Active suspension system · Damping adjustable shock absorber · Multi-mode switching · Full vehicle test · Ride comfort test

1 Introduction

Damper is used as damping device of active suspension control system, which plays an important role in attenuating vibration and improves ride comfort and handling stability [1]. In order to give full play to the vibration reduction performance of the suspension system, it is required that the damper has a variable damping coefficient to meet the requirements of damping in different driving conditions [2]. The main way to change the damping coefficient of the shock absorber is to change the viscosity of the damper and adjust the orifice area [3]. Variable damper and electrorheo logical damper are used as shock absorber for adjusting damping coefficient of damper [4, 5]. Mechanical damping adjustable damper, Pneumatic control damping adjustable shock absorber,

F. Xhafa et al. (eds.), *Advances in Intelligent Systems and Interactive Applications*, Advances in Intelligent Systems and Computing 686,
https://doi.org/10.1007/978-3-319-69096-4_71

Electromagnetic valve controlled damping adjustable shock absorber and Motor control damper are also used as Damper for regulating orifice area.

Some papers [6] have present a series of orifice area adjustable shock absorber and the damping force can be controlled by adjusting the throttle area of the throttle orifice. It is important theoretical research value application and prospects that how the tension damper and the compression stroke presents different damping characteristics [7], according to the driving conditions, the damping of the damper's stretching and compression stroke is multi-mode switching to meet the more complex conditions of the ride and handling and stability in the same mode [8].

This paper is based on a traditional hydraulic damper. A new type of multi-mode switching damping adjustable shock absorber is proposed. This paper also studies the working principle of damping multi-mode switching adjustable shock absorber and design damping multi-mode switching adjustable shock absorber, We test and verify the feasibility of the structural design of the multi-mode switching damping adjustable shock absorber through the random input test and pulse input test of linear ride comfort under different damping modes.

2 Logic Relationship and Control Strategy

The simplified structure and working principle of multi-mode switching damping adjustable shock absorber are shown in Fig. 1. According to the flow path of the shock absorber in compression stroke [9]. Qt is the flow of oil into the upper cavity, Qb is the flow of oil from the cavity of the shock absorber, Q1 is the flow of oil from the orifice of the piston to the upper cavity, Q2 flow rate of oil flow into the annular gap in the piston cavity, Qy is the flow of oil through the damping control valve, Qcrack is the damping liquid that flows into the cavity of the annular gap, Qc the flow of oil in a flow compensated cavity. According to the structural characteristics and working principle of the multi-mode switching damping adjustable shock absorber. As shown in Table 1, under different mode switching, the combinational logic relationship of Y1 and Y2 can switch each other [10] (Fig. 2).

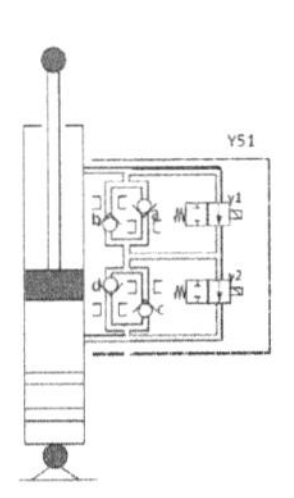

Fig. 1. Working principal of multi-mode absorber

Table 1. Combinational logic table of multi-mode switching damping adjustable shock absorber

State of damping force	Damper valve		Valve control	
	Tensile state	Squeezed state	Solenoid valvey1	Solenoid valvey2
1	Soft	Soft	Open	Open
2	Soft	Hard	Closed	Open
3	Hard	Soft	Open	Closed
4	Hard	Hard	Closed	Closed

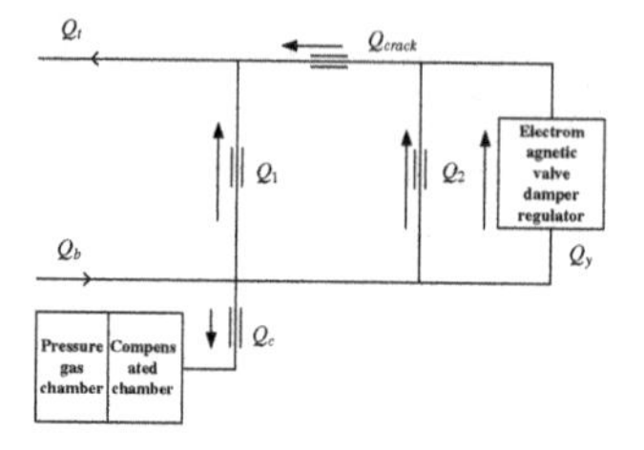

Fig. 2. Flow route of oil for stroke

The main function of the absorber is to attenuate the vibration of the body and improve the ride comfort and handling stability. The control strategy is the flowing [11], Y51 damper connects upper cavity and lower cavity of shock absorber, except for the Y51 fluid flow between the upper and lower cavity, the clearance between the throttle hole on the piston and the wall of the cylinder can also be circulated. Y1, Y2 are very small resistance solenoid valves [12]. Valve a and c are larger resistance one-way valves and Valve b and d are smaller. Shock absorber has four modes of operation by adjusting the two solenoid valve in the open and close combination: the soft compression and soft rebound mode, the hard compression and soft rebound mode, the soft compression and hard rebound mode and the hard compression and hard rebound mode.

3 Pulse Test and Analysis

The experiment mainly collects the acceleration value of the Z axis of the driver's seat, and the relationship between the maximum acceleration response and the speed of the driver's seat is shown in Fig. 3, the maximum acceleration of the driver's seat at the input of the pulse is shown in Table 2. The results of the acceleration value of the front part of the body on the pulse input are shown in Table 3.

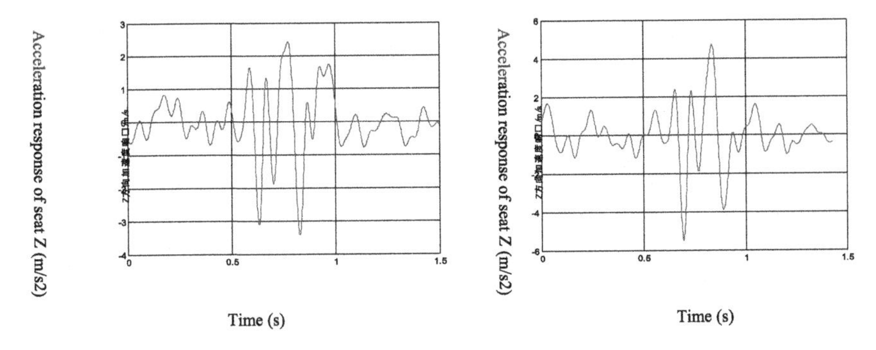

(a) Soft compression and soft rebound mode (b) Hard compression and soft rebound mode

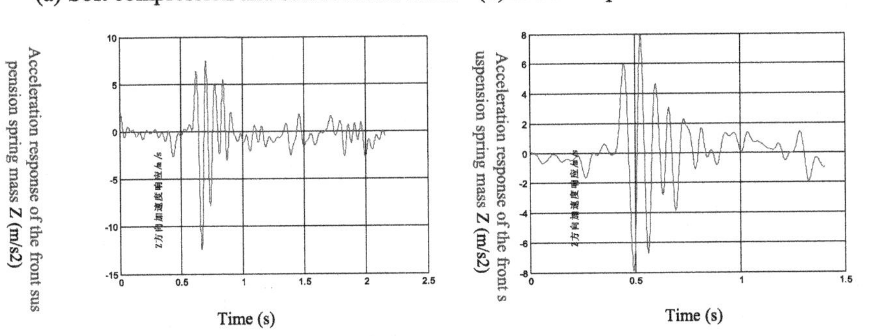

(c) Soft compression and hard rebound mode (d) Hard compression and hard rebound mode

Fig. 3. Acceleration response of the seat and sprung mass under pulse input (60 km/h)

Table 2. The maximum acceleration at driver's seat under pulse input

speed (km/h)	The maximum value of acceleration of seat (m/s^2)			
	Mode (a)	Mode (b)	Mode (c)	Mode (d)
20	2.60	2.49	3.13	3.24
40	3.67	2.85	3.87	3.70
60	2.64	3.47	2.81	4.95

Table 3. The sprung acceleration values on the front body under pulse input

speed (km/h)	The maximum value of acceleration of vehicle body (m/s^2)			
	Mode (a)	Mode (b)	Mode (c)	Mode (d)
20	8.580	6.950	9.210	7.460
40	10.94	19.22	13.53	19.08
60	11.91	20.10	11.97	16.74

Under different speeds, the suspension damping mode is different, which corresponds to the maximum acceleration of the seat; The damping mode corresponding to the maximum acceleration of the mass on the front part of the vehicle body is also different. Under the impulse excitation, the vertical acceleration response peak of the body is much larger than that of the vertical acceleration response of the seat, which reflects the important role of the seat for attenuating the vibration of the car body.

4 Random Input Test and Analysis

According to the national standard, the acceleration value of X, Y and Z axis is collected in the driver's seat, and calculates the total weighted acceleration RMS value, which is shown in Fig. 4. The driver seat three axial acceleration 1/3 frequency spectrum analysis are shown in Fig. 5, the results shows the relationship between the total weighted acceleration root mean square value of the driver's seat and the vehicle speed. The results of the total weighted acceleration RMS of the random road driver's seat are shown in Table 4.

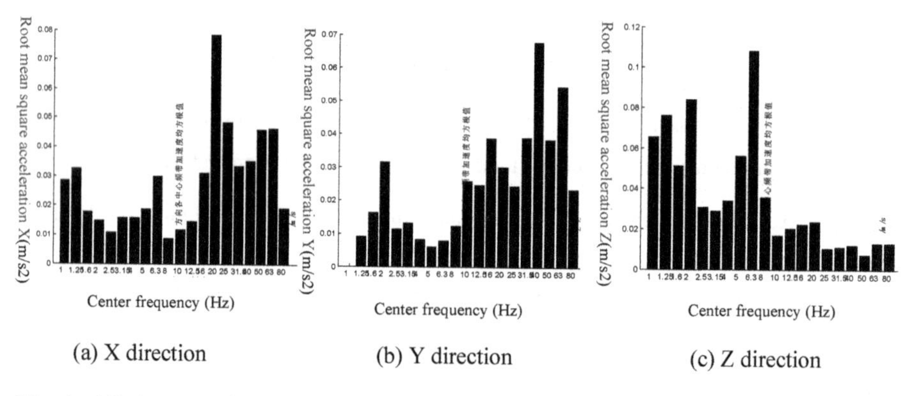

Fig. 4. 1/3 times acceleration spectrum analysis in three axial at the driver's seat (80 km/h)

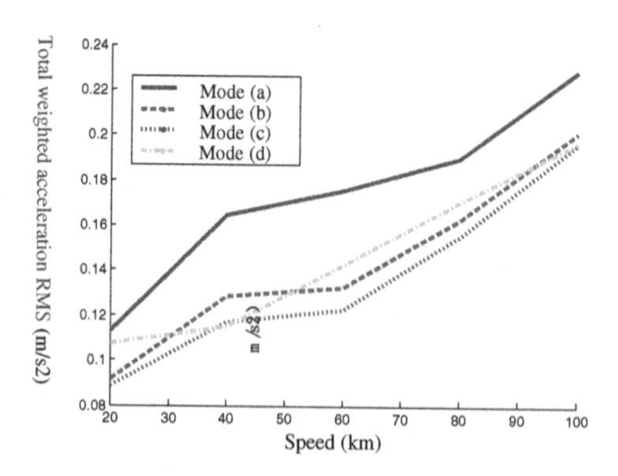

Fig. 5. The relationship between rms value of the total weighted acceleration speed and speed

Table 4. Total weighted acceleration RMS on random road at the driver's seat

Speed (km/h)	RMS value of the total weighted acceleration speed (m/s^2)			
	Mode (a)	Mode (b)	Mode (c)	Mode (d)
20	0.0886	0.0915	0.1074	0.1127
40	0.1171	0.1382	0.1150	0.1643
60	0.1223	0.1223	0.1423	0.1753
80	0.1552	0.1624	0.1702	0.1898
100	0.1961	0.2012	0.1969	0.2282

When the speed is 80 Km/h, weighted acceleration RMS value of the acceleration response of the driver's seat in three directions under different damping modes. In the full frequency range, weighted acceleration RMS, at two frequencies, which is near the natural frequency of the front suspension spring. The mean square root value of the weighted acceleration of the driver's seat had a significant increase trend with the increase of the speed, the acceleration under different damping modes may increase.

5 Conclusion

In this paper, the working principle of multi-mode switching damping adjustable shock absorber is proposed and designed. The ride comfort test of full vehicle under random input and pulse input were conducted respectively, which verify the feasibility of multi-mode switching adjustable shock absorber. Damper is used as damping device of active suspension control system. And the most important performance parameters are different under the model of damping force, the structure design of the shock absorber determines damping regulatory adjustment, according to requirements of damping force range under various driving conditions. Compared with the traditional damping adjustable shock absorber, the mode of multi-mode switching damping adjustable shock absorber is single, and the range is limited, Increase the damping adjustable working mode and adjusting range. It has important theory value and engineering application prospect for active suspension system and its control strategy.

Acknowledgments. Funding from the National Natural Science Foundation of China (Grant No: 61503163), the "333 project" of Jiangsu Province (Grant No: BRA2016440) and the six talent peaks project in Jiangsu Province (Grant No: ZBZZ-024) are gratefully acknowledged.

References

1. Tung, S.L., Juang, Y.T., Lee, W.H.: Optimization of the exponential stabilization problem in active suspension system using PSO. Expert Sys. Appl. **38**, 14044–14051 (2011)
2. Gysen, B.L.J., Janssen, J.L.G., Paulides, J.J.H.: Efficiency of a regenerative direct-drive electromagnetic active suspension. IEEE Trans. Veh. Tech. **60**, 1384–1392 (2011)

3. Kinagi, G.V., Pitchuka, S.P., Sonwane, D.: Hydro pneumatic suspension design for light military tracked vehicle. Commercial Veh. Eng. Congr. **2**, 111–120 (2012)
4. Zehsaz, M., Sadeghi, M.H., Ettefagh, M.M.: Tractor cabin's passive suspension parameters optimization via experimental and numerical methods. J. Terramech. **48**, 439–450 (2011)
5. Ruochen, W., Xiaoliang, Z., Long, C.: Integrated control system of vehicle body height and adjustable damping. Trans. Chin. Soc. Agric. Eng. **28**, 75–79 (2012)
6. Lin, Jinshan: Identification of road surface power spectrum density based on a new cubic spline weight neural network. Energy Procedia **7**, 534–549 (2012)
7. Wang, R., Jing, H., Yan, F., Karimi, H.R.: Optimization and finite-frequency H_∞ control of active suspensions in in-wheel motor driven electric ground vehicles. J. Franklin Inst. **352**, 468–484 (2015)
8. Zhongjie, L., Zuo, L., George, L.: Electromagnetic energy-harvesting shock absorbers: design, modeling, and road tests. IEEE Trans. Veh. Technol. **62**, 1065–1074 (2013)
9. Pan, H., Sun, W., Gao, H., Hayat, T.: Nonlinear tracking control based on extended state observer for vehicle active suspensions with performance constraints. Mechatronics **30**, 363–370 (2015)
10. Sun, W., Pan, H., Zhang, Y., Gao, H.: Multi-objective control for uncertain nonlinear active suspension systems. Mechatronics **24**, 318–327 (2014)
11. Wang, W., Song, Y., Xue, Y., Jin, H., Hou, J.: An optimal vibration control strategy for a vehicle's active suspension based on improved cultural algorithm. Appl. Soft Comput. **28**, 167–174 (2015)
12. Lin, J.-W, Shen, P.F., Wen, H.-P.: Repetitive control mechanism of disturbance cancellation using a hybrid regression and genetic algorithm. Mech. Sys. Sig. Process. **62–63**, 63 (2015)
13. Maciejewski, I., Glowinski, S., Krzyzynski, T.: Active control of a seat suspension with the system adaptation to varying load mass. Mechatronics **24**, 1242–1253 (2014)

Research on the System of Patrol Unmanned Aerial Vehicle (UAV) Docking on Charging Pile based on Autonomous Identification and Tracking

Zinan Qiu[1], Kai Zhang[1,2], and Yuhan Dong[1,2](✉)

[1] Graduate School at Shenzhen, Tsinghua University, Shenzhen, China
dongyuhan@sz.tsinghua.edu.cn
[2] Tsinghua-Berkeley Shenzhen Institute, Shenzhen, China

Abstract. This paper studies the system of patrol unmanned aerial vehicle (UAV) autonomous identifying, tracking the charging pile and distance calculation in the process of achieving innovative autonomous charging target. This paper firstly proposes an SRDCF-based (Spatially Regularized the Correlation Filters) identification and tracking algorithm. The algorithm extracts sift features within the scope of real-time image to match the existing template. Then, a minimum circumscribed rectangle around the target is created as the initial tracking box to determine target area. Afterwards, the target is tracked through training and detecting the location of area. Lastly, camera ranging module with monocular camera measures the distance between the camera and the target. Thus UAV body position relative to charging pile label target can be obtained. The real UAV experimental results shows that the target detection and tracking algorithm can accurately recognize and track the charging pile docking label under the condition of camera movement in the UAV normal flight. Compared with the STC (Spatio—Temporal Context) tracking algorithm, the improved SRDCF algorithm has improved accuracy and robustness obviously. In addition, camera ranging module can accurately measure the distance between camera to the target and the requirement of real-time performance and reliability is reached.

Keywords: Charging · Identification and tracking · Distance measurement

1 Introduction

In recent years, the unmanned aerial vehicles (UAV) has obtained rapid development and has been widely used in the fields of aerial photography, electric power, agriculture, mapping, military and so on. Many power companies at home and abroad are using rotorcraft unmanned aerial vehicle for electric power patrol. However, limited endurance capacity, short cruising range, small cruising radius restrict the further development of the rotorcraft unmanned aerial vehicle.

An effective way to solve this problem is autonomous charging of rotor UAV. The process of UAV independent charging is as follows. Firstly, UAV need determine

F. Xhafa et al. (eds.), *Advances in Intelligent Systems and Interactive Applications*, Advances in Intelligent Systems and Computing 686,
https://doi.org/10.1007/978-3-319-69096-4_72

the position within the scope of specified charging pile through the navigation and positioning module using GNSS (Global Navigation Satellite System) and INS (Inertial Navigation System). After getting the landing license through communicating with the charging pile system, UAV can be guided to land.

In the current Tracking—by—Detection algorithms, the STC(Spatio—Temporal Context) algorithm, space-time context visual tracking was a simple and rapid example [1]. STC algorithm had the strong robustness of illumination, scale changes, background interference, but poor performance with the dramatic change of target in the view or in the low-resolution video.

To achieve landing in charging pile label, it is not enough to just track the target, but the distance between the fuselage and the mark have also to be calculated. Camera ranging has a lot of advantages with low power consumption, high precision, good stability and so on.

To solve the above problem, this paper put forward a SRDCF-based (Spatially Regularized the Correlation Filters) [2] algorithm. In addition, we chose monocular camera that UAVs generally carry to measure the distance between the camera and charging pile label therefore we could provide achieve high-precision protection for the subsequent accurate flight and landing control.

2 Analysis of Algorithm

2.1 Target Detection Based on the SIFT Feature Matching and Target Tracking Based on the SRDCF

SIFT [3] is a detection algorithm of local characteristics. The shortcomings for using the SIFT to extract the image invariant features is hard to accurately extract the feature points for the target with the smooth edge. In order to improve the precision of matching, we introduced the RANSAC algorithm [4]. RANSAC can be used to filter the matching SIFT feature and eliminate false matching points (Fig. 1).

SRDCF (Spatially Regularized the Correlation Filters), is one of the notable algorithms which develops from KCF (Kernerlized Correlation Filter). It used sub-grid

Fig. 1. Result for locating the tracking target in the first frame. Our approach performed successfully at locating the target and generating the minimum circumscribed rectangle box around the target in the first frame.

to search generally after zoom processing and found the location area of target according to the scores. Finally the best position of target would be got for each scale in the frequency. The calculation of score is as follows:

$$s(u,v) = \frac{1}{MN}\sum_{m=0}^{M-1}\sum_{n=0}^{N-1}\hat{s}(m,n)e^{i2\pi(\frac{m}{M}u+\frac{n}{N}v)} \quad (1)$$

2.2 Distance Measurement System with Monocular Camera

UAVs generally carry their own camera therefore camera ranging will not introduce new devices and cost. The imaging geometry of the camera is similar to the principle of pinhole imaging model [5], as the Fig. 2 shows.

According to the principle of similar triangles, if H and F are known, we can find D by calculating the h in the camera image.

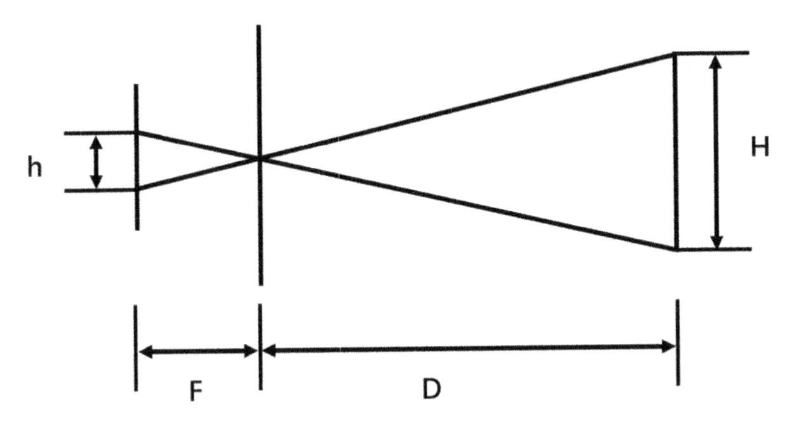

Fig. 2. Pinhole imaging model

$$D = \frac{F \cdot H}{\mathrm{h}} \quad (2)$$

Firstly, we read the video stream output from the camera, and divided it frame-by-frame. Then, the frame was applied gray, the median filter and edge detection. Ensured that the contour passed all our conditions and a min area rectangle was drew around the target. According to the triangle similarity and pinhole imaging model formula (2), the real distance could be calculated eventually.

3 Experimental Results and Analysis

3.1 Comparison of Two Video Tracking Algorithms

Experiment chose a video released by DJI UAV during a summer camp which filmed in a warehouse [6]. In addition, this paper also selected the STC tracking algorithm as contrast experiment and the experimental results of two tracking algorithms were analyzed and compared. The experimental environment is MATLAB R2014a, CPU Intel core i5, memory 4G, Window 7.

The evaluation criteria of target tracking is mainly the consistency of the tracking results and the actual real location. Evaluation criteria one: Center position error. Calculate the distance between the position of the tracking result center and the real target center.

$$E = \sqrt{(x - x_r)^2 + (y - y_r)^2} \tag{3}$$

Evaluation criterion two: calculate the overlap between the tracking box and the real target position.

$$R = \frac{A \cap B}{A \cup B} \tag{4}$$

The experiment selected a video of tracking the charge pile docking label (that is, the green box area). STC required manually choosing the target area in the first frame, but the method of this paper is automatically generating the target box. Since the result of RANSAC algorithm is random, the position of the target box was chosen generally consistent as the automatically generated box in the initial target tracking frame.

STC and SRDCF-based tracking results are shown in Figs. 3 and 4, respectively:

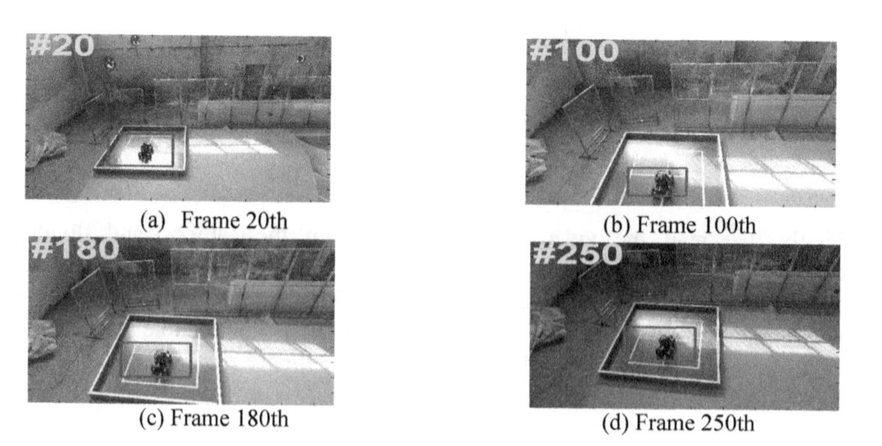

(a) Frame 20th

(b) Frame 100th

(c) Frame 180th

(d) Frame 250th

Fig. 3. Tracking result of the STC algorithm.

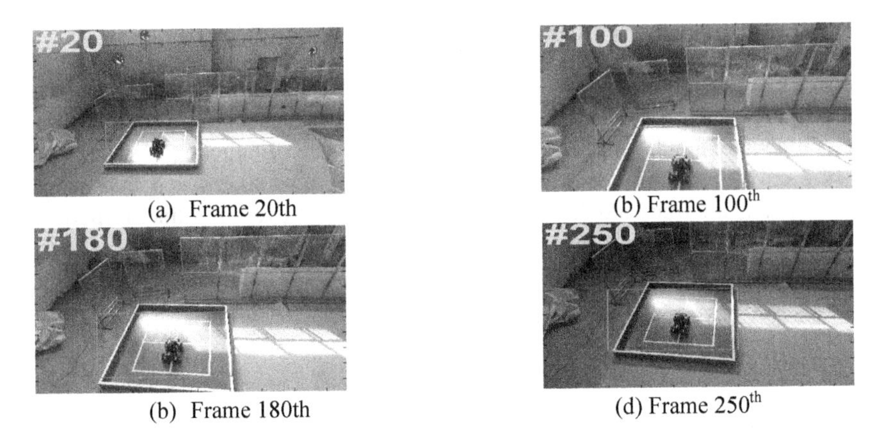

Fig. 4. Tracking result of the based SRDCF algorithm. Our approach achieved better and consistent results in challenging environment, such as occlusions, fast motion, target rotations compared with STC.

Figure 4 shows the result for the autonomous tracking method based on SRDCF adopted in this paper. It could automatically extract the feature and match the target from the beginning in the first frame and generate target tracking box as the objective for the follow-up tracking and detection.

The tracking position error of the target center and target box overlap of the two algorithms is shown in Fig. 5.

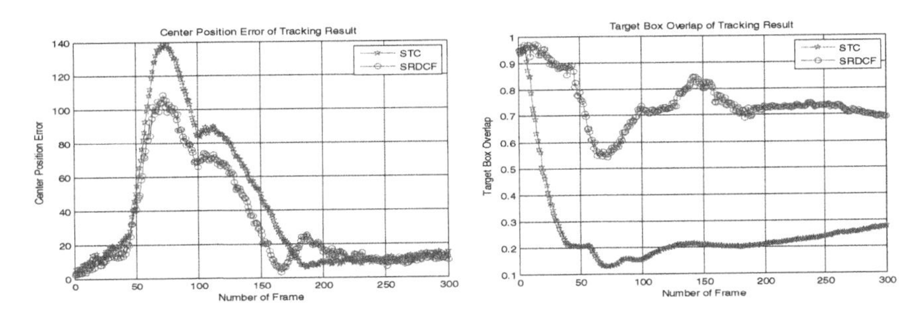

Fig. 5. Center position error and target box overlap of tracking result. Our approach clearly demonstrates robustness in both tracking evaluation criteria.

The comparison shows that the effect of SRDCF-based is better than STC significantly. Figure 5 showed that the SRDCF-based curve was almost under the STC curve and the maximum central position error of STC was close up to 140% of the SRDCF.

3.2 Monocular Camera Distance Measurement

To verify the effectiveness of our monocular camera distance measurement algorithm, we allow the camera to shoot the target within a distance of 80–180 cm from the target. Of course, the distance setting is based on the size of target and our target label is a 14.2 cm * 14.2 cm square area. In the range of 80–180 cm, red lines were set every 10 cm as the true value and we took 100 images at every red line. Experimental results are as the following Fig. 6:

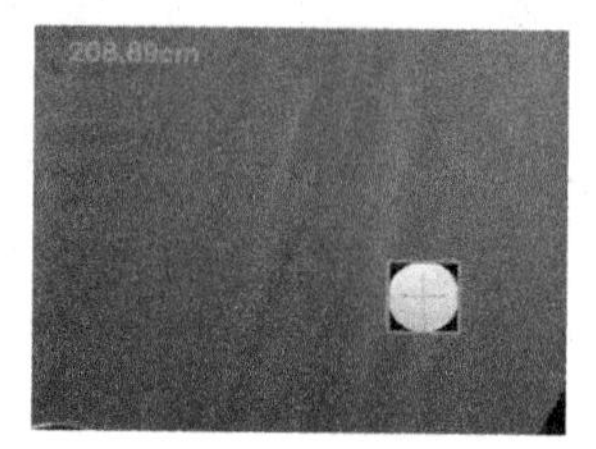

Fig. 6. Image processing process in the distance measurement

The distance calculated by our method were shown in the following Table 1:

Table 1. Experiment results for camera distance measurement. Every 100 images were took at every red line where we calibrated the true values. Then, average measuring distance and error were computed at every 10 cm. Error is defined as average absolute relative deviation.

True distance/cm	Average measuring distance/cm	Error/%
80	79.900	0.125
90	89.990	0.011
100	99.498	0.502
110	109.594	0.369
120	119.604	0.33
130	128.892	0.852
140	140.019	0.013
150	149.851	0.099
160	160.724	0.452
170	170.01	0.005
180	179.926	0.041

4 Conclusion

To achieve UAV independent landing on the charging pile docking area, the premise is that the UAV can always track the charge pile docking label and measure the distance between the body and label in the rapid movement. In this paper, we proposed an autonomous SRDCF-based tracking algorithm. Experiment results showed that, the SRDCF-based target tracking algorithm avoids the time hysteresis or error for manual choosing target area. At the same time, we applied a simple but effective method to measure the distance using the monocular camera that UAVs generally carry. The method we adopted has strong robustness and good adaptability and meets the reliability and accuracy requirements for UAV autonomous landing in charging pile in the wild and other harsh environments.

References

1. Zhang, K., Zhang, L., Yang, M.H., et al.: Fast tracking via spatio-temporal context learning. Computer Science (2013)
2. Danelljan, M., Ha'Ger, G., Khan, F.S., et al.: Learning spatially regularized correlation filters for visual tracking. IEEE international conference on computer vision. IEEE, 4310–4318 (2016)
3. Lowe, D.G.: Object recognition from local scale-invariant features. The proceedings of the seventh IEEE international conference on computer vision. IEEE, 1150 (1999)
4. Fischler, M.A., Bolles, R.C.: Random sample consensus: a paradigm for model fitting with applications to image analysis and automated cartography. Commun. Acm. **24**(6), 726–740 (1987)
5. Liu, Q., Pan, M., Li, Y.W.: Design of vehicle monocular ranging system based on FPGA. Chin. J. Liq. Cryst. Displays, **29**(3), 422–428 (2014)
6. https://pan.baidu.com/s/1dENxNxB

On Data Analysis in Forest Fire Induced Breakdown of the Transmission Line

Jiaqing Zhang[1,3,4](✉), Bosi Zhang[2], Hui Xie[1,3,4], Minghao Fan[1,3,4], and Liufang Wang[1,3,4]

[1] State Grid Anhui Electric Power Research Institute, Hefei 230601, China
dkyzjq@163.com
[2] Department of Safety Engineering, China University of Labor Relations, Beijing 100048, China
[3] Anhui Province Key Laboratory of Electric Fire and Safety Protection, Hefei 230022, China
[4] State Grid Laboratory of Fire Protection for Transmission and Distribution Facilities, Hefei 230022, China

Abstract. The forest fire is one of the great threats to the operation of transmission line. The reasons and prevent measures of forest fire induced breakdown of the transmission line were investigated, and the application of data analysis in this field was discussed in the present study. According to the data statistics on the forest fire induced breakdown of the transmission line, the occurrence of transmission line trip has apparent time and space rules. The flow chart of big data analysis on transmission line breakdown induced by forest fires with the data analysis was proposed. By using big data analysis, important information can be derived, which is helpful to the design of early-warning system, as well as the proper strategic decisions to control the damage of transmission line in forest fires.

Keywords: Data analysis · Big data · Forest fires · Transmission lines · Breakdown

1 Introduction

Breakdown of the transmission line induced by forest fires occurred frequently, and it usually caused huge losses [1–3]. Forest fire became one of the major threats to the safety of transmission lines in the wild. Compared with the transmission line faults caused by other severe disasters, such as strong wind, hail and so on, the trips of transmission line induced by forest fires had low success rate of reclosing and long outage time. Therefore, the breakdown of the transmission line induced by forest fires should receive enough attentions.

Data analysis is one of the most important technologies, which can show us the natural rule of things and provide useful information to help making decisions, and has applied on many industries [4]. However, it has not been introduced into the study on forest fire induced breakdown of the transmission line. The studies on forest fire

F. Xhafa et al. (eds.), *Advances in Intelligent Systems and Interactive Applications*, Advances in Intelligent Systems and Computing 686,
https://doi.org/10.1007/978-3-319-69096-4_73

induced breakdown of the transmission line were discussed, and the application of data analysis in this field was discussed in this study.

2 Overview of the Forest Fire Induced Breakdown of the Transmission Line

The breakdown of transmission line due to forest fire received much concerns [5, 6]. The main type of the breakdown of transmission line in forest fire is tripping operation. Hu et al. [7], Song [8] and Wu et al. [5] discussed the reasons of transmission line trip in their studies. The transmission lines usually locate at the height, which the flame cannot reach in most of forest fires. So the directly effect of the flame to the transmission line is usually not severe. However, the temperature of flame in the forest fires is higher than 1000 °C, so it may decrease the insulating properties of the insulating layer and air around the transmission line [7]. In addition, the ash particles and charge particles may float in the space between the transmission line and floor due to the entrainment of fire plume and environment wind [5]. Therefore, the transmission line discharges to the ground, and causes tripping operation finally. Due to the effect of forest fire, reclosing of the transmission line usually fails [5, 7].

3 Data Statistics and Analysis on Forest Fire Induced Breakdown of the Transmission Line

The principles of the occurrence and distribution of transmission line tripping due to forest fires are important to make proper strategy to remit the damage of transmission line in forest fires. These principles are usually obtained based upon the data recording and statistical analysis. The data of the wildfire induced transmission line tripping in different provinces, and the investigation the rule of transmission line accidents were summarized [9–12]. Some insights are summarized in Tables 1 and 2. It can be found that the transmission line wild fires usually had high risk in January to April. The different among the time period distributions of the fires was due to the region different that the data collected from in literature 1, 2 and 3 listed in Table 1. The average success rates of reclosing in wildfires were around 40% in the two studies in Table 2.

Except for the analysis data in Tables 1 and 2, the rules in different aspects about the transmission line wildfires were also discussed. Lu et al. [10] collected the hotspot data of Hunan province in 2001–2012 derived by using remote sensing satellites. The effect of time period on the occurrence of transmission line fault induced by wildfires was investigated. They claimed that there were six high risk periods in a year. According to the risk ranking, these periods were Tomb sweeping day, Spring ploughing, Spring festival, autumn harvest, the burning autumn harvest and Ghost Festival. Based upon the data of transmission line fires, they also claimed that the number of fires was negatively related to the precipitation. Zhang et al. [11] presented the statistics of some tripping operation of transmission lines induced by wildfires, and the influencing factors and evolution process of wildfire induced tripping of the transmission line were analyzed. Among kinds of wildfires, the ground fires and crown

Table 1. Distributions of transmission line wildfires in different months

Month	Literature		
	Literature 1 [16]	Literature 2 [11]	Literature 3 [17]*
January	0	103	1
February	11	32	23
March	15	49	3
April	3	67	0
May	0	3	0
June	0	0	0
July	0	0	0
August	1	11	1
September	0	7	0
October	0	7	0
November	0	19	0
December	0	35	0

*Data of 220 kV transmission line in 2010

Table 2. Success rate of reclosing in wildfires

Literature	Year					
	2009	2010	2011	2012	2013	2014
Literature 2 [11]*	40%					
Literature 3 [17]		25%	0%	–	33%	

*Data of 220 kV transmission line

fires are most likely to cause the tripping operation of transmission lines. Wu et al. [12] investigated the data of 220–500 kV lines trips in wildfires, as well as the behaviour of these wildfires. It was found that wildfires had high probability in February and March in the south region of China. Once the transmission line trip occurred in wildfires, the success rate of reclosing was usually low. According to the data in Wu's research, the rates that the outage time is longer than 30 minutes in the 220 and 500 kV transmission lines are 79 and 64%, respectively. Based upon the data of trip of 500 kV transmission lines due to the forest fire in Jiangxi Province, Peng et al. [9] indicated that the trip of transmission lines due to the forest fire had high risk in December to next April in a year, and high risk period appeared at 13:00 to 22:00 in a day. Huang and Shu [13] investigated the voltage levels of lines, failure time distribution, fault phase distribution, success rate of line reclosing, fault continuity and locality based upon the data of transmission line trips of China Southern Power Grid in the first quarter in 2010.

To sum, the data related to the breakdown of the transmission line induced by forest fire are valuable. From these data, some rules, such as time distribution of occurrence and fault probability of breakdown of the transmission line were obtained.

4 Insights on Application of Big Data Analysis on Forest Fire Induced Breakdown of the Transmission Line

The concept of "big data" was proposed in 1980, and the technology of big data analysis was applied into IT and financial industries firstly [14]. The big data refers to the data sets whose size is beyond the ability of typical database software tools to capture, store, manage and analyze [15]. Compared with the small-scale data, the big data can provide subtle patterns and heterogeneities [16].

The big data analysis has been paid great attentions by many countries, and applied into many industries [4]. In the field of fire protection engineering, the big data analysis also received great attentions. Based upon the big data and two-layered machine learning model, Dutta et al. [17] established an ensemble method, which provided highly accurate estimation of bush-fire incidence from the climatic data. Li et al. [18] discussed the application of big data technology in fire safety emergency operations and management. The proceedings of a workshop addressed "Big Data and Fire protection Systems" held on 2016 in America [19]. In this workshop, fire researchers identified the opportunities for big data being used in decision making for built-in of fire protection systems. There are also many discussion and exploration on the application of big data analysis in fire protection in China [20–22].

According to the above studies, the applications of big data analysis on fire protection are still limited, and most of these applications are situated at the exploration stage. Big data analysis has been applied on electrical safety analysis [14, 23]. Tajasekaran et al. [24] and Zhu et al. [25] investigated the forest fire prediction as well as the fire alert systems by using big data analysis. However, there little study focused on the breakdown of electrical transmission line induced by forest fire with big data analysis. There are studies [1] focused on the forest fire monitor systems for transmission line, which can provide big data for transmission line breakdown induced by forest fire. The studies discussed in Sect. 3 only developed basic analysis on these data, and showed some simple rules. The deeper data mining and analyses on these data need of urgent development. Because the results are helpful to the development of early-warning technology of transmission line faults, as well as the making of proper strategic decisions in forest fires.

The flow chart of analysis on forest fire induced breakdown of the transmission line though big data analysis is shown in Fig. 1. The big data for analyzing transmission line fault induced by forest fire can be acquire through the forest fire monitor system, State Grid GIS platform, the monitor systems of transmission line and meteorological information system. After data preprocessing, data analysis can be conducted through kinds of methods, such as correlation analysis, cluster analysis, data integration and so on. Finally, some valuable results can be derived to predict the occurrence of transmission line fault in forest fires, and make proper strategic decisions for disaster control.

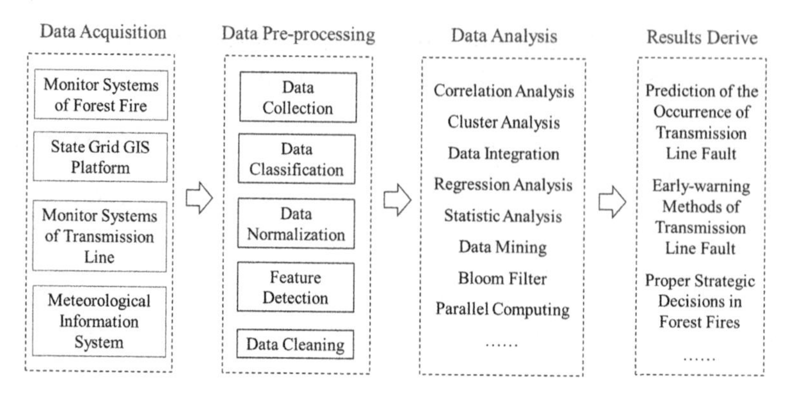

Fig. 1. Flow chart of analysis on transmission line breakdown though big data analysis

5 Conclusions

The studies on forest fire induced breakdown of the transmission line were analysis and the insights on applications of big data technology in this field were discussed in this study. According to the data statistics on forest fire induced breakdown of the transmission line, the occurrence of transmission line trip has high risk in January to April. The average success rates of reclosing in wildfires are only about 40%. The deeper data mining and analyses on the big data of transmission line forest fire need to be developed urgently. The flow chart of analysis on transmission line breakdown induced by forest fires by big data technology was proposed. With the big data analysis, important information can be derived, which contribute to the development of early-warning technology of transmission line faults, as well as the proper strategic decisions in forest fires.

References

1. Lu, J.Z., Wu, C.P., Li, Y., et al.: Research and application of forest fire monitor and early-warning system for transmission line. Power Sys. Prot. Control **42**(16), 89–95 (2014)
2. Li, Y., Tao, X.J., Chen, L.: The analysis and precautions of forest fire fault to transmission line. Hubei Electric Power **34**(6), 49–50 (2010)
3. Song, J.J., Guo, C.X., Zhang, J.J., et al.: A probabilistic model of overhead transmission line outage due to forest fire. Power Sys. Tech. **37**(1), 100–105 (2013)
4. Chen, M., Mao, S., Zhang, Y., et al.: Big Data: Related Technologies, Challenges and Future Prospects. Springer, Heidelberg (2014)
5. Wu, T., Ruan, J.J., Hu, Y., et al.: Study on forest fire induced breakdown of 500 kV transmission line in terms of characteristics and mechanism. Proc. CSEE **31**(34), 163–170 (2011)
6. Huang, D.C., Li, P., Ruan, J.J., et al.: Review on discharge mechanism and breakdown characteristics of transmission line gap under forest fire condition. High Voltage Eng. **41**(2), 622–632 (2015)

7. Hu, X., Lu, J.Z., Zeng, X.J., et al.: Analysis on transmission line trip caused by mountain fire and discussion on tripping preventing measures. J. Electric Power Sci. Tech. **25**(2), 73–78 (2010)
8. Song, J.J.: Time-varying outage models of overhead transmission line based on adverse disaster. Master Thesis, Zhejiang University (2013)
9. Peng, F.X., Xiong, X.: Study on tripping of 500 kV transmission line induced by wildfire and preventing measures. Power Sys. Tech. **2**, 58–61 (2014)
10. Lu, J.Z., Liu, Y., Yang, L., et al.: Rules of transmission line fire induced by wildfire. Fire Sci. Tech. **33**(12), 1448–1451 (2014)
11. Zhang, Y., You, F., Chen, H.X., et al.: Statistical analysis of wildfire accidents inducing flashover of transmission lines. Fire Sci. Tech. **30**(12), 1177–1180 (2011)
12. Wu, T., Ruan, J.J., Zhang, Y., et al.: Study on the statistic characteristics and identification of AC transmission line trips induced by forest fires. Power Sys. Prot. Control **40**(10), 138–148 (2012)
13. Huang, L., Shu, S.Y.: Analysis on fault trips caused by forest fire in csg in the first quarter of year 2010. Guangdong Electric Power **24**(3), 95–97 (2011)
14. Li, Z.P.: Research on transmission line management system and fault diagnosis based on big data analysis. Master Thesis, Zhejiang University of Technology (2015)
15. Trnka, A.: Big data analysis. Eur. J. Sci. Theol. **10**(1), 143–148 (2014)
16. Fan, J., Han, F., Liu, H.: Challenges of big data analysis. Nat. Sci. Rev. **1**(2), 293–314 (2014)
17. Dutta, R., Das, A., Aryal, J.: Big data integration shows Australian bush-fire frequency is increasing significantly. Royal Soc. Open Sci. **3**(2), 150–241 (2016)
18. Li, F., Reiss, M.: Application of "big data" for intelligent fire safety emergency operations and management. Council on Tall Buildings and Urban Habitat, Guangzhou (2016)
19. Big data and fire protection systems, Big data and fire protection systems workshop, San Antonio (2016)
20. Li, Q.: Application of big data on the field of fire protection. Telecom World, 10 (2014)
21. Yu, W., Li, H.T., Zhang, D.C.: Fire work is faced with opportunities and challenges in the age of big data. Fire Sci. Tech. **33**(9), 1061–1063 (2014)
22. Chen, H.Y.: Study on fire protection system by big data. China New Tech. Prod. **03**, 188 (2016)
23. Zhang, Z.C.: Electrical fire risk assessment based on data mining technology. Master Thesis, Capital University of Economics and Business (2016)
24. Rajasekaran, T., Sruthi, J., Revathi, S., et al.: Forest fire prediction and alert system using big data technology. International conference on information engineering, Management and security, London (2015)
25. Zhu, S.J., Zhang, J.Y., Xing, Z., et al.: A simulation model of big data analysis for fire alarm. International conference on advances in energy, Environment and chemical engineering, Changsha (2015)

Ultrasonic Guided Wave Testing Method of Gun Barrel Crack Defects Based on L (0, 2) Mode

Jin Zhang(✉), Xin Wang, Ying Wei, and Yang Shen

New Star Research Institute of Applied Technology, HeFei 230032, China
JGXYZhangJin@163.com

Abstract. Current detection means are faced with the single test parameters and are unable to detect the internal hidden defects; also, the point detection method adopted is of a low detection efficiency. In this paper, with the establishment of the three-dimensional model of barrel defects through the finite element simulation and the adoption of the line detection method based on the L (0, 2) mode guided wave, the quantitative relationship between the axial and circumferential size of crack defects and the echo reflection coefficient could be drawn, and a kind of nondestructive testing method of cannon gun tube crack defects based on ultrasonic guided wave will be put forward. Simulation experiments show that this method can realize high, quantitative detection of cannon gun tube crack defects.

Keywords: Nondestructive testing · L (0, 2) mode · Barrel crack · Reflection coefficient

1 Introduction

The gas of both high temperature and high pressure produced in weapon launch process makes the bore wall thin layer produce a sharp change of temperature through the wall bore forced heat transfer, resulting in changes and melting of wall materials, thus causing damages such as barrel ablation, cracks, grooves wear, corrosion, metal fouling etc., and then reducing the cannon firing accuracy, residual life and operational effectiveness [1–3]. Therefore, the effective and high precision detection of cannon gun tube defects is very important. Although the principle of the traditional barrel inspection method is relatively simple and the method is mature, there is a low accuracy and low efficiency of the detection [4]. Ultrasonic guided wave testing method, as a new nondestructive testing, shows incomparable advantages in underground pipeline and complex structure detection and other fields. Some scholars studied propagation mechanism of ultrasonic guided wave through complex structures such as the pipes, the thick beam, the pressure vessel and variable cross-section parts (such as high-speed rail wheels and axles), and analyzed the influence of different defect types to guided wave propagation [5], which provides theoretical basis for guided wave nondestructive testing [6, 7]. But the cannon gun tube defects ultrasonic guided wave nondestructive testing has not yet been reported. This article is based on

F. Xhafa et al. (eds.), *Advances in Intelligent Systems and Interactive Applications*, Advances in Intelligent Systems and Computing 686,
https://doi.org/10.1007/978-3-319-69096-4_74

the propagation characteristics of ultrasonic guided wave and the line detection instead of the conventional detection is adopted. With the establishment of three-dimensional model of barrel defects through the finite element simulation, the quantitative relationship between the axial and circumferential size of crack defects and L (0, 2) mode guided wave echo reflection coefficients could be drawn so as to realize the high efficiency and high precision detection of cannon gun tube crack defects.

2 Ultrasonic Guided Wave Nondestructive Testing Methods of Barrel Defects

2.1 Characteristics Analysis of Barrel Dispersion

The parameters of a certain type of cannon gun tube are set with inner diameter a = 62.5 mm, outside diameter b = 77.5 mm, and the length is 400 mm. Material parameters are shown in Table 1 [8]. With Disperse dispersion analysis software [9], From the longitudinal guided wave group velocity dispersion curve in barrel, it can be found that the dispersion phenomenon of low frequency L (0, 2) modes is not obvious with only two modes of L (0, 2) and L (0, 1). Also, the L (0, 2) mode guided wave propagation speeds faster than L (0, 1) mode, and L (0, 2) mode guided wave signals can be extracted from receiving echo signals through time window technology.

Table 1. Parameters of a certain type of cannon gun tube

Parameter	Value	Unit
Poisson's ratio	0.3	1
Density	7800	kg/m^3
Modulus of elasticity	203	Gpa

2.2 Principle of Ultrasonic Guided Wave Nondestructive Testing of Barrel Defects

The principle diagram of ultrasonic guided wave testing of cannon gun tube defects is shown in Fig. 1. A sensor thimble is fixed at the barrel side, and the piezoelectric transducer is evenly distributed, stimulating ultrasonic guided wave of longitudinal modes. Ultrasonic guided wave transmits within the tube wall, glutted with the whole body thickness direction. In the process of transmission, when ultrasonic guided wave meets the end face, it will produce face echo, continuing to spread along the body wall. However, when guided wave encounters defects in propagation process, guided wave mode converts due to the discontinuity of medium. The new mode resulting from the modal conversion is due to the existence of defects, so it carrys the barrel defect information. The location and size of the defect will be detected by comparing the new mode with reference values.

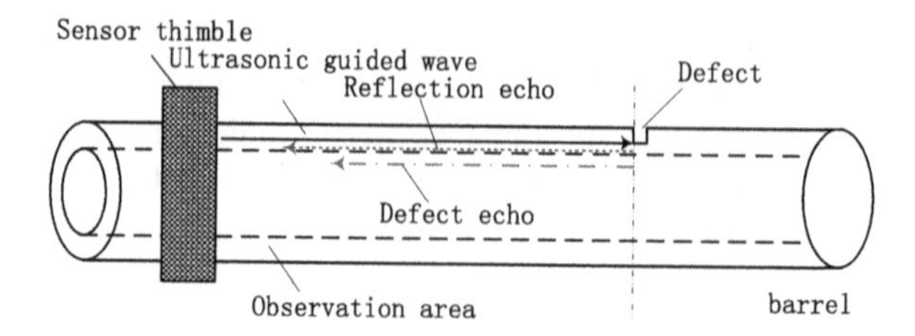

Fig. 1. The detection principle of ultrasonic guided wave

Defect location refers to the detection of the location of the defects in the barrel. To be specific, it refers to the calculation of the defect location away from the signal receiving end by using the echo time and the echo wave velocity as ultrasonic guided wave will produce the transformation mode echo when it encounters the barrel defects. Cracks with different depths, axial sizes and circumferential sizes are respectively set up at the inside surface of the tube, and the defect echo amplitudes, which are produced when guided wave converts the modes, are recorded under the circumstance of a defect size. The ratio of the maximum amplitude of the echo to the maximum amplitude of the excitation signal is defined as the echo coefficient, and the echo coefficient curves are drawn, respectively showing the relationship between echo coefficient and crack depth, axial size and circumferential size. Therefore, the relationship can be seen between the echo crack coefficient and the crack size, thus achieving the quantitative detection of the crack defects.

In this paper, the sine signal which is modulated by hanning window with its ultrasonic excitation frequency being 250 kHz is chosen as the excitation source. Its longitudinal transmission velocity is $C_{gL(0,2)} = 5388.5\,\mathrm{m/s}$.

3 Barrel Model and Meshing

100% circumferential crack defects are set on the 200 mm wall of the barrel middle with the depth being 50% of the thickness of the wall and the axial width being 1 mm. The model is shown in Fig. 2. The time step is set to $\Delta T < l/V_g$ and the minimum mesh size to l to guarantee the accuracy of the calculation. $T > (2l/V_g)$ is used as the transmission time for calculating, and l is the length of the pipe to ensure enough information. In this paper, $\Delta T = 0.5$ us, $T = 200$ us.

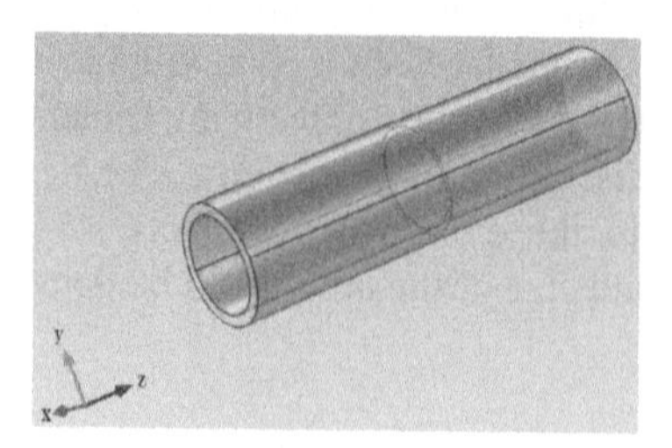

Fig. 2. Barrel model

After repeated simulation validation, the barrel model grid is set to 1/6–1/20 of the acoustic wavelength. The wavelength of longitudinal mode guided waves and the grid size are calculated as follows:

$$\lambda = \frac{c}{f} = \frac{5388.5}{250000} \approx 21.6\,\text{mm},\ \frac{1}{20}\lambda < n < \frac{1}{6}\lambda,\ n = 3\,\text{mm}$$

Λ refers to the guided wavelength; n is the maximum size of the grid, and the minimum size is set to be 1 mm. Barrel grid subdivision is shown in Fig. 3.

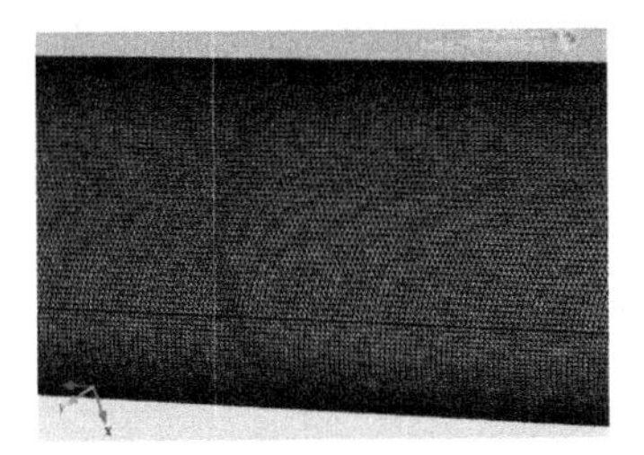

Fig. 3. Barrel grid model

4 The Simulation Experiment and Results

Under the effect of the axial excitation source, when t = 66 μs, the transmission process of L (0, 2) mode guided wave in the barrel is shown in Fig. 4.

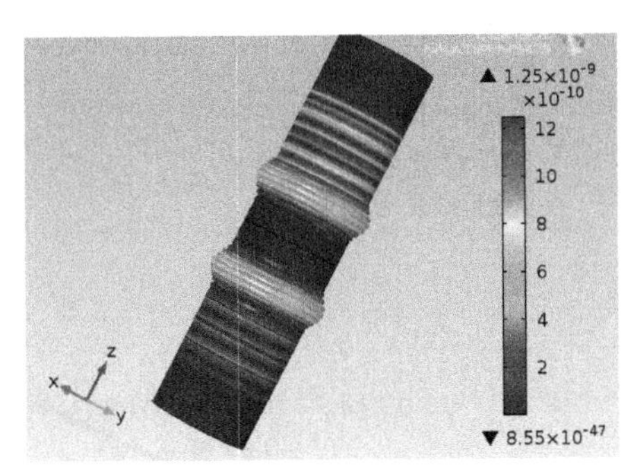

Fig. 4. L (0, 2) mode guided wave transmitting in the barrel

It is clearly seen from the picture that when L (0, 2) mode guided waves spread through the barrel wall crack defects, two modal guided waves come into being in tandem. This shows that the guided wave produces hereby a mode conversion phenomenon. Through the contrast of the lines' colors in the picture, most of the excitation guided waves going by defects convert into two modal guided waves through transmission, and a small number reflect into the same two modal guided waves. Differences drawn from the contrast with the propagation process schematic diagram of the L (0, 2) mode guided wave in the non-defect barrel indicate that the barrel defect detection can be achieved based on the L (0, 2) mode guided wave.

4.1 The Relationship Between Circular Crack Depth and Reflection Coefficient

A point on inner wall of the excitation end is selected as the observation point, and defect detection data can be obtained through the observation of the axial displacement change of the point as time passes by. Then the lining crack depth of the barrel is changed. Through the simulation experiment, an echo amplitude shown in Table 2 is obtained and the fitting curve is shown in Fig. 5.

Table 2. The experimental data of the relation between crack depth and reflection coefficient

The crack depth percentile (%)	The large amplitude of a flaw echo (mm)	Reflection coefficient (%)
1	8.5428*E−12	0.57
2	2.2153*E−11	1.48
3	3.1733*E−11	2.12
4	4.1108*E−11	2.74
5	5.0880*E−11	3.39
10	1.1277*E−10	7.52
15	1.8589*E−10	12.39
20	2.7619*E−10	18.41
25	3.7023*E−10	24.68
30	4.8085*E−10	32.06
35	5.9246*E−10	39.50
40	7.2240*E−10	48.16
45	8.5769*E−10	57.18
50	1.0026*E−9	66.84

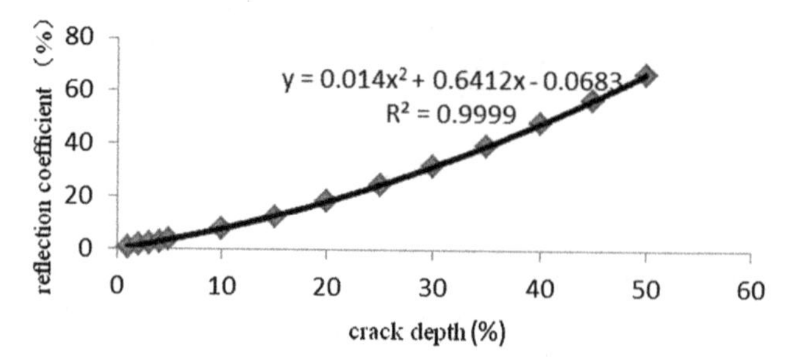

Fig. 5. The curve of the relation between crack depth and reflection coefficient

It can be seen from Fig. 5 that each data point meets the relationship of the quadratic polynomial equation, which is:

$$y = 0.014x^2 + 0.6412x - 0.0683 \quad (1)$$

Here, the horizontal axis represents the ratio of crack defect depth to barrel thickness, and the vertical axis represents reflection coefficient of a crack defect echo. It can be seen from the fitting results that the curve fitting degree R^2 is 0.9999, approximate to 1.

4.2 Crack Circumferential Position

The crack with the circumferential size Δθ being respectively 25 and 50% is set, as well as the location of the crack to be in the middle of the barrel, being only 200 mm far from the incentive side; also, its circumferential position θ is set to be 0. Then the simulation data obtained is shown in Table 3.

Table 3. Circumferential distribution of 25%, 50% circumferential crack echo

θ(°), $\Delta\theta$=25%	μ(mm), $\Delta\theta$=25%	θ(°), $\Delta\theta$=50%	μ(mm), $\Delta\theta$=50%
0	10.6455*E−10	0	9.9776*E−10
22.5	8.7601*E−10	7.5	10.1079*E−10
45	4.9803*E−10	15	10.4827*E−10
90	2.3115*E−10	30	11.1982*E−10
180	0.8208*E−10	60	8.3179*E−10
270	2.2988*E−10	90	4.9663*E−10
315	4.9693*E−10	135	2.2214*E−10
337.5	8.7577*E−10	180	2.3312*E−10

The maximum amplitude distribution in the direction of barrel circumferential of the defect echo in the Table 3 is drawn into the polar diagram as shown in Fig. 6.

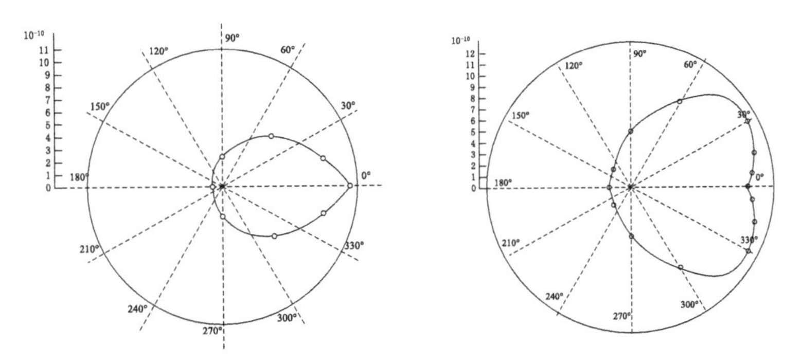

Fig. 6. Circumferential distribution of crack echo amplitudes with different circumferential sizes **a** 25%, **b** 50%

Through comparative analysis of the diagram 6 (a) and (b), it can be found that the echo amplitude is increased with the increment of crack circumferential size, and the ratio of the corresponding echo with the defect to the corresponding echo with the non-defect is also increased. When crack circumferential size increases to a certain extent, the maximum amplitude of the defect echo presents two extreme values in the circumferential direction as the reflection echo will expand along the circumferential. At that time the corresponding defect echo amplitude at the center position of the crack is no longer the maximum but in the middle of the two new extreme values. Therefore, when the transverse cracks in the barrel are circumferential located in the use of L (0, 2) mode guided waves, the maximum of the defect echo amplitude will correspond to the crack center if the crack circumferential size is small. When the crack circumferential size increases to the appearance of two echo amplitude extreme values, the center position of the two echoes corresponding to the circumferential position is the circumferential position of the crack center.

4.3 The Relationship between Crack Circumferential Size and Reflection Coefficient

In order to observe the relationship between crack circumferential size and reflection coefficient of the crack defect, the circumferential crack length is respectively set to 6.25, 12.5, 25, 37.5 and 50% with the crack width of 1 mm and the depth of 50% of the thickness of the barrel wall. Through the use of COMSOL post-processing function,

Table 4. The experiment results of crack circumferential size and reflection coefficient

Crack circumferential size (%)	The large amplitude of a defect echo (mm)	Reflection coefficient (%)
6.25	2.8932*E−10	19.29
12.5	5.6779*E−10	37.85
25	10.6455*E−10	70.97
37.5	12.6056*E−10	84.04
50	10.026*E−10	66.84

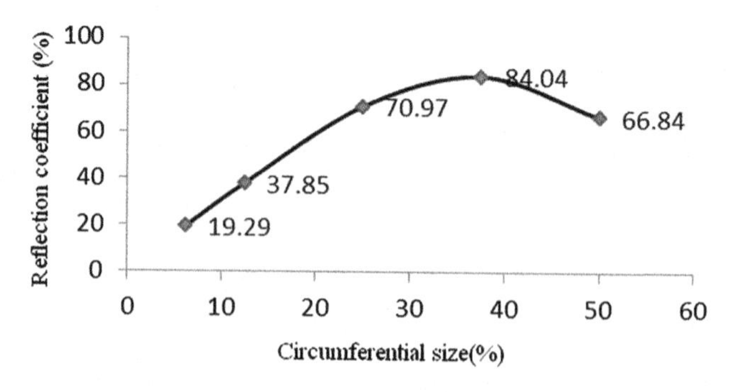

Fig. 7. The relation curve between crack circumferential size and reflection coefficient

the echo reflection coefficient of the crack defect corresponding to every crack circumferential length is respectively recorded as shown in Table 4, and the fitting curve is shown in Fig 7.

From the analysis of the variation trend of the curve in Fig 7, it can be found that with the increase of crack circumferential size, when crack circumferential size is less than 25% of the barrel length, the reflection coefficient of a guided wave defect echo basically keeps a linear growth with crack circumferential size. As the crack circumferential size continues to increase, the increasing trend of reflection coefficient of a defect echo slows down. When the circumferential size increases to 37.5% of the barrel circumference, the reflection coefficient of a defect echo is reduced instead of a monotone increasing trend.

This phenomenon coincides with the two maxima amplitude discussed above about the circumferential position of the transverse cracks. When the circumferential size of transverse cracks increases to a certain extent, the maximum defect echo appears two maxima at the circumferential position corresponding to the crack, and the defect echo amplitude corresponding to the crack center position will naturally decreases. Conclusions are consistent above and below [10].

5 Conclusion

On the basis of the three-dimensional finite element model of barrel crack, the influences of crack defects, which are of different sizes and directions, to the propagation of L (0, 2) mode guided wave in the pipe have been studied through simulation experiments. A kind of nondestructive testing method of the tube defect ultrasonic guided wave has been put forward, and can realize the positioning and quantitative detection of crack defects. The experimental results show that:

(1) With the increase of circumferential crack depth, guided wave reflection coefficient grows non-linearly, which meets the growth rule of the formula (1).
(2) As the crack circumferential size grows, the maximum amplitude of flaw echoes presents two extreme values. Therefore, when the transverse cracks circumferential located, the maximum of the defect echo amplitude will correspond to the crack center if the circumferential size is small. When the crack circumferential size continues to grow, the crack center has two extreme values corresponding to the center of the circumferential position.
(3) The circumferential spread of guided wave reflection signal will lead to the non-linear growth of echo reflection coefficient as the crack circumferential size increases. Turning point appears when the axial length of the defect is 37.5%.

Acknowledgments. This work was financially supported by the Equipment Advanced Research Foundation under (6140004030116JB91001).

References

1. Wang, G., Jin, Z., et al.: Cannon gun tube nondestructive testing methods. Fire Command Control. **23**(7), 18–20 (2015)
2. Yuhas, D.E., Mutton, M.J., Remiasz, J.R., et al.: Ultrasonic measurements of bore temperature in large caliber guns. Review of Progress in Quantitative NDE University of Illinois (Chicago), 20–25 July 2008
3. Jin, Zhi-ming: Gun interior ballistics. Beijing Institute Of Technology Press, Beijing (2004)
4. Yong, Jin, Zhi-jian, Su: Comparing to conventional NDT methods. Sci. Tech. Inf. **20**, 143 (2011)
5. Zheng, Ming-fang, Chao, Lu, Chen, Guo-zhu, Men, Ping: Modeling three-dimensional ultrasonic guided wave propagation and scattering in circular cylindrical structures using finite element approach. Phy. Procedia **22**, 112–118 (2011)
6. Gaul, L., Sprenger, H., Schaal, C., Bischoff, S.: Structural health monitoring of cylindrical structures using guided ultrasonic waves. Acta Mech. **223**, 1669–1680 (2012)
7. Ian, A.: Understanding and Predicting Gun Barrel Erosion. Technical Report DSTO, Australia (2005)
8. Bai, D.-Z.: Body tube failure and gun steel material. Weapons Industry Press, Beijing (1989)
9. Ta, D.-A., Wang, W.-Q., Wang, Y.-Y., et al.: Pipeline guided wave inspection in the choice of excitation frequency and sensitivity analysis. Nondestr. Test. **27**(2), 83–86 (2005)
10. Zheng, M.-F., Chao, L., Chen, G.-Z., Men, P.: Modeling three-dimensional ultrasonic guided wave propagation and scattering in circular cylindrical structures using finite element approach. Phy. Procedia **22**, 112–118 (2011)

Target Re-identification Based on Dictionary Learning

Gong Lianyou(✉) and Shi Guochuan

Army Officer Academy, Computer Center, Hefei 230031, China
2430401385@qq.com

Abstract. It is one the hot topics that how to improve the success rate of target recognition in the field of target recognition. Aiming at the change of the appearance of the target in the multi-camera surveillance videos, proposed a new algorithm of target re-identification that based on the sparse representation and LCC dictionary learning. Firstly, we constructs a joint learning dictionary, then training the joint dictionary by using the target data in the surveillance images. Then extracting the main features and carry on sparse coding of the re-identification target. Finally, using the features and sparse coding to match all the target data in another camera. In the simulation experiments, the results proved the superiority of ours algorithm, reduced the computational complexity, and improve the success rate of the re-identification.

Keywords: Dictionary learning · Re-identification · Sparse representation · Feature space · LCC

1 Introduction

In the field of target re-identification, how to improve the success rate of re-identification is a basic problem. Target re-identification especially difficult, because of the view angle and different external conditions may lead to the same target's appearance a significant change in different surveillance images. Early methods mainly depend on the face recognition technology [1]. These methods' recognition rate are very low when dealing with low resolution video and irregular images. In recent years, many researchers have proposed a series of new methods for the representation of appearance features, including weighted consistent region [2] and symmetry-driven accumulation [3]. At present, the re-identification algorithms of target can be divided into distance learning and local feature matching.

In the algorithms that based on distance learning, the target re-identification is described as an optimal similarity measure between two images. However, these algorithms usually need to label a large number of targets, which greatly increases the computational complexity. In the algorithms that based on local feature matching, the target re-identification is based on the local feature matching score. Compared with algorithms that based on distance learning, local feature matching algorithms can reduce the computational complexity to a certain extent, but the success rate is still low.

How to reduce the computational complexity and improve the re-identification rate is a difficult problem. Aiming at this, we proposed a new algorithm based on sparse

F. Xhafa et al. (eds.), *Advances in Intelligent Systems and Interactive Applications*, Advances in Intelligent Systems and Computing 686,
https://doi.org/10.1007/978-3-319-69096-4_75

representation and LCC dictionary learning. We assumes that the intrinsic structure of the same target is invariant, then constructed a proper over complete dictionary, then using the dictionary to describe the target's intrinsic structure features by sparse representation, finally, completed the re-identification process. Compared with traditional algorithms, the algorithm we proposed reduced the computational complexity and improves the success rate of target re-identification. The experimental results proved the superiority of our algorithm.

2 Dictionary Learning

2.1 LCC Dictionary Learning

Local Coordinate Coding (LCC) is a high dimensional nonlinear learning method based on manifold data. In the method of LCC, we given a set of unlabeled training data $\{x_1, \ldots, x_n\} \in R^{d \times n}$and dictionaries $D = \{d_1, \ldots d_n\} \in R^{d \times k}$, the dictionary learning is obtained by minimizing the reconstruction variance and the local penalty objective function.

$$\min_{D \in L, \alpha} \frac{1}{2} \|x - D\alpha\|^2 + \mu \sum_j |a^j| \|d_j - x\|^2 \tag{1}$$

Where α is the sparse coefficient of x, $a^j$and d_j indicate the jth component and the jth column elements of α and D respectively. $L = \{D \| di \leq 1, i = 1, \ldots, k\}$ is a convex set of D. Given a set of training samples, we can obtain an over complete dictionary by learning the set. The process can be completed by minimizing the objective function of the sum of all data samples to optimize D and α.

$$\min_{D \in L, \alpha} \sum_i \left(\frac{1}{2} \|x_i - D\alpha_i\|^2 + \mu \sum_j |a_i^j| \|d_j - x_i\|^2 \right) \tag{2}$$

Where α_i represents the coefficient of x_i. The above objective function is non-convex on D and α, so it is difficult to synchronize D and α. However, when we fixed α, the function of D is convex, and vice versa. Therefore, it is possible to optimize one variable when fixed the other variable, executed of this step to achieve the optimization process of D and α alternately. Specifically, when D is fixed, the different α_i can be separated into a single sparse coding problem. Definition$\beta = \Lambda\alpha$, Λ is a diagonal matrix, its elements are $\Lambda_{jj} = \|d_j - x\|^2$. While $d_j \neq x$, the inverse matrix of Λ exists, (otherwise, the x expressed by itself). Therefore, fixed D and x, the optimization of α can be converted to the optimization of β.

$$\min_{\beta} \frac{1}{2} \|x - D\Lambda^{-1}\beta\|^2 + \mu|\beta|_1 \tag{3}$$

Where $|\beta|_1 = \sum_j |\beta^j|$ represents L_1 norm. After optimizing of (3), can get α by $\alpha = \Lambda^{-1}\beta$, then we can use the constrained quadratic programming to optimization D.

$$\begin{aligned} &\min_{D \in L} \sum_i \left(\frac{1}{2} \|x_i - D\alpha_i\|^2 + \mu \sum_j |a_i^j| \|d_j - x_i\|^2 \right) \\ &= \min_{D \in L} \frac{1}{2} \operatorname{tr}\left[D^T D \left(\sum_i a_i a_i^T + 2\mu \sum\nolimits_i \right) \right] - \operatorname{tr}\left[D^T \left(\sum_i x_i a_i^T + 2\mu x_i \bar{a}_i^T \right) \right] \end{aligned} \tag{4}$$

Where $\bar{a}_i$ is the absolute value of α_i, $\sum_i$ is a diagonal matrix constructed by $\bar{a}_i$. Using iterative updates to solve the above minimization problem, we need to store two matrices $A = \sum_i a_i a_i^T + 2\mu \sum_i$ and $B = \sum_i x_i a_i^T + 2\mu x_i \bar{a}_i^T$, then using the block coordinate descent method to find the optimal dictionary, specifically, when updating the *j*th column elements of d_j^k, fixed other columns in the **K** iteration.

$$d_j^{k+1} = \Pi_L\left(d_j^k - \frac{1}{a_{jj}} \left(D^k a_j - b_j \right) \right) \tag{5}$$

Where a_j and b_j are represents the *j*th column of matrix A and B respectively, Π_L represents its projection on L.

In practice, we set **k** = 500, the balance coefficient is set to 0.15, and the dictionary learning optimization process iterates 15 times.

2.2 Joint Dictionary Learning

In the joint dictionary learning, given two sets of training data in joint feature space $X = \{x_1, x_2, \ldots, x_n\} \in R^{d \times n}$ and $Y = \{y_1, y_2, \ldots, y_m\} \in R^{h \times m}$. The goal of the joint dictionary learning is to obtain dictionary D_x and D_y, thus, the sparse representation coefficient $\alpha(x_i)$ of LCC should be the same as the sparse representation coefficient $\alpha(y_i)$. In all the given data, the labeled data are used to enhance the correlation between X and Y, the unlabeled data are used to obtain a better sparse representation for samples distribution. To integrate the above representation, a minimized objective function can be obtained as shown in formula (6):

$$Q(D_x, D_y, \alpha) = E_{labeled}\left(D_x, D_y, \alpha^{(s)}\right) + E_{unlabeled}\left(D_x, \alpha^{(x)}\right) + E_{unlabeled}\left(D_y, \alpha^{(y)}\right) \tag{6}$$

Where $\alpha^{(s)}$ represents the shared coefficient matrix of $\{x_i\}_{i=1\ldots t}$ and $\{y_i\}_{i=1\ldots t}$, $\alpha^{(x)}$ and $\alpha^{(y)}$ represents the sparse coefficient matrices of $\{x_i\}_{i=t+1\ldots n}$ and $\{y_i\}_{i=t+1\ldots m}$.

$$E_{labeled}\left(D_x, D_y, \alpha^{(s)}\right) = \sum_{i=1}^{t} \begin{pmatrix} \frac{1}{2}\left\|x_i - D_x\alpha_i^{(s)}\right\|^2 + \frac{1}{2}\left\|y_i - D_y\alpha_i^{(s)}\right\|^2 \\ +\mu \sum_j \left|\alpha_i^{(s),j}\right| \left\|d_j^x - x_i\right\| + \mu \sum_j \left|\alpha_i^{(s),j}\right| \left\|d_j^y - y_i\right\| \end{pmatrix} \tag{7}$$

(7) is a marked item, in order to obtain the result of the shared sparse coefficient matrix $\alpha^{(s)}$, it is necessary to reconstruct x_i and y_i, so there are:

$$E_{unlabeled}\left(D_x, \alpha^{(x)}\right) = \sum_{i=t+1}^{n} \left(\frac{1}{2}\|x_i - D\alpha_i^x\|^2 + \mu \sum_j \left|\alpha_i^{(x),j}\right| \left\|d_j^x - x_i\right\|^2\right) \quad (8)$$

$$E_{unlabeled}\left(D_y, \alpha^{(y)}\right) = \sum_{i=t+1}^{m} \left(\frac{1}{2}\|y_i - D\alpha_i^y\|^2 + \mu \sum_j \left|\alpha_i^{(y),j}\right| \left\|d_j^y - y_i\right\|^2\right) \quad (9)$$

(8) and (9) are unlabeled items to ensure the sparse representation and reconstruction of the unlabeled data. The optimization process of (6) is basically the same as (2). Specifically, fixed α, using (4) to optimize D_x and D_y, then fixed D, $\alpha_x$and α_y can be obtained by (2) directly. The joint coefficient matrix is combined with the description and the corresponding dictionary to learn the sparse representation coefficients. Using (10) represents the labeled data and the dictionary pairs.

$$z_i = \begin{bmatrix} x_i \\ y_i \end{bmatrix}_{i=1\ldots t} \qquad D_s = \begin{bmatrix} D_x \\ D_y \end{bmatrix} \quad (10)$$

The optimized objective function can be expressed by (11):

$$\min_{a^{(\delta)}} \sum_{i=1}^{t} \left(\frac{1}{2}\|z_i - D_s\alpha_i\|^2 + \mu \sum_j \left|\alpha_i^j\right| \left\|d_j^s - z_i\right\|^2\right) \quad (11)$$

Since (11) and (2) have the same form, we can solve it according to (3).

3 Target Re-identification based on LCC Dictionary Learning

3.1 Construction and Training of a Joint Dictionary

Assuming that $x \in X$ and $y \in Y$ are the visual features of the target captured by the two different cameras, the purpose of re-identification is to find a good matching measure $M : X \times Y \rightarrow \mathrm{R}$, the success rate of directly matching of the same target is very low. However, because the conditional of resolution and illumination dramatic changes, making M highly non-linear thus requiring a large number of labeled data, it is difficult to achieve in the actual situation. We use a transform function $F : X \rightarrow Y$ to describe the change of the target across cameras, because F is heterogeneous, it can't be obtained by directly learning, but an isomorphic $\hat{F}$ value can be estimated by minimizing the total score of the labelled data $\{(x_i, y_i)\}$ in the given metric matrix $\hat{M} : X \times Y \rightarrow \mathrm{R}$.

$$\hat{F}_{\hat{M}} = \arg\min_{F} \sum_{i} \left(\hat{M}(F(x_i), y_i)\right) + \mu\Omega(F) \tag{12}$$

Where $\sum_{i}(\hat{M}(F(x_i), y_i))$ is used to minimize the matching score, $\Omega(F)$ represents a regularization term, μ represents balance coefficient. Since $\hat{F}$ is a non-linear function, it required a large amount of training data, as shown in [6], any non-linear smoothing function $\sum_j \alpha^j(x)F(d_j^x)$ can be approximated by linear function $\alpha(x)$.

$$\left|F(x) - \sum\nolimits_j \alpha^j(x)F(d_j)\right| \le \beta\|x - D\alpha(x)\| + \gamma\sum_{j}|\alpha^j(x)|\|d_j - D\alpha(x)\|^2 \tag{13}$$

The quality of this approximate representation is bounded, simplified as (14):

$$\|x - D\alpha(x)\| + \gamma\sum_{j}|\alpha^j(x)|\|d_j - D\alpha(x)\|^2 \approx \|x - D\alpha(x)\|^2 + \gamma\sum_{j}|\alpha^j(x)|\|d_j - x\|^2 \tag{14}$$

By means of this approximation, the (12) can be transformed into (15):

$$\arg\min_{F} \sum_{i} \left(\hat{M}\left(\sum_{j} \alpha_i^j F\left(d_j^x\right), y_i\right)\right) + \mu\Omega(F) \tag{15}$$

Set $\hat{M}$ to L_2 norm distance, we can get (16):

$$\arg\min_{F(d_j^x)} \sum_{i} \left\|y_i - \sum_{j} \alpha_i^j F\left(d_j^x\right)\right\|^2 + \mu\Omega(F) \tag{16}$$

When fixed D_x and α, $F(d_j^x)$ becomes a constant projection vector, recorded the column vector projection matrix as D_y, using (1) as the regularization term [7], we can obtain another form of LCC dictionary learning as shown in (17):

$$\arg\min_{D_y} \sum_{i} \left(\|y_i - D_y\alpha_i\|^2 + \mu\sum_{j}|\alpha_i^j|\|d_j - x_i\|^2\right) \tag{17}$$

In practice, D_x and α are not fixed, need to minimize $E_{labeled}(D_x, D_y, \alpha^{(s)})$ of (7) to achieve the joint optimization of the three variables. In addition to the labeled data, we can use the generous unlabeled data of X to obtain a better sparse representation by minimizing $E_{unlabeled}(D_x, \alpha^{(x)})$, at the same time, by minimizing $E_{unlabeled}(D_y, \alpha^{(y)})$, the generous unlabeled data of Y can be regarded as a priori of $\hat{F}$.

3.2 Appearance Matching

In practice, given a target in a camera image, in order to avoid the change of image resolution and illumination conditions under different cameras, converted the feature representation x to $\hat{F}(x)$. However, the features of the local image blocks is highly dependent on the location of the image blocks, and image block angle and posture are changed, the direct matching is not robust, so we use a greedy nearest-neighbour patch matching strategy [8] to solve the feature transformation problem. During the experiments, a character image is generally divided into three regions: head, body, and legs, each image block can only be matched with the corresponding region. As shown in [5], the division of regions can be determined by the degree of asymmetry of the X axis, experimental results show that in the actual operation, using a fixed ratio (5:11:16) can get a better match effect.

4 Experimental Results

4.1 Target Re-identification Based on Single Camera

Under the condition of single camera, using the public dataset VIPeR to conduct experiments. Each image pair of VIPeR contains two images, the two images are captured by two cameras at different angles and illumination conditions, both of them have the same target person, each image is cropped to a size of 128 $\times$ 48 pixels. In the experiment, the images in the camera A are used for dictionary training, and the images in the camera B are used for the re-identification test. Repeat the experiment 10 times on the data set VIPeR, in each iteration, divided the data set into training data and test data randomly, 1/3 of the training data are labelled, the rest are unlabeled, and finally take the average of 10 results for performance evaluation.

We compared our algorithm with the most popular algorithms on the dataset VIPeR, including: SDALF [5], RDC [4], ELF [9], LF [10], ITML [11], LMNN [12], KISSME [13], and eLDFV [14]. The training and test data of all algorithms are the same, and using the cumulative matching characteristic (CMC) curve to describe the performance of each algorithm in Fig. 1.

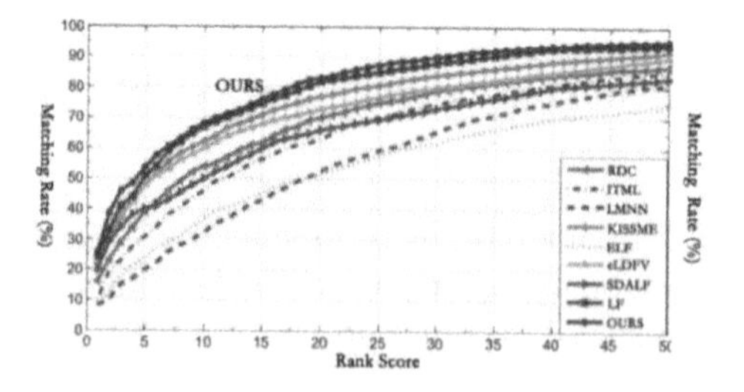

Fig. 1. Performance on the VIPeR dataset

As can be seen, compared with other recognition algorithms, in the case where the number of the most similar objects is the same, our algorithm has the highest probability contains the target to be re-identified. When returned objects is more than 30, the probability has exceeded 90%, and in a certain range, the more objects were returned, the probability of contains the re-identified target will be higher. It is not difficult to conclude that our algorithm has the highest success rate of target re-identification compared with the other eight kinds of re-identification algorithms in the same condition of single-camera.

4.2 Target Re-identification Based on Multi-camera

CAVIAR4REID is a dataset for multi-camera target re-identification, it consists of processing 26 sequence composition captured by two cameras in a shopping center. There are 72 persons, 50 of them appear in both camera images, the others 22 persons only appear in one of the camera image. We randomly selected 14 persons as labelled training data from the 50 persons appearing in two camera images, the remaining 36 people as test data. Set the 22 persons that only appeared in one of the cameras as the unlabeled training data. Iterative every algorithm for 10 times and take the average as the final result, it should be noted that, in each iteration process, the labelled data need to be randomly selected again.

In experiments, compared our algorithm with LF [10] and HPE [5] on the CAVIAR4REID dataset, the training data and test data used by the three algorithms are all the same. HPE is an unsupervised algorithm that does not require dictionary training, LF and our algorithm are required for dictionary training before matching test. The CMC curves of the three methods are shown in Fig. 2.

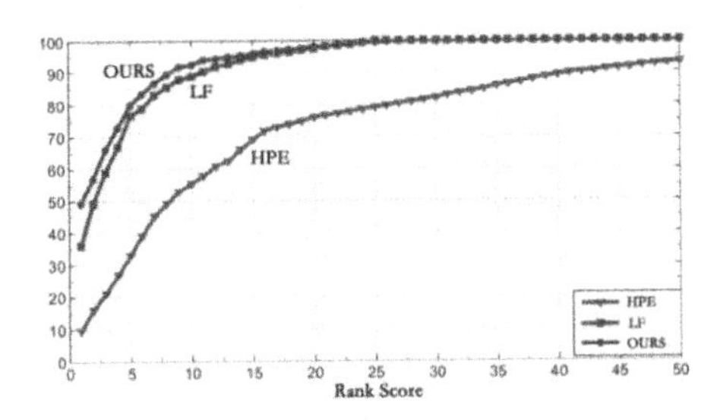

Fig. 2. Performance on the CAVIAR4REID dataset

As we can see, compared with LF and HPE algorithm, our algorithm has the highest probability contains the target to be re-identified under the same experimental conditions. When the returned objects in 10 or more, the probability contains the target to be re-identified has exceeded 90%, when returned objects is more than 25, the probability has reached 100%. It is not difficult to see that our algorithm is still the best in the three algorithms for multi-camera target re-identification experiments.

5 Conclusion

In order to solve handle the challenge of resolution and lighting condition changes in different cameras, we proposed a new algorithm for target re-identification that based on dictionary learning. By comparison with other re-identification algorithms in two publicly available datasets, the results proved that under the same conditions, the matching rate of our algorithm is higher than others re-identification algorithm both in single camera or multi-camera, its computational complexity and the amount of labelled data are both far less than the similar algorithms, which reflects the superiority of our algorithm in solving the target re-identification problem in the multi-camera images.

References

1. Sivic, J., Zitnick, C., Szeliski, R.: Finding people in repeated shots of the same scene. Proc. BMVC (2006)
2. Oreifej, O., Mehran, R., Shah, M.: Human identity recognition in aerial images. Proc. CVPR (2010)
3. Zheng, W., Gong, S., Xiang, T.: Reidentification by relative distance comparison. IEEE Trans. Pattern Anal. Mach. Intell. **35**(3), 653–668 (2013)
4. Dikmen, M., Akbas, E., Huang, T., Ahuja, N.: Pedestrian recognition with a learned metric. Proc. ACCV (2010)
5. Bazzani, L., Cristani, M., Perina, A., Murino, V.: Multiple-shot person reidentification by chromatic and epitomic analyses. Pattern Recogn. Letters **33**(7), 893–903 (2012)
6. Yu, K., Zhang, T., Gong, Y.: Nonlinear learning using local coordinate coding. Proc. NIPS (2009)
7. Liu, J., Cui, L., Liu, Z., et al.: Survey on the regularized sparse models. Chin. J. Comput **7**, 1307–1325 (2015)
8. Guan, X.: A method of object tracking based on feature point matching. Infrared Technol. **38** (7), 597–601 (2016)
9. Gray, D., Tao, H.: Viewpoint invariant pedestrian recognition with an ensemble of localized features. Proc. ECCV (2008)
10. Pedagadi, S., Orwell, J., Velastin, S., Boghossian, B.: Local fisher discriminant analysis for pedestrian re-identification. Proc. CVPR (2013)
11. Davis, J., Kulis, B., Jain, P., Sra, S., Dhillon, I.: Information theorietic metric learning. Proc. ICML (2007)
12. Weinberger, K., Blitzer, J., Saul, L.: Distance metric learning for large margin nearest neighbor classification. Proc. NIPS (2006)
13. Ostinger, M.K., Hirzer, M., Wohlhart, P., Roth, P., Bischof, H.: Large scale metric learning from equivalence constraints. Proc. CVPR (2012)
14. Ma, B., Su, Y., Jurie, F.: Local descriptors encoded by fisher vectors for person re-identification. Proc. ECCV (2012)

Research on High Impedance Fault Detection Method of Distribution Network Based on S Transform

Li Mengda(✉)

Shanghai Dianji University, Shanghai 201306, China
lmdlxblym@126.com

Abstract. Due to high impedance when the distribution network fault over current relay can't generate action, and therefore more difficult to detect its time and frequency information. At present, all kinds of methods in signal processing, S transformation is the most effective method used to extract the frequency distribution. In this paper, S transform the distribution network of high impedance fault detection and simulation results show that the distribution network fault detection method based on S transform high impedance can accurately identify the distribution network of high impedance fault event, the results can be of practical engineering has a certain value.

Keywords: S transform · High impedance fault · Detection method

1 Introduction

In the high-impedance fault detection, Sedighi and other scholars put forward a combination of wavelet transform and soft computing application classification [1]. However, due to wavelet transform on the noise suppression effect is poor, even if the signal-to-noise ratio reaches 30 Db, there will be great errors in the results. In the literature [2, 3], the author uses the method of artificial neuron group, although the system can accurately identify the standard signal [2, 3], but the fuzzy neural network on the system frequency changes require a lot of data and long training time, Frequency signal is difficult to get a better application. It can be seen in the above references used in the research methods are with some limitations [4].

In this paper, we deal with the current signal under the condition of fault and no fault by S-transform, we can extract the frequency information at higher frequency resolution and get the accurate frequency information at lower time resolution. And can extract the time information at a higher time resolution, and obtain accurate time information at a lower frequency resolution. The neural network can then be trained and tested using high-impedance fault current-signal feature points and be able to differentiate between faulted and fault-free events.

F. Xhafa et al. (eds.), *Advances in Intelligent Systems and Interactive Applications*, Advances in Intelligent Systems and Computing 686,
https://doi.org/10.1007/978-3-319-69096-4_76

2 Systematic Research

In this paper, three-phase radial distribution feeder model, for example, and in the MATLAB platform to build its schematic diagram shown in Fig. 1. A 6 kV generator with a capacity of 10 MVA is connected to a 6/10 kV and 10 MVA transformers and the distribution network runs at 10 kV. The simulation model used Blokset model of power system, sampling frequency selection 1.0 kHz. The high impedance fault current is shown in Fig. 2. From the data, high-impedance faults occur after 1/2 normal cycles. Therefore, if a high impedance fault occurs under a linear load condition, the current signal after the fault contains a higher harmonic component than the current signal before failure. Therefore, in the linear load conditions, the extraction of harmonic components can be easily distinguished from high impedance fault (HIF) and no fault (NF). However, if the non-linear load conditions, the fault current and fault-free current contains a higher harmonic component. Therefore, in the power grid, if there is a nonlinear load, it will be difficult to identify HIF and NF, which will also be a crucial issue. This paper will be under the conditions of non-linear load current signal analysis.

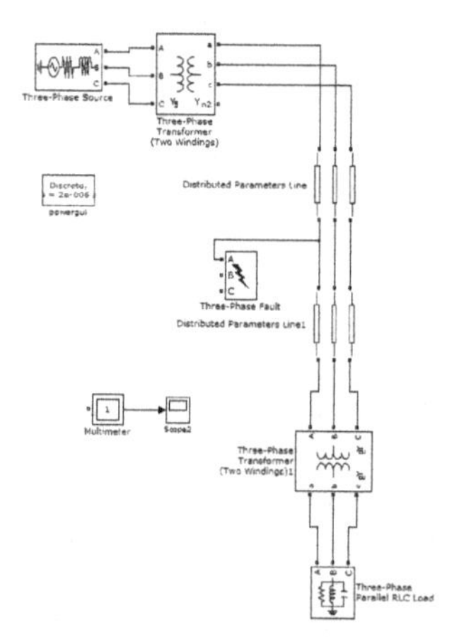

Fig. 1. High impedance fault simulation model

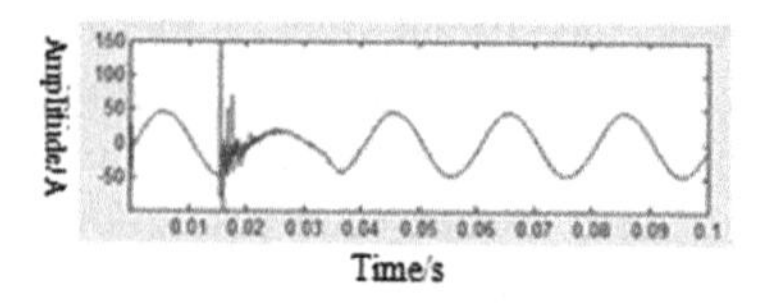

Fig. 2. High impedance fault current under a linear load (A phase)

3 S Transformation

S-transform is based on continuous wavelet transform and is studied by studying short-time Fourier transform. It is based on a movable and scalable Gaussian window function [5]. S transform is a multi-resolution Fourier transform and spectrum analysis, Since its standard deviation is the frequency of an inverse function, therefore, dimensionality reduction is reduced [6]. The Gaussian function g(t) is shown in Eq. (1):

$$g(t) = \frac{1}{\sigma\sqrt{2\pi}} \exp^{-[t^2 f^2/2\sigma^2]} \tag{1}$$

Among σ is the standard deviation, ST is defined as:

$$S(f, \tau, \sigma) = \int_{-\infty}^{\infty} h(t) g(\tau - t, \sigma) e^{-i2\pi ft} dt \tag{2}$$

S transform is a multi-resolution Fourier transform, The main parameter of the expansion is to increase the width of the window function, and then to reduce the frequency, and vice versa [7, 8]. In this paper, The reciprocal relation between Gauss window function and frequency absolute value is chosen, namely:

$$\sigma(f) = T = \frac{1}{|f|} \tag{3}$$

Among them, the cycle is T. The uniform selection is constant, and the Gaussian window in the formula (3) is within the narrowest time domain. S transformation can be written in the form of Eq. (4).

$$S(f, \tau) = \int_{-\infty}^{\infty} h(t) \times \frac{|f|}{\sqrt{2\pi}} e^{-((\tau - t)^2 f^2/2)} \times e^{-i2\pi ft} dt \tag{4}$$

It can be seen that in this definition, when the frequency is zero, the result of the S transform is zero, without any information, $S(f, \tau)$ is therefore defined as being time independent and is the average of h(t). As follows:

Here $f = 0$

$$S(o, \tau) = \lim_{T - \infty} \frac{1}{T} \int_{-T/2}^{T/2} h(t) dt \tag{5}$$

If the discrete S transform is performed, the h(t) can be represented in discrete form by h[pT]. At this point, the p ranges from 0 to N − 1 and serves as the discrete time sequence of the signal h(t). The time series h[pT] of the discrete Fourier transform can be expressed as:

$$H[\frac{n}{NT}] = \frac{1}{N}\sum_{p=0}^{N-1} h[pT]e^{-(i2\pi nk/N)} \tag{6}$$

Where n is from 0 to N − 1 and the discrete Fourier inverse transform is expressed in terms of expressions (7).

$$h[pT] = \sum_{n=0}^{N-1} H[\frac{n}{NT}]e^{2\pi ink/N} \tag{7}$$

The discrete S-transform is a projection of the set of h[pT] vectors of time series. The resulting vector set is non orthogonal, and the elements of the S matrix are uncorrelated, the basic vectors are divided into N-dimensional local vectors one by one to change the Gaussian distribution. Therefore, the sum of the local vectors of the N dimension is the primitive fundamental vector. The discrete time series H[pT] of the S transform is expressed as:

$$S[\frac{n}{NT}, jT] = \sum_{m=0}^{N-1} H[\frac{m+n}{NT}]e^{-(2\pi^2 m^2/n^2)}e^{i2\pi mj/N} \tag{8}$$

When n = 0

$$S[0, jT] = \frac{1}{N}\sum_{m=0}^{N-1} h[\frac{m}{NT}] \tag{9}$$

Among j, m, n $= 0, 1, 2 \ldots N-1$.

4 Feature Extraction Based on S Transform

In the power distribution network, the time-frequency transformation is considered to be S-transform to extract the HIF and NF characteristics of the current signal under different operating conditions, the characteristics of the fault is extracted half a cycle of the current signal. The current signals of the HIF and NF are generated through the design of the distribution model. The frequency and time information of the current signal is based on the previous S matrix data and is extracted by appropriate frequency and time resolution.

The expression (10) is the frequency information extracted in the S matrix.

$$a = \max(abs(S^T)) \tag{10}$$

In the expression, a represents the maximum absolute value of the transpose of the S matrix, which represents the amplitude frequency information of the signal.

The expression (11) is the time information extracted in the S matrix.

$$b = \max(abs(S)) \tag{11}$$

Here, b represents the maximum absolute value of the S transform matrix, which represents the time-frequency information of the signal. The following four expressions are the time and frequency information of electrical energy and standard deviation.

$P(f) = \text{sum}(a^2)$ $\sigma(f) = \text{std}(a)$ $P(t) = \text{sum}(b^2)$ $\sigma(t) = \text{std}(b)$

The frequency information can be extracted at higher frequency resolution and lower time resolution, which will provide different frequency components in the HIF and NF current signals. The time information, i.e., can be extracted at a higher temporal resolution and a lower frequency resolution; will provide different time samples in the HIF and NF current signals.

Figures 3 and 4, respectively, in the case of HIF and NF current signal time and frequency of the sample and the time-frequency signal of the aggregation were compared, where the characteristics of short-time amplitude reflect the time information, the frequency characteristic reflects the frequency information.

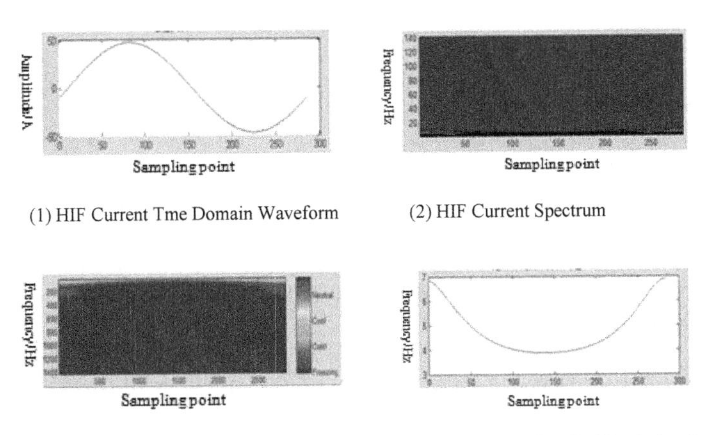

(1) HIF Current Tme Domain Waveform (2) HIF Current Spectrum

(3) Time - Frequency Clustering of HIF Current Signal (4) S Transformation Standard Deviation Curve

Fig. 3. S-transform in case of HIF. (1) HIF current time domain waveform, (2) HIF current spectrum, (3) Time-frequency clustering of HIF current signal, (4) S transformation standard deviation curve

As can be seen from the figure, there are some characteristics are different, some are natural overlap. The above graphs provide a great deal of information about the extraction ability. The classification of the above features is easy to be used in the design of probabilistic neural network and fuzzy neural network, and finally distinguish HIF and NF. Therefore, in the non-linear load conditions to extract features for HIF and NF distinction, usually based on neural classifier intelligent classification technology.

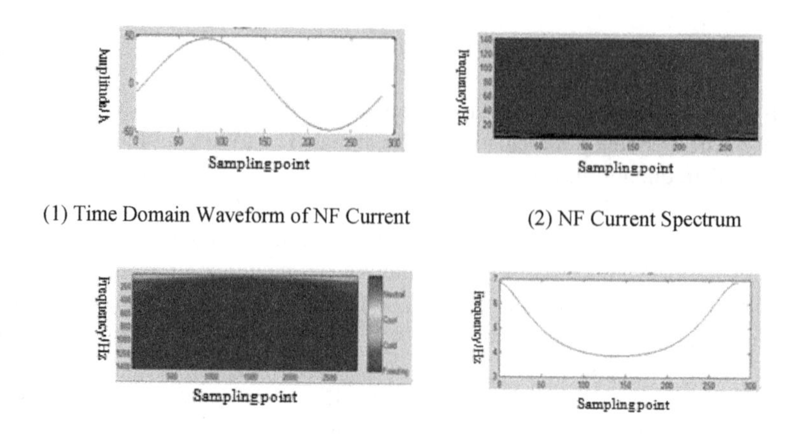

(1) Time Domain Waveform of NF Current (2) NF Current Spectrum

(3) Time - Frequency Clustering of NF Current Signal (4) S Transform Standard Deviation Curve

Fig. 4. S-transform in the case of NF. (1) Time domain waveform of NF current, (2) NF current spectrum, (3) Time-frequency clustering of NF current signal, (4) S transform standard deviation curve

5 Conclusion

In this paper, the intelligent detection technique of high impedance fault is introduced, which tries to distinguish HIF and NF under nonlinear load condition. In this paper, the time-frequency distribution of HIF and NF current signals is extracted by S-transform. The extracted signal characteristics can identify NF and HIF quickly and accurately, and can be extended to the fault detection and protection method of large-scale distribution network.

Acknowledgements. Heilongjiang Natural Science Foundation, E201410.

References

1. Xiangning, X., Yonghai, X.: Power quality analysis. Power Syst. Technol. **25**(3), 66–69 (2001)
2. Sedighi, A.-R., Haghifam, M.-R., Malik, O.P.: Soft computing applications in high impedance fault detection in distribution systems. Electr. Power Syst. Res. **76**(1), 136–144 (2010)
3. Sharaf, A.M., El-Sharkawy, R.M., Talaat, H.E.A.: Novel alpha-transform distance relaying scheme. Proceedings IEEE-CECE Conference, Calgary, Canada (2006)
4. Feng, Y., Li, L., Han, X.: Second-order nonsingular terminal sliding mode control of uncertain MIMO linear systems. In: IEEE. Proceedings of the 2006 1st International Symposium on Systems and Control in Aerospace and Astronautics Harbin, 1350–1355. China, 19–21 January 2012
5. Dai Yuxing, Q.H.: Detection and localization of power quality disturbance signals based on S-transform modulus matrix. Trans. China Electrotech. Soc. **22**(8), 120–125 (2013)

6. Shouliang, L., Xianyong, X., Honggeng, Y.: Short-time power quality disturbance classification based on S-transform time-frequency matrix similarity. Power Syst. Technol. **30**(5), 67–71 (2013)
7. Yang, H.T., Liao, C.C.: A de-noising scheme for enhancing wavelet-based power quality monitoring system. IEEE Trans. Power Deliv. **16**(3), 353–360 (2010)
8. Pinnegar, C.R., Mansinha, L.: The S-transform with windows of arbitrary and varying shape. Geophysics **68**(1), 381–385 (2013)

Research on the Protocols of VPN

Shuguang Zhang[✉], Ailan Li, Hongwei Zhu, Qiaoyun Sun, Min Wang, and Yu Zhang

Information Department, Beijing City University, Beijing, China
shugzhang@163.com

Abstract. Nowadays, the scale of enterprises expands continually, and its use of the network is constantly changing. Most enterprises in the country and even abroad are equipped with branches of the organization, want to share resources, work together to improve the efficiency by linking the various branches of the network. How can these branches communicate safely on the Internet? Virtual Private Network (VPN) can realize the secure data transmission equal to the traditional private network, while cutting great costs of network's establishment and maintenance. In this paper, the basic principles and several protocols of VPN technology are analyzed, and the IKE process is described in detail.

Keywords: VPN · Protocol · GRE · IPSec · IKE

1 Introduction

A virtual private network (VPN) extends a private network across a public network, and enables users to send and receive data across shared or public networks as if their computing devices were directly connected to the private network [1]. VPN refers to the erection of private security virtual logical link to establish the connection to form an expansion of the network. VPN can realize the function equal to the traditional private network, while cutting great costs of network's establishment and maintenance.

VPN now has grown up. It can provide users with a safe and reliable data communication network of business operation ignoring the actual geographical position. Applications running across the VPN may therefore benefit from the functionality, security, and management of the private network.

This paper analyses the basic principle and protocols of VPN technology. The IKE process is described in detail.

2 The Basic Principle of a VPN

VPN technology is a dedicated network of communications among user groups based on a public network. Any group of VPN users do not use a dedicated physical link, while through the Internet service provider's public network communication reflects its virtual nature. The external users can not access the internal resources, while VPN internal users are free to communicate securely reflecting its specificity.

F. Xhafa et al. (eds.), *Advances in Intelligent Systems and Interactive Applications*, Advances in Intelligent Systems and Computing 686,
https://doi.org/10.1007/978-3-319-69096-4_77

Packets are not delivered publicly, which are encrypted and then packaged as IP packets by the protocol, passed through the channel. The source and destination of the tunnel communicate each other through various protocols to negotiate a common plan, then the data is encrypted and encapsulated to verify the security through a variety of ways.

Packets are encrypted and encapsulated in IP packets, even if they are intercepted by users who are not the VPN's users, and there is no way to read the original data before encryption.

After the packets arrive at the destination tunnel address, the decryption of the packets is based on the scheme that has been negotiated before, and the original packets obtained after decryption are forwarded to the destination through the data communication device.

3 Key Protocols of VPN

3.1 The GRE Protocol

GRE is the generic routing encapsulation Protocol. GRE is a tunneling protocol developed by Cisco Systems that can encapsulate a wide variety of network layer protocols inside virtual point-to-point links over an Internet Protocol network. GRE is Cisco Company's proprietary tunnel protocol, which is a layer 3 protocol in TCP/IP model and the OSI model. GRE can encapsulate the packets formed by various network protocols into IP packets, encapsulate them into IP tunnels, and create a virtual point-to-point logical tunnel link transmission packets through the interconnection network between routers. GRE packets that are encapsulated within IP use IP protocol type 47.

The reason why GRE is called a lightweight tunneling protocol is that the number of bytes in the header is small and the speed of the packet encapsulation is fast. The data transmission for the GRE is transparent.

GRE encapsulated packets have four components. The load data is the original IP packet. GRE adds a GRE header to the original IP header and adds a new outer IP header before the GRE header to implement GRE technology encapsulation of the original IP packet.

GRE can encapsulate various network layer protocol packets, but doesn't set up any security mechanism, so the packet is delivered in plaintext and may be intercepted by non-VPN users.

GRE encapsulation structure is as follows.

new IP header	GRE header	Original IP header	TCP/UDP header	DATA

3.2 The IPSec Protocol

IPSec (IP Security), which is layer 3 protocol in TCP/IP model and the OSI model, can only support the IP protocol to guarantee the security of transmission.

The complete IPSec design is a framework for multiple services, algorithms, and granularities [2]. The framework structure does not specify hash and encryption, and is negotiated with specific algorithms in each session. The security of the communication is guaranteed after negotiation.

IP-level security encompasses three functional areas: authentication, confidentiality, and key management [3]. The framework combines several security technologies into one system, including the Internet key exchange (IKE), Encapsulating Security Payload (ESP) and IP authentication headers (AH):

(1) Internet Key Exchange protocol (IKE): negotiates of algorithm, key, peer-to-peer tunnel parameters, and then completes the encapsulation of data. The IKE protocol periodically performs dynamic key updates between two communicating parties.
(2) Encapsulating Security Payload protocol (ESP): IP protocol type number 50, authenticates, encapsulates, and encrypts packets.
(3) Authentication Headers (AH): IP protocol type number 51, which only authenticates and encapsulates packets but does not encrypt, transmitting data messages in plain text.

IPSec adds an IPSec header between the original IP header and the load, encrypts the original load, verifies IPSec and load, and ensures data integrity. It provides privacy, integrity, and source authentication for data traffic passing through the VPN.

The encapsulation structure of IPSec is as follows.

IP Header	IPSec Header	Transport Layer Header	Application Layer Header	Application Layer Data

3.3 IKE Protocol

Internet key exchange protocol (IKE) operates using pairs of messages called exchanges that are sent between an initiator and a responder [4]. The main task of this protocol is

(1) The way of authentication requires prior consultation, authentication and establishment of two communicating parties of IPSec;
(2) Key exchange, generating random HMAC and encryption keys, key updates;
(3) Negotiate encryption protocol, hash, encapsulation protocol, mode, key expiration date and other parameters;
(4) Authenticate both parties, manage the key.

IKE consists of three protocols

(1) the ISAKMP: determines the data exchange architecture, the essence of IKE
(2) SKEME: encryption authentication
(3) Oakley: IPSEC framework to support more protocols.

The source and destination UDP port number of the ISAKMP is 500.

Site-to-site IPSec VPN negotiation phase is divided into two sections.

Before secure site-to-site transmission of data, a series of parameters should be determined through consultation of the negotiation process, which is done by IKE. IKE negotiation run in two stages:

Stage 1: Establish a security management connection between communicating parties. There is no actual flow of data in this so-called connection, which is used to protect the second phase of the negotiation process. The concrete negotiation process is as follows.

(1) How to authenticate a peer
(2) The encryption algorithm
(3) The HMAC approach, MD5 or SHA
(4) The Diffie-Hellman key groups
(5) Main negotiating mode or aggressive negotiation mode
(6) Survival of SA.

Stage 2: When the peers establish a secure management connection, consult to build a secure data connection parameters. After the negotiation, the secure data connection will be established between the two sites. VPN users can securely communicate data through the established connection. The concrete negotiation process is as follows

(1) Packaging technology, AH or ESP
(2) The encryption algorithm
(3) The HMAC approach is MD5 or SHA
(4) Transfer mode
(5) Survival of SA

There are six messages used to negotiate Key's Exchange during stage 1. The first and second messages are used for the peer negotiation encryption mechanism, and the third message is used for the exchange of keys.

Both sides of the communicating parties will generate their own public and private keys at the same time. The third and fourth information is then exchanged for the public key and the private key.

After the public key is exchanged, the peer uses the DH algorithm to generate a shared secret based on the private key of the peer to associate with its own public key. The three SKEY is generated by calculating the shared secret, temporary value, and pre-shared key.

SKEYID_d Calculation of subsequent key resources
SKEYID_a Provides the integrity and authentication of subsequent messages
SKEYID_e Encrypts the following information

The contents of the ISAKMP packet of the third and fourth messages are ISAKMP header, key exchange load and temporary load.

The fifth and sixth messages are used to authenticate the peers by the information encrypted with SKEYID_e.

Thus IKE SA has been established and main mode authentication is accomplished.

The second stage of the negotiation process has three messages.

The first and second messages are used for the peer entities to consultant the encapsulation protocol, mode, encryption algorithm and other parameters of IPSec SA. They are also used to send each other the public key and temporary value calculated with the DH algorithm.

Before the third message is sent, the peer entity generates a new DH secret at both ends to generate the final key for encryption and decryption.

The third message is passed by the initiator to the responder to confirm and verify that the channel remains valid.

The security association (SA) is an agreement negotiated by the communication entities in the VPN, which determines the IPSec protocol, transcoding mode, key, and key validity of the protection packet security.

There are two types of security-related SA as follows

ISAKMP SA	negotiate the IKE traffic to verify the peer entity algorithm, only one
IPSec SA	the algorithm for negotiating the IP traffic of peer entity

Different protocols will produce different security associations (SAs), both parties use ESP and AH, for ESP and AH will generate different SA.

The process of keys' exchange in main mode is shown in Fig. 1. Alice is an initiator and Bob is a responder.

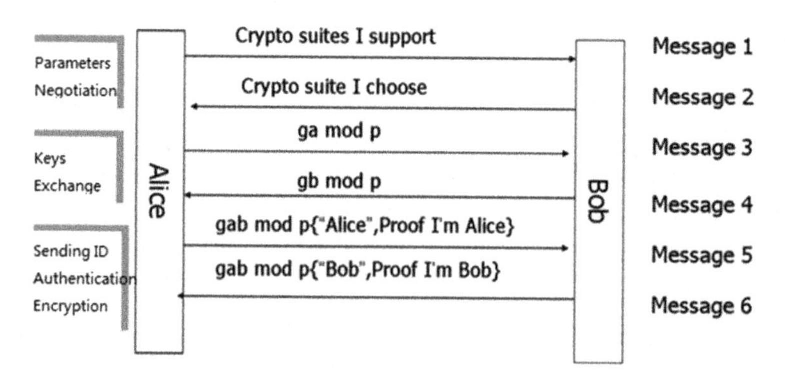

Fig. 1. The process of keys' exchange in main mode

Acknowledgements. This work was supported by the non-governmental education promotion project of Beijing: Comprehensive practical teaching base for modern communication technology in Beijing City University.

References

1. Wikipedia https://en.wikipedia.org/wiki/Virtual_private_network
2. Tanenbaum, A.S., Wetherall, D.J.: Computer Networks, pp. 832–835. China Machine Press, Beijing (2013)
3. Stallings, W., Brown, L.: Computer Security Principles and Practice, 2nd edn, pp. 716–724. China Machine Press, Beijing (2013)
4. Fall, K.R., Stevens, W.R.: TCP/IP Illustrated, vol. 1, The Protocols, 2nd edn. China Machine Press, Beijing (2013)

Penetration Level Permission of Small Hydropower Station in Distributed Network

Xin Su[1(✉)], Xiaotian Xu[1], Genghuang Yang[2], and Xiaiyi Hao[2]

[1] Chengxi Power Supply Branch, Tianjin Electrical Power Company, Tianjin, China
susin@126.com
[2] School of Automation and Electrical Engineering, Tianjin University of Technology and Education, Tianjin, China

Abstract. Small hydropower station, connecting with the power system through the distribution network, can lead to the changes of the detected current for protection relay and the protection coverage. This paper firstly analyzes the effects of the small hydropower station connecting with the IEEE7 power system in current protection theoretically, and calculates the corresponding parameters of current section I and section II protection of the original power distribution without small hydropower station in PSCAD/EMTDC environment. Lastly, it analyzes the maximum allowable capacity of the small hydropower station in the distribution network.

Keywords: Small hydropower station · Current protection · Allowable capacity · PSCAD/EMTDC

1 Introduction

Rural small hydropower as a typical synchronous motor type distributed power supply in China's southern rural areas has achieved good economic and environmental benefits [1]. Small hydropower stations are generally distributed in the mountainous areas of abundant water resources, widely distributed and weak connectivity with the large power grid, while the protection for the distribution network is relatively simple [2, 3]. When the line fails, especially for rural power grid, which protection of the program and the device is relatively simple, hydropower stations to maintain the operation of the grid must upgrade the existing relay protection system. Obviously, the cost of the program is too high compared to the small hydropower revenue.

At present, low-voltage distribution network of rural in China are mostly unilateral power supply, radiation network [4]. The relay protection of distribution network generally installs current protection of three-stage—no time limit current protection (stage I), time limiting current protection (stage II) and timing overcurrent protection (stage III), which the first two for the main line protection [5].

Because it is the single power supply to provide current for the fault point, so just complete the removal of the fault by simply breaking the circuit breaker beside the system. However, when the distributed power supply accesses into the distribution network, the distribution network is changed into a multi-power structure. If the failure

F. Xhafa et al. (eds.), *Advances in Intelligent Systems and Interactive Applications*, Advances in Intelligent Systems and Computing 686,
https://doi.org/10.1007/978-3-319-69096-4_78

happens, it may cause short-circuit current size and direction changes, which takes certain impact on reliability for relay protection [6]. Based on the access capacity of small hydropower stations, this paper analyzes the influence of the access of small hydropower stations on the protection of current I and II in rural low-voltage distribution networks, so as to obtain the maximum security capacity of the distribution system under the premise of accurate operation of the original protection.

2 Influence of Small Hydropower Access on Distribution Current Protection

The typical distribution network IEEE7 node system is shown in Fig. 1. When the three-phase ground fault occurs at F1 point, the influence of the access of small hydropower to the current protection of the distribution network is analyzed. In the figure, the voltage of SG is the system power of ES, and small hydropower is connected to bus C.

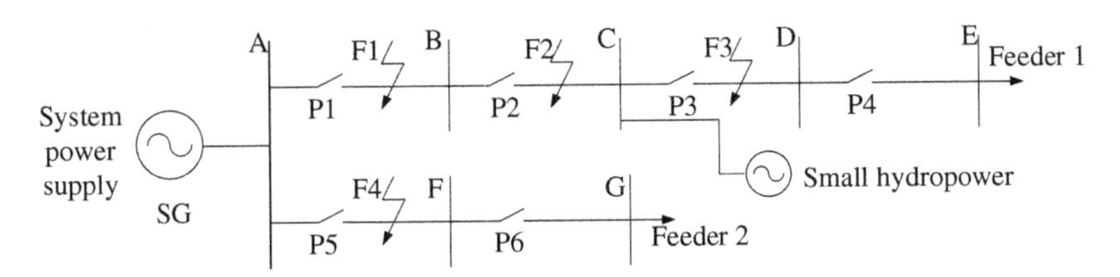

Fig. 1. Structural diagram of IEEE7 node distribution network with small hydropower

2.1 The End of AB Line Happens Short-circuit at F1 Point

When the small hydropower fails at F1 in the downstream, with ignoring the feeder 2, the system is simplified to Fig. 2.

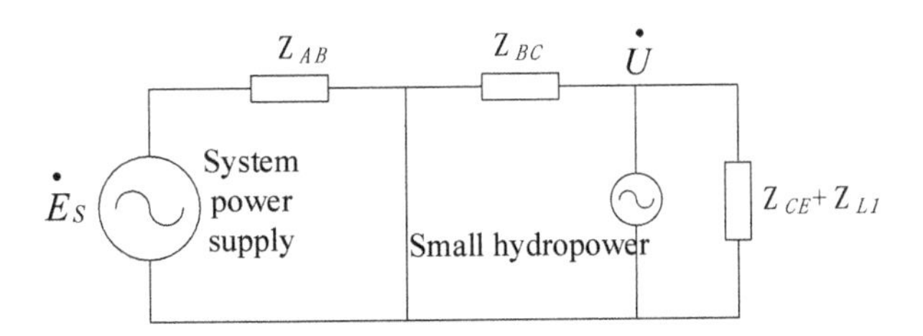

Fig. 2. The simplified diagram of three-phase fault system at F1 point

At this time, the short-circuit currents flowing through P1 and P2 respectively are these as follow.

$$I_{k1} = \frac{E_s}{Z_{AB}} \tag{1}$$

$$I_{k2} = \frac{U}{|Z_{BC}|} = \sqrt{\frac{P \cdot R_L}{|Z_{BC}|^2 + |Z_{CE} + Z_{L1}||Z_{BC}|}} \tag{2}$$

It can be obtained that the case F1 is shorted, the fault current flowing through P1 is only related to the system power supply and has no relationship with the accessed small hydropower. The magnitude of the reverse current I_{k2} flowing through P2 is proportional to the apparent power of the small hydropower, P2 can accurately remove the fault as long as I_{k2} exceeds the set value. But the action of P2 will make the downstream of small hydropower to be the island, it can have an impact on the operation of the system.

2.2 The End of BC Line Happens Short-circuit, Short-circuit Fault Occurs at F2 Point

When the small hydropower fails at F2 in the downstream, with ignoring the feeder 2, the system is simplified to Fig. 3.

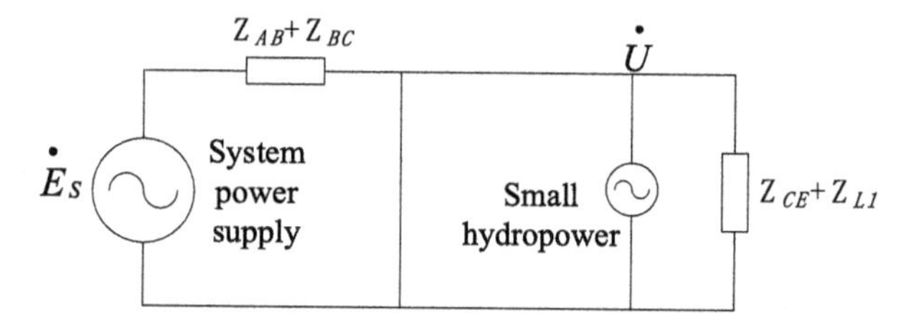

Fig. 3. The simplified diagram of three-phase fault system at F2 point

At this time, short-circuit current of P1 and P2 only supports by the system power supply and the size is the same.

$$I_{k1} = I_{k2} = \frac{E_s}{Z_{AB} + Z_{BC}} \tag{3}$$

Since the small hydropower has no effect on the fault current of P1 and P2, P1 and P2 can move normally. However, due to the presence of small hydropower, after the P2 moves, small hydropower will still inject the short-circuit current to the fault point, so that the fault point arc cannot be extinguished, the line reclosing can be failure and power outage time can be extended. Therefore, the small hydropower should timely remove from the system [7].

3 Determination of Protection Parameters of Distribution Networks Without Small Hydropower

As shown in Fig. 1, build an IEEE7 node system based on PSCAD/EMTDC. The three-phase short-circuit fault at the end of each line is simulated, and the corresponding parameters of time - limited protection with time - limited protection are given. The parameters of the distribution network are shown in Tables 1, 2 and 3. Among them, the positive sequence impedance of the line is 0.104 + j1.270, the negative sequence impedance is 0, the zero sequence impedance is 0.900 + j3.428, and a tower is installed every 60 m.

Table 1. Line parameters of power distribution network

Name of line	Length of line (km)	Name of line	Length of line (km)
AB	3	DE	10
BC	5	AF	2
CD	8	FG	8

Table 2. Load of end feeder

Name of feeder	Active power (MW)	Reactive power (MVAR)
Feeder 1	2	4
Feeder 2	0.9	1.8

Table 3. Power supply parameters of distribution network

Name of power supply	Reference voltage (kV)	capacity (MVA)	Minimum reactance (H)	Maximum reactance (H)
SG	10.5	150	0.0003	0.0004

4 Access Capacity Analysis of Small Hydropower Stations

From the above analyses we can see that the access of small hydropower will lead to the distribution of the short circuit current and the direction of the current flow in the corresponding line of the distribution network, and then make the protection malfunction and expand the fault range, and finally affect the stable operation of the power grid. Access capacity refers to the maximum capacity of the distributed power supply to the system when the distributed power supply is connected to the distribution network, without changing the protection of the original protection, and the protection can be reliably operated [3].

The following is the simulation and discussion of the fault in different locations when the access capacity of small hydropower to short-circuit current impact, and obtained the distribution network of small hydropower access capacity. The simulation topology diagram is shown in Fig. 1, the small hydro power access point is C point, the excitation system of small hydropower is selected as the AC1A module in PSCAD/EMTDC, the water turbine is TUR1 module, and the water turbine controller is GOV1 module.

4.1 A Short Circuit Fault Occurred at F1 Point

From the above analysis, we can see that when F1 fails, the small hydropower station injects the fault current through the P2 to the fault point. When it exceeds the setting value, it leads to the misoperation of the P2 protection in section II, which mainly affects the action of P2, and has little effect on the P1 (Table 4).

Table 4. Short-circuit current flowing through different protection when F1 point is faulty

S_{DG}/MA	I_{DG}/KA	I_1/KA	I_2/KA	I_3/KA
0	0	2.7583	0.0114	0.0114
4	0.6495	2.7720	0.6628	0.0140
8	1.0520	2.7320	1.0543	0.0220
12	1.3240	2.6970	1.3188	0.0283
14	1.4277	2.6839	1.4199	0.0302

(S_{DG} is the capacity of small hydropower, I_{DG} represents the current of small hydropower output, I_1, I_2, I_3 are P1, P2, P3 detected current)

It can be seen from Fig. 4, the access capacity should be less than 11 MVA in order to make the P2 reliable.

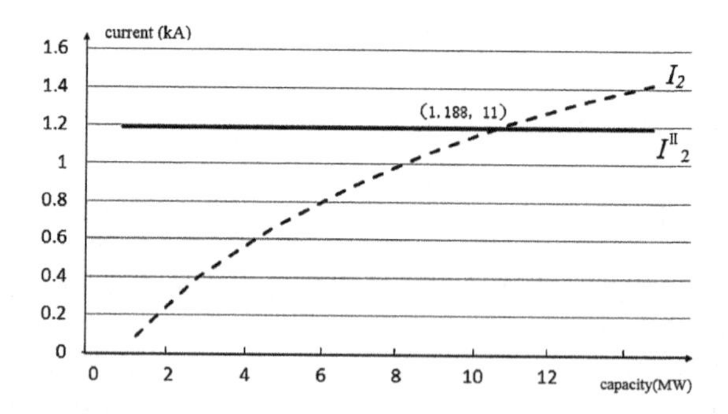

Fig. 4. Small hydropower access capacity when F1 point fails

4.2 A Short Circuit Fault Occurred at F2 Point

From the foregoing analysis, when the short-circuit fault occurs at the F2 point, the short-circuit current flowing through the protection 1 and the protection 2 is provided for the power supply of the system, and the small hydropower will not affect the line protection. But to take into account the previously mentioned in the protection action removal system power, the small hydropower is still connected with the fault point on the distribution network.

5 Conclusion

In summary, the access of hydropower stations in rural low-voltage distribution network, the impact of its relay protection can be summarized as:

Protection loses its directionality. When the fault occurs upstream of the P2, for P2, the small hydropower will inject the fault current, which may be greater than the set value in severe cases, resulting in P2 malfunction. When the adjacent line fails, the small hydropower will flow through the short circuit current through P1 and P2, and also may cause maloperation of protection.

At the same time, on the basis of this, simulation and analysis under the premise of ensuring the reliability of the original relay protection, access to small hydropower stations should be considered when the protection of the location, so as to get the maximum safe access to small hydropower capacity. This provides an effective basis for the site selection, permitted capacity and protection of the corresponding small-scale hydropower stations.

Acknowledgments. This paper is supported by the project of Tianjin University of Technology and Education (K-GD2014-1023).

References

1. Lu, Y., Gao, B.: Analysis of distribution network fault and protection range with distributed power supply. J. Shanghai Univ. Electr. Power **30**, 511–514 (2014)
2. Jiang, F., Du, X.: China's small hydropower development status and problems. China Rural Water Hydropower **1**, 82–86 (2004)
3. Xue, P.: The main problems and solutions of the improvement and expansion of small and medium hydropower stations. China Rural Water Power Hydropower **1**, 133–136 (2014)
4. Zhao, M., Yan, H., Liu, X.: Study on rural distribution network protection algorithm with distributed power supply. China Rural Water Hydropower **1**, 172–179 (2014)
5. Zhang, C., Ji, J.: Influence of distributed generation on feeder protection of distribution network. Relays **34**, 9–12 (2006)
6. Huo, L.: Power System Relay Protection. China Electric Power Press, Beijing (2008)
7. Chen, L.: Distribution Network Protection Scheme with Distributed Power supply. Nanjing University of Science and Technology, Nanjing (2014)

Research on Smooth Switching Method of Micro-Grid Operation

Wenbin Sun[1(✉)], Genghuang Yang[1], Xin Su[2], and Xiaotian Xu[2]

[1] School of Automation and Electrical Engineering, Tianjin University of Technology and Education, Tianjin, China
Sunwb2017@126.com

[2] Chengxi Power Supply Branch, Tianjin Electrical Power Company, Tianjin, China

Abstract. It is of great significance, especially for power quality, to realize the smooth switching and conversion between the two operation modes of Micro-Grid. In this paper, for the control of the two operation modes of Micro-Grid, constant power control strategy and Constant-Voltage Constant-Frequency control strategy are adopted, and analysis of these two kinds of control strategies. In order to solve the transient oscillation problem in the switching process between the two operation modes of Micro-Grid, a smooth switching control strategy is proposed. And then the strategy is applied to realize the smooth switching to ensure power quality in this paper. Lastly, a mode is constructed by PSCAD/EMTDC software to test whether the strategy is effective and practicable or not.

Keywords:: Micro-Grid · Grid-connected · Isolated grid · Constant power control · Constant-Voltage Constant-Frequency control · Smooth switching

1 Introduction

With the development and use of non-renewable energy, a series of environmental pollution problems have been brought up. However, the emergence of distributed generation such as hydro, wind, solar power and other renewable energy, not only to be environment-friendly, but also to improve power quality by making full use of their own advantages with large power grid. The Micro-Grid is mainly composed of distributed generation and energy storage device, and its operation mode can be switched between the grid-connected and isolated island according to the situation of the large power grid.

The control of Micro-Grid is usually based on the characteristics of distributed generation and its operation mode. Alternating current is converted from distributed generation by the inverter. For the control of Micro-Grid operation modes, it mainly refers to the control of inverter, reference [1–5] point out that the constant power control strategy can be chosen to stay given active power and reactive power in the grid-connected environment, and in order to obtain stable voltage and frequency, Constant-Voltage Constant-Frequency control strategy works well in the isolated grid

F. Xhafa et al. (eds.), *Advances in Intelligent Systems and Interactive Applications*, Advances in Intelligent Systems and Computing 686,
https://doi.org/10.1007/978-3-319-69096-4_79

environment. In this paper, these two kinds of control strategies are adopted for the control of the Micro-Grid operation mode.

There is a failure in the switching between the two Micro-Grid operation modes while Constant-Voltage Constant-Frequency control strategy with droop characteristic and constant power control strategy are adopted for the control of the Micro-Grid operation mode [6]. In this paper, a reasonable method of tracking state of inverter controller is brought out to solve this problem in the switching process between the two operation modes of Micro-Grid and realize the smooth switching and conversion between the two operation modes of Micro-Grid.

2 Micro-Grid Overall Structure

As shown in the following figure, distributed generation unit 1 (DG1), with step-up transformer, DG2 and load are connected to the bus which is connected to the 35 kV large power grid with switch K1 and step-down transformer (Fig. 1).

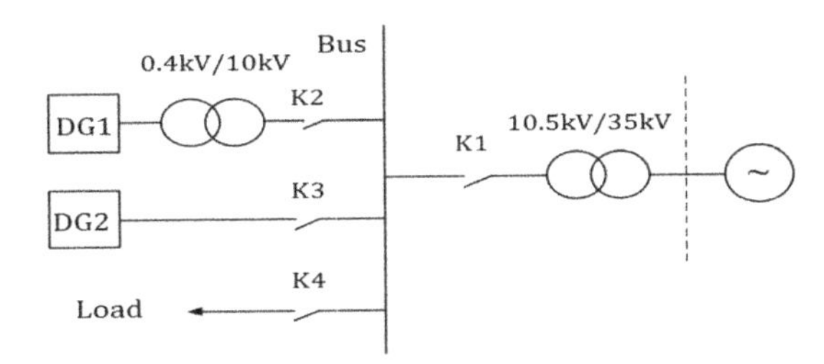

Fig. 1. Overall structure of Micro-Grid

3 DG Unit Structure

In Fig. 2, U_{dc} is DC voltage generated by distributed generation with converter or rectifier. The i_l, i_c, i_o and u_o represent filter inductor current, filter capacitor current, output current and output voltage.

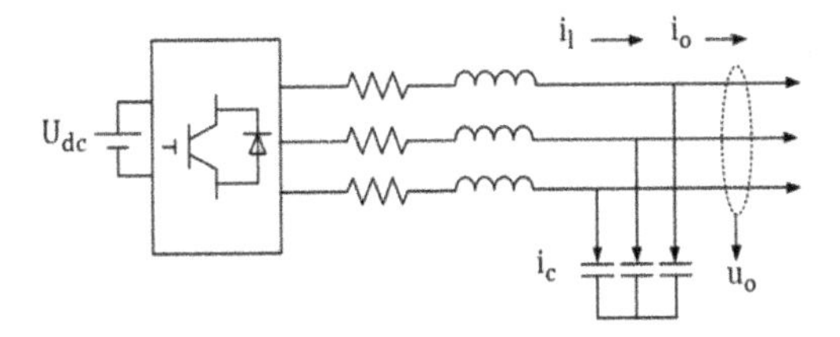

Fig. 2. Structure of DG unit

4 Inverter Control Strategy

4.1 Constant Power Control (PQ Control)

As Fig. 3 above shows, in the d–q rotating coordinate system, i_{ldq}, i_{odq} and u_{odq} respectively represent the *d* axis component and *q* axis component of filter inductor current, output current and output voltage. As for u^*_{dq} and $u^*_{\alpha\beta}$ used as voltage modulation signal of inverter, these two parameters, which are generated by double loop structure of the PQ control, are the *d*, *q* axis component and α, β axis component. And the P^*, Q^* are the reference value of active power and the reference value of reactive power, while P^0 and Q^0 are actual output active power and actual output reactive power. ω is the angular frequency and theta is coordinate transform angle.

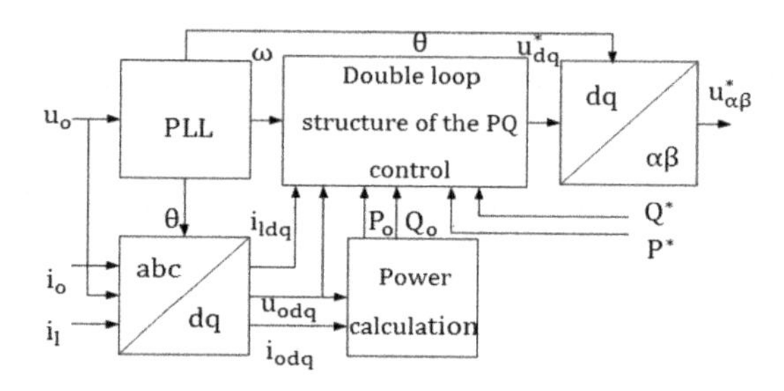

Fig. 3. Structure of the PQ control

Figure 4 describes the double loop structure of the PQ control. The i^*_{ld} and i^*_{lq} are reference values of the inductor current *d*, *q* axis component, respectively, i_{ld} and i_{lq} are values of the inductor current *d*, *q* axis component.

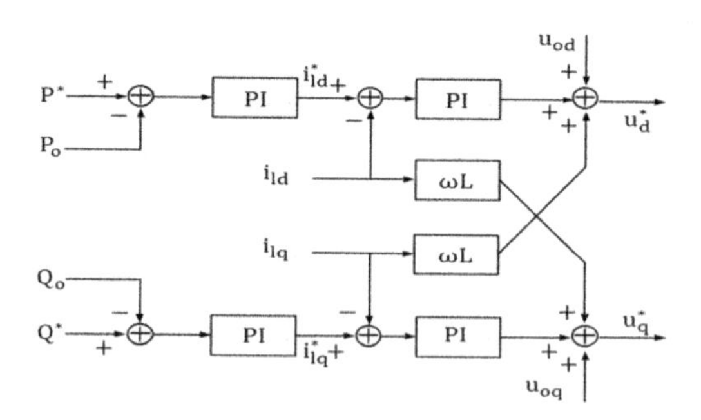

Fig. 4. Double loop structure of PQ control

4.2 Constant-Voltage Constant-Frequency Control (VF Control)

In Fig. 5, it is the structure of the VF controller. The structure is composed of three parts, which are coordinate transformation, three-phase voltage instruction and double loop structure of VF control.

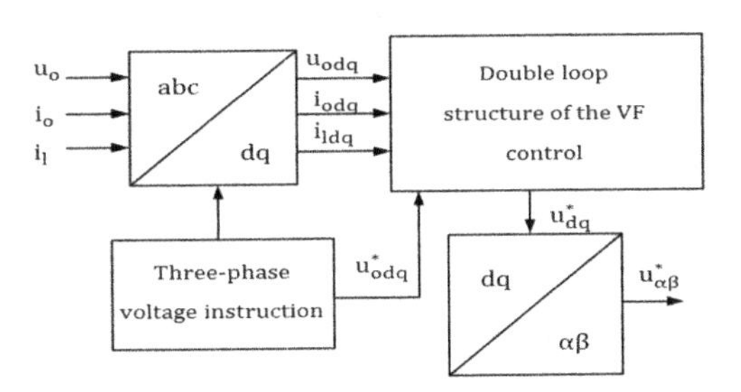

Fig. 5. Structure of VF control

As is indicated in the Fig. 6, double loop structure of VF control consists of two parts: voltage loop and current loop [7]. The inner loop is the instantaneous current regulating loop of the inductor current, which forms the current servo system, and it can speed up the dynamic process of anti-interference and improve the dynamic performance of the system. The outer loop composed of the instantaneous voltage control loop can improve the output voltage waveform and improve the output accuracy [8]. The process that the reference value of inductor current is generated by the difference between the reference value of output voltage and actual value of the output voltage after feed-forward decoupling and the PI controller is called as voltage outer loop. Similarly, the process of generation of the voltage modulation signal is called as current inner loop.

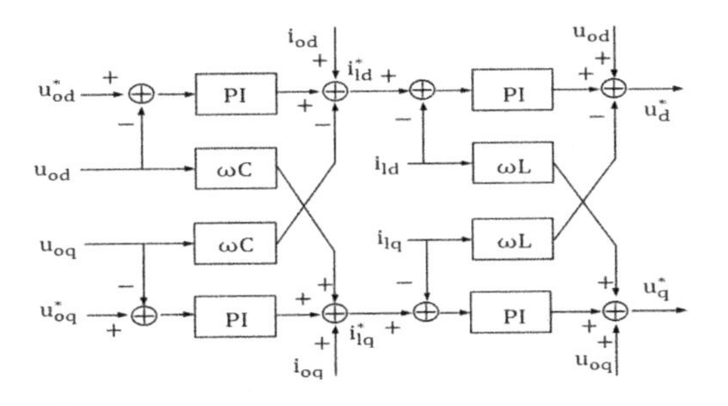

Fig. 6. Double loop structure of VF control

5 Switch Control Structure

When the fault occurs in the large grid, Micro-Grid receives the corresponding signal information, and then its work form switches from the grid-connected to the isolated grid. And the control of Micro-Grid will be changed. But in the switching process between the two operation modes of Micro-Grid, due to failing to get the state information of switcher in real time, as a result, the voltage and frequency are easy to produce a larger transient oscillation at the switching time, and even the failure of the mode switching. In order to solve the problem, this paper presents the method, tracking the inductor current reference value for the state information of inverter controller in real time, and uses it to realize smooth switching and conversion between the two operation modes of Micro-Grid.

As is shown in Fig. 7, in the grid connected environment, K8, K7 are closed and PQ control strategy is adopted. Then the outer loop of VF control will track the reference value of inductor current generated by the outer loop of PQ control in real time. When switching from the grid-connected to the isolated grid, VF control strategy is adopted, K8, K7 are opened and K5, K6 are closed at the same time, because of the tracking of the outer loop of VF control, the single received by current inner loop will not jump, therefore, the voltage modulation signal will also control the inverter stably. Lastly, the purpose to solve the transient oscillation problem is achieved. Similarly, when switching from the isolated grid to the grid-connected, there is also no problem.

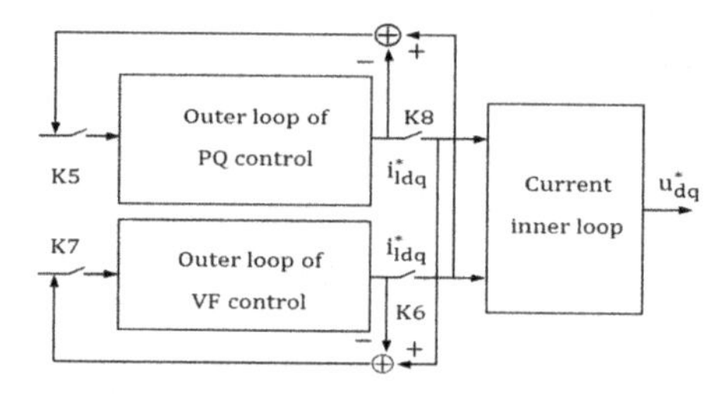

Fig. 7. Structure of switch controller

6 Simulation

For the smooth switching strategy proposed in this paper and according to Fig. 1, as is shown in Fig. 8, a mode is constructed by PSCAD / EMTDC software to test whether the strategy is effective and practicable or not. The PQ control method is adopted in the grid-connected. 150 kW, 0 kW, 0.25 MW, and 0.93 represent the values of reference active power, reference reactive power, resistance-inductance load and power factor. Simulation runs 3 s. In 0–1, the operation mode of Micro-Grid is grid-connected and DG1 works in the PQ control method. In 1 s moment fault occurs to the large grid, Micro-Grid is disconnected to the large grid and DG1 does not work. For the moment, K3 is closed and DG2 works in the VF control method providing electricity for the load continuously. After the fault recovery, in 2 s time, K3 is opened, K1 and K2 are closed. Then DG1 connected to the large grid and the large grid provide electricity for the load together. As we can see from the Fig. 9, at the moment of switching between the two operation modes of Micro-Grid, bus voltage and frequency values are within the allowable range and there is no larger oscillation.

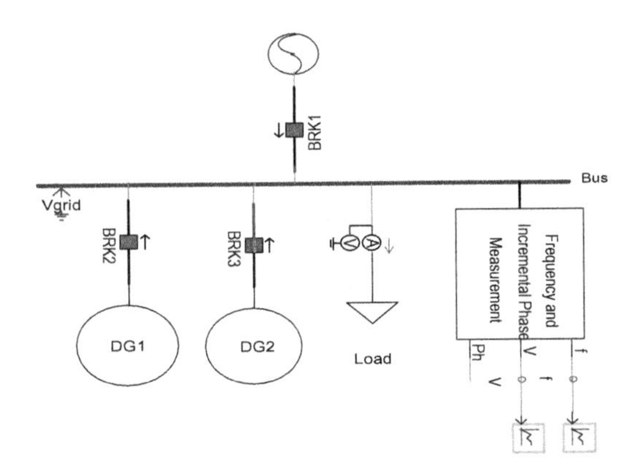

Fig. 8. PSCAD simulation model of Micro-Grid

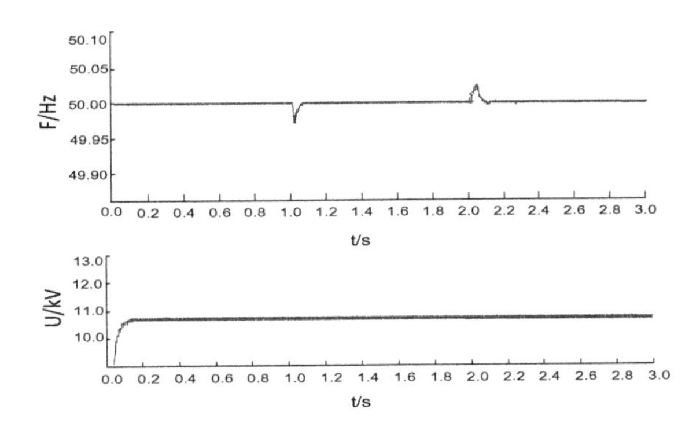

Fig. 9. Bus frequency and voltage

7 Conclusion

In this paper, the operation control strategies of the Micro-Grid work effectively. And this paper has dealt with the common problem of the larger transient oscillation in the switching process by the smooth switching strategy proposed in this paper. Simulation results show the feasibility and reliability of the strategy.

Acknowledgments. This paper is supported by the project of Tianjin University of Technology and Education (XJKC030155 and XJKC031112).

References

1. Schmelzer, C., Schweyen, R.J.: Generators Control Systems in Intentionally Islanded MV Micro-grids. Int. Symp. Power Electron. **46**(4), 557–565 (2008)
2. Guan, Y., Weiyang, W., Xiaoqiang, G.: Control technology of three phase inverter in microgrid. Proc. Chin. Soc. Electr. Eng. **31**(33), 52 (2011)
3. Wang, C., Xiao, X.: Integrated control and analysis of microgrid. Autom. Electr. Power Syst. **32**(67), 98–103 (2008)
4. Yu, X., Jiang, Z., Zhang, Y., et al.: Control of parallel inverter-interfaced distributed energy resources. Energy 2030 Conference, 1–8 (2008)
5. Su, L., Li, G., Jin, Z.: Modeling control and testing of a voltage-source-inverter-based microgrid, 4th International Conference on Electric Utility Deregulation and Restructuring and Power Technologies, pp. 724–729 (2011)
6. Wang, H., Li, G.: Control strategy of microgrid with multiple distributed power supply. Electr. Power Autom. Equip. **32**(5), 19–23 (2012)
7. Pogaku, N., Prodanović, M., Green, T.C.: Modeling, analysis and testing of autonomous operation of an inverter-based microgrid. IEEE Trans. Power Electron. **22**(2), 613–625 (2007)
8. Ju, H.: Research on grid connected control of multi inverter power supply in distributed micro grid power system. Doctoral Dissertation of He Fei University of Technology (2006)

A New BDD Algorithm for Fault Tree Analysis

Wei Liu[1(✉)], Yong Zhou[1], and Hongmei Xie[2]

[1] College of Computer Science and Technology,
Nanjing University of Aeronautics and Astronautics, Nanjing 211106, China
liuwei339@qq.com
[2] College of Continuing Education, Nanjing University of Aeronautics
and Astronautics, Nanjing 210016, China

Abstract. FTA is a NP-hard problem. When using the Binary Decision Diagram (BDD) to analyze a fault tree, the size of the BDD transformed by the fault tree increases as the number of basic events increases. When using the BDD method to analyze a large-scale fault tree, there is not enough memory to store the resulting BDD structure, and the process of obtaining the minimum cut set is time consuming. This paper introduces a simplified BDD (SimBDD) structure to solve these problems. SimBDD is a BDD whose scale linearly relates to the number of basic events in the fault tree. Therefore, when using this structure to deal with large fault tree, the structure is small and easy to store. The number of paths in SimBDD is small and the minimum cut set gets faster.

Keywords: BDD · Fault tree analysis · SimBDD · Minimum cut sets

1 Introduction

Fault tree analysis method is one of the main analytical methods of safety system engineering, which is widely used in the nuclear industry to obtain the response models for Probabilistic Safety Assessment (PSA). Binary decision diagram (BDD) analysis is one of the most effective methods to analyze fault trees [1–7]. In the BDD method, the fault tree is converted to a BDD structure to represent a logical Boolean expression for a particular system failure mode. The size of the BDD grows exponentially with the number of fault tree variables. When using the BDD algorithm to solve large-scale fault trees, a large amount of memory requirements will limit the use of the algorithm.

This paper presents a simplified form of BDD to solve the above problem. This new form of BDD method is a simplified BDD structure of the BDD method based on the BDD connection method [8]. In the SimBDD structure, only one BDD node corresponds to each basic event node of the fault tree. In fact, the number of BDD nodes (excluding nodeOnes and nodeZeros) in the SimBDD is the same as the number of basic events in the fault tree. Thus, the SimBDD structure does not require a lot of memory to store. By connecting BDD nodes, this method can quickly build the corresponding SimBDD structure, without a lot of calculation.

The structure of this paper is as follows. The next section describes the normal BDD algorithm. Section 2 provides the details of SimBDD, while Sect. 2.4 uses a case to explain the procedure of getting FT's MCSs using the SimBDD algorithm.

F. Xhafa et al. (eds.), *Advances in Intelligent Systems and Interactive Applications*, Advances in Intelligent Systems and Computing 686,
https://doi.org/10.1007/978-3-319-69096-4_80

Benchmark tests were performed to show the efficiency of the SimBDD algorithm for large fault trees and the results are described in Sect. 3. The conclusions of the study are provided in Sect. 4.

2 Simplified BDD

2.1 Structure of SimBDD

Figure 1 presents the general SimBDD structure. Its corresponding FT is T. We call the corresponding SimBDD of T as BDD_T. The nodes in rectangle are omitted in Fig. 1. We name the head of BDD_T as T_head.

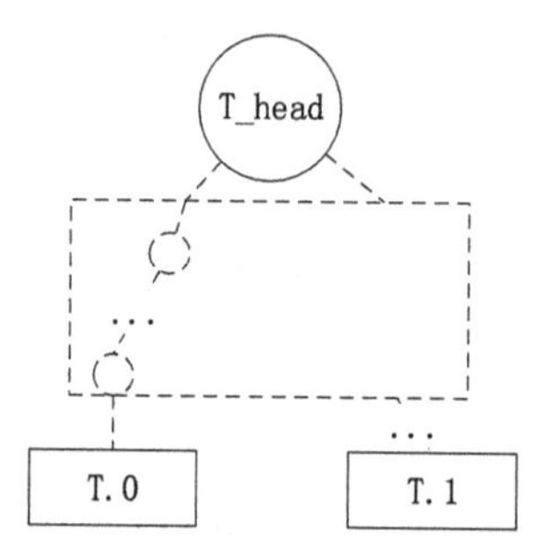

Fig. 1. Structure of SimBDD

Definition: leftmost node

Each leaf node in the SimBDD is connected to a different nodeZero. We call the edge which represents that the node occurs as edgeOne and the edge which represents that the node does not occur as edgeZero. Assume that the edgeZero of each node in a SimBDD is on the left side of its edgeOne, the leftmost node of a SimBDD represents the zero node that the leftmost path connects to.

The structure of SimBDD can be described as:

```
Struct SimBDD{
   String node                 //name of the node
   SimBDD zeroLink             //linked node when the node not occur
   SimBDD oneLink              //linked node when the node occur
   List<SimBDD> parents        //parent nodes of the node
   SimBDD zeroNode             //leftmost node of the SimBDD
   SimBDD oneNode              //nodeOne of the SimBDD
}
```

In Fig. 1, T.1 is the terminal nodeOne shared by all the nodes in BDD_T. That is, all paths that lead T true will eventually connect to T.1. T.0 in Fig. 1 represents the leftmost nodeZero of BDD_T. We can find that the leftmost path of SimBDD is consist of edgeZeros of nodes.

2.2 Operations of SimBDD

Suppose there are two SimBDD, BDD_T1 and BDD_T2, and the head nodes are T1_head and T2_head, respectively.

As shown in Fig. 2a, when $T = T1 \bullet T2$, we first connect the parent nodes of T1_head.oneNode to T2_head with their edgeOne and add these parent nodes to T2_head.parents. Then, we assign T2_head.oneNode to T1_head.oneNode. Finally, the SimBDD whose head is T1_head is the one that T corresponds.

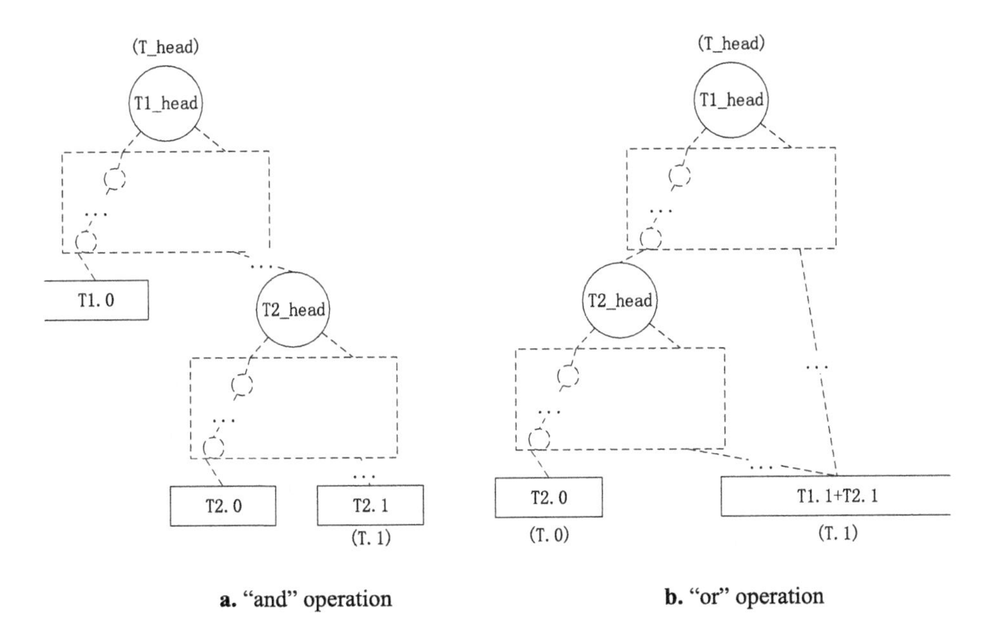

a. "and" operation **b.** "or" operation

Fig. 2. Operation of SimBDD

The procedure of getting the SimBDD of $T = T1 \bullet T2$ is as follows:

```
/* "and" operation of SimBDD*/
SimBDD_AND(T1_head,T2_head)
  parents ← T1_head.oneNode.parents
     for each parent in parents
    parent.oneLink ← T2_head
    T2_head.parents ← add(parent)
  T1_head.oneNode ← T2_head.oneNode
  return T1_head
```

As Fig. 2b shows, when $T = T1 + T2$, we first connect the parent node of T1_head.zeroNode to T2_head with their edgeZero and add the parent node to T2_head.parents. Then, we merge T1_head.oneNode and T2_head.oneNode. Finally, the SimBDD whose head is T1_head is the one that T corresponds.

The procedure of getting the SimBDD of $T = T1 + T2$ is as follows:

```
/* "or" operation of SimBDD*/
SimBDD_OR(T1_head,T2_head)
parents ← T1 _head.zeroNode.parents
for each parent in parents
      parent.zeroLink ← T2_head
      T2_head.parents ← add(parent)
parents ← T2 _head.oneNode.parents
for each parent in parents
      parent.oneLink ← T1_head.oneNode
      T1_head.oneNode.parents ← add(parent)
   T1_head.zeroNode ← T2_head.zeroNode
   return T1_head
```

2.3 Convert from FT to SimBDD

In this paper, FT is converted to SimBDD in depth-first order.

The following pseudocode shows the procedure of converting from FT to SimBDD, where T.root represents the root node of the fault tree T, ftnode represents a fault tree node, ftnode.name represents the name of fnode, ftnode.child represents the set of ftnode's child nodes, ftnode.child [i] represents the (i−1)th subnode of ftnode, ftnode.calc represents the corresponding gate of intermediate event node ftnode. The function TransToSimBDD uses the root node of the FT as its input and returns a SimBDD that corresponds to the FT.

```
/* The procedure of transforming from FT to SimBDD */
TransToSimBDD(T.root)
TransToSimBDD(ftnode)
  theBDD ← create a new SimBDD node
  if ftnode.childs is null
    theBDD.node ← ftnode.name
    theBDD.zeroLink ← theBDD.zeroNode
    theBDD.oneLink ← theBDD.zeroNode
    theBDD.zeroNode.parents ← add(theBDD)
    theBDD.oneNode ← add(theBDD)
  else
    for i ← 0 to the length of ftnode.childs
      child ← ftnode.childs[i]
      childBDD ← TransToSimBDD(child)
      if i=0
        theBDD ← childBDD
      else
        if ftnode.calc="AND"
          theBDD ← SimBDD_AND(theBDD,childBDD)
        else if ftnode.calc="OR"
          theBDD ← SimBDD_OR(theBDD,childBDD)
  return theBDD
```

2.4 Method of Getting MCSs

The pseudocode of getting the cut sets of a fault tree is as follows. The input of the function GetCSS is the head node of a FT's corresponding SimBDD and the output of the function is the cut sets of the fault tree.

```
/* Get cut sets*/
GetCSS(T_head)
  list ← create a empty list
  CSS ← create a empty list
  FindCS(T_head,list,CSS)
  return CSS
/*Traverse to find all cut sets*/
FindCS(bddNode,list1,CSS)
  if bddNode=T_head.oneNode
    CSS ← add(list1)
  else
    if bddNode.zeroLink is not null or a nodeZero
      list2 ← copy(list1)
      FindCS(bddNode.zeroLink,list2,CSS)
    if bddNode.oneLink is not null
      if list1 do not contains bddNode.node
        list1 ← add(bddNode.node)
      FindCS(bddNode.oneLink,list1,CSS)
```

From the Proposition 1, the cut sets obtained by the above method contain all the minimum cut sets of the fault tree. After all the cut sets are obtained, delete all cut sets which contain other cut sets. After the above procedure is completed, the remaining sets are the minimum cut sets of the fault tree.

3 Test Results

We have tested the open source tool OpenFTA and the SimBDD algorithm on six fault trees from a benchmark set [9] and the results are given in Table 1.

Table 1. FT test result through OpenFTA and SimBDD

FT	Size	Basic events	Solutions	Time (s)	
				OpenFTA	SimBDD
Das9201	204	122	14217	197.2	34.5
Isp9606	130	89	1776	3.6	0.1
Das9204	83	53	16704	320.7	76.4
Das9205	71	51	17280	120.4	14.6
Das9206	233	121	19518	735.4	54.0
ftr10	269	175	305	<0.1	<0.1
isp9603	186	91	3434	43.2	7.2

Table 1 lists the time to find the minimum cut sets through OpenFTA and SimBDD. The size in Table 1 means the total number of gates and basic events in the FT. Generally, we can get correct MCSs of FTs in both of these ways. Comparing the time of the two methods, it is found that it's much faster to obtain the MCSs through the SimBDD algorithm than the OpenFTA.

4 Conclusion

This paper introduces a simplified BDD structure and describes the transition from the fault tree to the simplified BDD structure. When the fault tree is converted to a normal BDD structure, the number of BDD nodes increases exponentially with the number of basic event nodes of the fault tree. Therefore, for large fault trees, sometimes the memory is not sufficient to store its corresponding BDD. If a large fault tree is converted to a SimBDD, the number of nodes will be less than three times the number of its basic event nodes. Therefore, the storage of the SimBDD structure requires only a small amount of memory. SimBDD is built very fast, and it is always smaller than the normal BDD of the same tree. It has nice speed to get all MCSs of a fault tree using the SimBDD.

Acknowledgment. This paper is funded by the Fundamental Research Funds for the Central Universities (NS2016088). And the paper is supported by China's 13th plan of key basic research project (JCKY2016206B001).

References

1. Deng, Y., Wang, H., Guo, B.: BDD algorithms based on modularization for fault tree analysis. Prog. Nucl. Energy **85**, 192–199 (2015)
2. Bryant, R.E.: Graph-based algorithms for boolean function manipulation. IEEE Trans. Comput. **35**(8), 677–691 (1986)
3. Brace, K.S., Rudell, R.L., Bryant, R.E.: Efficient implementation of a BDD package. In: ACM/IEEE Design Automation Conference, pp. 40–45. ACM (1991)
4. Rauzy, A.: A brief introduction to binary decision diagrams. J. Européen Des Systèmes Automatisés Hermes **4**(4), 206–207 (1996)
5. Rauzy, A.: New algorithms for fault trees analysis. Reliab. Eng. Syst. Saf. **40**(3), 203–211 (1993)
6. Andrews, J.D., Bartlett, L.M.: Efficient basic event orderings for binary decision diagrams. In: Reliability and Maintainability Symposium, 1998. Proceedings. IEEE Xplore, pp. 61–68. (1998)
7. Sinnamon, R.M., Andrews, J.D.: Improved efficiency in qualitative fault tree analysis. Qual. Reliab. Eng. Int. **13**(5), 293–298 (1997)
8. Remenyte-Prescott, R.: System Failure Modelling Using Binary Decision Diagrams. Loughborough University, Loughborough (2007)
9. A benchmark of boolean formulae http://iml.univ-mrs.fr/~arauzy/aralia/benchmark.html

A Sparse Nonlinear Signal Reconstruction Algorithm in the Wireless Sensor Network

Zhao Xia-Zhang(✉) and Yong-Jiang

Hunan Chemical Industry Vacation Technology College, Zhuzhou, China
997561071@qq.com

Abstract. For dealing with the problem of sparse nonlinear signal reconstruction of wireless sensor networks (WSNs), a distributed sparse nonlinear signal reconstruction algorithm which is on account of square root unscented kalman consensus filter (SRUKCF) with embedded pseudo-measurement (PM) is proposed in this paper. The pseudo-measurement embedded square root unscented kalman consensus filter (SRUKCF-PM) is reestablished in the information form. By introducing pseudo-measurement technology into SRUKCF, a kind of distributed reconstruction algorithm is exploited to integrate the random linear measurements from various nodes of the WSNs, so that all filters are able to achieve agreement about the calculation of sparse nonlinear signals. The simulation results demonstrate the effectiveness of the proposed algorithm.

Keywords: Unscented kalman filter · Consensus filter · Compressive sensing · Wireless sensor network · Signal reconstruction

1 Introduction

Compressive sensing (CS) has received considerable attention in the last decade [1]. It has shown that sparse signals can be reconstructed from far fewer observations using overwhelming probability settling the problem of a 1-norm minimization. However, many systems for which compressive sensing are used are dynamical. CS has been extensively applied in WSNs [2]. The Kalman Consensus Filter (KCF) is a state-of-the-art distributed algorithm for fusing multiple measurements from different sensors [3]. It is a distributed approach that combines the distributed Kalman filter and the control-theoretic consensus algorithms. In Ref [4], there proposed three kinds of distributed Kalman filtering algorithms. Based on the information form of Kalman filter, analysis of stability and performance of the Kalman-consensus filter provided in Ref [3], the algorithms were established.

The aim of the paper is to exploit a kind of distributed filtering method to reconstruct a sparse nonlinear signal with a few random linear measurements from WSN. The proposed method is built by the way of the square root unscented kalman consensus filter and the technology of pseudo-measurement. The rest of the paper is structured like this. Section 2 states the essential problems of compressed sensing and review the Kalman filtering embedded pseudo-measurement equation. And in Sect. 3, the Kalman filtering with pseudo-measurement equation is recommended into the

F. Xhafa et al. (eds.), *Advances in Intelligent Systems and Interactive Applications*, Advances in Intelligent Systems and Computing 686,
https://doi.org/10.1007/978-3-319-69096-4_81

square root unscented kalman consensus filter to exploit a kind of distributed sparse nonlinear signal reconstruction algorithm. Finally, in Sect. 4, experiment results are shown to prove the performance of this method.

2 Kalman Filter Sparse Signal Estimation

2.1 Recovery of sparse signal

Because of an R^n-valued random discrete-time process $\{x_k\}_{k=0}^{\infty}$ that is sparse, where $\|x_k\|_0 \ll n$, x_k is named s-sparse if $\|x\|_0 = s$. Suppose that x_k evolves on the basis of the dynamic model as follows

$$x_k = A_k x_{k-1} + q_{k-1} \tag{1}$$

where $A_k \in R^{n \times n}$ is the state transition matrix, $\{q_k\}_{k=0}^{\infty}$ is a zero-mean white Gaussian sequence by covariance $Q_k \geq 0$, and $x_k \sim N(x_{0|-1}, P_{0|-1})$. Equation (2) is the m-dimensional linear measurement of x_k

$$z_k = H_k x_k + \mathbf{r}_k \tag{2}$$

where the measurement matrix is shown by $H_k \in R^{m \times n}$, and $\{r_k\}_{k=0}^{\infty}$ represents a zero-mean white Gaussian sequence by $R_k \geq 0$. When $m < n$, recovering signal from measurement is an ill-posed problem.

In Ref [1], x_k is precisely recovered via optimally solving the below problem (3)

$$\min_{\hat{x}_k \in R^n} \|\hat{x}_k\|_0 \text{ s.t. } \|y_k - H_k \hat{x}_k\| \leq \varepsilon \tag{3}$$

However the problem (3) is NP-hard, and there is no available solution. Luckily, the Ref [1] states that assume the measurement matrix obeys the restricted isometry property, we can obtain the solution of problem (3) by overwhelming probability via solving the convex optimization of problem (4)

$$\min_{\hat{x}_k \in R^n} \|\hat{x}_k\|_1 \text{ s.t. } \|y_k - H_k \hat{x}_k\| \leq \varepsilon \tag{4}$$

2.2 Kalman Filtering Pseudo-measurement

As regards to the system presented in above (1) and (2), Kalman filtering is able to offer an estimate of x_k which solute the below unconstrained l_2 minimization problem

$$\min_{\hat{x}_k \in R^n} E_{x_k | y_1, \ldots, y_k} \left[\|x_k - \hat{x}_k\|_2^2 \right] \tag{5}$$

In Ref [6], the stochastic case of (4) and its dual problem is discussed

$$\min_{\hat{x}_k \in R^n} \|\hat{x}_k\|_1 \text{ s.t. } E_{x_k|y_1,\ldots,y_k}\left[\|x_k - \hat{x}_k\|_2^2\right] \tag{6}$$

$$\min_{\hat{x}_k \in R^n} E_{x_k|y_1,\ldots,y_k}\left[\|x_k - \hat{x}_k\|_2^2\right] \text{ s.t. } \|\hat{x}_k\|_1 \leq \varepsilon' \tag{7}$$

Through structuring a pseudo-measurement equation

$$0 = H_k x_k - \varepsilon_k' \tag{8}$$

3 Distributed SR-UKF for Sparse Signal Reconstruction

Kalman-consensus filter proposed in Ref [4] is a kind of distributed estimation algorithm for WSNs. Ref [3] analyzes its stability and performance detailedly. In this part, CSKF-1algorithm can be reestablished in the information from of SR-UKF, and embedding the pseudo-measurement technology in the KCF to exploit a kind of distributed sparse signal estimation algorithm for WSNs.

Because the topology of a WSN is stated by an undirected graph $G = (V, E, A)$ of the order N with the set of nodes $V = \{1, 2, \ldots, N\}$, the set of edges $E = V \times V$, and the adjacency matrix $A = [a_{i,j}]$ with nonnegative adjacency element $a_{i,j}$. An edge of G of the graph is positive; that is, $a_{i,j} > 0 \Leftrightarrow (i,j) \in E$. The node j is named a neighbor of node i which is denoted by N_i. Suppose that G is strongly connected. In a non-linear dynamic system, the signals, the dynamic system and measurement satisfy the dynamical model as follows

$$x_k = f(x_{k-1}) + q_{k-1} \tag{9}$$

$$z_k = H_k x_k + r_k \tag{10}$$

4 Experimental Evaluation

In Fig. 1, there is a WSN which has 6 nodes. Its topology is shown by an undirected graph $G = (V, E, A)$ with the nodes' set $V = \{1, 2, 3, 4, 5, 6\}$, the edges' set $E = \{(1,2), (1,3), (2,4), (3,5), (3,6), (4,5), (5,6)\}$, and the adjacency matrix:

$$A = \begin{bmatrix} 0 & 1 & 1 & 0 & 0 & 0 \\ 1 & 0 & 1 & 1 & 1 & 0 \\ 1 & 1 & 0 & 0 & 1 & 1 \\ 0 & 1 & 0 & 0 & 1 & 1 \\ 0 & 1 & 0 & 0 & 1 & 0 \\ 0 & 0 & 1 & 0 & 1 & 0 \end{bmatrix} \tag{11}$$

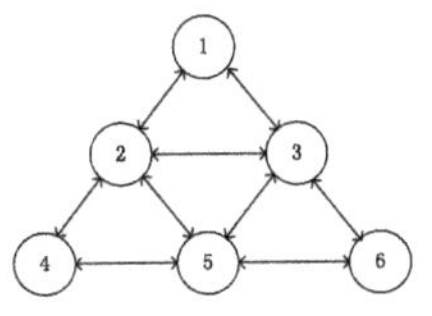

Fig. 1. Topology structure of WSN

In the first experiment, the simulation parameters are as follows: $m = 72$, $s = 10$, $n = 256$, $Q(i,i) = 3^2$, $R_k = 0.25^2 \times I_{12\times 12}$, $f(x) = 0.5x + 25\frac{x}{1+x^2}$. This process can be described by

$$x_k(i) = \begin{cases} f(x_{k-1}(i)) + q_{k-1}(i), q_{k-1}(i) \sim N(0, Q_k), if & x_k(i) \in \text{supp}(x_k) \\ 0, & otherwise \end{cases} \quad (12)$$

where $i \sim U_i[1, 256]$. According to the network, the 72-dimensional measurements are gained. We suppose that every node get a 12-dimensional measurement, it means that, $H_{i,k} \in R^{12\times 256}$ is obtained from a Gaussian distribution $N(0, 1/72)$. Set $\hat{x}_{0|-1} \sim N(0, 5)$, $\hat{P}_{0|-1} = 0.1^2 I_{256\times 256}$, and $\sigma_i^2 = 200^2$, $\gamma = 1$.

Figure 2 shows the estimates of time-varying actual signal of x_8, x_{73}, x_{127} and x_{157} from #1 sensor node. It is obvious that the #1 sensor node provides satisfactory estimations of the actual signal on the supports. In addition, NMSE (normalized mean squared error) is used to assess the error performance of all sensor nodes.

$$\text{NMSE} = \frac{||x - \hat{x}||}{||x||_2^2} \quad (13)$$

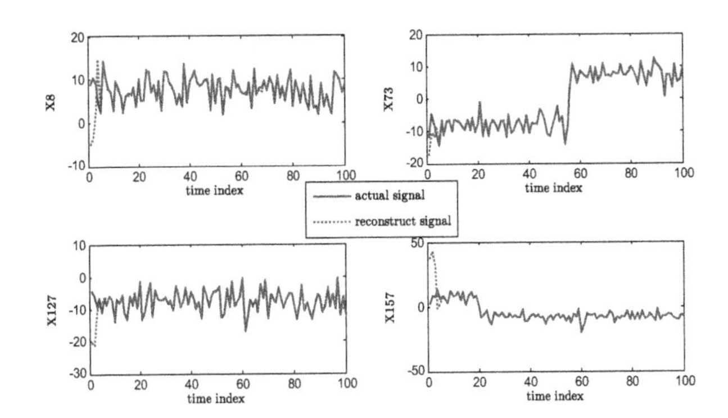

Fig. 2. The estimates $\hat{x}_i$ of sparse signal on its supports (i = 8, 73, 127, 157)

The normalized mean squared error is presented in Fig. 3. The result demonstrates that the NMSE of all sensor nodes gradually converge to zero which means the filter is stable, can achieve agreement about sparse signals estimations asymptotically (Fig. 4).

In particular, we design another experiment that the support x_4 is nonzero for the total measurement interval, x_{42} becomes zero at $n = 61$, x_{91} is nonzero between $n = 41$ and $n = 80$, x_{120} is zero between $n = 61$ and 80. Throughout the time interval, all the rest supports remain zero.

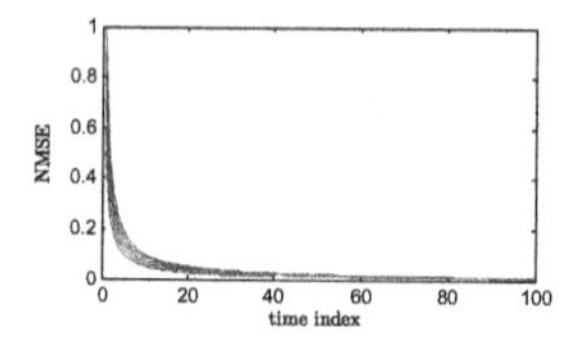

Fig. 3. NMSE of algorithm 1 versus time

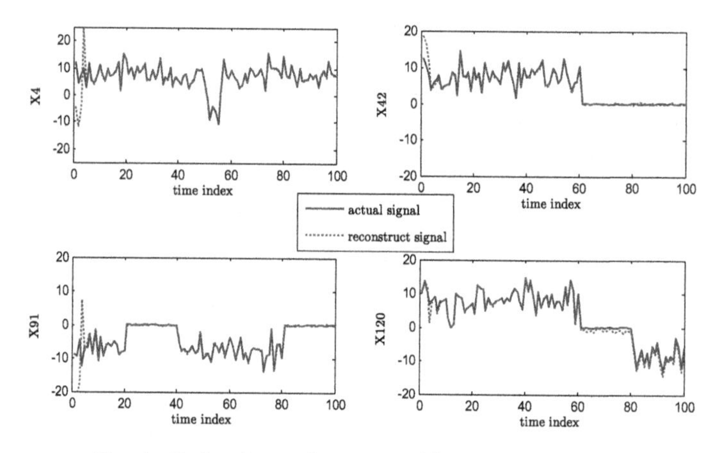

Fig. 4. Estimation performance with variational support

From the above, It can be inferred that even the number of the measurements is 20 times smaller than the signal's dimension for each sensor node in the distributed WSNs, the SRUKCF-PM consensus filter also can reconstruct the sparse nonlinear signal effectively.

5 Conclusion

We considered the signal reconstruction problem of sparse nonlinear signal in distributed WSNs with energy constraints. To deal with the nonlinearity of signal in distributed system, we combine the square root unscented kalman filter with consensus algorithm. Meanwhile, introducing the pseudo-measurement equation into the filter to deal with the sparsity of signal. The performance of presented algorithm is assessed by experiments. The experiment results demonstrate that SRUKCF-PM offers satisfied estimates of sparse nonlinear signal utilizing far fewer measurements than required traditionally. This is very promising in the distributed WSNs with energy constraint; the lifetime of WSNs can be prolonged. Moreover, we design the experiment in which the supports of sparse signal change with time. The results demonstrate that the SRUKCF-PM enables precisely track the variation of support. In the future, we will try to take the signal quantization into consideration and test algorithm performance in real-world applications.

Acknowledgments. This work was supported by scientific research project of Hunan Provincial Education Department (No. 17C0552).

References

1. Candes, E.J., Romberg, J., Tao, T.: Robust uncertainty principles: exact signal reconstructionfrom highly incomplete frequency information. IEEE Trans. Inf. Theory **52**(2), 489–509 (2006)
2. Yang, G., Tan, V.Y.F., Ho, C.K., et al.: Wireless compressive sensing for energy harvesting sensor nodes. IEEE Trans. Signal Process. **61**(18), 4491–4505 (2013)
3. Olfati-Saber, R.: Kalman-consensus filter: optimality, stability, and performance. In: Proceedings of the 48th IEEE Conference on Decision and Control (CDC), pp. 7036–7042. Shanghai, China (2009)
4. Olfati-Saber, R.: Distributed Kalman filtering for sensor networks. In: Proceedings of the 46th IEEE Conference on Decision and Control (CDC), pp. 5492–5498. Los Angeles, Calif., USA (Dec 2007)

Chinese POS Tagging with Attention-Based Long Short-Term Memory Network

Nianwen Si(✉), Hengjun Wang, Wei Li, and Yidong Shan

Department of Computer Science, Zhengzhou Institute of Information Science and Technology, Zhengzhou, China
snwl608@163.com

Abstract. Traditional Chinese Part-of-speech tagging (POS tagging) typically use statistical approach, which makes it rely on hand-crafted feature templates. In this paper, we propose an effective attention based bidirectional Long Short-Term Memory model for Chinese POS tagging. In our model, first, the word distributed embedding features are used as the input of the network to go through network transformation. Then, the bidirectional LSTM network is used to encode the input tokens. What's more, the attention mechanism is introduced into the hidden layer to make the model focus more on the related part. Finally, the state transition probability matric is explored and the log likelihood loss function is used for model training. In the experiment, we test the tagging model in Penn Chinese Treebank to observe its performance. The results show that the proposed model achieves competitive accuracies compared with state of the art works.

Keywords: Natural language processing · Part-of-speech tagging · Long short-term memory · Attention mechanism · Contextual information

1 Introduction

Part-of-speech tagging (POS tagging) is an essential component for Chinese natural language processing (NLP). Researches for Chinese POS tagging have been studied early. In previous works, they attempted to use statistical machine learning method such as hidden Markov model (HMM) [1], maximum entropy (ME) [2] and conditional random field (CRF) [3]. However, due to the heavily relying on the feature engineering, these traditional methods can only utilize limited and sparse features. What's more, large amounts of computations for features will put the model in the risk of over-fitting, which will reduce the generalize ability of the model.

In recent years, deep learning (DL) has been widely applied in NLP community. Collobert et al. [4] developed the unified neural network architecture for POS tagging and semantic role tagging (SRL). In their neural model, the designed neural network can learn rich features automatically, which alleviate the relying on the manually crafted features and achieved competitive tagging accuracies compared with traditional methods. Zheng et al. [5] further presented the fast and effective neural network model for Chinese word segmentation and POS tagging, which shows better performance using less feature set. The current researches show that high level contextual

F. Xhafa et al. (eds.), *Advances in Intelligent Systems and Interactive Applications*, Advances in Intelligent Systems and Computing 686,
https://doi.org/10.1007/978-3-319-69096-4_82

information is needed to solve the ambiguity of POS, which is the main problem in POS tagging. To exploit more contextual features, [6] presented a unified tagging solution using hierarchical Long Short-Term Memory network (LSTM) for multiple tasks in NLP. The LSTM based tagging model achieved high accuracies compared with previous works. What's more, attention mechanism has been introduced in neural networks, and shows great effectiveness in neural network machine translation [7]. It allocates the attention probability for the LSTM hidden vector at different time steps, which leads to the high attention been paid to the particular context.

In this paper, we propose an effective attention based LSTM POS tagging model. The proposed model utilizes LSTM to obtain the contextual information of current word, at the same time introduces the attention probability in LSTM hidden vector. By allocate different attention probability in importance, the model will better employ the contextual features around the target word and improve the focusing ability for important contextual features. We test the model on Penn Chinese Treebank dataset, the results show that the proposed model achieves better accuracies compared with traditional methods, and reach the similar accuracies compared with the existing best tagging systems.

2 Long Short-Term Memory

LSTM is the particular type of Recurrent Neural Networks (RNNs). It shows better ability in obtaining contextual information in sequential data. In common LSTM network structure, the information of sequential data is transferred from one direction. While in bidirectional LSTM network, it consists of the forward LSTM layer and backward LSTM layer. Figure 1 illustrates the structure of the bidirectional LSTM network.

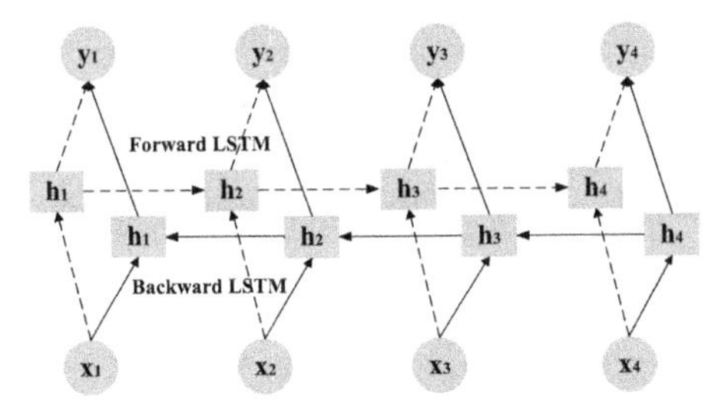

Fig. 1. The Bi-LSTM model structure

The LSTM network consists of three parts: the input layer, the hidden layer and the output layer. In the hidden layer, there are two separate parts which includes the forward LSTM and the backward LSTM layer. The forward LSTM is used to compute the forward hidden state while the backward LSTM is used for backward hidden state. Comparing with the uni-directional LSTM, the bi-LSTM could make use of richer features including the former and the later history information, which enables the bi-LSTM network to gain the better ability in feature learning and representation. Therefore, the bi-LSTM network has been applied into many NLP tasks and achieved great success in most of them.

3 POS Tagging with Attention-based LSTM

3.1 Model

In this paper, we propose an effective neural model for POS tagging based on the Bi-BLSTM, at the same time adding the attention mechanism to further improve the tagging accuracy. The model structure can be illustrated in Fig. 1. In this section, we will present the tagging model in detail from three parts: input layer, hidden layer, output layer.

(1) Input layer. we adopt the word distributed embedding trained by word2vec tools and develop the lookup table using the embedding matrices. By using the index of words, each word could find its corresponding word embedding in the lookup table, which will be utilized as the input of Bi-LSTM network.
(2) Hidden layer. In hidden layer, the computation process can be divided into three parts.

First, the forward and backward hidden state is calculated using the LSTM. For the uni-directional LSTM network, it computes the current hidden state $\boldsymbol{h_t}$ using the current input vector $\boldsymbol{x_t}$ and the previous hidden state $\boldsymbol{h_{t-1}}$. While for the bi-directional LSTM network, both the forward hidden state and backward hidden state are be computed at the same time:

$$\begin{aligned} \overrightarrow{\boldsymbol{h}}_t &= \mathrm{LSTM}(\boldsymbol{x}_t, \overrightarrow{\boldsymbol{h}}_{t-1}) \\ \overleftarrow{\boldsymbol{h}}_t &= \mathrm{LSTM}(\boldsymbol{x}_t, \overleftarrow{\boldsymbol{h}}_{t-1}) \end{aligned} \tag{4}$$

Where m denotes the dimensionality of the hidden layer. The function LSTM() represents the transformation of the LSTM network, which is described in Sect. 2.

Then, The hidden state $\boldsymbol{h_t} \in R^{m\times 1}$of Bi-LSTM can be obtained by using the forward hidden state $\overrightarrow{\boldsymbol{h}}_t$ and the backward hidden state $\overleftarrow{\boldsymbol{h}}_t$:

$$\boldsymbol{h}_t = \boldsymbol{W_1}\overrightarrow{\boldsymbol{h}}_t + \boldsymbol{V_1}\overleftarrow{\boldsymbol{h}}_t + \boldsymbol{b_1} \tag{5}$$

Where $\boldsymbol{W_1} \in R^{m\times m}$ and $\boldsymbol{V_1} \in R^{m\times m}$ are weight matrices and $\boldsymbol{b_1} \in R^{m\times 1}$ is the bias term. The compositional hidden state contains the history information from two directions, which will provide rich contextual features for tagging.

Last, the new hidden layer state is computed according to the attention mechanism. Where computing the new hidden layer energy e_t, the corresponding compositional hidden layer state $\boldsymbol{h_t}$ will be used which consists of the forward and backward history information.

$$e_t = \boldsymbol{V_2} \tanh(\boldsymbol{W_2 h_t} + \boldsymbol{b_2}) \tag{6}$$

Where $\boldsymbol{W_2} \in R^{l \times m}$ and $\boldsymbol{V_2} \in R^{1 \times l}$ are weight matrices and $\boldsymbol{b_2} \in R^{l \times 1}$ is the bias term.

$$\alpha_t = \frac{\exp(e_t)}{\sum_{t=1}^{T} \exp(e_t)} \tag{7}$$

$$\boldsymbol{s_t} = \sum_{t=1}^{T} \alpha_t \boldsymbol{h_t} \tag{8}$$

Where α_t is the attention weight corresponding to $\boldsymbol{h_t}$. The new hidden state vector $s_t \in R^{m \times 1}$ is obtained according to the previous hidden state at different time steps, whose dimensionality is the same with initial hidden state.

(3) Output layer. The output layer adopts the softmax as the activation function and calculate the probability distribution on the tagging set.

$$\boldsymbol{y_t} = \text{softmax}(\boldsymbol{W_3 s_t} + \boldsymbol{b_3}) \tag{9}$$

Where $\boldsymbol{W_3} \in R^{L \times m}$ denotes the weight matric between the attention hidden layer and the output layer. $\boldsymbol{b_3} \in R^{L \times 1}$ is the corresponding bias term.

3.2 Training

Given N training examples $T = \{(x^{(i)}, y^{(i)})\}_{i=1}^{N}$, where $x^{(i)} = [x_1^{(i)}, x_2^{(i)}, \ldots, x_n^{(i)}]$ denotes the ith sentence in the training set. $x_i^{(i)} (i = 1, 2, \ldots, n)$ denotes the ith word of the sentence and n is the sentence length. The corresponding POS tag sequence of the sentence is $y^{(i)} = [y_1^{(i)}, y_2^{(i)}, \ldots, y_n^{(i)}]$. In the model training, we use the log likelihood loss function. The final loss function is added with the L2-regulation term and formally defined as follows:

$$\mathrm{L}(\theta) = -\frac{1}{m} \sum_{i=1}^{m} \log(P(y^{(i)}|x^{(i)}; \theta)) + \frac{1}{2} ||\theta||^2 \tag{10}$$

Where the $P(y^{(i)}|x^{(i)}; \theta)$ is the score of the tagging sequence. θ is the model hyper-parameter set. In the training process, we use the stochastic gradient descent (SGD) with mini-batches. The parameter m in the (10) is the batch size. The dropout measure is also used in training process, which will randomly drop a part of the hidden state to avoid over-fitting.

3.3 Decoding

We adopt the similar decoding algorithm with [6] and introduce the transition matric $\boldsymbol{A}$ into the network. The score of each tagging sequence can be obtained as follows:

$$s(x_{1:n}^{(i)}, y_{1:n}^{(i)}) = \sum_{t=1}^{n} (\beta \boldsymbol{A}_{y_{t-1}y_t} + (\boldsymbol{y}_t)_{y_t}) \tag{11}$$

Where $\boldsymbol{A}_{y_{t-1}y_t}$ denotes the transition probability from tag y_{t-1} to tag y_t. Hypothesis that $\mathrm{Y}(\mathrm{x}^{(i)})$ denotes the set of tagging sequence containing all the tagging sequence of the sentence $x^{(i)}$, the final training objective is to find the sequence which obtains the highest score from the sequence set.

$$\hat{y}^{(i)} = \underset{y \in Y(x^{(i)})}{\arg\max}(s(x^{(i)}, y) + \Delta(\bar{y}^{(i)}, y)) \tag{12}$$

Where $\bar{y}^{(i)}$ denotes the gold standard tagging sequence and $\hat{y}^{(i)}$ denotes the predicted sequence. The structured loss function $\Delta(\bar{y}^{(i)}, y)$ is defined in the following formula:

$$\Delta(\bar{y}^{(i)}, y) = \sum_{j=1}^{n} \kappa l(\bar{y}_t^{(i)} \neq y_t) \tag{13}$$

Where κ is the loss parameter. As we can see, $\Delta(\bar{y}^{(i)}, y)$ grows proportionally with the number of incorrect tags in the tagging sequence.

4 Experiments and Analysis

4.1 Experiment Settings

To test the performance of the POS tagging, we conduct experiment on Penn Chinese Treebank (CTB5.0) dataset. Following previous works, the dataset is split into three parts: section 1–270, 400–931, 1001–1151 for training, 301–325 for development, 271–300 for testing. In the experiment evaluation, we adopt the tagging accuracy P to evaluate the performance. The hyper-parameters are tuned on the development set. We save the hyper-parameters for the comparison which obtains the best performance in the development set. The final setting of the parameters are listed as follows: Word embedding size 50, LSTM hidden layer size 120, learning rate 0.5, L2 regulation 10-4, dropout rate 0.4, mini-batch size 50, transition weight 0.75.

4.2 Results and Comparisons

In order observe the impact of the model component in the tagging process, we choose three schemes: LSTM, Bi-LSTM and Bi-LSTM+attention. The result is listed in Table 1.

Table 1. The results of the different models on ctb5

Model	CTB5.0	
	Dev	Test
LSTM	95.14	94.12
BLSTM	95.32	94.57
BLSTM+ attention	**95.76**	**95.63**

we can see that the Bi-LSTM model obtains higher tagging performance compared with LSTM model, because it can make use of the contextual information both from the forward and backward direction. What's more, when the attention layer is added in the Bi-LSTM model, the tagging model achieves the best tagging accuracy compared with other two models.

To make comparisons with other POS tagging model, we further list the recent works for POS tagging on CTB5 dataset in Table 1. Clearly, from Table 2 we can see that, comparing with the traditional tagging model, our model achieves better accuracies. While for other tagging system such as transition-based model, we also reach highly competitive accuracies. Particularly, In the joint model like [9] and [11], the model can make use of more features from other tasks. This will help to improve the POS tagging performance. While the model proposed in this work only the least features (word embeddings) but results in competitive tagging accuracies.

Table 2. Comparisons with previous works

Model	Method	Accuracy (P)
[1]	Semi-supervised HMM	95.48
[3]	CRF	92.07
[9]	Undirected graph	95.37
[10]	Easy-firs, beam search	94.01
[11]	Transition-based	**96.0-**
This work	Attention-based LSTM	95.63

5 Conclusion

In this work, we propose an effective Bi-LSTM network based model for Chinese POS tagging, which solves the problem of feature engineering in traditional tagging model. By introducing the attention mechanism into the hidden layer, the tagging accuracies are further improved and achieved competitive results compared with other recent works. Considering that he POS tagging will benefit from the syntactic information, in the further work, we will attempt to develop the joint model for POS tagging and syntactic parsing. By jointly make tagging and parsing tasks, the performance of two tasks will benefit from each other and improved together.

References

1. Han, X., Huang, D.: Research on Chinese part-of-speech tagging based on semi hidden Markov model. J. Chin. Comput. Syst. **36**(12), 2813–2816 (2015)
2. He, J., Wang, H.: Chinese word sense disambiguagtion based on maximum entropy model with feature selection. J. Softw. **21**(6), 1287–1295 (2010)
3. Yu, D., Ge, Y., et al.: Chinese part-of-speech tagging based on conditional random field. Microelectron. Comput. **28**(10), 63–66 (2011)
4. Collobert, R., Weston, J., Bottou, L., et al.: Natural language processing (almost) from scratch. J. Mach. Learn. Res. **12**(1), 2493–2537 (2011)
5. Zheng, X., Chen, H., Xu, T.: Deep learning for Chinese word segmentation and POS tagging. In: Conference on Empirical Methods in Natural Language Processing (2013)
6. Zhou, Q., Wen, L., Wang, X., et al.: A hierarchical LSTM model for joint tasks. In: Chinese Computational Linguistics and Natural Language Processing Based on Naturally Annotated Big Data. Springer International Publishing, Switzerland (2016)
7. Bahdanau, D., Cho, K., Bengio, Y.: Neural machine translation by jointly learning to align and translate. Comput. Sci. (2014)
8. Hinton, G.E., Srivastava, N., Krizhevsky, A., et al.: Improving neural networks by preventing co-adaptation of feature detectors. Comput. Sci. **3**(4), 212–223 (2012)
9. Zhu, C., Zao, T., et al.: Joint Chinese word segmentation and POS tagging system with undirected graphical models. J. Electron. Inf. Technol. **32**(3), 700–704 (2010)
10. Ma, J., Zhu, J., Xiao, T, et al.: Easy-first POS tagging and dependency parsing with beam search. In: Meeting of the Association for Computational Linguistics, pp. 110–114. (2013)
11. Wang, Z., Xue, N.: Joint POS tagging and transition-based constituent parsing in Chinese with non-local features. In: Meeting of the Association for Computational Linguistics, pp. 733–742. (2014)

A New Weighted Decision Making Method for Accurate Sound Source Localization

Yingxiang Sun and Jiajia Chen(✉)

Singapore University of Technology and Design, Singapore, Singapore
jiajia_chen@sutd.edu.sg

Abstract. Sound source localization is a challenging task in adverse environments with high reverberation and low signal-to-noise ratio. To accurately localize the source through classification methods, the number of sub-spaces for classification decision should be large. However, this also causes high misclassification rate, leading to larger localization error, especially in adverse environments. In this paper, we propose a new weighted decision making method (WDMM), which can effectively improve the localization accuracy of the likelihood based classification algorithms, by revisiting and accessing the probabilities of the adjacent sub-spaces. The synthetic experimental results have shown that the average mean and average standard deviation of the localization errors from the 20 different acoustic environments by the proposed WDMM are only 0.42 and 0.21 m respectively in a 4.0 m $\times$ 4.0 m $\times$ 4.0 m room. The 20 different acoustic environments include the high reverberation T_{60} up to 0.6 s and low signal-to-noise ratio to -10 dB. Compared to localization results without WDMM, the proposed method has reduced averages of mean and standard deviation localization errors by 35.8 and 55.2% respectively.

Keywords: Sound source localization · Machine learning · Classification

1 Introduction

In spatial audio processing, sound source localization (SSL) is a major topic with broad applications, including localizing sniper fire in urban battlefields, police identification of the location of criminals, automatically steering of the camera to the direction of the active talker and robot reaction for audio signal. For indoor SSL, a lot of efforts [1] have been made in the past decades. However, the SSL accuracy in the adverse environments with high reverberation and low signal-to-noise-ratio (SNR) is still very challenging which needs to be addressed.

Besides conventional methods [2, 3], machine learning method [4] has been applied to solve SSL problem. In this paper, we propose a new weighted decision making method (WDMM) for likelihood based classification algorithms to localize sound source in adverse environments. By revisiting and accessing the probabilities of the adjacent sub-spaces, the accuracy of the SSL can be effectively improved, compared with the results without WDMM.

F. Xhafa et al. (eds.), *Advances in Intelligent Systems and Interactive Applications*, Advances in Intelligent Systems and Computing 686,
https://doi.org/10.1007/978-3-319-69096-4_83

The remaining of this paper is organized as follows. In Sect. 2, the problem of SSL accuracy in terms of 3-dimensional distance error is presented. In Sect. 3, a new WDMM to improve the accuracy is proposed. The synthetic experimental results with and without the proposed WDMM in various acoustic environments are presented in Sect. 4, followed by the comparisons and discussions. Our conclusions are presented in Sect. 5.

2 Problem Formulation

In this section, we present the SSL accuracy problem to be addressed in this paper. It is assumed that a stationary single sound source is arbitrarily placed inside a 3-dimentional enclosed rectangular room. The microphone array which consists of K microphones inside the room is also stationary. Assume the source signal is $s(t)$, the signal received by the kth microphone can be expressed as

$$r_k(t) = h_k(t) * s(t) + n_k(t), \tag{1}$$

where * denotes convolution. $h_k(t)$ is the room impulse response between the sound source and the kth microphone, containing the multipath propagation and attenuation information. $n_k(t)$ is the noise at the kth microphone with $k \in [1, K]$.

We can rewrite (1) as

$$R = H * s(t) + N, \tag{2}$$

where $R = [r_1,, r_2, \ldots, r_K]^T$ denotes the received signal set. $H = [h_1, h_2, \ldots, h_K]^T$ denotes the room impulse response set. $N = [n_1, n_2, \ldots, n_K]^T$ denotes the noise set.

When a room is divided into Q sub-spaces, each sub-space can be approximately represented by a 3-dimensional Cartesian coordinate inside it. Therefore, if the volume of each sub-space is small, SSL inside a 3-dimensional room can be transformed into a likelihood based classification problem by deciding which particular sub-space the source belongs to. The sub-space c_{sc} which the sound source is classified into is the solution of the classification problem while its coordinate representative $[x_{sc}, y_{sc}, z_{sc}]$ is the solution of SSL.

Assume the actual source location is $[x_{ac}, y_{ac}, z_{ac}]$ inside the sub-space c_{actual}. Even if the classification is wrong when $c_{sc} \neq c_{\text{actual}}$, the localization error δ can always be evaluated by

$$\delta = \sqrt{(x_{sc} - x_{ac})^2 + (y_{sc} - y_{ac})^2 + (z_{sc} - z_{ac})^2}. \tag{3}$$

To minimize δ, therefore, volume of the sub-space should be small enough so that δ is bounded by the sub-space dimension when the classification is correct. Meanwhile, the classification correctness rate should be as high as possible. However, when the sub-space volume is smaller, it becomes more challenging to distinguish the different acoustic features of the adjacent sub-spaces and hence the classification correctness rate will drop. In addition, δ becomes even larger when the actual source location is close to the boundaries of two adjacent sub-spaces. To solve this problem, therefore, a new WDMM is proposed in the next section by accessing the classification information of not only the sub-space with the top likelihood, but a couple of adjacent sub-spaces as well.

3 Proposed Weighted Decision Making Method

Let's denote the likelihood based algorithm used for classification as **LBA**. Assume that the probability of the sound source being classified into the *j*th sub-space is p_j with $j \in [1, Q]$. Therefore, the probabilities of all sub-spaces can be obtained by **LBA**, where $\sum_{j=1}^{Q} p_j = 1$ is satisfied. Then, we select the β most possible sub-spaces with top probability following the descending order, whose probability sum is less than a threshold depending on sub-space size, i.e. $\sum_{j=1}^{\beta} p_j \leq p_{th}$.

After these β sub-spaces are selected, we perform the SSL through the following two steps, which are initial estimation and final estimation. Assume $L_i = (x_i, y_i, z_i)$ is the central point of the *i*th sub-space with $i \in [1, \beta]$ which represents the *i*th sub-space in Cartesian coordinate system, we initially estimate source location $L_{pe} = (x_{pe}, y_{pe}, z_{pe})$ by

$$x_{pe} = \sum_{i=1}^{\beta} p_i \cdot x_i; y_{pe} = \sum_{i=1}^{\beta} p_i \cdot y_i; z_{pe} = \sum_{i=1}^{\beta} p_i \cdot z_i. \tag{4}$$

We denote this procedure as **IE**(p_i, L_i). With (4), the distance between L_i and L_{pe} can be computed by

$$d_i = \sqrt{(x_i - x_{pe})^2 + (y_i - y_{pe})^2 + (z_i - z_{pe})^2}. \tag{5}$$

The smaller d_i indicates that the actual source location is more likely to be close to the *i*th sub-space, which means that its probability is supposed to be increased. Therefore, new weight assigned to the *i*th sub-space is inversely proportional to d_i, which can be derived by

$$w_i = \frac{(1/d_i)^{\mu}}{\sum_{i}^{\beta} (1/d_i)^{\mu}}, \tag{6}$$

where the controlling parameter μ is in the range of (0, 1). We denote this procedure as $\mathbf{NW}_{\mathbf{ss}}(d\ i,\ \mu)$.

In order to reduce the localization error caused by misclassification further, we adjust the initial estimation by more representative points in the second step. ψ representative points of each selected sub-space are utilized to represent the sub-space position more accurately. Similar to the initial estimation, weights of the representative points can be computed by

$$w_{i,t} = \frac{\left(1/d_{i,t}\right)^{v}}{\sum_{t=1}^{\psi}\left(1/d_{i,t}\right)^{v}}, \tag{7}$$

where $w_{i,t}$ is the weight for the tth representative point $rp_{i,t}$ in the ith sub-space with $t \in [1, \psi]$. d_t^i is the distance from $L\,pe$ to $rp_{i,t}$. The controlling parameter v is in the range of (0, 1). We denote this procedure as $\mathbf{NW_{rp}}(d_t^i, v)$.

Therefore, we can decide the location of sound source, i.e. $L_s = (x_s, y_s, z_s)$, as

$$x_s = \sum_{i=1}^{\beta} w_i\left(\sum_{t=1}^{\psi} w_{i,t}\cdot x_{i,t}\right); y_s = \sum_{i=1}^{\beta} w_i\left(\sum_{t=1}^{\psi} w_{i,t}\cdot y_{i,t}\right); z_s = \sum_{i=1}^{\beta} w_i\left(\sum_{t=1}^{\psi} w_{i,t}\cdot z_{i,t}\right), \tag{8}$$

Where x_t^i, y_t^i and z_t^i are Cartesian coordinates of $rp_{i,t}$. We denote this procedure as $\mathbf{FL}(w_i, w_{i,t}, L_{i,t})$.

The pseudo code of the proposed WDMM is summarized in Table 1.

Table 1. The pseudo code of the proposed WDMM

```
WDMM(LBA, IE, NW_ss, NW_rp, FL); // proposed weighted decision making method
begin
p_j=LBA; // j∈[1, Q], compute the probability by likelihood based classification algorithm
  for all i ∈ [1, β]
    L_pe=IE(p_i, L_i); // obtain initial estimation of source location
    w_i=NW_ss(d_i, μ); // assign new weights to sub-spaces
    for all t ∈ [1, ψ]
      w_i,t=NW_rp(d_i,t, v); // assign new weights to representative points
    end
  end
  L_s=FL(w_i, w_i,t, L_i,t); // obtain final source location estimation
return L_s = (x_s, y_s, z_s);
end
```

4 Synthetic Experimental Results and Discussion

In this section, synthetic experiments with and without the proposed WDMM are conducted to illustrate its effectiveness.

We simulated a typical medium size meeting room whose dimension is 4.0 m 4.0 m × 4.0 m. A microphone array consists of 6 elements and the elements are placed at K_1 = (1.8 m, 2.0 m, 2.0 m), K_2 = (2.2 m, 2.0 m, 2.0 m), K_3 = (2.0 m, 1.8 m, 2.0 m), K_4 = (2.0 m, 2.2 m, 2.0 m), K_5 = (2.0 m, 2.0 m, 1.8 m) and K_6 = (2.0 m, 2.0 m, 2.2 m). The sound source is placed on a spherical surface whose radius equals to 1.5 m, centred at the centroid of the room. On the spherical surface, the sound source is placed at 189 different positions, i.e. 21 different azimuth angles from −160° to +160° and 9 different elevation angles from −60° to +60°, both with even intervals.

The sound source adopts the clean speech files [5] sampled at 8 kHz. The reverberation time T_{60} in the room is set to be different levels as 0, 0.1, 0.2, 0.4 and 0.6 s while the SNR is set to be as −10, −5, 0 and 10 dB. The frame duration of the sound signal and the overlap rate between two frames are set to be 0.064 s and 62.5% respectively. The room is divided into 4096 equal-dimension rectangular sub-spaces whose dimensions are 0.25 m × 0.25 m × 0.25 m. The parameters p_{th}, μ, and v are set to be 0.004, 0.25, and 0.25 respectively. Moreover, 8 vertexes are chosen as representative points in each adjacent cluster. Naive Bayes [6] is employed as **LBA**. Room impulse responses are computed by fast image source method [7]. All the synthetic experiments are implemented in Matlab.

The localization error of sound source at the *n*th test position between the actual location and the estimated location is defined as ε_n, computed by (3) in Sect. 2. The mean error $\bar{\varepsilon}$ can be computed by

$$\bar{\varepsilon} = \frac{1}{N}\sum\nolimits_{n=1}^{N} \varepsilon_n, \tag{9}$$

where N = 189 presents the total number of testing sound source positions.

The standard deviation σ can be defined as

$$\sigma = \sqrt{\frac{1}{N-1}\sum\nolimits_{n=1}^{N} \varepsilon_n^2}. \tag{10}$$

The results are summarized in Table 2. A total of 20 different acoustic environments are simulated, with T_{60} varying from 0 to 0.6 s and SNR varying from –10 dB to 10 dB. The localization errors in terms of the mean error and standard deviation are collected from the 189 testing locations with each environment. The averages of mean and standard deviation localization errors in these 20 environments are summarized in Fig. 1.

Table 2. Localization results with and without WDMM

SNR/dB		−10					−5				
T_{60}/s		0	0.1	0.2	0.4	0.6	0	0.1	0.2	0.4	0.6
with WDMM	$\bar{\varepsilon}$/m	0.38	0.33	0.38	0.67	1.01	0.37	0.29	0.30	0.37	0.64
	σ/m	0.21	0.17	0.16	0.34	0.37	0.22	0.14	0.13	0.20	0.34
without WDMM	$\bar{\varepsilon}$/m	0.91	0.74	0.72	1.07	1.30	0.85	0.51	0.40	0.50	0.67
	σ/m	0.53	0.46	0.47	0.79	0.90	0.52	0.34	0.29	0.44	0.63
SNR/dB		0					10				
T_{60}/s		0	0.1	0.2	0.4	0.6	0	0.1	0.2	0.4	0.6
with WDMM	$\bar{\varepsilon}$/m	0.38	0.33	0.30	0.30	0.43	0.37	0.37	0.34	0.34	0.45
	σ /m	0.22	0.15	0.11	0.17	0.27	0.22	0.17	0.13	0.13	0.31
without WDMM	$\bar{\varepsilon}$/m	0.84	0.50	0.74	0.36	0.42	0.78	0.52	0.39	0.37	0.42
	σ/m	0.54	0.34	0.46	0.90	0.42	0.49	0.36	0.30	0.30	0.41

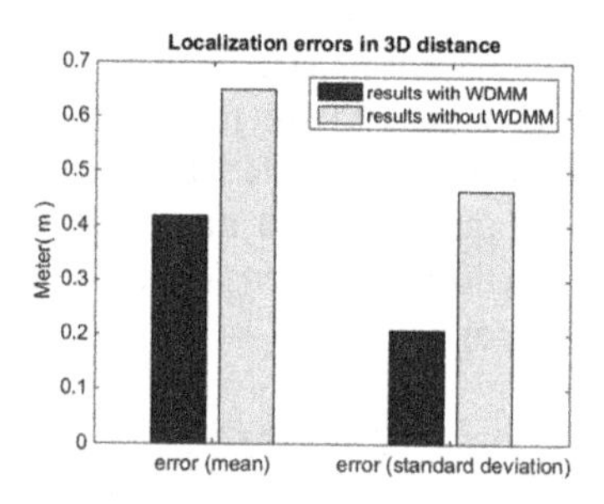

Fig. 1. Localization errors in terms of 3-dimentional distance

From the results of Table 2, when SNR is fixed and T_{60} varies from 0 to 0.6 s, the mean localization errors with WDMM generally increase. It can be seen that the mean error with WDMM increases significantly when T_{60} varying from 0.4 to 0.6 s with SNR = −10 and −5 dB, while it varies slightly and maintains a low level with other environments. This shows that the WDMM has good performances except very adverse environments. In contrast, generally, the mean localization error without WDMM can be observed a decreasing tendency followed by an increasing tendency, when SNR is fixed and T_{60} varies from 0 to 0.6 s. The mean errors without WDMM are generally larger than those with WDMM, especially in the environments with SNR = −10 dB. These results validate that the proposed WDMM can work effectively to address the misclassification problem and therefore improve SSL accuracy in high reverberation and low SNR environments. In terms of the standard deviation localization error, results with WDMM are also significantly better than those without WDMM, indicating better robustness.

From Fig. 1, it can be observed that WDMM can decrease average mean and average standard deviation localization errors by 35.8 and 55.2%, achieving only 0.42 and 0.21 m respectively.

5 Conclusions

In this paper, we proposed a likelihood based WDMM to improve the accuracy of SSL in the challenging high reverberation and low SNR environments. The proposed WDMM effectively reduced the SSL errors by utilizing probabilities of the adjacent sub-spaces and assigning new weights to these sub-spaces as well as the representative points inside them. The synthetic experimental results have shown that average mean and average standard deviation localization errors by WDMM are only 0.42 and 0.21 m respectively. Compared with the localization results without WDMM, averages of mean and standard deviation localization errors are reduced by 35.8 and 55.2% respectively.

References

1. Argentieri, S., Danès, P., Souères, P.: A survey on sound source localization in robotics: from binaural to array processing methods. Comput. Speech Lang. **34**(1), 87–112 (2015)
2. Canclini, A., Antonacci, E., Sarti, A., Tubaro, S.: Acoustic source localization with distributed asynchronous microphone networks. IEEE Trans. Audio Speech Lang. Process. **21**(2), 439–443 (2013)
3. Zhang, C., Florêncio, D., Ba, D., Zhang, Z.: Maximum likelihood sound source localization and beamforming for directional microphone arrays in distributed meetings. IEEE Trans. Multimedia **10**(3), 538–548 (2008)
4. Laufer-Goldshtein, B., Talmon, R., Gannot, S.: Semi-supervised sound source localization based on manifold regularization. IEEE/ACM Trans. Audio Speech Lang. Process. **24**(8), 1393–1407 (2016)
5. http://ecs.utdallas.edu/loizou/speech/noizeus/
6. Murphy, K.P.: Machine Learning: A Probabilistic Perspective. MIT press, Cambridge (2012)
7. Lehmann, E., Johansson, A.: Diffuse reverberation model for efficient image-source simulation of room impulse responses. IEEE Trans. Audio Speech Lang. Process. **18**(6), 1429–1439 (2010)

The Efficiency Factors of Point-to-Point Wireless Energy Transfer System in a Closed Satellite Cavity

Yue Yin, Xiaotong Zhang(✉), Fuqiang Ma, and Tingting Zou

USTB, Beijing, China
zxt@ies.ustb.edu.cn

Abstract. The research on the radio-frequency wireless energy transmission in the cavity has just started at present, which is limited to the feasibility and loss analysis of fixed size and working frequency. This paper deeply studies the point-to-point wireless energy transmission model in the cavity, analyzes the field and the energy loss energy of the transmission in the cavity. Based on the electromagnetic simulation software HFSS, the cavity simulation model is used to obtain the different influencing factors of energy transmission. The results show that the greater transmission efficiency and working bandwidth can be obtained when the transmitting and the receiving antenna are in the diagonal of the rectangular cavity when the TM220 wave is transmitted in the cavity. The longer the receiving antenna length is, the lower the transmission efficiency is. In addition, when the cavity size is larger and the resonant frequency is low, the loss is not great.

Keywords: Wireless energy transmission · Cavity · Energy loss

1 Introduction

In recent years, the wireless energy transmission has developed rapidly especially and set off a research revolution. Wireless energy transmission is divided into near-field coupling and far-field radiation in mainstream way [1]. Compared to near-field coupling, microwave wireless energy transmission is more suitable for space wireless energy transmission [2].

Satellites is closed metal cavity. At present, the wireless energy transmission research has just started due to the limited applications in the confined space. Sanchai Eardprab et al. used the analysis method of Dyadic Green's function to analyze the expression form of the field that is generated by the single antenna in a closed cavity [3, 4]. The 3D-cavity and the coil can transmit energy at the resonant frequency by coupling [5]. If the transmit and receive antennas are placed in the conductor cavity and only the inter-antenna coupling action is taken into account, the energy transmission efficiency can reach 80% [6]. There is a loss when the TM220 wave is transmitting in the cavity, which due to the cavity wall and antenna material [7].

This paper mainly studies the point-to-point wireless energy transmission system in the cavity, and analyzes the factors that affect the wireless energy transmission of the

F. Xhafa et al. (eds.), *Advances in Intelligent Systems and Interactive Applications*, Advances in Intelligent Systems and Computing 686,
https://doi.org/10.1007/978-3-319-69096-4_84

cavity. The second part mainly analyzes the cavity wireless energy transmission model in theory. The third part mainly introduces how to get the influence of different factors on the wireless energy transmission efficiency through simulation. Then conclusion is drawn in the last section.

2 Model Establishment

Figure 1 shows the simulation model that is a rectangular cavity structure with a size of $a \times b \times l$. Two identical antennas are put into the cavity. One of the antennas is used to shoot the electromagnetic waves and its location coordinate (x_t, y_t, z_t). The function of the other antenna is to receive the electromagnetic waves and its location coordinate is (x_r, y_r, z_r).

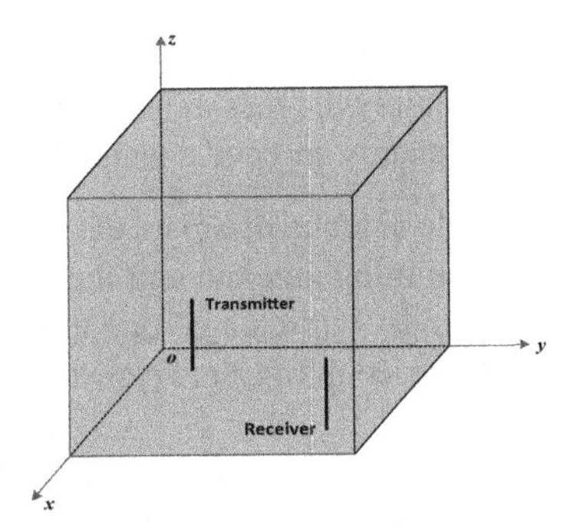

Fig. 1. Simulation mode

If we assume that there are ideal medium in the cavity and then the resonant frequency could be calculated according to the theory of resonant cavity.

$$(f_0)_{mnp} = \frac{c_0}{2}\sqrt{\left(\frac{m}{a}\right)^2 + \left(\frac{n}{b}\right)^2 + \left(\frac{p}{l}\right)^2} \tag{1}$$

where c_0 is the speed of light, m, n, p is the mode of electromagnetic waves propagating in the cavity.

The Green function in the [8] can be applied to describe the electric field in the cavity :

$$\begin{aligned} \vec{E}(\vec{r}) = -j\omega\mu_0 \iiint_v \left[\vec{\vec{G}}_{EJ}\left(\vec{r}, \vec{r}'\right) \cdot \vec{J}_t(\vec{r}') + \vec{\vec{G}}_{EJ}\left(\vec{r}, \vec{r}'\right) \cdot \vec{J}_r(\vec{r}')\right] dv' \\ - \iiint_v \left[\nabla \times \vec{\vec{G}}_{EM}\left(\vec{r}, \vec{r}'\right) \cdot M_t(\vec{r}') + \nabla \times \vec{\vec{G}}_{EM}\left(\vec{r}, \vec{r}'\right) \cdot M_r(\vec{r}')\right] dv' \end{aligned} \tag{2}$$

where $\vec{\vec{G}}_{EJ}(\vec{r},\vec{r}')$, $\vec{\vec{G}}_{EM}(\vec{r},\vec{r}')$ is Green function in the electric field[10]]. $\vec{J}_t(\vec{r}')$, $\vec{J}_r(\vec{r}')$is the current density flowing through the two antenna, $\vec{M}_t(\vec{r}')$, $\vec{M}_r(\vec{r}')$ is the flux source generated at the bottom of the transmitting/receiving antenna, $\omega = 2\pi f_0$ is the frequency of the resonant wave in the cavity, μ_0 is the vacuum permeability [6].

According to Max, the magnetic field at any point in the cavity can be obtained as follows:

$$\nabla \times \vec{H}(\vec{r}) = j\omega\vec{E}(\vec{r}) \tag{3}$$

So that the energy stored in the cavity can be:

$$W = W_e + W_m = \frac{\varepsilon_0}{4}\int_V |E_z|^2 dV + \frac{\mu_0}{4}\int_V \left(|H_x|^2 + |H_y|^2\right) dV \tag{4}$$

where ε_0 is vacuum dielectric constant.

The source of energy loss is mainly the loss of antenna coupling, the loss of antenna resistance and the absorption of energy in cavity wall, as follows:

(1) The omnidirectional monopole antenna dissipate energy around and receive energy from around. When the transmitting antenna and the receiving antenna in different locations, the mathematics model will change due to the reflection and absorption of the cavity wall. It affect the coupling between the antenna and thus affect the energy transmission efficiency.
(2) Because of its internal resistance, the antenna will lose part of the energy. For monopole antennas, the calculation of their internal resistance can be calculated according to the resistance of cylindrical conductors, $R_d = \rho\frac{l_0}{S_0} = \rho\frac{l_0}{\pi a_w^2}$, where ρ is conductivity that only related to antenna material and temperature, $l_0$is the antenna length, S_0 is cross sectional area of the antenna. The power loss on the transmit and receive antennas can be expressed as [6, 7].

$$\begin{cases} P_t = I_t(z)^2 \cdot R_d = 4\pi\rho l_0 J_t^2(z) \\ P_r = I_r(z)^2 \cdot R_d = 4\pi l_0 \rho J_r^2(z) \end{cases} \tag{5}$$

(3) Since the cavity wall cannot be an ideal conductor, the electromagnetic wave must be absorbed in the conductor wall and a current is formed in the cavity wall. The surface resistance of the wall is $R_s = \sqrt{\frac{\pi f_0 \mu}{\sigma}}$, thus, The loss of power per unit area on the wall of the cavity can be expressed as $P_{av} = \frac{1}{2}|J_s|^2 R_s$. According to the principle of electromagnetic analysis can be obtained the loss of energy on the cavity wall is:

$$P_L = \frac{R_S}{2}\oint_S |J_s|^2 dS = \frac{1}{2}\sqrt{\frac{\pi f_0 \mu}{\sigma}}\oint_S \left|\vec{e}_n \times \vec{H}(\vec{r})\right|^2 dS \tag{6}$$

where σ is conductor conductivity of cavity wall.

3 Simulation

The cavity structure model is extremely complex, and we use the numerical analysis tool HFSS to simulate the point-to-point wireless energy transmission model in the cavity.

3.1 Changing the Transmitting Antennaor Receiving Antenna Position

The simulation uses the cavity size of 1m × 1m × 1m, the transmitting antenna length is 0.17 m, and receiving antenna is 0.175 m, observing the energy transfer efficiency of TM220 wave in cavity.

3.1.1 Different Locations of the Transmitting Antenna

The receiving antenna position is fixed at (0.7, 0.7, 0) (meter) and the locations of the nine transmit antennas : (0.2, 0.2, 0), (0.2, 0.4925, 0), (0.2, 0.785, 0), (0.4925, 0.2, 0), (0.4925, 0.4925, 0) (0.2, 0.785, 0), (0.785, 0.2, 0), (0.785, 0.4925, 0), (0.785, 0.785, 0). When the antenna selected at different positions, the length of the receiving antenna is [0.168, 0.178]. In the figure, red expresses as the highest efficiency, and blue expresses as the lowest efficiency (Fig. 2).

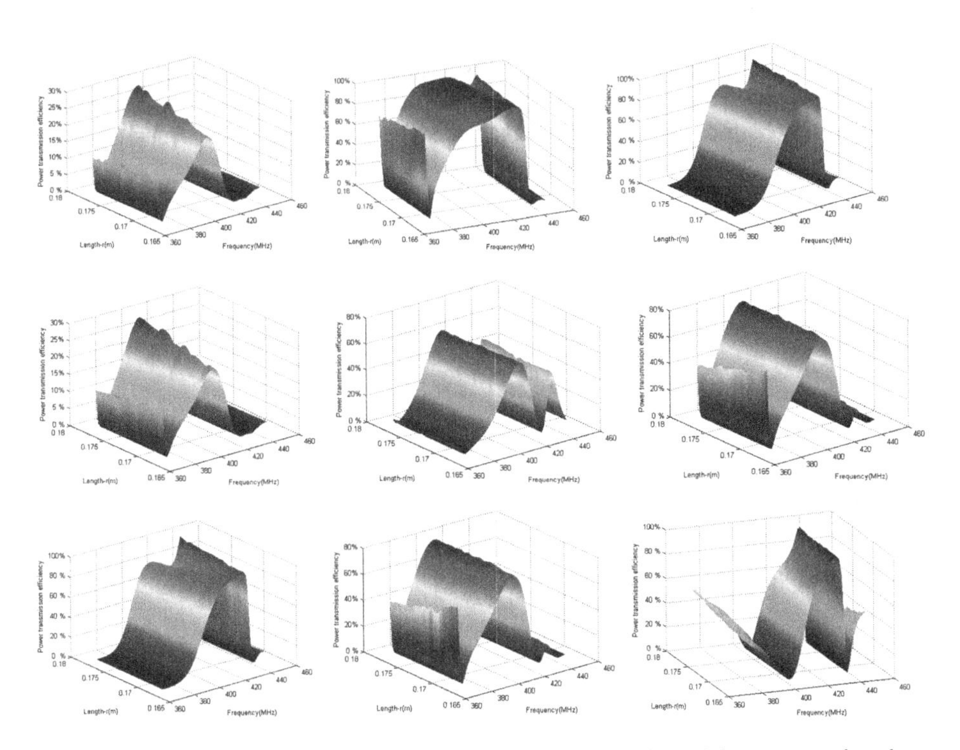

Fig. 2. Influence of different transmitting antenna position and receiving antenna length on Energy Transmission Efficiency (x-frequency, y-length_r, z-power transmission efficiency)

When the transmitting antenna and the receiving antenna in the same diagonal and the transmitting antenna in the cavity corner, the wireless energy transmission efficiency is relatively high, the bandwidth is relatively wide. When the antenna is in the center of the cavity, the highest value of wireless energy transmission efficiency is relatively low. The energy transfer efficiency near the resonant frequency will be recessed, the degree of depression increasing as the length of the receiving antenna increases. When the transmitting antenna is at the edge of the cavity, the energy received by the receiving antenna will affect by the frequency, and the transmission efficiency will decrease rapidly on both sides of the peak. The distance between the transmitting and the receiving antenna is close, and the energy received at the receiving end is more.

3.1.2 Different Location of the Receiving Antenna

The transmitting antenna is fixed at A (0.2, 0.2, 0). B (0.6, 0.6, 0), E (0.7, 0.7, 0), H (0.9, 0.9, 0) on the same diagonal as the transmitting antenna locations. Then take A as the origin, done the three points of the three concentric circles respectively, In the three concentric circles, points were selected and A point after the connection with the diagonal angle of 15°.

When the transmitting antenna and the receiving antenna are in the same diagonal, the energy transfer efficiency can achieve higher. Once the location of the receiving antenna from the diagonal, the energy reception efficiency will greatly reduce. Regardless of the distance between the transmitting antenna and the receiving antenna or near or far, the infinite energy transmission efficiency reduces. Receiving antenna in vicinity of the central location, the received energy will be very small, especially when not on the diagonal, the received energy is only about 2% (Fig. 3).

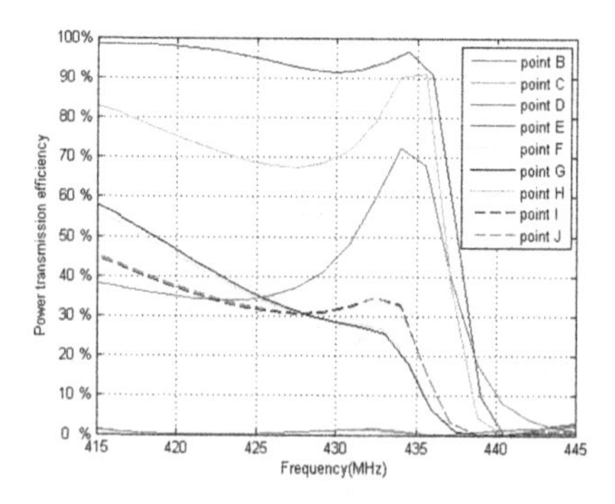

Fig. 3. Influence of receiving antenna position on antenna transmission efficiency

3.2 Changing the Transmission Efficiency of the Receiving Antenna Length

It can be seen from formula (5) that the length of the antenna contributes to the energy loss of the antenna itself. In general, the longer the length of the antenna, the greater the loss of the antenna resistance. The simulation shown in Fig. 4 uses the cavity size of 1m × 1m × 1m, the position of the two antennas are (0.2, 0.2, 0), (0.7, 0.7, 0).

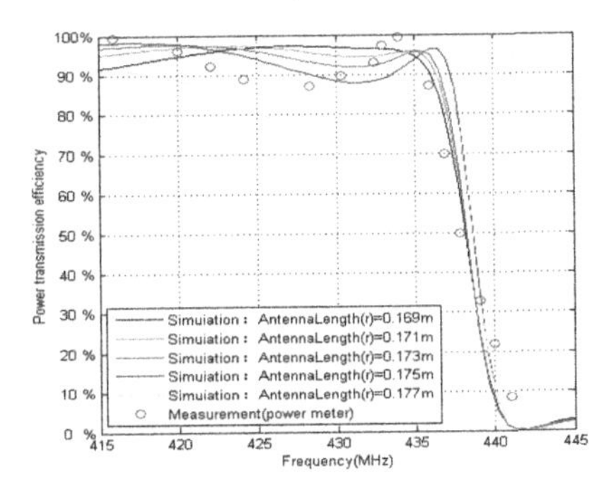

Fig. 4. Influence of receiving antenna length on energy transmission efficiency

3.3 Loss Analysis in Different Size Cavity Models

In this part of the simulation using 1m × 1m × 1m size cavity, the antenna position is (0.2, 0.2, 0), length of 17 cm, the receiving antenna position (0.8, 0.8, 0), the length of 17.5 cm for the original size. And then the original size of the cavity to fold two times, due to changes in the size of the cavity and the resonant frequency changes, the material loss and the degree of coupling between the antennas will be affected. The transmission efficiency of TM220 waves under different cavity sizes will show different characteristics.

Figures 5 and 6 shows the different losses when the size is 1/2 and 1/4 of the original cavity. It can be seen that the energy transmission efficiency can still reach > -1dB. Both the loss of the internal resistance of the antenna and the loss of the cavity wall begin to increase, and occupy a considerable proportion, especially in vicinity of TM220 wave resonant frequency, can reach -1.25dB.

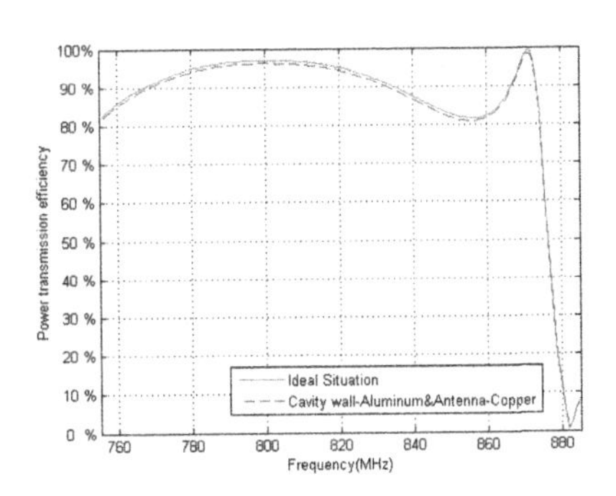

Fig. 5. Effect of energy transfer efficiency on the size of 1/2

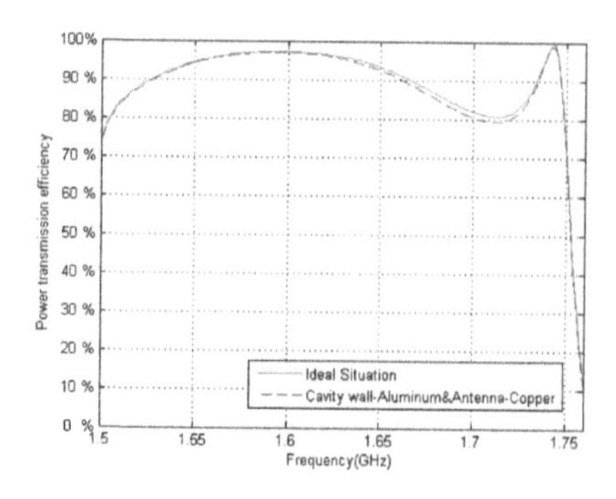

Fig. 6. Effect of energy transfer efficiency on the size of 1/4

4 Conclusion

This paper studies the point-to-point wireless energy transmission model in the cavity, analyzing energy of the transmission and the energy loss in it and establishing the influence of different influencing factors on the wireless energy transmission of the cavity based on the electromagnetic simulation software HFSS. In the rectangular cavity, when the transmitting antenna and the receiving antenna are in the rectangular diagonal position, the wireless energy transmission efficiency is the highest. When the length of the transmitting antenna is fixed, the longer the length of the receiving antenna, the greater the loss, the lower the wireless energy transmission efficiency. Without regarding to the effect of temperature, with the increase in the size of the cavity model, both the loss caused by the coupling of the antenna, the loss caused by the cavity wall or the loss caused by the internal resistance of the antenna will increase. The smaller the cavity is, the cavity wall and antenna resistance loss cannot ignore.

Acknowledgments. The paper is sponsored by the Database architecture, standard specification and database for material genetic engineering (No. 2016YFB0700503).

References

1. Ma, J., Zhang, X., Huang, Q., Cheng, L., Lu, M.: Experimental study on the impact of soil conductivity on underground magneto-inductive channel. IEEE Antennas Wireless Propagation Lett. **14**, 1782–1785 (2015)
2. Strassner, B., Chang, K.: Microwave power transmission: historical milestones and system components. Proc. IEEE. **101**(6), 1379–1396 (June 2013)
3. Eardprab, S., Phongcharoenpanich, C.: Electromagnetic field coupling between a probe and a rectangular aperture mounted on open-ended cavity. In: 2005 5th International Conference on Information Communications and Signal Processing, Bangkok (2005)
4. Lertwiriyaprapa, T., Phongcharoenpanich, C., Krairiksh, M.: Analysis of radiation characteristics of a probe fed rectangular cavity-backed slot antenna with finite-size ground plane. In: Antennas and Propagation Society International Symposium, Salt Lake City, UT, USA, vol. 2, pp. 714–717. IEEE (2000)

5. Chabalko, M.J., Sample, A.P.: Three-dimensional charging via multimode resonant cavity enabled wireless power transfer. IEEE Trans. Power Electron. **30**(11), 6163–6173 (2015)
6. Wu, J.: Feasibility study of efficient wireless power transmission in satellite interior. Microwave Opt. Technol. Lett. **58**, 2518–2522 (2016)
7. Liu, C., Zhang, X., Wang, P.: Feasibility study about the power loss of wireless power transmission in cavity. In: 2016 2nd IEEE International Conference on Computer and Communications, October 2016
8. Lu, M., Bredow, J.W., Jung, S., Tjuatja, S.: Evaluation of Green's functions of rectangular cavities around resonant frequencies in the method of moments. IEEE Antennas. Wirel. Propag. Lett. **8**, 204–208 (2009)

Personalized POI Recommendation Model in LBSNs

Zhong Guo and Ma Changyi(✉)

University of Macau, TAIPA Macau SAR, Zhuhai 999078, China
xhyhforever@sina.com

Abstract. The development of location-based social networks (LBSNs) generates large volume of check-in data. Point-of-interest recommendation (POI) is important for users to find some attractive venues, sometimes when users are in some places far away from their living cities. However, POI recommendation is so difficult compared to the classical recommender system. Users may access only a small portion of POIs, with a sparse user-POI matrix. The bulk of the POIs accessed by users located in a near close to users' residences, which leads it difficult to put in a good word for POIs when the user travels to a faraway region. Meanwhile, uses' preferences may be different in various geographical regions. Different users may prefer to go to different venues at different time. From our paper, we present a novel model represented as probabilistic graphical model to describe users' check-in behaviors, which can overcome the data sparsity for the users far away from their living cities. To demonstrate our proposed model can recommend effectively, we do experiments to calculate the precision. The results show our model can do recommendation effectively and efficiently.

Keywords: Recommendation system · LBSNs · Data sparsity · Joint modelling

1 Introduction

As computer network and the vast amounts of mobile devices with positioning function develop dramatically fast, location-based social networks (LBSNs), such as Foursquare, Go-Walla and Twitter have been generated [3, 4]. Therefore, developing recommendation systems for LBSNs to supply users with POIs have recently call attention to researchers [10]. We can use this application to help a user when he travels to a faraway city, especially he has little sense about the new environment. Some existing works shows that the recommendation system is presented as advice for out-of-town users in LBSNs [3]. The recent paper studies the users' check-in behaviours and do recommendation for the users whether they are in their hometown or out of their hometown. In addition, we call this two problems hometown recommendation and out-of-town recommendation.

F. Xhafa et al. (eds.), *Advances in Intelligent Systems and Interactive Applications*, Advances in Intelligent Systems and Computing 686,
https://doi.org/10.1007/978-3-319-69096-4_85

2 Preliminaries

2.1 SAGE Model

Whether some variable is affected by more than one variables, we cannot use the traditional topic model. Based on this observation, we can adopt Sparse Additive Generative Model (SAGE) to solve the above problem effectively. And the basic idea of the SAGE model can be described as Eq. 1.

$$P(v|\theta_u^{user}, \theta_l^{crowd}) = \lambda_u P(v|\theta_u^{user}) + (1 - \lambda)P(v|\theta_l^{crowd}) \quad (1)$$

In Eq. 1, λ_u is the "switching" variable, which should be estimated before recommendation. In addition, the SAGE model can combine two or more parameters as shown in Eq. 2.

$$P(v|\theta_u^{user}, \theta_l^{crowd}) = P(v|\theta_u^{user} + \theta_l^{crowd}) = \frac{\exp(\theta_{u,v}^{user} + \theta_{l,v}^{crowd})}{\sum_{v'} \exp(\theta_{u,v'}^{user} + \theta_{l,v'}^{crowd})} \quad (2)$$

2.2 Related Work

Geo-Social Influence. It implies people incline to search after close-by POIs of a POI which their friends or they have visited before. Some recent works [7–11] show that there exits a close relationship between users book-in activities and physical distances as well as social connections, thus a majority of recent POI suggestion work largely concentrate on applying the geographical and social impact to be helpful for the accuracy of recommendations. **Content Information.** Lately, Hu et al. [3] consider spatial aspect and textual aspect of user posts from Twitter, then present a spatial topic model for POI recommendation. **Crowds' Effect.** The region-level popularity of POIs also affects user visiting behaviours [5, 6]. In the real life, when a user first come to a region which means that the user is unfamiliar to the region. Meanwhile, the user do not have check-in histories in this region. Under this situation, crowds' comment is important for these users to decide to go to which kinds of POIs.

3 Proposed Model

3.1 POI Recommendation Problem Definition

Given a target user u_q, l_q and D denotes his/her current position and a user book-in record dataset, respectively. So the query is $q = (u_q, t_q, l_q)$. We want to advice a list of POIs of interest to the user u. And fix a threshold d, if the distance $|l_q - l_u|$ between the target user's current position and his/her home position is greater than d, then we call this problem is an out-of-town recommendation. Otherwise, it is a home-town recommendation.

3.2 Proposed Model

To model users' book-in behaviors, a probabilistic graphical model is proposed. Figure 1 demonstrates the graphical notation of the proposed model. Users' book-in behaviors data set is modelled and taken as observed stochastic variables, which is represented as shaded circles in the following Fig. 1. As the exited works show to us, [1, 2] the topic distribution of every user book-in record and the region index of every user book-in record are the latent random variables. We use z and r to denote them, respectively.

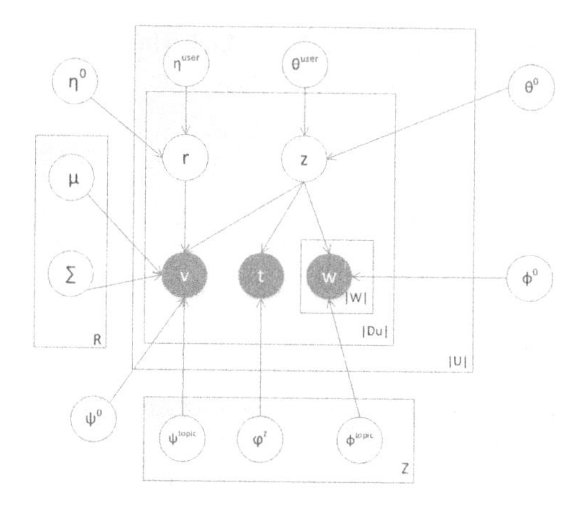

Fig. 1. Graphical representation of joint modelling

Following equation is the likelihood function in our proposed method. We use this equation to do get the maximum value of the equation and then do the recommendation step. For details, you can feel free to contact with us to find the Appendix.

$$
\begin{aligned}
& P(z, r, w, i, t|\Theta) \\
& \quad = P(z|u, \theta^0, \theta^{user}) \times P(r|\eta^0, \eta^{topic}) \times P(w|z, \phi^0, \phi^{topic}) \times P(i|r, z, \mu, \Sigma, \psi^0, \psi^{topic}) \times P(t|\varphi^z) \\
& \quad = \prod_{u=1}^{|U|} \prod_{i=1}^{|D_u|} \alpha_{u,z_{u,i}} \prod_{u=1}^{|U|} \prod_{i=1}^{|D_u|} \gamma_{z_{u,i},w_{u,i,n}} \prod_{u=1}^{|U|} \prod_{i=1}^{|D_u|} \prod_{n=1}^{|W_{V_{u,i}}|} \beta_{z_{u,i},w_{u,i,n}} \prod_{u=1}^{|U|} \prod_{i=1}^{|D_u|} \delta_{z_{u,i},i_{u,i}} N(l_{u,i}|r_{u,i}, \mu, \Sigma) B(t|\varphi^{z,1}, \varphi^{z.2})
\end{aligned} \tag{3}
$$

3.3 POI Recommendation

Our proposed method can be used to do personalized POI recommendation. For a document with a target query user, we want to recommend top-k POI set that he has not been visited. Also the problem can be described by using some notations, given the user of a document d and the words, the probability that the user u accesses POI v at the time of t is computed as in the following Equation.

$$
\begin{aligned}
P(W_v, V_v|u,t,\Theta) &= \sum_{z=1}^{K} P(v, W_v, t|u, l, \Theta) \\
&= \sum_{z=1}^{K} P(z|\theta^0, \theta^{user}) \times P(W_v|z, \phi^0, \phi^{topic}) \times P(r|\eta^0, \eta^{topic}) \times P(v|z, r, \mu, \Sigma, \psi^0, \psi^{topic}) \times P(t|\varphi^z) \\
&= \sum_{z=1}^{K} \alpha_{u,s,l,z} \times (\prod_{n=1}^{|W_V|} \beta_{z,w_{v,n}})^{|\frac{1}{W_v}|} \times \gamma_{z,v} \times \delta_{z,i} \times N(v|\mu, \Sigma) \times B(t|\varphi^{z,1}, \varphi^{z,2})
\end{aligned}
\tag{4}
$$

4 Performance Evaluation

4.1 Evaluation Process

For the experimental data sets, firstly, we chose seventy percent of the original data sets as the training data, so other 30% are identified as test data. We concentrate on POI recommendation for users who are far away from their hometown [2, 9].

Method for Evaluation. In most of the recommender system, precision@k is wisely able to assess the proposed model. The high-k accuracy usually calculated as $1/_k$. If the ground truth is in the high-k set, we mark hit number in top-k set as 1, and zero otherwise.

Competitors. In the experiment part, we select the following exited works as the competitors:

Probabilistic Matrix Factorization. Probabilistic matrix factorization is a very popular and useful model in recommendation system.

Latent Dirichlet Allocation. This is a kind of topic model, means LDA model, it is well-used for location recommendation [3].

Joint Model-S1. It is a simplified version of the proposed model. In this model, we remove the content words, and there exits only one latent variables topic z in this model, So this model is similar to the LDA model to some extent.

Joint Model-S2. It is another simplified version of the proposed model. This model don not consider the topic distribution z. In this model, we only consider the latent variable region r.

Joint Model-S3. It is another variation of the proposed model, which generates both the index and the coordinates of POIs, and potential variable are region r and the topic z, respectively.

Joint Model. It is the whole joint model proposed in Sect. 2.

4.2 Effectiveness

In this paragraph, we show the final experimental results of the competitors and our proposed model after tuning all the parameters. Figure 2a and b shows the advice work on the Twitter and the Foursquare, separately. We also analyze the contributions of different components in joint model, by comparing the performance of its simplified versions and joint model: Joint Model-S1, Joint Model-S2, and Joint Model-S3. Notice that modelling the index of POIs in Joint Model-S1 is even more accurate than modelling the coordinates of POIs in Joint Model-S2. Comparing Joint Model and Joint Model-S3, we see that the user interests interest representation in the documents really improve accurate of POI suggestion. Moreover, it is clear that Joint Model-S3 outperforms Joint Model-S2, demonstrating the contribution of exploiting the correlation of user movements.

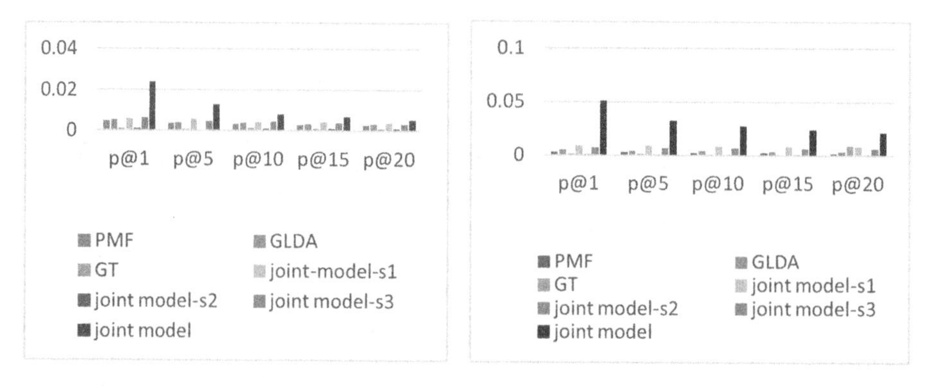

Fig. 2. (a) …(b) …

To show the impact of the hyper parameters, we do experiments to analyze the results of different numbers of regions, the results are as Fig. 3a and b. The results show that our proposed model is prefer to other competitors.

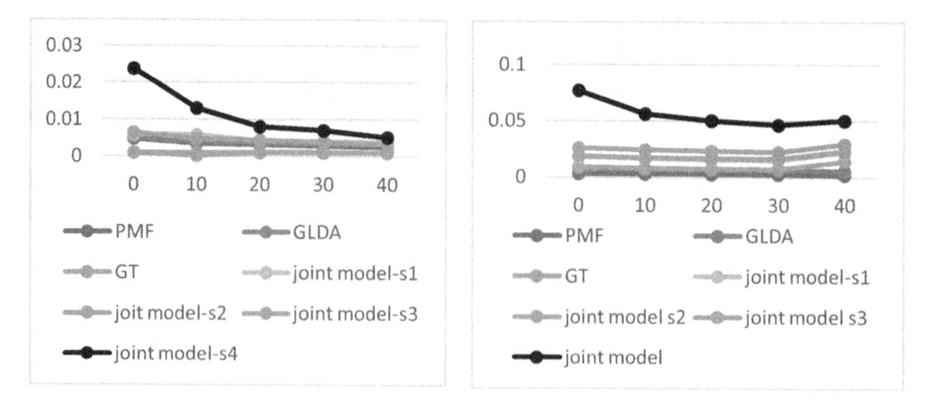

Fig. 3. **(a)** Precision on Foursquare dataset **(b)** Precision on Twitter data

5 Conclusion

From our paper, we present a novel joint probabilistic graphical model to simulate all the users' book-in behaviours in Location-Based Social Networks, which strategically integrates several factors of temporal effect, content, geographical influence and crowds effect. The results show that our model can overcome the issues of data sparsity and user preference drift across geographical regions, especially when users travel out-of-town. To demonstrate the adaptability and elasticity of joint modelling, we investigated how this model do effective recommendation on two datasets.

References

1. Ference, G., Ye, M., Lee, W.C.: Location recommendation for out-of-town users in location-based social networks. In: Proceedings of the 22nd ACM international conference on Information & Knowledge Management, pp. 721–726. ACM, New York, Oct 2013
2. Cheng, C., Yang, H., King, I., Lyu, M. R.: Fused matrix factorization with geographical and social influence in location-based social networks. In Aaai vol. 12, pp. 17–23. July 2012
3. Hu B., Ester M.: Spatial topic modeling in online social media for location recommendation. In: Proceedings of the 7th ACM conference on Recommender systems-RecSy, pp. 25–32. ACM, New York Nov 2013
4. Cui, B., Mei, H., Ooi, B.C.: Big data: the driver for innovation in databases. Natl. Sci. Rev. **1** (1), 27–30 (2014)
5. Wang, W., Yin, Z., Sadiq, S.,et al.: SPORE: A sequential personalized spatial item recommender system, 2016 IEEE 32nd International Conference on Data Engineering (ICDE), vol. 12, pp. 17–23. July 2016
6. Cheng, C., Yang, H., Lyu, M. R., King, I.: Where You Like to Go Next: Successive Point-of-Interest Recommendation. In: IJCAI, vol. 13, pp. 2605–2611. Aug 2013
7. Cheng, Z., Caverlee, J., Lee, K., Sui, D.Z.: Exploring millions of footprints in location sharing services. ICWSM **2011**, 81–88 (2011)
8. Hu, B., Ester M.: Social Topic Modeling for Point-of-Interest Recommendation in Location-Based Social Networks. 2014 IEEE International Conference on Data Mining, pp. 1082–1090. IEEE, USA, Aug 2014
9. Yin, H., Zhou, X., Cui, B. et al: Adapting to user interest drift for POI recommendation. IEEE Transactions on Knowledge and Data Engineering. IEEE, USA, Oct 2016
10. Zhang, C., Ke, W.: POI recommendation through cross-region collaborative filtering, Knowledge and Information Systems, pp. 369–387. April 2016
11. Wang, W., Yin, H., Chen, L.: Geo-SAGE: A geographical sparse additive generative model for spatial item recommendation. In: Proceedings of the 21th ACM SIGKDD International Conference on Knowledge Discovery and Data Mining, pp. 1255–1264. ACM, New York Mar 2015

Verification of CAN-BUS Communication on Robots Based on xMAS

Xiujuan Cao(✉)

College of Information Engineering, Capital Normal University,
100048 Beijing, China
godloveyou_cao@163.com

Abstract. With the development of robot systems, distributed control systems become more and more widespread, which leads to much more complex communication specifications and hardware structure designs. Since the communication structure has important influence on robot security and real-time property, we proposed using formal method to verify the repeater with xMAS and ACL2 in this paper. Giving out the data flow logic information can also keep the low level properties while avoiding state explosion. The formalization method and model logic can also be used in many areas.

Keywords: Robot Communication · CAN-bus · ACL2 · Formalization · xMAS

1 Introduction

Nowadays, in general there are two types of industrial robots' control systems: centralized and distributed. Since centralized control systems depend on Master computer a lot, once the master breaks down, the whole control system will paralysis, which may lead to low reliability and security [3]. The reasons why distributed control systems are better options are their good performances on scope of application, scalability, control speed, system decoupling, system maintainability and resistance to a single point of failure, etc [2]. In the future, robots will be widely used in most important areas like industry, education, agriculture, health care, military, and so on. Because robots will be closely related to our daily life, the importance of ensuring human security is self-evident [7]. Actually, although distributed control systems can improve the hardware reliability, communication systems are becoming more and more complex with systems integration components quantity increasing gradually, so there still exist a great of security risks.

Verification is using mathematical logic to create mathematical model of the system, proving the properties and finally proving the correctness of the design. So verification can be used as a dependable method to verify the communication network system [9].

F. Xhafa et al. (eds.), *Advances in Intelligent Systems and Interactive Applications*, Advances in Intelligent Systems and Computing 686,
https://doi.org/10.1007/978-3-319-69096-4_86

2 xMAS

2.1 Primitives

xMAS (eXecutable Micro-Architectural Specification) model is made by Intel company. It is a network created by connecting components with channels in order. All the properties were abstracted from the network.

There are eight primitives: Source, sink, fork, join, switch, merge, queue and function [11]. As shown in the following pictures (Fig. 1).

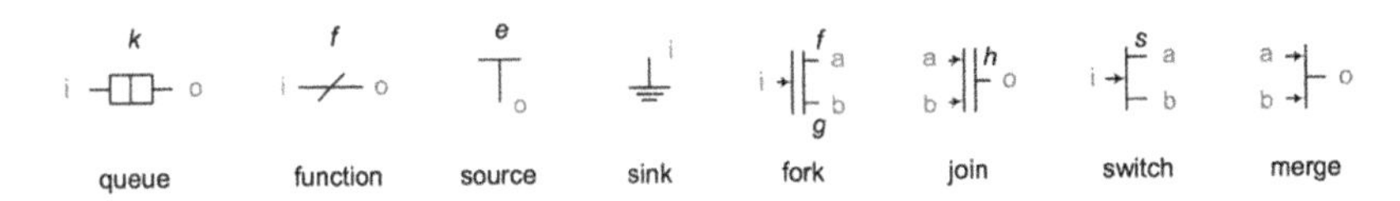

Fig. 1. xMAS model primitives [10]

These primitives are created by providing the synchronous equations that are generated for them. The synchronous equations have very strict rules on the inputs and outputs of the primitives. Switch is given here as an example, and further references can be learned [6, 10] for more details.

Switch: Switch is a primitive which can choose route for packets in the network. The input channel is ready only after the output channel is ready, which leads to the input channel state is lagged one cycle after the output channel.

We modified the definition of the equations to make sure both input and output channels can be connected at the same time, by which we can improve the speed of channel calculate as well as keep the primitive's function.

next(a.irdy):=pre(i.irdy) and s(pre(i.data)) next(a.data):=pre(i.data)
next(b.irdy):=pre(i.irdy) and not s(pre(i.data)) next(b.data):=pre(i.data)
next(i.trdy):=pre(a.trdy) and s(pre(i.data)) or pre(b.trdy) and not s(pre(i.data))

2.2 Channels

Channel is composed by two Boolean signals irdy (initiator is ready) and trdy (target is ready) for control and a data signal (data). There are two components connected with each channel, one is the initiator, the other is the target. Initiator sends data into the channel through its output port, it determines the signals of irdy and data. Target reads data from channel by the input port and determines the signal trdy. If both irdy and trdy are true, data will be transferred from initiator to the target. Channel is represented by a line in current model and it is just a triple and doesn't store state.

2.3 Networks

An xMAS network is composed by three parts: components, channels and sequential network. Components is the set of all the primitives used in the model. Component was defined by five parts: id_com, type_com, ins, ous and param_com. Channels is the set of all channels, each channel was defined by: id_chan, initiator, target, param_chan and data. Sequential network is the set of all sequential network which has delay because of the data transfer. So it always comes up with two of queue, source or sink. It was defined by : init, target and connect-channels.

For example, in Fig. 2, a simple micro-architectural model with a source that generated packet 'Red', a switch dependingon the packet was route to com_2-queue or com_4-sink, if it chose com_2, a queue could store ten elements and then send data to com_3-sink [5].

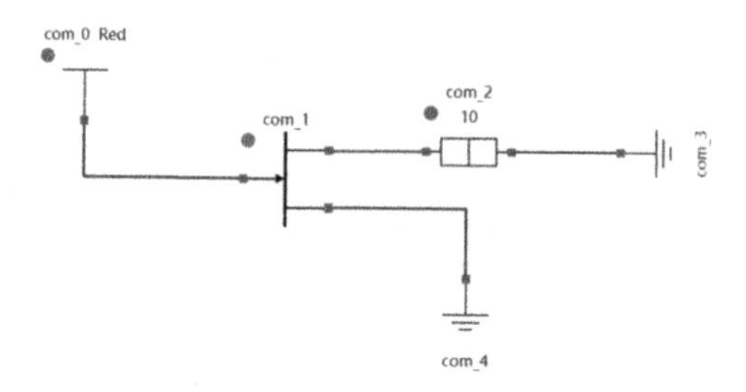

Fig. 2. Micro-architecture network example

3 ACL2

The ACL2 theorem prover has been developed at the university of Texas at Austin [1]. The programming language is developed from Common Lisp. It is designed to support automatic reasoning on inductive logical theories. The most two important principles are the definition principle and the induction principle [13].

To build ACL2 models of systems, the system must be understood by being modeled and ACL2 as a programming language. ACL2 will help to construct the proof and prevent logical mistakes. The user has the responsibility to abstract the properties. ACL2 objects have several data types. Numbers include integers, rational and complex rational items; Strings are sequences of characters; Symbols have a package name and symbol name. One of the most important data structures is list. Lists are represented by enclosing their elements in parentheses. It uses t and nil to express the Boolean values true or false. ACL2 is learned from reference [1]. Acl2 is a powerful system for integrated modeling, simulation and inductive reasoning, which has been used to verify various complex theorems of mechanical verification, such as microprocessors, floating point operators and many other structures.

4 The Communication Model

4.1 Communication

Signal requires time when it is transferred through transmission line and the length of the line has great influence on the characteristics of the signal. To be exact, the signal's reliability will decrease with the increase of transmission length [8]. To solve the problem between the maximum transmission distance and transmit rate, repeaters were put into the robot communication network to assure real-time and the reliability. In our model, distributed nodes are connected by CAN network, the repeater will route and conduct forward the datagram. Usually the number of nodes that CAN network could mount depends on the internal resistance of the drive chip, the distributed capacitance and resistance of transmission line, the driving voltage, the baud rate, the distance, etc. Take the robot in current study as an example, when the network mount more than fourteen nodes, it will report errors easily. CAN repeater could increase the maximum nodes number and increase the communicate distance (Fig. 3).

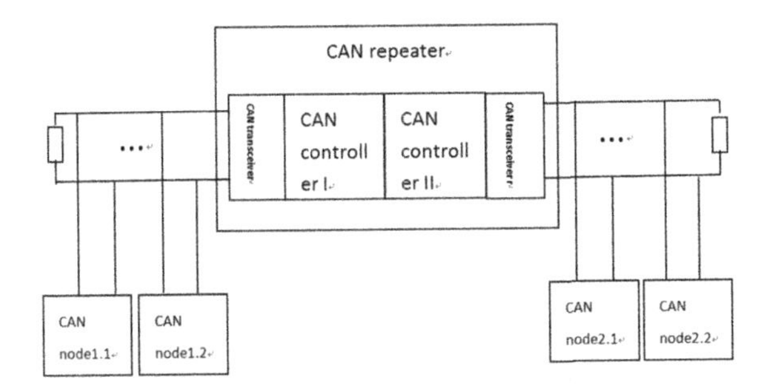

Fig. 3. The repeater architecture model [12]

4.2 The Sender Part

Sender first read the state register to get permission to the sender buffer. After making sure the sender buffer is ready, it will write the data which need to be transferred into sender buffer. Data have been written successfully, if not, try again. When data has been written successfully into the buffer, set send buffer address then transfer the send flag. Transfer eight bytes data through CAN bus and set send command, return the main program. The sender program flow chart is shown as Fig. 4.

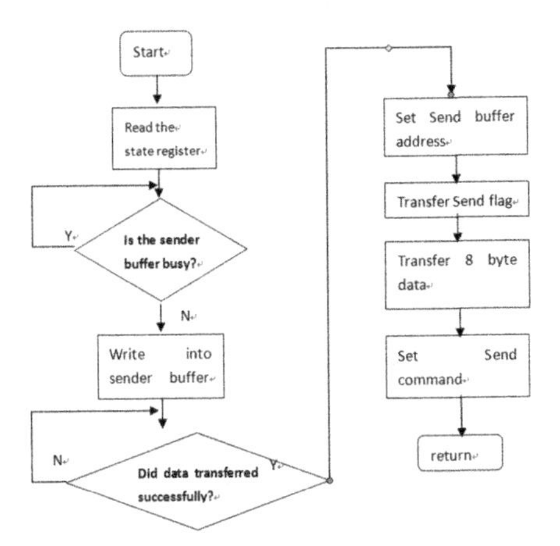

Fig. 4. The sender program flow chart

4.3 The Receiver Part

The receiver always read the state register firstly to get the highest message ID. Then the state flag is set accordingly. Quit the interrupt process as well as jump to different programs according to different message cases. If the message's destination is local subnet, compute the length of message and set the address of receive buffer. Transfer the data to be received. Finally release receive buffer and return the interrupt program. The receiver program flow chart is shown as Fig. 5.

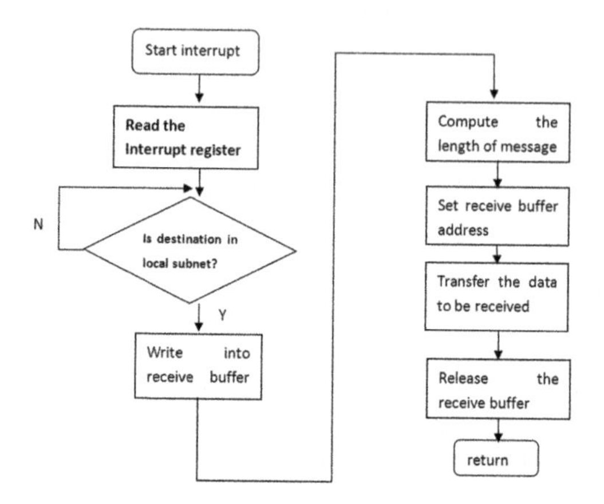

Fig. 5. The receiver program flow chart

4.4 The Repeater Model

This model is a simple communication network with three nodes. The node in the middle serves as a repeater, it receives data from node1 and transfers it to node3. This model shows the receiving and sending logic.

In this model, com_0 represents one can-bus, it transfer datagram to the repeater from com_0. Function com_1 will check if the datagram needs to be transferred through repeater to another can-bus, then by switch com_2, if not, the network will discard the datagram by sink com_3. Otherwise the join com_4 will fetch the signal created by com_14 of whether the receive buffer is null. If both buffer signal and data are ready, data will be written into the receive buffer com_5 and then achieve computing length and setting address using function com_6. After that giving a signal to join com_8 through merge com_7 while getting the signal of send buffer state. When both the data from receiver part and the state of send buffer are ready, join com_8 will give a signal to switch com_9 to write data into send buffer com_10. If there are problems during the transfer process, switch com_5 will give com_7 an error signal to start the transfer again. And if the data is transferred successfully, fork com_11 will give com_8 a buffer busy signal as well as set address and send flag by function com_12, then transfer data to another can-bus expressed by sink com_15, also release receive buffer and send buffer through function com_14 then generate a buffer ready signal. By this model, data in source com_0 can be transferred to sink com_15 correctly (Fig. 6, 7 and 8).

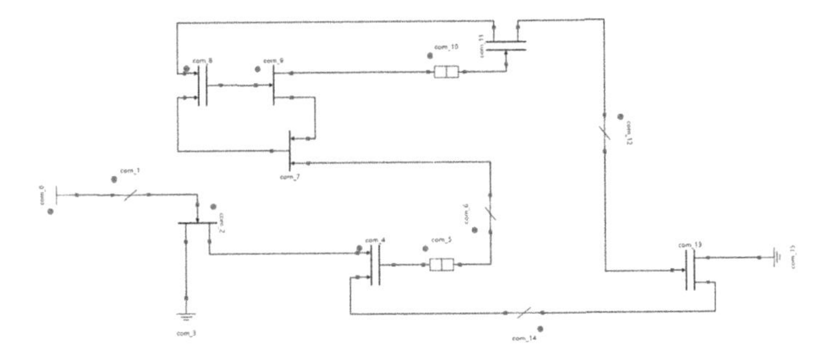

Fig. 6. Robert communication network repeater model

```
Summary
Form:  ( DEFCONST *REPEATER* ...)
Rules: NIL
Warnings:  None
Time:  0.01 seconds (prove: 0.00, print: 0.00, other: 0.01)
 *REPEATER*
ACL2 !>
```

Fig. 7. Robert communication network ACL2 model

```
Q.E.D.

The storage of PROPERTY-13 depends upon the :type-prescription rule
COMPONENT-PORTS-NUMBER=CHANNELS*2.

Summary
Form:  ( DEFTHM PROPERTY-13 ...)
Rules: ((:EXECUTABLE-COUNTERPART COMPONENT-PORTS-NUMBER=CHANNELS*2)
        (:EXECUTABLE-COUNTERPART NTH)
        (:TYPE-PRESCRIPTION COMPONENT-PORTS-NUMBER=CHANNELS*2))
Warnings:  Subsume and Non-rec
Time:  0.19 seconds (prove: 0.00, print: 0.00, other: 0.19)
 PROPERTY-13
```

Fig. 8. Robert communication network model property

5 The ACL2 Model

5.1 Express the Model in ACL2

Now we express the communication model in ACL2. Since the model code is in large amount, we only give the picture of the result showed in ACL2.

5.1.1 Verify the Model Connection

The number of component ports must equal to double channel number as the result of each channel connect to an input port and an output port [4].

(defthm property-13

(component-ports-number == channels*2 (car *transponder2*) (nth 1 *transponder2*)))

Acknowledgments. In this paper, we proposed a formal method to verify the communication of robot CAN network repeater and proved its validity. With this model, we can keep the property on hardware level and have clear logic of each part as well as data flow. Also the formalization method we used here can be expended into many areas. But there are also problems: the repeater takes some time when forwarding the data which increases the transmission delay. So it needs to be modified more carefully about the real-time control. Since we don't have flow control, if data are too much, the network buffer may overflows and leads to data lost.

References

1. Kaufmann, M., Manolios, P., Moore, J.: Computer aided reasoning: ACL2 Case Studies. Kluwer Acdemic Pub, Dordrecht (2002)
2. Ya, G., Xiaojuan, Li., Yong G.: Theorem Prover ACL2-based Verification of the node Communication in the robot operating system ROS. J. Chin. Comput. Syst. **35**(9), 2126–2130 (2014)
3. Li, Y., Li, X., Guan, Y.: xMAS-based Formal Verification of SpaceWire Credit Logic. Comput. Sci. **43**(2), 113–117 (2016)
4. Chatterjee, S., Kishinevsky, M.: Automatic generation of inductive invariants from high-level microarchitectural models of communication fabrics[J]. Formal Methods Syst. Des. **40**(2), 147–169 (2012)
5. Verbeek, F.: Formal verification of on-chip communication fabrics. UB, Nijmegen (2013)
6. Chatterjee, S., Kishinevsky, M., Ogras, U.Y.: Quick formal modeling of communication fabrics to enable verification. High level design validation & test workshop IEEE international **29**(3), 42–49 (2010)
7. Gotmanov, A., Chatterjee, S., Kishinevsky, M.: Verifying deadlock-freedom of communication fabrics, verification, model checking, and abstract interpretation, pp. 214–231. Springer, Berlin Heidelberg (2011)
8. Verbeek, F., Schmaltz, J.: Hunting deadlocks efficiently in microarchitectural models of communication fabrics. In: Bjesse, P., Slobodová, A. (ed.) Proceedings of the 11th Conference on Formal Methods in Computer Aided Design (FMCAD 2011). pp. 223–231. [s.l.], IEEE/ACM, (2011)
9. Joosten, S.J.C., Schmaltz, J.: Automatic extraction of micro-architectural models of communication fabrics from register transfer level designs, Design, Automation & Test in Europe Conference & Exhibition (DATE). IEEE **2015**, 1413–1418 (2015)
10. Joosten, S.S., Verbeek, F.F., Schmaltz, J.J.: WickedXmas : designing and verifying on-chip communication fabrics. S.l, S.n (2014)
11. Borrione, D., Helmy, A., Pierre, L., et al.: A generic model for formally verifying NoC communication architectures: A Case Study, International Symposium on Networks-on-chip, pp. 127–136. IEEE Computer Society, US (2007)
12. Gastel, B V., Schmaltz, J.: A formalisation of XMAS. Eprint Arxiv (2013)
13. ACL2Version7.0. http://www.cs.utexas.edu/users/moore/acl2/ (2015)

Network Traffic Classification Using Machine Learning Algorithms

Muhammad Shafiq[1], Xiangzhan Yu[1(✉)], and Dawei Wang[2]

[1] School of Computer Science and Technology, Harbin Institute of Technology, Harbin 150001, China
{muhammadshafiq, yuxiangzhan}@hit.edu.cn
[2] National Computer Network Emergency Response Technical Team Coordination Center of China, Beijing, China
stonetools@yeah.net

Abstract. Nowadays, Network Traffic Classification has got pivotal significance owing to high growth in the number of internet users. People use a variety of applications while browsing the pages of internet. It is very crucial for internet service providers (ISPs) to keep an eye on the network traffic. Most of the researches made on Network Traffic Classification, using Machine Learning Based Traffic Identification to collect data set from one campus network, don't provide far better results. In this paper, we attempt to achieve highly precise results using different kinds of data sets and Machine Learning (ML) algorithms. We use two data sets, HIT and NIMS data sets for this work. Firstly, we capture online internet traffic of seven different kinds of applications such as DNS, FTP, TELNET, P2P, WWW, IM and MAIL to make data sets. Then, we extract the features of captured packets using NetMate tool. Thereafter, we apply three ML algorithms Artificial Neural Network, C4.5 Decision Tree and Support Vector Machine to compare the results of each algorithm. Experimental results show that all the algorithms give highly accurate results. But C4.5 decision tree algorithm provides 97.57% highly precise results when compared to other two machine learning algorithms.

Keywords: Network traffic classification · Machine learning. IM application · WeChat traffic classification

1 Introduction

Nowadays many techniques are used for traffic classification, but the very first one is Port-Based Technique. In this technique, network applications were first register their ports in the Internet Assigned Number Authority (IANA) [1]. However, this technique remains ineffective on account of use of dynamic port number as an alternative of well-known ports numbers. The other one is Payload-Based Technique. It also remains ineffective due to encryption of applications data to evade from being detected. As many network applications use encryption technique, it is very challenging to inspect the encrypted packet data.

In the previous years, numerous supervised and unsupervised machine learning (ML) methods have efficiently been applied in traffic classification. Using Bayesian

F. Xhafa et al. (eds.), *Advances in Intelligent Systems and Interactive Applications*, Advances in Intelligent Systems and Computing 686,
https://doi.org/10.1007/978-3-319-69096-4_87

Analysis Technique [2] and Bayesian Neural Network [3], Moor [2] built 248 statistical features data sets for traffic classification in 2005. Using these techniques, they built data sets wherein they achieved high classification accuracies. However, in our previous work study in [4–8] we two different network trace traffic datasets and classify WeChat application traffic accurate and got very promising accuracy results using ML classifiers.

In this study, we have utilized three essential machine learning (ML) classifiers for network traffic identification taken broad categories of network application trained samples to classify unknown application classes, such as WWW, IM, MAIL, P2P, TELNET, FTP, and DNS. In this paper, we have used two combined data sets NIMS [9] and HIT data sets. The NIMS data sets are publically available on internet [10] and the HIT data sets are our own developed data sets. In this research study, our main aim to achieve high accuracy and also to increase the accuracy results from all the selected classifiers. To achieve high accuracy results, it is very important to have high quality data sets. In this research, we have used two combined data sets, NIMS and HIT data sets, which were collected from different location servers. We have used twenty two "22" extract features and three "3" different types of machine learning classifiers to classify DNS, FTP, TELNET, P2P, WWW, IM and MAIL applications. In this research work, we study the effectiveness of the three selected machine learning (ML) classifiers.

The rest of this paper study is organized as follows: Sect. 2 includes introductory information about machine learning (ML) algorithms. Section 3 throws light on Data sets. Section 4 gives information about feature extraction. Section 5 contains an experiment and result analysis. The last Sect. 6 is the conclusion that we derive from this research.

2 Machine Learning Algorithms

In this research study, we use three ML algorithms, which are used in many areas of computer science. However, in this paper, we use these algorithms for network traffic classification to classify unknown classes, which are already used for network traffic classification. The preliminary information of these three algorithms is given bellow:

The multilayer Perceptron algorithm is a feed forward multilayer artificial neural network, which is also called as Back Propagation Neural Network. The abbreviation ANN stands for Artificial Neural Network. Multilayer Perceptron algorithm is the most widely used algorithm. It is a supervised machine learning technique algorithm. The multilayers perceptron algorithm includes of many layers. The initial layer is an input layer and the last layer is an output layer, while the mid layer is a hidden layer. When the input data is put into this algorithm and processing procedure goes to output. Then the output result is compared to the desired output. If the errors occur in output, then this error is sent back input neural network to adjust the weight to decrease the error with such iteration. This process is the training process of ANN algorithm.

C4.5 is a decision tree ML algorithm [11], used to produce a decision tree. C4.5 is also known as statistical classifier as this algorithm can be used for classification. This algorithm uses training set as ID3. ID3 (Iterative Dichotomiser 3) is an algorithm utilized to produce DT (Decision Tree).

In machine learning, Support Vector Machine (SVM) is a supervised learning technique [12] applicable to both regression and classification. In other words, support vector machine uses machine learning tools to increase the predicting accuracy. It is widely applied for its discriminative classification, which separates hyper plane. In this research work, SVM gives highly precise result of 92.46% using 45,978 data sets.

3 Internet Traffic Dataset

In this study, we use two combined datasets, NIMS and HIT data sets for network traffic classification. We only take 50% of NIMS data sets and 50% of HIT data sets. Thereafter, we combine these data sets manually together with word processor application. It is a very crucial task and takes much time.

3.1 NIMS Data Set

The NIMS data sets are publically available on internet for the researchers [9]. These data sets include DNS, Telnet, FTP, HTTP and Limewrie etc. However, we take only HTTP, FTP, Telnet and DNS for our research work. It should be kept in mind that we have taken only 50% of packet for our work because these data sets have also Limewire and other types of data sets. These data sets were developed in 2007. The NIMS data sets are stored in .arff file format which is supported format of Weka toolkit [13]. There are two separate data sets 1 and 2. We have downloaded the first number of data sets. These data sets consist of 22 statistical features that are derived from the TCP/IP packet header.

3.2 HIT Data Set

The second data set is collected at laboratory of Harbin Institute of Technology (HIT) using Wire Shark, [14] application. Wire Shark is a packet tracing and network packet analyzer software that captures the packets and then displays all the information of captured packet in detail. In this process of capturing real time internet traffics, we have taken seven internet applications FTP, TELNET, WWW, DNS, IM, E-MAIL and P2P. In capturing process FTP, TELNET, WWW, DNS and P2P, packets are captured in 2015 and the other IM and MAIL data sets are captured in 2016.

4 Features Extraction

As we have discussed in Sect. 2 that we use two combined data sets collected from different location network, the NIMS and HIT data sets. The NIMS data sets are .CSV format, which is a readable format of Weka tools [13]. In the development of HIT data sets, we use NetMate tool [15] for feature extraction. Through this software application, we extract twenty two features. In the selection of features for traffic classification, we don't use flow feature like source ip, source port number, and destination ip and destination port number. We have developed the HIT data sets, then we marge the NIMS data sets and HIT data sets into a single file. After extraction and selection of feature, we save the file as a comma delimited .CSV format for Weka tool using MS Excel application. MS Excel is a software package of Microsoft Company, which is used for data analyzing and data calculation.

5 Experiments and Result Analysis

5.1 Methodology

In this research study, we use three machine learning (ML) algorithms Support Vector Machine (SVM), ANN and C4.5 decision tree in a data mining software Weka3.7 tool [13]. Weka is mostly used as data mining software for traffic classification. We use the classifier MLP for ANN and SMO for SVM and trees.j48 for C4.5 in Weka software. In this work, we use 45,978 data samples. These data samples are the combination of NIMS data sets and HIT data sets. Using these data sets, we implement the selected algorithms in Weka software. In our experiment for training and testing, we use 10-folder cross validation. In this work, the accuracy, precision; training time and recall are employed to calculate the performance of ANN, SVM and C4.5 algorithm.

5.2 Result and Analysis

We use ANN, SVM and C4.5 machine learning algorithm and then we compare these classifiers to find out which algorithm gives high performance accurate results. In Weka tool, we use function multilayer perception for ANN, Function SOM for Support Vector Machine and trees.j48 for C4.5 decision tree machine learning algorithms. We have shown the accuracy and training time of applied algorithms for network traffic classification to classify unknown classes (Fig. 1).

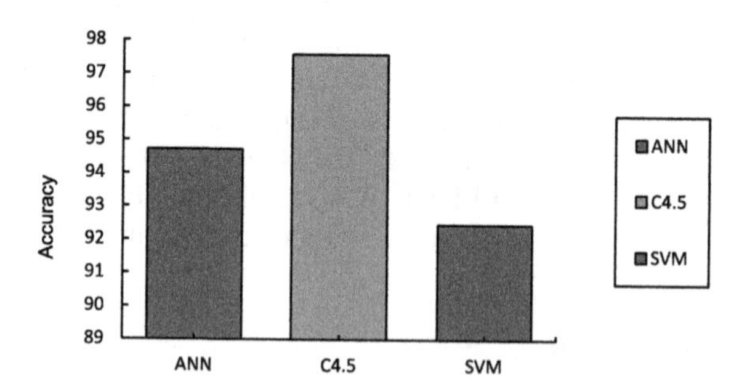

Fig. 1. Accuracy comparison by graph

Below is the comparison analysis chart of algorithms used in this research works. The classification accuracy of ANN is 94.72 is better than the SVM classification accuracy. However, to compare the training time of ANN to Support Vector Machine

(SVM), the SVM produce better results in terms of training time as shown in Table 1. The following Fig. 2 shows the recall comparison of selected internet traffic applications. The recall of C4.5 shows very high results in terms of recall as compared to ANN and SVM. The recall values of FTP, DNS, TELNET, P2P, WWW, MAIL and Instant Messaging is very high from the ANN and SVM recall values. So, it is evident again that C4.5 gives high performance in recall values. Figure 3 is the precise comparison of internet traffic applications, which shows that the C4.5 gives very high performance in internet traffic classification in terms of precise values. The C4.5 classifier precision value is greater than ANN and SVM classifiers. In this analytical process, all the algorithms ANN, C4.5 and SVM give very high classification results in terms of Classification Accuracy, Recall and Precision to classify unknown classes. But the overall result of C4.5 is very high, which is 97.57%.

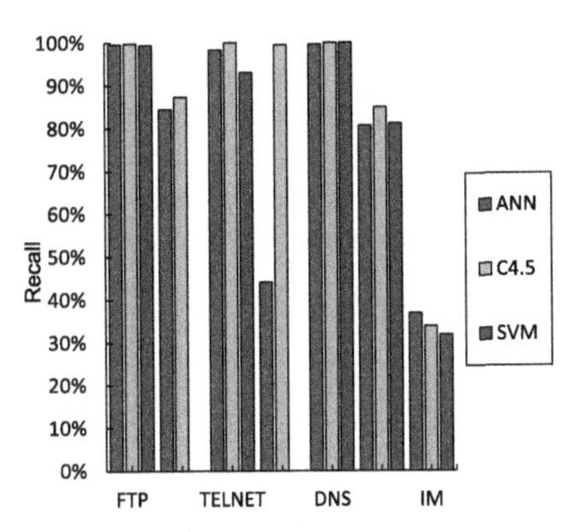

Fig. 2. Recall of selected internet applications

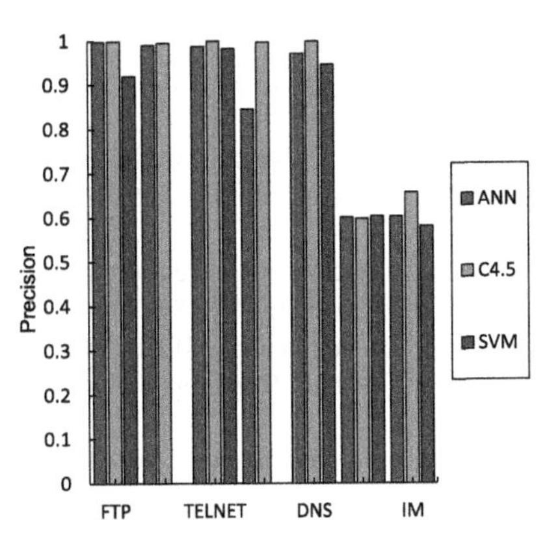

Fig. 3. Precision comparison

6 Conclusion

In this research study, we use two data sets, NIMS and HIT data sets. NIMS data set is publically available data set on internet. For HIT data sets, we captured the traffic using Wire Shark application, which is packet capture software. Thereafter, we used NetMate tool to extract the features from the captured packets, and then we combined both data sets using MS Excel software. Then, we classified the internet traffic applications using three machine learning algorithms to classify the internet traffic. In this experimental work, all the three algorithms gave very high accuracy result. But C4.5 decision tree give very high classification accuracy result as compared to other two algorithms. The C4.5 decision tree classification accuracy performance was 97.57% and training time was 1 s using 45,978 samples. So, our research work proved that C4.5 decision tree machine learning algorithm is a very good algorithm for network traffic classification. Also we concluded that data set perform very important role in network traffic classification. For effective network traffic classification pure data set is very important. Our next research will on data set and feature selection network traffic classification.

Acknowledgements. This work was supported by National Natural Science Foundation of China under Grant No. 61571144 and National Key Research and Development Plan of China under Grant 2016QY05X1000.

References

1. Nguyen, T.T.T., Armitage, G.: A survey of techniques for internet traffic classification using machine learning. IEEE Commun. Surv. Tutorials **10**(4), 56–76 (2008)
2. Moore, A.W., Zuev, D., Crogan, M.: Discriminators for Use in Flow Based Classification. Intel Research Tech. Rep. (2005)
3. Auld, T., Moore, A.W., Gull, S.F.: Bayesian neural networks for internet traffic classification. IEEE Trans. Neural Netw. **18**(1), 223–239 (2007)
4. Shafiq, M., Yu, X., Laghari, A.A.: WeChat text messages service flow traffic classification using machine learning technique. In: 6th International Conference on IT Convergence and Security (ICITCS). IEEE (2016)
5. Shafiq, M., Yu, X.: Effective packet number for 5G IM WeChat application at early stage traffic classification. Mobile Inf. Syst., Article ID 3146868, 22 pages, (2017). doi:10.1155/2017/3146868
6. Shafiq, M., et al.: WeChat text and picture messages service flow traffic classification using machine learning technique. In: 2016 IEEE 18th International Conference on High Performance Computing and Communications; IEEE 14th International Conference on Smart City; IEEE 2nd International Conference on Data Science and Systems (HPCC/SmartCity/DSS). IEEE (2016)
7. Shafiq, M., et al.: Effective feature selection for 5G IM applications traffic classification. Mobile Inf. Syst. (2017)
8. Shafiq, M., et al.: Network traffic classification techniques and comparative analysis using machine learning algorithms. In: 2nd IEEE International Conference on Computer and Communications (ICCC). IEEE (2016)

9. Alshammari, R., Zincir-Heywood, A.N.: A flow based approach for SSH traffic detection. In: IEEE International Conference on Systems, Man and Cybernetics, ISIC, pp. 296–301, 7–10 Oct 2007
10. To download NIMS data sets https://web.cs.dal.ca/~riyad/Site/Download.html
11. Korting, T.S.: C4.5 algorithm and multivariate decision trees. Image Processing Division, National Institute for Space Research—INPE, SP, Brazil
12. Kim, H., Claffy, K.C., Fomenkov, M.: Internet Traffic classification demystified: myths, caveats, and the best practices. In: Proceedings of the 2008 ACM CoNEXT Conference. Publisher: ACM, ISBN
13. Weka tools (Online). http://www.cs.wikato.ac.nz/ml/weka/
14. To capture online traffic, Wire shark tool, Application: http:// www.wireshark.org
15. Introduction to NetMate toll, download information https://dan.arndt.ca/nims/calculating-flow-statistics-using-netmate/comment-page-1/

Fault Analysis and Fault Diagnosis Method Review on Active Distribution Network

Zhang Tong[1(✉)], Liu Jianchang[1], Sun Lanxiang[2], and Yu Haibin[2]

[1] Department of Automation, College of Information Science and Engineering, Northeastern University, Shenyang 110819, China
Zhangsun6411@163.com

[2] Key Laboratory of Industrial Control Network, Shenyang Institute of Automation, Chinese Academy of Sciences, Shenyang 110016, China

Abstract. This paper analyses the current situation and researches the fault cause and the common fault type in active distribution network (ADN). The fault diagnosis methods are summarized and the advantage of the common methods are compared. The general problems consist in the current fault diagnosis methods of ADN are researched in this paper. Combining the development trend of ADN and the power supply reliability demand, the further research direction of ADN fault diagnosis method is proposed, which lays the foundation of the AND self healing and protection control.

Keywords: Active distribution network (ADN) · Fault analysis · Fault diagnosis

1 Introduction

The active distribution network is a distribution power system containing distribution generation (DG) resource, which has the ability to control and operation of distribution network. ADN provides power for the customers directly at the end of the power grid. The operation safety of ADN affects social production and customer normal life directly. Therefore, the distribution system should serve high quality energy to satisfy the customer demand. An abnormal equipment may cause cascading failure in ADN. After a fault occurs, the information signals such as measurement, protection, and wave record are exact for fault diagnosis, detection and location. The diagnosis mechanism is utilized to determine the fault feeder, fault location and isolate fault zones. The efficient fault diagnosis can minimize outage area and shorten the blackout time. When power failure occurs, the fault diagnosis is the basis of the ADN system protection strategy. So fault diagnosis is very important in the ADN self-healing restore and management. This paper research the comprehensive evaluation, comparison and prospect of fault diagnosis methods based on fault analysis. The research lays foundation for the further study of fault detection and achieves active power distribution network automation control.

F. Xhafa et al. (eds.), *Advances in Intelligent Systems and Interactive Applications*, Advances in Intelligent Systems and Computing 686,
https://doi.org/10.1007/978-3-319-69096-4_88

2 The Development Status

Active distribution network contains three stages: power generation, transmission and distribution network. The topological structure becomes complex, which consists of the feeder line, step-down transformer, circuit breaker, relay protection, automatic control device, measure instrument, communication network and other equipment. With the development of the distributed power resource technology, the DG installed around the industrial and appliance load access to AND in large scale. The single-generation-radiant traditional distribution network changes to multi-generation weak loop network structure [1]. The DG installation position, capacity and operation mode have a major impact on the power flow, feeder voltage and the transmission loss of the distribution system. The above changes caused by DGs brings difficulty on the fault diagnosis, location and self-healing in AND. Simple exact of voltage current amplitude and switch signals is not enough.

The distribution network research falls behind the European, USA and other developed countries, which the interruption time is far more than developed countries. Since the complex topological structure and various branches of the distribution network, the improvement of the fault diagnosis speed and the fault location accuracy is an important challenge.

3 The Fault Type and Reason Analysis

The AND fault types can be classified at the fault resistance aspect as follows: high resistance fault, earth fault, and intermittent fault. The fault causes are complex and disorder, which increases the difficulty of fault diagnosis.

The high resistance fault in ADN. The high-resistance fault is defined as that the transmission line contacts the high impedance ground or other objects, when the overhead line breaks. For example, after the high voltage transmission line breaks, the feeder line contacts the surrounding trees and causes a short-circuit fault.

Short-circuit earth fault. Short-circuit earth fault is one of the most common fault in power distribution network system. The single-phase to earth fault occurs most frequently and is difficult to monitor in the distribution circuit. Assuming that a short-circuit earth fault occurs in the customer side, the current drop is not obvious to induce tripping operation and the drop type fuse has no action. There the fault feature exact is the critical technical issue of fault diagnosis. The short circuit earth fault at night is easier to detect because of its spark phenomenon. The fault monitoring mainly relies the transient signal analysis in circuit system. The transient signal stored mass information about the single phase to earth fault. Another transient process feature is that it can reflect the fault character and avoid the earthing mode effect. Hence the deep analysis of transient process is beneficial for fault diagnosis. When a single phase to earth fault occurs, the transient signal analysis of the fault current and voltage can improve the efficiency of the distribution circuit fault diagnosis.

Intermittent fault: Intermittent fault refers to the intermittent discharge phenomenon in the ADN operation. Arc light is produced during the discharge progress The intermittent fault character is the instantaneity and repeatability. The fault time interval extends from second level to minute, hour and several days. The intermittent fault may cause inter-phase short circuit by hanging object, birds, lightning, and external force.

4 Traditional Fault Diagnosis

When an ADN occurs, the system normal operation recovery is the important condition to reduce the interruption time. The fault diagnosis is one of the prerequisites for fault restore. Therefore international researchers propose a variety of fault location methods. The traditional fault location methods contain impedance method, traveling wave method, S-injection method and so on. The advantage of impedance method is the low cost, In the concrete implement process, the resistance method is limited to path impedance, line load and the generation resource. The traveling wave method can locate fault accurately, but the fault diagnosis process takes long time. Sang proposes the S-injected signal method in 1993, The proposed method sends S signals to track the fault position and fault diagnose in ADN, The proposed has a high fault location precision.

5 Fault Diagnosis Method of Artificial Intelligence Technology

The ADN is a distributed complex system, which combines distribution, real-time and off-line. The artificial intelligent methods develop rapidly in recent years and show the high efficiency. Therefore the power researchers introduce the artificial intelligence concept to the distribution network fault location, diagnosis, and restore.

5.1 Expert System

Expert system (ES) is a mature subject of artificial intelligence. ES concludes the responding rules and sets up rulebase on the base of the expert knowledge and experience. The efficient inference mechanism applies the standard rules to the industrial system and computes the correct conclusion from the imprecise information. The ES is a relatively mature diagnosis method and is applied in the actual production. Due to the different knowledge representation and reasoning mechanism, the ES methods contain heuristic rule and positive negative reasoning system. Shao [2] proposes an ES for distribution network fault location based on GIS. The ES combines the geographic information of power distribution GIS, equipment management, network topology structure to the ES rule base. The fault section is located through dynamic search and back reasoning. The knowledge acquisition, organization, check and maintenance are difficult, which becomes the bottleneck of ES method application. The ES receives the final correct diagnosis based on the huge knowledge base research. The ADN research is at the start stage and can't provide comprehensive and systematic application database. Therefore the ES is limited in fault diagnosis, location and self-healing restore research of ADN.

5.2 The Artificial Neural Network

The artificial neural network (ANN) is the common method used in the distribution network fault diagnosis. ANN has strong self-learning ability compared with ES. Bi [3] proposes a fault location method based on RBF model, which increases the training speed and improves the fault diagnosis accuracy. The proposed method is efficient to diagnose the multiple fault and abnormal action of protection device. Ziolkowski [4] analyses the voltage and current signals, and constructs the fault feature vectors such as root mean square value, fault alarm rate constant, skewness and symmetrical component. Butler [5] detects low resistance and high resistance earth fault based on ANN method in the neutral point ground system and the neutral ineffectively ground system. The proposed method accuracy is above 90%.

5.3 Fuzzy Set Theory

The fuzzy set theory is widely used in the fault diagnosis. The fault membership is determined based on the equipment membership function. When the fault membership value exceeds the switch equipment threshold value, the fault section is detected. In the ADN fault diagnose, the relationship of the fault and omen is fuzzy, which the theoretical description is not clear. Therefore the fault diagnosis result is fuzzy. The fuzzy fault diagnosis constructs the fuzzy relationship matrix based on the expert experience, which is in the fault symptom and fault reason space. The common methods combine the fuzzy relationship matrix according to fuzzy inference rules. The advantages follows: excellent analysis ability of the uncertainty problems; it utilizes language variables to describe the expert experience and is easier to understand; it can get more result of different priority levels. The fuzzy theory has good fault-tolerance capability to improve other artificial intelligence methods. But the membership function determination and membership function modification of the network structure change are the research key issues.

5.4 Rough Set Theory

The rough set method is used to simplify the fault location decision table and exports minimum reduction form of the section position, which can locates fault rapidly and accurately. Z. Pawlak proposes rough set theory. The rough set theory can analyze and process the imprecise, inconsistent and incomplete information, which is based on the classification of the observation and measurement data. The rough set theory can discover the implicit knowledge, reveal the potential regularity and has a good fault tolerant performance. There are disadvantages in the rough set theory application. The more complete decision information table is, the credibility of the reduction rules is higher. But in the actual large scale network, the comprehensive collection fault samples will bring the amount of work. The rules produced by decision table method is much more than the rules produced by fault diagnosis expert knowledge of the traditional power grid. To ensure the diagnosis rapidity and accuracy, the rule search optimization is also a research issue in the future. In the power system fault diagnosis, considering the large-scale power system and multiple failure, the decision table is very complex, even occurs “combination explosion” problem. When fault occurs in the ADN, the existing SCADA system provides limited alarm signals to the dispatchers. The limited information makes the fault response system complicate and produces the incomplete alarm information, which brings great difficulties to the fault diagnosis research.

5.5 Bayes Method

The bayes method is utilized for fault section location. The synchronous fault information is uploaded to the control center. The control center analyzes the network structure and selects the fault information sequence. The fault position is located according to the maximum fault significant degree in the distribution brunch. The bayes method can shield fault information to a certain degree. The inexact reasoning method is utilized to exclude the negative impact of the error message and achieves the efficient section location. The bayes method is accurate and synchronized. But there may encounter the unique maximum value deficiency. Therefore the algorithm calculation will fail, it is necessary to process further.

5.6 Traveling Wave Method Combined with Impedance Method

Lu [6] proposes a single end fault location method combined traveling wave with impedance method based on wavelet transform. The simulation result shows: in the section near midpoint, the fault diagnosis based on impedance method may miscalculate. When the fault resistance increases, the location error is higher. The appropriate both-side fault location method can counteract the fault resistance effect and improve the location precision obviously.

5.7 PMU Combined with Computer Aided System

C.W. Liu researches the fault location methods and gets a series of research achievements based on PMU. Liu [7] proposes an adaptive fault location method using PMU. The basic thought is: the synchronous voltage and current signal is measured by PMU and the line parameter is identified online in ADN. The improved discrete Fourier transform method is utilized to exact the fundamental component from the transient electric parameter. The proposed method reduces the effects of line parameter change, measurement error and random disturbance. Jiang [8] proposes a fault location method of the transmission line with series compensation equipment based on the PMU. Different from the common location method, the proposed method doesn’t need exact mathematical model of the series compensation device and the priori knowledge of the operating state. The location method is adapt to different series compensation unit component with no additional phase shift. The simulation result indicates that the proposed method has a high fault location precision of 99.95%. The fault diagnosis method based on is the future development direction of ADN fault diagnosis.

6 Conclusion

This paper studies the fault diagnosis method application in power distribution network. Different fault diagnosis technology has its own characters and application scope. Due to the distribution network complex structure, fault type and fault resistance, the fault diagnosis research is a key technology and occupies an important position in power system restore. With the ADN development, the appropriate fault diagnosis scheme is designed according to the local distribution network structure and engineering field condition. The artificial intelligence method is the future development trend of the ADN fault diagnosis. The fault diagnosis method combining different technology can improve the accuracy and speed, which lays foundation for the ADN system protection and management.

References

1. Wu, Y.-K., Lee, C.-Y., Liu, L.-C.: Study of reconfiguration for the distribution system with distributed generators. IEEE Transm. Power Deliv. **25**(3), 1678–1685 (2010)
2. Shao, X., Tian, J.: The fault location application based on GIS expert system in distribution network. Zhejiang Electric Power **4**, 12–15 (2008)
3. Bi, T., Ni, Y., Fuli, W.: A novel neural network approach for fault section estimation. Proc. Chin. Soc. Electr. Eng. **22**(2), 73–78 (2002)
4. Ziolkowski, V., Nunes da Silva, I., Flauzino, R., Ulson, J.A.: Fault identification in distribution lines using intelligent systems and statistical methods. Circ. Syst. Sig. Process. Inf. Commun. Technol. Power Sour. Syst. **1**, 175–178 (2006)
5. Butler, K.L., Momoh, J.A.: A neural net based approach for fault diagnosis in distribution networks. IEEE Power Eng. Soc. Winter Meet. **1**, 353–356 (2000)
6. Lu, J., Li, Y., Li, J., Xue, Y.: A new single-end fault location method based on traveling wave and resistance method. Autom. Electr. Power Syst. **31**(23), 65–67 (2007)
7. Al-Mohammed, A.H., Abido, M.A.: An adaptive PMU based on fault location algorithm for series compensated lines. IEEE Trans. Power Syst. **29**, 2129–2137 (2014)
8. Jiang, J.A., Jang, J.Z., et al.: An adaptive PMU based fault detection/location technique for transmission lines part II. PMU implement and performance evaluation. IEEE Trans. Power Deliv. **15**(4), 1136–1146 (2000)

Extensive Survey on Networked Wireless Control

Li Lanlan[1,2(✉)], Wang Xianjv[3], Ci Wenyan[4], and Chalres Z. Liew[5]

[1] Chuzhou Vocational Technology College, Anhui, China
jlunco@126.com
[2] Nanjing University of Posts and Telecommunication, Jiangsu, China
[3] Fuyang Normal University, Anhui, China
[4] School of Electric Power Engineering, Nanjing Normal University Taizhou College, Taizhou, China
[5] Charles Laboratory, SmartSys Workshop, Hamilton, ON, Canada

Abstract. Networked control, especially wireless one, on account of the flexible and effective application it facilitates, has been increasing scholars with diverse background in various fields participate into such a hot issue. Due to the advance of new technologies and applications, relevant classic issues, such as delay, queuing and discussion on stability under a certain of effectiveness, have been brought up on a new height to research with new challenges. This survey paper presents an epitome of NWCS, which is short for Networked Wireless Control System, and related research in the past few years covered with the discussions range from characteristics to materializations.

Keywords: NWC · NWCS · Stability analysis · Sensor network

1 Introduction

Network is a kind of medium providing a exchanging links via relay method to perform communication. Networks enable remote data transfers and data exchanges among users, reduce the complexity in wiring connections and costs of Medias, and provide ease in maintenance [1]. But these facilities are at the cost on its elaborateness and complexity on administration, such as routing for instance.

With the popularization and extension for application of network, increasing nodes have been accessing into, this causes the raising of cost and complexity on both administration and computation particularly when resource is limited. Enormous and complicated progresses must have been negative affecting on system's stability and effectiveness. The professional manage to deal with that aggravated situation by continual modifying, not only in protocol in different layer but also in processing method.

Stability and efficiency are two primary factors in system. Regarding the whole network as an entirety including terminal, nodes and links, the performance of this comprehensive system would be more sophisticated. Coordinating, synchronizing, matching is all influence upon combined but distributed system. Protocols such as TCP/IP, UDP and CAN have been widely used for long duration of development, in which TCP/IP so far has been taken for granted as a effective protocol in Internet while

F. Xhafa et al. (eds.), *Advances in Intelligent Systems and Interactive Applications*, Advances in Intelligent Systems and Computing 686,
https://doi.org/10.1007/978-3-319-69096-4_89

UDP as for LAN and CAN for industries especially in car industries. Based on OSI seven-layer model, many new methods and ideas for improving network performance have been proposing by scholars and engineers to develop and fulfill the request of wireless communication. Bluetooth, Ad hoc, UWB (ultra wideband) and so on, there is a growing number of people who is inclined to developing network towards wireless direction. Wireless and wideband these hot keywords have been gradually grabbed the sight of researchers.

2 Related Tech

As for researching and developing, network module generally be abstracted as a 7 layer model according to its functions on account of to agree on a unified basic protocol that facilities interlink between different systems from different fields. So called the 7 layers of the OSI model is the basic description for general network framework. Some new models or special models proposed are on the base of that to extend. It is an Open Interconnection System in seven layers, which defines a networking framework for implementing protocols. Control information delivers from the bottom to the upper application.

Researches in different field do different work in different layer to improve the network performance. Even in the some research background, such as control for instance, unique characters exist respectively in link control, route control and terminal control. Most algorithms are often direct at application while hardware design physical layer. But some scholars research also account for the other layer such as data-link layer and transport one to improve or programming efficiency or for system adapted.

Based on such general regular model description, several extensions tech or model in wireless communication arise.

Bluetooth is a wireless technology designed by Ericsson. Open-stack is the character of its protocol. Including Intel, Microsoft, IBM, Motorola, Nokia, NEC, and SONY and so on, so many corporations adopt it that make this tech broadly used in diverse cyber production. It has been one of the most popular wireless techs in application. But its cost in some scene is too high to be afforded by some small-scale companies, especially when applied in limited-gains development. As for its chipset, firmware is embedded into RF hardware along with an antenna; developer can utilize its functions via HCI modification linked with equipment.

Ad hoc network combined by wireless mobile nodes completely, which is characterized by dynamic forming and organizing, limited bandwidth and multi-hop. It has been used in many communication circumstance such wireless mesh networks and sensor networks. Because there is no solid stationary router, any node within the net plays not only a sender or receiver role, but also transmitter for data relay. Dispersed structure contributes to a special way to arrange the tasks, distributing the task of whole net to each scattered nodes. This kind of communication method, to some extent, guarantees a certain of stability and robust in system. However, in most wireless ad hoc network applications, having a range of nodes will compete for access to shared wireless media that can lead to conflicts. Bandwidth limitation sometimes also aggravates such conflict. The use of cooperative wireless communication improves the anti-jamming by improving the decoding of the desired signal by causing the destination node to combine self-interference and other node interference. And it's opening channel and nodes information exchanging also gives rise to the research on its network security.

Another hot tech that usually be mentioned by researchers of information technology is UWB.

This is a radio technology that can be used at very low levels for short-range high-bandwidth communications by using most of the radio spectrum. The method uses a pulse carrier with a sharp carrier pulse, a logical convexity at a set of center frequencies. UWB has traditional applications in non-cooperative radar imaging. The latest applications for sensor data acquisition, precise positioning and tracking applications.

3 Theory and Methodology

In its general sense, delay refers to a lapse of time. Up to control, delay is one of the key factors that affects the stability and efficiency of the system. In network or networked circumstance, the random network delay can be analyzed with introduce of Queuing methodology. In a computer network, the queuing delay is a critical component of network latency, which is the time that the job waits until it can be executed in the queue. The tail packet discarding strategy is the traditional packet dropping policy used by the internet router. When the queue length is large to the maximum length, the packets arriving at the queue are discarded and the discarding behavior remains until the queue is reduced due to the packet being transmitted. When a packet loss occurs, a connection-based protocol such as TCP reduces its transmission rate to attempt to service the queued packets so that the queue is empty. Because the packet drops from the input (tail) of the queue, it is also called the tail drop. During network congestion, all forms of latency occur more than once, and the retransmission of such packets results in overall latency.

The stability of time delay systems can be divided as two primary results, which is of time independent stabilization and time-dependent stabilization. The former one can stabilize a system irrespective of delay. However, time dependent stabilization is in relation to the size of delay. An upper bound can be provide to render the system stable within the range of delay less than that upper limitation.

As for time domain, system can described as model in continuous time and discrete time to discuss. Because of the universal of the continuous time physical system, many theories are developed from or based on this common and natural circumstance. Even so, controllers implemented are usually digital one in fact; discrete time seems to be more reasonable approach to introduce. And digital processing often provides facilities on data collection and computation. The delayed system equation converts the delay of the known delay into a higher order delay. To some extent, this is an applicable method to deal. But the amounts of what the known delays exist in a system would give rise to large-dimensional systems, in which the scheme will complicate process considerably such as compute and store. Furthermore, this approach is not applicable for systems with unknown delay.

Krasovskii and Repin use the Lyapunov method for stability analysis of time delay systems. Recently, the time derivative has been introduced in the modified Lyapunov-Krasovskii function.

In Ref. [2], Mr. Yue, Han and Peng design a controller of network control systems with state feedback and perform asymptotic stability at the exponential speed of convergence. And in Ref. [7], feedback control was taken one step ahead to be introduced into uncertain system with time varying input delay and discussed as delay feedback.

Reference [17] proposes a networked control system that is suitable for use in switching systems where the last control signal is reused when the received data is lost, which is used to stabilize a networked control system with bounded interference and unreliable links.

Reference [9] solves the networked control systems under persistent disturbance rejection via state feedback. Such research provides a direction for enhancing system robust base on stability guarantee under some extreme noise circumstance. And several available conclusions are given in the paper. Mr. Tian, Yue and Peng work out a method of control for networked control systems, in Ref. [10], quantized output feedback was introduced into NCS framework. Via quantize state of input and output, this will provide facilities not only for analysis but also process in digital. Paper finally performs asymptotic stabilization of NCS.

Reference [5] gives a Lyapunov-Krasovskii method to solve the asymptotic stability of discrete-time systems. In Ref. [12], robust stabilization was introduced to analyze the uncertain system with unknown input delay and lower the conservative results so that this method is of that one applied approach and basis to calculate or design. It also provides an available condition for system that of any delay $\tau \in [\tau 0 - \delta, \tau 0 + \delta]$, where $\tau 0 \geq \delta$.

Combining NCS and Robust is a hot topic for researching. Reference [8] discusses the problem of system-based robust H∞ control uncertainties. The designed H∞ controller in the paper is of memory less type. It provides possibility and leaves room for the further step that works such as introduce of Markov chains into research. Reference [20] regards network-induced delays model as Markov chains. Through the analysis of the stochastic stability theory, a sufficient condition for the existence of strong H∞ perturbation attenuation level is obtained. A differential dropout compensator is designed between the sensor and the controller in the system, where the communication data and the signal interaction are performed by the output feedback with the state observer. Reference [13] gives an approach to sample in networked control system with scheduling method. The discussions primarily focus on the NCS with multi loops. Several schemes have been proposed and proved by given examples.

Intelligent control is one of the hot trends for research. Neural networks have been introduced in large areas.

In Ref. [16], paper uses Generalized Predictive Control (GPC) to overcome the influence of modeling error and uncertainty of system, an error predictive model is further built based on BP neural network. Discussion was made focus on characteristics of uncertain delay of networked control system, and the schemes are introduced and tested under random-delay-time characteristic data transmission.

Reference [11] gives an intelligent control in network control with neural network. It describes the disturbances by means of Brownian motion model in a stochastic view and abstract the delay that of time-varying as a Bernoulli stochastic model to analyze.

Through analysis and proof, paper gives some results of delay-distribution-dependent criteria for the discrete-time stochastic delayed neural networks (DSNNs) and gives sufficient condition and proof of robust global exponential stability of system in the mean square sense. Primarily, Lyapunov method was used in the works.

Reference [19] introduces a fuzzy control method for networked control system to evaluate feasibility and a networked control system for double inverted pendulum control is implemented in the paper.

Reference [6] go a step further to perform robust H∞ control in delay-dependent method for T–S fuzzy system with interval time-varying delay. Based on the parallel distributed compensation (PDC) method, Mr. Yue, Tian and Zhang give and prove several propositions to expound the asymptotic stability conditions for the existence of robust H∞ controller are obtained in the paper.

4 Conclusion and Prospect

Some main fruits and idea related to NWC have been introduced and discussed above. But there still be emerging in large numbers of new challenges and tech continuously.

To sensor network, people had ever taken for granted that modified Ad hoc tech and traditional NCS theory could fulfill the request of the application. Facts proved that those imaginations are far from practices. Besides the algorithm issues, operation system are also a problem needed solving. Several OS such as TinyOS, developed by Berkeley, and Linux now are regarded as available OS at hand. Protocol is also one key factor to perform. But until now, there is not a unified approach.

In the static configuration, mobile computing technology further expands the scope of the NWC implementation.

Another tech may lead a great revolution is quantum computation and quantum computer tech. Quantum computers use quantum mechanical phenomena to perform operations on data. Quantum computing is the use of quantum properties of these data to operate. This will prompt the rate of data process and bring about unlimited application in diverse field including NWC.

The mobile Internet and the Internet of Things are evolving at an unprecedented rate, causing the mobile data business to explode. In order to meet the challenges of massive flow, the mobile network is "no capacity to limit the wireless network", the so-called "big pipeline" direction, technology continues to make a breakthrough. In the future of wireless technology evolution, it'll adapt to the application scenarios to meet the user experience as a determinant.

Acknowledgements. The study is supported by Anhui provincial quality projects (2014jyxm516, 2015zy098, 2016ckjh137), college quality engineering projects (zlgc2014031, zlgc2015001), key project of domestic and foreign visiting and the training of excellent young and middle-aged talents in colleges and universities (gxfxZD2016343), key project of Natural Science Research in Anhui province (KJ2017A727).

References

1. Tipsuwan, Y., Chow, M-Y.: Control methodologies in networked control systems. Control Eng. Pract. **11**, 1099–1111 (2003)
2. Yue, D., Han, Q.-L., Peng, C.: State feedback controller design of network control systems. IEEE Trans. Circ. Syst.-II: Express Breiefs **51**(11), 640–644 (2004)
3. Vladimerou, V., Dullerud, G.: Wireless control with bluetooth
4. The 7 Layers of the OSI Model in Webopedia's
5. Stojanovi, S.B., Debeljkovi, D.L., Mladenovi, I.: A Lyapunov-Krasovskii. Methodology for asymptotic stability of discrete time delay systems. Serb. J. Electr. Eng. **4**(2), 109–117 (2007)
6. Tian, E., Yue, D., Zhang, Y.: Delay-dependent robust H∞ control for T–S fuzzy system with interval time-varying delay. Fuzzy Sets Syst. (2008)
7. Yue, D., Han, Q.-L.: Delayed feedback control of uncertain systems with time-varying input delay. Automatica **41**, 233–240 (2005)
8. DongYue, Qing-Long Han, Lam, James: Network-based robust ∞ control of systems with uncertainty. Automatica **41**, 999–1007 (2005)
9. Yue, D., Lam, J., Wang, Z.: Persistent disturbance rejection via state feedback for networked control systems. Chaos Solitons Fractals (2007)
10. Tian, E., Yue, D., Peng, C.: Quantized output feedback control for networked control systems. Inf. Sci. **178**, 2734–2749 (2008)
11. Zhang, Y., Yue, D., Tian, E.: Robust 9 delay-distribution-dependent stability of discrete-time stochastic neural networks with time-varying delay. Neurocomputing **72**, 1265–1273 (2009)
12. Yue, Dong: Robust stabilization of uncertain systems with unknown input delay. Automatica **40**, 331–336 (2004)
13. Peng, C., Yue, D., Gu, Z., Xia, F.: Sampling period scheduling of networked control systems with multiple-control loops. Math. Comput. Simul. **79**, 1502–1511 (2009)
14. Raymond Momut Director of Marketing ARESCOM, INC.Open Systems Interconnection (OSI) Reference Model 23 Jan 2000
15. Zimmermann, H.: OSI reference model-the ISO model of architecture for open systems. Interconnection IEEE Trans. Commun., Vol. Com-28, NO. 4 (1980)
16. Wu, A., Wang, D., Liang, J.: Application of generalized predictive control with neural network error correction in networked control system. In: Proceedings of the 5th World Congress on Intelligent Control
17. Ohori, A., Kogiso, K., Sugimoto, K.: Stabilization of a networked control system under bounded disturbances with unreliable communication links via common Lyapunov Function approach. In: SICE Annual Conference (2007)
18. Zhicheng, J., Lu, X., Linbo, X.: The integrated design of control and scheduling for networked control system. In: Proceedings of the 26 Chinese Control Conference (2007)
19. Sun, Z-y., Hou, C-Z.: Fuzzy control for networked control system. IEEE Xplore. Restrictions apply (2005)

Internet and Cloud Computing

Research and Design of Smart Home System Based on Cloud Computing

Huiyi Cao, Shigang Hu(✉), Qingyang Wu, Zhijun Tang, Jin Li, and Xiaofeng Wu

School of Information and Electrical Engineering, Hunan University of Science and Technology, Xiangtan 411201, China
hsg99528@126.com

Abstract. With the rapid development of Internet of Things (IoT) technology and cloud computing technology, smart home received more and more attention, which focuses on integrating with the home life-related facilities and building efficient residential facilities and family affairs management system to get safe, convenient, comfortable and artistic home life. Firstly, this paper introduces some problems of traditional smart home, which analyzes the advantages of 'Internet plus' and cloud computing combining with smart home and proposes ae new kind of smart home system based on cloud computing. At last, The simulation experiment proves that the new system can solve and improve the problems existing in traditional smart home system.

Keywords: Smart home · Cloud computing · Internet plus

1 Introduction

Currently, with the 'Internet Plus' era, people increasingly dependent on the basic necessities of the Internet, at the same time, the Internet has also brought to people's lives more convenient [1]. At this point, the traditional smart home reveals a lot of problems, which can no longer meet the requirements of consumers. 'Internet Plus' combined with the smart home has become a trend.

Now, there are many smart home systems on the market, but they will more or less bring bad influence to the user experience, which has been unable to meet consumer demand [2]. In recent years, with the Internet of Things technology and the rapid development of cloud computing, they provide technical support for the development of smart home [3]. And smart home development has also made great progress.

This paper first introduces some problems of traditional intelligent home, which analyzes the advantages of 'Internet plus' and cloud computing combining with smart home and proposes a new type of smart home system based on cloud computing. Finally, The simulation experiment proves that the new system can solve and improve the problems existing in traditional smart home system.

F. Xhafa et al. (eds.), *Advances in Intelligent Systems and Interactive Applications*, Advances in Intelligent Systems and Computing 686,
https://doi.org/10.1007/978-3-319-69096-4_90

2 Analysis of the Present Situation of Smart Home

Smart home system is the use of a variety of advanced technology, integration of individual needs, will be related to the home life of the various subsystems, such as security, lighting control, air conditioning control and other organic combination to achieve intelligent control and management purposes [4]. A typical smart home system is shown in Fig. 1.

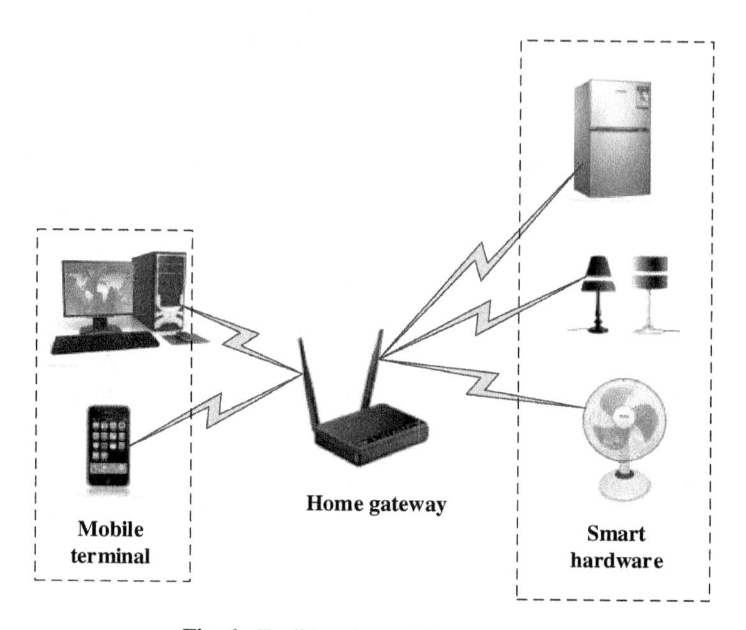

Fig. 1 Traditional smart home systems

Smart home now has entered the outbreak period. Data projections, China's smart home market will reach 180 billion yuan in 2018. However, there are data show that 87.5% of consumers are not satisfied with the smart home products. Although the smart home has great potential market, but the current sales market is limited to some high-end crowd or part of the pilot public buildings. Smart home is not universal in the market because of smart home system there are many problems [5].

- There is no business model.
  At present, including the smart home of the Internet of Things industry are exploring business models. Only by selling hardware equipment profitability can not be long, because a set of equipment sales may be used in a period of more than 10 years. If only the hardware update is not enough to support these hardware equipment manufacturers.
- Not really smart.
  Smart home users do not find the pain point, but for smart and smart. which does not give the user the opportunity to experience the real convenience and visible

economic benefits. The intelligent use is only dispensable, and users cannot rely on it. Users do not have enough economic benefits to drive, who develop long-term habits difficult to change.

- There are security risks.
  Manufacturers ignore the safety of smart home. Related reports show that a lot of people's lives and intelligent equipment is only an increase of network control functions, but to the hacker's attack has brought convenience, which making the traditional smart home systems exist security risks.
- Cannot handle large amounts of data.
  As the number of devices in the smart home system increases, the data generated during the operation of these intelligent devices grow exponentially. For the massive data generated, the traditional smart home system hardware conditions simply cannot meet such requirements. Massive data cannot be stored and calculated to limit the development of smart home.

Traditional smart home these problems seriously hindered its position in the market, it is necessary to propose a new intelligent home system to solve or improve these problems.

3 Design of Smart Home System Based on Cloud Computing

Cloud computing is a product of distributed computing, parallel computing, utility computing, network storage technologies, virtualization, load balancing, high available and other traditional computer technologies and network technologies [6]. Cloud computing is rapidly deployed in the Internet industry with its ultra-large scale, high scalability, virtualization, high reliability, on-demand services, fast services and extremely low cost. It has three service models including SaaS, PaaS and IaaS [7], as shown in Fig 2.

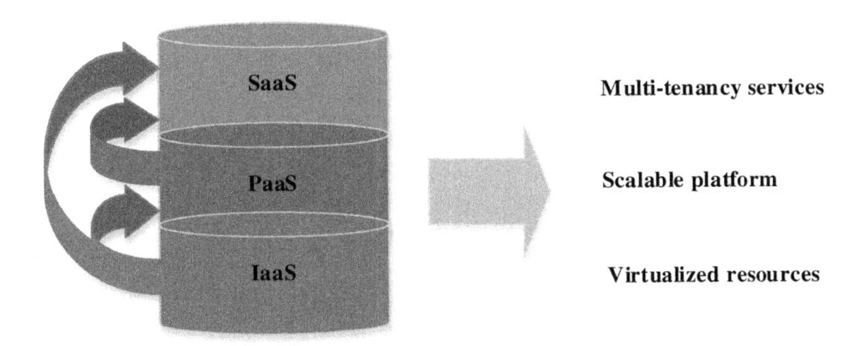

Fig. 2 Three service modes of cloud computing

In this paper, the design of smart home system is based on B/S architecture development, the system uses the IaaS cloud computing model. We will build on the basis of IaaS PaaS smart home system. We directly rented Aliyun's ECS (Elastic

Compute Service) and deployed it, including firewalls, load balancers, web servers, database servers, backup servers, and file servers. Specific cloud deployment architecture shown in Fig. 3.

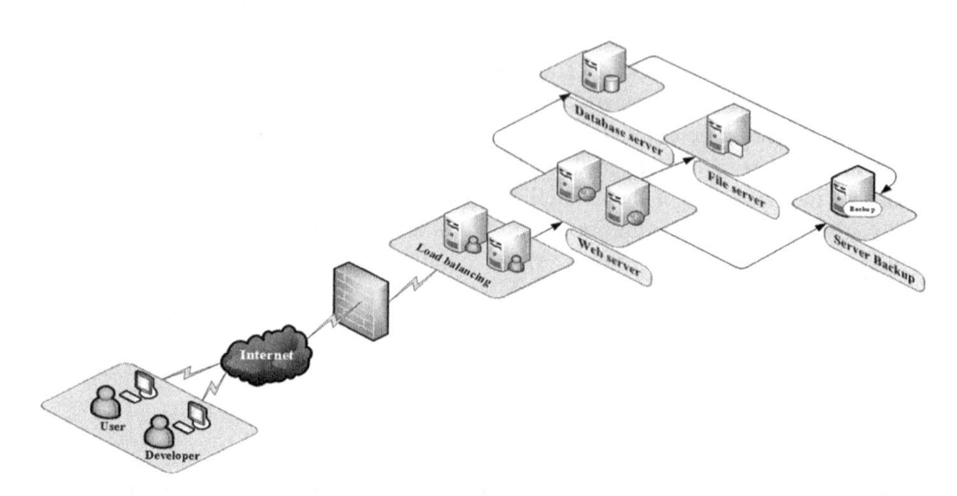

Fig. 3 Cloud deployment architecture

The application of cloud computing technology in the smart home system makes the design of the home gateway becomes very simple. ECS and home gateway use TCP/IP protocol for communication. In the course of communication, we adopt algorithm encryption to encrypt the data in order to enhance the security of the system. As long as the family of smart home devices connect home gateway by the lwIP, ZigBee, wifi and other wireless communication protocols, which can communicate with the cloud, so that the system to achieve online control and real-time detection purposes. The architecture of a complete cloud-based smart home system is shown in Fig. 4.

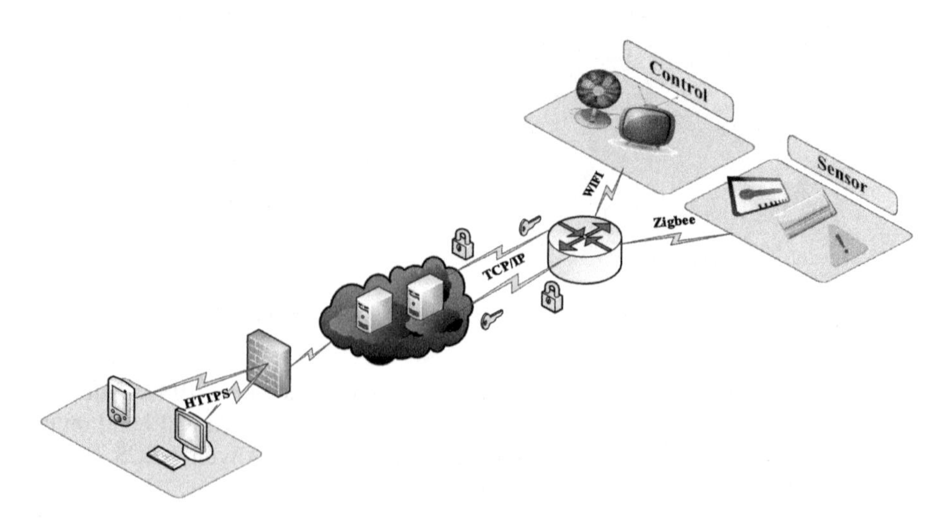

Fig. 4 The architecture of smart home based on cloud computing

4 Results and Discussion

After the system was built, we designed the software part of each function module by the asp.net technology. After the programming was completed, the target circuit board through the network cable to access the home gateway, the core board IP address was set to ECS IP address. Then entered the ECS IP address (https://118.178.230.195) on the client (pc, phone or pad), and entered the correct user name and password in the corresponding interface, we can log into the system's main control interface.

Then the client can control the target board by the simulation of household appliances, the system successfully completed the Web page and the underlying hardware interaction. So far this paper has completed the complete design of intelligent home control system. The complete smart home system is shown in the Fig. 5.

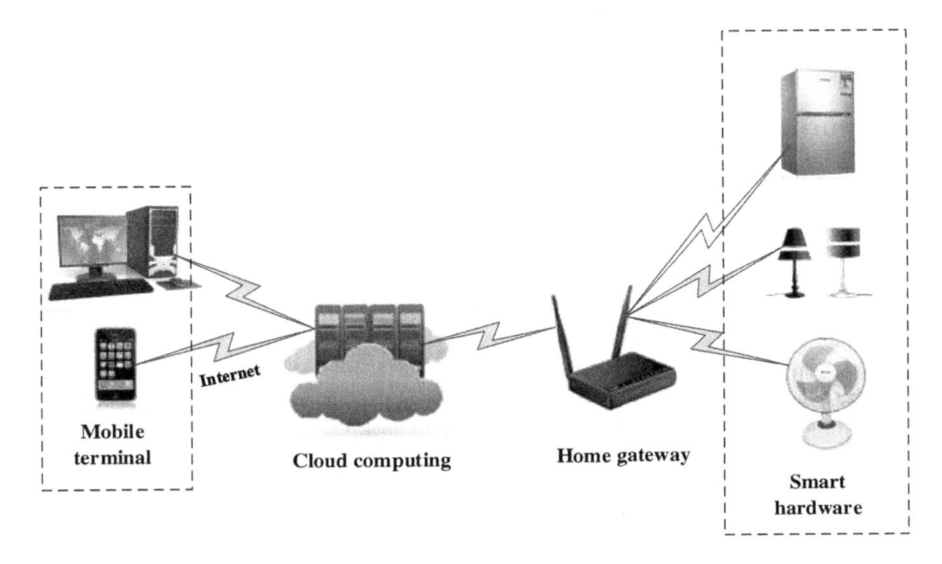

Fig. 5 Ideal reference image and two blurred images

The experimental results show that the system has the following advantages:

- The B/S architecture model developed by the system compatibility is very good, and closely integrated with the Internet.
- The system focused on the cloud, reducing and clarifying the task of the gateway, making the system standardization and generalization.
- The system can realize the dynamic access of massive intelligent devices.
- Cloud computing provides a basis for the storage and calculation of big data.
- The communication process by adding encryption algorithm to enhance the security of the system.

5 Conclusions

This paper describes the traditional intelligent home the shortcomings of existence, revealing the so-called smart home tepid status quo reasons. And a smart home based on cloud computing system is proposed by the study of cloud computing technology, the advantages of cloud computing and integration of 'Internet plus' new economic model. Finally, the experimental results show that the system is practical. Compared with the traditional intelligent home system, the system improves the flexibility and real-time of the user access, reduces the cost, and solves and improves some problems existing in the traditional smart home.

Acknowledgements. This research was financially supported by the National Natural Science Foundation of China (Grant Nos 61674056, 61675067, 61376076, 61377024 and 61474042); supported by the Scientific Research Fund of Hunan Provincial Education Department (Grant No. 16A072).

References

1. Suryadevara, N.K., Mukhopadhyay, S.C., Wang, R., Rayudu, R.K.: Forecasting the behavior of an elderly using wireless sensors data in a smart home. Eng. Appl. Artif. Intell. **26**(10), 2641–2652 (2013)
2. Puustjärvi, J., Puustjärvi, L.: The Role of Smart Data in Smart Home: Health Monitoring Case. Procedia Comput. Sci. **69**, 143–151 (2015)
3. Garcia-Valls, M., Cucinotta, T., Lu, C.: Challenges in real-time virtualization and predictable cloud computing. J. Syst. Architect. **60**(9), 726–740 (2014)
4. Lertlakkhanakul, J., Choi, J.W., Kim, M.Y.: Building data model and simulation platform for spatial interaction management in smart home. Automat. Constr. **17**(8), 948–957 (2008)
5. Catala, A., Pons, P., Jaen, J., Mocholi, J.A., Navarro, E.: A meta-model for dataflow-based rules in smart environments: Evaluating user comprehension and performance. Sci. Comput. Program. **78**(10), 1930–1950 (2013)
6. Khan, M.A.: A survey of security issues for cloud computing. J. Netw. Comput. Appl. **71**, 11–29 (2016)
7. Kavvadia, E., Sagiadinos, S., Oikonomou, K., Tsioutsiouliklis, G., Aïssa, S.: Elastic virtual machine placement in cloud computing network environments. Comput. Netw. **93**, 435–447 (2015)

Website Structure Optimization Based on BS Matrix

Yonghong Xie[1,2], Can Cui[1,2], Aziguli[1,2(✉)], and DaoLe Li[1,2]

[1] School of Computer and Communication Engineering, University of Science and Technology Beijing, 100083 Beijing, China
894547148@qq.com

[2] Beijing Key Laboratory of Knowledge Engineering for Materials Science, 100083 Beijing, China

Abstract. Applied the results of user's behaviour analysis to optimize website structure, this essay accordingly refines the session on behaviour cluster. Thus the solution model for optimizing website structure based upon BS (Behaviour-Service) matrix has been put forward. This model demonstrates application solutions respectively involving page navigator recommendation and link relations adjustment.

Keywords: User Behaviour Analysis · Website Structure Optimization · Page Navigation Recommendation · Link Relations Adjustment

1 Introduction

With the rapid and continuous development of Internet, the initial motivation that people acquiring information has gradually changed from the demand for enormous information to appropriate information. Website as one of the main information carriers involving multiple aspects like structure and service, is tending to be much more complicated and diverse than ever. On that premise, optimizing website structure constantly and providing personalized services are the key focus points when it comes to building brands and meeting customer requirements. By analyzing the behaviour data from a large amount of users, we can come up with good solutions to optimize website structure and provide personalized services.

In order to optimize website structure, scholars have done many researches. In 1997, Mike Perkowitz and Oren Etzioni defined adaptive web sites, which achieves website structure optimization by learning user access mode [1]. However, after PageGather algorithm used in adaptive websites to create clusters, the clusters still need to be selected by administrators. Besides, no consideration for page content while ascertaining similarity between two pages may damage the inherent structure of the website. In 2001, Yen and others used web accessibility and popularity into network linking, and they combined the expected number of links with access rate to adjust website structure [2]. But only using access rate to measure the popularity of web pages is inaccurate. In 2008, Yanhuan Huang proposed optimization scheme to solve the

F. Xhafa et al. (eds.), *Advances in Intelligent Systems and Interactive Applications*, Advances in Intelligent Systems and Computing 686,
https://doi.org/10.1007/978-3-319-69096-4_91

shortage of mutual information exchange in traditional algorithm by associated web page recommended method based upon ant-colony algorithms [3]. Since it is the variant of collaborative filtering mechanism, the optimization scheme has the inherent shortcomings of collaborative filtering mechanism. The scheme does not take the interaction between pages into account either, so it can not pay full attention to the characteristics of the website structure. In 2013, ETS (Enhanced Tabu Search) algorithm, proposed by Peng-Yeng, achieved the multi-constraint website structure optimization, but it is inapplicable to large websites and has some limitations [4]. Moreover, Hamed Qahri Saremi extended quadratic assignment problem to analyse the structure of site links [5]. Corin et al. [6], Lempel et al. [7] and Rafiei et al. [8] proposed site link or popularity analysis method based upon Markov chain. However, current researches on website structure optimization does not combine enhancing the rationality of website structure with providing users with better browsing experience very well.

To solute this problem, based on user behaviour analysis model and through structuring BS matrix, this paper considered the problem of page navigation recommendation and links relationship adjustment in website structure optimization from two dimensions and more fine-grained degrees to form website structure optimization model. To optimizing model of the website structure proposed in this paper, various experiments are done based on data set, to evaluate model's actual application effect. We have got relatively ideal assessment results.

2 BS Matrix and Its Extension Matrix

Before the introduction of BS matrix, a few concepts need to be explained:

(1) Service Set: pages collection of certain website services, which consists of a series of converging dynamic pages with certain keywords.
(2) Behaviour Set: the collection of user session clustering results through fuzzy

clustering. Each element in the collection represents a category of users, and according to the different membership, specific users can belong to more than one class.

A BS matrix is expressed as follow.

$$\mathrm{BS} = \begin{pmatrix} \mathrm{S}_{11}^{\mathrm{H}} & \cdots & \mathrm{S}_{1\mathrm{C}}^{\mathrm{H}} \\ \vdots & \ddots & \vdots \\ S_{S1}^{H} & \cdots & S_{SC}^{H} \end{pmatrix} \tag{1}$$

Where $\mathrm{S}_{\mathrm{ij}}^{\mathrm{H}}$, $i \in [1, \mathrm{s}]$, $j \in [1, c]$ is the collection of user session sequences. BS matrix consists of two components, namely Behaviour Set and Service Set. B dimension of the matrix (column vector), represents a certain type of user behaviours' distribution on all Service Sets; S dimension (row vector), represents all user behaviours' distribution on a certain type Service Set.

Through the process to construct the BS matrix mentioned above, we can see that a specific session may be discretely distributed in each location of a certain column in the matrix. This distribution structure vividly illustrates that a particular type of user's behaviours will involve multiple services of the website. So when recommending pages to users, different types of service pages should be provided. This will be explained later in page navigation of optimized model.

Two extension matrix, BSP (Behaviour-Service-Page) matrix and BSL (Behaviour-Service- Link) matrix, are derived from BS matrix.

BSP matrix, the deformation of BS matrix, is used to aid BS matrix learning and page navigation recommendations. BSP matrix breaks up each element S_{ij}^{H}, $i \in [1, s]$, $j \in [1, c]$ of the BS matrix to remove sequentiality, and only retains pages and the occurrence frequency of pages in the element. After conversion, each element in the BSP matrix is an extension pages collection of a number of pages and their appearing numbers.

BSL matrix, another deformation of BS matrix, is used to aid the link relationships adjustments. Based on pages in each element $S_{ij}^{H}, i \in [1, s], j \in [1, c]$ of BS matrix, BSL matrix forms adjacency matrix L representing link relationships.

$$\mathrm{L} = \begin{pmatrix} a_{11} & \cdots & a_{1n} \\ \vdots & \ddots & \vdots \\ a_{n1} & \cdots & a_{nn} \end{pmatrix} \tag{2}$$

In L matrix, aij is initialized to 0. When the session in S_{ij}^{H} has the access sequence from page i to page j, $a_{ij} = a_{ij} + 1$, or it remains the same. Each element in the BSL matrix is an adjacency matrix, but the value n of different elements is different.

3 Website Structure Optimization Model based on BS Matrix

Website structure optimization model of the paper is on the basis of BS matrix. Problems which are going to be solved mainly focused on how to recommend appropriate follow-up pages to the user, and how to adjust links between pages through the model. Model structure is as shown in Fig. 1.

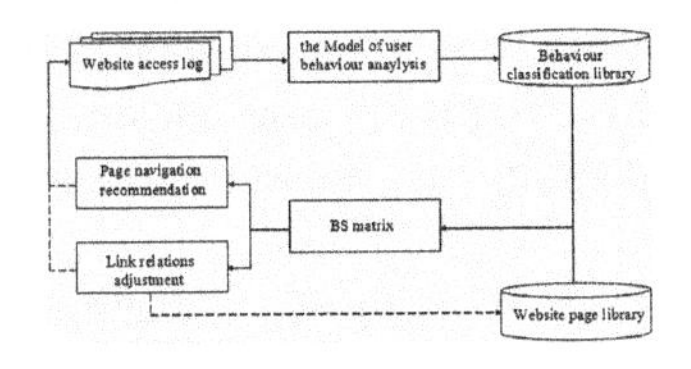

Fig. 1. Website structure optimization model

Model application is divided into two main aspects, recommending page navigation and adjusting link relations.

Recommending page navigation is divided into three steps.

First of all, for new user User, we assume that he has accessed to n pages, and the formed access path is $H = \{p_1, p_2, ..., p_n\}$. Calculate the sequence pages' dissimilarity between H and each type of user behaviour clustering prototype. Since clustering prototype is the representative sequences of such users, so we can sort the calculated sequence pages' dissimilarity in ascending order, and take the first two as vested in H's behaviour class, assuming as B_1 and B_2.

Then locate the column of B_1 and B_2 in the BS matrix, and calculate the intersected element number $|H \cap S|$ of H and each type of Service Set and sort it in descending order. Select two Service Sets whose $|H \cap S| > 2$ and have the most intersected elements as vested in H's Service Set, assuming as S_1 and S_2. At this point, we can get four corresponding element blocks as shown in Fig. 2.

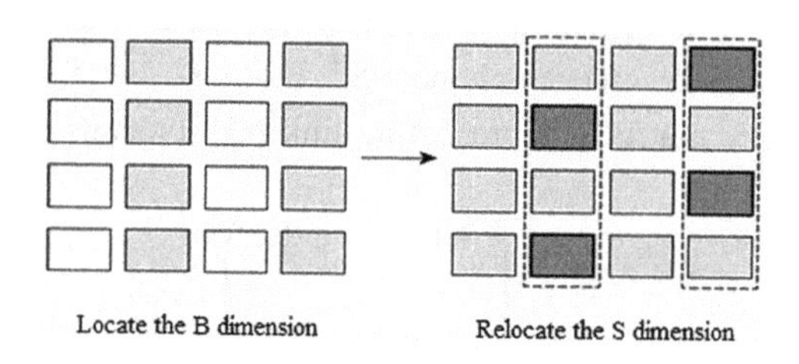

Fig. 2. Representation of BS matrix location

Finally, take the top 10% hot pages from the four corresponding element blocks in the BSP matrix, and merge them to form the initial candidate pages set. If the elements taken out of the four element blocks all include a page, then add the page access number. Then delete the page that already appeared in H, sort the rest of the pages according to the access time in descending order. Finally form the candidate pages set, and recommend some which have more visits to users.

Link relations adjustment is mainly divided into three steps. Assume we adjust the link relations within Service Set Support, specific adjustments are as follows.

First, calculate BSL matrix through the BS matrix. Each element of BSL matrix is a $n_i \times n_i, i \in [1, c]$ adjacency matrix L, and n_i indicates the number of pages contained in the column i of the Service Set.

Second, merge the BSL matrix rows where support is located and adjacency matrix $L_i, i \in [1, c]$ of all columns. Then we get the adjacency matrix $L_{surport}$ of the Service Set Support. The algorithm of merging two adjacency matrices L_i and L_j is shown as follow.

Step1 traverse each row of L_j in turn. If row_k is not in L_i, increase by a column and a row in L_i. Initialize elements in new added row and column to 0, repeat Step1 until all of the rows in L_j appear in L_i.

Step2 traverse each element of L_j in turn. Let $e_i = e_i + e_j$, where e_i is the element in the corresponding location of L_i matrix.

Step3 repeat Step2 until Service Set Support has only one adjacency matrix. We call the adjacency matrix $L_{surport}$. The value of each element e_j in $L_{surport}$ is the importance degree of the link relations between the pages i and pages j

Finally, sort the elements e_j of $L_{surport}$ matrix according to the values in reverse order. If $e_j$value is in the former 10%, increase the practical link relations between page i and page j; if the value is in the last 5%, remove the actual link relations between page i and page j.

4 Model Evaluations

Experimental environment includes: Windows 8.1, MySql, and Java. Experimental data set is the USTB dataset of the undergraduate teaching website in the University of Science & Technology Beijing, which contains 4,947,573 users'access records. Before evaluation, we have already finished the preparatory work such as data preprocessing and user behaviour clustering based on fuzzy clustering.

Page navigation recommendation is assessed by using AUC (Area Under ROC Curve) indicators to measure the accuracy of the results. Consider the depiction of the ROC Curve, for current featured pages p_{reco} and user browsing path $H = \{p_1, p_2, \ldots, p_n\}$. The experiment gives two methods to determine if p_{reco} is relevant to user.

Method one: Navigate to H's user class which has the highest degree of membership. Get the class's all pages in the BSP matrix and make them as the comparing page set SP_1. Calculate the average number of occurrences of SP_1 in all pages. If p_{reco} appears more than the average, it is considered that p_{reco} is relevant to H.

There are 18 different experiments based on different data groups and the ROC Curve drawing methods. Experimental data are randomly divided into 10 parts, and each experiment randomly selects some of them as the training set, and the rest as the test set. Each allocation and AUC assessment results are shown in Table 1.

Table 1. Evaluation result of page navigation recommendation AUC

Experiment	Number of training set sessions	The proportion of training set (%)	AUC1	AUC2
1	2302	30	0.6124	0.6528
2	2302	30	0.5973	0.6135
3	2302	30	0.6972	0.6836
4	3387	50	0.6835	0.7073
5	3387	50	0.6975	0.7168
6	3387	50	0.7087	0.7257
7	5419	80	0.7162	0.7319
8	5419	80	0.7273	0.7436
9	5419	80	0.7234	0.7440

Analyze the results in Table 1 from two aspects:

(1) Judging from the comparison of two methods, when the Service Set clustering method of this paper is used, the recommended results are generally better than that only using users'clustering method. And only number 3 experiment appears abnormal. Through the partition of Service Set, candidate selections in candidate recommendation lists have been more granular filtered, and are more in line with user expectations.
(2) Comparing the AUC values of the experiments coded as 1-3, 4-6, 7-9, it can be seen that, the 1-3 group with less training set data shows a certain degree of randomness and the remaining 6 groups have some stability in the group. And we can know the recommended results of this method are relatively stable without much fluctuation results in one data set.

The experiment takes the Number 9 data set in the accuracy experiment. Depending on the recommended set's length, five experiments are done, and results are shown in Table 2.

Table 2. Evaluation result of page navigation recommendation coverage

Experiment	Length of recommended set	Coverage (%)
1	3	66.53
2	4	71.32
3	5	76.46
4	6	81.24
5	7	85.76

By observing the evaluation results, we can see that as the length of recommended set increases, the coverage becomes higher, and the recommended set can basically cover most of the Service Sets in the site. In addition, after viewing specific Service Set page, we find out that there is a Service Set called "repair and buy", which does not have any pages. Since we have deleted the session whose length is less than 4, the Service Set of which page is less than 3 does not have any user session.

The assessment of link relations adjustment assessment uses the indicator of the site structure optimization based on target function put forward by Youwei Wang. The problem is described as calculating the minimum value of the objective function f(X), the calculation formula of f(X) as shown in Eq. (3).

$$\mathrm{f(X)} = \sum_{i=1}^{n}\sum_{j=1}^{n} c_{ij} * p_j * \min\{d_{ij}(X)\} \tag{3}$$

Where X is the adjacency matrix of the link structure of the website. The elements of Xmatrix are:

$$X_{ij} = \begin{cases} 0, \text{if thelinkdoesnotexist} \\ 1, \text{if thelink(i,j)exists} \end{cases} \quad (4)$$

Where c_{ij} is the correlation between the pages, p_j is a weighted value, and d_{ij} is the reachable distance between pages. In the experiment, we assume that $p_j = 1$, and c_{ij} is calculates as follow.

$$c_{ij} = 1 - d_{url}(p_{il}, p_{jk}) \quad (5)$$

After solving BSL matrix, make link adjustment suggestions, and suggestions that adding or removing links in the Service Set distribution are shown in Table 3.

Table 3. Distribution of Service Set link relations adjustment

Service Set	Number of links advised to increase	Number of links advised to remove
5	1	1
11	2	1
16	1	0
19	1	0
21	1	0
24	1	1

After getting adjustment proposals, we evaluate the link structure before and after optimization with objective function. Based on Eq. (3), Eq. (4), and Eq. (5), evaluate the objective function values f(X)of each Service Set after adjustment.

In view of the fact that the given method only does the link optimization within the Service Set, Service Sets which do not change before and after optimization are omitted. In this paper, we only provide 6 Service Sets'comparison results in Table 3 as shown in Fig. 3.

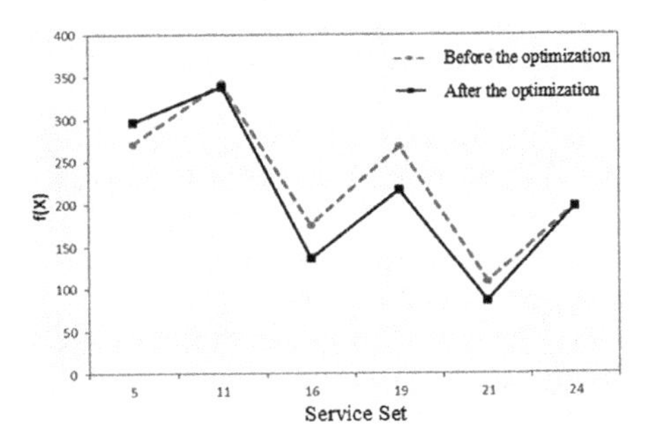

Fig. 3. Evaluation result before and after optimization

According to Fig. 3, we can see that after adjustment, the site structure has been improved significantly. But there is a problem that the total number of pages in USTB data set is relatively small, while the number of Service Set is large. Hence, the number of pages in each Service Set is relatively small, and the impact of increasing or removing 1 link is big. Due to the deletion of links, the optimization result is not obvious in number 5, 11 and 24 Service Set. So the proposed link relations adjustment method based on BS matrix does not have obvious effects for a website with small number of pages.

5 Conclusion

By partitioning Behaviour Set and Service Set, this paper presents the BS (Behaviour-Service) matrix, and on the basis website structure optimization model is formed. Specific optimization steps are given to recommend page navigation, and adjust link relations. Through experimental assessment, the model performs well in website structure optimization.

Acknowledgement. This work is supported by the National Key Technology R&D Program in 12th Five-year Plan of China (No. 2013BAI13B06).

References

1. Perkowitz, M., Etzioni, O.: Adaptive web sites: Automatically synthesizing web pages. Proceedings of the Fifteenth National Conference on Artificial Intelligence and Tenth Innovative Applications of Artificial Intelligence Conference, pp. 727–732 (1998)
2. Yen, B.P., Fu, K.: Accessibility on web navigation. International Conference on E-Commerce Engineering: New Challenges for Global Manufacturing in the 21st Century, Sept 2001
3. Huang, Y., He, Z.: A study on adaptive website based on ant colony algorithm with multi-agent cooperation and feedback mechanism. J. Fuzhou Univ. (Natural Science Edition) **36**(6), 814–818 (2008)
4. Yin, P.Y., Guo, Y.M.: Optimization of multi-criteria website structure based on enhanced tabu search and web usage mining. Appl. Math. Comput. **5**(11), 11082–11094 (2013)
5. Saremi, H.Q., Abedin, B.: Website structure improvement: Quadratic assignment problem approach and ant colony meta-heuristic technique. Appl. Math. Comput. **195**(1), 285–298 (2007)
6. Anderson, C.R., Domingos, P., Daniel, S.W.: Adaptive web navigation for wireless devices. In Proceedings of the Seventeenth International Joint Conference on Artificial Intelligence, 2001
7. Lempel, R., Moran, S.: The stochastic approach for link-structure analysis (SALSA) and the TKC effect. Comput. Netw. **33**, 387–401 (2000)
8. Rafiei, D., Mendelzon, A.O.: What is this page known for? Computing web page reputations. Comput. Netw. **33**, 823–835 (2000)

The Effects of Bank Employees' Information Security Awareness on Performance of Information Security Governance

Shiann Ming Wu, Dongqiang Guo, and Yung Chang Wu(✉)

College of Business Administration, National Huaqiao University, Fujian, China
Jasonwu988@gmail.com

Abstract. The transfer of huge amounts of funds over international financial markets from commercial activities involving the banking industry must provide a reliable operating environment to ensure data security. This study explores the relationship between bank employees' information security awareness and performance of information security governance by employing empirical methods and collected data through a questionnaire with the subjects being bank employees. The goal is to understand the effects of the subjects' information security awareness on the banks' performance of information security governance from the analysis results. The results demonstrate that information security awareness produces a significant positive impact on the performance of information security governance - that is, information security awareness significantly affects the performance of information security governance. Moreover, the better the bank employees understand information security, the more significantly the information security governance influences information security awareness.

Keywords: Information security awareness · Commercial banks · Performance of information security governance

1 Introduction

Research reports published by well-known information security and communications companies such as RSA (EMC), Symantec, and Verizon have noted various network and computer systems' attacks, types of intrusions, and possible future trends (Symantec, 2013) [1]. Although the Internet has no borders, the construction of information security and key information infrastructure protection networks in Taiwan still needs to be carried out according to local conditions by drawing on the analysis of relevant literature.

After the Personal Information Protection Act published detailed implementation rules in October 2010, enterprises in the future will assume more responsibilities for information security and need to conduct overall assessment and reviews related to information security during the transition of regulation orientation. However, the majority of enterprises and organizations consider information technology (IT) as a tool to achieve strategic goals and seize opportunities and only discuss how to combine organizational strategies and IT applications while paying little attention to the

F. Xhafa et al. (eds.), *Advances in Intelligent Systems and Interactive Applications*, Advances in Intelligent Systems and Computing 686,
https://doi.org/10.1007/978-3-319-69096-4_92

importance of information security governance. Therefore, this study probes into bank employees' information security awareness and performance of information security governance.

2 Literature Review

The objectives of information security awareness and education and training were established so that members of the organizations could be aware of their responsibilities for protecting corporate information assets and understand the value of information resources and potential threats [2]. Barker (2014) stressed the gap between actual and perceived security risks [3]. The Insurance Brokerage Association of Taiwan (2014) regulated that operators applying for information security assessment for computer systems need to meet the following relevant qualifications: A. They should acquire knowledge about information security management such as holding the certificate of Certified Information Security Manager (CISM) or passing the exam of Information Security Management System Lead Auditor, ISO 27001 LA. B. They should have knowledge of information security technology and ability like holding the certificate of Certified Information Systems Security Professional (CISSP). C. They should have the ability to simulate hacker attacks, such as holding the certificate of Council Ethical Hacking (CEH) and of the Certified Incident Handler (CIH). D. They should be familiar with vehicle applications of the financial field or system development or have audit experience [4], illustrating the importance of employees' knowledge about information security.

Posthumus and Von Solms (2004) maintained that information security governance is a process that describes how management copes with issues concerning information security and must be combined with the organizations' IT governance structure and strategy to provide strategic guidance. The board of directors and high-level executives are responsible for achievement of organizational objectives, appropriate risk management, resources used, and the supervision of implementing information security plans [5].

Computer literacy and information security literacy must be differentiated. Computer literacy refers to an individual's familiarity with a basic set of knowledge needed to use a computer. Information security literacy means the information security knowledge that individuals must be familiar with to protect information and systems and could be applied to any operating system and as the learning basis for future information security education (MA Alnatheer, 2014) [9]. NIST SP800-16 (1998) maintained that the foundation of information security is a group of information technology vocabulary and concepts and must be complied with by everyone in an organization. All personnel using computer technology or its products, regardless of their special job responsibilities, must know IT security basics and be able to apply them.

3 Research Framework and Methods

(1) Basic data on bank employees' organizations
In light of the control field specified by ISO27001: 2013, we interviewed five information security and banking professionals, and after discussions, summarized the 12 variables, including organization size, gender, age, service years, job position, nature of work and whether the company obtained ISO27001 certification into "personal differences", "position differences", and "organizational differences" to investigate the influences of these variables on information security awareness.

(2) Dimensions of information security awareness
This study quoted the nine courses under the foundation of information security knowledge and literacy curriculum framework set out in NIST SP800-16 (1998) as the measurement dimensions of "information security awareness", which were taken as the independent variables of this study.

(3) Dimensions of information security governance performance

This study cited the assessment tool for information security governance proposed by CGTF (2004) and divided "information security governance performance" into "organization's IT business dependency", "risk management assessment", "organization and people assessment", and "processes assessment" as dependent variables herein.

H1: The "individual differences" of bank employees have a significant impact on information security awareness.

H2: The "position differences" of bank employees exert a significant impact on information security awareness.

H3: The "organizational differences" of bank employees produce a significant impact on information security awareness.

H4: The "information security awareness" of bank employees produces a significant impact on information security governance performance of the banking industry

(4) Questionnaire design and sampling methods
The contents of the questionnaire consisted of three parts. Part 1 was "basic data on organizations", Part 2 was "information security awareness", and Part 3 was the scale of "information security governance performance" dimensions, with each part containing 12, 35, and 15 questions, respectively (totaling 62 questions). The population of this study was defined in "bank employees". Questionnaire pre-test: 30 peers were invited for a pre-test in the way of filling in the questionnaires without time limits. They were asked for their opinions and views about this questionnaire. The questionnaire was then modified based on their actual responses and drawbacks of the questionnaire design in order to meet "surface validity" and "content validity" to a certain degree. After the inappropriate questions were deleted, the Cronbach's α value of the "information security awareness" scale was 0.900, while that of the "information security governance performance" scale was 0.850.

4 Results and Analysis

This study explores the relationship between information security awareness and the direct and indirect effects of behavior patterns under the management regulation of ISO27001: 2013. The test of goodness-of-fit for the overall pattern was used to evaluate the goodness-of-fit between the model and the observed data; it was the external quality of the pattern in an attempt to figure out whether the empirical results were consistent with the theoretical pattern. The ideal numerical values and the evaluation results of each item are as follows. The chi-square value ratios of the theoretical model range from 0.002 to 2.746, which is in accordance with Bagozzi and Yi (1988) in which χ^2/df must be less than 3; GFI is more than 0.8, which is an acceptable threshold of an ideal value, while AGFI is greater than 0.8, and its goodness-of-fit also reaches an acceptable range. Moreover, NFI = 0.893 and meets the recommended value of Ullman (2001), whereas NNFI = 0.932 and IFI = 0.939 exceed the indicator of goodness-of-fit of over 0.9, as suggested by Bentler & Bonett (1980). CFI meets the ideal value of greater than 0.9, as indicated by Bagozzi and Yi (1988). RMSEA of all items (except that of the organization's protection regulations of intellectual property rights and records) is 0.096 and is slightly greater than the ideal value, which should be less than 0.08 as indicated by MacCallum et al. (1996). In summary, the goodness-of-fit of this study's overall pattern is sound.

Arbuckle (2008) pointed out that when C.R. > 1.96 or p-value < 0.05, they are deemed significant (C.R. is the t-value). Suppose that the estimated parameters of the positive direct influences of 1–4 are 0.842, 0.825, 0.865, and 0.913 (C. R. = 0.921/p-value = ***, C.R. = 0.926/p-value = *, C.R. = 0.931/p-value = *, C. R. = 0.915/p-value = ***), respectively. This establishes the hypothesis that four information security frameworks are significantly positively correlated with information security awareness.

This study regarded all dimensions of the "information security awareness" as the influence factors and analyzed the influences of each dimension of the information security performance in the hope of identifying the dependent variable combinations that can effectively affect the "information security governance performance".

(1) The impact of information security awareness on an organization's IT business dependency: The better the awareness of "technical control" or "system life cycle control" is, the higher the performance of an "organization's IT business dependency" will be.
(2) The impact of information security awareness on risk management assessment: The better the awareness of "technical control" or "risk management" is, the higher the performance of "risk management" will be.
(3) The impact of information security awareness on people assessment: The better the awareness of "technical control" is, the higher the performance of "organization and people assessment" will be.
(4) The impact of information security awareness on assessment of processes: The better the awareness of "laws and regulations" and "risk management" is, the higher the performance of "processes assessment" will be.

After the above statistical analyses were summarized, all verification results in this study H1–H4 are presented in the following table.

Research Hypotheses	Results
H1: The "individual differences" of bank employees have a significant impact on information security awareness	Partially significant
H2: The "position differences" of bank employees exert a significant impact on information security awareness	No significant impact
H3: The "organizational differences" of bank employees produce a significant impact on information security awareness	No significant impact
H4: The "information security awareness" of bank employees produces a significant impact on information security governance performance of the banking industry	Significant impact

5 Research Results' Discussion and Suggestions

5.1 Research Results' Discussion

In this study's statistical analysis the respondents' awareness about the dimension of "laws and regulations", especially for the item of "understanding that organizations should comply with laws and regulations to develop organizations' information security policy", is the lowest ($M = 3.52$, $SD = 0.96$). Consequently, the banking industry should enhance the promotion of laws and regulations.

Ke (2010) offered four information business departments of the Taipei City Government as an example to study the information security governance performance of government organizations. Findings show that organizations' IT business dependency is high and very high [8], which is similar to the results obtained by statistical analysis and calculations for the banking industry in this study.

Regarding internal risk management, in the "information security governance performance" scale of this study, the banking industry has the lowest score in the dimension of "risk management assessment" ($M = 3.70$, $SD = 0.77$). Although this score is greater than the average, the banking industry should still strengthen its risk management [10].

In terms of organization and people, by comparing the survey of this study with the study by Wu [6], Huang [7], and Ke [8], the banking industry has not assigned special personnel to be responsible for information security. Additionally, it has a low score in the item of "designating special information security personnel or units responsible for maintaining security plans and ensuring information security" ($M = 3.63$, $SD = 0.93$).

After the dependent variable combinations that can effectively affect "information security governance performance", through verification by regression analysis of Hypothesis 4 in this study on the impact of "information security awareness" on "information security governance performance", it is found that "laws and regulations", "system on-line and information sharing", "risk management", "acquisition/development/installation/implementation of control" and "technical control" can all effectively influence information security governance performance. Moreover, the respondents have a higher performance in the dimension of "information security awareness".

Bank employees naturally have a better awareness and a higher score in "technical control" of everyday information system operation and maintenance ($M = 4.16$, $SD = 0.61$), "system on-line and information sharing" ($M = 4.10$, $SD = 0.61$), "acquisition/development/installation/implementation of control", and "technical control" ($M = 4.17$, $SD = 0.61$) in respect of information security awareness, because these aspects are within the scope of daily work execution. Other dimensions of information security awareness, such as "organization and information security" ($M = 4.07$, $SD = 0.56$), "sensitivity" ($M = 4.04$, $SD = 0.60$), "management control" ($M = 3.91$, $SD = 0.64$), and "operation control" ($M = 3.73$, $SD = 0.75$), have certain influence on the "information security governance performance". Moreover, bank employees tend to perform well in these dimensions. Due to the different natures of businesses, employees are less familiar with these dimensions, hence contributing to their low awareness. In addition, under the circumstance where the same resources must be allocated, conducting business is generally more important that enacting information security, because organizations in the banking industry aim to make profits and their business is effectively conducted to earn profits for the whole organization. This is also a problem the banking industry needs to overcome in order to strengthen security awareness.

5.2 Future Improvement Projects of Information Security Governance

In the future the banking industry may first assess organizations' IT business dependency on a regular basis and then carry out an overall evaluation of information security governance performance. If there is a gap between an organization's IT business dependency and the overall evaluation of information security governance, like an organization's IT business dependency is "very high", but there is "continuous improvement" or "strengthening improvement" of the overall evaluation of information security governance, then improving risk management should be the top priority, because both corporate governance and IT governance involve the issues or areas of governance. Therefore, improving objectives and projects should be the top priority.

References

1. Symantec: Internet Security Threat Report 2013. Symantec Corp, CA, USA (2013)
2. Hong, K.S., Chao, L.Y.: A study of theory for information security management. MIS Rev. **12**(6), 17–47 (2003)
3. Barker, K.: The gap between real and perceived security risks. Comput. Fraud Secur. Bull. **2014**(4), 5–8 (2014)
4. The Insurance Brokerage Association of Taiwan: Regulations on Standardizing Assessment of Computer System Information Security (2014)
5. Posthumus, S., Von Solms, R.: A framework for the governance of information security. Comput. secur. **23**, 638–646 (2004)

6. Wu, C.P.: The study of governmental employees' personal cognition and behavior towards IT security. Master's Thesis of the Department of Public Administration and Policy, National Taipei University, 2005
7. Huang, K.C.: Information Security Governance of Public Sectors, Master's Thesis of the Department of Information Management of National Kaohsiung First University of Science and Technology, 2006
8. Ke, S.S.: Government agencies study of information security governance - A case study of Taipei City Government, Master's Thesis of Department of Information Management, Tamkang University, 2010
9. Alnatheer, M.A.: A conceptual model to understand information security culture. Int. J. Soc. Sci. Hum. **4**, 104–107 (2014)
10. Wu, S.M. et al.: Research into information security strategy practices for commercial banks in Taiwan. Adv. Intell. Syst. Comput. **541**, 182–187, Springer

A Comparative Study on Agglomeration Effects of the Central Cities of Three Urban Agglomerations in China– A Case Study of Producer Services

Qingmin Yuan, Xiao Luo(✉), and Jian Li

The Center of Recycling Economy and Enterprise Sustainable Development, Tianjin University of Technology, 300384 Tianjin, China
luoxiao3524@163.com

Abstract. The central city drives the development of the surrounding cities through the agglomeration effect of producer services. This paper measures the clustering abilities of producer services in central city of the three urban agglomerations through the location quotient coefficient. Conclusions can be drawn as follows: the agglomeration effect of the central cities of "Beijing-Tianjin-Hebei" is the largest, followed by the "Yangtze River Delta" and the "Pearl River Delta". Based on this, this paper puts forward some policy suggestions for the healthy development of the three major urban agglomerations.

Keywords: Urban agglomeration · The central city · Producer services · Industrial agglomeration

1 Introduction

As a growth pole of economic development of urban agglomeration, the central city outputs the resources and services to other city through the external service function, and makes the factor flow between cities, which is the main driving force to the development of cities [1]. Producer services, as an important economic form of the central city, is gradually replacing the manufacturing industry as the main driving force of economic growth. Industrial agglomeration has an important influence on the economic development of a region, and many scholars (Cui Yu-ming [2], Zhang Yunfei [3], Yu Binbin [4]) have done a deep research on it. The characteristics of spatial agglomeration of producer services are obvious, showing the trend of centralization and specialization in general [5–7]. "13th Five-Year Plan Proposal" requires that the urban agglomeration should play the leading role of radiation, optimize the development of "Beijing-Tianjin-Hebei", "Yangtze River Delta", "Pearl River Delta" three urban agglomeration, and strengthen regional service functions [8]. Therefore, this paper takes the productive service industry of the central cities of the three major urban agglomerations as the research object, estimates its agglomeration effect, discusses its impact on the development of urban agglomeration, and provides effective suggestions for guiding the positive and healthy development of urban agglomeration.

F. Xhafa et al. (eds.), *Advances in Intelligent Systems and Interactive Applications*, Advances in Intelligent Systems and Computing 686,
https://doi.org/10.1007/978-3-319-69096-4_93

2 Research Object and Model Establishment

2.1 Research Object

This paper selects the three major urban agglomerations as the research subject. Reference to the existing research results [9–11], this paper selects Beijing, Shanghai and Guangzhou, respectively, as the central city of the three major urban agglomerations. According to industry classification given by the National Bureau of statistics, the producer services (replaced by PSs in the below) is defined as transportation, storage and postal industry, information transmission, computer services and software industry, financial industry, leasing and business services, scientific research, technical services and geological prospecting industry five industries, hereinafter referred to as the logistics industry, information industry, the financial industry, business industry and technology industry (replaced by L, I, F, B and T in the below). The data of the relevant indicators are derived from "China Urban Statistical Yearbook (2006-2015)" and the relevant cities statistical yearbook.

2.2 Model Establishment - the Convergency Index of the Industry

This paper uses location quotient coefficient to measure the industrial agglomeration ability of the three major urban agglomerations in China. The formula is as follows:

$$LQ_{ij} = \frac{x_{ij}/x_i}{x_j/x} \tag{1}$$

Among them, x_{ij} represents the number of employees in city i of industry j, x_i represents the total number of employees in city i, x_j represents the total number of employees in the national industry j, x represents total number of employees in China. When $LQ_{ij} < 1$, it indicates that city i industry j is lack of concentration; when $1 < LQ_{ij} < 1.5$, it indicates that city i industry j is in a general agglomeration state; when $LQ_{ij} > 1.5$, it indicates that city i industry j is in a highly concentrated state.

3 The Agglomeration Effect of Producer Services in Central Cities

3.1 Analysis on the Agglomeration Ability of Producer Services in Central Cities

Location quotient (formula 1) measures the degree of specialization of an industry of a city in a country. Figure 1 is the location quotient of the producer services and the various industries in the central city of the three urban agglomerations from 2005 to 2014.

In Fig. 1, Beijing's overall agglomeration capacity of the producer services increased steadily, Shanghai and Guangzhou are both improved in volatility. Changes of the industrial agglomeration ability are different, the agglomeration ability of the various industries in Beijing is on the rise; In addition to the 2013 Shanghai's industrial agglomeration ability is basically a downward trend; Guangzhou's industrial agglomeration ability is relatively stable.

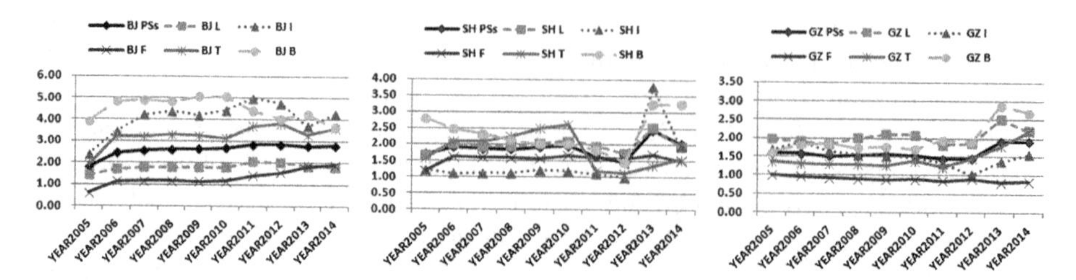

Fig. 1. The location quotient of producer services and internal industries in central city

Based on the vertical axis location quotient value, we can know that ① the producer services in Beijing have prominent regional advantages in the whole country, which have been rising continuously in the past 10 years. In addition to the financial industry, the rest of the industries are in a state of high agglomeration. ② In Shanghai, in addition to the information industry, the rest of the industries are in a state of high agglomeration, location quotient value fluctuates up and down in 2. The development of producer services in Shanghai is relatively balanced, and the location advantages are relatively concentrated. ③ The producer services in Guangzhou are more stable, and the change trend of each industry is basically the same and relatively concentrated. The agglomeration ability of logistics, business and information industry is stronger than the other two industries.

3.2 The Difference of the Agglomeration Ability of Producer Services

The results showed in Table 1 are the agglomeration ability differences. The greater the absolute value of the data in the table indicates that the difference is greater; the positive value indicates that the agglomeration ability of the surrounding city is not as good as that of the central city, negative and vice.

Seen from Table 1, In the three urban agglomerations, "Beijing, Tianjin and Hebei" producer services have the largest difference, followed by the "Yangtze River Delta" and "Pearl River Delta".

In "Beijing-Tianjin-Hebei", the difference of information, technology and business industries is obvious, the agglomeration ability of them in Beijing is higher than the surrounding cities; The differences of logistics and financial industries are relatively small, the agglomeration ability of surrounding cities are stronger than Beijing, as a whole the overall difference is smaller.

The difference of producer services in the "Yangtze River Delta" is mainly reflected in the logistics, science and technology and the business industries. There is still a gap in the agglomeration ability of each industry between the surrounding cities and

Table 1. The location quotient difference of each city and central city

	City	L	I	F	T	B	PSs	City	L	I	F	T	B	PSs
BTH	TJ	0.6	3.3	0.4	1.9	3.2	1.5	BD	1.2	3.4	0.2	1.8	4.2	1.8
	SJZ	0.3	3.3	−0.1	1.9	3.9	1.3	ZJK	0.9	3.2	0.2	2.6	4.0	1.7
	TS	0.8	3.5	0.1	3.0	3.9	1.8	CD	0.7	3.0	−0.3	2.7	3.7	1.5
	QHD	−0.8	3.3	−0.3	2.5	3.9	1.1	CZ	0.9	3.1	0.1	2.9	3.9	1.7
	HD	0.9	3.5	0.2	2.6	4.0	1.8	LF	1.2	3.1	0.2	1.9	3.9	1.7
	XT	1.2	3.2	0.0	2.7	4.2	1.9	HS	0.7	3.0	−0.4	2.8	4.2	1.6
	AVE	**0.9**	**3.2**	**0.2**	**2.4**	**3.9**	**1.6**							
YRD	NJ	0.3	−0.1	0.8	0.2	1.1	0.5	HZ	1.1	−0.4	0.5	0.4	0.7	0.6
	WX	1.3	0.6	0.6	1.1	1.6	1.1	NB	1.2	1.0	0.4	1.3	1.1	1.0
	ChZ	1.0	0.8	0.5	1.1	1.6	1.0	JX	1.6	1.1	0.8	1.4	1.3	1.3
	SZ	1.5	0.8	0.7	1.6	1.9	1.3	HuZ	1.4	0.8	0.5	1.4	1.7	1.2
	NT	1.3	0.8	0.5	1.5	1.8	1.2	SX	1.7	1.1	0.9	1.6	1.9	1.5
	YZ	1.4	0.6	0.7	1.2	1.8	1.2	ZS	0.2	0.6	0.3	1.1	0.6	0.5
	ZJ	1.1	0.9	0.3	1.1	1.7	1.0	TZ_2	1.5	0.9	0.2	1.3	1.5	1.1
	TZ_1	1.4	0.6	0.3	1.4	1.5	1.1	**AVE**	**1.2**	**0.8**	**0.5**	**1.2**	**1.5**	**1.0**
PRD	SZ	0.7	0.1	−0.2	0.5	−0.3	0.2	ZQ	1.2	0.5	0.0	1.0	1.4	0.9
	ZH	1.4	0.4	0.2	1.1	1.2	0.9	HZ	1.7	1.1	0.2	1.2	1.6	1.2
	FS	1.4	0.3	−0.5	1.0	1.4	0.8	DG	1.2	0.8	−1.5	1.1	1.5	0.5
	JM	1.5	0.9	−0.3	1.2	1.6	1.0	ZS	1.4	0.8	−0.3	1.0	1.4	0.8
	AVE	**1.3**	**0.6**	**0.4**	**1.0**	**1.3**	**0.8**							

Shanghai. Shanghai industrial development advantages are obvious, driving the development of surrounding cities.

In "Pearl River Delta", the differences of information and financial industries are relatively small. The small difference in information industry is due to the strong agglomeration ability of each city, and the convergence of Guangzhou development. The small difference in the financial industry is due to the "Pearl River Delta" urban agglomeration overall weak agglomeration ability, and lack of concentration caused.

4 Conclusion

As the growth pole of the economic development of the urban agglomeration, the central city's research on the agglomeration effect is very important to the development of the urban agglomeration.

(1) The degree of agglomeration of producer services in the central city of the "Beijing-Tianjin-Hebei" is the highest among the three major urban agglomerations, followed by the "Yangtze River Delta" and the "Pearl River Delta". The three urban agglomerations should make full use of the advantages of industrial agglomeration to make balanced development of various industries.
(2) The higher the degree of industrial agglomeration in the central city, and it is easy to make a bigger difference with the surrounding cities. We should further

strengthen the inter-city linkages, reduce the differences between cities, and achieve the coordinated development of urban agglomeration state.

(3) The development of urban agglomeration needs to pay attention to the agglomeration function of the central city. "Beijing-Tianjin-Hebei" should make full use of the prominent central city status of Beijing, focusing on the development of regional economic growth pole cities. "Yangtze River Delta" should improve the central city's radiation driven role of the surrounding cities. "Pearl River Delta" should make good use of the advantages of balanced development of cities.

References

1. Zhou, Y.: Urban Geography, pp. 171–186. The Commercial Press, Beijing (2003)
2. Cui, Y., Dai, B., Wang, P.: Research on urbanization, industrial agglomeration and TFP growth. Chin. J. Population Sci. **127**(04), 54–63 (2013)
3. Yunfei, Z.: Empirical studies on the relationship between industrial agglomeration and economic growth within the urban agglomeration—based on panel data analysis. Econ. Geogr. **34**(1), 108–113 (2014)
4. Yu, B., Yang, H., Jin, G.: Can industrial agglomeration improve regional economic efficiency? - A spatial econometric analysis based on Chinese urban data. J. Zhongnan University of Econ. Law **3**, 121–130 (2015)
5. Coffey, W.J.: The geographies of producer services. Urban Geogr. **21**(2), 170–183 (2000)
6. Harrington, J.W., Campbell, H.S.: The suburbanization of producer service employment. Growth and Change **28**(3), 335–359 (1997)
7. Searle, G.H.: Changes in produce services location, Sydney: Globalization, technology and labor. Asia Pac. Viewpoint **39**(2), 237–255 (1998)
8. The central committee of the communist party of China to develop the 13th five-year plan for national economic and social development proposal. People's Daily, 2015-11-04(001)
9. Niu, H.: A comparative analysis of the economic radiation of the central city to the surrounding economic circle – Based on the case of Beijing and Shanghai economic circle. J. Guangxi Univ. (Philos. Soc. Sci.) **31**(2), 29–34 (2009)
10. Shuqin, Y.: A study on improving the power of Guangzhou as a national central city. Urban Insight **6**, 166–177 (2014)
11. Qiu, L., Shen, Y., Ren, B.: China's service industry development and spatial structure. The Commercial Press, Beijing (2014)

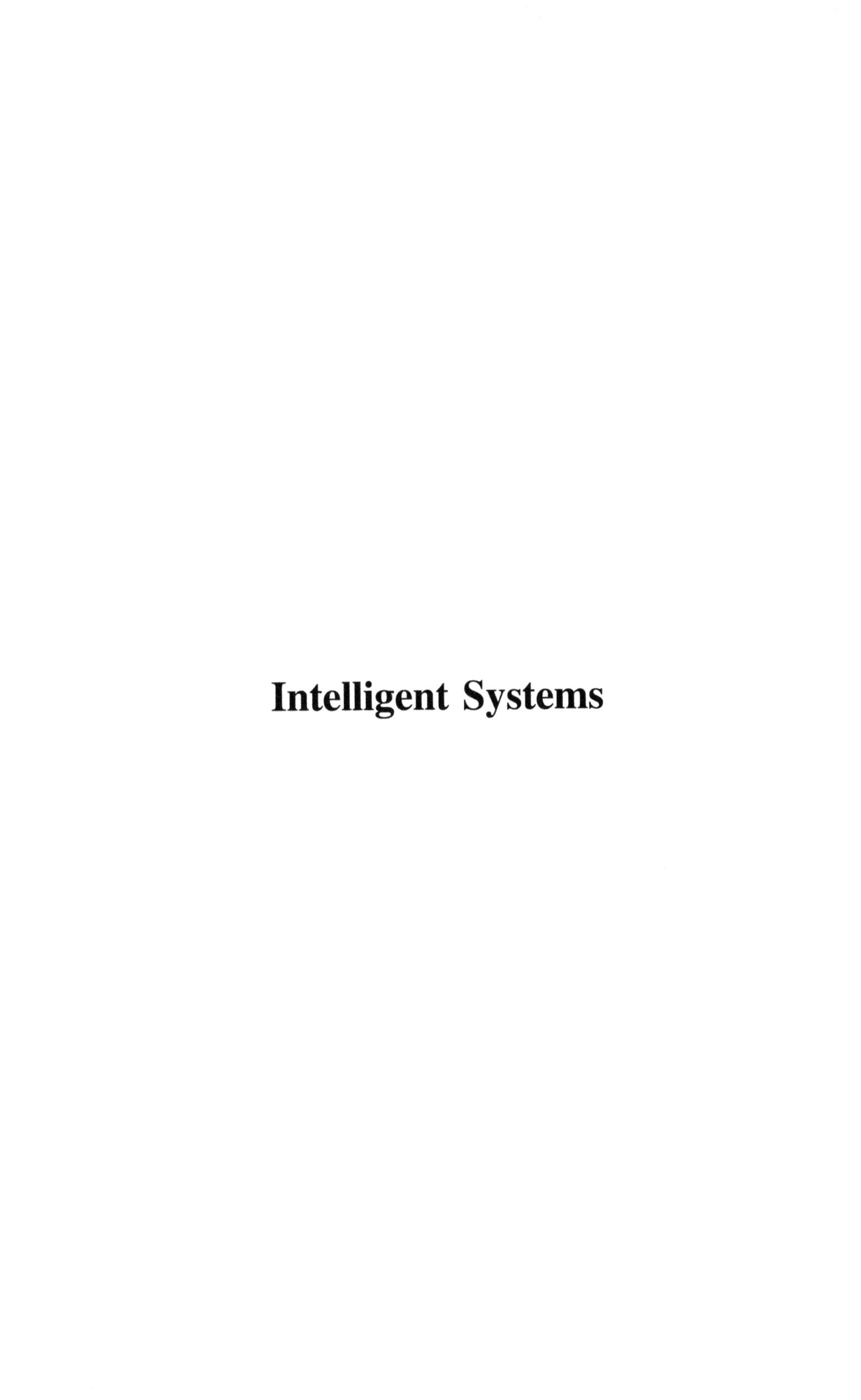

Intelligent Systems

A Dynamic Node's Trust Level Detection Scheme for Intelligent Transportation System

Ge-ning Zhang(✉)

International School, Beijing University of Posts and Telecommunications, Beijing, China
gening96@163.com

Abstract. As the basic unit of Intelligent Transportation System (ITS), Vehicular Ad Hoc Network (VANET) is responsible for transmitting traffic related messages between vehicle nodes, so it is important to ensure communication security of VANET. However, because of openness of wireless channels, self-organizing of network topology and mobility of vehicle nodes, detecting OBU's trust level in real time traffic environments is very difficult. This paper presents a dynamic node's trust level detection scheme, which is based on uncertain reasoning algorithm. Furthermore, the proposed scheme is integrated with asymmetric encryption to implement dynamic OBU's identity authentication.

Keywords: Intelligent transportation system · Trust level · Identity authentication · Uncertain reasoning algorithm · Communication security · Asymmetric encryption

1 Introduction

Intelligent Transportation System (ITS) is the development tendency of future transportation management. As the fundamental part of ITS, Vehicular Ad Hoc Network (VANET) is composed of plentiful On Board Units (OBUs), which are located in vehicles and equipped with various sensors. In ITS, VANET is responsible for providing OBU to OBU as well as OBU to Road-Side Unit (RSU) traffic related messages transmission, so it is important to ensure communication security between VANET nodes. However, finding a suitable mechanism to securely transmit a large amount of traffic messages in VANET is still an unresolved challenge, the reason of which can be summarized in following two reasons. First, licensed spectrum of VANET is at the free frequency band of 5.850–5.925 GHz, which is divided into several sub-channels. These channels are all opened to public users, the broadcast messages in channels can be easily acquired by all nodes, so that the wireless communications may unavoidably be interfered by malicious nodes. Further, rapid changes of network topology and high mobility of nodes make it difficult to perform OBU's identity recognition and trust level detection [1, 2].

F. Xhafa et al. (eds.), *Advances in Intelligent Systems and Interactive Applications*, Advances in Intelligent Systems and Computing 686,
https://doi.org/10.1007/978-3-319-69096-4_94

This paper presents an OBU's trust level detection model, which is based on uncertain reasoning algorithm. Further, the proposed model is integrated with asymmetric encryption communication to implement dynamic OBU's identity authentication.

2 OBU'S Trust Level Detection

2.1 OBU's Trust Level Detection Model

OBU's trust level detection model is based on uncertain reasoning algorithm [3], it has three aspects:

(1) Building OBU's trust level recognition frame

In ITS, major threats are attack behaviors to normal communication process by malicious nodes. Malicious node's attack behaviors can be divided into passive attacks and active attacks. Passive attacks include messages interception, communication interruption and traffic analysis. Active attacks include messages tampering, denial of service and counterfeiting communication connection. In general, the malicious nodes in VANET have typical communication traits, such as high packet loss rate, high packet sending frequency and abnormal communication traffic [4, 5]. Based on uncertain reasoning algorithm, OBU's trust level recognition rulers can be built as follows:

ruler1: if e_1 then $H\{CF(H, e_1)\}$
ruler2: if e_2 then $H\{CF(H, e_2)\}$
ruler3: if e_3 then $H\{CF(H, e_3)\}$

In above rulers, e_1 denotes the event of "OBU has high packet loss rate", e_2 denotes the event of "OBU has high packet sending frequency", e_3 denotes the event of "OBU has abnormal communication flow", H denotes the event of "OUB is a well-meaning node". The value range of $CF(H, e_i)$ $(i = 1, 2, 3)$ is between $[-1, 1]$, which denotes the supporting degree of condition e_i to conclusion H. Let $p(H)$ be the probability of H, $p(H/e_i)$ be the condition probability of e_i to H, these parameters can be obtained by analyzing OBU's previous communication behaviors. And then, $CF(H, e_i)$ can be calculated by

$$CF(H, e_i) = MB(H, e_i) - MD(H, e_i) \quad (i = 1, 2, 3) \tag{1}$$

In formula (1), we have

$$MB(H, e_i) = \frac{\max\{p(H/e_i), p(H)\} - p(H)}{1 - p(H)} \tag{2}$$

$$MD(H, e_i) = \frac{\min\{p(H/e_i), p(H)\} - p(H)}{-p(H)} \tag{3}$$

(2) Certainty factor analysis

Let $CF\ (e_i)$ be the certainty factor of event e_i, the value range of which is [− 1, +1]. $CF\ (e_i)$ can be calculated by formula (4) and formula (5).

$$\begin{aligned} CF(e_1) &= \frac{pl - pl_{ave}}{pl_{\max} - pl_{ave}} \\ CF(e_2) &= \frac{df - df_{ave}}{df_{\max} - df_{ave}} \end{aligned} \tag{4}$$

$$CF(e_3) = \begin{cases} 2 \times \dfrac{tf - tf_{ave}}{tf_{\max} - tf_{ave}} - 1 & if\,(tf \geq tf_{ave}) \\ 2 \times \dfrac{tf_{ave} - tf}{tf_{ave} - tf_{\min}} - 1 & if\,(tf < tf_{ave}) \end{cases} \tag{5}$$

In formula (4) and formula (5), pl is an OBU's packet loss rate, pl_{ave} is the OBU's average packet loss rate in ITS, pl_{max} is the maximum of OBU's packet loss rate in ITS, df is an OBU's packet sending frequency, df_{ave} is the OBU's average packet sending frequency, df_{max} is the maximum of OBU's packet sending frequency, tf is an OBU's average communication flow, tf_{ave} is the OBU's average communication flow, tf_{max} is the maximum of OBU's communication traffic, tf_{min} is the minimum of OBU's communication traffic. These parameters can be calculated through monitoring and analyzing OBU's previous communication behaviors.

(3) OBU's trust level analysis

The trust level of conclusion H under condition e_i is

$$CF_i(H) = CF(H, e_i) \times CF(e_i) \quad (i = 1, 2, 3) \tag{6}$$

At last, an OBU's integrated trust level is

$$CF_{i,j}(H) = \begin{cases} CF_i(H) + CF_j(H) - CF_i(H) \cdot CF_j(H) & [CF_i(H), CF_j(H) \geq 0] \\ CF_i(H) + CF_j(H) + CF_i(H) \cdot CF_j(H) & [CF_i(H), CF_j(H) < 0] \\ \dfrac{CF_i(H) + CF_j(H)}{1 - \min\{|CF_i(H), |CF_j(H)|\}} & [otherwise] \end{cases} \tag{7}$$
$$(i = 1;\ j = 2, 3)$$

2.2 Examples of OBU's Trust Level Analysis

Tables 1 and 2 are OBU's statistical parameters in ITS, Table 3 is the analysis results of OBU's trust level. It can be observed with high-reliability that "OBU 1 is malicious node, OBU 2 is neutral node, and OBU 3 is well-meaning node".

Table 1. The value of statistical parameters of OUBs

$p(H)$	0.70	df_{max}	80 packet/s	tf_{max}	20 M bit/s
$P(H/e_1)$	0.20	df_{ave}	12 packet/s	tf_{min}	200 K bit/s
$P(H/e_2)$	0.15	pl_{max}	0.60	tf_{ave}	8 M bit/s
$P(H/e_3)$	0.10	pl_{ave}	0.10	–	–
$CF(H, e_1)$	−0.7143	$CF(H, e_2)$	−0.7857	$CF(H,e_3)$	−0.8571

Table 2. The value of statistical parameters of OBU 1, OBU 2 and OBU 3

OBU number	*pl*	*df* (packet/s)	*tf* (bit/s)
OBU 1	0.35	20	300 K
OBU 2	0.05	10	5 M
OBU 3	0.005	3	8 M

Table 3. Trust level analysis results of OBUs

	OBU 1	OBU 2	OBU 3
$CF(e_1)$	0.5	−0.1	−0.19
$CF(e_2)$	0.1176	−0.0294	−0.1324
$CF(e_3)$	0.9744	−0.2308	−1
$CF_1(H)$	−0.3572	0.0714	0.1357
$CF_2(H)$	−0.0924	0.0231	0.1040
$CF_3(H)$	−0.8352	0.1978	0.8571
$CF\ (H)$	−0.9039	0.2723	0.8893

3 Dynamic OBU's Identity Authentication Scheme

In proposed scheme, OBU's trust level detection and identity authentication are centrally controlled by RSU and transportation management center, as shown in Fig. 1 [6]. RSU is responsible for dynamically monitoring OBU's communication behaviors, regularly calculating and noticing OBU's behavior parameters to transportation management center. At last, OBU's trust level is synthetically analyzed by transportation management center, and then, it is recorded into identity database.

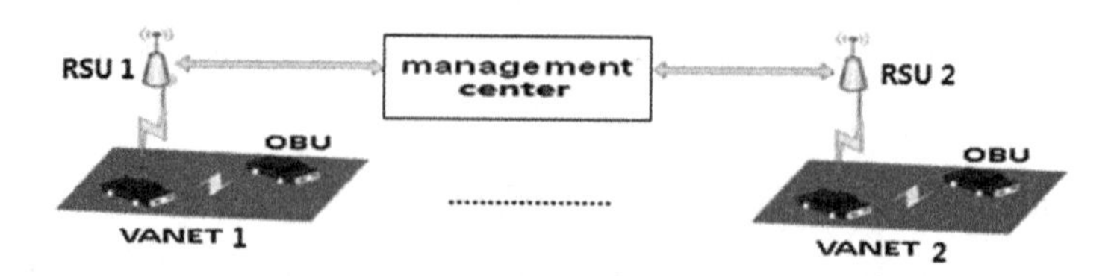

Fig. 1. The basic structure of ITS

During communications of VANET nodes, duple asymmetric encryption is implemented. Every rightful OBU and RSU is assigned a unique public key and private key by transportation management center. For secret communication, public key is used for encryption, and private key is used for decryption. For identity authentication, private key is used for encryption, and public key is used for decryption [7, 8].

Management center possess an identity database, which contains the information of public keys and trust levels of all rightful members in ITS. RSU also maintains a temporary OBU's identity database, which records the information of public keys and trust levels of current members in VANET. If an OBU doesn't belong to the member of VANET, the relevant record would be deleted from temporary identity database.

The process of packet's transmitting between RSU and OBU is shown in Figs. 2 and 3. RSU periodically broadcasts VPKP (VANET Public Key Packet), which contains public keys and trusts levels of current VANET numbers, so that each node can dynamically verifies other OBU's authentication and trust level. Meanwhile, OBU periodically sends VNDP (VANET Number Declare Packet), which is respectively encrypted by own private key and VANET's public key, so it can only be decrypted by RSU with OBU's public key and own private key. When receiving OBU's VNDP, RSU verifies OBU's identity and inquires about OBU's trust level from identity database in transportation management center, and then records it into own temporary identity database. If not receiving OBU's VNDP during expected time, the record would be removed from temporary identity database of VANET.

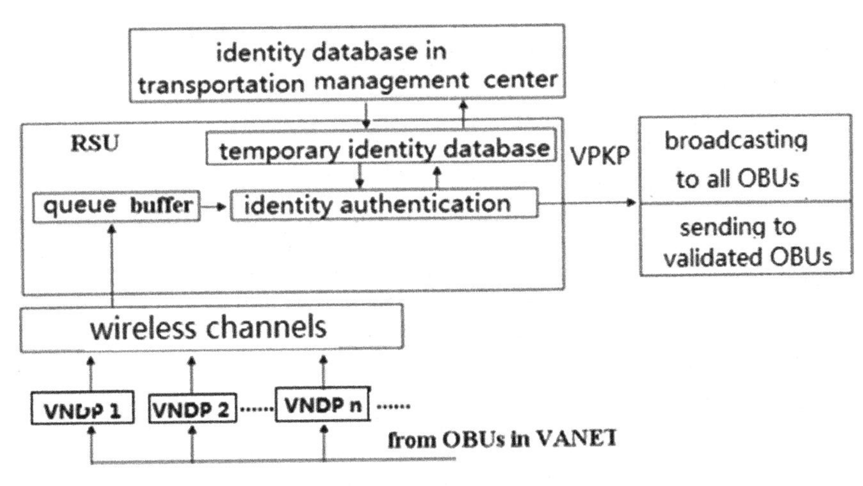

Fig. 2. The process of OBU's identity authentication and trust level detection

In addition, duple asymmetric encryption is used for packet's transmitting between VANET nodes, it has following three types:

- Broadcast packets are encrypted with node's private key. It can be decrypted by other nodes with its public key.

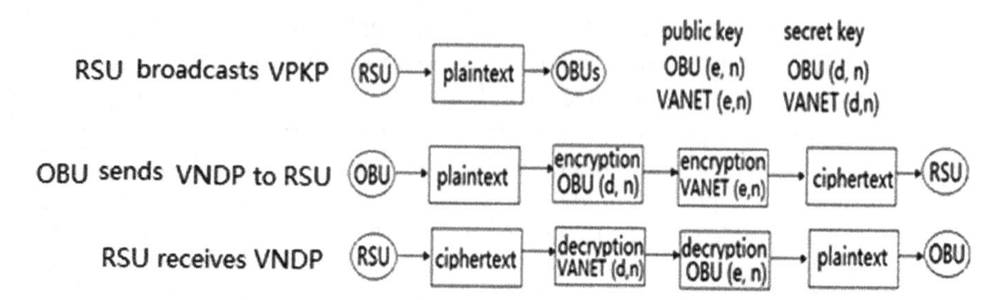

Fig. 3. The process of asymmetric encryption between RSU and OBUs

- The packets from RSU to OBU are encrypted with RSU's private key and OBU's public key. It can be decrypted by OBUs with RSU's public key and own private key.
- The packets from OBU to RSU are encrypted with OBU's private key and RSU's public key. It can be decrypted by RSU with OBU's public key and own private key.

4 Conclusions

This paper proposes a node's trust level detection model for ITS. Further, the proposed model is integrated with asymmetric encryption to perform OBU's identity authentication and security communications. As further work, we will extend research to more accurate trust level detection algorithm, complex VANET simulation model and high efficient secrecy communication system for physical layer of VANET.

References

1. Yano, A.: Featured topic: Intelligent transportation system technology has entered a new phase. SEI Tech. Rev. **78**(12), 6–7 (2014)
2. Chen, C., Lv, Z.Y., Fu, S.S., Peng, Q.: Overview of the development in cooperative vehicle-infrastructure system home and abroad. Transp. Inf. Secur. **29**(1), 102–105 (2011)
3. Wang, Y.Q.: Principles and Methods of Artificial Intelligence. Xian Jiaotong University Press, Xian (2001)
4. Luo, J.H.: Research on dynamic trust model for MANETs, Doctoral Dissertation of University of Electronic Science and Technology of China (2009)
5. Kong, S.J.: Research on authentication architecture and trust model in VANET, Master Dissertation of Shanghai Jiao Tong University of China (2010)
6. Jing, C., Xu, H.G., Wei, X., He, S.: Preliminary implementation of road-car network based on IEEE 802.11p/1609. Appl. Res. Comput. **28**(11), 4219–4221 (2011)
7. Cao, X.H., Zhang, Z.C.: Information Theory and Coding. Tsinghua University Press, Beijing (2012)
8. Stinson, D.R.: Cryptography: Theory and Practice, Publishing House of Electronics Industry, Beijing (2009)

An Application of Multi-Objective Genetic Algorithm Based on Crossover Limited

Shi Lianshuan(✉) and Chen YinMei

School of Information Technology Engineering, Tianjin University of Technology and Education, Tianjin, China
shilianshuan@sina.com

Abstract. Aimed at the problem of slow convergence of genetic algorithms, an improved genetic algorithm is given. The improved algorithm is based on the traditional NSGA, and uses crossover limited and elitist strategy to solve multiple objective network optimization. By these operations, the algorithm can effectively improve the speed of convergence. At the same times, the algorithm also uses objective layer method to select the non-dominated individuals. The current algorithm is used to solve the network optimum design problem with multiple objectives. Two objectives, the cost of path and delay of path are considered. A few examples for the network optimization generated by randomly are used test the algorithm. The result shows the algorithm can find better Pareto solutions.

Keywords: Multi-objective network optimization · Pareto optimal solution · Non-Dominated sorting · Crossover limited

1 Introduction

In many fields, such as engineering and management fields, a lot of design problems are multiple objectives optimum design problems. The solutions of these problems are different from single objective optimum design problem. Its solutions are a set of solutions, not one single optimal solution. Traditionally, a multi-objective optimum design problem firstly is transformed into a scalar single objective optimization design problem by a group of given numbers in advance. Then, the algorithm for solving the single objective optimization is used to solve MOP. These algorithms are based on a preference-based method. A preference vector is used to transform multiple objectives to a scalar objective [1]. Because a few heuristic algorithms are based on population-based search approach, it can obtain multiple pareto optimal solutions in one single run. They are naturally suited for solving a multi-objective optimum design problem [2, 3].

Genetic algorithm is a meta-heuristic algorithm that based on the evolutionary theory with the basic idea of survival of the fittest. In order to achieve the satisfied solution or optimal solution, it simulates the biological genetic behavior of genetic variation and survival competition in the nature and makes the solution of the problem improved in the competition. Since it can be achieved the satisfied solution or optimal solution during the multi-objective network optimization problem (MONP), so it is a better choice to use genetic algorithm to solve a MONP. Nowadays, more and more

F. Xhafa et al. (eds.), *Advances in Intelligent Systems and Interactive Applications*, Advances in Intelligent Systems and Computing 686,
https://doi.org/10.1007/978-3-319-69096-4_95

researchers have much enthusiasm to do a deep study on genetic algorithm because of its advantages such as having nothing to do with the problem domain, quickly random search capability, process simple, easy to combine with other algorithms, potential parallelism and a wide range of application.

Current work is introduced as follows. In Sect. 2, the basic concepts of the Pareto optimal solution and the mathematical model of MOP are introduced. In Sect. 3, the proposed approach is introduced. In Sect. 4, The network optimization design problem case with two objectives is given. Finally, the conclusion is given in Sect. 5.

2 The Mathematical Model of Multi-Objective Optimization

A multi-objective optimum design problem can be described as bellow: n decision variables, m objective functions, k equality constraints, l inequality constraints. The decision variable functions include the objective functions and the constraint functions. The mathematical model can be described as follow:

$$\text{Minmize}\mathbf{Y} = f(\mathbf{X}) = (f_1(\mathbf{X}), f_2(\mathbf{X}), \ldots, f_m(\mathbf{X}))^{\mathrm{T}}$$

$$\text{subject to } g(\mathbf{X}) = (g_1(\mathbf{X}), g_2(\mathbf{X}), \ldots, g_k(\mathbf{X}))^T \geq \mathbf{0}$$

$$h(\mathbf{X}) = (h_1(\mathbf{X}), h_2(\mathbf{X}), \ldots, h_l(\mathbf{X}))^T = \mathbf{0}$$

$$\mathbf{X} = (x_1, x_2, \ldots, x_n)^T \in \Omega, \quad \mathbf{Y} = (y_1, y_2, \ldots, y_m)^T \in \mathrm{Z}$$

Where, $\mathbf{X} = (x_1, x_2, \ldots, x_n)^T$ represents the decision variable vector, $\mathbf{Y} = (y_1, y_2, \ldots, y_m)^T$ represents the objective function vector, n and m are the number of the variables and the number of the objective functions, Ω represents the decision variable space, Z represents the objective function space, the constraints $g(\mathbf{X}) \geq 0$, $h(\mathbf{X}) = 0$ defines the feasible region S, $\mathrm{S} = \{\mathbf{X} | g(\mathbf{X}) \geq 0,\ h(\mathbf{X}) = \mathbf{0}\}$, l and k are the number equality constrains and the number of inequality constraints.

Definition 1 Let $X^* \in \mathrm{S}$ be a feasible solution, if S for all $X \in \mathrm{S}$, $f_k(\mathbf{X}^*) \leq f_k(\mathbf{X})$, $\mathrm{k} = 1, 2, \ldots, m$, then solution $X^*$is called a optimal solution of the multi-objective optimization problem.

In lot of cases, these objectives are always conflict with each other. The improvement of a objective often lead to becoming bad of another objective. In the most of cases, the optimal solution in which all objectives obtain optimum is almost not existed, or it is not necessary to find the optimum objective value for each objective. In many cases, people consider how to find the pareto optimal solutions of the MOP. In past years, a lot of methods to solve the multi-objective optimum design problems have been proposed.

Definition 2 Let $X_1, X_2 \in$ S are feasible solutions, if $f_j(X_1) \leq f_j(X_2)$ for $j = 1, 2, \ldots, m.$, and there exists a k, such that $f_k(\mathbf{X}_1) \leq f_k(\mathbf{X}_2)$, then it is said $\mathbf{X}_1$ dominates $\mathbf{X}_2$. If a feasible solution $\mathbf{X}'$ does not be dominated by any $\mathbf{X}(\mathbf{X} \in$ S), then $\mathbf{X}'$ is said a Pareto optimal solution of MOP.

The MOP solution is different the single objective optimal problem, its solution is a set of non-dominated solutions, that form the Pareto Curve, also known as Pareto Optimal. Srinivas and Deb [4] developed the non-dominated Sorting Genetic Algorithms (NSGA). The non-dominated chromosome is removed one by one at every layer from the population. Before removing these individuals, the rank value of every individual is assigned in each layer. The individual in the first layer has the best fitness. In this contribution, the Pareto solutions are considered for MOP.

There, for the MONP, two objectives are considered, one is communication cost, another is transmission delay. It can be described as a problem: Suppose that a network has n nodes, E arcs, find a path from the source point s to the end point t, minimize the length of the path, as well as the delay. A multi-objective network optimization problem is a typical combination optimization design problem, and is a NP hard problem or NP-complete [5].

3 Algorithm Description

In the process of selection, this contribution adopts the algorithm of filtering and crossover-limited. In the process of crossover, select randomly an individual as the first parent, then the second parent individual is selected based on the first individual fitness value. If the fitness value of the second individual is more than the first parent, then allow crossover between the selected parents. If no individual which fitness is more than the first parent being selected, the first parent is directly put into next group.

3.1 Chromosome Decoding Method

Due to a network optimization is to find optimal paths, and a path is composed of multiple nodes. If each node is represented in binary, the length of each chromosome will be very long. It is not conducive to the genetic operation, and also it will bring the problem of storage. For the convenience of operation, the algorithm use node sequence encoding. Because the length of a path from the source node to the end node are different, so we use a node sequence encoding with dynamic length.

3.2 Generate the Initial Population

Assuming that the nodes number of network is m, size of population is n. Firstly, set node number for all nodes of network with natural number 1, 2, …, m.

The maximum length of individuals from the source node to the end node t is m, in this case, the path pass all nodes of network. The minimum length is 2, that is, directly from the source node s to the end node t, no the middle points. To generate an individual, firstly, the starting node s is taken as first gene value of individual. For other genes, use the random function to generate. Firstly, a random integer $p(p \neq$ s and

p belongs [1, m]) is generated, if k = t, then the end node have been obtained, an individual have been generated; Otherwise, continue generating.

Suppose that the size of population is N. Firstly, 2N individuals are generated randomly, then N individuals are selected by as initial population to participate in the genetic operation.

3.3 Choose the Non-Inferior Solution

In order to sort the population with N individuals by the non-dominated levels, each individual is compared with other individuals of the population to check whether this individual is dominated by other individuals. That need more computational times. Here used the objective layered algorithm [6] to produce the first layer individuals (non-inferior solutions). Due to the following to the fitness of the first layer of the solution, so the first layer of pareto solutions first transferred to the population of the back, just in front of the first individual fitness calculation.

3.4 Crowding-Distance Calculation for the Non-Inferior Solutions

Firstly, the chromosomes in the population are ordered in descending order according to the k-th objective values. In order to preserve the region end points, set the crowding-distance value of these individuals with the maximum or the minimum objective values to maximum crowding-distance value (set a large number). For other individuals in the population, the crowding distances can be calculated as follow method 1. Calculate the deference between the (i−1)-th and the (i + 1)-th objective function values; 2. Each individual's crowding-distance value equal to the sum of the normalized differences of its each dimensional objective value as follow [7]:

$$dist\_crow(x_i) = \sum_{j=1}^{m} \frac{f_j(x_{i+1}) - f_j(x_{i-1})}{f_{j,\max} - f_{j,\min}},$$

Where, the f_j is the j-th objective function value.

In the population updating operation, the individuals with larger crowding-distance value will have priority to be selected. This method is simple, and has a good effect on the 2-dimensional problems. When the crowding distance of all the individuals have been calculation, these individuals are sorted in ascending order according to the crowd-distance of individuals. After that, m individuals can be selected according to the crowding distance from larger value to smaller value.

3.5 The Fitness Calculation of the Remaining Individuals

The method for calculating the fitness of inferior solution by TongJing [8] is used. For any other individual (inferior solution) x, firstly calculate the Euclidean distances between the individual x and non-inferior individuals. Then calculate minimum value of the Euclidean distances. For the individual x, the fitness is defined as:

$$F(x) = 1 / (1 + \|x - y(x)\|2)$$

Where, y (x) is the inferior individual in inferior individuals of population with the minimum Euclidean distance between x, $\|X - Y(X)\|2$ represents the Euclidean distance between x and y (x).

3.6 The Crossover Restrictions Mechanism

Its advantage is that can accelerate the convergence of the algorithm. firstly, randomly select an individual as the first parent of crossover, then second parent for crossover can be selected based on their fitness. Only when the fitness of an individual is greater than the first parent, this individual can be selected as second parent to cross. If no individual with bigger fitness than the first parent, then the first parent is directly put into the next population.

4 Experiments and Results

In order to test the algorithm, we generate randomly a network with 8 nodes. The adjacency matrix, the length matrix of path and the delay matrix, are given in Figs. 1 and 2.

$$\begin{bmatrix} 0 & 6 & 3 & 5 & 15 & 20 & 50 & 90 \\ 6 & 0 & 8 & 17 & 20 & 30 & 12 & 16 \\ 3 & 8 & 0 & 50 & 70 & 8 & 2 & 35 \\ 5 & 17 & 50 & 0 & 60 & 30 & 24 & 10 \\ 15 & 20 & 70 & 60 & 0 & 20 & 10 & 5 \\ 20 & 30 & 8 & 30 & 20 & 0 & 3 & 50 \\ 50 & 12 & 2 & 24 & 10 & 3 & 0 & 20 \\ 90 & 16 & 35 & 10 & 5 & 50 & 20 & 0 \end{bmatrix}$$

Fig. 1. The length matrix

Genetic algorithm parameter settings: the node 1 is the source point, the node 8 is the end point, the number of objective functions m is 2, the population size N is 50, the maximum number of non-inferior solutions is m = 20, the mutation probability p_u is 0.1, the termination conditions is the maximum generation number maxgen = 100. When the algorithm is performed 100 times, the probability of obtaining two optimal paths is 95%. Two optimum paths are as follow, 1-2-8, the objective value is (22,4), 1-4-8, the objective values is (15,6).

$$\begin{bmatrix} 0 & 3 & 5 & 4 & 8 & 10 & 15 & 13 \\ 3 & 0 & 11 & 6 & 7 & 5 & 12 & 1 \\ 5 & 11 & 0 & 4 & 2 & 8 & 30 & 27 \\ 4 & 6 & 4 & 0 & 20 & 17 & 11 & 2 \\ 8 & 7 & 2 & 20 & 0 & 13 & 16 & 3 \\ 10 & 5 & 8 & 17 & 13 & 0 & 7 & 15 \\ 15 & 12 & 30 & 11 & 16 & 7 & 0 & 8 \\ 13 & 1 & 27 & 2 & 3 & 15 & 8 & 0 \end{bmatrix}$$

Fig. 2. The delay matrix

5 Conclusions

When genetic algorithm is used to optimization problem, that it has slow convergence speed, an improved algorithm is proposed. Based on non-dominated Sorting Genetic Algorithm with Elite strategy, a proposed algorithm with strategy of crossover limit is given. Through the use of the restriction mechanism in the process of crossover, can effectively improve the convergence speed of the evolution. The improved genetic algorithm is used to solve the MONP, the experimental results show that the improved algorithm can quickly find the optimal solution set.

Acknowledgements. This work was supported by Tianjin Research Program of Application Foundation and Advanced Technology (14JCYBJC15400).

References

1. Lobato, F.S., Steffen, J.V.: Multi-objective optimization firefly algorithm applied to (bio) chemical engineering system design. Am. J. Appl. Math. Stat. **1**(6), 10–116 (2013)
2. Omkar, S.N., Khandelwal, R., Yathindra, S., Naik, N.G., Gopalakrishnan, S.: Artificial immune system for multi-objective design optimization of composite structures. Eng. Appl. Artif. Intell. **2**(21), 1416–1429 (2008)
3. Wong, E.Y.C., Yeung, H.S.C., Lau, H.Y.K.: Immunity-based hybrid evolutionary algorithm for multi-objective optimization in global container repositioning. Eng. Appl. Artif. Intell. **22** (2), 842–854 (2009)
4. Srinivas, N., Deb, K.: Multi-objective optimization using non-dominated sorting genetic algorithms. Evol. Comput. **2**(3), 221–248 (1994)
5. Huang, R., Ma, J., Hsu, D.F.: A genetic algorithm for optimal 3-connected telecommunication network designs. In: Proceedings of the 1997 International Symposium on Parallel Architectures, Algorithms and Networks (ISPAN '97) pp. 344–350

6. Shi, L.S., Chen, Y.M.: A layered approach based on objectives to multi-objective optimization. Int. J. Intell. Eng. Syst. **5**(1), 11–19 (2012)
7. Deb, K., Agrawal, S., Pratap, A.: A fast and elitist multi-objective genetic algorithm: NSGA-II. IEEE Trans. Evol. Comput. **6**(2), 182–197 (2002)
8. Tong, J., Zhao, M.W.: A multi-objective evolutionary algorithm for efficiently solving pareto optimal front. Comput. Simul. **26**(6), 216–218 (2009)

A Method for Extracting Objects in Physics Using Unit Semantic Model

Yanli Wang[1(✉)] and Pengpeng Jian[2]

[1] Henan University of Economics and Law, Zhengzhou, China
dayanmei@huel.edu.cn
[2] Central China Normal University, Wuhan, China

Abstract. Physics problem understanding in natural language (NL) is always a challenge for machine. To address this challenge, this paper proposes a general approach to extract physics objects of middle school physics problems in Chinese. According to the appearing of object name and unit in a given physics problem, this approach use unit semantic model to extract physics objects. In the process of physics object extraction, an unit semantic model described by the components of semantics, unit and output is proposed to identify physics objects. The effectiveness of our proposed approach was examined in a dataset collected from the textbook. Experimental results demonstrated that the proposed method is very effective in identifying physics objects.

Keywords: Objects · Extraction · Unit semantic model

1 Introduction

Extracting physics objects is a key step of understanding physics problems, which is a challenge for machine in natural language processing (NLP). Problem understanding is to extract information from the content of problem text, diagrams and formulas, which can be used to find the solution. The method of problem understanding is composed of automatic and non-automatic understanding by machine. In non-automatic area, [1] and [2] achieved some applications by using Lisp language to manually input the known physics conditions, such as physics objects, physics relations between objects and so on. [1] designed and developed a physics computer aided instruction (CAI) system, whose main work was automatic computer reasoning of physics knowledge and producing of traditional style solution based on the known conditions.

The existing algorithms on automatic problem understanding can be divided into three categories: formal action, formal language and formal equation. The main idea of formal action is that regarding the process of reducing from known conditions to unknown conditions as the process of problem solving, and it's easy to extract the known conditions in a problem, but it's hard to confirm the derivative action of solutions. So, the formal action algorithm focuses on the process of derivative action, which is required in the problem solving. [3] proposed an algorithm of

F. Xhafa et al. (eds.), *Advances in Intelligent Systems and Interactive Applications*, Advances in Intelligent Systems and Computing 686,
https://doi.org/10.1007/978-3-319-69096-4_96

automatically solving arithmetic word problems, whose goal of problem understanding was to form knowledge frames and solution frames. In one hand, the algorithm used the method of sentence template matching to form the knowledge frames and filled with elements in the corresponding knowledge frames. In another hand, the algorithm used the method of problems classifier to select the solution frames for a given problem. Because of the knowledge frame designed by Kintsch can only express one-step arithmetic word problems, so [4] extended Kintsch's method by adding sentence templates and increasing the elements of knowledge frames that multi-step arithmetic word problems can be solved. The insufficient of formal action algorithm is failed to give a complete classification system of problems.

The main idea belongs to formal language is that using formalized language which is more simple than NL to represent a given problem, and it's easy to extract solution information from formalized language. So, there are three steps for the algorithm of problems understanding. First of all, defining a formal language model. Second, setting up a method to transform NL into formal language. Third, creating a method to extract solution information in the formal language. [5] developed a system of Dolphin to automatically solving the semantic analysis and inference in arithmetic word problems. The DOL method of problem understanding is to set up DOL language with structured semantic to represent problem text, and using a semantic analyzer to realize transformation from math problem text to DOL tree. Because of the complexity of NL expression, formal language method has not yet to create a algorithm of formalized language for all the math problems.

The main idea in formal equation is to use a set of equations to denote word problems, which is the embodiment form of formal equation set. So, the assignment of this problem understanding algorithm is to select formal equations and confirm the corresponding relations between the formal equation elements and the entities of problem text. [6] proposed an algorithm for arithmetic word problems understanding based on machine learning. Assuming formal equation is linear, the algorithm can extract a set of linear equations and record the correspondences between variables and formal equation by using machine learning algorithm. [7] improved the problem understanding method of [6] that they designed the effective feature templates to describe the relations between knowns and unknowns, and in this way the process of complex derivation will be reduced by filling template slots. [8] developed another form of formal equation algorithm, whose method was to analyze the text and transform both body and question parts into tag-based logic forms and then performing inference on them.

The three categories of problem understanding are all lack of the ability to extract implicit relations. They are all almost using the way of semantic analysis, which has a variety of semantic expression, so that the analysis algorithms are very complex.

The remainder of this paper is organized as follows. Section 2 describes the unit semantic models for physics objects extraction. Section 3 presents the implement of extracting physics objects. In Sect. 4, experiments have been done to evaluate the proposed method. Finally, we conclude this paper in Sect. 5.

2 Unit Semantic Model

This section presents the unit semantic models for extracting physics objects. An unit semantic model consists of three parts, which are semantics portion, unit portion and output relation portion. In an unit semantic model, the semantics portion is the structure of keyword, the unit portion is the unit of keyword and the output relation is a relation set of physics objects among the involved problems.

Definition 1: An unit semantic model for physics problem is defined as a triple $M = (K, U, O)$, where K represents keyword element, U is the unit of keyword K, and O is the object relation. Let $\Sigma = \{M_i = (K_i, U_i, O_i)|i = 1, 2, \ldots, m\}$ denote all the prepared models, where K is included by a set of object names, such as the object names of length unit can be Distance (represented by symbol s), Gourney (represented by symbol s) and Height (represented by symbol h); U is included by a set of units themselves, for example the units of length can be Kilometer (represented by symbol km), Meter (represented by symbol m), Centimeter (represented by symbol cm) and so on. So M is a set of models which is the combination between set K and set U. It is also called as a pool of unit semantic models for physics problem.

With the pool of the prepared unit semantic models, a procedure (**Procedure I**) is proposed to identify physics object relations. The main point of this procedure is how to use the unit semantic model to match with object name and unit together in the problem text.

3 Physics Objects Extraction

This section presents the algorithm to extract a set of object relations of a given physics problem, which can be used to record the basic units of problems. This algorithm is built on the method of extracting object relations by using unit-semantics models. The algorithm including the combination of unit semantic models, a model matching method and an instantial method, is used to transform physics objects into equations and creating a corresponding triple to record the detail properties of objects.

The goal of annotating the problem text of a physics problem is to transform the problem text into a new form by doing text annotation. During the physics problem text annotation, ICTCLAS [9] is used to parse the text into phrases and to annotate these phrases with POS (part-of-speech) labels.

3.1 Physics Object Representation

Definition 4 (Physics object): A physics object is an identifiable collection of matter that expresses a physics concept. The physics object which is extracted by our approach has three properties, including object name, object value and object unit.

There are three categories of physics object, whose names are general object (only contains name), normal object (contains name, value, unit) and common sense object.

The general object is a basic unit, which indicates the object of physics entity such as the resistance R, the voltage U and so on. The normal object includes object name property, object value property and object unit property, which will be converted into equations equivalently in the next process. The common sense object is the object which is defined and accepted by some authorities in physics, such as the speed of light in a vacuum is 2.99 * 108 m/s. These three categories of physics objects are mainly indicated by unit semantic models.

Definition 5 (Physics object representation): A physics property representation is a triple $o = (s, v, w)$ in which s is the physics object symbol, v is the value of physics object and w is the unit of s.

From the expression of unit semantic model $M = (K, U, O)$, the physics object relation O can be extracted by **Procedure I** and the detail properties of O will be recorded by the triple $o = (s, v, w)$. With the storage content of O and o, the physics object will be parsed out clearly.

3.2 Unit Semantic Model Matching

Procedure I: Extraction of Physics Objects
The input is a physics problem in NL with POS annotation, denoted as Θ.
The output is a set of physics objects, denoted as Δ.
Load $\Sigma = \{M_i \mid i = 1, 2, \ldots, m\}$;
Initialize Θ as $\Theta(0)$, Δ as empty;
While TRUE
  For i from 1 to m
  If matching K_i of M_i in order with $\Theta(i - 1)$ is FALSE then Continue;
    If matching U_i of M_i
      Put O_i (normal object) of M_i into Δ;
    Else If matching K_i' (common sense object name) of M_i
      Put O_i' (common sense object) of M_i into Δ;
    Else
  Put O_i'' (general object) of M_i into Δ;
  Mark the matched portion of $\Theta(i - 1)$ and denote it as $\Theta(i)$. Break For loop; If the whole $\Theta(i)$ is marked break WHILE loop;

3.3 Implementation of Physics Objects Extraction

Extracting physics objects is a basic and important work for physics problems understanding, and it has three separate steps. First of all, the problem text should be uniformed by parsing into phrase and annotating with POS. Second, the list of object names and object units that may appear in the physics problem should be formed in model preparation. Third, physics objects will be extracted by using Procedure I and an object table will also be created to record the correspondences between properties and variables. Then all the properties of the extracted objects are replaced by the assigned variables. As a result, a set of linear equations is formed. The pseudo-code of the algorithm for extracting physics objects in physics problem is formed below (Algorithm I) based on the above discussion.

Algorithm I: Physics Objects Extraction in physics problems

Input: a physics problem.

Output: a system of equations ζ and the corresponding table between properties and variables Γ.

Step 1:(Parsing and Annotation) Uniform the problem text; parse the problem text into phrases and annotate the phrases with POS;

Step 2:(Model preparation) Prepare object name set and object unit set for models;

Step 3:(Unit semantic Model Matching) Extract all the physics objects of the given problem using **Procedure I** and form a set of the physics objects Δ;

Step 4:(Equation Instantiation) Forming the corresponding table between properties and variables Γ, instantiate the physics objects in Δ and form a set of linear equations ζ.

4 Experiments

4.1 Experimental Setup

Datasets: The dataset contains 63 physics application problems in Chinese, which is collected from the textbook of middle school published by the people's education press in 2013. At the same time, other basic datasets are prepared for the experiment, including units dataset, objects dataset and common sense dataset. Table 1 gives the statistics of these datasets.

Table 1. Statistics on the datasets of experiment

	#Ps	#Us	#Os	#Cs
Total	63	52	32	138

Note *#Ps* Problems, *#Us* Units, *#Os* Objects, *#Cs* Common sense.

4.2 Experimental Results

Evaluation metrics: Evaluation is performed in the setting that a system can choose NOT to achieve all experimental results in the test set. For evaluating the proposed algorithm, we follow the setting that three metrics are adopted in reporting evaluation results. Assuming that n is the size of test set, m is a method generates answers of n problems a. The three metrics for evaluation are listed: (1) Precision = k/m, (2) Recall = k/n, (3) F1 (measure) = 2k/(m + n).

All the physics objects extraction by Algorithm I are all feasible. As shown in Table 2, our method records the results of general objects, normal objects and common sense objects in terms of precision, recall and F1 score.

Table 2. physics objects extraction results

	Method	Precision (%)	Recall (%)	F1 (%)
#GOs	USM	97.2	61.2	75.1
#NOs	USM	92.1	56.5	69.4
#COs	USM	88.1	55.4	67.9

Note #GOs General objects, *#NOs* Normal objects, *#COs* Common sense objects, *USM* Unit semantic model

5 Conclusions and Future Work

This paper presents a new method to extracting physics objects in junior middle school. Following this method, the unit semantic model is used to extract physics objects. And the method gets good performance according to the high precision and a reasonable recall in the experiment.

In the future, we want to extend the research in multiple directions of physics problems solving. First, the algorithm to extract associated relations between physics objects is worth to study. Second, the method of physics diagram recognition in physics problem is another research point. Third, achieving fully automated physics problems solving is one of the ultimate goals.

References

1. Li, CG., Wang, C., Xiaojing., Wei, H.: A physics ICAI system with traditional style solution. Comput. Appl. pp. 16–19 (2001. 02)
2. Shan, F., Shan, G.: Research on automatic reasoning system of physics intelligent education platform of junior high school. Graduate student degree thesis (2014)
3. Kintsch, W., Greeno, J.G.: Understanding and solving word arithmetic problems. Psychol. Rev. **92**(1), 109–129 (1985)
4. Ma, Y.: The study of automatically solving math word problems based on the cognitive model. Central Compilation and Transl. Press (2012)

5. Shi, S., Wang, Y., Lin, YC., Liu, X., Rui, Y.: Automatically solving number word problems by semantic parsing and reasoning. In Conference on Empirical Methods in Natural Language Processing (2015)
6. Kushman, N., Zettlemoyer, L., Barzilay, R., Artzi, Y.: Learning to automatically solve algebra word problems. In Meeting of the Association for Computational Linguistics, pp. 271–281 (2014)
7. Zhou, L., Dai, S., Chen, L.: Learn to solve algebra word problems using quadratic programming. In Conference on Empirical Methods in Natural Language Processing, pp. 817–822 (2015)
8. Liang, CC., Hsu, KY., Huang, CT., Li, CM., Miao, SY., Su, KY.: A tag-based english math word problem solver with understanding, reasoning and explanation. In Conference of the North American Chapter of the Association for Computational Linguistics: Demonstrations pp. 67–71 (2016)
9. Zhang, HP., Liu, Q.: ICTCLAS. Institute of Computing Technology, Chinese Academy of Sciences, (2002) http://www.ict.ac.cn/freeware/003ictclas.as.asp

A Novel Matching Technique for Two-Sided Paper Fragments Reassembly

Yi Wei[1(✉)], Lumeng Cao[1], Wen Yu[1], and Hao Wu[2]

[1] School of Automation, Wuhan University of Technology, 122 Luoshi Road, Wuhan, People's Republic of China
weiyi@whut.edu.cn

[2] School of Resources and Environmental Engineering, Wuhan University of Technology, 122 Luoshi Road, Wuhan, People's Republic of China

Abstract. Paper fragments reassembly has been playing an important role in many places such as public security and even archaeology. Combined with the travelling salesman problem, a novel approach based on the matching of greyscale difference matrix is adopted. Experimental results demonstrate its potential in speed, accuracy and less human intervention for double-sided paper fragments reassembly. The study may provide a new direction for the automatic stitching or image mosaic technique.

Keywords: Greyscale difference matrix · Traveling salesman problem · Paper fragments reassembly · Automatic stitching

1 Introduction

Shredded paper, especially two-sided shredded paper recovery has been a difficult task yet wildly used in many aspects, such as historical documents restoration, judicial evidence recovery and so on. When the number of text fragments increases to hundreds of thousands of pieces, an automatic stitching or image mosaic technique with fast speed, high accuracy and less human intervention becomes a big challenge.

In this paper, studies have been carried out on the reassembly of two-sided shredded paper cut by a shredder. Firstly, each paper fragment is represented by its greyscale matrix, from which two columns on its left and right sides are extracted respectively to form a new $n \times 2$ matrix. Due to the high similarity of gray value along the broken border of the originally connected part, any two fragments which have the maximum matching score are considered previously connected, and the sorting order of a group of fragments which has the smallest degree of difference is the corresponding stitching result.

The Traveling Salesman Problem (TSP), aims to choose the shortest route to visit N different cities once and only once. To solve this problem, the objective function, which is the sum of matrix rows and columns of the different elements, should reach the minimum. By substituting the distance matrix with the greyscale difference matrix, TSP is introduced and thus the calculation of the minimum value of greyscale difference matrix becomes the searching of the shortest path.

F. Xhafa et al. (eds.), *Advances in Intelligent Systems and Interactive Applications*, Advances in Intelligent Systems and Computing 686,
https://doi.org/10.1007/978-3-319-69096-4_97

2 Mathematical Modeling

If both transverse and longitudinal cut are considered, it comes huge quantities of fragments and makes the following study too complicated. Therefore the cutting of paper is conducted only along the longitudinal direction in the paper. Besides, assumptions are also made to simplify the task and they include: plain text only without overlapping, characters with same size, spacing and style, fragments under recovering with rectangular shape, from the same piece of paper and smooth cut.

Given the above conditions, edges of each fragment are extracted and built an n × 2 matrix. Two matrices are then matched. A total difference by summing all possible combinations of fragments is obtained, and the preliminary result is the one with the smallest matching price. It is worthwhile to mention that, human intervention must be taken to rule out the irrational splicing.

Take any scrap of paper and built its grayscale matrix. Two columns of the left and right edges of the matrix are extracted to form an n × 2 matrix.

$$A_i = \begin{bmatrix} a_{i1}(t) & a_{i2}(t) \\ a_{i3}(t) & a_{i4}(t) \end{bmatrix}, \; i = 1, 2\ldots, n \tag{1}$$

Where $ai1$ (t) and $ai3$ (t) are the grayscale values of pixels on the left edge of the fragment, $ai2$ (t) and $ai4$ (t) are the gray values of the right edge. n is the total number of edge pixels, and the number of fragments is z + 1.

Let k represents the degree of difference, the matching rules are that the first column of one matrix match the second column of the other matrix, $ai1$ (t) matches $aj1$ (t), where i and j are the number of fragments, and t is the number of the edge pixel.

$$\begin{cases} k_{ij}(t) = 0; & if\; a_{i1}(t) = a_{j1}(t) \\ k_{ij}(t) = 1; & Otherwise \end{cases} \tag{2}$$

The difference matrix can be defined as:

$$Q = \begin{bmatrix} q_{00} & \cdots & q_{pz} \\ \vdots & \ddots & \vdots \\ q_{z0} & \cdots & q_{zz} \end{bmatrix} \tag{3}$$

Where

$$qij = \sum_{i=1}^{n} kij(t), t = 1, 2, \cdots \cdots, n \tag{4}$$

is the matrix element. The minimum degree of difference is written as

$$k \min = \min \sum_{i=0}^{z} qij \tag{5}$$

Adding the following constraints:

(1) $i \neq j$, means that the left and right edges of a fragment cannot match each other.
(2) The cumulative elements of matrix cannot be the same row or column.

3 Algorithm Design and Experimental Results

Step 1: Read the scraps of paper image and extract gray values of each fragment on their edges. A 1980×2 matrix A_i of each fragment is obtained.
Step 2: Match the corresponding columns of the matrix according to the established model and yield the grayscale difference matrix as shown in Figs. 1 and 2.

$$Q=\begin{bmatrix}
1980 & 596 & 498 & 603 & 471 & 545 & 75 & 513 & 458 & 492 & 457 & 542 & 427 & 532 & 554 & 484 & 521 & 462 & 440 \\
571 & 1980 & 595 & 528 & 156 & 486 & 474 & 594 & 531 & 525 & 460 & 571 & 480 & 547 & 547 & 521 & 470 & 527 & 543 \\
579 & 497 & 1980 & 444 & 468 & 466 & 526 & 404 & 409 & 455 & 476 & 457 & 464 & 489 & 543 & 395 & 56 & 499 & 549 \\
502 & 582 & 412 & 1980 & 491 & 519 & 479 & 481 & 430 & 508 & 127 & 536 & 385 & 504 & 496 & 500 & 493 & 436 & 566 \\
536 & 518 & 494 & 525 & 1980 & 113 & 519 & 523 & 310 & 508 & 501 & 402 & 465 & 540 & 358 & 412 & 489 & 582 & 436 \\
569 & 503 & 475 & 572 & 532 & 1980 & 566 & 416 & 451 & 115 & 538 & 557 & 444 & 497 & 531 & 463 & 472 & 531 & 507 \\
442 & 532 & 362 & 573 & 445 & 295 & 1980 & 343 & 1980 & 412 & 483 & 318 & 281 & 448 & 322 & 184 & 405 & 478 & 358 \\
489 & 521 & 489 & 610 & 454 & 544 & 426 & 1980 & 453 & 529 & 502 & 547 & 438 & 515 & 521 & 465 & 466 & 105 & 527 \\
472 & 504 & 570 & 585 & 521 & 399 & 541 & 495 & 1980 & 558 & 525 & 462 & 505 & 594 & 152 & 462 & 549 & 544 & 412 \\
481 & 529 & 469 & 482 & 516 & 526 & 534 & 454 & 457 & 1980 & 544 & 433 & 510 & 63 & 591 & 455 & 498 & 511 & 537 \\
579 & 537 & 65 & 534 & 574 & 474 & 522 & 440 & 409 & 437 & 1980 & 487 & 420 & 469 & 549 & 499 & 516 & 535 & 537 \\
474 & 490 & 480 & 565 & 531 & 487 & 521 & 103 & 354 & 442 & 541 & 1980 & 451 & 470 & 492 & 406 & 431 & 466 & 454 \\
409 & 579 & 483 & 568 & 490 & 422 & 528 & 422 & 207 & 393 & 482 & 377 & 1980 & 469 & 421 & 75 & 440 & 459 & 437 \\
456 & 518 & 544 & 551 & 567 & 473 & 477 & 469 & 358 & 508 & 589 & 408 & 539 & 1980 & 470 & 384 & 519 & 546 & 92 \\
479 & 577 & 345 & 606 & 406 & 424 & 450 & 430 & 225 & 497 & 412 & 461 & 90 & 467 & 1980 & 389 & 466 & 437 & 513 \\
547 & 529 & 497 & 128 & 574 & 542 & 630 & 572 & 617 & 593 & 564 & 519 & 586 & 503 & 629 & 1980 & 504 & 607 & 581 \\
584 & 94 & 572 & 525 & 583 & 511 & 623 & 473 & 564 & 504 & 591 & 554 & 597 & 548 & 524 & 586 & 1980 & 562 & 518 \\
110 & 544 & 516 & 589 & 571 & 513 & 559 & 447 & 450 & 536 & 553 & 450 & 495 & 452 & 450 & 432 & 583 & 1980 & 486 \\
471 & 535 & 495 & 546 & 542 & 474 & 500 & 512 & 439 & 499 & 540 & 161 & 536 & 473 & 525 & 403 & 472 & 551 & 1980
\end{bmatrix}$$

Fig. 1. Difference matrix of the front side

$$Q=\begin{bmatrix}
1980 & 404 & 385 & 392 & 362 & 86 & 437 & 451 & 426 & 410 & 425 & 405 & 382 & 419 & 377 & 383 & 411 & 418 & 425 \\
329 & 1980 & 407 & 268 & 354 & 450 & 393 & 333 & 320 & 46 & 399 & 401 & 394 & 387 & 389 & 315 & 349 & 302 & 377 \\
316 & 327 & 1980 & 233 & 321 & 435 & 378 & 90 & 347 & 355 & 342 & 330 & 319 & 392 & 380 & 374 & 294 & 339 & 308 \\
359 & 336 & 349 & 1980 & 376 & 440 & 125 & 403 & 418 & 372 & 399 & 341 & 368 & 393 & 421 & 347 & 397 & 364 & 327 \\
231 & 232 & 313 & 1980 & 1980 & 416 & 321 & 273 & 290 & 272 & 287 & 287 & 242 & 287 & 301 & 295 & 241 & 256 & 263 \\
280 & 67 & 386 & 243 & 349 & 1980 & 394 & 354 & 377 & 337 & 342 & 376 & 313 & 340 & 338 & 372 & 338 & 317 & 284 \\
359 & 384 & 119 & 274 & 320 & 404 & 1980 & 369 & 362 & 386 & 345 & 369 & 290 & 399 & 365 & 331 & 405 & 330 & 369 \\
336 & 363 & 358 & 289 & 381 & 405 & 326 & 1980 & 343 & 323 & 380 & 350 & 305 & 430 & 382 & 80 & 368 & 395 & 370 \\
326 & 345 & 320 & 237 & 339 & 379 & 368 & 368 & 1980 & 397 & 372 & 310 & 77 & 400 & 376 & 328 & 368 & 343 & 332 \\
352 & 363 & 376 & 301 & 379 & 421 & 418 & 438 & 399 & 1980 & 388 & 404 & 419 & 98 & 380 & 420 & 374 & 305 & 394 \\
425 & 354 & 349 & 286 & 360 & 432 & 437 & 329 & 62 & 308 & 1980 & 331 & 366 & 377 & 335 & 319 & 399 & 308 & 377 \\
139 & 334 & 399 & 238 & 320 & 456 & 377 & 329 & 428 & 362 & 315 & 1980 & 326 & 379 & 417 & 347 & 255 & 354 & 339 \\
439 & 382 & 373 & 326 & 346 & 384 & 457 & 353 & 326 & 392 & 359 & 431 & 1980 & 373 & 39 & 399 & 417 & 384 & 379 \\
346 & 341 & 354 & 261 & 319 & 433 & 392 & 360 & 343 & 405 & 124 & 316 & 361 & 1980 & 340 & 354 & 370 & 367 & 332 \\
351 & 322 & 349 & 276 & 294 & 418 & 411 & 343 & 384 & 374 & 357 & 407 & 346 & 307 & 1980 & 447 & 391 & 104 & 307 \\
340 & 345 & 374 & 287 & 343 & 455 & 378 & 284 & 391 & 343 & 400 & 360 & 335 & 386 & 386 & 1980 & 354 & 331 & 96 \\
336 & 371 & 390 & 309 & 95 & 403 & 394 & 340 & 399 & 371 & 318 & 404 & 375 & 424 & 422 & 394 & 1980 & 353 & 374 \\
299 & 340 & 421 & 230 & 318 & 446 & 399 & 309 & 390 & 356 & 361 & 371 & 368 & 359 & 385 & 391 & 41 & 1980 & 337 \\
399 & 398 & 355 & 298 & 376 & 422 & 357 & 381 & 356 & 402 & 353 & 83 & 344 & 411 & 427 & 363 & 391 & 392 & 1980
\end{bmatrix}$$

Fig. 2. Difference matrix of the reverse side

Table 1. Front side recovery order after vertical slice

008	014	012	015	003	010	002	016	001	004	005	009	013	018	011	007	0017	000	006

Table 2. Front side recovery order after vertical slice

003	006	002	007	015	018	011	000	005	001	009	013	010	008	012	014	017	016	004

Step 3: Exclude elements on the major diagonal direction of the difference matrix, and choose a group of 19 different elements in different rows and columns.
Step 4: Accumulate matching scores and get the best match with the smallest degree of difference. In Fig. 1, the minimum degree of difference among front side fragment is 2592, and 2107 for reverse side fragment in Fig. 2. The output is the matching sequence as shown in Tables 1 and 2:
Step 5: Splicing fragments in the order, the recovery results are shown as follows: (Fig. 3)

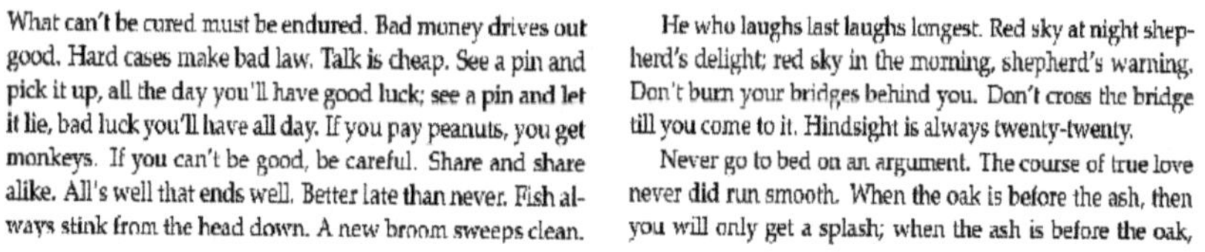

What can't be cured must be endured. Bad money drives out good. Hard cases make bad law. Talk is cheap. See a pin and pick it up, all the day you'll have good luck; see a pin and let it lie, bad luck you'll have all day. If you pay peanuts, you get monkeys. If you can't be good, be careful. Share and share alike. All's well that ends well. Better late than never. Fish always stink from the head down. A new broom sweeps clean.

(a) front side fragment recovery

He who laughs last laughs longest. Red sky at night shepherd's delight; red sky in the morning, shepherd's warning. Don't burn your bridges behind you. Don't cross the bridge till you come to it. Hindsight is always twenty-twenty.
Never go to bed on an argument. The course of true love never did run smooth. When the oak is before the ash, then you will only get a splash; when the ash is before the oak,

(b) reverse side fragment recovery

Fig. 3. Two-sided paper fragment stitching results

4 Conclusions

A mathematical model based on the grayscale difference matrix approach in conjunction with the TSP technique is studied in the paper to solve the problem of paper fragment reassembly. The major advantage is the extraction of the dominating features from fragment edges, and thus less information is used to fulfill the splicing task. However due to its nature of dealing with large amount of calculation, speed becomes its shortcoming when processing huge data set. In addition, manual intervention is required.

By introducing other methods such as genetic algorithm, particle swarm optimization algorithm and so on is our future research direction to improve above two existing problems.

Acknowledgements. This research is financially supported by the National Science Foundation of China 41671406, and Hubei Provincial Natural Science Foundation 2016CFA013.

References

1. Li, X., Liu, Z.Y., Tan, F.X.: Novel quantum ant colony algorithm for TSP. Comput. Eng. Appl. **47**(32), 42–44 (2011)
2. Hou, S.F., Zhang, Y.S., Jiang, T.: Application of Patch- Based texture synthesis in image creating. Comput. Technol. Dev. **17**(7), 247–249 (2007)
3. Li, Q., Zhang, B.: A fast matching algorithm based on image gray value. J. Softw. **17**(2), 216–222 (2006)

4. Jia, H.Y.: Research on the Key Technologies of Computer-aided Paper Fragments Reassembly. National University of Defense Technology, Changsha (2005)
5. Justino, E., Oliveira, L.S., Freitas, C.: Reconstructing shredded documents through feature matching. Int. Forensic Sci. **160**(2), 140–147 (2005)
6. Sun, J.: Research on ant colony: algorithm for solving traveling salesman problem, Doctoral Thesis of Wuhan University of Technology (2005)

An Improved Memetic Algorithm with Novel Level Comparison for Constrained Optimization

Xinghua Qu[1], Wei Zhao[2], Xiaoyi Feng[4], Liang Bai[3(✉)], and Bo Liu[4]

[1] School of Computer Science & Engineering, Nanyang Technological University, Singapore, Singapore
[2] Beijing Institute of Astronautical System Engineering, 100076 Beijing, China
[3] National Computer Network Emergency Response Technical Team/Coordination Center of China, Beijing, China
bailiang@cert.org.cn
[4] Academy of Mathematics and Systems Science, Chinese Academy of Sciences, 100190 Beijing, China

Abstract. Memetic algorithms combining with effective constraints handling techniques have been becoming the focus of concern as well as the subject of substantial research issue in decision-making in complex systems. As one of the most effective techniques for constraints handling, level comparison based method has been used to solve various constraint optimization problems. However, in most of the existing research, the constraint satisfaction level keeps constant or follows a fixed regulation during search, which may weaken the efficiency of relaxation of constraint violation. In this paper, we propose a TLBO based constrained optimization method, labeled as TLBO–LCSQP by incorporating a novel level comparison technique to effectively handle the constraints, and the sequential quadratic program (SQP) to enhance the searching performance. According to the simulation on a well-known constrained benchmark, the proposed TLBO–LCSQP could effectively enhance the constraints handling efficiency and greatly improve the searching ability.

Keywords: Memetic algorithm · Teaching-Learning based optimization · Level comparison method · Sequential quadratic program

1 Introduction

Designing an algorithm that can not only pick out the feasible region quickly but also locate the optimal solutions precisely has been becoming an urgent research topic. Without loss of generality, we can define the constrained optimization problems (COPs) as follows:

F. Xhafa et al. (eds.), *Advances in Intelligent Systems and Interactive Applications*, Advances in Intelligent Systems and Computing 686,
https://doi.org/10.1007/978-3-319-69096-4_98

$$\min f(x)$$
$$subject\ to \tag{1}$$
$$g_i(x) \leq 0,\ i = 1,\ 2, \ldots,\ m; h_j(x) = 0,\ j = 1,\ 2, \ldots,\ n$$

where $x = [x_1, x_2, \ldots, x_n]$ represents the design variables, n indicates the size of x, $g_i(x)$ denotes the inequality constraints, $h_j(x)$ represents equality constraints, m and n denote the number of constraints, respectively. To solve the constrained optimization problems, various kinds of optimization algorithms [1, 2] and constraints handling techniques [3–6] have been proposed.

As a universal problem solver [7–10], Teaching-Learning based Optimization (TLBO) has been introduced into various kinds of COPs [11, 12]. As an effective constraints handling technique, α constrained method [13] can transform a constrained optimization problem into an unconstrained one with two unique components: (1) relaxation of the limit to consider a solution as feasible, and (2) a lexicographical ordering mechanism.

In this study, we propose a TLBO based constrained optimization method, labeled as TLBO–LCSQP by incorporating a novel level comparison technique to effectively handle the constraints, as well as the Sequential Quadratic Program (SQP) method to strengthen the searching quality.

2 The Proposed TLBO–LCSQP

TLBO is a population based algorithm which imitates the teaching and learning process among a group of students in the classroom [11, 14, 15]. The detail of TLBO can refer to [11]. In this section, we propose the TLBO based constrained optimization method, namely TLBO–LCSQP by presenting the novel level comparison technique and the Sequential Quadratic Program (SQP).

2.1 Level Comparison Technique

Level comparison (LC) is a fuzzy control theory inspired constraint-handling technique [2, 6] which utilizes feasibility relaxation and lexicographical ordering mechanism to convert the COP into an unconstrained problem.

The definition of satisfaction level $\mu(x)$ and the ranking principle of LC method can refer to [2]. Wang and Li [2] proposed a linear changed rule of defined α level to exploit the advantage of constraint violation. In this paper, we propose a novel and simple α iterated level rule, which works as the following rules.

$$\alpha_{g+1} = \alpha_g + [1 - \alpha_g] \cdot \frac{1}{4} \tag{10}$$

where α_{g+1} denotes the α level value of the g + 1 generation, α_g represents the α level value of the g generation. If we define $\alpha_1 = 0.1$, the curve of α level value can be depicted as follows: (Fig. 1)

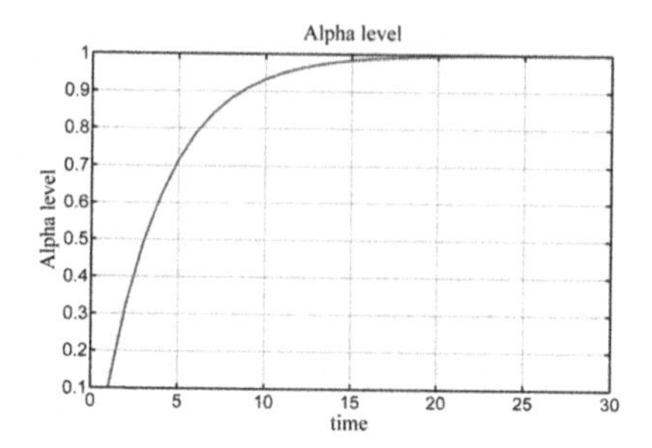

Fig. 1. Novel alpha level value

According to this novel α level, the α value can gradually get close to 1. With the increase of generation, the value of α will get infinitely approximate to 1, which enables the algorithm with more flexible to exploit the boundary between feasibility and infeasibility regions.

2.2 TLBO–LCSQP

To effectively deal with the constraints and balance the global exploration and local exploitation of the optimization algorithm, a new level comparison based constraints handling method and a local search based on Sequential Quadratic Program are introduced in the original TLBO respectively. In the proposed TLBO–LCSQP, the TLBO would find its optimal results near the true global optimum. Then, the TLBO results (i.e., the teacher in the current population) will be used as an initial point for the SQP algorithm. Furthermore, the principle of sorting the individuals is based on the novel α level rule. The pseudo-code of the proposed TLBO–LCSQP is depicted in Fig. 2.

Initialization
While the maximum iteration time is reached
Teaching and *Learning phase* [11]; define the one with best cost value as $X_m^{gnew}(t)$.
For all the inequality constraints $g_i(x)$ and equality constraints $h_j(x)$, calculate the satisfaction level of each constraint (i.e $\mu_{g_i}(x)$ and $\mu_{h_j}(x)$) [2].
Calculate the $\mu(x)$ (i.e., constraints satisfaction level of the entire problem [2].
Reorder the individuals according to the level comparison method [2].
Get the updated $X_m^{gnew}(t)$.
Define x_{best}^k=$X_m^{gnew}(t)$ as the initial point of SQP; Apply SQP as the local search and update $X_m^{gnew}(t)$.
End while

Fig. 2. The flowchart of the proposed TLBO–LCSQP

3 Numerical Results

In this section, simulation results based on a well-known constrained benchmark (i.e., the welded beam design problem and tension) and comparisons with other methodologies. The defination of the welded beam design problem can refer to reference [1]. To identify the effectiveness and efficiency of the proposed TLBO–LCSQP, the simulation is applied on the mentioned above welded beam design problem and compared with other algorithms (including co-evolutionary particle swarm optimization [1], geometric programming [16], genetic algorithm with penalty function [17], genetic algorithm based co-evolution [3], and multi-objective optimization with feasibility rule [18]. The simulation and comparison are listed as below.

From Table 1, it can be found that TLBO–LCSQP has superior performance than other algorithms. The solution of TLBO–LCSQP is closer to the boundary between constrained region and unconstrained region, which demonstrates that the constraints handling ability of the TLBO–LCSQP has been enhanced. In Table 2, TLBO–LCSQP performs better than any other algorithm listed in the table with regarding to best result, mean value, worst solution and standard deviation.

Table 1. 30 times independent runs' results and comparisons

Design variables	TLBO–LCSQP	He and Wang [1]	Ragsdell and Phillips [16]	Deb [17]	Coello [3]	Coello and Montes [18]
$x_1(h)$	0.2057296	0.202369	0.245500	0.248900	0.208800	0.205986
$x_2(l)$	3.470489	3.544214	6.196000	6.173000	3.420500	3.471328
$x_3(t)$	9.036623	9.048210	8.27300	8.178900	8.997500	9.020224
$x_4(b)$	0.2057296	0.205723	0.245500	0.253300	0.210000	0.206480
$g_1(x)$	−7.0395e-10	−12.839796	−5743.826517	−5758.60378	−0.337812	−0.74092
$g_2(x)$	−2.1828e-11	−1.247467	−4.715097	−255.576901	−353.902604	−0.266227
$g_3(x)$	−2.7756e-17	−0.001498	0.0000000	−0.004400	−0.001200	−0.000495
$g_4(x)$	−3.4330	−3.429347	−3.020289	−2.982866	−3.411865	−3.430043
$g_5(x)$	−0.0807	−0.079381	−0.120500	−0.123900	−0.083800	−0.080968
$g_6(x)$	−0.2355	−0.235536	−0.234208	−0.234160	−0.235649	−0.235514
$g_7(x)$	−3.6380e-12	−11.681355	−3604.275002	−4465.270928	−363.232384	−58.666440
$f(x)$	1.7248523	1.728024	2.385937	2.433116	1.748309	1.728226

Table 2. Statistical results and comparisons

Method	Best	Mean	Worst	Std dev
TLBO–LCSQP	1.7248523	1.7248523	1.7248523	6.207e-27
He and Wang [1]	1.728024	1.748831	1.782143	0.012926
Coello [3]	1.748309	1.771973	1.785835	0.011220
Ragsdell and Phillips [16]	2.385937	–	–	–
Deb [17]	2.433116	–	–	–
Coello and Montes [18]	1.728226	1.792654	1.993408	0.074713

4 Conclusion

This research presents an effective constrained memetic algorithm (i.e. TLBO–LCSQP), in which a novel level comparison based constraints handling technique and a local search refinement of Sequential Quadratic Program (SQP) are introduced. Computational results based on the welded beam design problem prove that TLBO–LCSQP could enhance both the searching efficiency and constraints satisfaction.

Acknowledgements. This work was supported in part by National Natural Science Foundation of China (Grant Nos.71390331, 61673058), National Science Fund for Distinguished Young Scholars of China (Grant No. 61525304), Defense Industrial Technology Development Program, and Key Research Program of Frontier Sciences, Chinese Academy of Sciences (QYZDB-SSW-SYS020).

References

1. He, Q., Wang, L.: An effective co-evolutionary particle swarm optimization for constrained engineering design problems. Eng. Appl. Artif. Intell. **20**, 89–99 (2007)
2. Wang, L., Li, L.P.: Fixed-structure H-infinity controller synthesis based on differential evolution with level comparison. IEEE Trans. Evol. Comput. **15**, 120–129 (2011)
3. Coello, C.A.C.: Use of a self-adaptive penalty approach for engineering optimization problems. Comput. Ind. **41**, 113–127 (2000)
4. Deb, K.: An efficient constraint handling method for genetic algorithms. Comput. Methods Appl. Mech. Eng. **186**, 311–338 (2000)
5. He, Q., Wang, L.: A hybrid particle swarm optimization with a feasibility-based rule for constrained optimization. Appl. Math. Comput. **186**, 1407–1422 (2007)
6. Mezura-Montes, E., Coello, C.A.C.: Constraint-handling in nature-inspired numerical optimization: past, present and future. Swarm Evol. Comput. **1**, 173–194 (2011)
7. Feng, L., Ong, Y.S., Tan, A.H., Tsang, I.W.: Memes as building blocks: a case study on evolutionary optimization plus transfer learning for routing problems. Memetic Comput. **7**, 159–180 (2015)
8. Bai, L., Jiang, Y., Huang, D., Liu, X.: A novel scheduling strategy for crude oil blending. Chin. J. Chem. Eng. **18**, 777–786 (2010)
9. Bai, L., Jiang, Y.H., Huang, D.X.: A novel two-level Optimization framework based on constrained ordinal optimization and evolutionary algorithms for scheduling of multipipeline crude oil blending. Ind. Eng. Chem. Res. **51**, 9078–9093 (2012)

10. Bai, L., Wang, J., Jiang, Y., Huang, D.: Improved hybrid differential evolution-estimation of distribution algorithm with feasibility rules for NLP/MINLP engineering optimization problems. Chin. J. Chem. Eng. **20**, 1074–1080 (2012)
11. Qu, X., Zhang, R., Liu, B., Li, H.: An improved TLBO based memetic algorithm for aerodynamic shape optimization. Eng. Appl. Artif. Intell. **57**, 1–15 (2017)
12. Yu, K., Wang, X., Wang, Z.: Constrained optimization based on improved teaching—learning-based optimization algorithm. Inform. Sci. **352–353**, 61–78 (2016)
13. Takahama, T., Sakai, S.: Constrained optimization by applying the alpha constrained method to the nonlinear simplex method with mutations. IEEE Trans. Evol. Comput. **9**, 437–451 (2005)
14. Qu, X., Li, H., Zhang, R., Liu, B.: An effective TLBO-based memetic algorithm for hypersonic reentry trajectory optimization. IEEE Congr. Evol. Comput. (CEC) **2016**, 3178–3185 (2016)
15. Qu, X., Liu, B., Li, Z., Duan, W., Zhang, R., Zhang, W., et al.: A novel improved teaching-learning based optimization for functional optimization. In 2016 12th IEEE International Conference on Control and Automation (ICCA), pp. 939–943 (2016)
16. Ragsdell, K.M., Phillips, D.T.: Optimal Design of a Class of Welded Structures Using Geometric Programming. J. Eng. Ind. **98**, 1021–1025 (1976)
17. Deb, K.: Optimal design of a welded beam via genetic algorithms. AIAA J. **29**, 2013–2015 (1991)
18. Coello, C.A.C., Montes, E.M.: Constraint-handling in genetic algorithms through the use of dominance-based tournament selection. Adv. Eng. Inform. **16**, 193–203 (2002)

FPGA Based Real-Time Processing Architecture for Recurrent Neural Network

Yongbo Liao[1,2(✉)], Hongmei Li[2], and Zongbo Wang[3]

[1] State Key Laboratory of Electronic Thin Films and Integrated Devices, Chengdu, China
lyb@uestc.edu.cn

[2] School of Energy Science and Engineering, University of Electronic Science and Technology of China, Chengdu, China

[3] Electrical Engineering and Computer Science (EECS), University of Kansas (KU), Kansas, USA

Abstract. A field programmable gate array (FPGA)-based real-time processing architecture for recurrent neural network (RNN) is proposed and presented; the proposed FPGA processing architecture is based on echo state network (ESN) and can get the output weights of RNN in real-time. The proposed architecture and the performance have been verified on an Altera FPGA chip. Experimental results show that the real-time hardware RNN can be trained to recognize different duty cycles of the input signal. We also performed experiments to investigate the ESN demand for resources and systems convergence in FPGA.

Keywords: FPGA · RNN · ESN · Recognize · Duty cycle · Verify · Systems convergence

1 Introduction

Most of the popular training methods for RNN are computationally intensive and have extremely slow convergences [1]. Echo state network (ESN) as an architecture with supervised learning principle in RNN is conceptually simple, computationally inexpensive, and has fast convergence [2]. Recently, there have been a few papers which start research about realization of hardware implementation in FPGA. A software framework which simulates a RNN circuit is introduced in [3], and a FPGA/Software framework is proposed in [4]. But both the frameworks in [3] and [4] train and obtain output weight in the software and by offline computation.

The paper proposes an FPGA based ESN processing architecture for realizing the RNN in real-time. It is conceptually simple, resources inexpensive and has fast convergence.

F. Xhafa et al. (eds.), *Advances in Intelligent Systems and Interactive Applications*, Advances in Intelligent Systems and Computing 686,
https://doi.org/10.1007/978-3-319-69096-4_99

2 Formalism and Theory

The essential echo state network is a discrete-time system, formed by sigmoid units, which contains N reservoir units, K inputs and L outputs, already applied to machine learning tasks in supervised manner. There are a training input vector $\mathbf{u}(n) \in \mathbb{R}^K$ and a desired target output vector $\mathbf{y}_{\mathbf{target}}(n) \in \mathbb{R}^L$ which are given before training process start. Here n = 1, …, T is the discrete time index indicating training period, where the last index T is determined by the number of data elements which form the training dataset [1, 2]. The task of ESN is to generate output vector $\mathbf{y}(n) \in \mathbb{R}^L$ which matches the $\mathbf{y}_{target}(n)$ through supervised learning. The reservoir can be interpreted as a non-linear high-dimensional state vector $\mathbf{x}(n) \in \mathbb{R}^N$ of the feeding input vector $\mathbf{u}(n)$. The reservoir state update Eq. is [1]

$$\mathbf{x}(n+1) = f(\mathbf{W}^{\mathbf{in}}\mathbf{u}(n+1) + \mathbf{W}\mathbf{x}(n) + \mathbf{W}^{\mathbf{back}}\mathbf{y}_{\mathbf{target}}(n)) \quad (1)$$

where $f(\cdot)$ is a sigmoid function, $\mathbf{W}^{\mathbf{in}} \in \mathbb{R}^{N \times K}$ is the input weight matrix, $\mathbf{W} \in \mathbb{R}^{N \times N}$ is the internal reservoir weight matrix, $\mathbf{W}^{\mathbf{back}} \in \mathbb{R}^{N \times L}$ is the output feedback weight matrix.

The linear readout is shown as

$$\mathbf{y}(n+1) = f_{out}(\mathbf{W}^{\mathbf{out}}(\mathbf{u}(n+1), \mathbf{x}(n+1))) \quad (2)$$

where $f_{out}(\cdot)$ is a sigmoid function, or identity function, $\mathbf{W}^{\mathbf{out}} \in \mathbb{R}^{L \times (K+N)}$ is the trained output weights.

Learning of the output weights $\mathbf{W}^{\mathbf{out}}$ can be solved in (2) by ridge regression with Tikhonov regularization [2]:

$$\mathbf{W}^{\mathbf{out}} = \mathbf{Y}_{\mathbf{target}}\mathbf{X}^T(\mathbf{X}\mathbf{X}^T + \alpha^2\mathbf{I})^{-1} \quad (3)$$

where $\mathbf{Y}_{\mathbf{target}} \in \mathbb{R}^{L \times T}$ including all $\mathbf{y}_{\mathbf{target}}(n)$, and $\mathbf{X} \in \mathbb{R}^{N \times T}$ containing all $\mathbf{x}(n)$, both collected into matrices during the training period n = 1,…,T, respectively. Every $\mathbf{x}(n)$ embodies the reservoir with $\mathbf{u}(n)$ during the training period n = 1,…,T. $\mathbf{X}^{\mathbf{T}}$ denotes $\mathbf{X}$ transposed, $\mathbf{Y}_{\mathbf{target}}\mathbf{X}^{\mathbf{T}} \in \mathbb{R}^{L \times N}$ and $\mathbf{X}\mathbf{X}^{\mathbf{T}} \in \mathbb{R}^{N \times N}$ are independent of the index T denoting the length of the training sequence, and can be computed incrementally when the training data pass through the reservoir. $\mathbf{I} \in \mathbb{R}^{N \times N}$ denotes the identity matrix, α denotes a regularization factor, $(\cdot)^{-1}$ denotes matrix inversion.

3 Proposal FPGA Structure Implementable

The proposed FPGA implantation structure mapped the (1, 2, and 3) into six modules. As shown in Fig. 1, Input module, reservoir module, and random weights generator realize the (1). Equation (2) includes input and output module. The training module realized the (3). The system switch module is designed to judge and stop training, and then continue to normal ESN system function after the machine learning is finished.

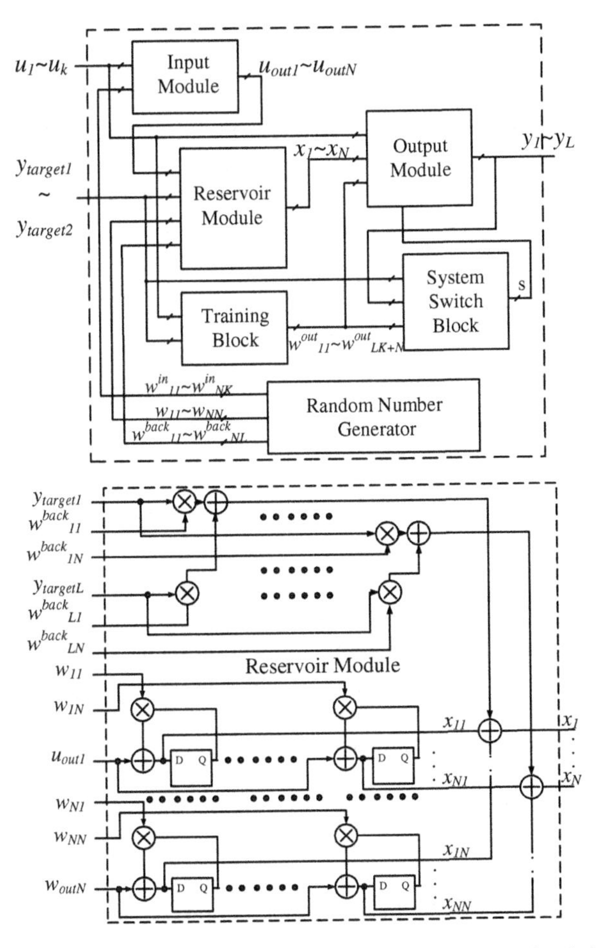

Fig. 1 The proposed ESN architecture and reservoir module circuit in FPGA

The data flow and processing steps of achieving real-time ESN training are explained below:

Given: A training input and targeting output sequence $\mathbf{u}$ (n) and $\mathbf{y}_{\mathbf{target}}$ (n), n = 1… T.

Wanted: Apply the harvestable trained network on anew feeding input data $\mathbf{u}$ (n) to computing $\mathbf{y}$ (n) through employing the trained output weights $\mathbf{W}^{\text{out}}$ following the relationship of teaching input/output.

Step 1. Random weights generator procure an untrained dynamical reservoir network weights ($\mathbf{W}^{\text{in}}$, $\mathbf{W}$, $\mathbf{W}^{\text{back}}$) and set up connection of in input module and reservoir module.

Step 2. The network is driven by entering the given training input $\mathbf{u}$ (n) from the input module, and by teacher-forcing the teacher output $\mathbf{y}_{\mathbf{target}}$ (n) from the reservoir module, per each clock cycle.

Step 3. Collecting the network state **x** (n) as a state collecting matrix **X**, and the desired output $\mathbf{y}_{\mathbf{target}}$ (n) as a collecting matrix $\mathbf{Y}_{\mathbf{target;}}$ training $\mathbf{W}^{\mathbf{out}}$ with stochastic gradient descent [5] computed by (3), per each clock cycle.
Step 4. The network ($\mathbf{W}^{\mathbf{in}}$, **W**, $\mathbf{W}^{\mathbf{back}}$, $\mathbf{W}^{\mathbf{out}}$) with definite values is now ready to feeding new input data and computing the wanted output. The update **y** (n+1) in the output module is calculated by the novel input sequences **u** (n+1), and so does the update **x** (n+1) in the reservoir module.
Step 5. Repeat step 3 to 4 until match the stop condition. It is same for **y** (n) and $\mathbf{y}_{\mathbf{target}}$ (n) when the output signal is just "1" or "0" and using stochastic gradient descent way.
Step 6. After getting the final $\mathbf{W}^{\mathbf{out}}$, the training block with network is turned off, and the $\mathbf{W}^{\mathbf{out}}$ value of training block is set into $\mathbf{W}^{\mathbf{out}}$ registers in output module by system switch block.

4 Experimental and Result

The proposed FPGA structure has been implemented on an Altera Stratix III FPGA using VHDL. We used a different duty cycle pattern recognition benchmark task similar to that demonstrated by Kulkarni et al. [3] for memristior-based RC. As shown in Fig. 2, the whole processing is composed of two periods: training period and normal period. In the training period, the signal $\mathrm{y}_{\mathrm{target1}}$ shows the expected response of signal u_1 which is a different duty cycle pattern. The signal $\mathrm{y}_{\mathrm{target1}}$ should converge to 0 for the duty cycle less than 50%, and 1 for the duty cycle more than 50%. The output signals are y_1 which is real output signal. We train the ESN in FPGA and get $w_{11}^{\mathbf{out}}$, $w_{21}^{\mathbf{out}}$ and $w_{32}^{\mathbf{out}}$ online. When we get the final output weights, the system switch module changes the signal S from 0 to 1,and then sets $w_{11}^{\mathbf{out}}$, $w_{21}^{\mathbf{out}}$ and $w_{32}^{\mathbf{out}}$ to the output module registers of $w_{1}^{\mathbf{out}}$, $w_{2}^{\mathbf{out}}$, and $w_{3}^{\mathbf{out}}$. In the normal period, we import different duty cycle pattern signals into the trained ENS. The output signal (y_1) will change between the input signal (u_1) duty cycle is more than and less than 50%.

Figure 3 shows the percentage of FPGA logic utilization and error rate in relationship with the overall number of neural cells implemented in the FPGA. It can be noticed that the logic utilization is less than 60% until the neural number is 512 units.

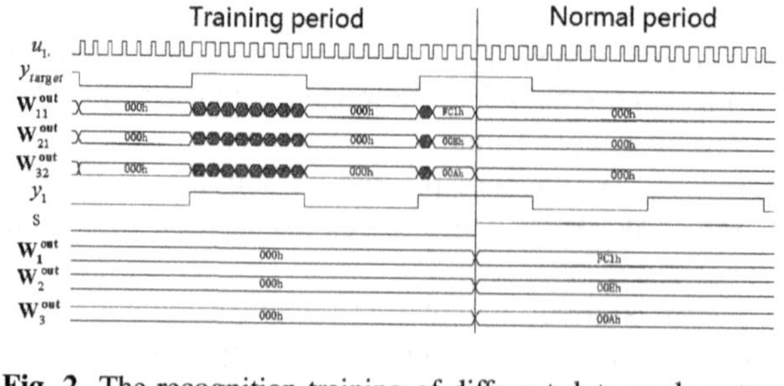

Fig. 2 The recognition training of different duty cycle pattern

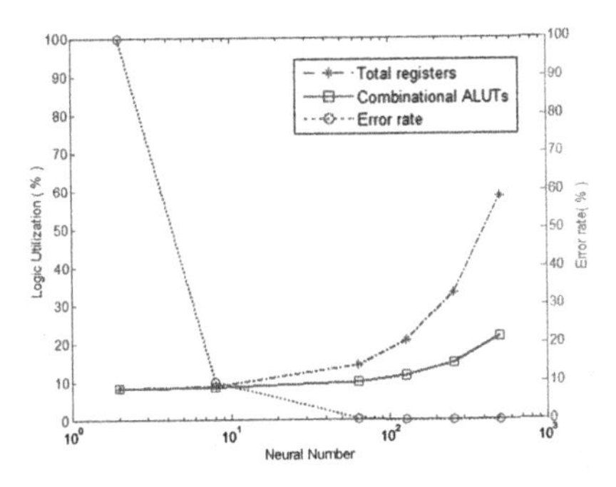

Fig. 3 The relationship of ENS number vs logical utilizations (in FPGA Stratix III EP35L70F780C4) and error rate

And the error rate between the real output signal and the desired output is zero when the neural number is over 16 units. Thus, the proposed circuit is inexpensive in resource consumption and features fast convergence.

5 Conclusions

A real-time ESN architecture for RNN is proposed and implemented in FPGA. The output weights have been calculated online and in real-time when the ESN is training. The experiment, which investigates the RNN ability of remember and recognize, was setup by two different frequency signals.

References

1. Lukoševičius, M.: 'A practical guide to applying echo state networks', neural networks: tricks of the trade, pp. 659–686, 7700. Springer Berlin Heidelberg, (2012). doi:10.1007/978-3-642-35289-8_36
2. LukošEvičIus, M., Jaeger, H.: Reservoir computing approach- es to recurrent neural network training. Comput. Sci. Rev. **3**(3), 127–149 (2009). doi:10.1016/j.cosrev.2009.03.005
3. Kulkarni, M.S., Teuscher, C.: 'Memristor-based reservoir computing', Nanoscale Architectures (NANOARCH), 2012 IEEE/ACM International Symposium on. IEEE, pp. 226–232 (2012)
4. Alomar, M.L., Canals, V., Martinez-Moll, V., et al.: 'Low-cost hardware implementation of reservoir computers', International Workshop on Power and Timing Modeling, Optimization and Simulation (PATMOS), pp. 1–5 (2014). doi: 10.1109/ PATMOS.2014.6951899
5. Bach, F.: Adaptivity of averaged stochastic gradient descent to local strong convexity for logistic regression. J Mach. Learn. Res. **15**(1), 595–627 (2014)

Publication Recommendations of Manuscripts Using Improved C4.5 Decision Tree Algorithm

Didi Jia[1], Wu Xie[2,3(✉)], Zhiyong Chen[1], and Baohua Qiang[3,4]

[1] School of Computer Science and Information Security, Guilin University of Electronic Technology, 541004 Guilin, China
[2] Guangxi Colleges and Universities Key Laboratory of Cloud Computing and Complex System, 541004 Guilin, China
xiesixchannels@126.com
[3] Guangxi Key Laboratory of Trusted Software, 541004 Guilin, China
[4] Guangxi Cooperative Innovation Center of Cloud Computing and Big Data, 541004 Guilin, China

Abstract. In view of the problem that the result accuracy of the research publication recommendations of manuscripts is often low. It is very difficult for many contributors to find the suitable journals from many recommendation results. To solve this problem, a novel method of publication recommendations for authors' manuscripts is proposed using our improved C4.5 decision tree algorithm. For the shortcomings of too many values of data samples during dealing with the traditional C4.5 methods, the five-value logic ideas are adopted to improve this C4.5 algorithm to apply in the research fields of publication recommendations. The experimental results show that the related publications are easier to obtain for authors via the decision trees with the improved C4.5 method than before, and this method has higher accuracy than ever. It is of great significance to help the researchers, especially for those who have no enough experience to choose the most suitable journals from a lot of publications.

Keywords: Manuscript · Publication recommendation · Data mining · C4.5 algorithm · Five-value logic

1 Introduction

Publication recommendation of manuscript is to recommend the related journals for the contributors who have the needs to submit their manuscripts. The relationships between the publication scopes of journals and the direction of the manuscript need to be relatively consistent. Authors also often concern about the factors of journal impact and manuscript innovation level. Good recommendations results are not only to meet the needs of contributors, but also the time and effort for contributors are often reduced to find their desired journals which are related to their manuscript. There are two main aspects researches on current publications recommendations of manuscripts. One is to recommend literatures which are consistent with the research fields of researchers [1, 2]. The other is to find the recommended publications in accord with the research fields of manuscripts [3–8]. Through social networking, Hiep, Klamma [3, 4] explored

F. Xhafa et al. (eds.), *Advances in Intelligent Systems and Interactive Applications*, Advances in Intelligent Systems and Computing 686,
https://doi.org/10.1007/978-3-319-69096-4_100

journals which were published in the network by other authors of the manuscripts in the same research fields. In 2016, Nghiep, Tin [7] utilized the publications citations by the researchers to build data samples for publication recommendations. The existing publication recommendation methods have provided some convenient ways for the submission users to submit manuscripts. However, there is a problem that the accuracy results of publication recommendation for manuscripts are still low. It is very difficult for many authors to search the most suitable publications from too many potential recommendation results for their research fruits.

In this paper, the traditional C4.5 algorithm is improved using the five-value logic theory, and this improved method is applied to recommend publications for authors' manuscripts. The SCI (Science Citation Index) journals in JCR (Journal Citation Reports) are selected as the data resource examples to build a large enough sample for the authors who want to submit their high-level manuscripts from scientific research fruits. The remainder of this paper is arranged as follows.

In Sect. 2, the C4.5 algorithm is introduced and improved for the applications of the publication recommendations of scientific manuscripts. The recommendations experimental results of data samples of SCI journal are obtained through data mining with the decision trees via the improved C4.5 algorithm in Sect. 3. Then in Sect. 4, this new method to recommend appropriate publication of manuscripts is discussed. Finally, a brief conclusion is followed in Sect. 5.

2 Publication Recommendations of Manuscript Via the Improved C4.5 Algorithm

The C4.5 algorithm is improved using five-value logic. For the binary logic, the two values of '0' and '1' are used to characterize the true value and false values with two deterministic states. In order to make use of the C4.5 decision tree algorithm which plays an important role in the classification tasks of manuscript-publication recommendations, it is necessary to control the numbers of attribute values of the training data samples. So, in our work, the five-value logic theory is used to improve the C4.5 decision tree algorithm for data mining. The attribute value is limited to the fixed value on the interval of [0, 1]. Here, m is the values of five numbers, that is, v_1, v_2, …, v_5.. The C4.5 algorithms are improved with the entropy theory and five-value logic method to select the appropriate attributes as the root node. The steps to improve the C4.5 algorithm using five-value logic are as follows.

(1) The information entropy of the category attributes is calculated. If there are n samples in the journal data sample set S. The five-value logic theory is used to divide the category attribute of the relevance information of the journal. The manuscript has m values between 0 and 1, that is, v_1, v_2, . . ., v_5. The category attribute C has m different values, i. e. C_i ($i = 1, 2, \ldots, 5$). Let s_i be the number of samples in C_i, and we can get the information entropy of sample set S with the class attribute C as follows:

$$I(s_1,s_2,\ldots s_5) = -\sum_{i=1}^{5} p_i \log_2(p_i) \tag{1}$$

Among them, P_i represents the probability of any sample of the classification C_i.

(2) The information entropy of decision attributes is operated. The decision attribute A is selected in sample set S. If A has k different values, then the decision attribute A is divided into k categories for sample set S. Thus, the information entropy that sample set S is divided by A:

$$E(A) = -\sum_{j=1}^{k} A_{1j} + A_{2j} + \cdots + A_{5j}/nI(s_j) \tag{2}$$

A_{ij} ($i = 1, 2, \ldots, 5; j = 1, 2, \ldots, k$) represents the number of samples of the class C_i in a subset of S_j. For the decision attributes of continuous variables, the five-valued logic theory is used to discretize the value of the decision attributes.

(3) The information gain is obtained. The information gain of A can be calculated according to the above information as follows.

$$Gain(A) = I(S) - E(A) \tag{3}$$

(4) The information entropy of the decision attribute value is calculated.

$$SplitInfo(A) = -\sum_{j=1}^{k} p_j \log_2(p_j) \tag{4}$$

P_j represents the probability that any attributes value of the total samples.

(5) The information gain rate is achieved. The information gain rate is chosen the test attribute in the C4.5 algorithm. The formula of information gain rate is as follows:

$$GainRatio(A) = Gain(A)/SplitInfo(A) \tag{5}$$

The decision attribute with the largest information gain is selected as the root node of the improved C4.5 decision tree. The new tree model can be used to generate classification rules, which can be utilized to recommend appropriate publications. Publications are recommended for authors' manuscripts. The data resources in this paper are from the JCR in 2015. The collected SCI journals data in the origin JCR table are adopted. The impact factor, 5-year impact factor, immediacy index, the amount of information, cited half-life index are utilized to evaluate the publications. To facilitate the needs of scientific researchers, the tool of network is often used to obtain journals information. It takes a lot of time to divide the research direction of journals in the JCR publications. For instance, the divisions of journals in the computer fields are classified according to the requirements of the research direction classifications in CCF (China Computer Federation). Then, the relationship between journal research fields and manuscript themes are established. The relationships as data attributes of the data sample are focused on. The direction correlation degree attribute of journals and manuscripts is researched by data mining of decision trees.

3 Experiment Results

The improved C4.5 algorithm is used for JCR journal recommendations of manuscripts by an example. The decision tree model of the C4.5 method of using JCR table is taken as the example. In our work, the influence factor ranges from 0.76 to 2.76. The key words are data mining, Bayesian network, Bayesian probability, prior probability and posterior probability. The result is shown as Fig. 1 after data mining.

4 Discussion

In comparison with current methods, it can be seen from Fig. 1 that our publication recommendation methods via the improved C4.5 algorithm has two advantages.

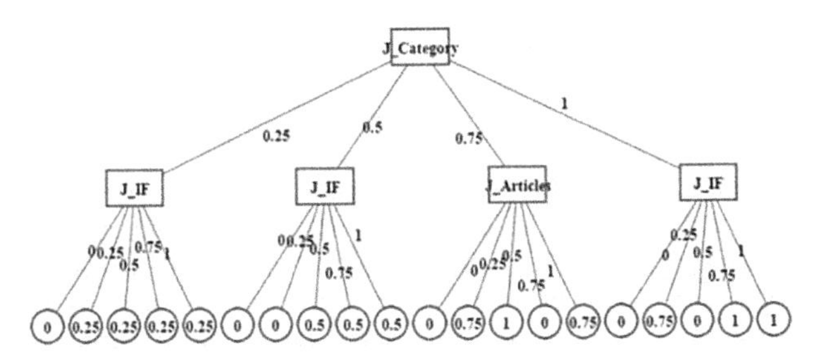

Fig. 1. The decision tree results of data mining via the improved C4.5 algorithm

Firstly, the scales of the decision trees of data mining with the improved C4.5 algorithm are smaller than before. The main reason is that the C4.5 algorithms are improved via the five-value logic methods instead of uncertain numbers of subtrees of decision trees for the traditional C4.5 algorithm. The original irregular attribute values become regular finite thresholds after the improvements. The classification results are uniform and reasonable as shown in Fig. 1. The decision trees which are generated via the improved C4.5 algorithm have fixed numbers of tree branches.

Secondly, the results of publication recommendations of manuscripts using our new methods are more accurate than ever. After data mining with the improved C4.5 algorithm, it is also easier for authors to find the suitable publication sets from lots of candidate journals. The reasons mainly include two aspects. One is the five-value logic theory is utilized in the processes of data mining with the improved C4.5 algorithm. The other is the unified division method is used in the preprocess steps before data mining. The classification rules generated by the new intuitive decision tree model are easy to be understood, and the accuracy of the classification results using our methods is higher than other current recommended methods.

5 Conclusion

In this paper, the C4.5 algorithm of decision tree has been improved with five-value logic theory. This new method has been applied in the fields of publication recommendations of manuscripts. The results show that the new recommendations methods of publications not only have the advantages of small-scale decision trees, but also it has higher accuracy than before. It is easier to be understood intuitively for manuscript submissions than ever. For next work, more journals are added to expand the sample spaces to provide more choices for contributors, and the data records of diverse journals are increased to improve the recommended accuracy of manuscripts.

Acknowledgments. This work is supported by Guangxi Key Laboratory of Trusted Software (Nos. kx201510, kx201413), Guangxi Colleges and Universities Key Laboratory of Cloud Computing and Complex System (No. 14106), Guangxi Cooperative Innovation Center of Cloud Computing and Big Data (Nos. YD16510, YD16802, YD16509).

References

1. Pera, M.S., Ng, Y.K.: A personalized recommendation system on scholarly publications. Proceedings of the 2011 ACM International Conference on Information and Knowledge Management, CIKM'11, Glasgow, United Kingdom, 2133–2136 (2011)
2. Anh, V.L., Hai, V.H., Tran, H.N., Jung, J.J.: A recommendation system for scientific publication by discovering keyword relationships. Lect. Notes Comput. Sci. **8733**, 72–82 (2014)
3. Klamma, R., Cuong, P.M., Cao, Y.: You never walk alone: recommending academic events based on social network analysis, pp. 657–670. Complex Sciences. Springer, Heidelberg (2009)
4. Hiep, L., Tin, H., Susan, G., Kiem, H.: Exploiting social networks for publication venue recommendations. Proceedings of the International Conference on Knowledge Discovery and Information Retrieval, KDIR, Barcelona, 239–245 (2012)
5. Erict, M., Alberto, B., Giulio, P.: Publication venue recommendation based on paper abstract. Proceedings of International Conference on Tools with Artificial Intelligence, ICTAI, Limassol Cyprus, 1004–1010 (2014)
6. Alzoghbi, A., Ayala, V., Peter, M., Fischer., Lausen, G.: PubRec: recommending publications based on publicly available meta-data. Proceedings of the LWA 2015 Workshops: KDML, FGWM, IR, and FGDB, Trier, Germany, 11–18 (2015)
7. Tran, H.N., Huynh, T., Hoang, K.: A potential approach to overcome in scientific publication recommendation. Proceedings of 2015 IEEE International Conference on Knowledge and Systems Engineering, Ho Chi Minh City, 310–313 (2015)
8. Tin, H., Trac-Thuc, N., Hung-Nghiep, T.: Exploiting social relations to recommend scientific publications. Lect. Notes Comput. Sci. **9795**, 182–192 (2016)

Design Replica Consistency Maintenance Policy for the Oil and Gas Pipeline Clouding SCADA Multiple Data Centers Storage System

Miao Liu[1(✉)], Jia Shimin[2], and Yuan Mancang[1]

[1] China Petroleum Longhui Automation Engineering Co. Ltd, China Petroleum Pipeline Engineering Co.Ltd, Tianjin, China
lhbj_liumiaol@cnpc.com.cn

[2] Science and Technology Center, China Petroleum Pipeline Engineering Co.Ltd, Tianjin, China

Abstract. The multiple data centers storage system of clouding SCADA is proposed and the replica consistency maintenance policy is designed in paper. The oil and gas pipeline clouding SCADA multiple data centers storage system can support storage enormous capacity data, expanding in data center lever and independent reading and writing for each data center. The two stage lock replica consistency maintenance policy can ensure the consistency of data copy of multiple data centers.

Keywords: Clouding SCADA · Multiple data centers · Storage system · Replica consistency maintenance

1 Introduction

Supervisory Control and Data Acquisition system (SCADA) realizes data acquisition, device control, measurement, parameter adjustment, all signal alarming function etc. by data communication network to monitor and control operating equipment of remote stations. The SCADA system is developed for the production process distributed in large space span. Recently, the SCADA system is used in long distance transportation pipeline network widely [1, 2]. It is used in CNPC Beijing Oil & Gas Transportation Center and each regional Oil & Gas Transportation Center to monitor the pipeline operation process and control every pipeline metal instruments.

The SCADA system includes remote terminal equipment, main station computers, monitor terminal, peripheral device and so on. The system is a hierarchical control system [3]. It is a distributed computer control system with high reliability. The computers of pipeline control center periodically queries the oil and gas pipeline stations or remote control valve chamber, collects operation data and status information of each station continuously and sent operating instructions by data transmission system [4, 5]. Cloud computing is a computing model that the internet is utilized to realize accessing sharing resource pooling (computing facilities, storage devices, applications and so on) anytime and anywhere, on-demand and conveniently [6, 7]. Computer

F. Xhafa et al. (eds.), *Advances in Intelligent Systems and Interactive Applications*, Advances in Intelligent Systems and Computing 686,
https://doi.org/10.1007/978-3-319-69096-4_101

resources as a service is the important form of expression. It masks the questions of data center management, large scale data processing, applications deployment and so on. The users can fast apply for or release resources according to users' traffic load and pay for the used resource by on-demand payment [8]. The service quality is improved and at the same time, the operation and maintenance costs is reduced. The cloud computing has caught more attention of the industry and academia since it is proposed as a great innovation of information industry [9]. The application framework of oil and gas pipeline SCADA system based on cloud computing (clouding SCADA) has been designed to improve the reliability and resource utilization of the traditional SCADA system [10].

In order to realize storage enormous capacity data, expanding in data center lever and independent reading and writing for each data center, the multiple data centers storage system of oil and gas pipeline is proposed and the two stage lock replica consistency maintenance policy is proposed in the paper.

2 The Application Framework of Clouding SCADA System

The application framework of clouding SCADA system can be designed to improve the reliability and resource utilization of the traditional SCADA system. It is shown in Fig. 1.

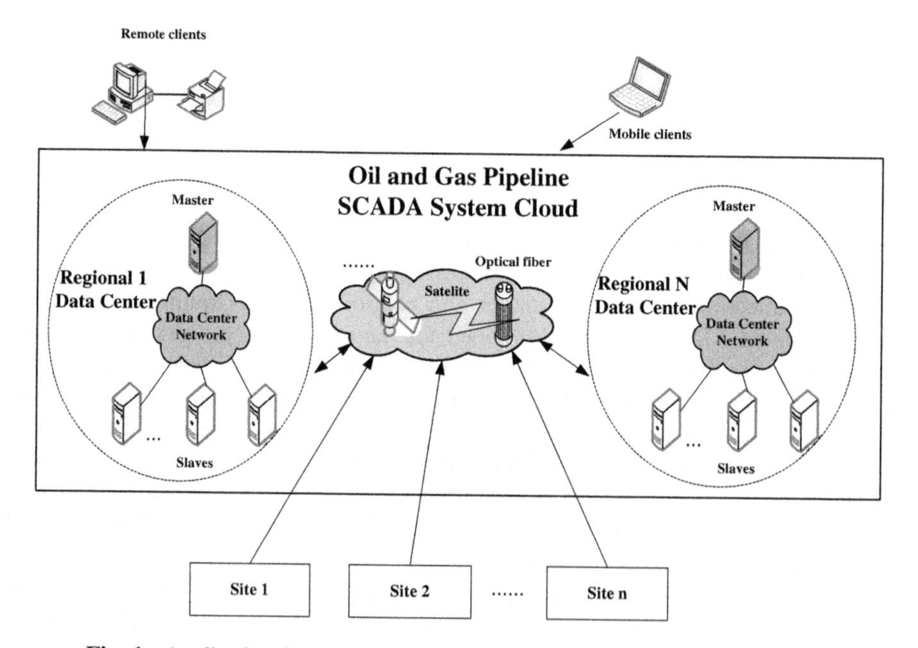

Fig. 1. Application framework of SCADA system based on cloud computing

We can see from Fig. 1. The Master-Slave structure is used in the clouding SCADA system. Master server is in charge of maintaining raw data. In addition, it is responsible for system resource scheduling. In the data center of clouding SCADA system, the distributed clouding architecture is used.

3 The Multiple Data Centers Storage System Design for Clouding SCADA

3.1 Oil and Gas Pipeline Clouding SCADA System Multiple Data Center Storage Model

The oil and gas pipeline multiple data center storage system is composed by multiple distributed data center. These distributed data center set can be expressed as $MDS = \bigcup_{i=1,2,\ldots,|MDS|} \{dc_i\}$. There, dc_i expresses No. i data center. The oil and gas pipeline multiple data center storage system is abbreviated to SCADA-MCMSS (SCADA Mutiple Cascading Master-Slave Storage System).

$$\text{Set SCADA-MCMSS} = \{\text{MDS, MM}\} \tag{1}$$

$$\begin{aligned} \text{Set MM} = (&\text{mm\_ID, mm\_IP, StroageInfo, Io\_info,} \\ &\text{Cpu\_Info, Mem\_Info, GS, Cm\_Info, other\_Info}) \end{aligned} \tag{2}$$

There, MM expresses Main Master of system. It is the main server of storage system. mm_ID, mm_IP express the ID and IP address of Main Master, respectively.

$$\text{Set Storage\_Info} = (\text{StorageTotal, StorageUsed, StorageUsage}) \tag{3}$$

$$\text{Io\_Info} = (Io\_Total, Io\_Used, Io\_Usage) \tag{4}$$

$$Cpu\_Info = (Cpu\_Total, Cpu\_Used, Cpu\_Usage) \tag{5}$$

$$Mem\_Info = (Mem\_Total, Mem\_Used, Mem\_Usage) \tag{6}$$

$$CmInfo = \bigcup_{i=1,2,\ldots,|MDS|} \{CM_i\} \tag{7}$$

There, Storage_Info is the storage space of Main Master, Io_Info is IO throughput, Cpu_Info is CPU information, Mem_Info is memory information, and Cm_Info is Client Master information. GS (Global File Name Space) expresses the file and data object naming space. Other_Info is the reserved bits of Main Master. Set $S_i = \bigcup_{j=1,2,\ldots,|dc_i|} \{s_{ij}\}$ expresses storage nodes set of data center dc_i.

$$\text{Set } dc_i = \{S_i, CM_i\} \tag{8}$$

dc_i is a local HDFS system. It's system structure, data reading and writing mechanism, heart beating mechanism is same to HDFS.

$$\text{Set } s_{i,j} = (i, j, Storage\_Info, Io\_Info, Cpu\_Info, Mem\_Info, Blocks) \tag{9}$$

There, $s_{i,j}$ is the j storage node of data center dc_i, $Blocks = \bigcup_{m=1,2\ldots} \{block_m\}$ is data block of storage nodes.

$$\text{Set } CM_i = (i, cm\_ID, cm\_IP, Storage\_Info, Io\_Info, Cpu\_Info, Mem\_Info, meta\_Info, other\_Info) \tag{10}$$

CM_i is Client Master of the i data center dc_i. It is the NameNode of dc_i. cmID and cm_IP are the ID and IP address of CM_i.

$$\text{Set meta\_Info} = (\text{BandWidth, Storage\_Info, FailureTime, dc\_FNS, Si\_Info, other\_Info}) \tag{11}$$

Meta_Info is the metadata information storage in CM_i, it is metadata information of dc_i. Bandwidth is network delay of data center. Storage_Info is the storage space information. FailureTime is average MTBF. $dcFNS_i = (Files, FileNum, TotalAccess, TotalAnswered)$ is the namespace of file and data object. $Files = \bigcup_{j=1,2,\ldots,|FileName|} \{File_i\}$ is the file or data object set of dc_i. FileNum expresses the files and data object number of dc_i. TotalAccess and TotalAnswered is visited number and response request number of dc_i

$$\text{set } File_i = (FileGUID_i, BlockSize, FileBlocks_i, FileReplicaNum_i, FileMeta\_Info_i) \tag{12}$$

There $File_i$ is the i files. FileGUID is the only GUID. BlockSize is default block size. FileReplicaNum is the copy number of file block. FileMeta_Info is metadata information of file, including file owner, user group, file size, read and write permissions, visited number, response request number and so on.$FileBlocks_i$ is data block list of file, $FileBlocks_i = \bigcup_{n=1,2,\ldots} \{block_n\}$. $SiInfo = \bigcup_{j=1,2,\ldots,|dc_i|} \{s_i\}$ is the metadata information set of each storage nodes in dc_i.

In the cloud storage system, MM and CM_i is master and slave relationship. The storage status of data center in each CM_i is reported to MM by heart beating information. MM manages the CM_i just like NameNode manages DataNode in HDFS. So, in the system, MM and CM_i, CM_i and each storage nodes in dc_i organize a cascade master slave architecture. It is called Multiple Cascading Master-Slave Storage System (MCMSSS).

In the storage system, when users visit some file or object, the data center dc_i will search $GUID(obj)$ information in namespace of local CM_i. If the $GUID(obj)$ information is not find in local CM_i, it is searched in other name space. *MM* will return the information of dc_i which has $GUID(obj)$ data. Users visit and operate data *Obj* according to related information and related mechanism of MCMSS.

4 Design Replica Consistency Maintenance Policy

The replica consistency maintenance policy based on two stage lock and clouding SCADA the multiple data center storage system can be shown in Figs. 2 and 3.

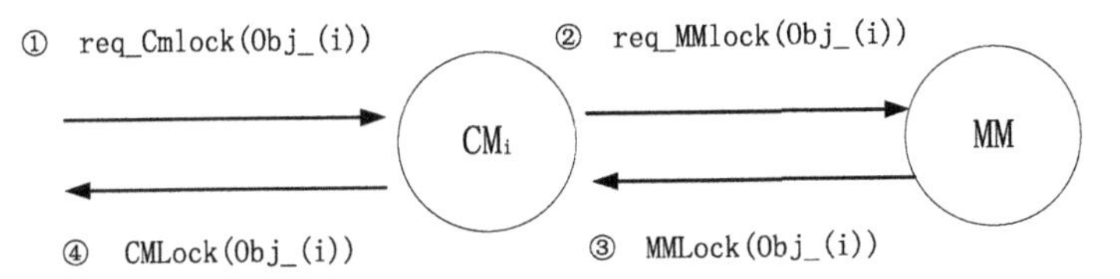

Fig. 2. The procedure of requesting a lock

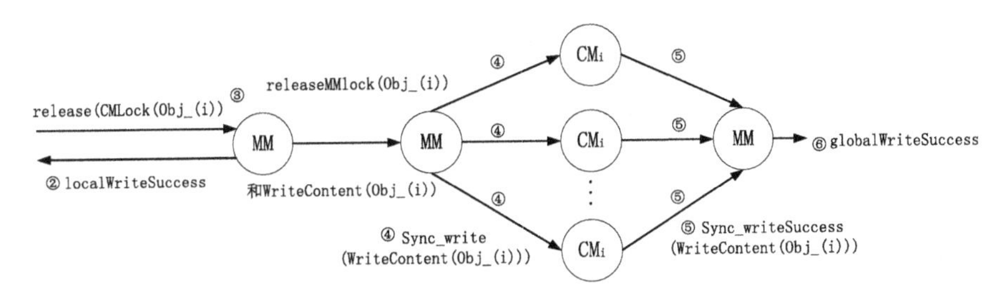

Fig. 3. The procedure of release the locks

In this replica consistency maintenance policy, if users want to update the *i* copy *Object_(i)* of data object *Obj*, the phase of obtaining the two lock must be gone through.

(1) The procedure of requesting a lock (Fig. 2)

① Before *Object_(i)* is updated, users must get a lock for permitting Client Master i (*Object_(i)*) to operate *Object_(i)*. It means the users must apply to *Object_(i)* for a lock to operate *Object_(i)*. This process can be described as *req_CMlock(Object_(i))*.
② Before *Object_(i)* locking the operation of *Object_(i)*, the lock of permitting Main Master (MM) to operate *Object_(i)* must be obtained. It means *Object_(i)* must apply to MM for a lock to operate *Object_(i)*. This process can be described as *req_MMlock(Object_(i))*.

③ If Main Master responses lock request of *Obj*, MM locks the operation to *Obj* in system global namespace and return the lock of operating $Object\_(i)$ to $MMLock(Object\_(i))$. This procedure can be described as $MMLock(Object\_(i))$;
④ When $Object\_(i)$ receives the permission operation lock to $Object\_(i)$ from Main Master, the news of $req\_CMlock(Object\_(i))$ success is returned at once. This procedure can be described as $CMLock(Object\_(i))$.

(2) The procedure of release the locks (Fig. 3).

① Users apply to CM_i for releasing the lock to operate $Object\_(i)$. This procedure can be described as $CMLock(Object\_(i))$.
② If releasing $CMLock(Object\_(i))$ is responded, the news of successful releasing lock is returned to User and the updated *Obj* can be seen in the data center with CM_i. This procedure can be described as $localwriteSucess$.
③ If ② happens, CM_i apply for releasing the lock to operate $Object\_(i)$, at the same time, the updated content is sent to MM. This procedure is described as $realeaseMMLock(Object\_(i))$.
④ MMlaunches the update work of synchronizing $Object\_(i)$ to CM_i with *Obj* copy. This procedure can be described as $Sync - write(WriteContent(Object\_(i)))$.
⑤ Each CM_i with *Obj* copy returns synchronization updating complete news to MM after it's $Object\_(i)$ updating is synchronized completely. This procedure is described as $SyncwriteSuccess(WriteContent(Object\_(i)))$.
⑥ If all procedures $SyncwriteSuccess(WriteContent(Object\_(i)))$ are finished, MM releases the lock to operate $Object\_(i)$ in the system global namespace. This procedure can be described as $globalWriteSuccess$.

5 Summary

The cloud storage system for oil and gas pipeline SCADA system based on cloud computing and the replica consistency maintenance policy are designed in the paper. This is a Multiple Cascading Master-Slave Storage System. In the system, the metadata information is saved in Client Master of each data center. Then Client Master reports the metadata information to Main Master. Because of two layer metadata extraction, the storage pressure of Main Master is less. So, the SCADA-MCMSSS can storage ultra large capacity data content and support expansibility on data center level. In additional, each data center has internal name space. In most situation, each data center can provide independent decision and data reading and writing visit. In addition, the two stage lock replica consistency maintenance policy can ensure the consistency of data copy of multiple data centers.

References

1. Bailey, D., Wright, E.: Practical SCADA for Industry, pp. 236–243 (2015)
2. Cherdantseva, Y., Burnap, P., Blyth, A., et al.: A review of cyber security risk assessment methods for SCADA systems. Comput Secur. **56**, 1–27 (2016)
3. Sun, P., Li, J., Wang, C., et al.: A generalized model for wind turbine anomaly identification based on SCADA data. Appl. Energy **168**, 550–567 (2016)
4. Castellani, F., Astolfi, D., Sdringola, P., et al.: Analyzing wind turbine directional behavior: SCADA data mining techniques for efficiency and power assessment. Appl. Energy, 510–523 (2017)
5. Rezai, A., Keshavarzi, P., Moravej, Z.: Key management issue in SCADA networks: a review. Eng. Sci. Technol. Int. J., 367–377 (2016)
6. Jain, A., Mahajan, N.: Introduction to cloud computing, The cloud DBA-Oracle. Apress, 143–150 (2017)
7. Packer, R.G.: The ethics of cloud computing. Sci. Eng. Ethics **23**(1), 21–39 (2017)
8. Liu, M., Wang, X., Yang, C., et al.: An efficient secure internet of things data storage auditing protocol with adjustable parameter in cloud computing. **13**(1), 155014771668657 (2017)
9. Niki, K., Maragoudakis, M., Loukis, E., et al.: Prediction of propensity for enterprise cloud Computing adoption, Hawaii International Conference on System Sciences, **10**(3), 44–51 (2017)
10. Miao, L., Yuan, M.C., Wang, C.Q., et al.: The oil and gas pipeline SCADA system based on cloud computing and the scheduling algorithm. Advanced Materials Research, pp. 1406–1410 (2014)

SPSS-Based Research on Language Learning Strategy Use

Yang Xu(✉)

School of Foreign Language, Shenyang Ligong University, Liaoning, China
xuxu29@Tom.com

Abstract. This research aims to investigating the language learning strategies use of foreign language learners. The computer program SPSS (Statistical Package for Social Science) was adopted, and a questionnaire with 24 items about learning strategies was made to collect data from 32 Chinese university freshmen. The study of the collected data shows a positive relationship between learning strategy use and efficiency, which also verifies that the deep influence of the conventional learning strategies prevents those students from employing many other new ones to improve their foreign language skills.

Keywords: SPSS · Learning strategies · Efficiency · Language skills

1 Introduction

People usually wonder why some people learn a foreign language much more efficiently than others. Experts prefer to explain them in this way, because there are lots of factors influence the learners' learning outcomes, such as age, sex, previous experience with language learning, personal factors, language aptitude, attitude, motivation, intelligence and cognitive styles [1]. But can we professors help our foreign language learners to find a suitable learning strategy to learn more effectively? We do have a lot of reasons to give those questions affirmative answers. Many researchers gradually find the significant difference between the teacher-center and the student-center classroom teachings, so they spare no efforts to convert teaching methods into learning strategies, to arouse students' interests and improve their involvement in class. The application of learning strategies will increase their effective involvement and thus to improve their language skills. What I try to do in this research is to find out what learning strategy they employed, and whether they are efficient strategies for their foreign language learning.

1.1 Language Learning Strategy (LLS) and Related Concepts

Reid defined LLS in 1995 as a learner's "natural, habitual, and preferred way(s) of absorbing, processing, and retaining new information and skills" [2], which is different from learning styles. Language Learning is given the definition of "best reserved for general tendencies or overall characteristics of the approach employed by the language learner, leaving techniques as the term to refer to particular forms of observable learning behavior" [3]. Wenden and Rubin define LS as "…any sets of operations,

F. Xhafa et al. (eds.), *Advances in Intelligent Systems and Interactive Applications*, Advances in Intelligent Systems and Computing 686,
https://doi.org/10.1007/978-3-319-69096-4_102

steps, plans, routines used by the learner to facilitate the obtaining, storage, retrieval, and use of information" [4]. Whereas Rubin later in 1987 wrote that LS are "strategies which contribute to the development of the language system which the learner constructs and affect learning directly" [5]. Oxford [6] provides lots of examples of LLS, such as learning ESL, watching U.S. TV soap operas, guessing the meaning of new expressions and predicting what will come next and thus the following definition: "… language learning strategies—specific actions, behaviors, steps, or techniques that students use to improve their progress in developing L2 skills. These strategies can facilitate the internalization, storage, retrieval, or use of the new language. Strategies are tools for the self-directed involvement necessary for developing communicative ability" [6].

1.2 The Classification of Language Learning Strategies

Many scholars attempted to classify Language Learning Strategies, and their classification will be listed in time sequence. In 1985, O'Malley got three main subcategories: Metacognitive, Cognitive and Socio affective strategies [7]; In 1987, Rubin divided learning strategies into three types: learning (cognitive strategies and metacognitive learning strategies), communication and social strategies [5]; In 1990, Oxford classified those strategies into direct strategies (memory, cognitive and compensation strategies) and indirect strategies (metacognitive strategies, affective strategies and social strategies) [8]; In 1992, Stern's classification goes like this: management and planning strategies, cognitive strategies, communicative-experiential strategies, interpersonal strategies and affective strategies [9].

2 Research Methods

2.1 Objects

There are 32 Chinese university freshmen involved in this experiment, and all of them are non-English major students in automotive technology department.

2.2 Tools

Questionnaires: At the beginning of the experiment, I invited a group of experienced foreign language teachers to discuss and design a questionnaire for investigating the condition of the language learning strategies use by the students. Oxford's classification in 1990 is employed. There are totally 24 items in the questionnaires, which were divided into six categories with memory strategies, cognitive strategies, compensatory strategies, metacognitive strategies, affective strategies, and social strategies. The students are asked to mark their corresponding levels with numbers 1–5 for never using this strategy, rarely using the strategy, sometimes, often and always using this strategy. The 24 items are as follows,

I. Memory strategies: I review the text to memorize the usage of the new words; I link new words to some visual images such as pictures, familiar people; I relate some words to the information I already learned; I use the handouts for class lectures to memorize the information.
II. Cognitive strategies: I practice speaking by imitate others' speech; I read the vocabulary several times a day; I read English newspapers to improve my reading comprehension; I separate the new words into parts I am familiar with.
III. Compensatory strategies: I guess the part of speech of a word based on its formation; I use other words to express the word which I can not remember; I guess the meaning of a word based on the context while I am reading; I use body languages to express my meaning when I don't know the proper word.
IV. Metacognitive strategies: I review what I learned before the class; I talk about my feeling of learning English with others; I force myself to speak more even when I am afraid of using English; I make some learning plans to improve English.
V. Affective strategies: I listen to some English songs when I feel nervous; I try to relax myself when I am learning English; I see English movies to amuse myself even though I don't understand everything; I study some customs to arouse my interest in English learning.
VI. Social strategies: I discuss a confusing point with my classmates; I speak to my roommates in English; I ask the speaker to slow down when I can't keep up with him; I ask others to correct my faults when I am speaking English.

2.3 Data Collection and Analysis Method

We gathered their university entrance scores for reference to evaluate the efficiency of those language learning strategies. Before the survey was carried out, the students were told that these questionnaire were anonymous survey and therefore would do no effect on their final grades. They had 20 min to finish it in class, and were asked to think about their own real conditions carefully. 32 questionnaires were sent to the students, and all of them are available ones. Statistical Package for Social Science (SPSS) was adopted to analyze these collected data.

3 Results and Analysis

First, we divided all the students' university entrance scores into 3 groups: relatively lower score, medium scores, and higher scores groups, and then the sum of the language learning strategies figures into 3 groups as well to analyze the relationship between them. After carefully comparison, we found that the university entrance scores were quite consistent with the learning strategies. So it could be concluded that those learning strategies play an effective role in their language learning, and thus have direct relation to their final scores. However, the analysis of the higher scores group shows the relationship between the strategies and the scores were not as close as the other two groups, which means the effective adaption of those learning strategies help to increase their final scores, but it seems there are some other factors effect the scores positively.

The 24 items are not enough for us to get to know all the factors influencing their foreign language learning, so we could not tell what the other reasons are, which needs a quite different questionnaire and survey, and therefore a further research to get a complete conclusion.

With SPSS, we sum up all the figures of each strategy categories in the questionnaires, and made Table 1. In this table, the average value of each strategy category is around 3, which proves that all these learning strategies are used frequently by those students. And obviously the social strategies are used a little bit more widely than others. In these 6 categories, the social strategies' mean value is comparatively higher, while its Std. deviation is relatively lower. Social strategies refer to those methods such as seeking out friends who are native speakers of the target language, working with peers in a classroom settings, which are strongly recommended by all the foreign language instructors [10]. We also find that the affective strategies group has medium mean value but a comparatively higher std. deviation value. Affective strategies involve a lot of emotional factors, such as one's attitude, motivation and value view, etc., which usually make an significant influence on language learning process. For example, anxiety is the hot topic a lot of linguistic experts are quite interested in, and it has a great impact on learners' performance. The higher std. deviation value of the affective strategies in Table 1 tells that the students do use different strategies to relax themselves when studying English, but since we didn't list all the possible items in the questionnaire, we could only take a guess that those students need to find methods which fit to themselves to arouse one's interest in language learning, even to improve their performance level.

Table 1. The statistics of six categories of learning strategy

Category	Number	Minimum	Maximum	Mean	Std. deviation	Average
Memory strategies	32	6	16	11.58	2.402	2.97
Cognitive strategies	32	7	18	12.06	2.287	3.02
Compensatory strategies	32	8	17	12.36	2.603	3.03
Metacognitive strategies	32	7	17	12.19	2.562	3.05
Affective strategies	32	8	18	12.41	2.987	3.12
Social strategies	32	9	19	12.78	2.396	3.23

We picked up the most popular and representative learning strategy in each categories to form the Table 2. The Sig. value of all the strategies are higher than 0.05, which shows lower significance probability of those learning strategies. All these strategies are used widely by this group of students. Half of the strategies' Std deviation is higher than 1, while the other half is lower than 1. When people begin to learn a foreign language, they are suggested to read more, and get to know the new words in text, which made it easy to use them again in the same situation. So this learning strategy became the most efficient one with the highest R value. The language instructors would tell their students that they could separate some polysyllabic words into several familiar parts according to their formation, that guessing the meaning of the words on the context is a reading skill they need to get, that for better performance, the students need to review what they learned before next class, and in the leisure time, listening to English songs and seeing English movies will not only arouse one's interest in English, but also learn different traditions and customs. Though all the language learners are encouraged to speak more and ask whenever they meet difficulties in communication, sometimes they are too shy to ask questions. So social strategies represented by this is not as efficient as others strategies. Actually, all the above six learning strategies are the most conventional language learning strategies used in our traditional classroom teaching and self-learning process. Those strategies are well known among the foreign language learners and they do help learners get higher scores in all kinds of examinations. By analyzing the collected data and their university entrance scores, the efficient use of those conventional learning strategies do contribute to the higher scores. Between these two aspects exists a close correlation.

Table 2. The most efficient strategy of each categories

Categories	Items	R	Sig.	Std. deviation
Memory strategies	I review the text to memorize the usage of the new words	0.458	0.010	1.002
Cognitive strategies	I separate the new words into parts I am familiar with	0.402	0.020	0.0981
Compensatory strategies	I guess the meaning of a word based on the context while I am reading	0.389	0.008	0.0862
Metacognitive strategies	I review what I learned before the class	0.358	0.052	1.008
Affective strategies	I see English movies to amuse myself even though I don't understand everything	0.364	0.048	1.009
Social strategies	I ask the speaker to slow down when I can't keep up with him	0.348	0.034	0.0965

4 Conclusion

The analysis shows that the 6 categories of learning strategies are widely used by these university freshmen in their language learning. But different learners may adopt different learning strategies, even the same learner would prefer a quite different strategy in some particular conditions. From this experiment, we found that the conventional strategies are more popular than others. It is probably the traditional examination tendency that leads to the current situation. Language instructors usually design the final exam to test how much knowledge points in the book have been mastered by the students, so they prefer to choose some topics and questions out of the texts. To review the texts repeatedly and even to completely recite the contents of the book may contribute to their higher final grades. Thus, those conventional learning strategies gain their popularity among students. However, the further detailed analysis tells us that to the low scores group and medium scores group, there is a close correlation between the efficient use of these learning strategies and their final scores. The more strategies they used, the higher scores they got. But to the higher scores group, the learning strategies use are not so closely correlated to the final grades, which means the use of those learning strategies is not the only reason to help them get the higher scores, there must be some other factors to effect their language learning and thus improve their language capability.

References

1. Ellis, R.: The study of second language acquisition. Shanghai Foreign Language Teaching Press (1999)
2. Reid, J.: Learning styles in the ESL/EFL classroom. Heinle & Heinle, Boston (1995)
3. Richard, J.C., Renandya, W.A. (eds.): Methodology in language teaching: an anthology of current practice. Cambridge, Cambridge (2002)
4. Wenden, A., Rubin, J. (eds.): Learner strategies in language learning Englewood Cliffs. N.J, Prentice-Hall International (1987)
5. Rubin, J.: Learner strategies: theoretical assumptions, research history and typology. In: Wenden, Rubin (eds.) (1987)
6. Oxford, R.L: Language learning strategies in a nutshell: update and ESL suggestions, Cambridge (2002)
7. O'Mally, J., Chamot, A.: Learning strategies in second language acquisition. Cambridge University Press (1990)
8. Oxford, R.: Language learning strategies: what every teacher should know. Rowley, Mass, Newbury House (1990)
9. Stern, H.H.: Issues and options in language teaching. OUP, Oxford (1992)
10. Richards, J.C., Schmidt, R., Kendrick, H., Kim, Y.: Longman dictionary of language teaching and applied linguistics. Foreign Language Teaching and Research Press (2005)

Study on Intelligent Monitoring of Power Transformer Based on FPGA

Fu-Sheng Li(✉), Xin-Dong Li, Hong-Xue Bi, and Guang Jin

Locomotive Vehicle Department, Zhengzhou Railway Vocational & Technical College, Zhengzhou 450052, China
lifusheng666@126.com

Abstract. This paper discusses the design and implementation of a transformer intelligent monitoring system based on the FPGA Stratix IV. The monitoring system utilizes the chip resources of FPGA, and uses precise rectification circuit and amplifying circuit to collect data of voltage and current. Pulse Value is converted from the electric energy with the sensor and the pulse conversion circuit, and is sent into the FPGA for processing. The software part of the system can complete the monitoring of power transformer parameters and running state, including the record of voltage, current, reactive power, apparent power, frequency, power factor, and automatic formation of various power curve analysis report, monthly management curve report, etc. The system adopts double anti-interference measures, and can be used on transformers of self closing railway lines and various substations.

Keywords: Transformer · Intelligent monitoring · FPGA

1 Introduction

Power transformer is an important equipment of power system, its intelligent monitoring is an important component of modern power management, and its original data acquisition is very critical [1, 2]. However, some data collection work is done after the power outage, which greatly affects efficiency. Some automatic monitoring systems are relatively large, with high equipment investment and poor geographical adaptability [3–5]. If the fault occurs in the power transformer and is not handled promptly, it will cause serious economic losses, and the consequences are often disastrous [6]. This paper discusses a kind of transformer intelligent monitoring system which can make full use of the original equipment, and read all kinds of data of the power transformer with modern electronic technology, and send the information into the core of FPGA. The system can handle and record voltage, current, power, electric energy, power factor, frequency, phase and other parameter of the single-phase, three-phase three wire and three-phase four wire AC circuit. In this way, it can realize the safe and reliable operation of the power transformer.

F. Xhafa et al. (eds.), *Advances in Intelligent Systems and Interactive Applications*, Advances in Intelligent Systems and Computing 686,
https://doi.org/10.1007/978-3-319-69096-4_103

2 Experiment Setup and Procedures

The hardware components of this system are shown in Fig. 1. It is mainly composed of FPGA data processing module, data acquisition module, current and voltage conversion module, keyboard and display control module, zero crossing detection module, real-time clock and USB communication interface module.

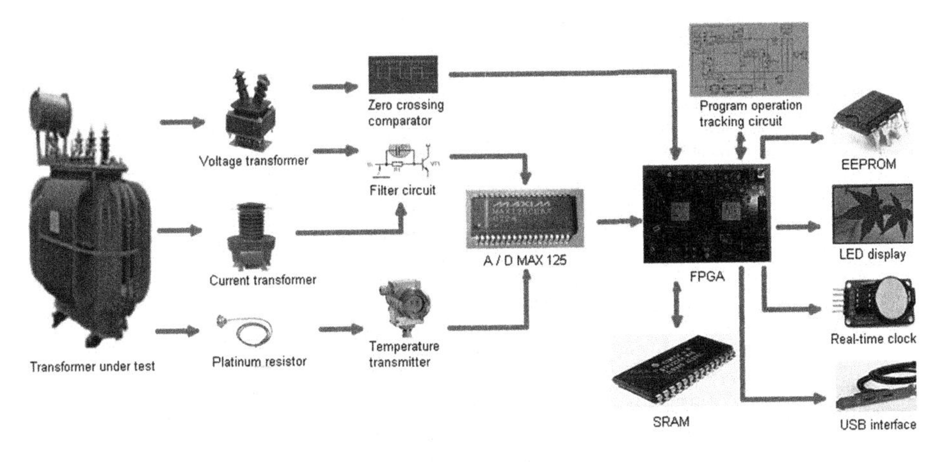

Fig. 1. Hardware block diagram of monitoring system

Among them, FPGA chose the Stratix IV model introduced by Altera Company, the data read card chose the EEPROM of 16 K capacity, which can read more than 60 transformer data. We can insert the data reading card into the transformer intelligent monitoring system, and transmit the measured data to the PC computer with the microcomputer communication through the USB port for calculation, then generate various charts. In addition, the system has a low pass filter in the power source in order to isolate the digital channel.

The main program flow chart of the transformer intelligent monitoring system is shown in Fig. 2. The command redundancy and software trap method are adopted to solve the flying off problem of the program. If the program forms a dead loop, this is a problem that command redundancy and software traps can not solve, but we can reset the circuit in a timely manner and let the program restart [7].

The program adopts digital filtering technique, which collects the data continuously for five times and then takes the average value. In order to reduce the interference of instantaneous ripple signal, the digital filter technology is adopted in the program. In the process of data collection, the measured data is collected continuously 5 times, and the average value is taken. The initial value of the minimum is the full range value, and the initial value in the minimum data memory is zero.

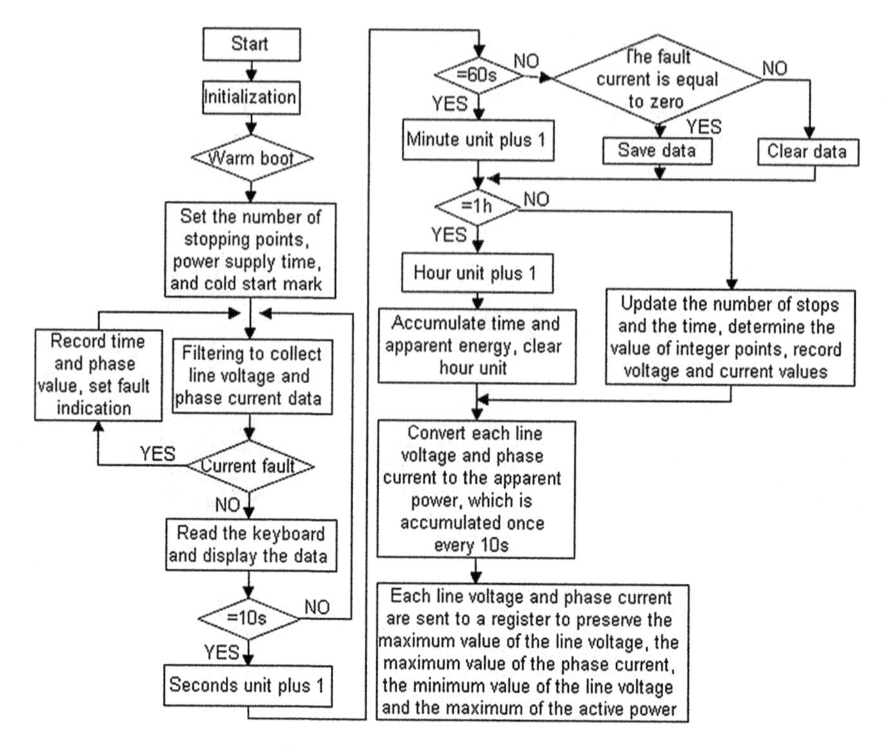

Fig. 2. Master program flow chart

3 Data Detection and Calculation Methods

We can input the square wave signals of three-phase AC shaping through the voltage comparator to the capture port of FPGA respectively. The clock frequency of the counter T1 is 1 MHz, and the counter is incremented every 1 μs. The value of counter T1 is automatically latched into the register and a signal is issued. When the capture port occurs a positive jump, the difference between the value of the two input capture signals is the signal cycle. So we can calculate the frequency based on this cycle. Assuming that m_1 and m_2 are the count values of the corresponding one of the first and the n + 1th signals detected by the capture port, the number of counter overflows is N, so the measured frequency is:

$$f = \frac{n}{(m_1 - m_2 + N \times 65536) \times 10^{-6}} \tag{1}$$

Suppose the frequency signal period is T, Ix(n) and U(n) is the discrete sampling sequence obtained by uniformly sampling N points of current and voltage in one cycle. When the sampling theorem is satisfied, the relevant parameters of the frequency signal can be calculated according to the numerical integration method. Let N be large enough and select the appropriate one, the effective value of the AC voltage is:

$$U = \sqrt{\frac{1}{T}\int_{0}^{T} u^2(t)dt} \tag{2}$$

The value of U has the characteristics of discretization. If the continuously varying voltage values are replaced by a sampling sequence in one cycle, we can derive the following formula according to the rectangular algorithm of numerical integration:

$$U_m = \sqrt{\frac{1}{T}\sum_{m=0}^{N-1} u_m^2 \Delta T} \tag{3}$$

In the formula, U_m is the voltage sampling value, ΔT is the sampling period. Since the two adjacent sampling intervals are equal, there is a constant N:

$$N = \frac{T}{\Delta T} \tag{4}$$

If we bring Eq. (4) into Eq. (3), we can calculate the effective value of AC sampling voltage:

$$U = \sqrt{\frac{1}{N}\sum_{m=0}^{N-1} U_x^2(n)} \tag{5}$$

In the same way, the effective value of the current can be calculated:

$$I = \sqrt{\frac{1}{N}\sum_{m=0}^{N-1} I_x^2(n)} \tag{6}$$

Single-phase average power is calculated as:

$$P = \frac{1}{T}\int_{0}^{T} u(t)i(t)dt \tag{7}$$

After the formula (7) is discretized, we can be obtained:

$$P = \frac{1}{N}\sum_{n=0}^{N-1} U_x(n)I_x(n) \tag{8}$$

The Ix(n) and Ux(n) of the formula (8) are the sampling values of the current and voltage at the same time.

Therefore, according to effective value of AC voltage and AC current as well as the AC active power, we can find the following three parameter values:

$$S = UI \tag{9}$$

$$Q = \sqrt{S^2 - P^2} \tag{10}$$

$$\cos\varphi = \frac{P}{S} \tag{11}$$

Among them, S is the apparent power, Q is the reactive power, and $\cos\varphi$ is the power factor.

Equal periodic sampling or synchronous sampling is the measurement method used in this monitoring system. The measurement must be met:

$$\mathrm{T} = \mathrm{N} \times \mathrm{TS} \tag{12}$$

In the formula, T is the period of the signal to be measured, TS is the sampling interval, and N is the number of points sampled every cycle.

4 Results and Discussion

This paper uses the S12 series power transformer as the test object. The main parameters of the S12 transformer include a high voltage of 10.00 kV, a low voltage of 400.00 V, a high voltage resistance of 4.9850 Ω, a low voltage resistance of 9.3207 mΩ, and a nominal capacity of 200 kVA. The main parameters of the three-phase electrical energy tested are shown in Table 1.

Table 1. The test results

Category parameter	A phase	B phase	C phase	Three-phase average value
Voltage effective value	396. 55 V	399. 92 V	400. 10 V	398. 86 V
Absolute mean of the voltage	395. 20 V	398. 58 V	399. 26 V	397. 69 V
Current effective value	0.5780 A	0.6011 A	0.5638 A	0. 5810 A
Active power loss	0.2243 kW	0.1838 kW	0.1238 kW	0.1773 kW

According to the test results, it can be calculated that the apparent power S is 231.7377 kVA, the power factor $\cos\varphi$ is 0.8675. Thus, the voltage deviation rate is −0.2953%, so the system can fully meet the monitoring requirements.

5 Conclusions

This transformer intelligent monitoring system takes full advantage of the internal resources of FPGA and 14 bit synchronous sampling A/D converter MAX125. Its software system is written by Verilog HDL language and C language, so that it can be upgraded easily. It is widely used, especially for systems that require long automatic monitoring of power transformers, and has been successfully used in railway power systems.

Acknowledgements. The work was supported by Henan Science and Technology Development Project under award 172102210557.

References

1. Xu, K.L., Zhou, M., Ren, J.W.: An object-oriented power system fault diagnosis expert system. International conference on electrical engineering 1999 (ICEE'99), Hong Kong, pp. 223–234, 1999
2. Mcarthurs, D.J., Davsion, M., Hossack, J.A., et al.: Automating power system fault diagnosis through multi-agent system. In: Proceeding of the 37th annual Hawaii international conference on system sciences, Big Island, pp. 947–954, 2004
3. Liu, Q., Jiang, F., Deng, D.: Design and implement for diagnosis system of hemorheology on blood viscosity syndrome based GrC. In: Proceedings of 9th international conference on rough sets, fuzzy sets, data mining, and franular computing, pp. 413–420, 2003
4. Yang, G.P.: The application of artificial intelligence (AI) in power transformer fault diagnosis. In: CI CED proceedings, pp. 561–563, 2000
5. Michel, D., Alonso, D.P.: Interpretation of gas-in-oil analysis using new IEC publication 60599 and IEC TC10 databases. IEEE Electr. Insul. Mag. **17**, 31–32 (2001)
6. Zhang, L.Q., Guo, Q.W., Liu, J.D.: Discussion on solving the problem of temporary low voltage of power supply by wide distribution voltage regulating transformer. Electr. Eng. **10**, 49–50 (2011)
7. Liu, Y.Q., Nie, W.Q.: Application of embedded system in power transformer monitoring. Appl. Electron. Technol. **10**, 65–67 (2010)

Mutual Fund Performance Analysis Using Nature Inspired Optimization Techniques: A Critical Review

Zeenat Afroz, Smruti Rekha Das, Debahuti Mishra, and Srikanta Patnaik(✉)

Department of Computer Science and Engineering, Siksha 'O' Anusandhan University, Bhubaneswar, Odisha, India
{zafroz132431, das.smrutirekha13}@gmail.com, {debahutimishra, srikantapatnaik}@soauniversity.ac.in

Abstract. Successful prediction of mutual fund with maximum accuracy is a great challenge because of highly fluctuating behaviour of the financial market. The prediction of Net Asset Value (NAV) of mutual fund helps investors to wisely plunge money into profitable mutual funds. This survey covers more than 40 related published articles in the field of mutual fund and gone through a systematic review of the various nature inspired techniques, used for NAV prediction with its performance analysis. The performance of mutual fund is highly correlated with the stock market, hence some of the stock market prediction analysis is also well throughout. Through this survey it is found that very few works is done in the field of NAV prediction using optimization techniques while for performance analysis of mutual fund many nature inspired optimization (NIO) techniques have been employed. On the whole, this paper gives a comprehensive review of the literature in the field of mutual fund.

Keywords: Mutual fund · Net asset value (NAV) · Nature inspired optimization (NIO)

1 Introduction

The term 'Market' can be described as any platform where people can trade goods and services in exchange of money, which is influenced by the demand and supply measure. Technically, financial market is a marketplace where sale and purchase of financial instruments takes place such as securities, currencies, and similar assets. Securities [1] can be like shares, mutual funds, bonds and so on. Mutual Fund can be considered as an intermediary tool in the financial market for raising money [2]. It is basically an investment scheme made up by pooling money from several investors, professionally handled by Asset Management Company (AMC). AMC makes use of the fund by spreading it over numerous securities such as bond, stock, debentures, money market instruments and similar assets of various companies. All the AMC's are regulated under Security Exchange Board of India (SEBI). The fund is managed by the fund manager appointed by AMC; who are professional analysts and have detailed knowledge of the financial market for strategic investment. Unlike stock market,

F. Xhafa et al. (eds.), *Advances in Intelligent Systems and Interactive Applications*, Advances in Intelligent Systems and Computing 686,
https://doi.org/10.1007/978-3-319-69096-4_104

mutual funds are highly diversified which minimizes the risk in investment. Mutual fund can be bought in terms of units whose value per unit is generally known to be Net Asset Value (NAV) of Mutual Fund. Every mutual fund scheme has a unique investment objective, on accordance of which the portfolio is decided by the fund manager. Mutual fund portfolio gives complete idea about the investment strategy i.e., how much is invested in what.

To predict the NAV accurately and to enhance the performance of mutual funds various prediction model and optimization techniques, has been proposed by researchers. Several methods have been suggested for calculating the portfolio return and performance. The vast domain of nature inspired algorithms is used since long for the performance evaluation and prediction of mutual funds. The essence of nature inspired optimization (NIO) algorithm [3] lies in the way, how beautifully they mimic nature's processes, to find optimal solutions to complex optimization problem. These NIO techniques have often been applied to numerous financial problems [4] which in turn help in decision making and strategizing investment by determining trading rules for the specific market condition. Nature inspired computing is gradually earning importance [5] with time. The optimized investment decision maximizes the expected return gained from investment over the planning of the decision maker [6].

Researchers have explored many nature based optimization techniques in the field of finance for better prediction of future indices as well as for portfolio optimization. Particle Swarm Optimization (PSO) is a well known NIO technique, which Hsu et al. proposed for investigating trading strategy [7] for mutual fund. Similarly a Genetic Algorithm (GA) technique is given by Tsai for profitable portfolio selection [8] for yielding high return. Another NIO method named Fruit fly Optimization (FOA) is used by Hua et al. for the prediction of NAV [9] mutual fund using General Regression Neural Network (GRNN). A novel co-variance guided Artificial Bee Colony (ABC) algorithm is efficiently used by Kumar et al. for solving the portfolio optimization [10] problem, which is also addressed by Nebojsa Bacanin using the firefly algorithm [11]. Similar kinds of many NIO algorithms are taken up and modified by researchers and it is obvious from the results found that these algorithms are very suitable for solving financial time series problems.

For exploring the financial market and investing in it, depending only on analysis of historical data does not prove to be efficient. This fact motivates us to review and analyse various methods that have been applied to mutual fund schemes for prediction and performance evaluation. In this paper, we survey numerous nature inspired methodologies used for various mutual fund prediction and their performance evaluation that have been proposed since decades. The approach of the survey is theme wise, where variety of evolutionary algorithm used, are analysed and their findings and limitation are stated.

The paper is further organised as follow: Firstly we analyse how mutual funds prediction and performance analysis are made without any optimization techniques. Secondly, we explore the nature inspired methodologies used for optimization of mutual fund portfolio and performance. Lastly, in the conclusion we discuss which approach is better on comparison and we also state our future.

2 Literature Survey

2.1 Performance Analysis and Prediction Without Optimization

Financial market requires fast and instant response for continually changing dynamics of market. The financial market is a highly complicated area for investment where prediction often flops since the numbers of entity that cause a change are too many. Prediction of financial market value helps investor to wisely manage fund and make profit out of it. Mutual fund NAV is one of the important attribute for predicting since its value enables investors to know whether to buy, sell or hold securities. Various intelligent algorithms have been explored by researchers for analysing and predicting mutual fund behavior and NAV. However, researchers may be unaware [5] of the different approaches used by other researchers, since the growth in this vast domain is very fast.

Numerous models in various domains are proposed for mutual fund performance evaluation and NAV prediction which are very briefly stated below. ANN is also used with [12–14] other techniques such as Regression analysis and ARIMA for prediction. Methodologies such as FLANN, MLP and RBF are used for prediction of NAV [15–17] and also combinations of these methods are used on different datasets for NAV prediction of mutual fund. We find in some other papers where Auto Regressive models [18, 19] are proposed for prediction of various schemes of mutual fund. Another model combining k-means with boosting/bagging based classifier ensembles is proposed by researchers [20], which perform much better than the other methods in terms of prediction accuracy.

2.2 Performance Analysis and Prediction with NIO Technique

In this section we overview and discuss various NIO techniques taken up by researchers for mutual fund NAV prediction as well as its portfolio optimization. Since the performance of mutual fund also highly depends upon the stock market, some of them are also considered for our review.

2.2.1 PSO

PSO is an evolutionary optimization technique [21] originally developed by Kennedy and Eberhart in 1995. Since its development, it has been utilized as an optimization tool in numerous applications ranging from medical applications to computer graphics, music composition, etc. [22] and also in solving complex real time optimization problem. It is a well known population based stochastic optimization technique. Zhu et al. [23] have adopted a meta-heuristic approach for effectively solving portfolio optimization problem using PSO algorithm. According to Benninga [24], there are two types of risky portfolios, restricted and unrestricted. The compositions of the optimal risky portfolios developed by PSO optimize 6 portfolios in this paper, whose values of Sharpe Ratio are determined. These values are compared to other optimization techniques such as GA and Visual Basic Application solver (VBA) and it is observed that PSO performs better. Similarly, Xiong et al. further explores Discrete PSO (DPSO) and

PSO algorithm [25] for optimizing fully complex-valued RBF neural network (FCRBFNN) for prediction of synthetic and also real time data. The formation of the complex valued interval is model using FCRBFNN, which is further evolved with PSO and DPSO for combined optimization of structure and parameters. The interval Average Relative Variance (ARV) is used for error measurement, the lower is the ARV value, the better is the forecast. Prediction of the everyday interval stock price series in a one-step-ahead fashion is efficiently achieved using the proposed model.

In similar way, Turbulent Particle Swarm Optimization (TPSO) algorithm is one which makes use of minimum velocity threshold in order to control the velocity of particles [26] and overcomes premature convergence problem of PSO. Using this concept, Hsu et al. proposed a fund trading strategy [7] that combines TPSO and two variants of general moving average technique; one being the fixed length moving average (FLMA) and the other is the variable length moving average (VLMA) techniques to find good set of technical indicator parameters in order to achieve profit and minimize risk in mutual fund investment. To plan the trading strategy, a total of eight moving averages are used which determines the buy and sell point. The proposed model based on mixed moving averages and TPSO (MMPSO) observes that performance on average of each fund has improved and the model can achieve high profit when considering the transaction fee of mutual fund.

A hybrid method based on PSO is proposed by Pulido et al. for modelling an ensemble neural network [27] with fuzzy aggregation of responses for predicting the complex time series. Ensembling of neural network models [28] for prediction purpose behave remarkably well, in recent times. The PSO algorithm is implemented for optimization of the neural network architectures, in which for each of the modules, number of neurons and number of layers and thus the design architecture is chosen such that the neural network produces the best results for the considered time series. Results from each module are desegregated with an integration method based on type-1 and type-2 fuzzy logic, from which the prediction error is obtained. It is found that the proposed algorithm is good enough in terms of speed when compared to other optimization methods and is a good choice for predicting financial time series. In this paper, the basic PSO is used for optimization whereas it is enhanced to a Two-level PSO (TLPSO) [29] by Lu et al. who percepts the credit portfolio management as a min–max optimization problem. Author proposes a TLPSO technique to serve the objective of credit portfolio optimization by minimizing the loss of the portfolio in a given consulting budget constraint. In this paper, the two level of searching of TLPSO are namely, “external-search” which explores specific proportion of wealth invested, the searching process of the minimization problem and “internal-search” to find the best combination consulting firms, the searching process of the maximization problem. The particles in the external and internal population can evolve by original PSO or its variants [30]. The simulation results show that, TLPSO gives best solutions in less computational time as compared to GA and PSO. (Table 1)

Table 1. PSO

Sl. no.	Author	Methodology	Datasets used	Findings	Limitations
1	Hanhong Zhu, Yi Wang, Kesheng Wanga, Yun Chen	PSO	Shanghai Stock Exchange 50 Index	Experimental result shows the proposed model has good efficiency in choosing optimal portfolio. It is seen that PSO performs better than GA and VBA solver and clearly has greater efficiency in solving constrained optimization problems.	Some derived model of PSO can be used for handling larger portfolios in the investment.
2	Xiong Tao, Yukun Bao, Zhongyi Hu, Raymond Chiong	PSO and DPSO	New York Stock Exchange	The DPSO/PSO-FCRBF model is a promising tool for real time interval time series forecasting.	However computational time of the model could be reduced further.
3	Ling-Yuan Hsua, Shi-Jinn Hornga, Mingxing Heb, Pingzhi Fanc, Tzong-Wann Kaod, Muhammad Khurram Khan, Ray-Shine Run, Jui-Lin Laif, Rong-Jian Chenf	TPSO, FLMA and VLMA	Taiwan Stock	The proposed trading model provides good buy and sells points which infers that the model can achieve high profit and the minimize risk the in investment	As the output of the model is compared with the original performance of a fund, it should also be compared with few more well known techniques like GA
4	Martha Pulido, Patricia Melin, Oscar Castillo.	PSO with ensemble NN and fuzzy aggregation	Mexican Stock Exchange	The ensemble neural networks with type-2 fuzzy integration is a better choice as compared to type-1, in complex time series prediction	Type-1 and type-2 fuzzy systems could be further optimized with other variants of PSO.
5	Fu-Qiang Lu, Min Huang, Wai-Ki Ching, Tak Kuen Siu	TLPSO	Numerical Example	Comparing with GA and PSO, TLPSO computation time is less and it is efficient enough to solve typical optimization problems	Model could have been implemented with real time dataset.

2.2.2 GA

GA is a well known evolutionary nature inspired metaheuristic which mimics the process of natural selection. Following the bio-inspired operators such as mutation, crossover and selection, GA is used to generate high-quality solutions to solve optimization problems. The simplest form of Darwinian evolution is GA [31]. Using this simple technique of GA, Yu et al. approaches an evolutionary fuzzy neural network [32] using neural networks (NN), fuzzy logic and GA for financial prediction in different financial domains. Genetic fuzzy NN uses basic GA to initialize parameters [33] and then uses the gradient based learning algorithm to discover the fuzzy rules. To enhance the old genetic fuzzy NN, author merges GA with the gradient-descent learning algorithm in an iterative manner. GA generate optimized parameters for the fuzzy NN with knowledge discovery (FNNKD) [33], which trained to generate new parameter values, which will again be optimized by GA. Simulation result shows, the integration of GA and FNNKD incorporates advantages of each technique for better financial prediction. The prediction accuracy is further analyzed basing upon different simulations, and is more effective as compared to the old genetic fuzzy neural learning algorithm. In similar way, a multiobjective GA technique proposed by Kaucic et al. to generate technically most beneficial technical indicators [34], from a set of almost 5000 signals. In general, technical analysis is widely used to forecast future movements of market by analyzing past prices and volumes [35] and deals with the application of trend following methodologies to financial markets. In this paper, learning methods like boosting, Bayesian model averaging and committee are compared in different market trends. Experimental results show that both statistical and economical objectives are handled efficiently at a same time and the algorithm selects optimal set of technical indicators for the artificial trading system and also their parameterization.

For achieving economical objectives, strategizing the investment is very important. A GA based approach is given by Tsai et al. to strategize the investment [8] for achieving high annual return after considering the transaction fee in real time investment. To select potentially profitable funds, author uses GA and defines a scoring function that considers the performance indicators of all funds. Monitoring indicator is derived from global trend indicator which indicates buying and selling points of the funds. Author also defines various weights and obtains the relationship between the replacement threshold and cumulative Rate of Return (ROI). Simulation results suggest investors to take this investment strategy with dynamic weights based on GA, which makes an excellent annual return. The annual return of any investment very much depends on the risk involved in it and the issues relating management of risk. Two such issues of portfolio risk management are addressed [36], by Lin et al., namely efficient set selection and volatility forecasting. Portfolio Value-at-Risk (PVaR) [37] concept is used here for handling risks and returns for a multiple-asset portfolio investment. For addressing these issues, author proposes the evolutionary GA for portfolio volatility forecasting model as well as efficient set selection based on Extreme Value Theory (EVT) to optimize portfolios, maximize returns and minimize risk. Experimental results proves that the proposed mechanism brings out an efficient set of portfolios, and also estimates more appropriate threshold values for all assets in order to forecast PVaR more accurately. Furthermore interestingly, Chen et al. proposes a combination genetic algorithm (CGA) [38] for solving the investment strategy

Table 2. GA

Sl. no.	Author	Methodology adopted	Dataset Used	Findings	Limitations
1	Lixin Yu and Yan-Qing Zhang	Genetic Fuzzy NN	Bank prime loan rate, federal funds rate and discount rate from January 1955 to December 2000.	Learning accuracy and prediction accuracy of the model proposed is much better as compared to the base models.	The model can be enhanced for multidomain financial system
2	Massimiliano Kaucic	Multi objective GA	S&P 500	The nature and quality of the result depicts that the model works profitably, both in up and down trend of market.	The learning method has not been applied for long term investment.
3	Tsung-Jung Tsai, Chang-Biau Yang, Yung-Hsing Peng	GA	Type A Equity fund and US dollar.	Simulation result shows, the model provides excellent buy and sell points which brings in high annual return.	The proposed model is based only on basic GA approach which can might be enhanced by some derived GA model.
4	Ping-Chen Lin	GA and EVT	Taiwan Stock Exchange	The success rate of GA based PVaR using EVT model is much more than that of other evolutionary technique such as Harmony Search.	Even more efficient portfolio can be selected for by using some hybrid optimization techniques
5	Jiah-Shing Chen	CGA	Top 10 priced stocks of DJIA	Using CGA, the model efficiently derives long trading rules which proves to be helpful in optimizing portfolio of the selected fund.	Model may further be modified to derive short trading rules.

portfolio problem. The classical portfolio problem [39] constitutes of issue in distributing capital to a group of assets. Author compares the returns of CGA with uniform allocation on the multiple assets strategy portfolio. The average accumulated returns of CGA are better than the accumulated returns of uniform allocation. (Table 2)

2.2.3 ACO

In the domain of swarm intelligence, ACO is a probabilistic technique for solving computational problems following the behaviour of ants. As we know, financial decision making process often depends on classification models which involve the selection of the appropriate independent features, relevant for the problem. An automated feature selection technique [40] could greatly help in decision making in financial market. For doing so, Marinakis et al. used two nature-inspired algorithms [40], namely ACO and PSO, for feature selection. Author focuses on two financial classification problems being, credit risk assessment and audit qualifications. Three nearest neighbour (nn) classification based classifier were used for the classification. The performance of the proposed method was tested upon financial data. The experimental result indicates high performance in extracting a reduced set of features with great accuracy. PSO k-nn was found to be the most efficient results in terms of accuracy rates using only half of total available features. Index Fund Problem (IFP) is another type of financial optimization problem in the field of Portfolio management. IFP is addressed by Nigam et al. [41] by using an effective algorithm based on the ACO. In order to approximate the efficiency of the proposed ACO based algorithm, GA was taken for comparison. The experimental result demonstrates the strength of the ACO over the GA approach. Author also states that the proposed algorithm can also easily adapt other attributes of the IFP model such as transaction cost, residual return, etc., only by modifying the objective function definition as per requirement. (Table 3)

2.2.4 ABC, FOA, DE, Fireworks Algorithm, Modified Firefly Algorithm

A bee-inspired algorithm namely cOptBees was originally designed to solve problem related to data clustering [42]. The basic artificial bee colony algorithm [5] can be efficiently used in collective decision making for multi-criteria selection problems. ABC is one of the efficient and widely used optimization techniques of swarm intelligence. The portfolio optimization comprises of finding an optimal way to distribute funds among various available securities so as to maximize the return and minimize the risk. Kumar et al. [10] presents a novel co-variance guided Artificial Bee Colony algorithm to solve the portfolio optimization problem, named as Multi-objective Co-variance based ABC (M-CABC). The performance analysis of the proposed model M-CABC is done by verifying it on the set of global portfolio optimization test problems from the OR-library [43]. Five different experiments on this test suite is done using the model which proves it to be promising, powerful and capable of handling real life portfolio management tasks.

A novel optimization FOA is one which is inspired by the foraging behavior of fruit flies. An FOA optimized GRNN [9] is proposed by Huang et al. for NAV prediction of mutual fund. Author also focuses on constructing an efficient portfolio on the basis of DEA, Sharpe and Treynor indices [44] and ROR of nine mutual funds. The dataset

Table 3. ACO

Sl. no.	Author	Methodology	Dataset Used	Findings	Limitations
1	Yannis Marinakis	ACO and PSO	Doumpos and Pasiouras	It is observed that PSO k-nn gives the best results on both stated problem followed by ACO k-nn	The stated problem can be addressed using some other classifier also like SVM etc.
2	Ashutosh Nigam and Yogesh K. Agarwal	ACO	S&P100 index	Proposed ACO algorithm outperforms the well known GA	Model could have been compared with some other metaheuristics also.

collected was efficiently used in training and testing the model and experimental results shows that the portfolio selected by the Sharpe's index outperformed those selected by other indices and the NAV predicted by Fruit Fly optimized GRNN offered highest accumulated return rate compared to other models. In the field of evolutionary computing, Differential Evolution (DE) is a remarkable optimization technique that iteratively improves a candidate solution of an optimization problem, with regard to a given measure of quality. This global optimization technique converges faster [45] and with more certainty than most of the other acclaimed global optimization techniques. A Fireworks Algorithm [46] is given by Nebojsa Bacanin, for solving constrained portfolio optimization problem and author states that proposed algorithm has strong potential for as it performed better than some variants of GA. The same problem is also addressed by Bacanin et al. with nature inspired modified Firefly Algorithm (mFA) [11] which performs almost uniformly better when compared to techniques like Tabu Search, Simulated Annealing, GA, and PSO.

3 Conclusion

In this survey, we first tried to learn about financial market, mutual fund investment and portfolio selection. We observed various techniques explored by researchers to predict the NAV and optimize portfolio of various mutual funds in the domain of artificial intelligence. Variety of nature inspired methodologies are grouped theme wise which are efficiently used to optimize the prediction results as well as for portfolio optimization. NIO algorithms have recently gained a lot of attention from researchers and investment managers, because of the efficiency it holds. We observed, maximum work with NIO techniques are done for selecting profitable portfolio of the security in which one invests the money, and very few work is done for optimizing the predicted NAV value. The well known NIO algorithm must be implemented for optimizing the predicted NAV of the mutual funds, which might give very good results. Further exploration of nature inspired computing algorithms will surely be beneficial for financial prediction especially in mutual fund.

References

1. Drake, P.P., Frank, F.J.: The basics of finance: an introduction to financial markets, business finance, and portfolio management, **192**, Wiley, 2010
2. Poojara, J.G., Christian, S.R.: Funds mobilization in Indian financial market—a mutual fund perspective. IJEMR **2**(4), 1–14 (2012)
3. Binitha, S., Siva Sathya, S.: A survey of bio inspired optimization algorithms. Int. J. Soft Comput. Eng. **2**(2), 137–151 (2012)
4. Michalak, K., Lipinski, P.: Prediction of high increases in stock prices using neural networks. Neural Netw. World **15**(4), 359 (2005)
5. Kar, A.K.: Bio inspired computing—a review of algorithms and scope of applications. Expert Syst. Appl. **59**, 20–32 (2016)
6. Copeland, T.E., Weston, J.F., Shastri, K.: Financial theory and corporate policy. Reading, Mass.: Addison-Wesley, **3** (1983)
7. Hsu, L.Y., Horng, S.J., He, M., Fan, P., Kao, T.W., Khan, M.K., Run, R.S., Lai, J.L., Chen, R.J.: Mutual funds trading strategy based on particle swarm optimization. Expert Sys. Appl. **38**(6), 7582–7602 (2011)
8. Tsai, T.J., Yang, C.B., Peng, Y.H.: Genetic algorithms for the investment of the mutual fund with global trend indicator. Expert Syst. Appl. **38**(3), 1697–1701 (2011)
9. Huang, T.H., Leu, Y.H.: A mutual fund investment method using fruit fly optimization algorithm and neural network. Applied Mechanics and Materials, **571** Trans Tech Publications, (2014)
10. Kumar, D., Mishra, K.K.: Portfolio optimization using novel co-variance guided artificial bee colony algorithm. Swarm Evol. Comput. **33**, 119–130 (2017)
11. Bacanin, N., Tuba, M.: Firefly algorithm for cardinality constrained mean-variance portfolio optimization problem with entropy diversity constraint. Sci. World J. 2014, (2014)
12. Priyadarshini, E., Babu, A.C.: A comparative analysis for forecasting the NAV's of Indian mutual fund using multiple regression analysis and artificial neural networks. Int. J. Trade, Econ. Finan. **3**(5), 347 (2012)
13. Priyadarshini, E.: A comparitive analysis of prediction using artificial neural network and auto regressive integrated moving. ARPN J. Eng. Appl. Sci. **10**(7), 3078–3081 (2015)
14. Chiang, W.C., Urban, T.L., Baldridge, G.W.: A neural network approach to mutual fund net asset value forecasting. Omega **24**(2), 205–215 (1996)
15. Anish, C.M., Majhi, Babita: Net asset value prediction using FLANN model. Int. J. Sci. Res. IJSR **4**(2), 2222–2227 (2015)
16. Tsai, C.F., Lu, Y.H., Yen, D.C.: Determinants of intangible assets value: the data mining approach. Knowl.-Based Syst. **31**, 67 (2012)
17. Narula, A., Jha, C.B., Panda, G.: Development and performance evaluation of three novel prediction models for mutual fund NAV prediction. Annual Res. J. Symbiosis Centre for Manag. Stud. **3**, 227–238 (2015)
18. Roy, S., Ghosh, S.K.: Can mutual fund predict the future? an empirical study. J. Finan. Econ. **2**(1), 1 (2013)
19. Gowri, M., Deo, M.: Prediction model of the net asset returns of indian equity fund of mutual funds with application of ARMA. Indian J. Appl. Res. **5**(10), (2016)
20. Tsai, C.F., Lu, Y.H., Hung, Y.C., Yen, D.C.: Intangible assets evaluation: the machine learning perspective. Neurocomputing **175**, 110–120 (2016)
21. Poli, R., Kennedy, J., Blackwell, T.: Particle swarm optimization. Swarm Intell **1**(1), 33–57 (2007)

22. Sedighizadeh, D., Masehian, E.: Particle swarm optimization methods, taxonomy and applications. Int. J. Comput. Theory Eng. **1**(5), 486 (2009)
23. Zhu, H., Wang, Y., Wang, K., Chen, Y.: Particle Swarm Optimization (PSO) for the constrained portfolio optimization problem. Expert Syst. Appl. **38**(8), 10161–10169 (2011)
24. Benninga, S.: Financial modeling. MIT Press Books **1** (2008)
25. Xiong, T., Bao, Y., Hu, Z., Chiong, R.: Forecasting interval time series using a fully complex-valued RBF neural network with DPSO and PSO algorithms. Inf. Sci. **305**, 77–92 (2015)
26. Abraham, A., Liu, H.: Turbulent particle swarm optimization using fuzzy parameter tuning. Foundations of Computational Intelligence Volume 3. Springer, Berlin Heidelberg, pp. 291–312 (2009)
27. Pulido, M., Melin, P., Castillo, O.: Particle swarm optimization of ensemble neural networks with fuzzy aggregation for time series prediction of the Mexican stock exchange. Inf. Sci. **280**, 188–204 (2014)
28. Hansen, L.K., Salamon, P.: Neural network ensembles. IEEE Trans. Pattern Anal. Mach. Intell. **12**(10), 993–1001 (1990)
29. Lu, F.Q., et al.: Credit portfolio management using two-level particle swarm optimization. Info. Sci. **237**, 162–175 (2013)
30. Zhang, J., Avasarala, V., Subbu, R.: Evolutionary optimization of transition probability matrices for credit decision-making. Eur. J. Oper. Res. **200**(2), 557–567 (2010)
31. Aguilar-Rivera, R., Valenzuela-Rendon, M., Rodríguez-Ortiz, J.J.: Genetic algorithms and Darwinian approaches in financial applications: a survey. Expert Syst. Appl. **42**(21), 7684–7697 (2015)
32. Yu, L., Zhang, Y.Q.: Evolutionary fuzzy neural networks for hybrid financial prediction. IEEE Trans. Syst., Man, Cybernetics, Part C (Appl. Rev.) **35**(2), 244–249 (2005)
33. Zhang, Y., Kandel, A.: Compensatory genetic fuzzy neural networks and their applications. **30**. World Scientific, (1998)
34. Kaucic, M.: Investment using evolutionary learning methods and technical rules. Eur. J. Oper. Res. **207**(3), 1717–1727 (2010)
35. Faber, M.T.: A quantitative approach to tactical asset allocation. J. Wealth Manag. **9**(4), 69–79 (2007)
36. Lin, P.C., Ko, P.C.: Portfolio value-at-risk forecasting with GA-based extreme value theory. Expert Syst. Appl. **36**(2), 2503–2512 (2009)
37. Byström, H.N.E.: Managing extreme risks in tranquil and volatile markets using conditional extreme value theory. Int. Rev. Finan. Anal. **13**(2), 133–152 (2004)
38. Chen, J.S., Hou, J.L., Wu, S.M., Chan-Chien, Y.W.: Constructing investment strategy portfolios by combination genetic algorithms. Expert Syst. Appl. **36**(2), 3824–3828 (2009)
39. Gondzio, J., Grothey, A.: Solving non-linear portfolio optimization problems with the primal-dual interior point method. Eur. J. Oper. Res. **181**(3), 1019–1029 (2007)
40. Marinakis, Y., Marinaki, M., Doumpos, M., Zopounidis, C.: Ant colony and particle swarm optimization for financial classification problems. Expert Syst. Appl. **36**(7), 10604–10611 (2009)
41. Nigam, A., Agarwal, Y.K.: Ant colony optimization for index fund problem. J. Appl. Oper. Res. **5**(3), 96–104 (2013)
42. Cruz, D.P.F., Maia, R.D., Szabo, A., de Castro, L.N.: A bee-inspired algorithm for optimal data clustering. Evolutionary Computation (CEC), 2013 IEEE Congress on IEEE (2013)
43. Beasley, J.E.: OR-library: distributing test problems by electronic mail. J. Oper. Res. Soc. **41**(11), 1069–1072 (1990)
44. Basso, A., Funari, S.: A data envelopment analysis approach to measure the mutual fund performance. Eur. J. Oper. Res. **135**(3), 477–492 (2001)

45. Storn, R., Price, K.: Differential evolution—a simple and efficient heuristic for global optimization over continuous spaces. J. Global Optim. **11**(4), 341–359 (1997)
46. Bacanin, N., Tuba, M.: Fireworks algorithm applied to constrained portfolio optimization problem. Evolutionary Computation (CEC), 2015 IEEE Congress on IEEE, pp. 1242–1249, (2015)

Others

Synthesis Algorithm based on the Pre-evaluation of Quantum Circuits for Linear Nearest Neighbor Architectures

Dejun Wang, Zhijin Guan(✉), Yingying Tan, and YiZhen Wang

College of Computer Science and Technology, Nantong University, Nantong 226019, China
guan_g617@163.com

Abstract. In order to design quantum circuits for LNN, in this work, we propose a kind of synthesis algorithm based on the pre-evaluation of quantum circuits for LNN. Through pre-evaluation, not only can the algorithm accurately calculate if there are deletable redundant SWAP gates and remove them, but also convert every non-adjacent quantum gate to adjacent quantum gate, with inserting a minimal number of SWAP gates, and therefore get quantum circuits of minimal quantum cost (qc). As for quantum circuits of n lines and m quantum gates, the time and space complexity of the algorithm and optimized algorithm is $O(m\text{^}3)$ and $O(n\text{^}n+m)$, respectively. The results present that, with fewer average gates of quantum circuits and higher improvement efficiency of quantum cost, the algorithm has a wider range of application compared to the existing algorithm.

Keywords: Linear Nearest Neighbor · Quantum circuits · SWAP gates · Quantum cost · Logic synthesis

1 Introduction

Quantum logic synthesis [1] is what we use to achieve corresponding quantum circuits with given quantum logical gates according to constraint conditions and limitations of no fan-out, no feedback and meeting technological requirements achieving quantum circuits, and also optimizes quantum circuits under a certain cost model, with qc being as little as possible.

Quantum circuits synthesis originate in the research for quantum computer. In order to achieve the LNN constraint of quantum technology and construct quantum circuits for LNN, so far, several approaches converting quantum circuits of arbitrary architecture to quantum circuits of LNN architecture have been come up with [2–5].

The time of quantum computation depends of the number of gates in quantum circuits and physical operations achieving every gate, which is called quantum cost [6]. Decoherence of quantum system will be caused by the interference of the external environment, as a result, quantum computation must be finished within limited and coherent time, which requires that qc in quantum circuits should be minimized.

F. Xhafa et al. (eds.), *Advances in Intelligent Systems and Interactive Applications*, Advances in Intelligent Systems and Computing 686,
https://doi.org/10.1007/978-3-319-69096-4_105

The circuit, which is of minimal cost, used to achieve a certain specific function of minimal qc is called optimal circuit.

For the sake of constructing optimized quantum circuits for LNN architecture, following work will be finished in this paper: (1) Find out quantum circuits of minimal NNC by full permutation of the whole circuit line with optimized evaluation algorithm for quantum circuits; (2) In order to achieve the convert to LNN architecture from the original quantum circuits, the optimized approach of adding swap gates should be finished based on (1) to realize the convert to LNN architecture, which will make physical realization easier; (3) In order to decrease qc of quantum circuits and make the minimal number of additive swap gates, related algorithms solving the problems of optimization of quantum circuits will be come up with.

2 Reversible Logical Algorithm of Lnn Based on Pre-Evaluation

2.1 Comprehensive Algorithm

Theorem 1. In the quantum circuits constructed by NCV gate library, the minimal number of additive swap gates when a single arbitrary quantum gate converts to an adjacent quantum gate is equal to NNC value of the arbitrary quantum gate.

The relationship between the minimal number of additive swap gates and NNC value of arbitrary quantum gates is as follows:

$$sc = g\_nnc \tag{1}$$

There into sc is the minimal number of additive swap gates; g_nnc is NNC value of a single quantum gate.

From Theorem 1, we can know that the number of additive swap gates is certainly little. On the other hand, when adding swap gates to arbitrary quantum gates, we should take the circumstance of the qubit in quantum circuits into account to obtain as far as possible pairs of redundant swap gates with newly additive swap gates and existing swap gates in quantum circuits. Then, the pairs of redundant swap gates are removed to achieve the goal of decreasing swap gates in quantum circuits.

As for a know arbitrary quantum circuit, form left to right, the first arbitrary quantum gate is middle axle. The quantum circuits cascading system on the left side of this arbitrary quantum gate is N_l (excluding the arbitrary quantum gate), the rest of quantum circuits on the right side of this arbitrary quantum gate is N_r (including the arbitrary quantum gate). N_m is a group of quantum circuits including only swap gates, which is required when cascading N_l and N_r, the two partial quantum circuits.

Detailed algorithm is as follows:

Step1: Initialize N_l to null, N_m null and $N_r = N$.

Step2: Scan. Search the first non-adjacent quantum gate of quantum circuit from the input of the quantum circuit. If there is a non-adjacent quantum gate, we set it as g_1 and carry out Step3. If not, we carry out Step7.

Step3: Bounded by g_1, the left side of the quantum circuit is N_l (excluding g_1) and the right side together with g_1 is N_r.

Step4: Ordering. We undergo global reordering on N_r and obtain quantum circuit set, N_r (i) and swap gate group set, N_m (i).

Step5: Adjacence. We convert g_1 (i) of the N_r (i) to adjacent gate applying the algorithm of adding swap gate. (N_m (i) and N_r (i) can be cascaded to construct a quantum circuit)

Step6: The optimal evaluation of quantum circuit. After every operation of adjacence, we finally choose a minimal qc of N_r (i) as N_r applying the optimal evaluation of quantum circuit. Then, we turn to Step2.

Step7: Check up on redundant swap gates in established quantum circuit for LNN and remove those redundant.

Step8: End.

2.2 Optimal Evaluation Algorithm of Quantum Circuits

In quantum circuits constructed by NCV gate library, on the premise of input/output truth-value staying the same, we evaluate NNC and chaos value of quantum circuits by the heuristic algorithm and determine minimal set of results.

Detailed algorithm is as follows:

Step1: Calculate the high/low qubit of the non-adjacent quantum gate g_1. We suppose l_i as its low qubit, l_j high qubit and NNC n_l.

Step2: Judge the quantum status of low qubit. If it is close, we carry out Step5.

Step3: Calculate the number (i) of removed qubits, add correspondent number (i) of swap gates, remove redundant swap gate group and update the value of n_l of g_l.

Step4: Judge the value of n_l. If it is 0, turning to Step7.

Step5: Judge the quantum status of high qubit. If it is closed, we carry out Step6; otherwise, we carry out Step3.

Step6: According to the value of n_l, we add in minimal but necessary swap gates and then convert non-adjacent quantum gates to adjacent ones.

Step7: End.

2.3 The Algorithm of Computing Minimal Chaos Value

In order to find out a proper N_m structure cascading N_l and N_r, solve the problems of reordering quantum circuits, and determine chaos value of N_m, this paper proposes an algorithm to compute minimal chaos value.

Detailed algorithm is as follows:

Step1: Starting with $i = 0$, we read the element of origin.

Step2: Judge whether the element read is the last element of origin or not. If so, we carry out Step5; if not, we carry out Step3.

Step3: Store the location information of origin in target array in t[i].

Step4: Remove the element in the location in target array (forward elements in follow-up locations lead), and then carry out Step2.

Step5: Remove all of the elements in target array. Calculate the sum marked by t of elements in t[i].

Step6: End.

2.4 The Algorithm of Adding Swap Gates

In the algorithm of adding swap gates, this paper presents an approach of adding swap gates to arbitrary quantum gates, which can precisely compute whether every arbitrary quantum gate can get redundant swap gates pairs which could be deleted or not as well as compute how many pairs of redundant swap gates can be removed. We can obtain swap gates with minimal addition making use of the algorithm of adding swap gates, thus obtain quantum circuits of minimal qc.

The swap gates in N_m structure and additive swap gates in the algorithm of adding swap gates can form some pairs of redundant swap gates, which can be removed from quantum circuits to achieve the goal of decreasing qc.

Detailed algorithm is as follows:

Step 1: Compute the high/low qubit of arbitrary quantum circuit g_1 (l_i, l_j, k). Suppose its low qubit is l_i, high qubit is l_j, its NNC is n_1.

Step 2: Judge the quantum state of low qubit. If it is close, carry out Step 5.

Step 3: Compute cancellation layers of the qubit, add i corresponding swap gates, remove these i pairs of redundant swap gates, and update the NNC value, n_1 of quantum gate, g_1.

Step 4: Judge whether the value of n_1 is 0 or not. If so, turn to Step 7.

Step 5: Judge the quantum state of high qubit. If is close, carry out Step 6, otherwise carry out Step 3.

Step 6: According to the value of n_1, add minimal yet necessary swap gates to make arbitrary quantum gates adjacent.

Step 7: End.

3 Experimental Results and Analysis

The Experimental data in this paper all come from revlib, from which 31 groups of data are tested and 3 to 8 quantum circuits are covered. The number of quantum gates from test data is 0 to 50 (Table 1).

Experimental results are compared with representative reference [5], which is shown in Table 1. From the comparison and analysis of number of additive swap gates to quantum circuits, we find that in the same test data 23, 4 groups of experimental data present less number of additive swap gates than that in reference [5]; 9 groups of experimental data present equal number of additive swap gates to that in reference [5]; 7 groups of experimental data present more number of additive swap gates than that in reference [5].

We find from Fig. 1 that the optimization efficiency of additive swap gates (qc) in the algorithm proposed in this paper is stable and fruitful in the process of dealing with quantum circuits within the scope of 4 to 8 circuits. By contrast, the algorithm in reference [5] can only deal with more simple quantum circuits within 5 circuits, but the

Table 1. Experimental results

Benchmark	n	Gate	S	Ref.[15]			Ours				
				s_1	Time_1	%s_1	s_2	Time_2	%s_2	qc	%t
3_17_13	3	13	6	2	0.1	66.67	5	0	16.67	32	100
decod24-v0_38	4	17	18	4	23.6	77.78	7	0	61.11	42	100
decod24-v0_40	4	8	12	3	0.4	75	4	0	66.67	21	100
decod24-v1_42	4	8	12	2	0.4	83.33	2	0	83.33	21	100
decod24-v2_43	4	16	20	3	7.6	85	5	0	75	36	100
decod24-v3_46	4	9	18	3	0.4	83.33	4	0	77.78	21	100
rd32-v0_66	4	12	10	4	1.6	60	4	0	60	30	100
rd32-v1_68	4	12	10	4	1.6	60	6	0	40	37	100
rd32-v1_69	4	12	10	2	0.3	80	4	0	60	31	100
4_49_17	4	30	42	O	O	O	22	0	47.62	104	O
hwb4_52	4	23	28	O	O	O	16	0	42.86	71	O
4gt11_83	5	12	20	6	630.1	70.00	6	0.001	70	36	100
4gt11_84	5	7	14	1	8.9	92.86	1	0.001	92.86	13	99.99
4gt11-v1_85	5	7	14	1	16.6	92.86	1	0.001	92.86	14	99.99
4gt13-v1_93	5	15	24	6	9808.5	75	9	0.001	62.5	46	100
4mod5-v0_19	5	12	18	6	489.2	66.67	6	0.002	66.67	37	100
4mod5-v0_20	5	8	12	2	55.3	83.33	3	0.001	75	21	100
4mod5-v1_22	5	9	6	3	45.5	50	3	0.001	50	21	100
4mod5-v1_24	5	12	34	10	548.7	70.59	7	0.001	79.41	37	100
4mod5-v1_25	5	9	6	1	11.9	83.33	3	0.001	50	21	99.99
alu-v4_37	5	14	26	14	3669.9	46.15	8	0.002	69.23	51	100
mod5d1_63	5	11	32	10	745.5	68.75	4	0.001	87.5	23	100
mod5d2_70	5	14	12	18	3047.5	-50	7	0	41.67	46	100
mod5mils_71	5	12	18	7	735.7	61.11	7	0.001	61.11	37	100
4mod5-v1_23	5	24	50	O	O	O	19	0.001	62.00	90	O
mod5d1_63	5	11	32	O	O	O	4	0.001	87.5	23	O
graycode6_48	6	5	0	0	4.5	0.00	0	0.015	0	5	99.67
decod24-enable_125	6	21	44	O	O	O	16	0.016	63.64	75	O
xor5_254	6	5	20	O	O	O	3	0.01	85	25	O
ham7_106	7	49	100	O	O	O	52	1.324	48	217	O
rd53_138	8	44	88	O	O	O	40	88.994	54.55	164	O

comprehensive algorithm in this paper can deal with quantum gates at a higher quantity level. With the increasing of quantum circuits, the optimization efficiency of the algorithm in this paper has obvious advantages. The optimization efficiency of NNC of quantum circuits for LNN architecture applying the algorithm of optimized evaluation of quantum circuits is 47.22%. This algorithm also has advantages in performance period of CPU on the premise that searched space grow exponentially.

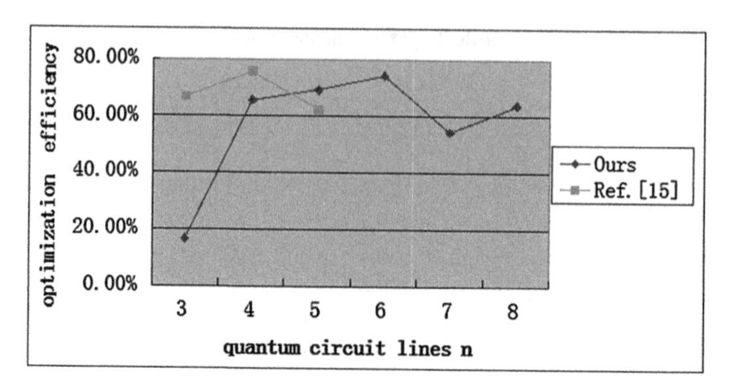

Fig. 1. optimization efficiency comparison chart

4 Conclusion

This paper presents an algorithm of adding swap gates to arbitrary quantum gates. While adding swap gates to arbitrary swam gates, the algorithm can produce redundant "swap gates pair" by newly-additive swap gates and existing ones in quantum circuits. Not only can this algorithm reduce the add of swap gates in the process of adding swap gates to arbitrary quantum gates, but also remove an original swap gate in quantum circuits. The algorithm can also accurately compute whether there is any chance that every arbitrary quantum gate can form redundant swap gates pair which is deletable as well as the number of redundant swap gates. We can obtain the number of minimal additive swap gates, that is quantum circuits of minimal qc, within a shorter time applying the algorithm.

Acknowledgments. This work was financially supported by the Nantong University Graduate Research and Innovation Project funded projects (No.YKC16087), Natural Science Foundation of Jiangsu Province (BK20151274) and National Natural Science Foundation(61403216).

References

1. D.Cheung, D.Maslov, and S.Severini. Translation techniques between quantum circuit architectures. Quantum Information Processing, 2007. available online at http://citeseerx.ist.psu.edu/viewdoc/download;jsessionid=C437A1E24389757B8BDC0D5C3D1DADAB?doi=10.1.1.133.7479&rep=rep1&type=pdf
2. Saeedi, M., Wille, R., Drechsler, R.: Synthesis of quantum circuits for linear nearest neighbor architectures. Quantum Information Processing **10**(3), 355–377 (2011)
3. Hollenberg, L.C.L., Greentree, A.D., Fowler, A.G., Wellard, C.J.: Two-dimensional architectures for donor-based quantum computing. Physical Review B **74**(4), 045–311 (2006)
4. Amlan Chakrabarti, Susmita Sur-Kolay, Senior Member, IEEE and Ayan Chaudhury. Linear Nearest Neighbor Synthesis of Reversible Circuits by Graph Partitioning. 2011. available online at http://arxiv.org/pdf/1112.0564v2.pdf

5. Robert Wille, Aaron Lye, Rolf Drechsler. Optimal SWAP Gate Insertion for Nearest Neighbor Quantum Circuits. In 19th Asia and South Pacific Design Automation Conference (ASP-DAC), 2014
6. Lee, S., Lee, S.J., Kim, T.: The cost of quantum gate primitives. Multiple Valued Logic and Soft Computing **12**(5/6), 561 (2006)

Study of an SIR Epidemic Disease Model with Special Nonlinear Incidence and Computer Simulation

Xiuchao Song(✉), Miaohua Liu, Hao Song, and Guohong Liang

Science College, Air Force Engineering University, Xi'an 710051, China
xiuchaosong@163.com

Abstract. An SIR epidemic disease model with special nonlinear incidence was dealt with in this paper. By constructing the Lyapunov function, the global stability of the infection free equilibrium is proved when one is greater than or equal to the basic reproduction number; and the global stability of the unique infection equilibrium of the system is also proved when one is less than the basic reproduction number. At last, the results were verified by computer simulation.

Keywords: Basic reproduction number · Nonlinear incidence · Globally stable

1 Introduction

One of the important mathematical models for epidemiological dynamics is the compartmental model, which was first proposed by Kermack and Mckendrick in 1927 [1]. Up to now the compartmental model is still widely used and continuously improved. Most compartmental models assume that the incidence is bilinear or standard. In recent years, the epidemic models with nonlinear incidence have become an extremely active area of research [2–6]. In this paper, we considered epidemic models with the special nonlinear incidence $g(S)h(I)$

$$\begin{cases} S' = A - \mu S - g(S)h(I) \\ I' = g(S)h(I) - (\mu + \alpha)I \\ R' = \gamma I - \mu R \end{cases}$$

where $S(t)$ is the susceptible compartment, $I(t)$ is the infectious compartment, and $R(t)$ is the recovery compartment. The recruitment rate of $S(t)$ is a constant A; μ is the per capita natural death rate; γ is the rate of recovery; α is the sum of the disease induced mortality and the recovery rate of the infected individuals. Considering the following equivalent system of the model:

F. Xhafa et al. (eds.), *Advances in Intelligent Systems and Interactive Applications*, Advances in Intelligent Systems and Computing 686,
https://doi.org/10.1007/978-3-319-69096-4_106

$$\begin{cases} S' = A - \mu S - g(S)h(I) \\ I' = -(\mu + \alpha)I + g(S)h(I) \end{cases} \tag{1}$$

where the functions $g(S)$ and $h(I)$ have the following properties:

(a) $g(0) = 0$ $h(0) = 0$; (b) $\lim_{I \to 0^+} \frac{h(I)}{I} = \beta$, β is positive.

(c) $g'(S)$ and $h'(I)$ are positive, $\left(\frac{h(I)}{I}\right)'$ is negative and the positively invariant set to system (1) is

$$\Omega = \{(S,\ I) \in R_+^2\ :\ (S + I) \le A/\mu\}$$

2 Existence of Equilibria

Obviously, system (1) has a disease free equilibrium $P_0(\frac{A}{\mu},\ 0)$. Using the next generation matrix [7], we get $R_0 = \frac{g(S_0)\beta}{\mu + \alpha}$. From the following equations, we can find the endemic equilibrium

$$\{A - \mu S = g(S)h(I) \tag{2}$$

$$\{(\mu + \alpha)I = g(S)h(I) \tag{3}$$

(2)–(3) yields

$$S = \frac{A - (\mu + \alpha)I}{\mu} \tag{4}$$

In order to ensure $S > 0$, it is required that $I < \frac{A}{\mu + \alpha}$. Substituting (4) into (3) yields

$$g\left(\frac{A - (\mu + \alpha)I}{\mu}\right) = \frac{(\mu + \alpha)I}{h(I)} \tag{5}$$

Let $G(I) := g\left(\frac{A - (\mu + \alpha)I}{\mu}\right)$ and $H(I) := (\mu + \alpha)\frac{I}{h(I)}$. Obviously, $G(0) = G(S_0)$ and $G(\frac{A}{\alpha + \mu}) = g(0) = 0$. As we know $g'(S) > 0$ so function $G(I)$ is decreasing. From the assumption $\left(\frac{h(I)}{I}\right)' < 0$ we know $H(I)$ is increasing and $\lim_{I \to 0} H(I) = \frac{\mu + \alpha}{\beta}$. So when $H(0) < G(0)$, that is to say when $R_0 > 1$, Eq. (5) has a positive root I^* in $(0, \frac{A}{\alpha + \mu})$. Therefore, $P(S^*, I^*)$ is the endemic equilibrium, where $S^* = \frac{A - (\mu + \alpha)I^*}{\mu}$.

3 Stability of Equilibria

Theorem 1 *When one is greater than or equal to the basic reproduction number, P_0 is a globally stable equilibrium.*

Proof Let: $L_1 = \int_{S_0}^{S} \frac{g(u)-g(S_0)}{g(u)} du + I$ then

$$\begin{aligned} L_1' &= \left(1 - \frac{g(S_0)}{g(S)}\right)(A - \mu S - g(S)h(I)) + g(S)h(I) - (\mu+\alpha)I \\ &= A - \frac{Ag(S_0)}{g(S)} + \mu S\frac{g(S_0)}{g(S)} + g(S_0)h(I) - (\mu+\alpha)I - \mu S \\ &= (A - \mu S)\left(\frac{g(S) - g(S_0)}{g(S)}\right) + g(S_0)h(I) - (I\mu + \alpha I) \\ &= (A - \mu S)\left(\frac{g(S) - g(S_0)}{g(S)}\right) + \frac{1}{(\mu I + \alpha I)}\left(\frac{g(S_0)h(I)}{\mu I + \alpha I} - 1\right) \end{aligned}$$

We know $\left(\frac{h(I)}{I}\right)' < 0$, $\lim\limits_{I\to 0}\frac{h(I)}{I} = \beta > 0$, so

$$L_1' \le \mu(S_0 - S)\left(1 - \frac{g(S_0)}{g(S)}\right) + \frac{1}{(\mu+\alpha)I}(R_0 - 1)$$

When $R_0 < 1$, L'_1 is negative definite, and when $R_0 = 1$, using the LaSalle's Invariance Principle [8] we can get the same conclusion. So P_0 is globally stable.

Theorem 2 *$P(S^*, I^*)$ is globally stable when one is less than the basic reproduction number.*

Proof Clearly, S^*, I^* satisfy equations

$$\begin{cases} A - \mu S^* = g(S^*)h(I^*) \\ (\mu+\alpha)I^* = g(S^*)h(I^*) \end{cases} \tag{6}$$

Define a Lyapunov function as following

$$L_2 = \int_{S^*}^{S} \frac{g(u) - g(S^*)}{g(u)} du + \int_{I^*}^{I} \frac{u - I^*}{u} du,$$

then

$$L_2' = \frac{g(S) - g(S^*)}{g(S)}(A - \mu S - g(S)h(I)) - \frac{I - I^*}{I}((\mu I + \alpha I) - g(S)h(I))$$

From (6), it follows that

$$L_2' = \frac{g(S) - g(S^*)}{g(S)}(\mu S^* + g(S^*)h(I^*) - \mu S - g(S)h(I)) + \frac{I - I^*}{I}\left(\frac{g(S)h(I)I^* - g(S^*)h(I^*)I}{I^*}\right)$$

For simplicity, dnote

$$x = \frac{S}{S^*} \quad y = \frac{I}{I^*} \quad u = \frac{f(S)}{f(S^*)} \quad v = \frac{g(I)}{g(I^*)}$$

then

$$L_2' = \mu S^*(1 - \frac{1}{u})(1 - x) + h(I^*)(v - y)\left(1 - \frac{1}{v}\right)g(S^*) + h(I^*)\left(3 - \frac{1}{u} - \frac{y}{v} - \frac{uv}{y}\right)g(S^*)$$

From assumption (c), we get $\mu S^*(1 - x)(1 - \frac{1}{u}) \leq 0$ and $(y - v)g(S^*)h(I^*)(1 - \frac{1}{v}) \geq 0$. It is easily to prove that $f(S^*)g(I^*)\left(3 - \frac{1}{u} - \frac{y}{v} - \frac{uv}{y}\right) \leq 0$, and so L'_2 is negative definite. Therefore, the endemic equilibrium $P(S^*, I^*)$ is globally stable when $R_0 > 1$.

4 Computer Simulation

Let $f(S) = S^3$, $g(I) = \frac{0.02I}{1+0.08I^{1/3}}$, and taking the corresponding parameters are: $A = 0.01$, $\mu = 0.01$, $\alpha = 0.06$, respectively. We can get $R_0 = 0.286 < 1$. The computer simulations were carried out with three sets of initial values of (0.7, 0.3), (0.5, 0.1) and (0.3, 0.6), respectively. We see that the three orbits tend to $P_0(1, 0)$ (Fig. 1) as t $+\infty$. When $\beta = 0.2$ and other parameters' values and initial values remain the same, we can get $R_0 = 2.86 > 1$. Do computer simulation again (Fig. 2), we see that the three orbits tend to infection equilibrium $P^*(0.602, 0.094)$, when $t \to +\infty$.

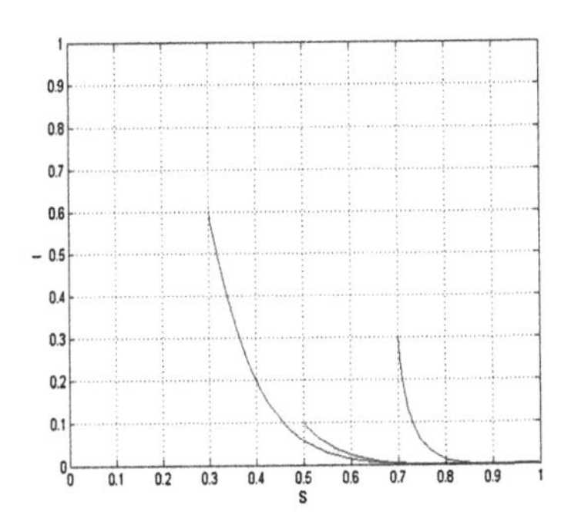

Fig. 1. The stability of P_0 when $R_0 < 1$

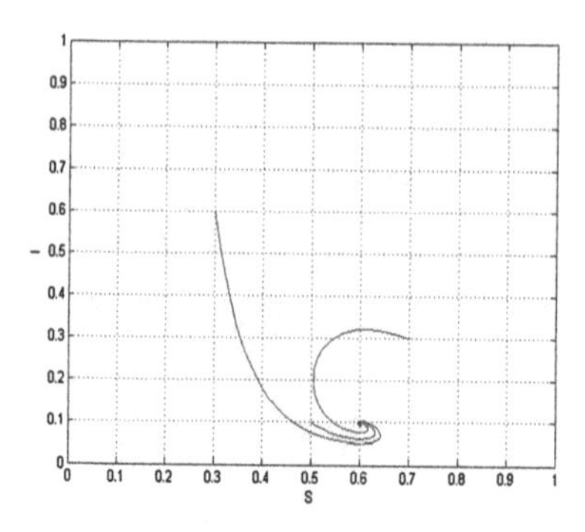

Fig. 2. The stability of P^* when $R_0 > 1$

References

1. Kermack, W.O., Mckendrick, A.G.: Contributions to the mathematical theory of epidemics. Proc. Roy. Soc. **A115**, 700–721 (1927)
2. Andrei, K., Philip, K.M.: Non-linear incidence and stability of infectious disease models. Math. Med. Biol. **22**(2), 113–128 (2005)
3. Andrei, K.: Global properties of infectious disease models with nonlinear incidence. Bull. Math. Biol. **69**(6), 1871–1886 (2007)
4. Andrei, K.: Lyapunov functions and global stability for SIR and SIRS epidemiological models with nonlinear transmission. Bull. Math. Biol. **30**(3), 615–626 (2006)
5. Andrei, K.: Global asymptotic properties of virus dynamic models with dose-dependent parasite reproduction and virulence and nonlinear incidence rate. Math. Med. Biol. **26**(2), 225–239 (2009)
6. Song, X., Ren, J., Song, H., et al.: Global stability of an SI epidemic model with spontaneous cure rate. China Science Paper **10**(5), 552–554 (2015). (in chinese)
7. van den Driessche, P., James, W.: Reproduction numbers and sub-threshold endemic equilibria for compartmental models of disease transmission. Math. Biosci. **180**(1), 29–48 (2002)
8. LaSalle, J.P.: The stability of dynamical systems. Regional conference series in Applied Mathematics, SIAM, Philadelphia (1976)

An Improvement Response Surface Method Based on Weighted Regression for Reliability Analysis

Xingchen Yu[1] and Zhangxue Gang[2(✉)]

[1] China Electronic Technology Group Corporation Thirty-Second Research Institute, Shanghai 201800, China
yu178844264@126.com

[2] Xi'an Institute of Space Radio Technology, Xi'an 710100, China
arif15@foxmail.com

Abstract. For structural reliability analysis, the response surface method is popularly used to reduce the computational efforts of numerical analysis. The general method in the response surface method is to use the least square regression method. To give higher weight to the points closer to the failure curve, an improvement response surface method based on weighted regression for reliability analysis is presented, the new weight function is built based on parallel circuit theory, because the closer sample points are to the failure curve (the smaller the branch resistance is), the higher weights are given (the greater the branch current is). Numerical applications are provided to indicate the significance of the presented method.

Keywords: Response surface · Weight function · Reliability · Parallel circuit

1 Introduction

To take into account uncertainties in structure design is a hot topic because the valuable information can be obtained. Therefore, reliability analysis is proposed, which is carried out by estimating the probability of failure. A vector of M random parameters $X = X_1, X_2, \ldots, X_M$ is introduced, and its joint probability density function are called as $f_X(x)$. Give that one researches a structural system, charactered by a performance function, say $G(X)$, a realization x of vector X is defined, $G(X) < 0$ is called failure domain, $G(X) = 0$ is called the limit state function(LSF) and $G(X) > 0$ is called safety domain. The probability of failure is given by function G reads

$$P_f = \int_{G(X) \leq 0} f_X(x) dx_1 dx_2 \cdots dx_M \tag{1}$$

A simple and accurate calculation of the probability of failure is to use the Monte Carlo method. But, the LSF is usually implicit in most of engineering problems, it needs to be calculated by a finite element method, Monte Carlo method can result in prohibitively large computational efforts. In order to improve the calculation efficiency

F. Xhafa et al. (eds.), *Advances in Intelligent Systems and Interactive Applications*, Advances in Intelligent Systems and Computing 686,
https://doi.org/10.1007/978-3-319-69096-4_107

and ensure the calculation accuracy, alternative methods have been proposed, such as quadratic response surfaces, polynomial chaos [1], kriging [2] and neural network.

By a response surface (RS), Bucher and Bourgund suggested a new adaptive interpolation scheme which make fast and accurate representation of the system behavior. The RS is utilized together with advanced Monte Carlo simulation techniques to obtain the desired reliability estimates. Based on weighted regression, Kaymaz and McMahon presented a response surface method (RSM), when the sample points are closer to the failure curve, the higher weights are given, then the response surface function is constructed. Nguyen et al. gave an adaptive RSM, by using the weighted regression method, the RS is fitted. Li et al. proposed a doubly weighted moving least squares for structural reliability analysis, a surrogate model based on a doubly-weighted moving least squares is presented.

In this paper, an improvement RSM is presented, the new weight function is built based on parallel circuit theory, because the closer sample points are to the failure curve (the smaller the branch resistance is), the higher weights are given (the greater the branch current is).

2 Experimental Design

Experimental design is introduced by Bucher and Bourgund, the sample points are selected along the coordinate axes x_i as the points at the mean vector $\bar{X}$ of the variables x_i and the sample points with coordinates $x_i = \mu_i \pm f_i\sigma_i$. For the first iteration, a constant value $f_1 = 3$ were employed and a constant value of $f_1 = 1$ for the following iteration were employed. Figure 1 shows the typical experimental design method in a two parameters example.

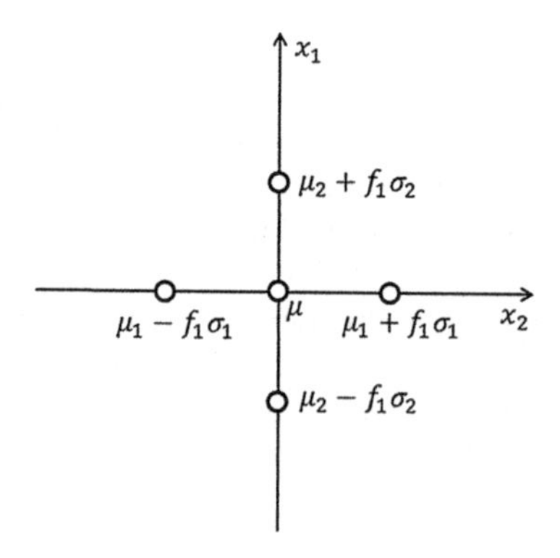

Fig. 1. Location of the sample points for a two parameters example

3 Response Surface Method

A linear response surface is given as

$$\hat{y}(x) = b_0 + \sum_{j=1}^{n} b_j x_j \tag{2}$$

where n is the number of basic random variables, b_j is the coefficients of polynomial function, the number of coefficients of polynomial function to be determined is $(n + 1)$.

In the common RSM, by using least square method the coefficients of the polynomial function are as

$$b = (X^T X)^{-1} X^T y \tag{3}$$

where $\boldsymbol{b}$ represents the coefficient matrix, $\boldsymbol{X}$ is the design matrix, and $\boldsymbol{y}$ is the response matrix.

The weighted regression approach based on least square method is presented to obtained the coefficients of the response surface function

$$b = (X^T W X)^{-1} X^T W y \tag{4}$$

where $\boldsymbol{W}$ is a diagonal matrix of weights. To give higher weights to the sample points closer to the failure curve, a new weight function is built based on parallel circuit theory, Fig. 2 shows the typical parallel circuit, the total resistance is shown as follow

$$R = \frac{1}{\sum_{i=1}^{n} \frac{1}{R_i}} \tag{5}$$

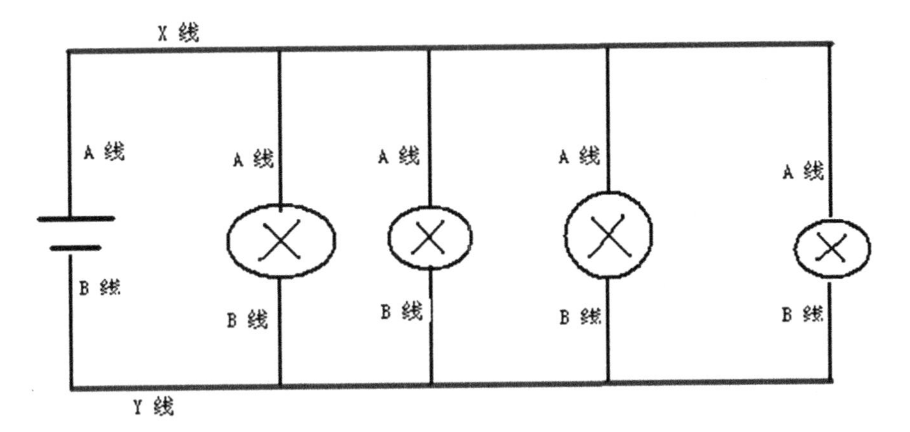

Fig. 2. Typical parallel circuit

If the total current I = 1A is employed (the sum of the weights is 1), the total circuit voltage is as follow

$$U = I \times R = 1 \times \frac{1}{\sum_{i=1}^{n} \frac{1}{R_i}} = \frac{1}{\sum_{i=1}^{n} \frac{1}{R_i}} \tag{6}$$

so the branch current (weight) is described

$$I_i = \frac{U}{R_i} = \frac{\frac{1}{\sum_{i=1}^{n} \frac{1}{R_i}}}{R_i} = \frac{1}{R_i \times \sum_{i=1}^{n} \frac{1}{R_i}} \tag{7}$$

Because the responses can be positive and negative, and to adjust the magnitude of the effect of responses on the computation results, the new weight function is built as follow

$$\begin{cases} w_{ii} = \frac{1}{|\hat{y}_i|^m \times \sum_{i=1}^{k} \frac{1}{|\hat{y}_i|^m}}, & \hat{y}_i \neq 0 \\ w_{ii} = 1, & \hat{y}_i = 0 \end{cases} \tag{8}$$

where k represents the number of experimental points, and m represents an arbitrary factor.

The presented RSM can be illustrated based on the following steps:

Choose 2n + 1 sample points based on μ_i and σ_i, $x_i = \mu_i \pm f_i\sigma_i$, where $f = 3$.

1. Compute the values of the performance function and weights at experimental points.
2. Get the linear response surface using Eq. (4).
3. Utilize the FORM to get the new center point and compute R by Monte Carlo simulation.
4. If the convergence is not satisfied, for a new iteration proceed to step 1, the convergence criteria is as follows

$$\left| R^r - R^{r-1} < \varepsilon \right| \tag{9}$$

4 Numerical Applications

By applying to applications chosen from various reference papers, the method proposed is tested. For some applications, the results obtained based on the method proposed are compared with the results obtained by Monte Carlo simulation. 10^8 experimental points are sampled when the reliability is estimated by using Monte Carlo simulation.

Example 1 This application is selected from Ref. [3], the LSF is described as follows

$$g(X) = e^{[0.4(x_1+2)+6.2]} - e^{(0.3x_2+5.0)} - 200$$

Where x_1 and x_2 are supposed to be uncorrelated and a standard normal distribution. $\varepsilon = 0.005$ and $m = 1$ are employed, the number of iterations to achieve convergence is two, iterative procedure and experimental design for the RSM proposed are seen from Fig. 3. The results are shown in Table 1. In terms of the results in Table 1, the presented method can yield better results as seen from reliability.

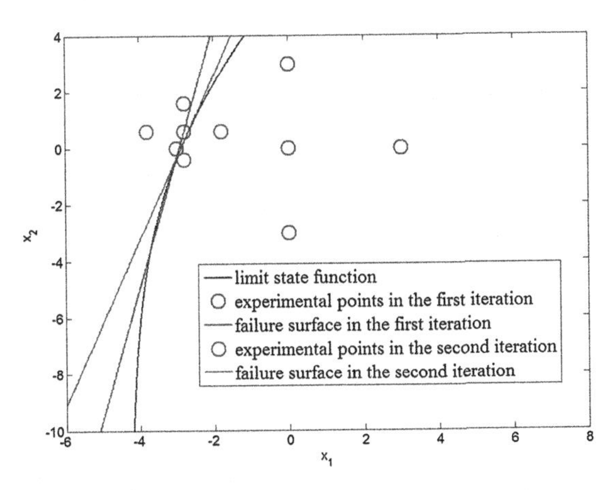

Fig. 3. Iterative procedure and experimental design for the general RSM

Table 1. The comparing results of the reliability analysis

Method	Reliability	Relative error	Experimental points
Monte Carlo simulation	0.99638	/	10^8
The RSM proposed	0.99676	+0.04%	10

Example 2 This application is selected from Ref. [3], the LSF is described as follows

$$g(X) = (5x_1+10)^3 + (5x_1+10)^2(5x_2+9.9) + (5x_2+9.9)^3 - 18$$

where x_1 and x_2 are supposed to be uncorrelated and a standard normal distribution. $\varepsilon = 0.005$ and $m = 1$ are employed, the number of iterations to achieve convergence is three, iterative procedure and experimental design for the RSM proposed are seen from Fig. 4, the results are shown in Table 2. The proposed approach can yield better results as seen from reliability.

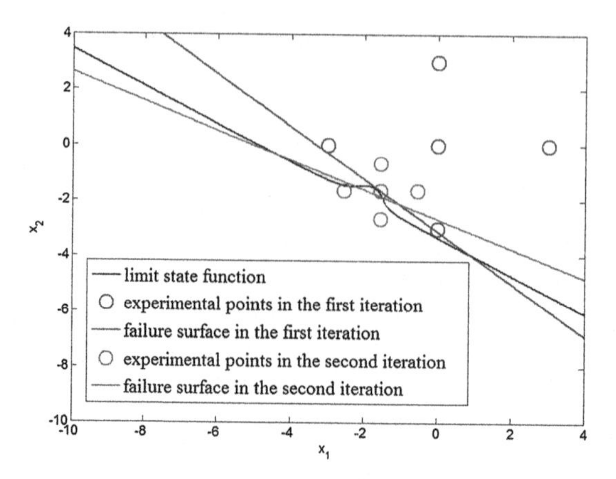

Fig. 4. Iterative procedure and experimental design for the general RSM

Table 2. The comparing results of the reliability analysis

Method	Reliability	Relative error	Experimental points
Monte Carlo simulation	0.99418	/	10^8
The RSM proposed	0.99080	−0.34%	10

5 Conclusion

In this study, to give higher weights to the sample points closer to the failure curve, an improvement approach for reliability analysis is presented, the new weight function is built based on parallel circuit theory, Numerical applications are provided to indicate the benefits of the proposed method, from the point of view of the results, the presented method can yield better results in terms of reliability to compare with the results obtained on account of Monte Carlo simulation.

References

1. Sacks, J., Welch, W., Mitchell, T., Wynn, H.: Design and analysis of computer experiments. Stat. Sci. **4**(4), 409–423 (1989)
2. Kaymaz, I.: Application of kriging method to structural reliability problems. Struct. Saf. **2**(7), 133–151 (2005)
3. Kaymaz, I., McMahon, C.A.: A response surface method based on weighted regression for structural reliability analysis. Prob. Eng. Mech. **20**, 11–17 (2005)

The Greedy Algorithm and Its Performance Guarantees for Solving Maximization of Submodular Function

Yunxia Guo(✉), Guohong Liang, and Jia Liu

School of Science, Air Force Engineering University, Xi'an 710051, China
1752954325@qq.com

Abstract. Maximizing or minimizing submodular function is widely used in combinatorial optimization problems. In this paper; we present an approximation algorithm for maximizing submodular function subject to independence system that be represented as the intersection of a limited number of matroids, and discuss its performance guarantee.

Keywords: Combinatorial optimization problems · Submodular function · Performance guarantee · Approximation algorithm

1 Introduction

Submodular function frequently appear in the analysis of combinatorial systems such as graphs networks, and algebraic systems. The theory and application of submodular functions has extensive attention in combinatorial optimization and other fields of combinatorial analysis in recent years.

The theory of submodular functions is now becoming mature, but a large number of fundamental and useful results on submodular functions are scattered in the literature.

A set of $E = \{1, 2, \cdots, n\}$, $2^E = \{X|X \subseteq E\}$ is power set of E, function $\varphi : 2^E \to R$ is submodular set function [1], we have for $A, B \in 2^E$,

$$\varphi(A \cap B) + \varphi(A \cup B) \leq \varphi(A) + \varphi(B)$$

In this paper, we present agreedy algorithm for maximizing submodular function subject to independence system that can be expressed as the intersection of matroids. A set of $E = \{1, 2, \cdots, n\}$,(E, F) is a uniformed matroid, as $F = \{S \subseteq E||S| \leq k\}$, considering $Z_{opt} = \max\{f(S)|S \in F\}$,$(E, F)$ is independence system that can be expressed as the intersection of M matroids. This algorithm is the extension of Fisher algorithm [2]. In the reference 2, the number of independent system is limited with $M + 1$ performance guarantee. The performance guarantee of greedy algorithm is $\alpha M + 1$. Maximizing submodular function is $f_S|F$ subject to independence system that can be expressed as the intersection of M matroids.

F. Xhafa et al. (eds.), *Advances in Intelligent Systems and Interactive Applications*, Advances in Intelligent Systems and Computing 686,
https://doi.org/10.1007/978-3-319-69096-4_108

2 Main Lemma and Proof

Lemma 1 (Fisher)

$$U_t \subseteq \bigcup_{m=1}^{M} SP_m(S_t), \quad t = 0, 1, \cdots$$

U_t is acquired from $t+1$, $r_m(S)$ is rank of set S in matroidm. SP_m is the closure of S inmatroidm, so [2]

$$SP_m(S) = \{e \in E | r_m(S \cup \{e\}) = r_m(S)\}.$$

Lemma 2 (Fisher), *if*

$$\sum_{\omega=0}^{t-1} x_\omega \le t, t = 1, 2, \ldots, k, \rho_{\omega-1} \ge \rho_\omega, \rho_\omega, x_\omega \ge 0, \omega = 1, \ldots, k-1, \rho_k = 0$$

we have

$$\sum_{i=0}^{n-1} \rho_i x_i \le \sum_{i=0}^{n-1} \rho_i.$$

Proof Considering the follow linear programming [3]

$$V = \max_x \left\{ \sum_{\eta=0}^{\vartheta-1} \rho_\eta x_\eta \middle| \sum_{\eta=0}^{\delta-1} x_\eta \le \delta, \delta = 1, \ldots, \vartheta, x_\eta \ge 0, \eta = 0, 1, 2, \cdots, \vartheta - 1 \right\}.$$

Its dual programming

$$W = \min_Z \left\{ \sum_{t=1}^{\vartheta} tZ \middle| \sum_{t=i}^{\vartheta-1} Z_t \ge \rho_i, i = 0, 1, \ldots, \vartheta - 1, Z_t \ge 0, t = 0, 1, 2, \ldots, \vartheta - 1 \right\}.$$

For

$$V = \max_x \left\{ \sum_{\tau=0}^{k-1} \rho_\tau x_\tau \middle| \sum_{\tau=0}^{\varepsilon-1} x_\tau \le \varepsilon, \varepsilon = 1, \cdots, k, x_\tau \ge 0, \tau = 0, 1, 2, \cdots, k-1 \right\}$$

Because

$$\rho_\xi \ge \rho_{\xi+1}, Z_\xi = \rho_\xi - \rho_{\xi+1}, \xi = 1, 2, \cdots, k-1, \rho_k = 0$$

is dual feasible solution, its optimal value is [4]

$$\sum_{t=1}^{k} t(\rho_{i-1} - \rho_i) = \sum_{i=0}^{k-1} \rho_i.$$

The weak duality theorem of linear programming, Lemma found, Finish.

Lemma 3 Suppose $f: D \rightarrow R$ be an arbitrary submodular on a allocation lattice $D \subseteq 2^E$. Then the set of all the minimizers of f given by

$$D_0 = \{X | X \in D, f(X) = \min\{f(X_2) | X_2 \in D\}\}$$

Forms a sublattice of D, i.e., for $\forall X_1, X_2 \in D_0$, $X_1 \cup X_2, X_1 \cap X_2 \in D$.

Proof For any $X_1, X_2 \in D_0$,

$$f(X_1 \cap X_2) + f(X_1 \cup X_2) \leq f(X_1) + f(X_2),$$

where
$\min\{f(X_1 \cup X_2), f(X_1 \cap X_2)\} \geq f(X_1) = f(X_2)$.
Hence we must have

$$\begin{aligned} f(X_1 \cap X_2) &= f(X_1 \cup X_2)\} \\ &= f(X_1) \\ &= f(X_2), \end{aligned}$$

i.e., $X_1 \cap X_2, X_1 \cup X_2 \in D$.

3 Greedy Algorithms and Performance Guarantee

Solving the greedy algorithm of $f_S | F$
Step 1 $i = 1, S_0 = \phi, E_0 = E$;
Step 2 If $E_{i-1} = \phi$, stop.
Step 3 Solving $e_i \in E_{i-1}$, if $\alpha \cdot \rho_{e_i}(S_{i-1}) \geq \max_{e \in E_{i-1}} \rho_e(S_{i-1}), \alpha > 0$
Step 4 Check $S_{i-1} \cup \{e_i\}$ belong to F or not;
Step 5a if $S_{i-1} \cup \{e_i\}$ not belong to F, $E_{i-1} = E_{i-1} \backslash \{e_i\}$, back to step 2;
Step 5b if $S_i = S_{i-1} \cup \{e_i\}, \rho_{i-1} = \rho_{e_i}(S_{i-1}), E_i = E_{i-1} \backslash \{e_i\}$;
Step 6 $i = i + 1$, back to Step 2 [5, 6].

Theorem, Assuming (E, F) is independence system that can be expressed as the intersectionof M matroids, f is nonnegative reduced modulus function defending at Ω. If Z_g, Z_{opt} are the objective function value and the optimal objective function value, $\frac{Z_{opt}}{Z_g} \leq \alpha M + 1$.

Proofing Assuming S, T are solution and optimal solution $f_S | F$ greedy algorithm, $|S| \leq k$, as (E, F) is the independent system, and also the matroid, $|T|$ cannot be k.

As $t = 1, \cdots, k$, $S_{t-1} = |T \cap (U_t \backslash U_{t-1})|$, U_t is the set of the t iteration. Without loss of generality, Assuming

$$U_0 = \Phi, U_k = E, \rho^*(S_i) = \max_{e \in E_i} \rho_e(S_i), i = 0, 1, \ldots, k-1.$$

As f is nonnegative reduced modulus function, subject to Lemma 1,

$$Z_{opt} = f(T) \leq f(S) + \sum_{e \in T \backslash S} \rho_e(S) \tag{1}$$

Assuming $t \in \{1, 2, \ldots, k\}$, $\rho_{q(t)} = \min\{\rho_i | i = 0, \ldots, t-1\}$,
For any $e \in T \cap (U_t \backslash U_{t-1})$, as the submodular of f,$\rho_e(S) \leq \rho_e(S_t)$.
The definition and algorithm of ρ^* is

$$\rho_e(S) \leq \rho_e(S_t) \leq \rho^*(S_t) \leq \rho^*\left(S_{\eta(t)+1}\right) \leq \alpha\rho_{\eta(t)} \tag{2}$$

If

$$\rho'_{\phi-1} = \alpha\rho_{q(\phi)}, \rho_e(S) \leq \rho'_{\phi-1}, \forall e \in \left(U_\phi \backslash U_{\phi-1}\right) \cap T, \phi = 1, 2, \ldots, k \tag{3}$$

As non hydrophobicity of $q(t)$, ρ'_{t-1} is non hydrophobicity, $\rho'_{t-1} \geq \rho'_t$.

$$\rho'_t = \alpha\rho_{q(t+1)} \leq \alpha\rho_t$$

From inequality (1),

$$\begin{aligned} f(T) &\leq \sum_{\beta \in T \backslash S} \rho_\beta(S) + f(S) \\ &\leq f(S) + \sum_{\beta \in T} \rho_\beta(S) \\ &= \sum_{s=1}^{k} \sum_{\beta \in T \cap (U_T \backslash U_{s-1})} \rho_\beta(S) + f(S) \\ &\leq f(S) + \sum_{\lambda=1}^{k} \rho'_{\lambda-1} S_{\lambda-1} \end{aligned} \tag{4}$$

As $S_{t-1} = |(U_t \backslash U_{t-1}) \cap T|$, $\sum_{j=1}^{r} S_{j-1} = |U_r \cap T|$,

From Lemma 2, as $U_t \subseteq \bigcup_{m=1}^{M} SP_m(S_t)$

$$|T \cap U_t| \leq \sum_{m=1}^{M} |T \cap SP_m(S_t)|$$

Also, T is Independent set of matroid, $r_m(SP_m(S_t)) = t, |T \cap SP_m(S_t)| \leq t$, Hence

$$\sum_{i=1}^{t} S_{i-1} \leq \sum_{m=1}^{M} |T \cap SP_m(S_t)| \leq Mt, t = 1, \ldots, k \tag{5}$$

As Non hydrophobicity of $\forall t, \rho'_t, S_t \geq 0, \rho'_t$, In Lemma 2, as $x_i = \frac{S_i}{M}, \rho_i = \rho'_i$. From inequality (4), $\sum_{\lambda=0}^{n-1} \rho'_\lambda S_\lambda \leq M \sum_{\lambda=0}^{n-1} \rho'_\lambda$, also from inequality (5).

$$\begin{aligned} f(T) &\leq M \sum_{l=0}^{m-1} \rho'_l + f(S) \\ &\leq M\alpha \sum_{l=0}^{m-1} \rho_l + f(S) \\ &= (\alpha M + 1) f(S). \end{aligned}$$

For example: Set $E = E_1 \cup E_2, E_1 = \{\upsilon_1, \upsilon_2\}, E_2 = \{\upsilon_3\}$, select at most one element, from E_1 and E_2, f is submodular function $f(\Phi) = 0, f(\{\upsilon_1\}) = \upsilon_1$ $f(\{\upsilon_2\}) = f(\{\upsilon_3\}) = 1, f(\{\upsilon_1, \upsilon_3\}) = f(\{\upsilon_1, \upsilon_2\}) = \alpha + 1$, $f(\{\upsilon_2, \upsilon_3\}) = 1$, $f(\{\upsilon_1, \upsilon_2, \upsilon_3\}) = \alpha + 1$, Visibility, The optimal solution is $f(\{\upsilon_1, \upsilon_3\}) = \alpha + 1$, and the solution of greedy algorithm and approximate greedy algorithm is $f(\{\upsilon_2, \upsilon_3\}) = 1$, so the performance guarantee of greedy algorithm and approximate greedy algorithm is $\alpha + 1$.

References

1. M Suiridendo A.1.5282-approximation algorithm for metric uncapacitated facility location problem. Proceeding Integer Programming and Combinatorial Optimization. Cambridge, MA, USA, 458 (2002)
2. Fisher, M.L., Nemhauer, G.L., Wolsey, L.A.: An analysis of approximations for maxing sub Modular set functions programming study, **8**, 73–87 (1978)
3. Gao, Y., Hao, Z., He, S.: An approximate algorithm for solving the maximum value of the function of non-increasing modulus function and its performance guarantee. Pract. Underst. Mathe. **12**(38), 148–149 (2008)
4. Narayanan, H.: Submodular functions and electrical networks. Vehicles: North Holland (1997)
5. Wang, W., Zhang, F., TuoXiaoli, He, S.: Local search algorithm for solving maximizing submodular set function. J. Wenzhou University (3) 2008

Model of Evaluation on Quality of Graduates from Agricultural University Based on Factor Analysis

Xuemei Zhao, Xiang Gao, Ke Meng, Xiaojing Zhou(✉), Xiaoqiu Yu, Jinhua Ye, Yan Xu, Hong Tian, Yufen Wei, and Xiaojuan Yu

Department of Information and Computing Science, Heilongjiang Bayi Agriculture University, Daqing 163319, People's Republic of China
zhouxiaojing7924@126.com

Abstract. According to the factors of affecting the employment quality for the university graduates, we established the index system of evaluation on the employment quality for the university graduates. The factor analysis was used to establish the model for evaluating the employment quality for the university graduates. Several data sets were analyzed to illustrate the proposed method with the SPSS software. The results not only included the employment quality for the graduates from different colleges, but also included the employment quality for the graduates from different departments in one university. The results coincide with the statistical data about the employment on the official website. Some advice and countermeasures were given on the personnel training and the employment promotion to the universities and education authorities, which can provide the regulatory on employment quality and services of the graduate employment.

Keywords: Index system · Model · Factor analysis · SPSS

1 Introduction

Although there are some theoretical studies on the standards of the employment of college graduates quality and the quality assessment of the current graduates, there are few studies on the method for evaluating of graduates' employment quality. The several existing evaluation methods mainly aimed at the graduates of the well-known colleges or the vocational and technical schools [1–3].

The comparative studies between different majors in the same universities and the same majors in different universities have not been reported in literature. If the employment quality of the university graduates is evaluated scientifically and objectively, a reference will be provided for the career development to the students, also an important basis for decision making will be provided on the university teaching reform and improve the quality of personnel training, along with the employment policy. Thereby the quality of the graduates will be improved.

F. Xhafa et al. (eds.), *Advances in Intelligent Systems and Interactive Applications*, Advances in Intelligent Systems and Computing 686,
https://doi.org/10.1007/978-3-319-69096-4_109

2 Index System of Employment Quality Evaluation for College Graduates

According to the characteristics of the graduates evaluated and the actual conditions, the index system of the employment quality evaluation for the university graduates was constructed based on four dimensions composed of the graduate employment structure, the personal satisfaction of the graduates, the social satisfaction and university factors.

Next, on the basis of experience, the graduate employment quality evaluation was selected by empirical in accordance with the concept of the employment quality evaluation, the indicators selected principles and the scope. Last, the index system employment quality evaluation for the university graduates was determined, in which, four level indicators and 13 secondary indicators were identified. The 13 secondary indicators were the working conditions, the average monthly wages, the social security benefits, the job stability and promotion opportunities, the professional counterparts, the ability-job fit, the career prospects, the evaluation of employers on graduates, the evaluation of the parents, the hardware and the software facilities of school, the curriculum, the teaching content and the practice, the daily teaching management and the service and the employment system, measures, employment guidance.

3 Model on Employment Quality Evaluation of University Graduates

3.1 Data Resources

The data analyzed in this paper is from statistics data about employment on the official website of career center. The data obtained in this way is more objective and credible. In addition, the data also includes the acquired data on graduates in the university, which is in the way of survey. The content of the questionnaire involved personal information, the average monthly wages, the social security benefits, the professional counterparts, the ability-job fit, the evaluation of the employers on the graduates and the evaluation of parents and so on. The process of survey takes advantage of the modern means of communication, including wechat, letters, surveys, e-mail, online surveys and telephone interviews. We try to avoid the subjective factors and randomly select a considerable number of respondents, which ensured the randomness and the fairness of the subject of the investigation.

3.2 Model on Employment Quality Evaluation of Graduates in B University and the Results

Take the graduates in the year of 2015 from XX colleges in B university as an example, there are totally 50 questionnaires handed out and 41 valid questionnaires returned, for which the effective rate is 80.01%. The time for distributing the questionnaires starts from May 1, 2015 to December 1, 2015. The data was analyzed with the SPSS software, and the results are listed in detail as follows.

From the results of Table 1, the first two principal components explained 93.609% of all variance, that is containing the amount of information of the original data reached 93.609%.

Table 1. Variance contribution rate

Principal component	Eigenvalue	Square root	Contribution rate	Cumulative contribution rate
1	10.884	3.299091	83.726	83.726
2	1.285	1.133578	9.884	93.609
3	0.776		5.97	99.58
4	0.055		0.42	100
5	6.71E-16		5.16E-15	100
6	2.15E-16		1.65E-15	100
7	1.13E-16		8.70E-16	100
8	9.95E-17		7.65E-16	100
9	6.95E-18		5.34E-17	100
10	−5.62E-17		−4.33E-16	100
11	−2.42E-16		−1.87E-15	100
12	−3.94E-16		−3.03E-15	100
13	−5.18E-16		−3.98E-15	100

This demonstrated that the first and second principal components can represent the original 13 index (Table 2).

Table 2. Factor loading matrix

Evaluation index	Eigenvalue of the first principal component	Eigenvalue of the second principal component
V1	0.797	−0.495
V2	0.966	−0.208
V3	0.934	−0.342
V4	0.931	−0.317
V5	0.849	−0.09
V6	0.957	−0.088
V7	0.881	0.273
V8	0.661	0.729
V9	0.986	0.146
V10	0.981	0.139
V11	0.954	0.171
V12	0.948	0.287
V13	0.992	−0.069

In accordance with the method of the index coefficient of principal component analysis results, each element of the eigenvectors of the first and the second principal component divided the square root of their eigenvalues, respectively.

So the formula for the coefficient of the first principal component was constructed in the following

$$\begin{aligned} y_1 = {} & 0.24158x_1^* + 0.29280x_2^* + 0.28310x_3^* + 0.28219x_4^* + 0.25734x_5^* \\ & + 0.29008x_6^* + 0.26704x_7^* + 0.20035x_8^* + 0.29887x_9^* + 0.29735x_{10}^* \\ & + 0.28917x_{11}^* + 0.28735x_{12}^* + 0.30068x_{13}^* \end{aligned} \quad (1)$$

The formula for the coefficient of the second principal component was constructed in the following:

$$\begin{aligned} y_2 = {} & -0.43667x_1^* - 0.18349x_2^* - 0.3017x_3^* - 0.27965x_4^* - 0.07939x_5^* \\ & - 0.07763x_6^* + 0.24083x_7^* + 0.64309x_8^* + 0.12879x_9^* + 0.12262x_{10}^* \\ & + 0.15085x_{11}^* + 0.25318x_{12}^* - 0.06087x_{13}^* \end{aligned} \quad (2)$$

where, x_1^* is the working conditions, x_2^* is the average monthly wages, x_3^* is the social security benefits, x_4^* is the job stability and the promotion opportunities, x_5^* is the professional counterparts, x_6^* is the ability-job fit, x_7^* is the career prospects, x_8^* is the evaluation of the employers on the graduates, x_9^* is the evaluation of the parents, x_{10}^* is the hardware and software facilities of school, x_{11}^* is the curriculum, the teaching content and the practice, x_{12}^* is the daily teaching management and the service, x_{13}^* is the employment system, the measures, the employment guidance.

On the sort problems using score of the principal component for the sample, the most popular method is using the linear combination with the main componenty_1, y_2, …, y_m. The variance contribution rate of each principal component was chosen as weights to construct comprehensive evaluation function [1–3]

$$F = \alpha_1 y_1 + \alpha_2 y_2 + \cdots + \alpha_m y_m \quad (3)$$

Sort or classify based on the value calculated of F.

Therefore, the model for the employment quality evaluation of the graduates was constructed as follows

$$F = 0.83726y_1 + 0.09884y_2 \quad (4)$$

If the results were divided into five levels, excellent, good, fair, poor and very poor, the rank of the score was listed in Table 3.

Table 3. The rank of the score for the employment quality evaluation of graduates from B university

Level	y_1	y_2	F	Percentage	Rank
Excellent	657.2278	36.4940	693.7218	0.2567	2
Good	1224.336	1.4752	1225.8112	0.4536	1
Fair	620.6082	−14.8822	605.7260	0.2249	3
Poor	152.5352	−9.8953	142.6399	0.0528	4
Very poor	36.7331	−2.5493	34.1838	0.0127	5

The results show the score of the good level is the most highest, the score of the excellent level is the second highest, the score of the fair level is the third highest, the score of the poor level is the fourth highest. So we can obtain the level of the employment quality evaluation of graduates from XX colleges is the second rank, that is good. The results coincide with the statistical data about the employment on the official website.

3.3 Longitudinal Comparison on Quality of Employment of Graduates from the Colleges in B University

The results obtained by the proposed method demonstrated that the level of the employment quality evaluation of graduates from SM college is the second rank. The results coincide with the data obtained from the statistical data about the employment on the official website and the survey (Table 4).

Table 4. The rank of the score for the employment quality evaluation of graduates from SM college

Level	y_1	y_2	F	Percentage	Rank
Excellent	98.3503	12.8484	111.1987	0.2776	2
Good	172.1731	1.3605	173.5336	0.4333	1
Fair	91.3289	−4.5736	86.7554	0.2166	3
Poor	25.9016	−2.5484	23.3532	0.0583	4
Very poor	5.8816	−0.1966	5.6849	0.0142	5

The results obtained by the proposed method demonstrated that the level of the employment quality evaluation of graduates from LX college is the second rank, which is almost the same percentage of the third rank, the "fair" one. Although the XJ major only set up 13 years, the teachers of this major continue to carry out reform and enhance the level of teaching and research. Especially in the last five years, on the basis of research teaching, actively study and explore the way to comprehensively integrate the entrepreneurship education to the professional training system. Construct the course system of innovation and entrepreneurship with "discipline integration, overall optimization" and the diversified practice system and the regular security system. They try to explore constructing the practical teaching mode composed by experiment, practice and training. The "Business-school" and "school-enterprise cooperative" were added to the new education mechanism. High employment rate gradually increased (Table 5).

Table 5. The rank of the score for the employment quality evaluation of graduates from LX college

Level	y_1	y_2	F	Percentage	Rank
Excellent	14.8972	−0.0634	14.8338	0.1274	3
Good	43.2849	−0.1635	43.1215	0.3705	1
Fair	39.9439	0.1046	40.0485	0.3440	2
Poor	13.6341	0.0967	13.7309	0.1180	4
Very poor	4.6158	0.0479	4.6638	0.0400	5

4 Conclusion

In this paper, first, we established the index system of evaluation on the employment quality for the university graduates. The index system contained 13 indexes. Then, by the principal component analysis, we established the model for the employment quality for the university students. Several data sets were analyzed to illustrate the proposed method with the SPSS software. The results not only include the employment quality for the graduates from different colleges and different departments in one university, but also include the employment quality for the graduates from the same major in different universities. The results coincide with the statistical data about the employment on the official website. This evaluation method is simple, which can provide an operable method to assess the quality of employment education for authorities.

Some advice and countermeasures were given on the personnel training and the employment promotion to the universities and education authorities. From the employers' perspective, they should provide a fair chance to the graduates from different universities to compete. From the school perspective, they should strengthen the school characteristics, improve personnel training system and increase the social practice base. The universities should insist on carrying out joint educational school-enterprise cooperation. In addition, the universities should strengthen the employment guidance, broaden the employment approaches and open up the job market. From the students' perspective, they should strengthen the professional knowledge, improve the professional skills and prepare the career planning. Moreover, they should actively participate in social practice, accumulate work experience and exercise good personal character.

Acknowledgements. The preparation of this manuscript is partially supported by the program of Innovative Entrepreneurship Training for University students of Heilongjiang province: Research on the evaluation for the employment quality of the college graduates from Heilongjiang Province and the real dataset analysis (201610223073).

References

1. Wei, Wand, Jian, Liu: Analysis of the application of pca in the quality of college graduates' employment. Res. Higher. Educ. Heilongjiang Province **5**(2), 84–87 (2016)
2. Yanwei, Wang: Six dimensions of the evaluation index system of the employment quality for college graduates should be considered. J. Yunnan Agric. Univ. **9**(2), 87–90 (2015)
3. Stavrovsky, E., Krasilnikova, M, Pryadko, S.: AHP and ANP as particular cases of Markov chains, IFAC Proc. Volumes **46**(9), 531–536 (2013)
4. Qiaoqiao, Li.: Study on evaluating index system of the employment quality for college graduates, Northeast Normal University (2012)
5. European Parliament, Employment And Social Affairs, Directorate General For Internal Policies Policy Department A: Economic and Scientific Policy, 124 (2009)

Distributed Collaborative Control Model Based on Improved Contract Net

Zhanjie Wang(✉) and Sumei Wang

Dalian University of Technology, Dalian, China
wangzhj@dlut.edu.cn

Abstract. The overall efficiency of multi-robot system is closely related to the cooperative control algorithm, so the research on cooperative control algorithm has been a hot topic in the field of multi-robot. In order to improve the overall efficiency of the multi-robots cooperative system, this paper proposes a Distributed Collaborative Control Model based on improved contract net (DCCM). According to the characteristics of the robot cooperative system, the negotiation and evaluation rules of the multi-robots cooperative system are proposed. This model combines the traditional contract net model with the broadcast algorithm to solve the shortcomings of the traditional contract net. This model solves the bottleneck problem of the traditional contract net model in the bidding evaluation stage, and also reduces the waiting time of the robot. With the help of the supervisory mechanism, the problem of the loss of the traditional contract net model in the task execution stage has been solved, and the stability of the system has been improved. Through the experiment of multiple trackless robots, it proves that this model can effectively reduce the execution time of the whole multi-robots system.

Keywords: Distributed · Contract net · Multi-robot

1 Introduction

Multi-robot cooperative control system is an important development content in the field of robot. The cooperation between multiple intelligent robots will greatly improve the overall intelligence of the robot, and make it possible to complete a lot of functions that single robots can't work [1, 2]. It is important to speed up the practical application of the robot. At present, the main control algorithms for the multi-robot coordination system include the contract net model, the Markov decision process model, the node planning method, the set covering theory and the market agreement method, among which the contract net in the dynamic task allocation has achieved good results [3, 4].

The contract net model was jointly funded by DARPA and the US National Science Foundation in the late 1970s. In 1980, Davis and Smith proposed the contract net model in accordance with the tender-bid-bid mechanism in the market. At present it has been widely used in robots, formation coordination, satellite and other fields [5, 6]. However, the traditional contract net model still exist the following two deficiencies [7, 8]:

F. Xhafa et al. (eds.), *Advances in Intelligent Systems and Interactive Applications*, Advances in Intelligent Systems and Computing 686,
https://doi.org/10.1007/978-3-319-69096-4_110

(1) There is a bottleneck in the tender evaluation of the tenderers
In the traditional contract net model, the tender information is collected by the bidding party, and then the tenders are evaluated according to the evaluation function. When the system scale is large, it is easy to cause bidding parties to slow down and cause the system bottleneck. As the traditional contract net model, intelligent robots in the bidding process, can only participate in the current tender. So the bottleneck time increased the waste of system resources.

2) Task Loss
In the traditional contract net model, the intelligent robot starts to carry out the task after the contract is established. Because the traditional model has no feedback message, when the intelligent robot can't perform the task normally due to the change of the environment situation, the task can't be canceled and executed, which results in the loss of the task.
Based on the current load of the robot, this paper proposed a DCCM for the shortcomings of the traditional contract net model. By introducing the broadcasting mechanism and the supervisory mechanism, this model solves the bottleneck, the resources waste and the task loss of the traditional contract network model.

2 Distributed Collaborative Control Model Establishment

This chapter mainly aims at improving the bottleneck problem, resource idle problem and task loss problem in the traditional contract net model, and establishes a control model suitable for multi-robot cooperative control system.

In order to solve the problem of the bottleneck of the bidding process, the evaluation process is carried out by the intelligent robot which participates in bidding. When the robot receives the tender information and generates its own tender, it passes the broadcast mechanism to other intelligent robots, and also receives the tender information of other intelligent robot broadcasts. The robot compares its own tenders with the received tender information, and continues to receive other tenders if the rating is higher than the other tenders received. If the assessment is over and the robot finds itself to be a winner, it becomes a contractor and is responsible for notifying the tenderer.

In a traditional contract net model, if two or more tenders are simultaneously issued in a system, the intelligent robot will not accept the tender from other tenderers while waiting for a tender to respond to the assessment results. In this bidding method, if the intelligent robot is not selected, it will lose another chance of bidding, which will result in idle resources. Since the system's evaluation process is transparent, intelligent robots can immediately begin to bid others when it can't bid successfully. This way reduces the time the robot waits for the bid results, so it can alleviate the idle resources and improve the resource utilization to a certain extent. In this study, we establish supervisory feedback mechanism to solve the problem of task loss. After the intelligent robot establishes the contract, the bidding party supervises the execution of the contractor's task and periodically asks the execution information of the task. Once some unforeseen circumstances occur and contractor can't complete the task, the tender side has to launch a new bidding process, and the tenderer elects a new contractor to carry out the task, which is repeated until the task is completed.

Through the improvement of the traditional contract net model, the main process of the DCCM in this system is shown in Fig. 1.

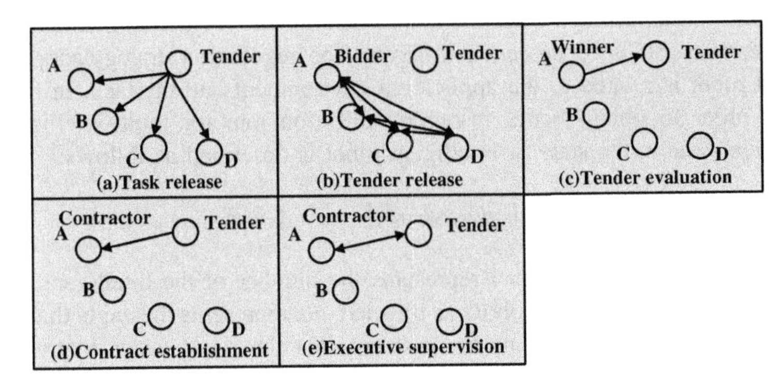

Fig. 1. The model of collaborative control

As shown in Fig. 1, the DCCM consists of the following steps:

(1) Task release as shown in Fig. 1(a). The direction of the tender intelligent robot release task information.
(2) Tender release as shown in Fig. 1(b). Intelligent robot bid after obtaining job information, according to their ability to calculate the value of the bid, and broadcast to other intelligent robots.
(3) Tender evaluation as shown in Fig. 1(c). The intelligent robot compares its own bid value with the bid value of the other intelligent robot broadcasts. If win, it will continue to compare. If failed, it gives up this bid, waiting for other tasks. The winner of the bidding process is responsible for notifying the tender.
(4) Contract establishment as shown in Fig. 1(d). The tenderer verifies the information of the winning bidder, and if he receives multiple wins at the same time, it indicates that the assessment process has caused an error due to net instability or other environmental factors. Secondary evaluated by the tender side, which selected the best tender person. Send confirmation information to the intelligent robot, and formally establish the agreement.
(5) Executive supervision as shown in Fig. 1(e). After the agreement is established, the contractor begins to execute the task. The tenderer periodic receives feedback on the task execution information. If a contractor is found to be unable to perform the task normally, or if the contractor has received the information from the task, return to step (a), and the tender will be re-launched.

The improved contract ne model effectively alleviates the bottleneck problem in the evaluation of tenders, the loss of tasks, and also relieves the idle situation of resources, which can better realize the cooperative control between multiple robots.

3 Negotiation Evaluation Rules

After intelligent robot receives the tender's information, if itself can meet the needs of the task, it will calculate the bidding value of this task and participate the bidding process based on their own situation. The negotiation process is a process that a group of intelligent robots achieve a common satisfaction state. Negotiation strategy in different types of consultation process is different. The negotiation strategy adopted by the intelligent robot is related to the application environment and the task. In this experiment, in order to obtain better resource utilization rate for multi-intelligent robot coordination system, the state of intelligent robot is described as follows:

$$S_i = \{dd_i, da_i, rd_i, ra_i\} \quad i = 1, 2, 3, \ldots \tag{1}$$

In the Eq. (1), the parameter i represents the number of the intelligent robot, dd_i represents the distance that the robot has traveled, da_i represents the angle that the robot has already turned, rd_i represents the distance that the robot needs to travel at the current time, and ra_i represents the angle that needs to turn in the remaining task of the robot. Due to the different magnitude of each attribute value, the attribute value is required to be normalized. The normalized formula is as follows:

$$y = \frac{x - MinValue}{MaxValue - MinValue} \tag{2}$$

In Eq. (2), x represents the parameter value before normalization, y represents the parameter value after normalization. $MaxValue$ and $MinValue$ respectively represent the maximum and minimum values in the same type of parameters. In this paper, we describe four parameter types of the robot state, such as dd_i, da_i, rd_i and ra_i. The four parameters are denoted by x_{ij} after normalization. Where i represents the robot number and j represents the task type number.

The calculation rules for each smart robot bid are as follows:

$$f_{ik} = \sum_{j=1}^{4} y_j x_{ij} \qquad (\sum_{j=1}^{4} y_j = 1) \tag{3}$$

In Eq. (3), f_{ik} represents the bid value of robot i for Task k, and y_j represents the weight of attribute x_{ij}. The robot calculates the bid value, and then informs the other robot by broadcast. Through the comparison of the bid value, the winner becomes a contractor, and signs an agreement with the tender. In this experiment, the rule of evaluation is simple, and the robot with small bidding value becomes a winner. By negotiation and cooperation, the load of each robot is balanced, and the executive time of the entire system is least.

4 Model Analysis

4.1 Performance Analysis

The finishing time of tasks depends on the execution time of tasks, it is necessary to shorten finishing time of tasks by reducing the execution time of tasks. In the process of resource allocation of the traditional contract network model, each robot generates its own bid value and sends it to the bidder. With the increase in the number of robots in the system and the increase in the number of tasks, the presentation of the tender process will lead to serious bottlenecks which will lead to a decline in system performance. In the DCCM, the evaluation of the tender by each robot, for the smaller bidder robot, can quickly get the assessment results. In DCCM, whenever there are new tasks that need to be competed, each robot root bids the rules to produce their own tenders. The bidding rules in this model introduce the amount of tasks and the amount of outstanding tasks. By introducing the completed task, all the robots in the whole system realize load balancing. Because the contract net model in the allocation of tasks takes intelligent robots' uncompleted tasks into account, so intelligent robots with fewer tasks are given priority, the overall execution time is minimal.

4.2 Experimental Verification

In this part of the experiment, the robot model is used to experiment with the proposed control model and sequence model to verify that the proposed model can effectively reduce the execution time of the task. Because the task in this article refers to a randomly generated task set, the execution time of the task can be converted to the final completion time of the task set. The experimental environment is set as follows: The multi-robot coordination system contains five intelligent robots, and each robot's speed is set to 15 cm/s. In the course of turning, the speed is equivalent to the angular velocity 14.3°/s. The tenderer obtains the information (path length, angle of rotation) required by the contract net according to the path information of each task. The path length range for each task is [0, 1000 cm], and the angle range is [0°, 1000°]. The number of tasks set in the experiment is $N \in \{25, 30, 35, 40, 45, 50\}$. The weights of the four attributes are set to 0.15, 0.15, 0.35 and 0.35. The comparison results of the execution time of the two scheduling algorithms are shown in Fig. 2.

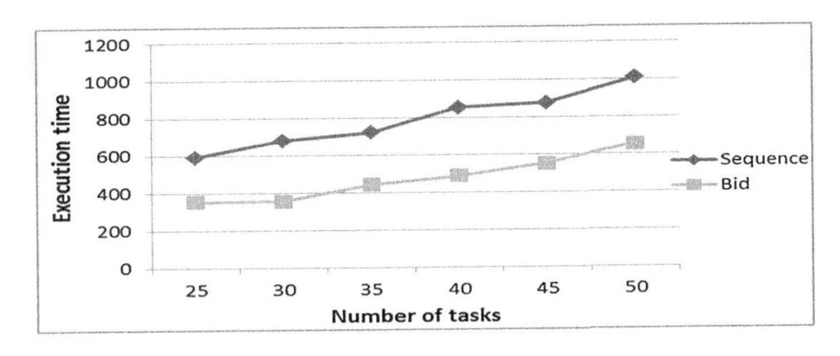

Fig. 2. The comparison of completion time

As shown in Fig. 2, the black curve represents the execution time of the task in the sequential model, and the gray curve represents the execution time of the task in the DCCM. It is obvious from the trend of the curve in the graph that with the increasing number of resources, the execution time of the contract net algorithm is generally less than the sequential algorithm.

5 Conclusion

In view of the existing problems of the traditional contract network model, DCCM is proposed to improve the overall efficiency of multi-robot cooperative system. By introducing the broadcasting mechanism and the supervisory mechanism, this model solves the bottleneck, the resources waste and the task loss of the traditional contract network model. The model shortens the execution time of the task set and improves the system stability of the system. Based on multiple robot entities, experiments were performed on DCCM and sequential models. In the same experimental environment, the execution time of the two tasks is compared. The experimental results show that the model can effectively reduce the execution time of the task. At the same time, the model improves the stability and reliability of the system and its effectiveness.

References

1. Chen, X., Hao, F., Ma, B.: Periodic event-triggered cooperative control of multiple non-holonomic wheeled mobile robots. IET Control Theory Appl. **11**(6), 890–899 (2017)
2. Wang, G., Wang, C., Du, Q., et al.: Distributed cooperative control of multiple nonholonomic mobile robots. J. Intell. Rob. Syst. **83**(3–4), 1–17 (2016)
3. Moreno, L., Mart, Fernando N., et al.: Differential evolution Markov chain filter for global localization. J. Intell. Robot. Syst. **82**(3), 513–536 (2016)
4. Li, G., Sha, J.C.: Research on task optimal allocation for distributed satellites system based on contract net protocol. J. Astronaut. **30**(2), 815–820 (2009)
5. Yuan, L., Dang, J., Zhao, S.: Emergency rescue strategy research of urban road accident based on contract net. Comput. Eng. Appl. (2016)
6. Li, Y.W., Li, B.A.: Research of multiple UAVs task allocation based on improved contract net. Adv. Mater. Res. **823**(823), 439–444 (2013)
7. Zhou, P.C., Han, Y.S., Xue, M.G.: Extended contract net protocol for multi-robot dynamic task allocation. Inf. Technol. J. **6**(5), 733–738 (2007)
8. Gupta, A.K., Gallasch, G.E.: Equivalence class verification of the contract net protocol-extension. Int. J. Soft. Tools Technol. Transfer **18**, 1–22 (2015)

K Distribution Clutter Modeling Based on Chebyshev Filter

Bin Wang(✉) and Fengming Xin

Northeastern University at Qinhuangdao, Qinhuangdao, China
wangbinneu@qq.com

Abstract. As technology continues to progress and develop, the electromagnetic environment becomes more and more complex. So in the research field of modern radar, it is very important to model clutter accurately. In this paper, in order to overcome the spectral broadening matter, we set up a new architecture based on Chebyshev filter and propose a novel clutter modeling method for K distribution clutter. In simulations, curves demonstrate that the estimated value of the improved Chebyshev filter method is more close to the theoretical value than the original zero memory nonlinearity method in the aspects of probability density and power spectral density, and the improved Chebyshev filter method is valid. In the end, the conclusions are given.

Keywords: Clutter modeling · Chebyshev filter · K distribution

1 Introduction

With the continuous development of modern technology, the demands for radar simulation system are increasing. Clutter modeling and simulation is the basis of echo analysis, which is very important in radar simulation. Accurate modeling in complex environment has important significance, and it attracts attention of researchers worldwide.

Many researchers have done much work in clutter simulation, which is an important part of radar echo simulator. In [1], the authors add different clutter distributions into the framework of track-ability, and significant differences are obtained. In [2], a new model is proposed, which is comprised by functions of radar frequency, polarization, sea state, and grazing angle. In [3], it is shown that when looking up or down wind local spectrum intensity is strongly linked to mean Doppler shift. In [4], with the environment of Saudi Arabian urban, South African urban and Saudi Arabian date farms, the statistical analysis of three types of land clutter are presented. In [5], the authors research on bistatic radar. Bistatic clutter scenario is defined and two land-clutter models are proposed. In [6], the authors propose a novel radar waveform optimization method, and the criterion is information theory. In [7], different from traditional signal-dependent stochastic models, the authors propose a new approach of stochastic transfer function. In [8], the authors describe X-band radar-based measurements of urban ground clutter, which is made from the CSIR campus in Pretoria. In [9], artificial neural network is used to classify four common radar clutter models.

F. Xhafa et al. (eds.), *Advances in Intelligent Systems and Interactive Applications*, Advances in Intelligent Systems and Computing 686,
https://doi.org/10.1007/978-3-319-69096-4_111

After simulating clutter models and extracting important features, the authors make neural network with two hidden layers.

In this paper, after analyzing characteristics of K distribution clutter, we establish a new architecture based on Chebyshev filter. Then we design Chebyshev filter and propose a novel clutter modeling method for K distribution clutter. In simulations, we will compare the improved Chebyshev filter method with the original zero memory nonlinearity (ZMNL) method.

2 Characteristics and Architecture of K Distribution Clutter

Rayleigh distribution, Log-Normal distribution and Weibull distribution are all based on a single statistic, and they are only suitable for single pulse detection. The main shortcoming is lack of time and spatial correlation of simulated clutter. K distribution overcomes the shortcomings of the above three distributions.

Probability distribution density function of K distribution is

$$f(x|a,v) = \frac{2}{a\Gamma(v+1)}\left(\frac{x}{2a}\right)^{v+1}K_v\left(\frac{x}{a}\right)(x>0, v>-1, a>0) \tag{1}$$

where $K_v(\bullet)$ is second type of modified Bessel function, a is scale parameter which is only related to the average value of clutter, and v is shape parameter which controls the shape of the distribution tail.

The mean value and variance of K distribution is

$$E(x) = \frac{2a\Gamma(v+3/2)\Gamma(3/2)}{\Gamma(v+1)} \tag{2}$$

$$\mathrm{var}(x) = 4a^2\left[v+1-\frac{\Gamma^2(v+3/2)\Gamma^2(3/2)}{\Gamma^2(v+1)}\right] \tag{3}$$

To seek integrals for probability density function, we can approximately get distribution function of K distribution

$$F(x) = 1-\frac{2}{\Gamma(v+1)}\left(\frac{x}{2a}\right)^{v+1}K_{v+1}\left(\frac{x}{a}\right) \tag{4}$$

The principle block diagram of ZMNL method is in Fig. 1. After the action of the filter $H(z)$, correlated Gaussian process can be obtained, the input of which is white Gaussian random process. Then we make nonlinear transformation, and the required clutter sequence is generated.

However, ZMNL method seems to be powerless in analyzing K distribution clutter. In order to generate K distribution clutter with ZMNL method, nonlinear relationship of clutter correlation coefficient before and after nonlinear transformation should be known. In fact, it is difficult to find such a nonlinear transformation in K distribution

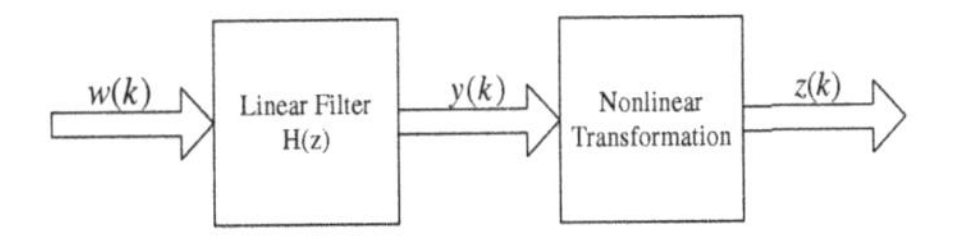

Fig. 1. Principle block diagram of ZMNL method

clutter. Generally we can use curve fitting method, but it will produce a certain error. So we consider improving the structure of ZMNL to reduce error.

In the K distribution Eq. (1), v is the shape parameter and a is the scale parameter. The relationship between v and a can be expressed as

$$a^2 = \frac{\sigma^2}{2v} \tag{5}$$

In most cases, the range of v is $0 < v < \infty$. v can be large or small. Figure 2 is the improved architecture with Chebyshev filter. The key problem is the design of Chebyshev filter.

In Fig. 2, $w_1(k)$ is a complex Gaussian white noise, $w_2(k)$ is a real Gaussian noise which is independent to $w_1(k)$.

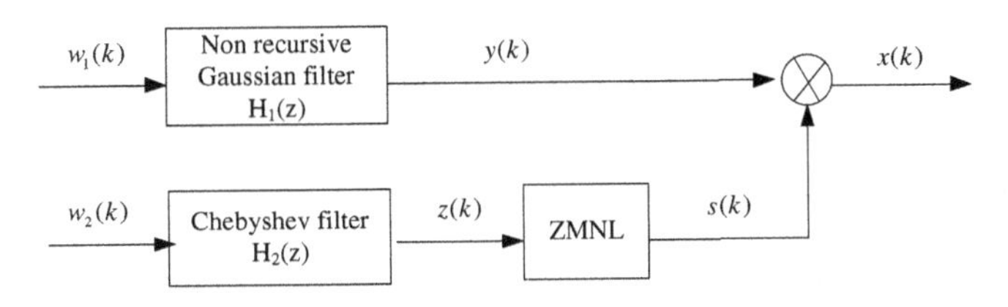

Fig. 2. Improved architecture with Chebyshev filter

Chebyshev filters have excellent property. Over the broad filter range, they can minimize the error. As there is pass-band ripple in Chebyshev filters, in some application situation we prefer to use more suitable filter. The filter should have a smoother response in the pass-band. However, it should have a more irregular response in the stop-band. So we can design Chebyshev filters to reduce the error when using curve fitting method with ZMNL.

3 Design of Chebyshev Filter

After designing filter $H_1(z)$ and $H_2(z)$ in Chebyshev filter, we can obtain K distribution clutter with our new model.

Chebyshev filter design is a key step in the improved ZMNL method. Amplitude characteristic of Chebyshev filter has monotone property in pass-band, while has equal ripple characteristic in the stop band. Its amplitude squared function can be expressed as

$$|H_a(j\Omega)|^2 = \frac{1}{1+\varepsilon^2[C_N(\Omega_{st})/C_N(\Omega_{st}/\Omega)]^2} \quad (6)$$

where Ω_{st} is the minimum frequency when the resistance band attenuation reaching the prescribed value. ε is positive number less than 1, which represents the size of pass band ripple. Generalized distributed random variable should be generated when using this model, and it can be obtained by taking a square root of Gamma distribution random variables. The expression of nonlinear transformation is

$$\frac{\gamma(v,s^2)}{\Gamma(v)}\mu(s) = 1 - Q(z) \quad (7)$$

where $\gamma(q, p)$ is incomplete Euler function.

It can be changed into

$$g(v, 2vs^2/\pi) = 1 - Q(z) \quad (8)$$

where $Q(z)$ is tail area of standard normal random variable, then

$$Q(z) = \int_z^{\infty} \frac{1}{\sqrt{2\pi}} \exp(-\frac{u^2}{2}) du \quad (9)$$

So the problem of generating $s(k)$ converts into the problem of solving formula (9). This is a nonlinear equation which can be solved using dichotomy method.

4 Simulations

In this section, we will compare the improved Chebyshev filter method with the original ZMNL method. Radar echo frames is 2000, sampling frequency is 1000 Hz, wavelength λ_0 is 0.05, root mean square value of σ_v (velocity distribution) is 1.0. Simulations results are in Figs. 3, 4, 5 and 6.

Figures 3 and 4 is real and imaginary part of K distribution clutter waveform. This is the basis of the experiment. The characteristics of K distribution clutter waveform can be obtained in Figs. 3 and 4.

Figure 5 is clutter amplitude distribution. It can be seen in this figure the estimated value with improved ZMNL is closer to the theoretical value than the estimated value with ZMNL. Figure 6 is clutter spectrum. It can be seen in this figure that in low frequency part (<50 Hz), the estimated value with improved ZMNL and ZMNL have little difference to the theoretical value, while in high frequency part (>50 Hz), the performance of our improved ZMNL is significantly better than ZMNL.

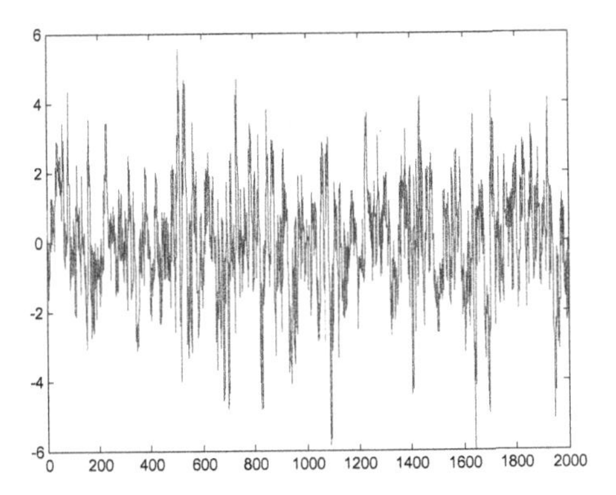

Fig. 3. Real part

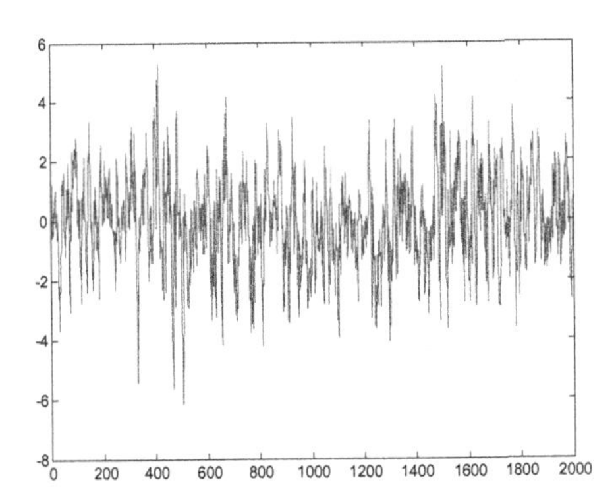

Fig. 4. Imaginary part

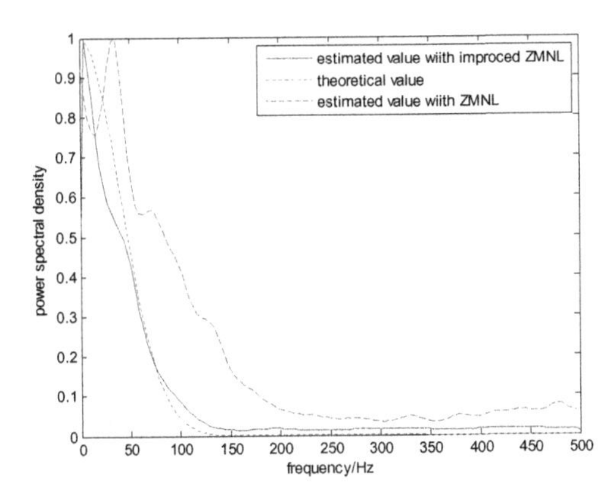

Fig. 5. Distribution of clutter amplitude

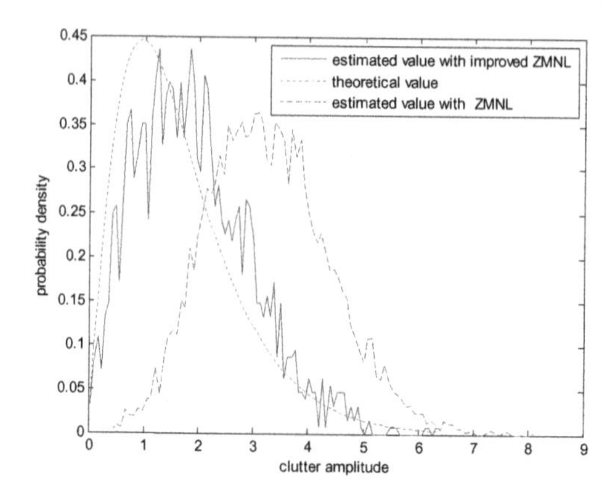

Fig. 6. Clutter spectrum

5 Conclusions

In modern research, it is very important to model and analyze clutter accurately in the modern radar field. The paper focuses on K distribution clutter modeling. As a result of difficult to find nonlinear transformation in K distribution, traditional ZMNL method seems to be powerless, and curve fitting method will produce a certain error. Our improved ZMNL model introduces Chebyshev filter, and it can minimize the error over the range of the filter, which lies between the characteristics of idealized and actual filter. So we can design Chebyshev filters to reduce the error when using curve fitting method with ZMNL. Simulation results show that the improved Chebyshev filter method is valid.

Acknowledgements. This work was supported by the NSFC (No. 61403067).

References

1. Schoenecker, S., Willett, P., Bar-Shalom, Y.: The effect of K-distributed clutter on trackability. IEEE Trans. Signal Process. **64**(2), 475–484 (2016)
2. Gregers-Hansen, V., Mital, R.: An improved empirical model for radar sea clutter reflectivity. IEEE Trans. Aerosp. Electron. Syst. **48**(4), 3512–3524 (2012)
3. Watts, S., Rosenberg, L., Bocquet, S., Ritchie, M.: Doppler spectra of medium grazing angle sea clutter; part 1: characterization. IET Radar Sonar Navig. **10**(1), 24–31 (2016)
4. Melebari, A., Abdul Gaffar, M.Y., Strydom, J.J.: Analysis of high resolution land clutter using an X-band radar. In: IEEE Radar Conference Proceedings, pp. 139–144. IEEE, Johannesburg, 2015
5. Pola, M., Bezousek, P., Pidanic, J.: Model comparison of bistatic radar clutter. In 13th Conference on Microwave Techniques Proceedings, pp. 182–185. IEEE, Pardubice, 2013
6. Leshem, A., Naparstek, O., Nehorai, A.: Information theoretic adaptive radar waveform design for multiple extended targets. IEEE J. Sel. Top. Signal Process. **1**(1), 42–55 (2007)

7. Guerci, J.R., Bergin, J.S., Guerci, R.J., Khanin, M., Rangaswamy, M.: A new MIMO clutter model for cognitive radar. In: IEEE Radar Conference Proceedings, pp. 1–6. IEEE, Philadelphia, 2016
8. Strydom, J.J., De Witt, J.J., Cilliers, J.E.: High range resolution X-band urban radar clutter model for a DRFM-based hardware in the loop radar environment simulator. In: International Radar Conference Proceedings, pp. 1–6. IEEE, Lille, 2014
9. Darzikolaei, M.A., Ebrahimzade, A.A., Gholami, E.: Classification of radar clutters with artificial neural network. In: 2nd International Conference on Knowledge-Based Engineering and Innovation Proceedings, pp. 577–581. IEEE, Tehran, 2015

Influence of Liquid Film Thickness on Dynamic Property of Magnetic-Liquid Double Suspension Bearing

Zhao Jianhua(✉), Wang Qiang, Zhang Bin, and Chen Tao

Hebei Provincial Key Laboratory of Heavy Machinery Fluid Power Transmission and Control, Yanshan University, Qinhuangdao 066004, China
zhaojianhua@ysu.edu.cn

Abstract. Due to the low viscosity of seawater, it is difficult to form the seawater-lubricated film, and the bearing capacity and stiffness of the seawater-lubricated film is very small. It is easily to cause the "overload" and "burning" phenomenon of the seawater-lubrication sliding bearing, and the operation stability and service life can be shorted. The magnetic bearing takes the bearing form of non-contact suspension, and it is more suitable as an auxiliary support to the seawater-lubrication sliding bearing. Therefore, the paper introduces the electromagnetic suspension support into the seawater-lubricated sliding bearing. And then a novel Magnetic-Liquid Double Suspension Bearing with the advantages of electromagnetic suspension and hydrostatic supporting can be formed. The structural characteristics, supporting mechanism, hydrostatic self-adjustment and electromagnetic-adjustment processes of Magnetic-Liquid Double Suspension Bearing can be analyzed in the paper. Based on force balance equation, electromagnetic equation and flow equation, the transfer function of different adjustment processes under constant-flow supply model are deduced. Then adjusting time, dynamic stiffness and phase margin are selected as dynamic indexes. The influence rule of the liquid film on capacity property of single degree freedom bearing system of Magnetic-Liquid Double Suspension Bearing can be analyzed. The results show that as liquid film thickness increases, dynamic stiffness decrease and adjusting time increase, and phase margin remain the same during the hydrostatic self-adjustment process. The proposed research provided a basis for the design of Magnetic-Liquid Double Suspension Bearing in the engineering practice.

Keywords: Liquid film thickness · Magnetic-Liquid double suspension bearing · Dynamic index

1 Introduction

Due to the low viscosity and special performances of seawater, makes the lubricating film of seawater-lubricated sliding bearing is difficult to be formed. As the electromagnetic bearing in the form of non-contact support, and with zero mechanical friction, high precision, that is very suitable used as an auxiliary supporting form for the seawater lubrication bearing support.

F. Xhafa et al. (eds.), *Advances in Intelligent Systems and Interactive Applications*, Advances in Intelligent Systems and Computing 686,
https://doi.org/10.1007/978-3-319-69096-4_112

Magnetic-Liquid Double Suspension Bearing System with seawater lubricated, mainly composed of the base, shell, magnetic lining, coil, stator, spindle and other components, as shown in Figs. 1 and 2.

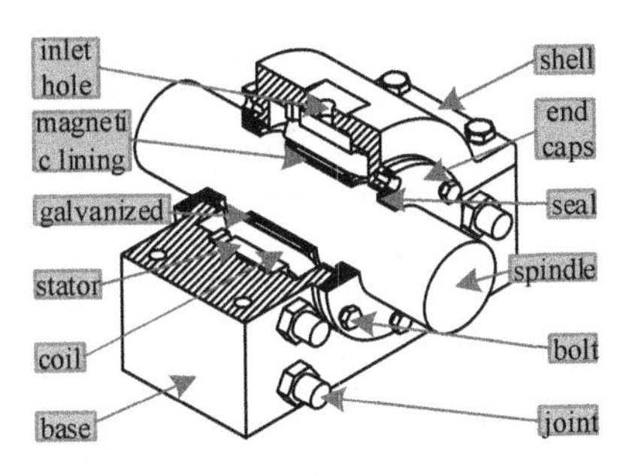

Fig. 1. Magnetic-liquid double suspension bearing semi-isometric view

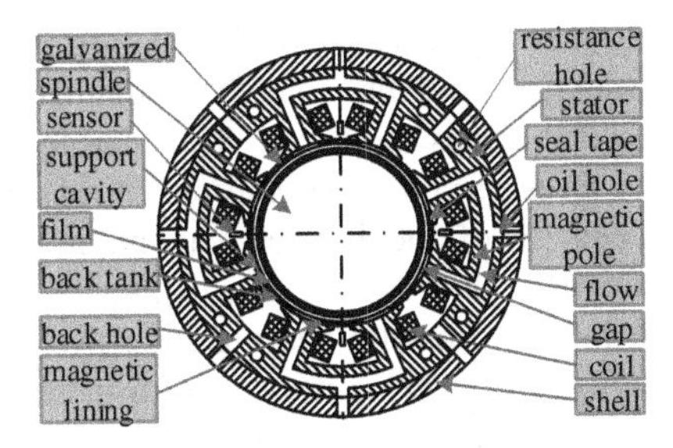

Fig. 2. Magnetic-liquid double suspension bearing full profile

Magnetic-Liquid Double Suspension Bearing System includes magnetic suspension and hydrostatic two support system (as shown in Fig. 3), and combines magnetic suspension support and hydrostatic support in one, so that the bearing carrying capacity, stiffness and operational stability can be significantly increased. The working process of the Magnetic-Liquid Double Suspension Bearing System is divided into two processes: hydrostatic self-adjustment and magnetic-regulation (as shown in Fig. 4).

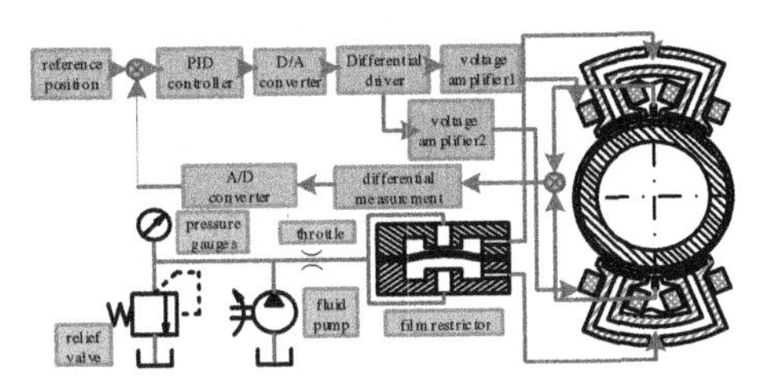

Fig. 3. Single degree freedom support regulation principle

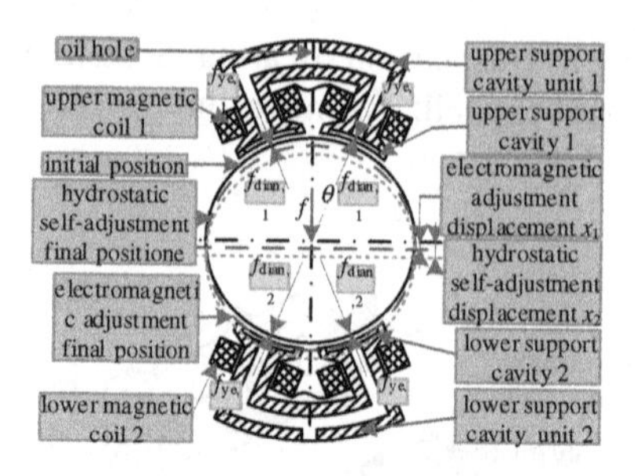

Fig. 4. Force diagram of magnetic-liquid double suspension bearing

In recent years, many scholars have carried on further research on liquid-hydrostatic bearing and electromagnetic bearing, and also acquired large achievements. In paper [1], the mathematical model of the hydrostatic thrust bearing is established, and the influence of water film thickness, water cavity thickness and inlet hole diameter on bearing capacity are analyzed. The results show that with the water film thickness increase, the bearing capacity of the bearing is significantly decreased. With the water cavity thickness and the inlet hole diameter increase, the bearing capacity increases first and then keeps no change. The bearing capacity and influencing factors of water-lubricated thrust bearing are analyzed in reference [2], and the results show that the bearing capacity of the bearing can be improved by choosing the appropriate friction pair and PV value. In reference [3], the basic characteristics of high speed spindle water-lubricated hydrostatic bearing rotor system are studied. And the results show that static stiffness of bearing-rotor system increases as the water supply pressure and load increased. Reference [4], a new type of magnetic suspension precision linear motion platform is proposed, and establish its three-dimensional finite element model, and analyze the magnetic suspension platform structural modal under the different working conditions. And the results show that the platform has the lowest rigidity when suspended, and the position of the suspension has no effect on the overall mode of the platform.

2 Kinetic Equation of Single DOF Supporting System

2.1 The Initial State of a Single Degree of Freedom System

In initial state, film thickness and bias current of the electromagnet are equal, and the electromagnetic attraction of the upper and lower supporting units is equal.

(1) Voltage equation Voltage amplifier output is:

$$e_{pian,1,0} = e_{pian,2,0} = i_0 R_{quan} \tag{1}$$

(2) Flow equation Magnetic fluid double suspension bearing adopts constant flow supply mode, and the upper and lower support cavity 1and 2 flow equation are:

$$\begin{cases} q_{1,0} = q_{fa,1,0} \\ q_{2,0} = q_{fa,2,0} \end{cases} \tag{2}$$

(3) Hydrostatic bearing force According to Navier-Stokes equation, the upper and lower $f_{\mathrm{ye},1,0}$ and $f_{\mathrm{ye},2,0}$ of support cavities 1 and 2 are:

$$\begin{cases} f_{ye,1,0} = 2p_{1,0}A_e\cos\theta = 2q_{1,0}R_0A_e\cos\theta \\ f_{ye,2,0} = 2p_{2,0}A_e\cos\theta = 2q_{2,0}R_0A_e\cos\theta \end{cases} \tag{3}$$

(4) Electromagnetic suspension bearing force. According to the Maxwell attractive formula, get up and down the pole 1, 2 $f_{\mathrm{dian},1,0}$ 、$f_{\mathrm{dian},2,0}$:

$$f_{dian,1,0} = f_{dian,2,0} = 2k\frac{i_0^2}{(h_0+l)^2}\cos\theta \tag{4}$$

(5) Spin axis mechanical balance equation Based on Newton's second law, the mechanical equilibrium equation is obtained:

$$f_{dian,1,0} + f_{ye,2,0} - f_{dian,2,0} - f_{ye,1,0} = mg \tag{5}$$

2.2 Working State of a Single Degree of Freedom System

Under the external load, the bearing shaft displacement is x, and the liquid film thickness of the upper and lower support cavities h_1, h_2 are:

$$\begin{cases} h_1 = h_0 + x\cos\theta \\ h_2 = h_0 - x\cos\theta \end{cases} \tag{6}$$

(1) Voltage equation As magnetic system is not regulated, output voltage is unchanged:

$$e_{pian,1,0} = e_{pian,2,0} = i_0R_{quan} \tag{7}$$

(2) Flow equation The change of the liquid film thickness causes the change of the liquid resistance so, the upper and lower support cavities flow rate is:

$$\begin{cases} q_1 = q_{fa,1,0} - A_b\dot{h}_1 - \dfrac{V_{oa}}{E}\dot{p}_1 \\ q_2 = q_{fa,2,0} - A_b\dot{h}_2 - \dfrac{V_{oa}}{E}\dot{p}_2 \end{cases} \tag{8}$$

(3) Hydrostatic bearing force Similarly, the working state of the upper and lower support cavity 1, 2 hydrostatic pressure bearing force is:

$$\begin{cases} f_{ye,1} = 2q_1 R_1 A_e \cos\theta \\ f_{ye,2} = 2q_2 R_2 A_e \cos\theta \end{cases} \tag{9}$$

(4) Electromagnetic suspension bearing force Similarly, the working state of the upper and lower poles electromagnetic levitation force is:

$$\begin{cases} f_{dian,1} = 2k\dfrac{i_0^2}{(h_1 + l)^2}\cos\theta \\ f_{dian,2} = 2k\dfrac{i_0^2}{(h_2 + l)^2}\cos\theta \end{cases} \tag{10}$$

(5) Shaft mechanical balance equation Similarly, the mechanical equilibrium equation of the rotating shaft is:

$$f_{dian,1} + f_{ye,2} - f_{dian,2} - f_{ye,1} - mg - f = -m\ddot{x} \tag{11}$$

2.3 Transfer Function of Single Degree of Freedom Bearing Unit

According to the formula from (1) to (10), ignoring the sensitive volume of Voa, and the linearization and Laplace transform, the transfer function is obtained:

$$G(s) = \frac{X(s)}{F(s)} = \frac{1}{T_2 s^2 + T_1 s + T_0} \tag{12}$$

3 Influence of Film Thickness on Static Performance

3.1 Static Performance Indexes

(1) Adjust time t_s After a period of time, the displacement of the bearing shaft for the first time to reach and always remain within the allowable error, the time required is known as the adjustment time t_s, and mathematical expression is:

$$t_s = \frac{8m}{T_1} \tag{13}$$

(2) Dynamic stiffness j_s U nder dynamic load, the capacity of the every displacement of the static hydrostatic guideway can resist the dynamic load f is j_s, that is:

$$j_s = \sqrt{\left(T_0 - m\omega^2\right)^2 + T_1^2\omega^2} \tag{14}$$

(3) Phase margin γ Owning appropriate relative stability, and it is necessary to have a certain stability margin from the critical point of the system. and mathematical expression is:

$$\gamma = 180^{\circ} + \omega_c = 180^{\circ} + \arctan\frac{-T_1\omega}{T_0 - m\omega^2} \tag{15}$$

3.2 Magnetic-Liquid Double Suspension Bearing Parameters

Design parameters of Magnetic-Liquid Bearing are shown in Table 1.

Table 1. Design parameters of magnetic-liquid double suspension bearing

Bearing quality m/kg	Dynamic viscosity μ/Pa s	Elastic modulus E/MPa	Zinc coating thickness l/mm	Pole area A/mm^2
66.82	1.3077×10^{-3}	2.4×10^3	0.5	1000
Coil number *N*/ Dimension-less	Liquid cavity width *A*/m	Liquid cavity length *B*/m	Axial fluid sealing tape width *b*/m	Sealing tape width *a*/m
633	0.1	0.02	0.004	0.006
Pump pressure p_s/MPa	Fluid chamber pressure p_1/MPa	Fluid chamber pressure p_2/MPa	Bias current Size i_0/A	Load *f*/N
3	1	1.3	1.2	2000

3.3 Influence of Liquid Film Thickness on Static Performance of Magnetic-Liquid Double Suspension Bearing by Hydrostatic Self-Adjustment

With the increase of liquid film thickness, the liquid resistance of the upper and lower two supports decreases, it is concluded that the damping term decreases enables the adjustment time of the system increases monotonically. As shown in Fig. 5.

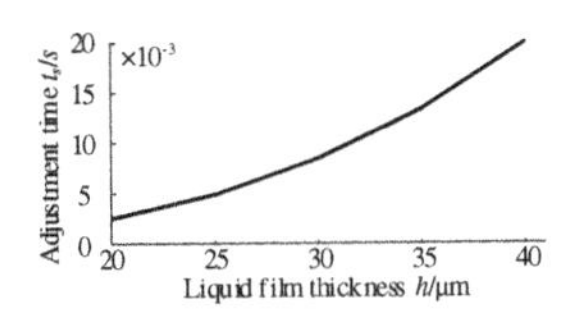

Fig. 5. Relationship between adjustment time and the thickness of liquid film

With the increase of liquid film thickness, the damping and stiffness of the two order system decrease, and based on the formula (14), the monotonic decreasing trend of the dynamic stiffness of the bearing system, as is shown in Fig. 6.

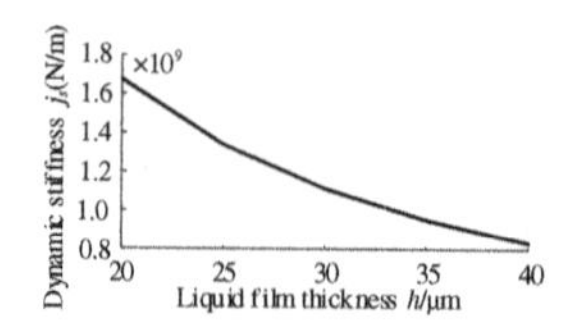

Fig. 6. Relationship between dynamic stiffness and liquid film thickness

With the increase of liquid film thickness, the stiffness and damping of the two order system decrease, the bearing phase margin increases, but the increase is not obvious, as shown in Fig. 7.

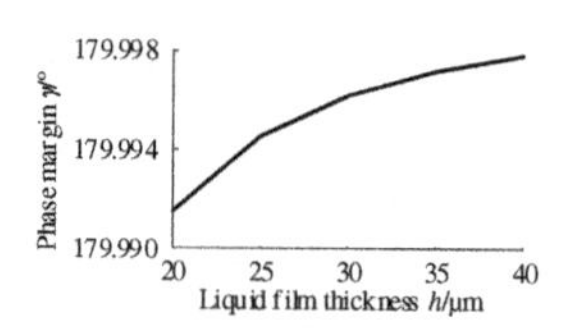

Fig. 7. Relationship between phase margin and liquid film thickness

4 Conclusion

(1) Transfer function of the single degree of freedom bearing system of the fluid and magnetic suspension bearing is two order system in the process of self-regulating.
(2) First term of transfer function is a damping term, which is mainly affected by dynamic extrusion effect of hydrostatic support cavity, and is related to the liquid resistance and the support cavity area; zero order term is stiffness term, which is related to liquid film thickness, supporting cavity area, initial current and coil turns and so on.
(3) Dynamic stiffness of the bearing decreases with the increase of the thickness of the liquid film in the process of self-adjustment, but the adjustment time is increase.

Acknowledgements. The project was financially supported by Natural Science Foundation of Hebei Province supported (E2016203324).

References

1. WANG, YZ., JIANG, D., YIN, ZW., et al.: Load capacity analysis of water lubricated hydrostatic thrust bearing based on CFD. J. Donghua Univ. **41**(4), 428–432 (2015)
2. Zhang, X., Zhou, Y., Wang, X.R., et al.: Research the influences on load capacity of water-lubricated thrust bearing. Machinery **49**(4), 16–19 (2011)
3. DONG, SX., MA, QS.: Basic performance analysis of high-speed spindle system supported by hybrid bearings. Manuf. Technol. Machine Tool **9**, 136–139 (2011)
4. GUO, NP., ZHOU, HB., et al.: Modal analysis of Maglev linear motion platform. J. Eng. Des. **18**(3), 183–190 (2011)
5. Zhao, J.H.: The Oretical Analysis and Experimental Research of Liquid Hydrostatic Slide's Performance of Gantry Turning and Milling Center. Yanshan University, Hebei (2013)
6. Ding, Z.Q.: Design of Liquid Hydrostatic Bearing. Shanghai Scientific & Technical Publishers, Shanghai (1989)
7. SHARMA, SC., PHALLE, VM., Jain, SC.: Influence of wear on the performance of a multirecess conical hybrid journal bearing compensated with orifice restrictor. Tribology Int. **44**(12), 754–764 (2011)
8. Liang, S.P.: Research on Nonlinear Dynamics of a Rotor-Active Magnetic Bearing System. Shanghai University, Shanghai (2009)
9. Formica, F.: Heart ware LVAD is a promising device for patients with end-stage heart failure. Ann. Thorac. Surg. **94**(6), 2180 (2012)
10. ZHANG, GY., YUAN, XY., MIAO, XS, et al.: Experiment for water-lubricated high-speed hydrostatic journal bearings. Tribology **26**(3), 238–240 (2006)
11. Pruijsten, RV., Lok, SI., Kirkels, HH., et al.: Functional and haemodynamic recovery after implantation of continuous-flow left ventricular assist devices in comparison with pulsatile left ventricular assist devices in patients with end-stage heart failure. Eur J Heart Fail. 14(3), 319–325 (2012)

Characteristics Analysis on Open-Type Liquid Hydrostatic Guideway with Unequal Area Oil Pocket

Zhao Jianhua(✉), Wang Qiang, Zhang Bin, and Chen Tao

Hebei Provincial Key Laboratory of Heavy Machinery Fluid Power Transmission and Control, Yanshan University, Qinhuangdao 066004, China
zhaojianhua@ysu.edu.cn

Abstract. Design of unequal-area oil pocket can decrease required pressure and flow of primary oil-supply oil pocket in open-type self-adaption oil-supply liquid hydrostatic slide. But anti-vertical-loads and anti-overturning-loads ability is quite different in composition from hydrostatic slide with equal-area oil pocket design. Taking single-row oil pocket group as research subject, ma matics relationship expression between length of primary oil-supply oil pocket and bearing capacity and stiffness of oil pocket group under vertical and overturning load is presented. Results indicate that with length of rimary oil-pocket increases, vertical bearing capacity and stiffness of unequal area oil pocket group is greater than of which equal area oil pocket group. But overturning bearing capacity and stiffness is smaller and anti overturning ability gets worse.

Keywords: Open-type liquid hydrostatic guideway · Self-adaption oil-supply scheme · Unequal-area design · Bearing capacity static stiffness

1 Introduction

Liquid-hydrostatic guide is a kind of hydrostatic bearing, with advantage of superior carrying capacity, high static and dynamic stiffness, and good antivibration performance [1, 2]. Open type hydrostatic guideway can only withstand unidirectional load, and n its static performance was limited [3–5]. In order to reduce operation range of oil film, and improve precision. Generally, ratio of allowable load and self-weight of guideway is small (Fig. 1).

At present, some scholars were designed and improved open-type liquid hydrostatic guideway, and developed a new adaptive oil supply system. According to Dai Huiliang [6] proposed a way that constant pump + proportional pressure valve to adjust oil cavity flow; Chen Peijiang [7]. With a company's milling center, for example, milling center has two gantry skateboards, and each skateboard is mainly composed of driving motor, transmission gear and hydrostatic cavity, etc. (shown in Fig. 2).

Gantry skateboards are subjected to vertical and overturning loads, two columns open-type hydrostatic oil cavity, and layout of each column of oil cavity with equidistant, equal-area, as shown in Fig. 3.

F. Xhafa et al. (eds.), *Advances in Intelligent Systems and Interactive Applications*, Advances in Intelligent Systems and Computing 686,
https://doi.org/10.1007/978-3-319-69096-4_113

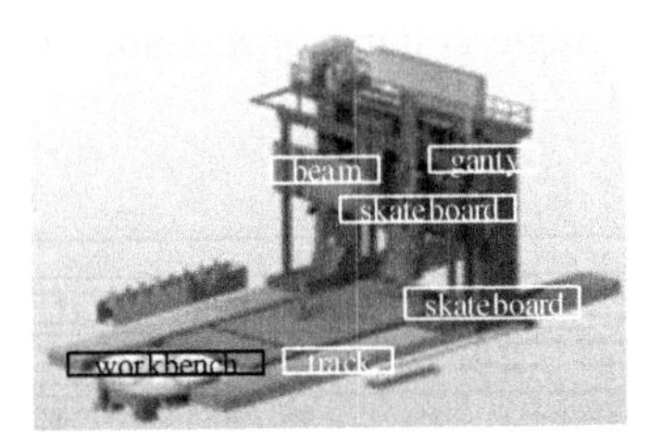

Fig. 1. Sketch of milling center

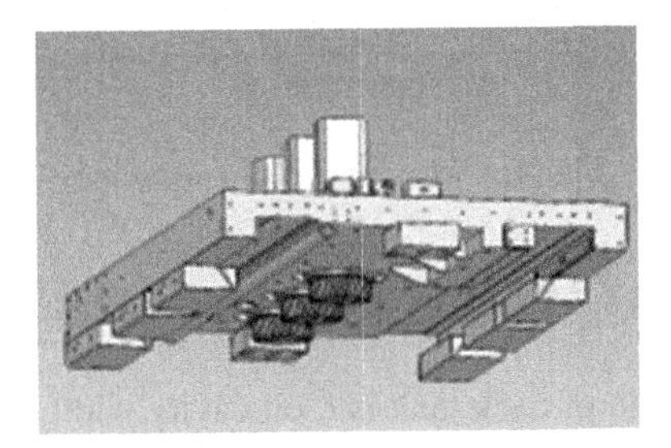

Fig. 2. Gantry skateboard of milling center

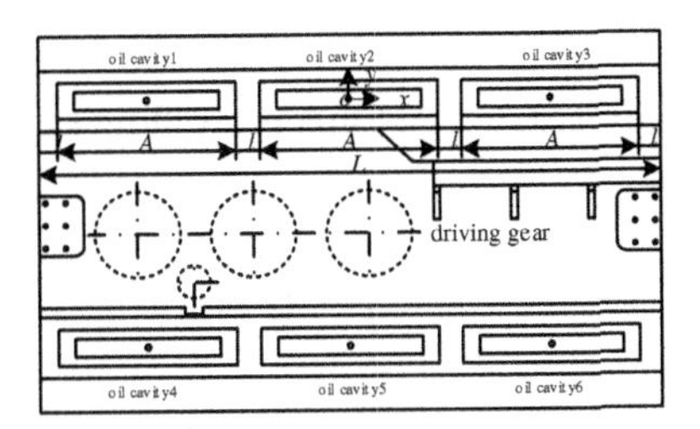

Fig. 3. Arrangement of oil cavities

With gantry skateboard single column open static chamber, for example, preuse layout form shown in Fig. 4.

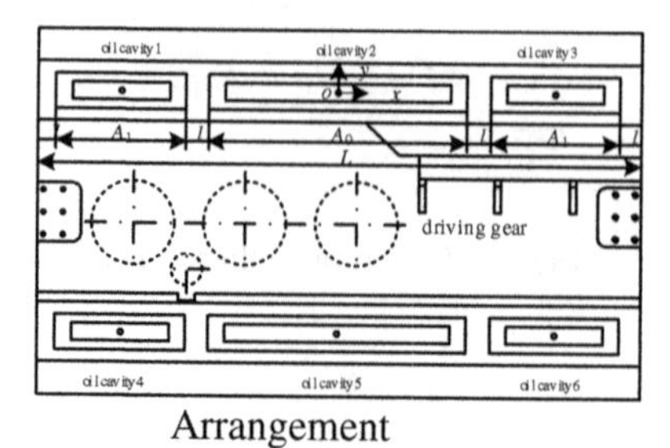

Fig. 4. Arrangement of oil cavities

2 Initial Parameters of Open-Type Hydrostatic Guideway

Length of slider is L, width of back tank is l, oil cavity width is B, and oil seal width is a, b, position x and bearing area A_e of each oil cavity are:

$$x_i = \frac{A_1}{2} i + li + \frac{A_0}{2}(i - 1) \tag{1}$$

effective length of oil cavity is L-4 l and guideway flow rate is q_0, n flow rate of oil cavity with length A_i is q_i:

$$q_i = \frac{q_0}{L - 4l} A_i \tag{3}$$

oil hydrostatic guideway oil film thickness is h_0, get bearing capacity f is:

$$f = p_r A_e \tag{4}$$

self-weight of center milling machine gantry is 140 t, and supported by two hydrostatic skateboard. Skateboard design parameters are shown in Table 1.

Table 1. SkateboFard design parameters

Oil cavity number	Oil cavity form	B/m	b/m	a/m	L/m
6	rectangle	0.25	0.05	0.05	2
l/m	film thickness h_0/μm	skateboard total flow q_0/(L/min)	oil type	dynamic resistance μ/(Pa·s)	oil density ρ/(kg/m^3)
0.05	70	1.76	VG15	0.04136	880

3 Static Performance of Guideway Under Vertical Load

3.1 Equilibrium Equation of Guideway

When F acts on geometric center of skateboard, and skateboard is translated from initial position to new equilibrium position along direction of F, as shown in Fig. 5.

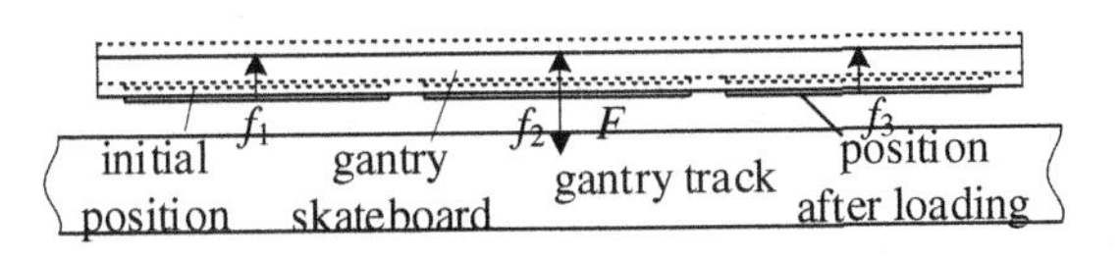

Fig. 5. Sketch of vertical load F subjects on skateboard

Oil film thickness changes same, from initial value h_0 to final value $h = h_0 + \Delta h$, and oil resistance of oil cavity R_i, pressure p_i are:

$$\begin{cases} R_i = \mu / \bar{B}_i h^3 \\ p_i = q_i R_i \end{cases} \tag{5}$$

each oil cavity load Δf_i is balanced with load F. And force analysis of skateboard, and n establish mechanical equilibrium equation:

$$\sum_{i=1}^{3} \Delta f_i + F = 0 \tag{6}$$

Introducing Eqs. (3)–(5) into Eq. (6), relationship between load F and oil film thickness variation Δh are obtained:

$$F = f_1(\Delta h) \tag{7}$$

Limit value of film thickness h_0, carrying capacity and static stiffness are deduced:

$$\begin{cases} F_0 = \lim\limits_{\Delta h \to h_0} f_1(\Delta h) = \dfrac{3\mu}{h_0^3} \sum\limits_{i=1}^{3} \dfrac{A_{e,i} q_i}{\bar{B}_i} \\ j_0 = -\dfrac{\partial F}{\partial(\Delta h)}\bigg|_{h=h_0} = \dfrac{3\mu}{h_0^4} \sum\limits_{i=1}^{3} \dfrac{A_{e,i} q_i}{\bar{B}_i} \end{cases} \tag{8}$$

3.2 Carrying Capacity of Guideway

(1) Guideway carrying capacity with equal area oil cavity When open-type hydrostatic oil cavity of gantry skateboard with equal area design, oil cavity area A_e, flow q and support flow coefficient are:

$$\begin{cases} A_e = (L - 4l - 3a)(B - b) \\ q = q_0/3 \\ \bar{B} = (L - 4l - 3a)/18b + (B - b)/6a \end{cases} \tag{9}$$

Introducing Eq. (8) into (9), and carrying capacity F_{deng} of guideway under vertical load is obtained:

$$F_{deng} = 18\mu q_0 \Phi_1 / h_0^3 \tag{10}$$

(2) Data in Table 2 is brought into Eq. (8) to obtain carrying capacity F_0 and static stiffness j_0 of unequal area oil cavity under vertical load are:

$$F_{fei} = 18\mu q_0 \Phi_2 / h_0^3 \tag{11}$$

3.3 Static Stiffness of Guideway

Static stiffness j_{deng}, j_{fei} of oil cavity design with equal and unequal area under vertical load are:

$$\begin{cases} j_{deng} = 18\mu q_0 \Phi_1 / h_0^4 \\ j_{fei} = 18\mu q_0 \Phi_2 / h_0^4 \end{cases} \tag{12}$$

3.4 Comparison of Static Performance

From Figs. 6 and 7 it can be seen that carrying capacity and static stiffness of gantry skateboard are decreased and n increased when meddle oil cavity is increased from 0.5 to 1.0 m. And carrying capacity and static stiffness of gantry skateboard are increased as oil flow rate increases from 1.74 to 1.78 L/min.

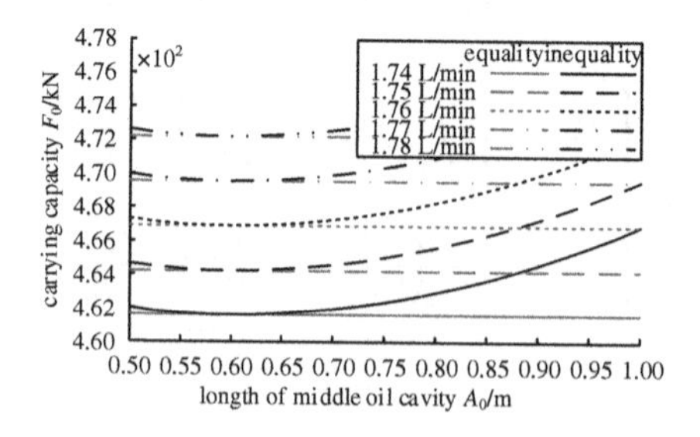

Fig. 6. Sketch of sliding plate bearing capacity and middle oil cavity length

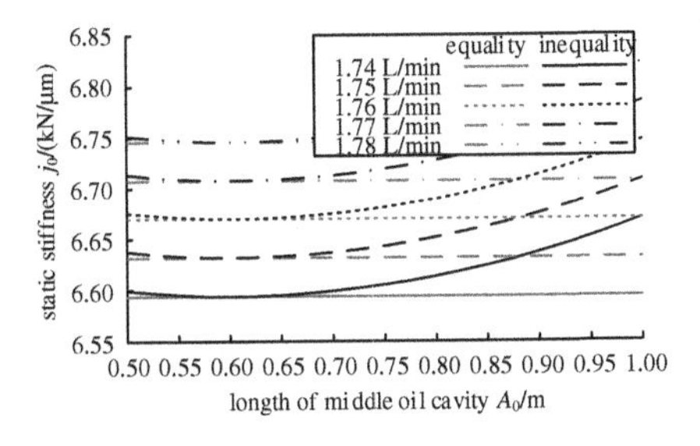

Fig. 7. Sketch of slide plate static stiffness and middle oil cavity length

4 Static Performance of Guideway Under Overturning Load

4.1 Equilibrium Equation of Guideway

When overturning load M acts on geometric center of slider, plate is shifted from initial position to new equilibrium position in direction of load M, As shown in Fig. 8.

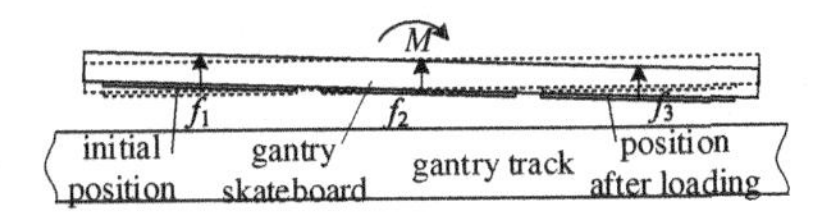

Fig. 8. Schematic diagram of a skateboard subjected to overturning loads M

According to equation of position change [8], change of oil film thickness Δh_i is:

$$\Delta h_i = (\mathbf{T} - \mathbf{I})[x_i \quad 0 \quad 0]^T \cdot [0 \quad 0 \quad 1]^T \tag{13}$$

According to formula (4), load under each oil chamber is Δf_i:

$$\Delta f_i = \frac{\mu q A_e}{\bar{B}(h_0 + \Delta h_i)^3} - \frac{\mu q A_e}{\bar{B} h_0^3} \tag{15}$$

Formula (15) is transformed into Taylor series, so it become:

$$\Delta f_i = -3 \frac{\mu q A_e}{\bar{B} h_0^4} \Delta h_i \tag{16}$$

Mechanical equilibrium equation is obtained by force analysis of slide plate:

$$-\sum_{i=1}^{3} x_i \Delta f_i + M = 0 \tag{17}$$

Relation between load M and rotation angle of slide plate is obtained:

$$M = f_2(\beta) \tag{18}$$

Limit of oil film is H0, and maximum angle of rotation of Longmen frame is:

$$\beta_{\max} = \lim_{\delta x \to 0} \arcsin(h_1/x_1) = \arcsin(h_1/x_1) \tag{19}$$

According to formula (18), carrying capacity M0 and static stiffness j_M are:

$$\begin{cases} M_0 = \dfrac{3\mu}{h_0^3 x_1} \displaystyle\sum_{i=1}^{3} \dfrac{qA_e x^2}{\bar{B}} \\ j_M = \dfrac{3\mu}{h_0^4 x_1} \displaystyle\sum_{i=1}^{3} \dfrac{qA_e x^2}{\bar{B}} \end{cases} \tag{20}$$

4.2 Carrying Capacity of Rail Under Overturning Load

(1) Rail bearing capacity of equal area design of oil cavity Loading capacity of guideway of equal area oil cavity under overturning load is M_{etc} respectively:

$$M_{\mathrm{etc}} = 4\mu q_0 \Phi_3 / h_0^3 \tag{21}$$

(2) Rail bearing capacity of oil cavity of inequal area Under overturning load, bearing capacity of inqual area oil cavity is M_{no}:

$$M_{\mathrm{no}} = 9\mu q_0 \Phi_4 / 4h_0^3 \tag{22}$$

4.3 Static Stiffness of Rail Under Overturning Load

Static stiffness of guideway with equal area and of inequal area is $j_{M\mathrm{deng}}$ 、$j_{M\mathrm{fei}}$:

$$\begin{cases} j_{M\mathrm{deng}} = 4\mu q_0 \Phi_3 / h_0^4 \\ j_{M\mathrm{fei}} = 9\mu q_0 \Phi_4 / 4h_0^4 \end{cases} \tag{23}$$

4.4 Comparison of Equal Area and Inequal Area Guideways

As shown in Figs. 9 and 10, with middle oil cavity increasing from 0.5 to 1 m, carrying capacity and static stiffness of Gantry frame slide under overturning load are decreasing in turn.

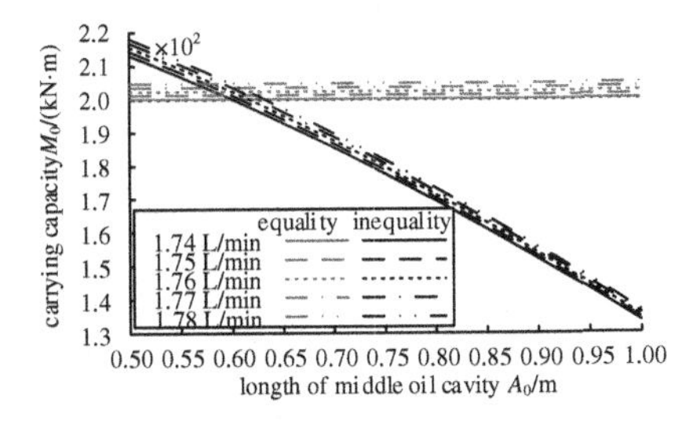

Fig. 9. Sketch of sliding plate bearing capacity and middle oil cavity length

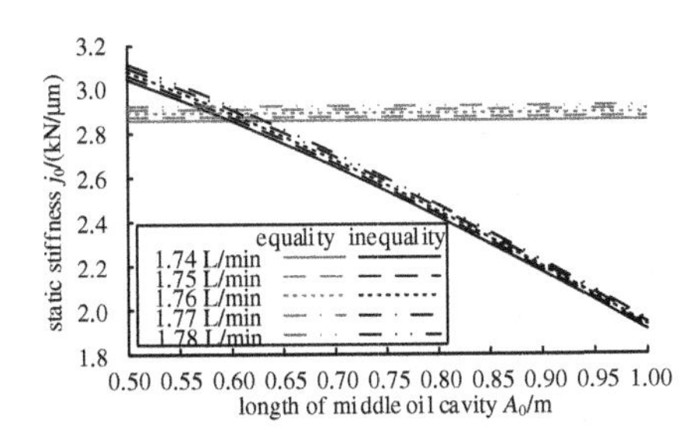

Fig. 10. Sketch of slide plate static stiffness and middle oil cavity length

5 Conclusion

(1) With increase of length of intermediate oil cavity, vertical bearing capacity and vertical static stiffness of slide plate are better than equal area design scheme, but lifting range is limited.

(2) With increase of length of intermediate oil cavity, overturning load capacity and static stiffness of slide plate are smaller than equal area design scheme, and capacity to withstand overturning load becomes worse.

Acknowledgements. The project was financially supported by Natural Science Foundation of Hebei Province supported (E2016203324).

References

1. SHAO, J.P., LI, H.M., YANG, X.D., et al.: Study on flow ability of gap oil film of multi-oil pad hydrostatic bearing with variable viscosity. In: 1st International Conference on Intelligent Human-Machine Systems and Cybernetics, Hangzhou, China, 2009, 1
2. MENG, Z.Y., MENG, X.Z., CHEN, S.Z.: Static characteristics of open hydrostatic sliding way and optimum parameters. J. Luoyang Inst. Technol. **21**(4):43–47 (2000)
3. ZHANG, Wei.: Research on oil film thickness of constant flow open circular hydrostatic slide. Mechanical Eng. (7):34–35 (2011)

4. Shuyan, Y.A.N.G., Haifeng, W.A.N.G., Feng, G.U.O.: Study of characteristics of oil film thickness in step bearing. Lubr. Eng. **37**(6), 15–19 (2012)
5. Shuyan, Y.A.N.G., Haifeng, W.A.N.G., Feng, G.U.O.: Influence of grooved surfaces on film thickness of hydrodynamic lubrication. Tribology **31**(3), 283–288 (2011)
6. Zhang, L., Dai, H.L., Liu, S.R.: Study on self-adaptive oil supply system of hydrostatic slide based on AMESim. Manuf. Technol. Machine Tool **12**, 25–28 (2010)
7. Peijiang, C.H.E.N.: Control schemes of oil film thickness for hydrostatic slide [J]. Manuf. Technol. Machine Tool **5**, 46–49 (2008)
8. SHAO, J., ZHANG, X.: Control schemes of oil film thickness for hydrostatic slide. Energy Conservation Technol. **24**(6):558–561 (2006)

Influence of Liquid Film Thickness on Static Property of Magnetic-Liquid Double Suspension Bearing

Zhao Jianhua(✉), Wang Qiang, Zhang Bin, and Chen Tao

Hebei Provincial Key Laboratory of Heavy Machinery Fluid Power Transmission and Control, Yanshan University, Qinhuangdao 066004, China
zhaojianhua@ysu.edu.cn

Abstract. Due to the low viscosity of seawater, it is difficult to form the seawater-lubricated film, and the bearing capacity and stiffness of the seawater-lubricated film is very small. It is easily to cause the "overload" and "burning" phenomenon of the seawater-lubrication sliding bearing, and the operation stability and service life can be shorted. The magnetic bearing takes the bearing form of non-contact suspension, and it is more suitable as an auxiliary support to the seawater-lubrication sliding bearing. Therefore, the paper introduces the electromagnetic suspension support into the seawater-lubricated sliding bearing. And then a novel Magnetic-Liquid Double Suspension Bearing with the advantages of electromagnetic suspension and hydrostatic supporting can be formed. The structural characteristics, supporting mechanism, hydrostatic self-adjustment and electromagnetic-adjustment processes of Magnetic-Liquid Double Suspension Bearing can be analyzed in the paper. Based on force balance equation, electromagnetic equation and flow equation, the transfer function of different adjustment processes under constant-flow supply model are deduced. Then bearing capacity, static stiffness and total power loss are selected as static indexes. The influence rule of the liquid film on capacity property of single degree freedom bearing system of Magnetic-Liquid Double Suspension Bearing can be analyzed. The results show that as liquid film thickness increases, static stiffness decrease, carrying capacity and total power loss remain the same during the hydrostatic self-adjustment process. The proposed research provided a basis for the design of Magnetic-Liquid Double Suspension Bearing in the engineering practice.

Keywords: Liquid film thickness · Magnetic-Liquid double suspension bearing · Static index

1 Introduction

Due to the low viscosity and special performances of seawater, makes the lubricating film of seawater-lubricated sliding bearing is difficult to be formed. As the electromagnetic bearing in the form of non-contact support, and with zero mechanical friction, high precision, that is very suitable used as an auxiliary supporting form for the seawater lubrication bearing support. Therefore, this paper introduces the magnetic

F. Xhafa et al. (eds.), *Advances in Intelligent Systems and Interactive Applications*, Advances in Intelligent Systems and Computing 686,
https://doi.org/10.1007/978-3-319-69096-4_114

suspension support into the seawater lubricated sliding bearing to form a new type of Magnetic-Liquid Double Suspension Bearing.

Magnetic-Liquid Double Suspension Bearing System with seawater lubricated, mainly composed of the base, shell, magnetic lining, coil, stator, spindle and other components, as shown in Figs. 1 and 2.

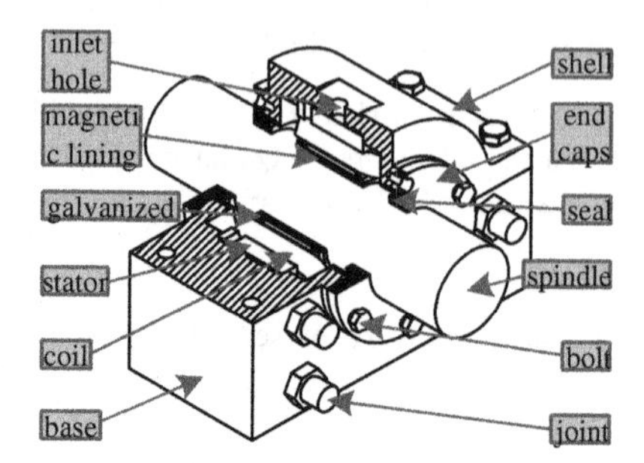

Fig. 1. Magnetic-liquid double suspension bearing semi-isometric view

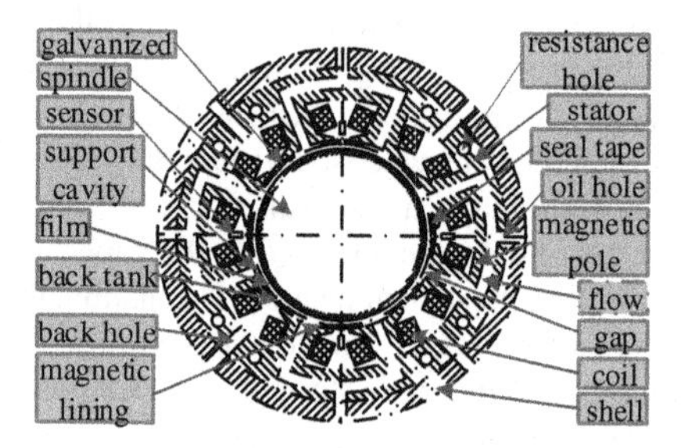

Fig. 2. Magnetic-liquid double suspension bearing full profile

Magnetic-Liquid Double Suspension Bearing System includes magnetic suspension and hydrostatic two support system (as shown in Fig. 3), and combines magnetic suspension support and hydrostatic support in one, so that the bearing carrying capacity, stiffness and operational stability can be significantly increased. The working process of the Magnetic-Liquid Double Suspension Bearing System is divided into two processes: hydrostatic self-adjustment and magnetic-regulation (as shown in Fig. 4).

In recent years, many scholars have carried on further research on liquid-hydrostatic bearing and electromagnetic bearing, and also acquired large achievements. In paper [1], the mathematical model of the hydrostatic thrust bearing is established, and the influence of water film thickness, water cavity thickness and inlet hole diameter on bearing capacity are analyzed. The results show that with the water film thickness increase, the bearing capacity of the bearing is significantly decreased. With the water cavity thickness and the inlet hole diameter increase, the bearing capacity increases first and then keeps no change. The bearing capacity and influencing factors of water-lubricated thrust bearing are analyzed in reference [2], and the results show that the bearing capacity of the bearing can be improved by choosing the appropriate friction pair and PV value. In reference [3], the basic characteristics of high speed spindle water-lubricated hydrostatic bearing rotor system are studied. And the results show that static stiffness of bearing-rotor system increases as the water supply pressure and load increased. Reference [4], a new type of magnetic suspension precision linear motion platform is proposed, and establish its three-dimensional finite element model, and analyze the magnetic suspension platform structural modal under the different working conditions. And the results show that the platform has the lowest rigidity when suspended, and the position of the suspension has no effect on the overall mode of the platform.

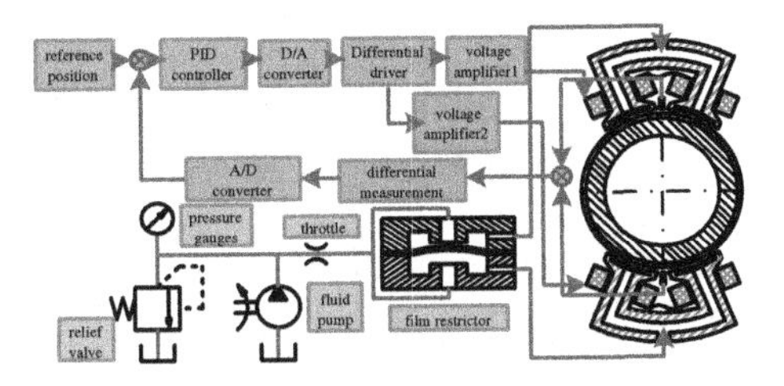

Fig. 3. Single degree freedom support regulation principle

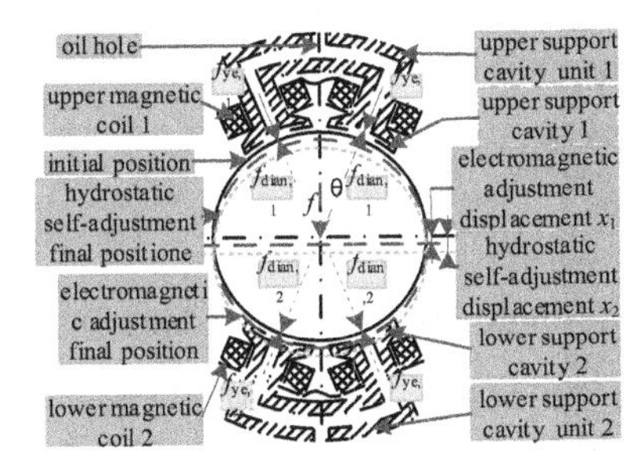

Fig. 4. Force diagram of magnetic-liquid double suspension bearing

2 Kinetic Equation of Single DOF Supporting System

2.1 The Initial State of a Single Degree of Freedom System

In initial state, film thickness and bias current of the electromagnet are equal, and the electromagnetic attraction of the upper and lower supporting units is equal.

(1) Voltage equation Voltage amplifier output is:

$$e_{pian,1,0} = e_{pian,2,0} = i_0 R_{quan} \tag{1}$$

(2) Flow equation Magnetic fluid double suspension bearing adopts constant flow supply mode,and the upper and lower support cavity 1 and 2 flow equation are:

$$\begin{cases} q_{1,0} = q_{fa,1,0} \\ q_{2,0} = q_{fa,2,0} \end{cases} \tag{2}$$

(3) Hydrostatic bearing force According to Navier- Stokes equation, the upper and lower $f_{ye,1,0}$ and $f_{ye,2,0}$ of support cavities 1 and 2 are:

$$\begin{cases} f_{ye,1,0} = 2p_{1,0}A_e \cos\theta = 2q_{1,0}R_0A_e \cos\theta \\ f_{ye,2,0} = 2p_{2,0}A_e \cos\theta = 2q_{2,0}R_0A_e \cos\theta \end{cases} \tag{3}$$

(4) Electromagnetic suspension bearing force. According to the Maxwell attractive formula, get up and down the pole 1, 2 $f_{dian,1,0}$ 、 $f_{dian,2,0}$:

$$f_{dian,1,0} = f_{dian,2,0} = 2k\frac{i_0^2}{(h_0 + l)^2}\cos\theta \tag{4}$$

(5) Spin axis mechanical balance equation Based on Newton's second law, the mechanical equilibrium equation is obtained:

$$f_{dian,1,0} + f_{ye,2,0} - f_{dian,2,0} - f_{ye,1,0} = mg \tag{5}$$

2.2 Working State of a Single Degree of Freedom System

Under the external load, the bearing shaft displacement is x, and the liquid film thickness of the upper and lower support cavities h_1, h_2 are:

$$\begin{cases} h_1 = h_0 + x\cos\theta \\ h_2 = h_0 - x\cos\theta \end{cases} \tag{6}$$

(1) Voltage equation As magnetic system is not regulated, output voltage is unchanged:

$$e_{pian,1,0} = e_{pian,2,0} = i_0 R_{quan} \tag{7}$$

(2) Flow equation The change of the liquid film thickness causes the change of the liquid resistance so, the upper and lower support cavities flow rate is:

$$\begin{cases} q_1 = q_{fa,1,0} - A_b \dot{h}_1 - \frac{V_{oa}}{E} \dot{p}_1 \\ q_2 = q_{fa,2,0} - A_b \dot{h}_2 - \frac{V_{oa}}{E} \dot{p}_2 \end{cases} \tag{8}$$

(3) Hydrostatic bearing force Similarly, the working state of the upper and lower support cavity 1, 2 hydrostatic pressure bearing force is:

$$\begin{cases} f_{ye,1} = 2q_1 R_1 A_e \cos\theta \\ f_{ye,2} = 2q_2 R_2 A_e \cos\theta \end{cases} \tag{9}$$

(4) Electromagnetic suspension bearing force Similarly, the working state of the upper and lower poles electromagnetic levitation force is:

$$\begin{cases} f_{dian,1} = 2k \frac{i_0^2}{(h_1 + l)^2} \cos\theta \\ f_{dian,2} = 2k \frac{i_0^2}{(h_2 + l)^2} \cos\theta \end{cases} \tag{10}$$

(5) Shaft mechanical balance equation Similarly, the mechanical equilibrium equation of the rotating shaft is:

$$f_{dian,1} + f_{ye,2} - f_{dian,2} - f_{ye,1} - mg - f = -m\ddot{x} \tag{11}$$

2.3 Transfer Function of Single Degree of Freedom Bearing Unit

According to the formula from (1) to (10), ignoring the sensitive volume of Voa, and the linearization and Laplace transform, the transfer function is obtained:

$$G(s) = \frac{X(s)}{F(s)} = \frac{1}{T_2 s^2 + T_1 s + T_0} \tag{12}$$

3 Influence of Film Thickness on Static Performance

3.1 Static Performance Indexes

(1) Carrying capacity F_0 When spindle displacement reached its maximum, liquid film thickness of one liquid cavity is zero, and then the external load f reaches the limit, which is called bearing carrying capacity F_0. The expression is:

$$F_0 = 2\cos\theta \sum_{i=1}^{2} (-1)^i \left(p_i A_e - \frac{k i_0^2}{(h_i + l)^2} \right) - mg \tag{13}$$

(2) Static stiffness j_0 Representation ability caused by liquid film deformation to resist the external load. The expression is:

$$j_0 = 2\cos\theta \sum_{i=1}^{2} \frac{2 k i_0^2}{(h_i + l)^3} + \frac{3 \mu A_e q_i}{\bar{B} h_0^4} \tag{14}$$

(3) Total power loss W_{zong} Total power loss of Magnetic Liquid Double Suspension Bearing is sum of support cavity power loss and electromagnetic coil heat loss. The expression is:

$$W_{zong} = 2\bar{B} h_0^2 \left(\frac{mg}{A_e} + 2 p_{1,0} \right) \frac{A_s}{A_t} + 4 i_0^2 R_{quan} \tag{15}$$

3.2 Magnetic-Liquid Double Suspension Bearing Parameters

Design parameters of Magnetic-Liquid Bearing are shown in Table 1.

Table 1. Design parameters of magnetic-liquid double suspension bearing

Bearing quality m/kg	Dynamic viscosity μ/Pa•s	Elastic modulus E/MPa	Zinc coating thickness l/mm	Pole area A/mm^2
66.82	1.3077×10^{-3}	2.4×10^3	0.5	1000
Coil number *N*/ Dimension-less	Liquid cavity width *A*/m	Liquid cavity length *B*/m	Axial fluid sealing tape width *b*/m	Sealing tape width *a*/m
633	0.1	0.02	0.004	0.006
Pump pressure p_s/MPa	Fluid chamber pressure p_1/MPa	Fluid chamber pressure p_2/MPa	Bias current Size i_0/A	Load *f*/N
3	1	1.3	1.2	2000

3.3 Influence of Liquid Film Thickness on Static Performance of Magnetic-Liquid Double Suspension Bearing by Hydrostatic Self-Adjustment

With the liquid film thickness increase, and due to the larger zinc coating thickness, the variation range of the air gap is small, so that the reduce magnitude are smaller, basically can be ignored, as shown in Fig. 5.

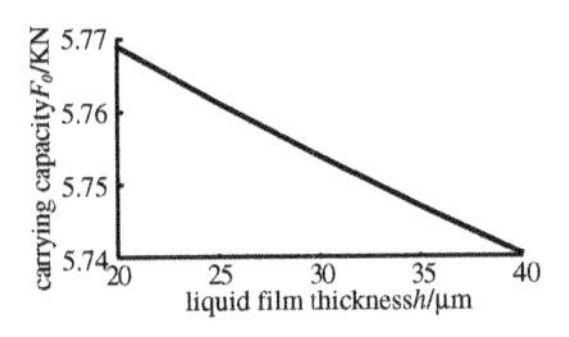

Fig. 5. Relationship between carrying capacity and liquid film thickness

The liquid film thickness is inversely proportional to the static stiffness. As liquid film thickness increases, the stiffness of bearing are decreased, as shown in Fig. 6.

As the bias current does not change, the magnetic coil heat loss remains unchanged; the bearing total power loss increases with the film thickness increase, but the amplitude is small and can be neglected. As shown in Fig. 7.

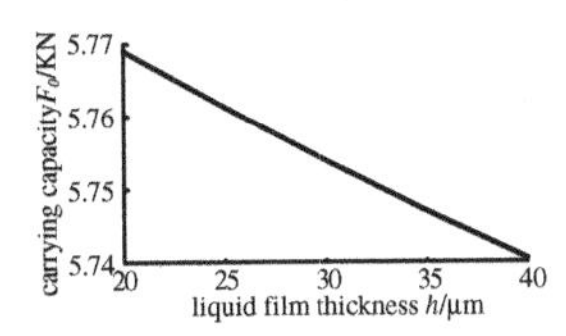

Fig. 6. Relationship between static stiffness and liquid film thickness

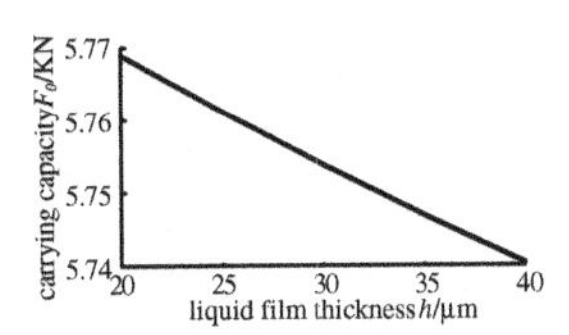

Fig. 7. Relationship between total power loss and liquid film thickness

4 Conclusion

(1) Transfer function of the single degree of freedom bearing system of the fluid and magnetic suspension bearing is two order system in the process of self-regulating.
(2) First term of transfer function is a damping term, which is mainly affected by dynamic extrusion effect of hydrostatic support cavity, and is related to liquid resistance and support cavity area. Zero order term is stiffness term, which is related to liquid film thickness, supporting cavity area, initial current and coil turns and so on.
(3) During self-regulation process, bearing carrying capacity and static stiffness decreased with liquid film thickness increases, and total power loss is almost unchanged.

Acknowledgements. The project was financially supported by Natural Science Foundation of Hebei Province supported (E2016203324).

References

1. WANG, YZ., JIANG, D., YIN, ZW., et al.: Load capacity analysis of water lubricated hydrostatic thrust bearing based on CFD. J. Donghua Univ. **41**(4), 428–432 (2015)
2. Zhang, X., Zhou, Y., Wang, X.R., et al.: Research the influences on load capacity of water-lubricated thrust bearing. Machinery **49**(4), 16–19 (2011)
3. DONG, SX., MA, QS.: Basic performance analysis of high-speed spindle system supported by hybrid bearings. Manuf. Technol. Machine Tool **9**, 136–139 (2011)
4. GUO, NP., ZHOU, HB., et al.: Modal analysis of maglev linear motion platform. J. Eng. Design **18**(3), 183–190 (2011)
5. Wu, Q., Qian, Y.M.: The stiffness optimization analysis of magnetically suspended table. Manuf. Autom. **14**, 60–63 (2013)
6. Harigaya, Y., Suzuki, M., Toda, F., et al.: Analysis of oil film thickness and heat transfer on a piston ring of a diesel engine: effect of lubricant viscosity. J. Eng. Gas Turbines Power **128** (3), 685–693 (2006)
7. ZHAO, JH., GAO, DR., ZHANG, ZC., et al.: Indeterminate mechanics model of bearing capacity of constant pressure oil pockets in hydrostatic slide. Chinese J. Mechanical Eng. **48** (22),167–176 (2012)
8. SHAO, JP., ZHANG, XT.: Control schemes of oil film thickness for hydrostatic slide. Energy Conservation Technol. **24**(6), 558–561 (2006)
9. Wu, G.Q.: Magnetic Suspension Supporting System and Its Control Technology Used for Numerical Control Machine. Shanghai University, Shanghai (2006)
10. KONG, XD., WANG, YQ.: Control Engineering Fundamentals. China Machine Press, Beijing (2007)
11. ZHAO, JH.: The Analysis and Experimental Research of Liquid Hydrostatic Slide's Performance of Gantry Turning and Milling Center. Yanshan University, Hebei (2013)

The Research on the Thinking of Large Data for Agricultural Products Market Consumption in Beijing

Chen Xiangyu, Gong Jing(✉), Yu Feng, and Chen Junhong

Key Laboratory of Urban Agriculture (North China), Ministry of Agriculture, Institute of Information on Science and Technology of Agriculture, Beijing Academy of Agriculture and Forestry Sciences, Beijing 100097, People's Republic of China
Cxy8132@sohu.com

Abstract. Deep excavation of large data provides immeasurable value in agricultural application, such as agricultural market consumption. This paper initially outlines the application status of consumption data for Beijing agricultural products market. Then, the analytical ideas of agricultural market consumption data are explored and the difficulties for building application architecture are investigated. Ultimately, the corresponding countermeasures are proposed based on the aforementioned analysis.

Keywords: Large data · Agricultural market consumption · Data analysis

1 Introduction

Effective agricultural large data significantly enhances the scientific decision-making ability. Agricultural market consumption data, as an important part for agricultural data, can effectively improve agricultural production decision-making, agricultural market management. Therefore, it is essential to collect and further analyze the agricultural market consumption data.

At present, many researchers have focused on large data. Domestic researches on large agricultural data are mainly limited to the macro level and rarely consider the application of lage data in the specific aspects of agriculture. Based on these reasons, this paper presents the investigation ideas for market consumption data of Beijing agricultural products by combining specific agricultural link, which provides beneficial reference for policy research.

Project source: Study on agricultural informatization strategy of Ministry of Agriculture in 2015.

F. Xhafa et al. (eds.), *Advances in Intelligent Systems and Interactive Applications*, Advances in Intelligent Systems and Computing 686,
https://doi.org/10.1007/978-3-319-69096-4_115

2 Application Status of Market Consumption Data of Beijing Agricultural Products

2.1 The Data Source

2.1.1 Statistical Data

The statistical data mainly includes main index and characteristic index related to Beijing agricultural products market consumption in some annuals. These data are normally obtained from the authority figures of relevant departments, such as the State Council, national or provincial statistical offices and so on. The figures not in the national or provincial (autonomous regions and municipalities) statistical scope should be verified and published by business department.

2.1.2 Monitoring Data

The monitoring data is mainly collected by Beijing municipal agriculture bureau through monitoring the agricultural products market of Beijing. Such all-round monitoring covers a complete business chain including the origin, the wholesale and the retail.

2.1.3 Survey Data

The survey data mainly refers to the market research data. Such data is mainly concerned with agricultural market or household food consumption. Generally, the survey data is purposefully and systematically collected, recorded, and sorted out by the relevant institutions and scholars through scientific methods.

2.1.4 Electricity Supplier Data

The electricity supplier data is mainly derived from the statistical results of indicators behind-the-scenes of major electricity supplier websites. The electricity providers can accurately grasp the behavior patterns of user groups or individual networks through analyzing and interpreting the data of the store.

2.2 Application Status

2.2.1 Consumption Analysis in "the Long-Term Beijing Agricultural Products Price Analysis Report"

In August 2014, "The analysis report about price change law of main agricultural products in medium and long term (2004–2013)" was released in Beijing. In this report, the data related to dozens of large-scale wholesale market are compared with each other. Market volatility and trend of people's "vegetable basket" products in recent 10 years were respectively presented, which contributes to the decision-making adjustment.

2.2.2 Consumption Analysis in Renwo Online Agel Ecommerce Ltd of Beijing

Renwo online Agel Ecommerce Ltd of Beijing creates the internet direct sales support system by combining internet, mobile, and things. It provides three major functions including data sharing, remote control and network data acquisition. Then, producers determine the number of products according to the requirement, which effectively avoid waste backlog caused by overcapacity. Moreover, the requirement for diversification, personalization, high-end consumer demand can be fully detected.

2.2.3 Data Analysis of Beijing Agricultural Information Technology Co Ltd

Beijing Agricultural Information Technology Co Ltd developed series of large data products. Such large data products mainly serve Information Center of Ministry of Agriculture, Township Enterprises Bureau of Ministry of Agriculture and so on. The government can guide agricultural production and make management decision by analyzing large data derived from these systems.

3 Agricultural Market Consumption Data Analysis

According to the existing research results, we propose the application framework of agricultural products market consumption data from 5 aspects, as shown in Fig. 1.

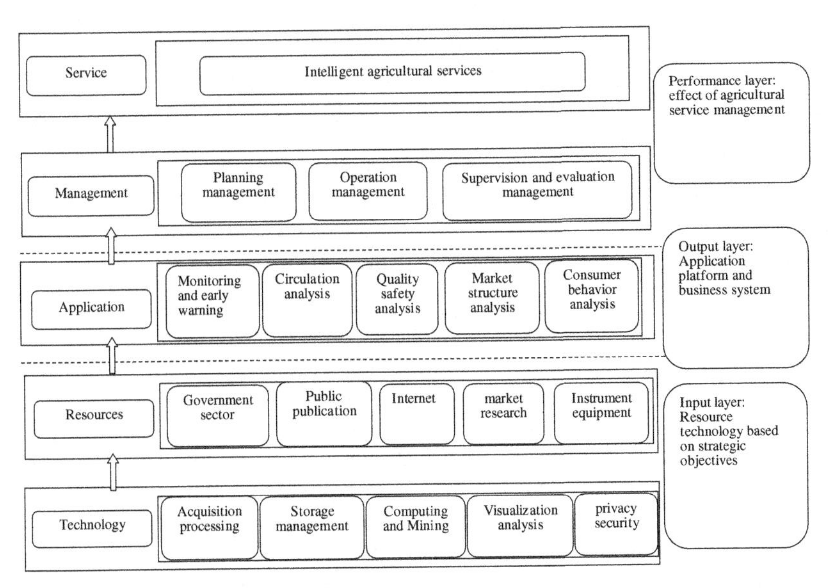

Fig. 1. Application architecture of agricultural products market consumption data

3.1 Service

The development of agricultural market consumption data is applied in service. Farmers, agricultural materials company, cooperative organizations of farmers and agricultural enterprises are the principal parts of agricultural activities. Government mainly provides the agricultural public service, such as infrastructure construction, technology promotion and information supply, to these agricultural activities.

3.2 Management

Government and main agricultural activities should participate in the whole process of agricultural activity, including project planning, construction, management and supervision. Specifically, their management function involves the intelligent agriculture planning management, operation management, as well as supervision and evaluation management.

3.3 Application

3.3.1 Monitoring and Early Warning of Agricultural Products

The monitoring and early warning of agricultural products needs real-time information capture of whole industry chain. The capture process includes data collection, information analysis, early warning and information dissemination. Then the information flow was formed and its rate and direction are revealed. The key nodes are analyzed subsequently and the dynamic early warning and precise control are finally realized.

3.3.2 Analysis of Agricultural Products Circulation

Nowadays, agricultural products safety has been difficult to guarantee for government. Therefore, the mode of production and circulation of agricultural products should be modified. Meanwhile storage resources allocation and distribution line of agricultural products needs to be optimized. Furthermore, the agricultural production and logistics standard system should be established and improved.

3.3.3 Quality Safety Analysis of Agricultural Products

Essentially, the quality and safety problems of agricultural products are mainly caused by information asymmetry. Large data analysis ensures the effective supply of information and the traceability of agricultural products quality and safety, and reduces the safety risk of agricultural product quality.

3.3.4 Analysis of Agricultural Market Pattern

Since the current agricultural market pattern is formed spontaneously, it is not synchronized with the production layout. Such imbalance will fundamentally shake the stability of agricultural industry chain and enlarge market volatility. Therefore, it is necessary to achieve top-level design of market system through large data analysis. Moreover, we can guide farmers to reasonably arrange production scale according to the relationship between production and marketing.

3.3.5 Analysis of Consumption Behavior of Agricultural Products

Large data technology promotes the development and innovation of data marketing for agricultural products. The agricultural enterprises can adjust marketing strategies by analyzing the behavior of such data.

3.4 Resources

The development and utilization of resources is the foundation for large data application. Data resources can be obtained from government acquisition, public publications, the Internet, market research, and so on.

3.5 Technology

The software ecosystem based on Hadoop distributed processing is adopted in the technical architecture of its applications. The decision analysis reports in PDF or DOC format (containing text, numbers, charts and statements) can be automatically generated through the report output function.

4 Problems in Lage Data Analysis for Agricultural Products Market of Beijing

4.1 Input Layer

4.1.1 Relatively Scarce Data Collection Channels

The existing acquisition channels for lage data of agricultural product market consumption in Beijing are unable to meet the actual needs. Meanwhile, many problems, such as insufficient data, inadequate coverage, discontinuity, poor timeliness, insufficient utilization, and so on remain to be solved.

4.1.2 Insufficient Data Resource Sharing

Due to the poor data integration and associative complementarity on the function, the information can't be shared and swapped effectively. The information is even out of touch with business processes and applications.

4.1.3 Data Quality Improvement Needs

On the one hand, part of large data for agricultural product market consumption is out of order and doesn't match the facts at present. On the other hand, due to the limitation of existing theory and technology level, the data source cannot be detected automatically and effectively, which lead to low availability of the data.

4.1.4 Unified Standards Requirements for Lage Data

Since the research institution normally analyzes the agricultural product market consumption in different ways, it results in the lack of comparability and the docking difficulty between the collected data and analyzed result.

4.2 Output Layer

4.2.1 Lack of Professional Data Analysis Tools

China is weak in data processing technology and difficult to satisfy the large-scale application needs of large data for agricultural product market consumption.

4.2.2 Inadequate Professionals for Large Data Analysis

Currently, the awareness of large data mining and utilization is inadequate in the agricultural production and management field of Beijing. Meanwhile, the spread of information technology education is not enough and the professional talents are still lacking.

4.3 Performance Layer

4.3.1 Requirement for Management Philosophy and Operation Mode Update

The traditional management ideas and decision-making methods cannot meet the current decision-making requirements for agricultural market consumption. On the one hand, the awareness of "large data" is inadequent. On the other hand, the establishment of relevant legal system cannot meet the urgent social needs of "large data" application.

4.3.2 Lack of Special Management Coordination Agencies

Currently, Beijing is still lacking of permanent and unified management coordination agencies for agricultural large data. With many barriers and relatively chaotic management, Beijing does not have unified policies, standards and objectives on the large data of agricultural products market consumption.

5 Suggestions on Promoting Lage Data Analysis of Agricultural Products Market in Beijing

5.1 Multi-Channel Data Collection Management System Establishment

Firstly, we should strengthen departmental joint, reduce the duplication of data collection, expand collection channels and adjust the data reporting mode. Thus, the acquisition mode can give priority to independent acquisition, meanwhile considering data purchase and exchange. Second, the public sharing mechanism from government as well as other relevant departments should be gradually improved. To effectively expand the degree of large data public sharing, we should give full play to the role of assessment and supervision.

5.2 Large Data Analysis Platform Construction for Agricultural Products Market Consumption

Firstly, the construction standards of agricultural products market consumption should be built. Therefore, the network interconnection, information exchange and resource sharing of all levels of information systems can be achieved. Meanwhile, we should establish effective platform for data acquisition, analysis and sharing. Then large data collection, analysis and resources sharing should be provided for supporting government decision-making.

5.3 Talent Cultivation for Lage Data Analysis of Agricultural Products Market Consumption

We can make full use of corporate and private large data resources. On one hand, large data related courses, professional direction and degrees in agricultural colleges and universities can be established. Meanwhile, support and evaluation systems for entrepreneurship and innovation of large data talents can be constructed. On the other hand, to cultivate large data talents with comprehensive ability, the cooperation between agricultural enterprises and universities or training institutions should be encouraged. Finally, we should investigate in decision-making methods, techniques and analytical tools to promote the development of core technologies.

5.4 Dedicated Management Agencies Establishment for the Large Data of Agricultural Products Market Consumption

To coordinate the data distributed in different departments, the specialized management agencies with higher level and higher coordination should be established. Firstly, unified data security management standard and standardized format for information data should be developed. Second, we should obtain the distribution of agricultural products market consumption data. Ultimately, the usage, security and release of large data should be regulated.

5.5 Effective Operating Mechanism for Supporting Policies of Platform Construction

The development of laws and regulations on the ownership, protection, collection, storage, processing, transmission, retrieval and authorization of all kinds of data should be improved. To use the data legally, use of user data, it is necessary to establish data confidentiality and risk classification management mechanism and clarify the boundaries of citizen's privacy and the right to learn the truth. Finally, we must strengthen law supervision and illegal strike for information security to reduce information security vulnerabilities. Furthermore, corresponding measures should be implemented to improve talents cultivation, relevant policies, funding supports and other guiding mechanisms.

References

1. STACB, S.: Knowledge processes and ontologies. IEEE Intell. Syst. Special Issue on Knowledge Manage. **16**(1), 26–34 (2001)
2. Zhang, QY., Zhao, J.: Agricultural information heterogeneous resource integration method. Agric. Lib. Inf. Sci. **27**(2), 11–12 (2015)
3. Xiong, CL.: Rural agricultural information service ability construction. Hunan Agricultural University, Changsha (2014)
4. DIETZ, JLG.: Enterprise ontology: Theory and methodology, pp. 35–80. Springer, Berlin, (2010)

The Application of Decision Tree in Workflow

Lisong Wang(✉), Yifan Chu, and Min Xu

College of Computer Science and Technology, Nanjing
University of Aeronautics and Astronautics, Nanjing 210000, China
wangls@nuaa.edu.cn

Abstract. The issue of path selection in workflow can be resolved by establishing a model of decision tree. The solution presented in this paper is based on analysis of workflow application in real world and probability theory. The solution can provide accurate and reliable runtime data for workflow engine to determine the future execution paths. The data provided by our solution has significant impact on the decision making of workflow, which contributes to optimization of workflow in many aspects. Finally, the efficiency of business process in real world can be greatly improved.

Keywords: Probability · Workflow · Decision tree

1 Introduction

Workflow technology has been integrated in many commercial applications. In many complex business processes, workflow plays an important role. So the efficiency of workflow execution is highly related to the quality of applications. Branching is a common pattern of workflow model. Branching pattern is used in workflow model to decide which path will be chosen based on runtime data provided by applications and the structure of workflow process model. Path selection has great impact on efficiency of workflow process instance. A model of decision tree is established in this paper and the runtime data can be calculated as application data with help of the probability theory which provides important data for the selection of path in runtime workflow process instance.

In recent years, focus has been largely put on model verification of workflow structure. The famous author Van der Aalst puts forward many publications on workflow verification with YAWL, a formalized language based on perti-net [1–3]. Currently, YAWL has been put into practice. In [4], features of YAWL(Yet Another Workflow Language) are elaborated for supporting the workflow management system development. The perti-net based modeling methods are widely applied in workflow study and become norms in many other aspects of workflow issues. The theory of stochastic process is also an useful tool of workflow verification. In [5], authors present a framework for the automated restructuring of stochastic workflows to reduce the impact of faults.

There are also some research on the issue of workflow efficiency. In [6], the author proposes a optimization formulation of workflow from the perspective of resources distribution. The author elaborates on the issue of distribution of resources under many

F. Xhafa et al. (eds.), *Advances in Intelligent Systems and Interactive Applications*, Advances in Intelligent Systems and Computing 686,
https://doi.org/10.1007/978-3-319-69096-4_116

kinds constraints in workflow. This paper aims to resolve the issue of raising the efficiency of workflow from perspective of path selection and find a better application data provider for workflow engine.

The structure of this paper is as follows: the second part introduces some definition related to our solution. The third part elaborates on the detail of resolving the issue with decision tree and probability theory. The last part concludes our solution.

2 Definition

2.1 Activities, Activity Instance and Transitions in Workflow

Activities in workflow process models represent logic steps in real business processes. In this paper, the set P represent the set of all activities defined in a workflow process. Each element of set P represents an activity in a workflow process. Here, a function is defined to describe the types of activities in workflow. *CM(E∈ P) = GENERAL, OR-SPLIT, AND-SPLIT, XOR-SPLIT ANDJOIN, OR-JOIN, XOR-JOIN*. The detailed definitions of types of activities in workflow are specified in [7]. The *XOR-SPLIT* and *OR-SPLIT* activities will be main focus in this paper.

Instances of activities in a workflow represent instances of activities defined in processes of workflow which are activated by workflow engine according to pre-specified rules. When an activity is activated, the process is running to the step represented by this activity and the work items generated by the activity instance should be operated. $\forall E \in P$, a function *AI(E)* is defined to represent the current runtime instance of activity *E*. Every instance of activities has its runtime status. For an instance I. here a function is defined to represent statuses of instances: *State(I)* ∈ *initated, active, completed, none*. *State(I)* = *initated* means the instance is initiated and waiting for being activated. *State(I)* = *active* means the instance is activated. *State(I)* = *completed* means the instance is completed and is qualified for activating the next instance of activity or has already activated the next instance of activity. *State(I)* = *none* means the instance cannot be initiated due to path selection.

Activities in workflow are connected by transitions which indicate the relationship and sequences between different activities in a workflow process. When tasks generated by an instance of activity are completed, the instance of next activity defined by transitions may be activated or the state of instance of next activity will be changed. $\forall t \in T$, two functions are defined: *from(t)*∈ *P*, *to(t)* ∈ *P*. *from(t)* denotes the source activity of *t*, while *to(t)* denotes target activity of *t*.

2.2 Process Instance, Decision Tree and Process Algebra CCS

An instance of process in workflow is created according to the definition of workflow process when a process in workflow starts to run. An instance of process in workflow consists of several instances of activities.

Decision tree [8, 9] is a model of anticipation. Each branch represents the outcome of the test and each leaf node represents a class label (decision taken after computing all attributes). The paths from root to leaf represent classification rules.

Some concepts in CCS [10] are used to describe workflow models. There are two basic elements in CCS: action and process. Action is used to describe transitions and process is used to describe activity in workflow model. Some operators in CCS are also introduced here. The prefix operator is defined by ".".

In workflow area, $A = a.B$ represents an instance of activity A should be able to activates the transition a then an instance of activity B is initiated. The selection operator is defined as "+". $C = a.A + b.B$ means that an instance of activity C should be able to be initiated by a transition a which is fired by an instance of activity A or by a transition b which is fired by an instance of activity B.

3 Application of Probability Theory in Workflow

3.1 Application of Decision Tree

There is a kind of activity called route activity. Here two types of route activity are discussed: *XOR-SPLIT* and *OR-SPLIT* activities. Figure 1 shows an *XOR-SPLIT* activity A. Only one of its successors, B or C, will be activated after an instance of A is completed. We formalize this type of activity as $A = b.B + c.C$. *OR-SPLIT* activity showed in Fig. 2 is complicated than *XOR-SPLIT* activity. When an instance of A is completed, then the number of instances of its successors N which will be activated satisfies $1 \leq N \leq M$, M is the number of its successors. The two types of activity act as path decision points of execution in runtime workflow. The selection result is calculated based on data provided by applications. The two types of activity can be modeled as decision points in a decision tree and the whole workflow process can be converted to a decision tree. The detailed algorithm is presented here.

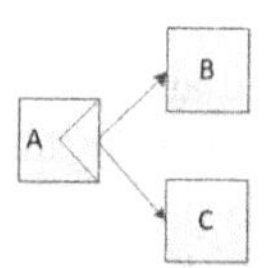

Fig. 1. XOR-SPLIT activity

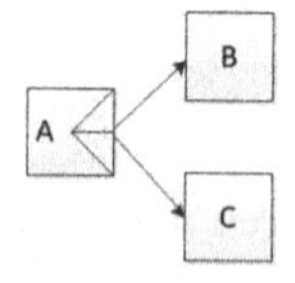

Fig. 2. OR-SPLIT activity

Algorithm 1 Convert a workflow process model to a decision tree

Input: a set of activity in a workflow process model P.

Output: a set of nodes make up of a decision tree converted from a workflow process model (deepest first)

Function: *activityNode(E∈ P)* represents the corresponding activity which is converted to a node in a decision tree in workflow process model.

Function: addTreeNode(E∈ P,S) each element in set *S* is added to a decision tree and become the child node of *E*.

Function: choice(E∈ P) represents a set of all possible activities whose instances can be activated by an instance of *E*.

Here *♯(S)* denotes the size of a set *S*.

decisionTree(P)

{root = the first activity in P P = P∪ root; call addNode(root, P);}

Sub addNode(P,P){S = {E|∀F∈ activityNode(P),F → E,CM(E) = XOR-SPLIT∨ CM(E) = OR-SPLIT}

if (♯(S) ==1) {E∈ S, P = P∪ E; ∀ST∈ choice(E), addTreeNode(E,ST); call addNode(E, P); }

else if (♯(S) > 1) { addTreeNode(E,S); P = P∪ E; S = A1,A2,A3.....;

X = choice(A1) × *choice(A2) choice(An)*//Cartesian product of all possible selection of activities

∀N∈ X, addTreeNode(E,N); call addNode(E, P);} }

Here explains the main idea of Algorithm 1. The start activity is the root node of a decision tree. The function addNode is recursively called to convert *XOR-SPLIT* and *OR-SPLIT* activities to tree nodes and add them and their all possible successors to a decision tree. The decision tree generated by Algorithm 1 only consists of decision nodes which are presented by squares while event nodes of decision tree presented by rounds will be added depending on different scenarios.

Take Fig. 3 as an example, Fig. 3 reveals a workflow process model. Figure 4 shows the corresponding decision tree converted with Algorithm 1.

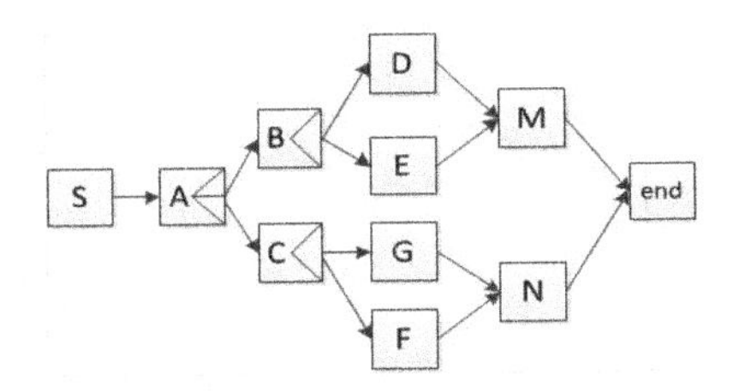

Fig. 3. A workflow process model

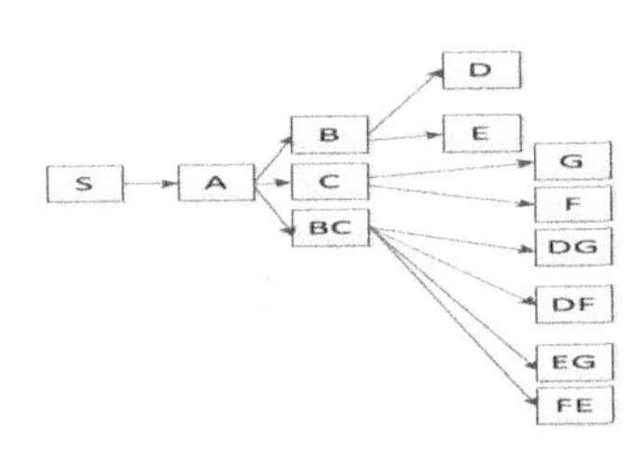

Fig. 4. The corresponding decision tree of Fig. 3

3.2 Path Selection Based on Probability and Decision Tree

As workflow process models can be converted to a decision tree, many issues of workflow such as path selection in the runtime can be resolved by some concepts in probability and decision tree. The application data can be calculated by probability theories, which provides important information for picking up the most efficient running path in a workflow process instance. Figure 5 shows a workflow process model representing a articles publishing process in a website. There are several route activities in the workflow process model. The activities *B* and *C* represent the tasks of initial verification. The activities *D, E, G, F* represent the tasks of reviewing and checking by senior editors. The efficiency of publishing process depends on many factors such as the efficiency of editors, the content of articles, the priority of articles. These factors can be viewed as providers of application data in workflow process instances. The activities should be published on website as quickly as possible. On the other hand, every piece of news or articles should be stringently verified by website administrators and goes through some other procedures. Let t be the time units the author want to get result of submitted articles. Based on statistics, the probability of completing its task in $t1$ $(t1 < t)$ for activity B is P_b. The probability of completing its task in $t1$ $(t1 < t)$ for activity C is p_c. The probability of completing its task in t for activity D is $P_t(T|D)$. The probability of completing its task in t for activity E is $P_t\ (T|E)$. The probability of completing its task in t for activity G is $P_t\ (T|G)$. The probability of completing its task in t for activity F is $P_t(T|F)$. We can take algorithm 1 to establish a decision tree. Base on the probability data, some event node is added to the decision tree. Figure 6 reveals the final decision tree. It is not difficult to figure out the probabilities of completing publishing news or articles in t time units for all running paths. The probability of finishing article publishing for choosing path from B to D is $P_b \times P_t\ (T|D)$. We can summarize the algorithm of getting the probability of achieving a goal in a specific running path in a workflow process.

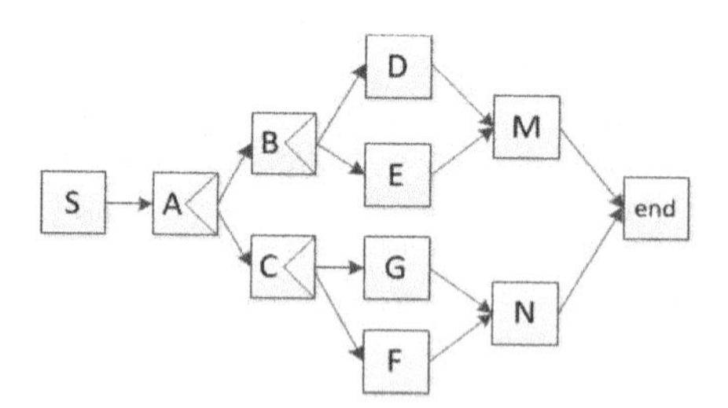

Fig. 5. Workflow process model for article submission

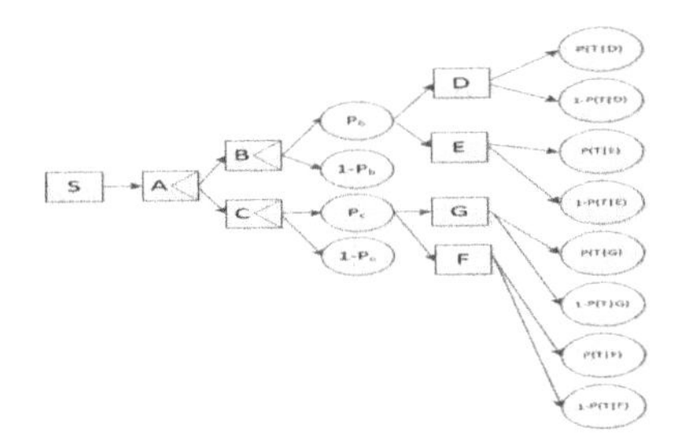

Fig. 6. Corresponding decision tree of Fig. 5

Algorithm 2 Getting probability for a specific running path
Input: decisionPath; *Output:* the probability of achieving a goal
pathProbability(decisionP ath)
{ result = 1; ∀m,n∊ decisionPath; if(isEventNode(m) == true∧ isDecisionNode (n) == true∧ parentNode(n) == m) result = P(m)*P(et|n); return result; }

3.3 Application of Bayes Theory

Sometimes the probability of some events which occur in front part of workflow process are relatively difficult to calculated compared with events occur in rear part of workflow process. For example, in the process of article submission revision, it is relatively easy to get some information of error omission rate in the final phase of the process. Based on the probability of error omission, it is convenient to figure out who are prone to make mistake during the process and what kind of resource configuration will lead to a high quality revision process. In this situation, the Bayes theory can be fully applied to provide useful data. For an activity M in a rear part of workflow process model, based on the statistical data and the anomaly analysis, calculate the probability P_x denoting the possibility of anomaly occurrence, calculate the probability $P_{ex}(D|X)$ denoting the anomaly occurrence possibility of the execution path from node D to node M. Consequently joint probability is given by $P(D\cap) = Pex(D|X) \times Px$. Meanwhile, the probability $(D\cap \overline{X}) = (1 - Px) \times Pex(D|\overline{X})$ denotes that no anomaly occurs in the execution path. Obviously, $P(D) = P(D\cap X) + P(D\cap \overline{X})$. Finally, based on the Bayes formula, $P(X|D) = P(D\cap X) \div P(D)$, the probability can be obtained, which denotes the probability of anomaly occurs at node M when node D is previously selected.

4 Summary

With probability theory and decision tree introduced, a solution is put forward to act as a new kind of provider of application data for workflow. The solution can provide very important data from the perspective of probability so that a proper recommendation of running path selection can be instantly figured out and be provided to the user of workflow. The solution makes contribution to the efficiency of workflow engine to some extent. On the other hands, the current solution is a relatively light-weight tool and can be functional in a small part of workflow engine. The solution should be extended to be more powerful. There are still many similar issues of worklfow should be addressed. In the future work, many useful basic knowledge such as probability theory, graph theory can be combined to be applied on those unaddressed issues.

Acknowledgements. The work of this paper was supported by the National Key Basic Research Program of China (Grant No.2014CB744903, 2014CB744904).

References

1. Van Der Aalst, W., Van Hee, Kahm., Hofstede, T, et al.: Soundness of workflow nets: classification, decidability, and analysis. Formal Aspects of Computing pp. 333–363 (2010)
2. Van Der Aalst, W.: Workflow verification: finding control-flow errors using petri-net-based techniques. Lecture Notes in Computer Science. Business Process Manage. pp. 19–128 (2000)
3. Van Der Aalst, W.: Verification of workflow nets, computer science, application and theory of petri nets, pp. 407–426 (1997)
4. Van Der Aalst, W.: Supporting the workflow management system development process with YAWL, CEUR Workshop Proceedings, pp. 33–40 (2013)
5. Herbert, LT., Hansen, ZNL.: Restructuring of workflows to minimise errors via stochastic model checking: An automated evolutionary approach, Reliability Engineering and System Safety, pp. 351–365 (2016)
6. Xiong, P.C., Fan, Y.S.: An optimization solution of the number of workflow resources under cost constraint. Comput. Integrated Manuf. Syst. **09**, 1–6 (2007)
7. Workflow Standards and Associated Documents [EB/OL], www.wfmc.org, 2008
8. Kennedy, HC., Chinniah, C., Bradbeer, P., Morss, L.: The construction and evaluation of decision trees: a comparison of evolutionary and concept learning methods, pp. 147–161 (1997)
9. Chen, Z., Lan, S., Han, H.: Multi-valued attribute and single-sampled data decision tree algorithm for entity identification. J. Comput. Inf. Syst. pp. 2927–2935 (2014)
10. Stirling, C.: Temporal logics for CCS, pp. 660–672 (2006)

Constructing and Analysis of the State Spaces of Workflow Process Models

Lisong Wang(✉), Yifan Chu, Min Xu, Yongchao Yin, and Ping Zhou

College of Computer Science and Technology,
Nanjing University of Aeronautics and Astronautics, Nanjing,
Jiangsu 210016, China
wangls@nuaa.edu.com

Abstract. In this paper, a novel formal language Z is adopted to describe the state spaces of workflow process models. We construct the state space for two aspects of workflow process: control flow and resource management. The formulation proposed in this paper can guarantee the correctness of process models to some extent. At the same time, it provides a useful tool for checking the consistence of workflow process models and real processes.

Keywords: Workflow · State space · Process models

1 Introduction

So far, the technology of workflow has been widely applied in many complex commercial projects. So designing workflow process models which can properly describe real business processes is an important topic in the community of workflow technology. Modelling workflow process models are becoming more challenging. Modelers of workflow process need a tool to check the correctness of designed models.

In the research area of workflow, many verification formulations for process models have been put forward. In [1–3], the author proposes the petri-net based modelling language. In this paper, the author takes full advantage of the concepts of live property and safety property to verify the soundness of workflow process models. In [4], the author uses a classical algorithm to transfer cyclic workflow process models to acyclic models. The complexity of workflow process models can be reduced with this algorithm applied so that it is relatively easy to check all kinds of features in process models. In [5], the author presents a solution of verifying process models with graph transformation. Traditional workflow nets can also be modified with reset arcs for soundness verification [6].

Some verification tools have been put into practice. In [7], features of YAWL (Yet Another Workflow Language) are elaborated for supporting the workflow management system development.

Currently, focus is mainly paid on checking soundness of process models. Sound workflow process models can avoid deadlock and livelock. But on the other hand, verifying the consistency between workflow process models and real business process

F. Xhafa et al. (eds.), *Advances in Intelligent Systems and Interactive Applications*, Advances in Intelligent Systems and Computing 686,
https://doi.org/10.1007/978-3-319-69096-4_117

is also an indispensable part of workflow technology. In this study, we propose a verifying method by constructing the state space of workflow process models. The state space of process models is described with Z notation. The concept of schema in Z notation [8, 9] can be used to describe not only the state of instances of activities in workflow process models but also the state of resource management in activities. State space of a process instance can be acquired and the corresponding information such as constraints of activities, behaviour of activities can be reveal in the form of schema.

The structure of this paper is as follows: the second part introduces some definition related to our solution; the third part elaborates the construction of state space of workflow process models with Z language. The fourth part concludes the whole paper.

2 Definition

2.1 Notations, Schema and Function

In our study, some notations in Z are introduced and integrated into the description of workflow process models. Table 1 lists some basic notations.

Table 1. Notations

Type	Example	Remark
Set	A, B, C	A = {a, b}
Relation	$X \leftrightarrow Y$	$X \leftrightarrow Y = 2^{X \times Y}$
Partial function	$X \mapsto Y$	$\{R: X \mapsto Y \mid \forall x \in X, y1, y2 \in Y, x\ R\ y1 \wedge x\ R\ y2 \Rightarrow y1 = y2\}$
Total function	$X \rightarrow Y$	$\{f{:}X \rightarrow Y \mid X \mapsto Y\ dom(f) = X$

Schema is the basic structure of Z notation. Schema in Z notation can be divided into two different types: state schema and operation schema [6]. The notation of schema is defined in Fig. 1:

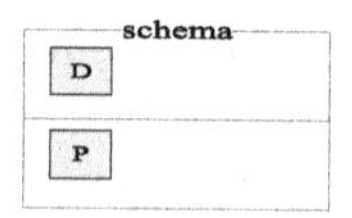

Fig. 1. Notation of schema

Functions can be described formally with schema. Figure 2 show the formalized description of a function. The declaration part shows the declaration of the function and the assertion part shows the detail definition of the function.

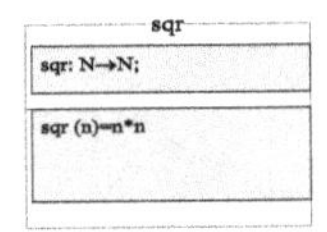

Fig. 2. Presentation of function

2.2 Activities in Workflow Process Models

Activities in workflow process models represent logic steps in real business process. In our study, the set *P* represents the set of all activities in a workflow process model. Each element in *P* represents an activity. Each activity instance represents a single invocation of an activity within an enactment of process instance. Here we use *AI* (*E*) to represents the current instance of activity *E*

2.3 Transitions in Workflow Process Models

The transitions in a workflow process model describe the relations between different activities [9]. A set *T* represents all transitions in a workflow process model. Here for $\forall$ $t \in$ T, there is a function defined: from (t):T→P, to (t):T→P. from (t) represents the source activity of the transition *t*. *to* (*t*) represents the target activity of the transition *t*.

2.4 The State of Instance of Activities and the Types of Activities

In runtime, every instance of an activity has a state. Here, for all activities in *P*, a function is introduced, {*AI* (*E*)| *E* ∈ *P*}→{*initiated*, *active*, *completed*, *none*}, *AI* (*E*) = *initiated* denotes that the instance of *E* has been initiated and waiting for being activated. *AI* (*E*) = *active* denotes the instance of *E* has been activated. *AI* (*E*) = *completed* denotes the instance of *E* has finished the operation and are ready to fire or has fired some other transitions.

A set is defined here to describe types of activities in workflow process models: *AT* = *{NORMAL, ANDSPLIT, ANDJOIN, ORSPLIT, ORJOIN, XORSPLIT, XOR-JOIN}*. At the same time, two functions should be introduced here to construct the mapping between activities and their types: stype: *P*→*AT*, *jtype: P*→*AT*. The definitions of types of activities are described in [10].

3 Constructing the State Space of Workflow Process Models

3.1 The State Space of Control Flow Pespective

Figure 3 is a workflow process model. In the model of Fig. 3, an instance of activity S should fire two transitions to activate activity A and B after it finishes its operation while the instance of activity C can not be activated until both A and B finish their operation

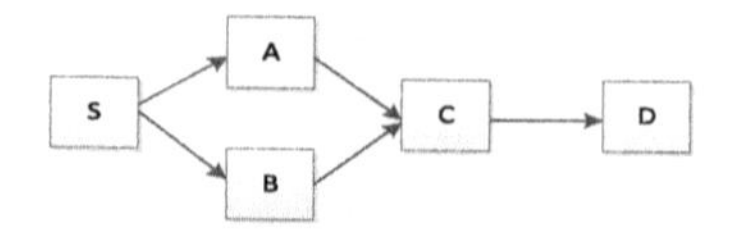

Fig. 3. A process model

(i) Definition schema of activity
Figure 4 is the definition schema of activity S which denotes the activity S is start activity of this process and is an ANDSPLIT-typed activity.

(ii) State schema of activity
State schema of activity describes the runtime abstraction of information of activity instances. Let's take the activity C in Fig. 3 as an example. First, the definition schema of activity C should be defined. Figure 5 is the definition schema of activity C, which denotes that C is an ANDJOIN-typed activity. Then the state schemas of activity C can be defined. Figure 6 is pre-active state schema of activity C. The schema illustrates that the instance of activity C can not be activated until the instance of its predecessor activities finish their operation. In declaration part of this schema, the variable *direcpre* is a subset of *P* and it denotes all direct predecessors of activity C. Figure 7 is post-active schema of activity C. The schema reflects that the instance of activity C is activated.

(iii) Operation schema of activity
The operation schema of activity can describe the behavior feature of activity instances. In control flow aspect of workflow process model, the behavior of activities mainly refers to firing transitions. Once transitions are fired, the state of some activity instances will change immediately.
Here we still take activity C in Fig. 3 as an example. Figure 8 is the operation schema of activity of C. The schema describes the state transition of activity instances after an instance of C fires transitions. In declaration part, *direcpost* is a subset of *P* and denotes all direct successors of C. When an instance of activity C fires transitions, the instance of C will be in state of completed and instances of all direct successors will be in state of either initialized or active.

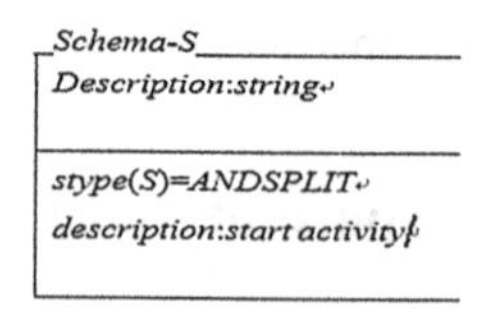

Fig. 4. Definition of Schema S

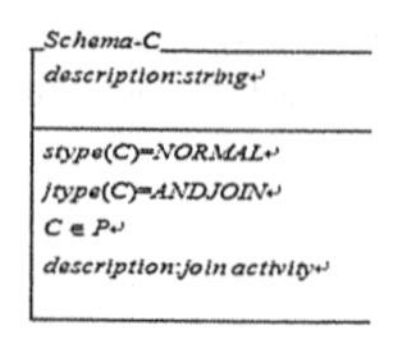

Fig. 5. Definition of Schema C

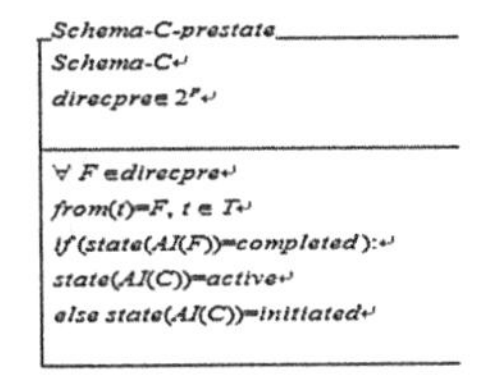

Fig. 6. Pre-active schema of C

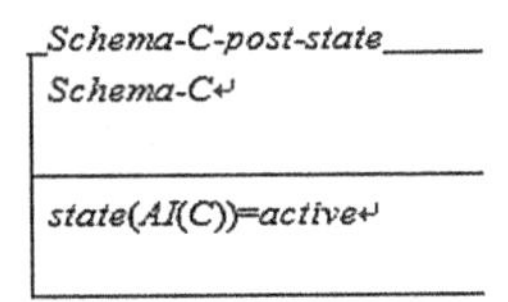

Fig. 7. Post-active schema of C

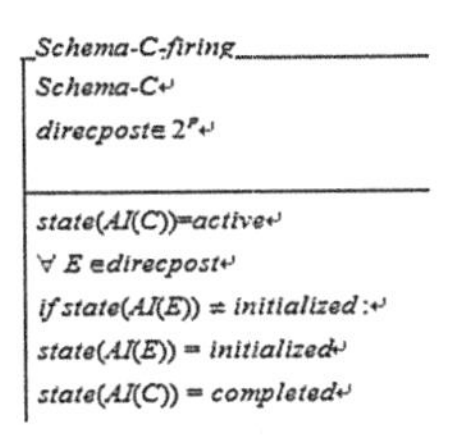

Fig. 8. Post-active schema of C

(iv) State space of control flow

So far, three kinds of schemas in control flow aspect of workflow process models are introduced. With those schemas mentioned above, we can construct the state space of workflow process model. The state space of process model in Fig. 4 can be formalized as follows:

WFStateSpace = schema-s-pre $\vee$ *schema-s-post* $\vee$ *((schema-A-pre* $\vee$ *schema-A-post)* $\wedge$ *(schema-B-pre* $\vee$ *schema-B-post))* $\vee$ *schema-C-pre* $\vee$ *schema-C-post* $\vee$ *schema-D-pre* $\vee$ *schema-D-post.*

3.2 State Space of Resource Perspective

A workflow process consists of many activities which represent logic steps in real business process. Every logic step should be operated by human or non-human objects such as machine in real world. The resource configuration of workflow is dedicated to dealing with allocation of human or non-human resources to execute tasks which are generated by activities in workflow process. Resource configuration can be divided into four phases: task creation, offering task to resource, allocating tasks to resource, finishing executing tasks.

Before introducing the state space of resource configuration, some global sets should be defined here. *Task* is the set of all tasks in a process instance. *Resource* is the set of all candidate operators of tasks in workflow process. *Created* is the set of created tasks. *Offered* is the set of all offered tasks. *Allocated* is the set of all allocated tasks. *Finished* is the set of finished tasks.

(i) Definition schema of task
In resource configuration perspective, a task is generated by an active instance of an activity. Figure 9 is the definition schema of task A. Figure 10 is a definition schema of resource A.

(ii) Operation schema of resource configuration
In task offering phase, the workflow engine pushes the task into the work lists of some resources, which reminds the resources responsible for the tasks that tasks are available in their work lists. The common strategies of task offering are shortest queue solutions, cyclic solutions and random solutions. In our study, we describe shortest queue solution with schema.

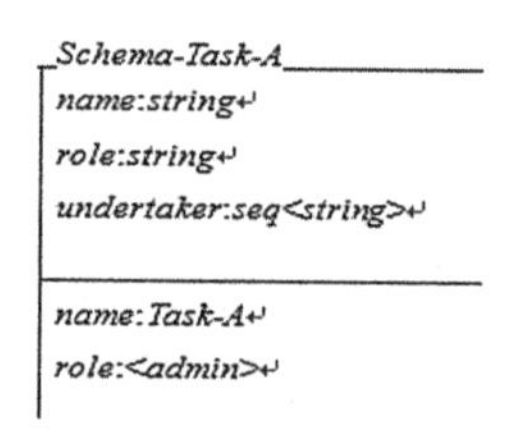

Fig. 9. Definition schema of task A

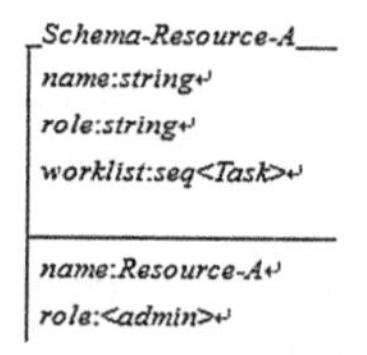

Fig. 10. Definition schema of resource A

First, we define a schema describing the state before a certain task is offered to some resources. Figure 11 is a schema which denotes the task should be created before being offering to resources.

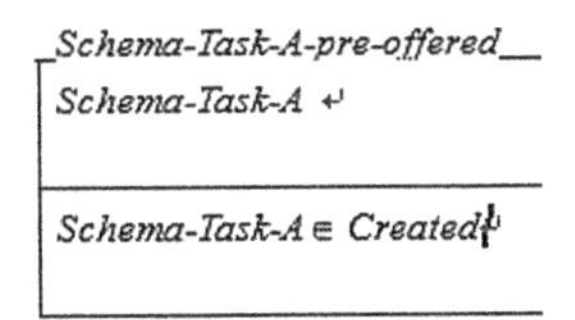

Fig. 11. Schema of pre-Offered

Figure 12 is the schema of shortest queue solution. In this schema, the input variable *h* represents the set of resource which are possible to be offered with task A. The workflow engine checks the number of work items in work lists of all resources whose roles satisfy the requirement defined in definition schema of task-A and offers the task A to resources with minimum work items in their work lists.

Task-A-shortest-queue-offering

Schema-Task-A
$h? \in Resource$
$min: N \longrightarrow N$

$h.role \in Schema\text{-}Task\text{-}A.role$
$\forall\, r \in h$
$m = min(\#r.worklist)$
$if\ \#(r.worklist) == m$
$receiver! = r$

Fig. 12. Schema of shortest queue solution

The function *min* declared in the schema will calculate the minimum number of workitems in work lists of all resources.

In the resource configuration perspective of workflow process, allocating tasks refers to that the resources proactively begin to execute the tasks which have been offered in their work lists or the workflow engine specifies the undertakers of tasks and the corresponding resources begin to execute tasks. The schema of allocating tasks is similar to the schema of offering tasks.

Now we can construct the state space of resource configuration with schemas. For example, in the process of configuring resource for task *A*, the state space ResourceSpace-task-A is as follows: ResourceSpace-A = Task-A-Created ∨ Task-A-shortest-queue-Offering ∨ Task-A-Allocating ∨ Task-A-Completed.

4 Conclusions

This paper makes full use of Z notation to provide a method to construct the state space of workflow. Our solution describes the state of workflow process with a formation of schema from perspectives of control flow and resource configuration. The formulation presented in our study can viewed as a useful tool for checking consistency between the process models of workflow and the real business process. But in some complex process, some side effects such as the state space is too large will emerge. So the modification on current research or a better solution should be our top priority in the future work.

Acknowledgements. The work of this paper was supported by the National Key Basic Research Program of China (Grant No.2014CB744903, 2014CB744904).

References

1. van der Aalst, WMP., van Hee, KM., ter Hofstede, AHM., Sidorova, N., Verbeek, HMW, et al.: Soundness of workflow nets: classification, decidability, and analysis. Formal Aspects of Computing pp. 333–363 (2010)
2. van der Aalst, WMP.: Workflow verification: finding control-flow errors using petri-net-based techniques. Lect. Notes in Comput. Sci. pp. 19–128 (2000)
3. van der Aalst, WMP: Verification of workflow nets, Application and Theory of Petri Nets, pp. 407–426 (1997)
4. Choi, Y., Zhao, JL.: Feedback Partitioning and Decomposition-Oriented Analysis of Cyclic Workflow Graphs, On the Move to Meaningful Internet Systems 2004: OTM 2004 Workshops, pp. 17–18 (2004)
5. Rafe, V., Adel, T.: A graph transformation-based approach to formal modeling and verification of workflows. Rahmani, Communications in Computer and Information Science, pp. 291–298 (2009)
6. Clempner, JB.: Classical workflow nets and workflow nets with reset arcs: using Lyapunov stability for soundness verification. J. Exp. Theor. Artif. Intell. p. 15 (2015)
7. van der Aalst, WMP.: Supporting the workflow management system development process with YAWL. CEUR Workshop Proc. pp. 33–40 (2013)
8. Jonathan, PB.: The Z Formal Specification Notation, The Z Formal Specification Notation, pp. 15–43 (1997)
9. Derrick, J., Boiten, E.: Combining component specifications in object-Z and CSP. Formal Aspects Comput. **13**(2), 111–127 (2001)
10. Workflow Standards and Associated Documents www.wfmc.org (2008)

Analysis on the Causes of Bad PCBA Heavy Tin

Junjie Lv(✉)

School of Electronic Information Engineering, Wuhan Polytechnic College, Wuhan, Hubei, People's Republic of China
593316640@qq.com

Abstract. The emergence of PCBA heavy disc soldering tin bad phenomena in two times of furnace process, the failure of the pad, a furnace a pad, not a furnace of solder surface observation, analysis of the FIB sample preparation section, the reason to search the AES failure surface composition analysis. The results showed that: the the failure pad at second times in front of the furnace has been oxidized, and the pad surface tin thickness dramatically thinned, resulting in a pad of tin.

Keywords: Tin precipitation · FIB profile preparation · AES component analysis · Tin plating

1 Case Background

The failure samples for a certain type double chip PCBA board, the PCB board after two SMT, found B small pads on the tin surface appear undesirable phenomenon, the sample failure rate of about 3/1000. The PCB board pad surface treatment process of electroless tin, tin appear pads are located in the second side patch.

2 Brief Analysis Method

Using SEM for failure pad, a furnace a pad, not a furnace pad surface morphology was observed, as shown in Fig. 1. The results show that no furnace pad surface tin layer forming good furnace a pad and the pad surface uplift failure of tiny particles. Show that the tin layer in the furnace after the surface of whiskers.

Through the FIB of failure pad, a pad and a furnace furnace without pad to create profiles, again through the EDS component line scanning on the profile of the surface, see Fig. 2 results: the results showed that the specific failure pad on the surface has Cu elements, pure tin layer in Sn has been basically complete with Cu the formation of alloy; furnace a pad surface at about 0.3 m depth Cu elements appeared, explained in a furnace after welding, pure tin thickness is about 0.3 m; not a furnace pad surface at about 0.8 m depth of Cu elements, that did not have the thickness of pure tin layer furnace the pad is about 0.8 M.

F. Xhafa et al. (eds.), *Advances in Intelligent Systems and Interactive Applications*, Advances in Intelligent Systems and Computing 686,
https://doi.org/10.1007/978-3-319-69096-4_118

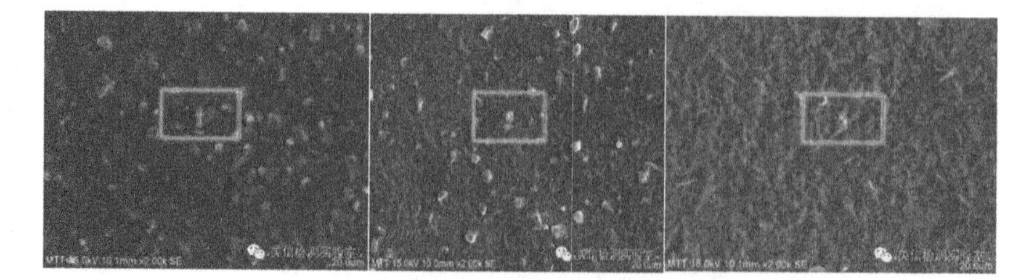

Fig. 1. Surface topography of different pads (3000X)

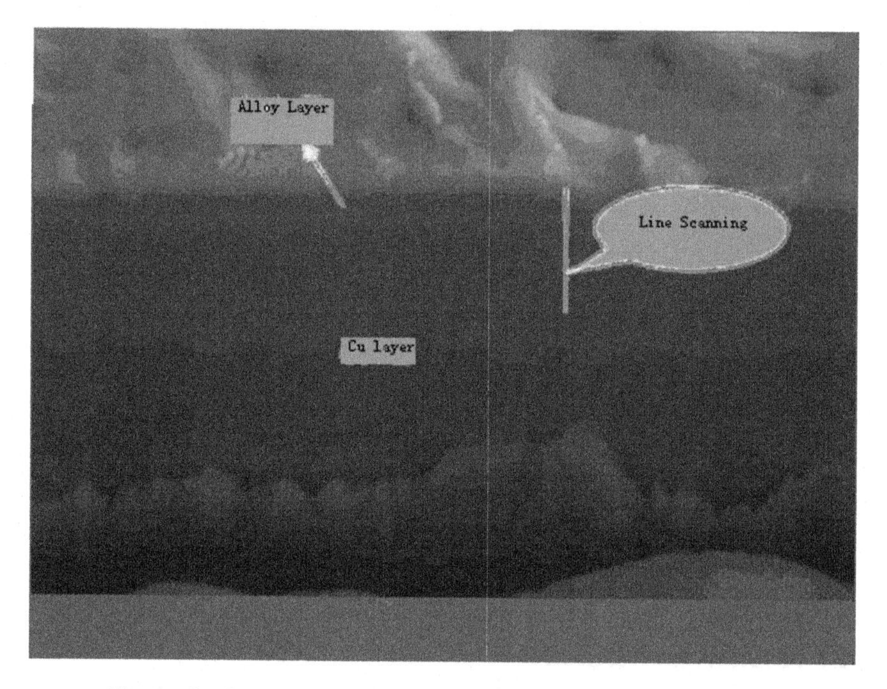

Fig. 2. Sectional Sect. (10000X) of the failed solder pad FIB sample

Thus with the increase of furnace number, surface tin pad thickness decreased significantly, tin thickness can hardly meet the requirements of the welding furnace again once. Because the precision of the EDS analysis of reasons, need to use AES on the surface of the high precision analysis.

Analysis of AES composition on 2.3 pad surface

Since the depth of the EDS is more than the thickness of the deposited tin, AES (Auger electron spectroscopy (5 nm) analysis) is used to analyze the composition of the failed solder pad and the surface of the once through pad.

Figure 3 is a map of the distribution of furnace pad in the depth of 0–220 nm composition curve, it can be seen from the figure, a furnace a pad surface in the range of 0–187 nm is C, Sn; Cu elements appeared at about 180 nm in the depth of tin and copper compounds have emerged. Thus, a welding furnace the disc surface tin layer

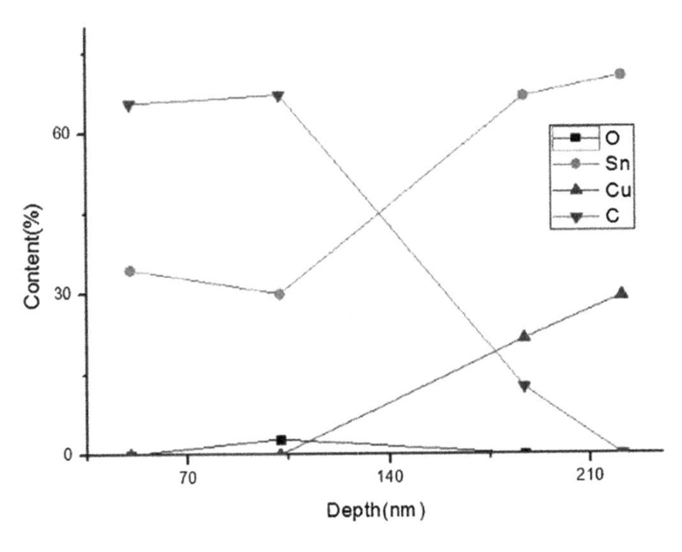

Fig. 3. Distribution of components in the depth range of 0–220 nm in the cross welding pad

had no serious oxidation, but metal compounds have grown, pad surface residual tin layer thickness.

3 Analysis and Discussion

Sink structure tin pad consists of copper layer, copper tin alloy layer, pure tin tin oxide layer and the surface layer, which can ensure the pure tin layer has good wetting pad surface during reflow process, the tin oxide layer on the surface of the material will be activated in the removal of solder paste flux, solder paste with the pure tin melting layer, make the pads of tin is good, in general, the pad surface at least pure tin layer are 0.2 m, in order to ensure the pad good weldability. The failure phenomenon appeared in the second patch, NG sample in the first furnace process, due to the high temperature accelerated diffusion between copper and tin, copper tin alloy layer becomes thicker, and pure tin oxide in pure tin layer is increased, serious consumption; in the second furnace welding flux, the inadequate activity of surface oxide is not completely removed (from the analysis of the AES component), lack of pure tin thickness, pad wettability Difference causes the pad to not solder. PCB solder alloy connection. solder paste in the metal content increases, the solder paste viscosity increase, can effectively resist the force created by the preheating process of vaporization, the increase of the content of metal to metal powder are close, in the melt when combined with without being blown, the increase of metal content may also reduce the solder paste printing after collapsing. The amount of flux in solder paste if too much, can cause local cave solder paste, result in tin beads is easy to produce. active hours solder paste, flux, flux oxidation ability is reduced, it will make the solder ball easy to produce (flux activity levels into R inactive, RMA moderate activity levels, RA, fully active, RSA super activity). In the solder paste, the higher the degree of metal oxide in welding metal powder combined with the resistance, the greater the solder paste and solder and components between the less invasion, thus reduce solderability. Metal powder particle size is smaller, the greater the total surface area of solder paste, so that a higher degree of oxidation of fine powder, lead to solder ball phenomenon. Most of the solder paste for short periods of time can endure 26.6 DHS C is the highest temperature, temperature can lead to high flux and solder paste from ontology, and transform the liquidity of the solder paste, further lead to bad printing. Generally do not recommend freezing solder paste, because can lead to precipitation of catalysts, reduce the welding performance. All the solder paste will absorb moisture, so to avoid in high humidity. If inhaled too much moisture, tin in use will cause explosion, residue, tin ball, in the reflow soldering components will shift, and poor welding, etc. According to the above situation, in the process of teaching should make students clear, training workshop of solder paste should be stored in an airtight form in constant temperature, and humidity of the air-conditioner, save the temperature of 0, 10 DHS C (in such conditions can be kept up to 6 months). If the high temperature alloy powder in the solder paste and flux react will reduce viscosity activity affect performance; Resin will produce crystallization in the low temperature solder, solder paste is form bad; If a short period of time the solder paste repeatedly appear different in different environment temperature change can make the solder paste, flux performance change and influence welding quality. Solder paste, solder paste storage environment influence on process and use requirements, solder paste and control decision-making aspects in this paper, the defects caused by the importance of solder paste in SMT production printing, the solder paste is in use and welding problems arising from the targeted control decisions are put forward.

4 Conclusion

NG samples were found in the second patch, note pad in the first furnace process, due to high temperature accelerated diffusion between copper and tin, copper tin alloy layer thickness, and increased layers of pure tin is oxidized in pure tin layer becomes thinner; in the second furnace welding flux. The activity is not strong enough, the surface of tin oxide is not completely removed, and the insufficient thickness of pure tin, lead solder wettability, does not appear on the tin failure.

5 Suggest

(1) Use more active flux to enhance the ability of solder paste to remove oxidation layer;
(2) Increasing the thickness of the PCB plate to ensure that the thickness of the tin layer can meet the solderability requirement once the furnace has passed;
(3) Increase nitrogen protection and reduce the oxidation of pad surface.

References

1. Zhangsi, S.: The electroless tin process of printed circuit information, 5 (2011)
2. Fu, L., Bin, L.: Do you understand tin? Printed circuit information, 9 (2012)
3. Fu, L., Bin, L., Xiaojin, G.: Welding failure analysis of printed circuit information method. PCB sink tin, 3 (2014)
4. Al-Sakran, HO.: Framework architecture for improving healthcare information systems using agent technology. Int. J. Managing Inf. Technol. (IJMIT) **7**(1) (Feb 2015)
5. Devi, CS., Ramani, GG., Pandian, JA.: Intelligent E-Healthcare Management System in Medicinal Science. Int. J. PharmTech Res. **6**(6), pp 1838–1845 (Oct-Nov 2014)

Characteristics of Solder Paste and Reflow Process Analysis

Junjie Lv(✉) and Xu Li

School of Electronic Information Engineering, Wuhan Polytechnic College, Wuhan, Hubei, People's Republic of China
593316640@qq.com

Abstract. No matter what the welding technology, should ensure that meet the basic requirements of welding, welding to ensure good results. High quality welding should have the following 5 basic requirements: 1. appropriate heat; 2. good wetting; 3. appropriate solder joint size and shape; 4. controlled tin flow direction in the welding process; 5. the welding surface does not move to have enough solder joint life, we must ensure that the shape and size of solder joint with welding end structure. The mechanical strength of the solder joint is too small, unable to withstand the stress in use, even after the welding stress is unable to bear. But once in use began to appear fatigue or creep cracking, the fracture speed is rapid. The shape of the solder joints will cause bad homes from the phenomenon of light, life expectancy shortened the solder joint.

Keywords: Appropriate heat · Welding process · Solder paste · Solder joints

1 Solder Paste

Solder paste by flux and solder powder composition, its quality is directly related to the quality of the product. Printing speed, adhesive force, after the return of bridge, built, lack of wettability, solder, solder false problems were associated with the production of solder paste on quality. First, according to the equipment and process conditions to choose the appropriate solder paste type. According to the melting temperature of the solder alloy can be divided into normal temperature (183 degrees C), high temperature, low temperature. According to the flux type, can be divided into rosin, disposable type and water soluble. According to the type of alloy can be divided into lead and lead-free. From the point of view of environmental protection, in the absence of heat sensitive components, the choice of non clean lead-free (high temperature) Han Xigao. According to the specific process requirements from the viscosity, particle size and other indicators to refine the selection of solder paste. The degree of viscosity of solder paste unit for the "Pa - S", fully automatic printing machine generally choose 200–600 Pa - S solder paste, and general manual and semi automatic selection for printing solder paste viscosity in 600–1200 Pa - S. Solder paste particle size according to the distance between the minimum distance of solder joints on the PCB board to determine if a larger spacing, can choose the particle size of the solder paste, and when

F. Xhafa et al. (eds.), *Advances in Intelligent Systems and Interactive Applications*, Advances in Intelligent Systems and Computing 686,
https://doi.org/10.1007/978-3-319-69096-4_119

the distance between each point is small, should choose a small number of solder paste particles; particle diameter generally less than the opening of the 1/5 template. But it is not as small as possible, small particles of solder paste, solder paste printing more clearly, but also more prone to collapse, the degree of oxidation and the opportunity is also high. Paste sealed at 0–10 °C when a period of six months. Precautions must be 24 h powered refrigerator, strict control of temperature 0–10 °C, need to check and record the temperature every day, do not paste the wall close to the fridge. New solder paste in a good state before the fridge tags, date paste and fill out sheets. Paste before use, preservation sealed state (20–25 °C Humidity 45–75%) in the back at room temperature for 4 h or more, and in the state of solder paste bottle label paper states defrosting time and fill solder cream out of the sheets. Must not use the method of heating solder paste to room temperature, rapid heating causes the solder paste flux deterioration in performance, thus affecting the welding results. When used in accordance with the requirements of this training solder paste, select the appropriate packaging specifications, generally 250, 500 g packaging, in order to avoid the loss caused by solder paste failure. Strict implementation of advanced first principles, and priority to the use of recycling (old) solder paste, but only once, and then do the remaining scrap treatment. The use of old solder paste must be mixed with the new solder paste, the new and old solder paste mixture ratio control in 4:1–3:1, and the requirements of the same model with the same batch. stir by hand, stir the 5–10 min in the same direction in the same direction, in order to achieve uniform mixing of alloy powder and flux. Performance: the use of a scraper scraper part of solder paste, tilt, solder paste can smoothly slide. Pay special attention to printing, if the solder paste is dry when the same direction can be manually stirred 1 min.

2 Manual Printing

Stencil before using, or check the leak up to light with a magnifying glass. Steel net examinations in the printing stage, tighten the bolts. In hand-printed circuit board positioning commonly mechanical positioning and orientation. Mechanical positioning is to take a piece of circuit board to the printing surface, move the circuit board, circuit board pads and align the stencil openings, at the rate of up to 90%, fixed on the selected hole with a PIN, cut off excess nail with pliers, nail hammering positioning, and printing fine tuning screw alignment of the civil service. Positioning is the circuit board to the right position and fixed with double-sided tape two pieces of circuit board clip critical of printed circuit board scrap. Special attention when installing steel plane must be contact with the circuit board was flat, otherwise, will result in a collapse of solder paste printing and stencil life reduction. To note is printed circuit boards prior to installation should check for warpage of the surface and surface cleanliness, because the circuit board surface is not clean or severely reduced oxide solder adhesion, and have an impact on quality of welding.

Take part of solder paste on the mixing knife steel mesh front, evenly placed as far as possible, be careful not to add in the leak, solder paste volume not too much to ensure steel line when printing solder paste into rolling a cylinder, with a diameter of 1–1.5 cm. Principle of solder paste to add small amounts several times, can be added at any time during the operation. Hand-printed, blade angle, pressure, and speed is hard to control, so printing at the beginning, be sure to observe, experience fine, find the most suitable angle. With blade backwards from the front of solder paste evenly scraped and blade angle is 45°–60°, the angle is too large, and easy to solder paste graphics not full, the angle is too small, easy solder paste graphics to stains, scrape out excess solder paste back in the front of the template. Too much pressure on the blade can easily stain solder paste graphics (minimum) pressure is too small, stay in the solder paste on the surface of the template easier to solder leak brought up together, causing printing and easy to solder block templates printed holes. But ensuring a clear edge of solder, surface roughness and thickness of blade-light as possible under the right conditions. Printing speed is too fast, too fast cause solder paste graphics full of printing defects, under normal circumstances, 10–20 mm/s. Printing should ensure solder is relative to the blade for rolling instead of sliding.

The molten solder must flow to the required direction, to ensure the controlled formation of the solder joint. In the wave soldering process of "stealing solder and solder mask plate" (green) is used, and the reflow soldering tin absorption phenomenon in the process, and is a tin flow direction control technology details. If welding process according to the mobile end mobile welding, and time, will not only affect the shape and size of solder joints, may cause the weld and the inner hole. This will affect the quality of life of solder joints. So the product design and process, must be taken into account in the process of welding welding end stays in reflow soldering state. In the process, in addition to the above general welding conditions, there is a little special, is the need to put through the printing process without chemical composition of solder paste in timely dispose of volatile.

Finished printing solder circuit boards, usually within half an hour of complete SMD, reflow soldering done in 2 h. If over time, must be washed, reprinting and patch. Each print completed a circuit board, should carry out checks, printing results were determined based on the reason for printing defects, when printing a piece of circuit board, may be appropriate to change the squeegee angle, pressure and speed, until you're satisfied. Check stain solder paste graphics (even), when leak plug or a template, feel free to use the fiber-free paper moistened with ethanol wipe the bottom surface of the template. Especially when printing narrow spacing products, due to its high requirements on the solder joints forming effects of commonly used high viscosity paste, pay special attention to cleaning work on the steel, printing out a circuit board you must wipe the bottom surface of the template. Printing double-sided SMT circuit board, should first be Indian components lighter, less components and when this patch after the welding is completed, further components or devices printed. Printing second surface Shi need in printing table Shang placed pad article, put circuit board frame up (pad article put in has completed posted tablets and welding this side no components of location Shang), pad article of height is slightly higher than circuit board Shang has welding of highest of components, at to note fixed template at gasket of height and printing stage circuit board positioning pin of height also to corresponding improve.

3 Results and Discussion

Careful observation of welding PCBA under the microscope observation of the solder joint shape and surface conditions, wetting degree, tin flow direction, residues and PCBA on the solder ball and so on. Especially for the above second points recorded at the welding difficulty should pay more attention. In general, after the above adjustment does not appear after what welding fault. But if a failure has occurred, according to the analysis of the fault model, then the mechanism with the upper and lower temperature zone control to adjust. If there is no fault, from the curve and plate joints decides whether to fine tune the optimization. The aim is to make the process set the most stable and minimum risk. Consider the problem of load and the production line speed adjustment, in order to get a good balance between quality and yield.

PCB is fed along a conveyor belt solder paste printing, automatically find the main mechanical side of the PCB, and positioning. Z-shaped frame to move up to the position of the vacuum plate, adding vacuum, firmly fixed in a specific location PCB. Visual axis (lens) and move it slowly to the first target Mark PCB (reference point), the machine can move the print network to make it align PCB., The machine can print and mobile networks in the Q axis X, Y-axis direction direction. 64G U disk welding, soldering temperature curve when the temperature setting is not the same, resulting in Fig. 1.

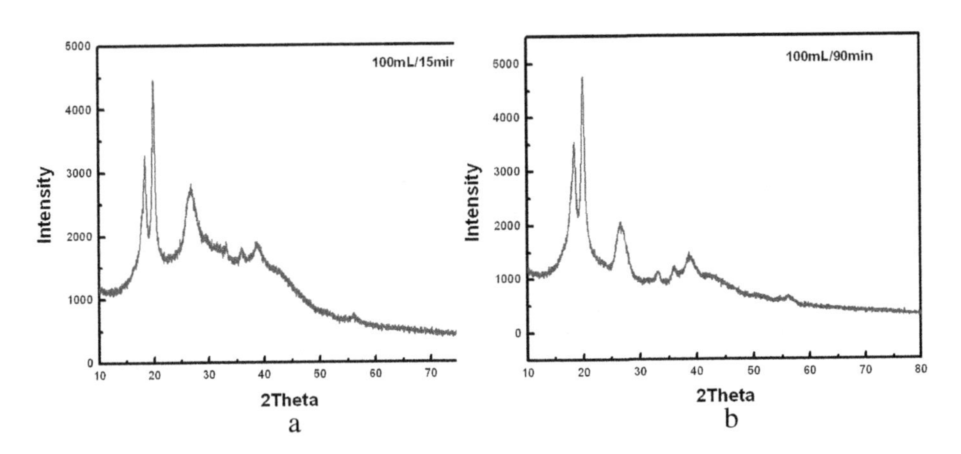

Fig. 1.

References

1. Hicks, D.: Phys. Rev. B **47**(24), 16631 (1993)
2. Chung, D.Y., Hogan, T., Brazis, P., et al.: Science **287**, 1024 (2000)
3. Harman, T.C., Taylor, P.J., Walsh, M.P., et al.: Science **297**, 2229 (2002)
4. Mateeva, N., Niculescu, H., Schlenoff, J., et al.: Appl. Phys. **83**, 3111 (1998)
5. Langiano, S.C.: Mater. Res. **13**, 465 (2010)

Design and Implementation of ARM7 Instruction Set Based on GDB Framework

Tao Yongchao(✉), Wu Xianghu, and Qu Mingcheng

Shenzhen Academy of Aerospace Technology, Shenzhen, China
taoyongchao652@163.com

Abstract. This paper alms to raise a instruction set simulator prototype based on ARM7. The simulator complete the application-simulate through execute the application instructions' fetch, decode and perform. In this paper, we realize the simulation of the ARM7 3-stage pipeline, decoding logic, the implementation of logic, a variety of abnormal patterns of switching and other basic functions.

Keywords: Instruction set simulator · ARM7 instruction set · GDB

1 Introduction

Now the simulator can be divided into two categories, one is the hardware emulator (emulator), one is the software simulator (simulator). The latter can be divided into the following schemes according to the different simulation levels of the microprocessor: processor full function model, instruction set simulator (ISS), embedded software compilation simulation and processor hardware simulation [1]. Compared with other schemes, the cooperative simulation technology of ISS has the advantages of fast speed, easy debugging of software and low cost of verification [2]. Especially for the development of a high-performance system, a software simulator near real-time simulation speed is essential.

This article is mainly to achieve the ARM7 instruction set simulator, In-depth understanding of ARM7 architecture. It mainly includes the simulation of the three-stage pipeline of ARM7, the simulation of decoding logic, the simulation of execution logic, the simulation of various abnormal modes and mode switching, and the simulation of ARM7 register group. At the same time, this paper designs a unified interface between instruction set simulator and GDB communication, so that the ARM7 simulator can be used as an example to communicate with GDB through the realization of unified interface. In addition, this article implements some basic debugging functions, such as breakpoint setting, program pause, resume, and stop. What's more, external functions such as external interruption response through communication with GDB.

F. Xhafa et al. (eds.), *Advances in Intelligent Systems and Interactive Applications*, Advances in Intelligent Systems and Computing 686,
https://doi.org/10.1007/978-3-319-69096-4_120

2 ARM7 Microprocessor Architecture

2.1 ARM7TDMI Microprocessors

The ARM7 family of processors is the mainstream embedded processor designed by ARM Company [3]. ARM7 series includes ARM7TDMI, ARM7TDMI-S and ARM720T. Among them, ARM7TDMI is shown in Fig. 1.

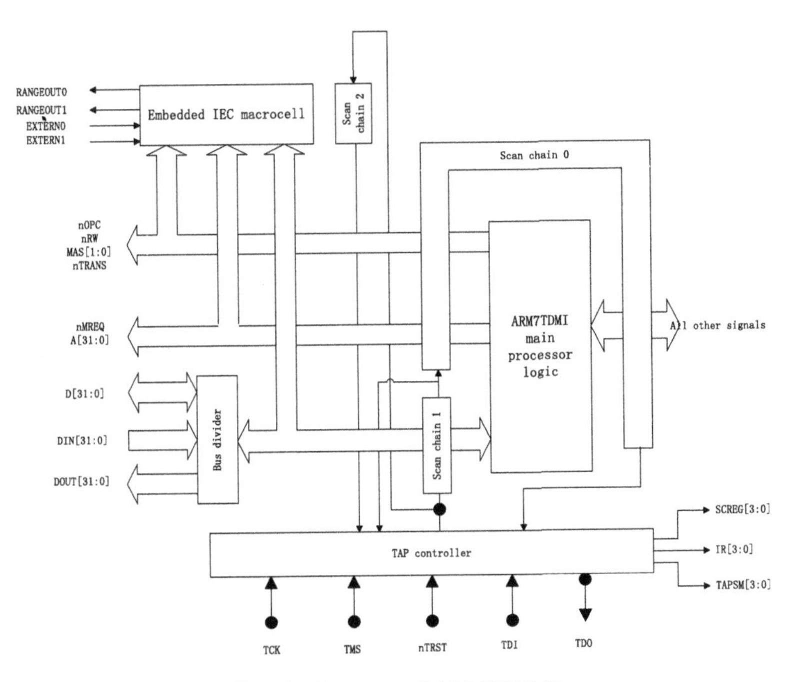

Fig. 1. Structure of ARM7TDMI

As shown, ARM7TDMI consists of three parts: (1) Processor Core; (2) TAP controller; (3) ICEBreaker. The processor contains 31 general-purpose registers and 6 status registers which are 32-bit.

2.2 ARM7 Instruction Set

ARM7 instruction length is 32 bits, and the ARM7 instruction space has a good structure, but each instruction still has a lot of details that need to be mastered [4]. The ARM7 instruction is a fixed 32-bit binary code, shown in Table 1:

Table 1. ARM7 typical instruction code table

31 28	27 25	24 21	20	19 16	15 12	11 8	7 0
Condition	Default	Operating	Flag	Target R	Operand R	Second operand	

1. [condition]: Execution constraint
2. < Operating > : Specific operation command
3. [Flag] : CPSR impact domain
4. < Target R > : The register to which the operation target belongs
5. < Operand R > : A register with operand data
6. < Second operand > : another operand data.

There are several types of ARM instruction sets, include Jump operation, data operation, Memory access instruction and so on. ARM7 instruction set is Load/Store type, so only through the Load/Store instruction to achieve access to memory. The addressing mode is based on the address code segment given in the instruction code to find the way to find the real operand address [5].

3 Simulation of the ARM7 Instruction Set

3.1 The Frame Design

The main function of this emulator is to simulate the processing power of the ARM7 processor on the instruction set. When the emulator obtains the executable program .elf file, through the analysis of GDB, the application code snippet, data segment, etc. loaded into the virtual memory of the emulator, then the simulator can be in memory to fetch, decode, Implementation, which can simulate the operation of the real ARM7 processor.

The simulator runs on a host computer that is based on a Windows system. The top program that loads the emulator DLL is GDB. Therefore, GDB is required to have a standard interface with the emulator to communicate and coordinate control. In other words, the simulator needs to provide a set of data interfaces with GDB standard interface, through this interface GDB can successfully load the emulator DLL, and call its export function on demand, simulation. At the same time, it should be pointed out that the standard interface between GDB and DLL needs to be versatile so that the simulation kernel based on other processor architectures can interact with GDB as long as it is implemented in accordance with the standard interface definition.

3.2 Processing Process Design

Firstly, GDB will executable file symbol analysis, and then load the code segment, data segment and other information to the simulation of virtual memory. The simulation core will be fetched in memory and decoded. After the executable file is loaded into memory by GDB parsing, the emulator can perform the simulation. The general process is shown in Fig. 2.

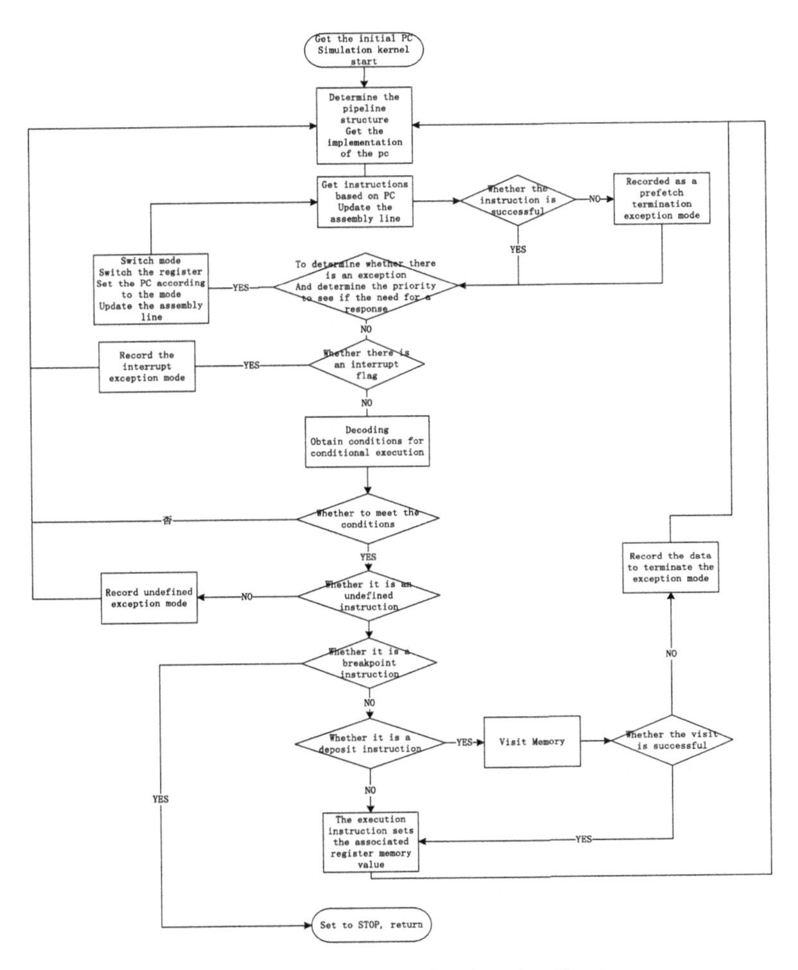

Fig. 2. Kernel simulation function flow

3.3 Instruction Decode Scheme

According to the implementation of ARM conditions, before the implementation of the instruction, the simulator need to determine whether the implementation of the conditions is met. If satisfied, it can perform further decoding, if not satisfied, it must implement the next instruction. If you want to determine an ARM instruction execution type, up to 12 bits need to be determined.

ARM instruction format can be divided into several components, as shown in Fig. 3:

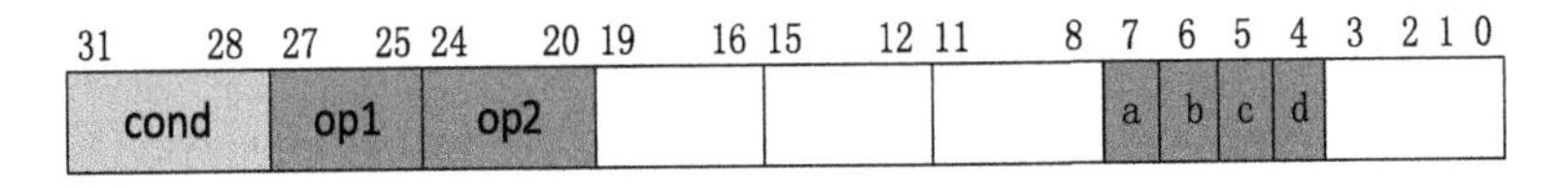

Fig. 3. ARM instruction format

Here the program used, not directly to the 12 to read the judge, but the level of layers of judgment. Mainly divided into three levels:

Level 1: The hierarchy is configured as op1. According to the value of op1 different, from 000 to 111 can be divided into seven categories. For the categories represented by 000 and 001, it is necessary to enter the level 2 to judge. While the other 010–111 representative of the category, you can directly to the level of 3 to judge.

Level 2: The hierarchy is configured as op2 and a, b, c, and d bits. If op1 is 000, then first judge a and d. If a = d = 1, then the need to determine the value of b and c, to be able to determine the multiplication instruction or Load/Store class instructions, and then into the level 3. If you do not meet a = d = 1, you also need to determine the op2 field, to determine whether the data processing class instructions or miscellaneous instructions, and then into the level 3. If op1 is 001, then directly determine the op2 field, to determine the data processing class instruction, the undefined instruction or the immediate transfer status register instruction, and then enter the level 3.

Level 3: The hierarchy is configured as op2. Into this level, it has been proved that what type of instruction has been determined, then only need to be prepared according to different types of different processing functions. Most of the op2 field is based on the specific operation code, and according to the operation code to extract the instruction information, and complete the operation.

3.4 Design of Emulator Exception

An exception is a random event that is randomly generated internally and externally. Handling exceptions need to keep the state, then deal with abnormal, so that it will not affect the original program can be implemented.

When an exception occurs, the R14 and SPSR of the exception pattern group are used to save the status. When processing exception returns, transfer SPSR to CPSR and transfer R14 to PC. As long as the normal program flow is temporarily stopped, the exception occurs. The state of the processor must be saved before exception handling. If several exceptions occur at the same time, you need to refer to the exception priority.

4 Instruction Set Simulator Function and Performance Test

Host is based on the windows system x86 machines, directly run by the VC development of the front-end procedures, through the link emulator operation, you can complete the GDB and simulation of the nuclear link, then, you can test validation. The test program contains most of the common instructions, including the type of assembly instructions and debugging statistics as shown in Table 2.

Table 2. Test program instruction type, number and execution

Instruction type	Number	Correct execution times	Correct rate (%)
MOV	122	122	100
Data processing	157	157	100
Store/Load	252	252	100
Jump Ins.	44	44	100
Multi-register transfer Ins.	8	8	100

Through the test, the simulator can handle the program correctly. The statistics are as follows Table 3

Table 3. Instruction set simulator simulation efficiency

Number of instruction executions	Execution time (s)	Speed (/s)
1,000,000	About 2	500,000
5,000,000	About 9	550,000
About 9,375,000	About 20	468,700

Acknowledgements. First of all, I must thank my parents, is them who gave me life, is them who care for me in every possible way to have my today! I would like to thank Professor Wu Xianghu for his kindness, guidance and guidance. Through the exchange with Pro. Wu, I understand the embedded system and the processor architecture with a great ascension, that will enable me to benefit for life.

References

1. Wang, H., Kuo, C.C.J.: Image protection via watermarking on perceptually significant wavelet coefficients. In: Proceedings of the IEEE Multimedia Signal Processing Workshop, Redondo Beach, California, pp. 278—284 1998
2. Reshadi, M., Bansal, N., Mishra, P., Dutt, N.: An efficient retargetable framework for instruction—set simulation. In: International Symposium on Hardware/Sotfware Codesign and System Synthesis (CODES + 1555), October 2003
3. Zhu, J., Daniel, D.: Gajski. AnUltra—fast instruction set simulator. IEEE Transactions on Very Large Scale Integration (VLSI) Systems, 10(3), (2002)
4. Zhu, D., Mosse, D., Melhem,R.: Multiple—resource periodic problem: how much fairness is necessary. In: Presented at International Real—Time Systems Symposium, (2003)
5. Cmelik, B., Keppel, D.: Shade: a fast instruction—set simulator for execution profiling. ACM Sigmetrics Perform **22**(1), 128–137 (1994)

Simulation and Research on the Rotor Flux Linkage Model of Asynchronous Motor

Xiayi Hao[1(✉)], Genghuang Yang[2], Xin Su[2], and Xiaotian Xu[2]

[1] School of Automation and Electrical Engineering, Tianjin University of Technology and Education, Tianjin, China
tutezdhl203@163.com
[2] Chengxi Power Supply Branch, Tianjin Electrical Power Company, Tianjin, China

Abstract. This paper based on the Vector Control theory uses the MATLAB/SIMULINK simulation software to simulate the rotor flux current model and rotor flux voltage model in the asynchronous motor. It comes to a conclusion that the two waveforms produced by simulating the rotor flux current model and voltage model are credible under the ideal conditions through comparing the two simulation waveforms. After the waveform simulated by asynchronous motor model, which has been transformation in different degree, is compared with the other simulated by original asynchronous motor model, it finds that the changes of motor parameters have a great influence on the rotor flux current model and that the integral part of the motor model affects the rotor flux voltage model. Thus, the following conclusion can be drawn that the current model is suitable for low speed operation of the motor and the voltage model for high speed operation.

Keywords: Vector control · Rotor flux linkage model · MATLAB/SIMULINK simulation · The asynchronous motor speed adaptability

1 Introduction

Since the Vector Control is applied in the asynchronous motor, the speed performance of motor is as good as that of DC motor. The Vector Control, which is also called Flux Orientation Control (FOC), rules the direction of the flux vector that is rotated at the synchronous speed as the reference axis direction and builds the mathematical model of the motor with the inputs, voltage and current, and outputs, the flux linkage and rotational speed. Therefore, Vector Control has the high performance of static-dynamic speed control of asynchronous motor [1–3].

In this paper, the mathematical model of rotor flux linkage, which is based on the Vector Control theory and the dynamic model of three-phase asynchronous motor speed control system, is designed and built. The waveforms are simulated by MATLAB/SIMULINK simulation software. It can make some reasonable assumptions with a further study of the model. To verify the hypotheses, the model is transformed

F. Xhafa et al. (eds.), *Advances in Intelligent Systems and Interactive Applications*, Advances in Intelligent Systems and Computing 686,
https://doi.org/10.1007/978-3-319-69096-4_121

and re-simulated. By comparing the differences between the former and latter, it can get some results and do evaluation for the built rotor flux linkage model [4, 5].

2 Model Construction and Simulation

2.1 Structuring the Simulation Model of Rotor Flux Linkage of Asynchronous Motor

The parameters of the rotor flux model are shown in the Table 1. Figure 1 shows the whole simulation model of the rotor flux linkage of asynchronous motor.

Table 1. The parameters of asynchronous motor

Rated voltage/U	380 V	DC power/U	510 V
Rated frequency/f	50 Hz	Stator self-inductance/L_s	0.071H
Stator resistance/R_s	0.435 Ω	Rotor self-inductance/L_r	0.071H
Stator mutual inductance/L_{ls}	0.002H	Leakage coefficient/σ	0.056
Rotor resistance/R_r	0.816 Ω	Time constant/Tr	0.087 s
Rotor mutual inductance/L_{lr}	0.002H	Moment of inertia/J	0.19 kg m^2
Inter-group interaction/L_m	0.069H		

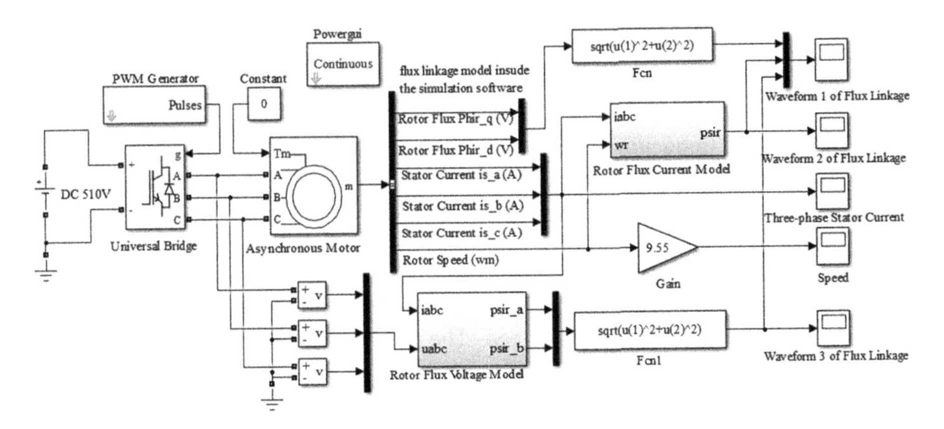

Fig. 1. The simulation model of rotor flux linkage

The data are used as the operating parameters of the simulation model [6]. The simulation waveforms are shown in Fig. 2.

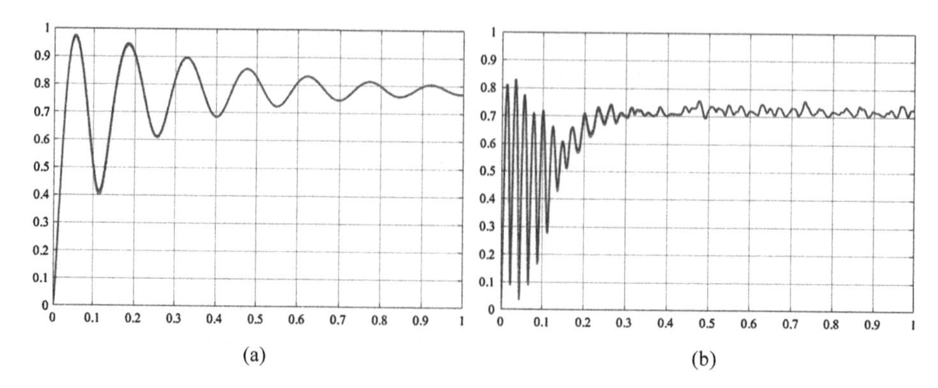

Fig. 2. **a** Frequency is 10 Hz and modulation degree is 0.2. **b** Frequency is 50 Hz and modulation degree is 0.9

3 Research on Two Flux Linkage Models

In order to research the two flux linkage models, they are compared with the flux linkage model inside the simulation software. By observing the coincidence of simulation waveforms produced by the three flux linkage models, it can make a performance evaluation between the rotor flux voltage model and rotor flux current model. Based on the evaluation, it helps to select the appropriate model to realize speed control at high precision for asynchronous motor. The reliability of the rotor flux voltage and current models is studied in following sections.

3.1 Research on the Rotor Flux Current Model

The changes of the frequency of the motor input voltage have directly influence on the magnetic saturation, which in turn affects the inductance of the windings. The following assumption is made: The changes of the inductance cause the flux linkage to become unstable.

In order to research whether the changes of the inductance in the process of start-up and working operation exert an influence on the flux linkage, the rotor flux voltage model and current model are transformed. The new models are shown in Fig. 3.

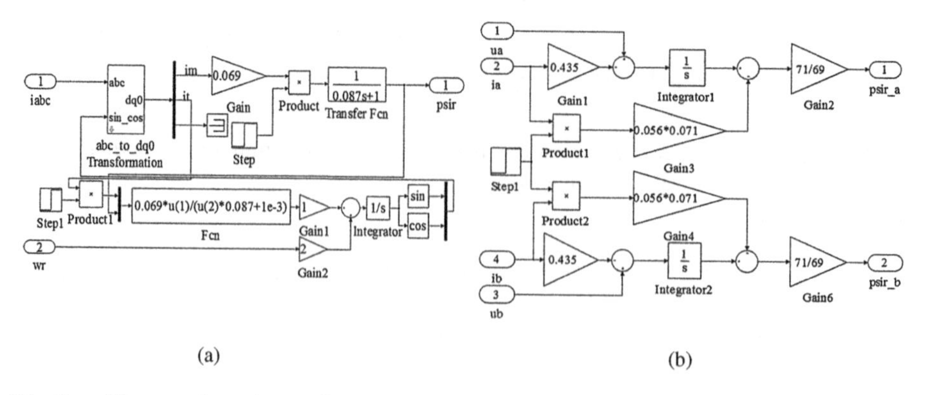

Fig. 3. **a** The transformed rotor flux current model. **b** The transformed rotor flux voltage model

Assuming that the motor is under different speeds, the inductance value changes. This paper selects the extreme condition to simulate, that is, the inductance value is set to be step signal at a point, for example, L_m and L_r are both increased to 1.5 times than original value at a time.

Take 0–0.02 s as the low-speed operating state of the motor, 0.02–0.1 s as the medium-speed operating state and 0.1–0.5 s as high-speed operating state. Run the simulation and obtain the flux linkage waveform shown in Fig. 4. The modulation of the PWM modulator is 0.2 and the output frequency is 10 Hz in the below simulation. From the waveform, it can be seen that if the inductance value changes, the waveform of rotor current model also changes accordingly.

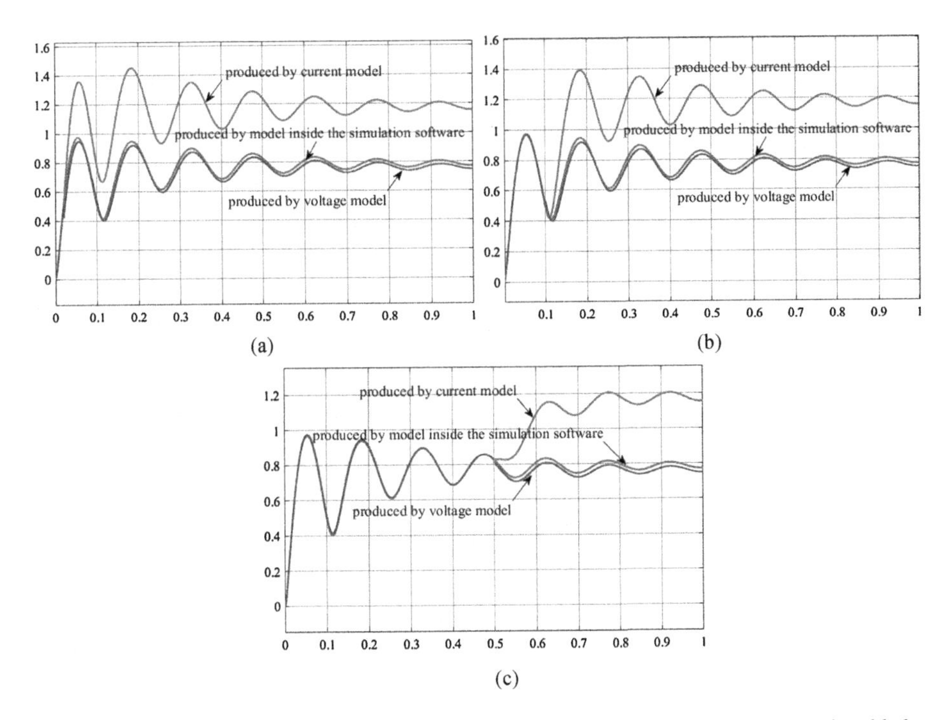

Fig. 4. **a** The step is added at t = 0.02. **b** The step is added at t = 0.1. **c** The step is added at t = 0.5

In summary, when the motor is at different speeds, there is a great impact on the waveform of flux linkage produced by current model as soon as the change of the inductance of the winding appears. Therefore, the above assumption is correct.

It can be obtained the same results for the current model by running the simulation if the modulation is 0.9 and the output frequency is 50 Hz.

Taking into account the actual situation, the frequency of motor input voltage can not be static. Therefore, the waveform of flux linkage produced by current model should be fluctuated irregularly in the vicinity of that of flux linkage produced by the flux linkage model inside the software, but the overall trend of the waveform tends to be stable. For the characteristic of stability and followability, the voltage model is better than the current model.

3.2 Research on the Rotor Flux Voltage Model

In Fig. 4c, the waveform between 0.5 and 0.6 s is magnified and the picture is shown as Fig. 5.

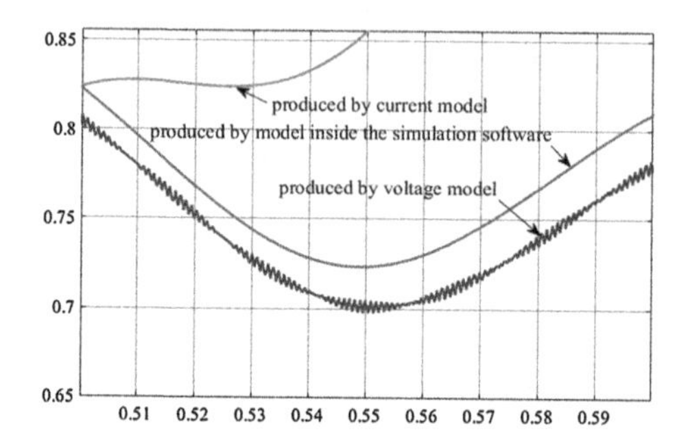

Fig. 5. The magnified waveform with step added at t = 0.5

At this time, the waveform of flux linkage produced by voltage model has a high similarity with that produced by model inside the simulation software, which can be considered to coincide with each other. And it can be seen that the voltage model is insensitive to changes of inductance value.

Based on the above conclusion, the voltage model is studied when the motor is under working operation. From the rotor flux linkage voltage model, it is found that there is an integrator inside the model. According to the characteristic of the integrator, the following assumption is made: The reliability of the rotor flux voltage model would be affected when the motor input voltage is not stable.

There is a large start-up current for motor in the process of starting. It is shown in Fig. 6.

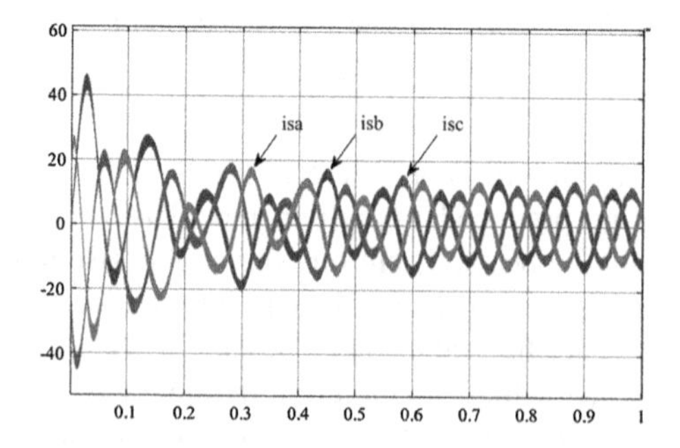

Fig. 6. The waveform of three-phase current

At that time, the stator resistance would withstand a large pressure drop, which can result in a large calculating error. From Fig. 6, it indicates that the integrator plays a bad role at the beginning in the voltage model.

In Fig. 2a, for example, magnify the waveform between 0 and 0.02 s and 0.5 and 0.6 s. The figures are shown in Fig. 7. As it can be seen from Fig. 7, the reliability of the voltage model is not as good as that of the current model when the motor is at start-up. This proves that the above assumption is correct.

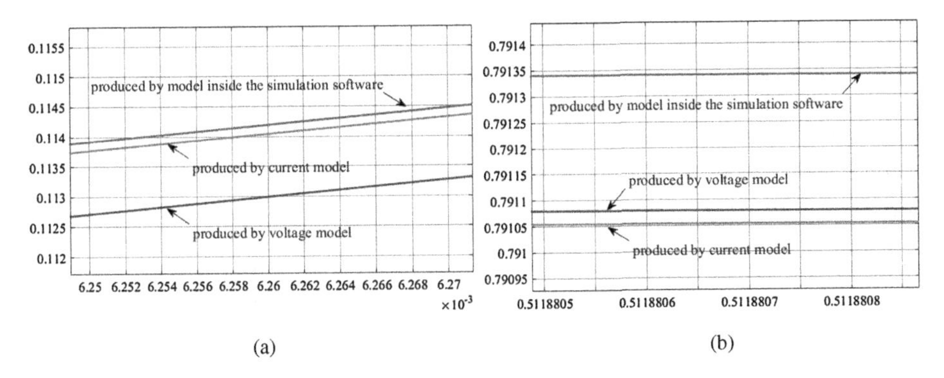

Fig. 7. **a** The magnified waveform during t = 0–0.02. **b** The magnified waveform during t = 0.5–0.6

In the reality, the error of flux linkage produced by voltage model is greater than that produced by current model caused by the fluctuations of input voltage frequency. Therefore, in the actual engineering application, the current model has better reliability if the motor is not in normal operation.

Considering the skin effect that exists during the operation of the motor, it is necessary to study whether the changes of the stator resistance caused by the temperature rise affect the flux linkage model. By checking the papers, it can be found that the variation range of the stator resistance is less than 0.2 Ω when the motor is working [7]. And this changes have little influence on the simulation model.

In the process of simulation, the waveform of flux linkage produced by voltage model has a slight difference with that produced by flux linkage model inside the simulation software. It is because the flux linkage model inside the simulation software can not change the stator resistance at any time. This difference has little effect on the simulation, so it can be ignored. After the value of stator resistance of voltage model is increased at a certain time, it is found that: The changes do not work on the waveform of flux linkage produced by voltage model.

4 Conclusion

For the rotor flux current model, the value of rotor flux linkage calculated by the rotor current model is absolutely affected by the stator inductance and rotor inductance whose values are changed when the motor is at start-up, which is independent of the input voltage frequency and PWM degree of the motor. As a whole, the rotor flux linkage is in a dynamic-steady state. For the rotor flux voltage model, the rotor flux linkage calculated by the voltage model is affected by the integrator existed in the model. There is a large start-up current in the process of starting for the motor, so it can generate a great integral accumulation. In conclusion, it is not suitable for observation with the flux linkage of voltage model at this time. In this paper, the best method of observing the flux linkage of asynchronous motor in different periods is obtained by simulation analysis: For the advantages and disadvantages of the rotor flux voltage and current model, it can be seen that when the motor is in the low-speed state, it is suitable to use the rotor flux current model to observe the flux linkage, and in the medium and high speed state, it is appropriate to choose voltage model for observation; Ideally, the flux linkage of the motor will eventually reach a steady state. It is proved that the speed control of asynchronous motor can be realized by using the Vector Control.

Acknowledgements. This paper is supported by the project of Tianjin University of Technology and Education (XJKC030155 and XJKC031112).

References

1. Wang, J.: Method of rotor coordinate system observation for rotor flux of asynchronous motor. S&M Electr. Machines, vol. **27** (2000)
2. Leonhard, W.: Control of Electrical Drives. Springer, Heidelberg (2001)
3. Park, S.S., Cho, G.H.: Rotor flux controlled induction motor drive for high performance applications, Electron. Lett. **26**, 439–440 (1990)
4. Liusev, P., Stumberger, G., Dolinar, D.: Electr. Power Appl. **152**, 112–118 (2005)
5. Chen, B.: The electricity pull automation control system—the motion control system, Beijing, (2003)
6. Hong, N.: Modeling and simulation of power electronics and motor control system, Beijing, (2010)
7. Chen, B: Asynchronous motor vector control system oriented by rotor flux. Power Electr. Technol. **1**, 42–45 (2008)

Application of Fuzzy PID Control for Oil Temperature in Overvoltage Withstand Test

Qiang Fang[1], Xin Su[2], Xiaotian Xu[2], and Genghuang Yang[1](✉)

[1] School of Electrical Engineering, Tianjin University of Technology and Education, Tianjin, China
ygenghuang@126.com

[2] Chengxi Power Supply Branch, Tianjin Electrical Power Company, Tianjin, China

Abstract. A fuzzy PID algorithm is proposed and implemented for the temperature control of the insulating oil of the test cup in the overvoltage withstand tester. Due to the specific heating of the insulating oil and the high viscosity, it is difficult to control the temperature. It is necessary to design other method to control the temperature rather than the conventional control method. Combined with some characteristics of temperature control, a method of fuzzy PID control is proposed and validated by MATLAB for the adaptive control ability. The simulation results show that the control strategy has good robustness and stability, and the control precision is higher than conventional PID control.

Keywords: Temperature control · Fuzzy control · Control algorithm · MATLAB simulation

1 Introduction

The traditional pressure tester does not consider the control of the ambient temperature of the tested sample. In this paper, the temperature control of the insulating oil of the test cup in the pressure tester is studied. The specific practice is to intercept part of the XLPE (cross-linked polyethylene) into the pressure test cup and fix it, and the cup is filled with insulating oil. The temperature of insulating oil is the ambient temperature of XLPE. This paper uses the 25# transformer oil as insulating oil. Through regulating the temperature of the insulating oil, it can achieve the purpose of controlling the ambient temperature around the XLPE. The heating is carried out mainly by heater strip. The heat generated by heater strip at the time of operation causes the oil around it to heat up and expand. Through the oil up and down convection, the heat spread to the upper cup. The balance of heating is achieved through the combination of heater strip and stirrer. The temperature of the insulating oil is controlled by the algorithm studied in this paper to ensure the stability of the ambient temperature of XLPE.

1.1 PID Control

PID control is simple, easy to implement, wide application and strong stability, so it is widely used in industrial control. According to statistics, in the field of industrial

F. Xhafa et al. (eds.), *Advances in Intelligent Systems and Interactive Applications*, Advances in Intelligent Systems and Computing 686,
https://doi.org/10.1007/978-3-319-69096-4_122

control, the proportion of PID control and its associated optimization control up to 90%. Even in the rapid development of advanced control technology, PID control technology is still the first to consider the method and occupy the most important position in the application. The performance of PID control mainly depends on the tuning of three parameters of P, I and D. only to set the appropriate parameters, PID controller can achieve optimal control to meet the control speed and accuracy.

2 Fuzzy Self—Tuning PID Control Algorithm

In this design, the design steps of the fuzzy module are as follows:

(1) Determine input and output variables. When the control quantity is adjusted in the field, it is mainly controlled according to the actual output and output rate of change. Therefore, the error e and the error rate of change ec are taken as the input of the fuzzy module in this system. The fuzzy controller has three output controls, namely "ΔK_P", "ΔK_I" and "ΔK_D". In this system, set the temperature T, then $e = T$ set value-T current value, $ec = d$ (e current value-e on the moment)/dt. Combined with the actual situation of field control, the range of positive and negative direction of the error e can not be symmetrical in the system design, otherwise the overshoot will be too large, Therefore, the value of the input value e is in the range of [−4, 12], the range of the value of ec is [−1, 1], the output value ΔK_P is in the range of [−900, 900], and the range of ΔK_I is [−12, 12], ΔK_D is in the range [−6000, 6000].

(2) Design language variable domain. In this paper, we define the language variables "error E" and "error change EC" respectively on the domain of e and ec, and define the language variable "control quantity ΔK_P" on the domain of ΔK_P, ΔK_I, ΔK_D, "Control amount ΔK_I", "control amount ΔK_D". In the fuzzy controller, the domain of the language variable is usually a finite discrete integer.
In the fuzzy control system, the quantization factor $Ke = 1/2$, $Kec = 6$, the scale factor $K\,\Delta K_P = 150$, $K\,\Delta K_I = 2$, $K\,\Delta K_D = 1000$. Fuzzy controller input as follow.

$$E = <k_e \cdot (e - \frac{e_H + e_L}{2}) > \quad (1)$$

$$EC = <k_{ec} \cdot (ec - \frac{ec_H + ec_L}{2}) > \quad (2)$$

< > means the represents rounding.

(3) Define the language value of the variable. The fuzzy subset of E is negative, ZO (zero), PS (positive), PM (median), PB (positive)}, EC and ΔK_P, ΔK_I, ΔK_D {NB (negative), NM (negative), NS (negative), ZO (zero), PS (positive), PM (median), PB (positive)}.

(4) Determine membership function. The two inputs of the fuzzy controller, the input range of the error E is moderate, and the input range of EC is too small. Therefore, both of them are suitable for the triangular membership functions with higher resolution. Since the value of the three outputs is magnified by a factor of 1000, the

effect of the actual temperature on the need for rapid sensitivity, so the same use of three-star membership function. Triangle function is characterized by sharp shape, so the resolution is high, the output caused by the output changes are relatively large, with high control sensitivity.

(5) Establish fuzzy control rules. PID parameter tuning, which takes into account the entire control process, K_P, K_I, K_D three parameters each play a role in the control process, and the effect of each in control. The key to fuzzy PID self-tuning control is to rely on the practical experience of the relevant control process, through the practice of continuous testing, modification, summary, and thus establish a good control performance and meet the control requirements of the control rules.
The control rules of ΔK_P. ΔK_P value directly related to the controller in the system response speed, can make the system bias smaller. In the early stages of adjustment, ΔK_P should take a large value to speed up the system response. In the middle of the adjustment, ΔK_P should be appropriately reduced, while ensuring the response speed to prevent excessive overshoot. In the late adjustment, ΔK_P value should be moderate or smaller, it is necessary to maintain the system stability should also prevent excessive overshoot.
The control rules of ΔK_I. ΔK_I in the control of the role of the system to reduce the steady-state error. In the pre-adjustment period, in order to effectively prevent the integral saturation and a large overshoot, therefore, ΔK_I value is smaller, usually take 0. In the middle of the regulation, in order to play the role of integral, should be appropriate to increase the value of ΔK_I. In the later stage of regulation, the value of ΔK_I is large, which can effectively reduce the static error of the system.
The control rules of ΔK_D. ΔK_D is mainly used to control the dynamic operating characteristics of the system. In the early stage of adjustment, ΔK_D value is larger, in order to increase the differential effect, thereby reducing the final overshoot. In the middle of the adjustment, ΔK_D changes on the dynamic characteristics of relatively large, so ΔK_D value should be appropriate to reduce. In the later stages of regulation, the value of ΔK_D is small in order to effectively reduce the effect of disturbance.

(6) Solve the fuzzification. The fuzzy output value is obtained by querying the fuzzy control rule table, and then the fuzzy control process is transformed into the precise control quantity to realize the control of the actuator. The fuzzy language variable of the output is obtained by querying the rule table, and then the fuzzy control can be obtained by determining the membership degree of the fuzzy language variable. In the process of solving the fuzzy process, the system uses the maximum membership method.

3 The Effect of Fuzzy PID Control in Temperature Control

In the experiment, two methods of conventional PID control and fuzzy self-tuning PID control are taken respectively, and the final control effect is recorded respectively. Statistics the thermometer to display the data, recorded every 30 s, and according to this data to draw out the control effect map.

Figures 1 and 2 show the temperature control effect of the conventional PID and fuzzy self-tuning PID control when the temperature is 30 °C. It is found that fuzzy self—tuning PID control can effectively improve the oversampling problem of conventional PID control. This control method also shortens the time to stabilize, and the final result satisfies the requirement that the error be within ±2 °C.

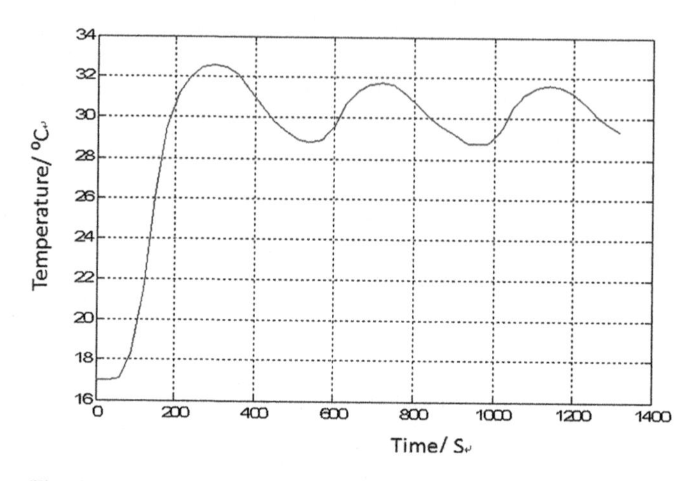

Fig. 1. Effect drawing m of PID control setting temperature 30 °C

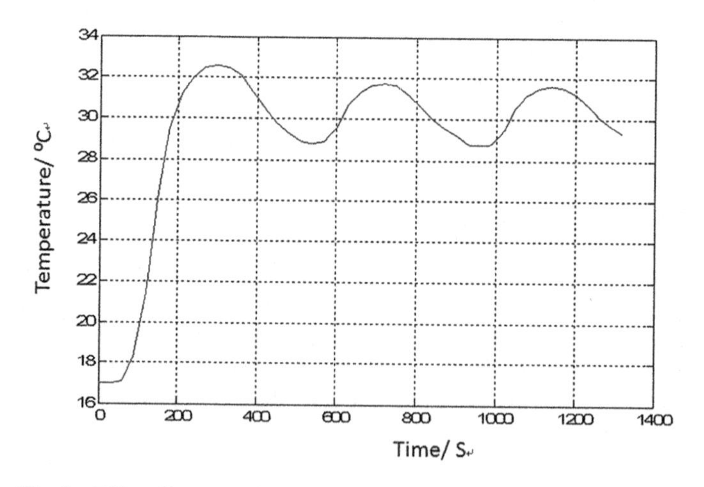

Fig. 2. Effect diagram of fuzzy PID control setting temperature 30 °C

Figures 3 and 4 show the temperature control effect of the conventional PID and fuzzy self-tuning PID control when the temperature is 90 °C. By comparing Figs. 3 and 4, it can be seen that fuzzy self-tuning PID control can appropriately reduce overshoot and improve dynamic characteristics over conventional PID.

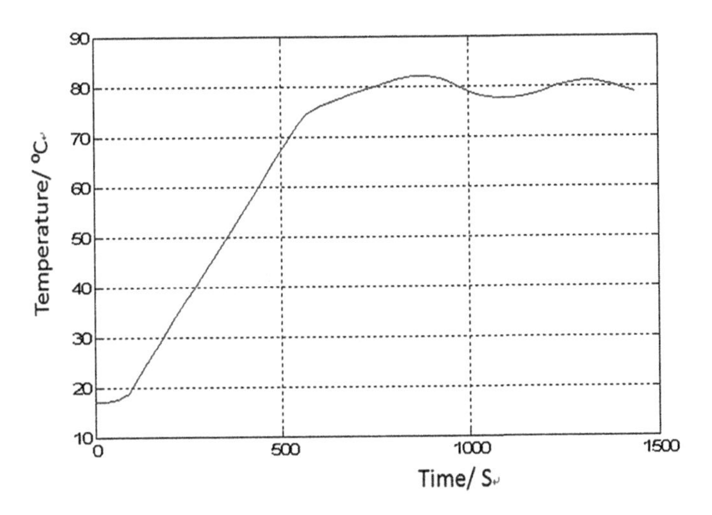

Fig. 3. Effect drawing m of PID control setting temperature 90 °C

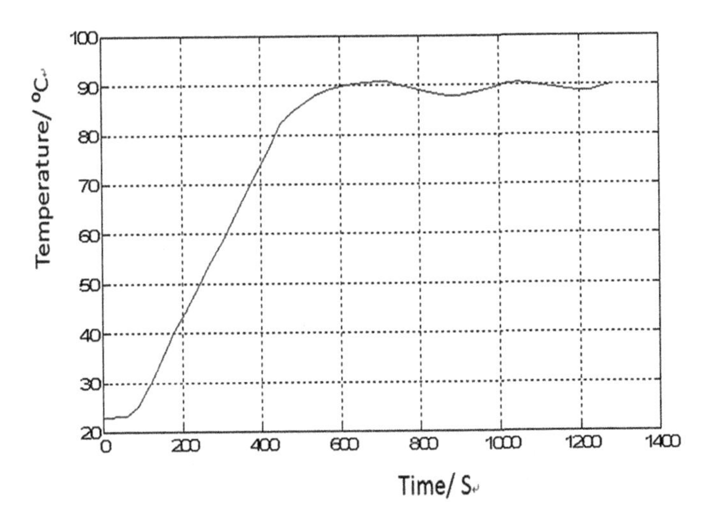

Fig. 4. Effect diagram of fuzzy PID control setting temperature 90 °C

4 Conclusion

Through the simulation analysis, the fuzzy control with fuzzy self-tuning PID control compared with the conventional PID control, with better robustness and reliability. It has a very important role in the effective control of temperature in the process of agricultural production and bio-fermentation.

Acknowledgements. This paper is supported by the project of Tianjin University of Technology and Education (XJKC030155 and XJKC031112).

References

1. Li, K.: Temperature control system intelligent PID control algorithm. Huazhong University of Science and Technology, Wuhan (2006)
2. Wen, K.: Research on intelligent PID algorithm and its application in temperature control. Donghua University, Shanghai (2009)
3. Meng, X.: PID parameter self-tuning method research and controller development. Dalian University of Technology, Dalian (2010)
4. Mao, Y., Luo, H., Zhang, J.: Design and Simulation of a fuzzy self-tuning controller with PID parameters. Autom. Instrum. **1**, 37–39 (2001)
5. Liu, Z., Jiang, X.: Overview of PID controller parameter tuning method. Power Syst. Autom. **21**, 79–83 (1997)
6. Wu, Z., Zhang, L., Zhang, B.: PID parameter tuning and optimization. China High-tech Enterp. **1**, 16–17 (2010)
7. Tang, K.S., Man, K.F., Chen, G.: An optimal fuzzy PID Controller. IEEE Trans. Ind. Electron. **1**, 757–765 (2007)

Study on High Overload Characteristics of Ultrasonic Electric Actuator

Xu Xuerong[1,2](✉), Tian Xiu[1], Fu Hongwei[1], Wang Yanli[1], and Hao Yongqin[1]

[1] Beijing Institute of Aerospace Control Devices, Beijing, China
1028626740@qq.com
[2] North University of China, Taiyuan, China

Abstract. According to the impact principle, the response characteristics of the newly designed anti-high overloaded steering gear under the half sine wave impulse excitation are studied, and the effect of the buffer structure is verified by the combination of the finite element simulation. At the same time, the establishment of a finite element under impact load model of the ultrasonic actuator, using the explicit algorithm in the simulation environment in the transient process of impact load.

Keywords: Ultrasonic electric actuator · High overload · Response characteristics · Finite element simulation

1 Introduction

Traveling wave type ultrasonic motor is a new type of special motor, which makes use of the converse piezoelectric effect of piezoelectric ceramics to realize the conversion of electric energy to mechanical energy [1–3]. The ultrasonic motor has the advantages of performance can be achieved with low speed and high torque, small volume, fast response self-locking, without reducing mechanism, and has good application prospect in the extreme environment of aerospace high vacuum and low temperature, strong radiation [4].

2 Single-Degree-of-Freedom Dynamic Model of Stator Structure at the Time of Impact

The actual impact time of the buffer structure is 3 ms, so the response time is short. When analyzing the shock response of the damping structure, the influence of damping is not considered [5]. It is mainly considered that in a very short time, the absorption energy of the damping system is very limited, and has no effect on the maximum response value [6].

F. Xhafa et al. (eds.), *Advances in Intelligent Systems and Interactive Applications*, Advances in Intelligent Systems and Computing 686,
https://doi.org/10.1007/978-3-319-69096-4_123

$$\ddot{y}(t) = \begin{cases} \ddot{Y}_0 \sin \omega t & 0 \leq t \leq \tau \\ 0 & t \geq \tau \end{cases} \quad (1)$$

in this formulation, $\omega = \pi/\tau$ .The shock response of the structure:

$$\ddot{x}(t) = \begin{cases} \frac{\ddot{Y}_0}{1-\lambda^2}(\sin \omega t - \lambda \sin \omega t) & 0 \leq t \leq \tau \\ \frac{\ddot{Y}_0 \lambda}{\lambda^2 - 1}\left[(1 + \cos\frac{\pi}{\lambda}) \sin \omega_n - (\sin\frac{\pi}{\lambda}) \cos \omega_n t\right] & t \geq \tau \end{cases} \quad (2)$$

in this formulation: $\lambda = \omega/(k/m)^{1/2}$; $\omega_n = (k/m)^{1/2}$.

System response can be divided into two stages: the excitation period for the first stage $0 \leq t \leq \tau$, after the excitation for the second stage $t \geq \tau$. The maximum value of the shock response should be analyzed according to the situation.

According to the formula (1) and formula (2), the maximum ratio of shock response of the system can be obtained. The results are as follows:

(1) When $0 < \tau f < 0.5$, $\omega > \omega_n$, the maximum response amplitude of the system appears in the second stage:

$$\eta_{\max} = \frac{4\tau f}{1 - 4(\pi \tau f)^2} \cos(\pi \tau f) \quad (3)$$

in this formulation: $f = \omega_n/2\pi$.

(2) When $f = \omega_n/2\pi$, $\omega < \omega_n$, the maximum response amplitude of the system appears in the first stage:

$$\eta_{\max} = \frac{2\tau f}{2\tau f - 1} \sin\left(\frac{2\pi n}{1 + 2\tau f}\right) \quad (4)$$

in this formulation: $n = \frac{1}{4} + \frac{1}{2}\tau f$, rounded, and then take the integer.

(3) When $\tau f = 0.5$, the maximum response amplitude of the system occurs at the end of the impact excitation:

$$\eta_{\max} = \frac{\pi}{2} \quad (5)$$

The natural frequency of the stator buffer structure is about 30 kHz and the excitation frequency is 167 Hz. The maximum response amplitude of the system appears in the first stage: $\eta_{\max} = 1$.

The simulated response of the simulation is shown in Fig. 1, and the maximum response of the system is about 1 when the maximum load moment is 3.5 ms. So the simulation results are reasonable.

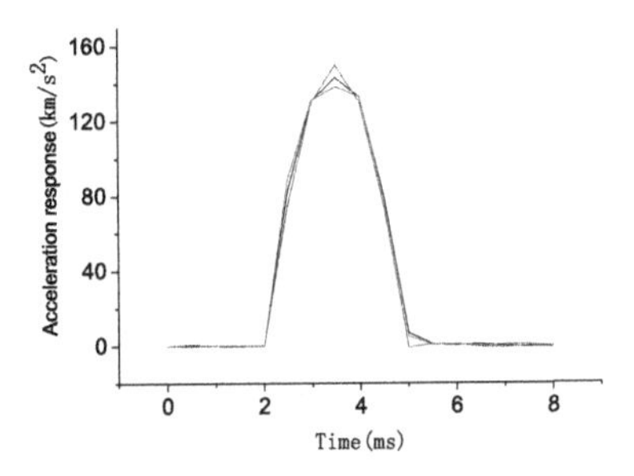

Fig. 1. Response of shock acceleration

3 Dynamic Response of Ultrasonic Motor Under Impact Load

3.1 Simulation Model of Motor

As shown in Fig. 2, 36 spring elements are used to simulate the wave spring, to simulate the transient state of the stator and rotor. In order to simplify the simulation model, the rotor, thrust bearing and angular contact bearing are made into a whole.

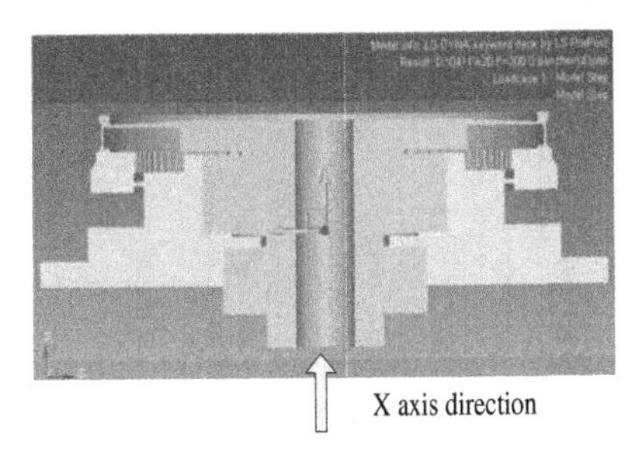

Fig. 2. Simulation model of motor

3.2 Shock Spectrum and Force Diagram of Stator before Impact

The acceleration (as shown in Fig. 3) direction of the impact overload is along the X axis, and the stress wave is transmitted to the base of the motor. At this time, the shock wave transmission in the motor is:

From the middle part of the base through the stator and rotor clamped transfer impact overload, the outer stator inner web clamped part relatively close to the tooth along the positive X axis motion trend, while the rotor outer ring relative to the rotor output shaft along the X axis negative to the movement.

Both the stator and the rotor have a thin wall structure, which is prone to large stress and deformation when subjected to high impact overload. Especially when the stiffness of the rotor, the smaller the peak stress exceeds the yield stress, the web will produce plastic deformation and permanent effects of stator and rotor contact, the output

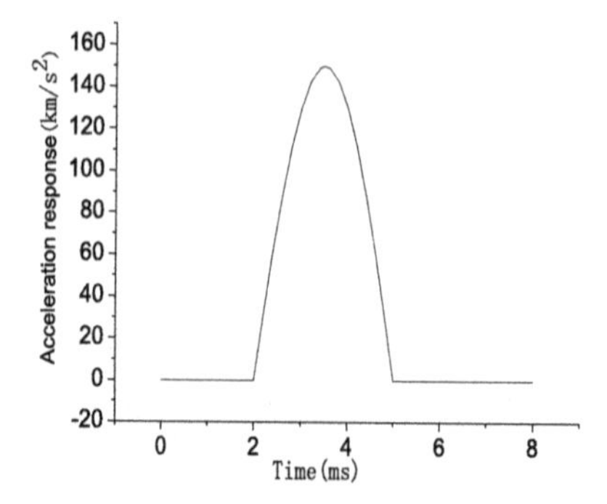

Fig. 3. Impact load curve

performance of the ultrasonic motor will be affected, even does not work because the rotor fault. The analysis shows that the stiffness of the stator is large and the deformation is very small.

The displacement and stress time history curves of the nodes along the radial direction of the rotor and the rotor are selected. Figures 4 and 5 is the displacement time history curve of the radial direction of the stator and rotor web of the ultrasonic motor under the condition of 15 thousand g high impact.

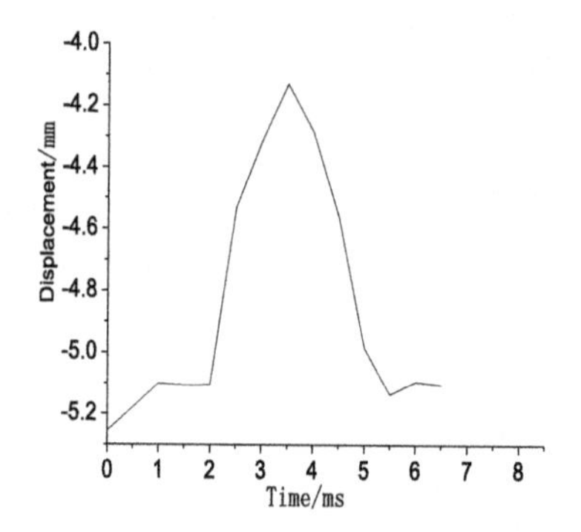

Fig. 4. Displacement of bottom and base

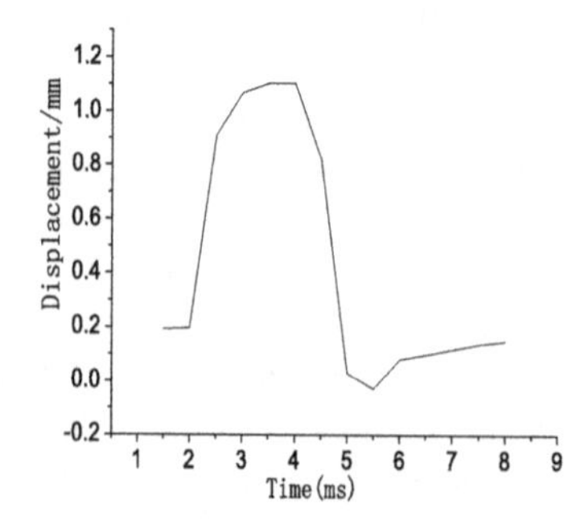

Fig. 5. Displacement variation of rotor

Because the stator relative deformation is small, so only in the inner and the outer edge of the web along the selected observation points, finally reached the same displacement (displacement caused by mobile base). The displacement of each observation point on the stator and rotor reaches the peak value at different time, and keeps a certain position at about t = 8 ms. At each point and the initial displacement difference reflects the permanent deformation and the distribution of the rotor in the stator and subjected to impact. Obviously, there is no obvious deformation of the stator web, the stiffness of the rotor is small, the displacement time history curve of each point along the radial direction, the displacement difference between the inner and outer two nodes can reach 1.2 mm.

In the process of impact load, the yield stress of the material in the stator is not up to the material (phosphor bronze) 345 MPa, as shown in Fig. 6, in the elastic deformation range. The stress produced by the rotor structure does not exceed the yield stress of the material (hard) 325 MPa, as shown in Fig. 7.

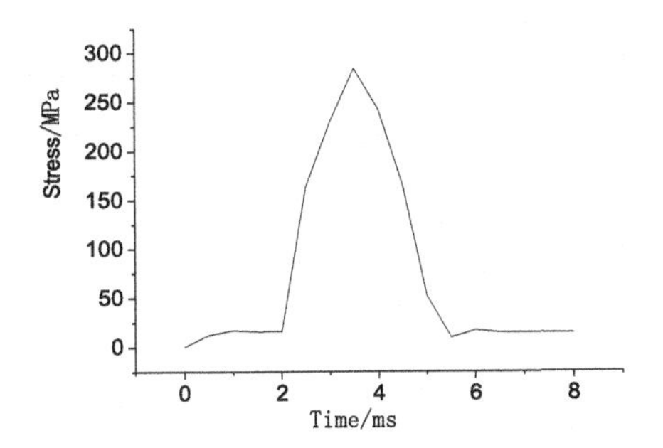

Fig. 6. Element stress diagram of cantilever beam

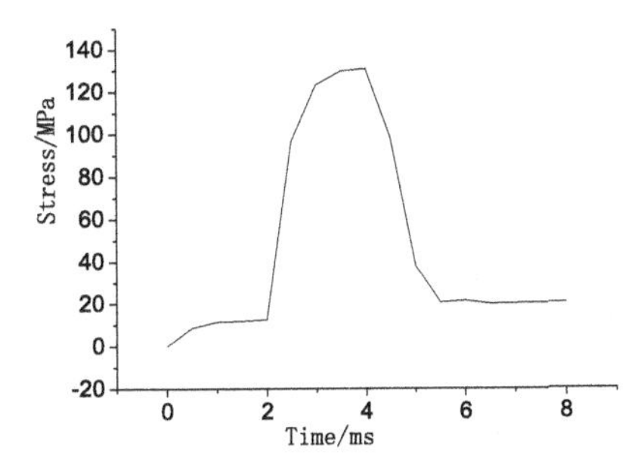

Fig. 7. Rotor element stress diagram

4 Conclusion

In this paper, the dynamic model of the stator buffer structure is established, and the response characteristics of the structure are analyzed. Combined with the finite element analysis method, the transient process under high overload impact environment is simulated, and the stress and strain of the components such as the rotor and the rotor are calculated.

Acknowledgments. We are grateful to the National Major Scientific Instruments and Equipment Development Project (2013YQ470765).

References

1. Zhao, C., Xiong, Z.: Development of piezoelectric ultrasonic motors in China. J. Vibr. Measur. Diagn. **17**(2), 7 (1997). (in Chinese)
2. Senjyu, T., Miyazato, H., Uezato, K.: Quick and precise control of an ultrasonic motor with dual mode control. Int. J. Electron. **80**(2), 191–200 (1996)
3. Li, X., Zhou, S., Yao, Z., et al.: Application of ultrasonic motor in locking device for magnetic bearing flywheel. J. Vibr. Measur. g- Diagn. **33**(4), 555–559 (2013). (in Chinese)
4. Chen, C., Zhao, C.: Modeling of the stator of the traveling wave rotary ultrasonic motor based onsubstructural modal synthesis method. Vibr. Eng. **2**, 238–242 (2005). (in Chinese)
5. Wang, J., Zhang, J., Zhang, Y.: Numerical simulation on stress of projectile materials under high overload. Ordnance Mater. Sci. Eng. **32**(1), 31–33 (2009). (in Chinese)
6. Yao, Z., Wu, X., Zhao, C.: Test of contact interface properties of stator and rotor in traveling wave ultrasonic motor. J. Vibr. Measur. g- Diagn. **29**(4), 388–391 (2009). (in Chinese)

Design and Analysis of Resonator for the Resonant Accelerometer

Yan Li(✉), Xi Chen, and Yunjiu Zhang

School of Mechanical Electronic & Information Engineering, China University of Mining and Technology, Beijing 100083, China
yanli.83@163.com

Abstract. This paper investigates the design and analysis of resonator for the resonant accelerometer. The difference in the resonator resonant frequency in driving and sensing mode and the resonator sensitivity were found to be key factors in determining the performance of the resonant accelerometer. The methodology suggests a simple way of designing and analyzing the resonator. Firstly, the three structural parameters that the width b, the thickness h and the length L of the resonant beam were optimized. And the sensitivity of the resonator to the axial force is related to the three structural parameters. Moreover, simulations of resonator performance are achieved using ANSYS finite element software. The simulation results show that the sensitivity of the resonator 1 is 3.35 Hz/μN, the sensitivity of the resonator 2 is 4.06 Hz/μN. At the same time, the linear degree of sensitivity of the resonator 1 is better than that of the resonator 2.

Keywords: Design · Resonator · Resonant accelerometer · Sensitivity

1 Introduction

Accelerometers are applied in a variety of motion sensing ranging from inertial navigation to vibration monitoring. Many kinds of accelerometers have been designed based on a n variety of different techniques [1]. A resonant accelerometer benefit from output frequency signal, high resolution and large dynamic range. Here, the input acceleration is measured in terms of a shift in the resonant characteristics of a sensing device coupled to the proof mass.

Resonant accelerometers have been previously demonstrated [2–5]. Resonator is the key sensor of the accelerometer, its performance determines the accelerometer scale factor, resolution and other performance indicators. Resonator design should meet the following requirements, such as great sensitivity, small vibration damping, small energy coupled with the external, and easy to excite and pick up. The design of the resonator includes two parts: the design on parameters of resonator and driving and testing of the resonator, this paper only analyzed the design on parameters.

F. Xhafa et al. (eds.), *Advances in Intelligent Systems and Interactive Applications*, Advances in Intelligent Systems and Computing 686,
https://doi.org/10.1007/978-3-319-69096-4_124

2 The Working Principle of Resonant Accelerometer

Resonant accelerometer is based on the principle of resonance measurement to achieve the measured acceleration measurement, with Fig. 1 for the schematic structure.

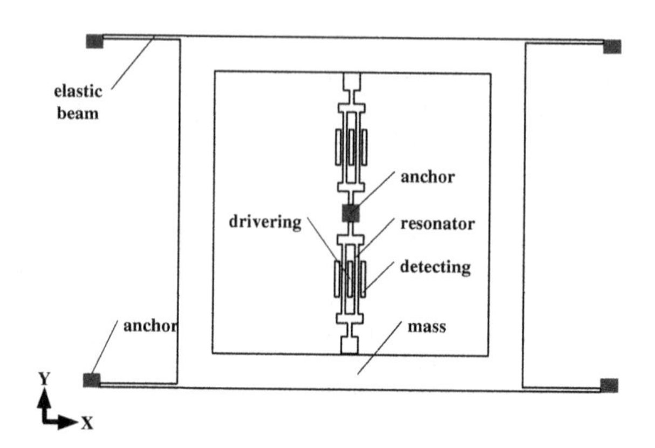

Fig. 1. Schematic of the resonant accelerometer

Resonant accelerometer mainly includes mass, elastic beam, resonator, drive unit and detection unit. When the resonator natural frequency changes under the inertial force in the Y-axis direction, the driving unit excites the resonator to vibrate and maintains the resonator in a resonant state, and the detecting unit picks up the vibration signal of the resonator, and the pick-up signal is fed back to the drive unit through control circuit of the closed loop, so that the frequency of the excitation force is always consistent with the natural frequency of the resonator to achieve tracking of the resonator natural frequency. The measured acceleration value could be converted by the amount of change in the natural vibration frequency based on the vibration signal picked up by the detection unit.

3 The Parameter Design of Resonator

Sensitivity is the most important indicator of the resonator, is also an important basis for selecting resonator structure parameters. In order to have the accelerometer with high scale factor, the resonator must have high sensitivity. According to the mechanical analysis on resonant beam of the resonator, the natural frequency change Δfcaused by the axial force F can be obtained:

$$\frac{\Delta f}{F} = \frac{0.298}{(2\pi)Ebh^2}\sqrt{\frac{Ebh}{0.397\rho bh + {}^{m_a}/_L}} \tag{1}$$

Where E, ρ, b, h, and L denote the elastic modulus, density, width, thickness and length of the resonant beam, respectively, and m_a is the resonator additional mass.

If the effect of additional mass is neglected, the resonator sensitivity to the axial force is related to the resonant beam width b and thickness h. The two structural parameters were optimized as following.

3.1 Resonant Beam Thickness *h*

From (1), it can be seen that the sensitivity of the resonator to the axial force is inversely proportional to the square of the resonant beam thickness h. The h from 2 to 8 μm is substituted into the Eq. (1), calculating the $\Delta f/F$. At the same time, use the finite element software to simulate and calculate the corresponding $\Delta f/F$, and show the theoretical results and simulation results in Fig. 2. The results are very close, when the h from 2 μm increase to 8 μm, the sensitivity from 15 Hz/μN reduced quickly to 1 Hz/μN or so. In addition, the selection of h is bound by the processing technology, the smaller the h, the harder the process. Therefore, in order to ensure that the resonator has a high sensitivity, we should select the minimum thickness of 4 μm can be processed by the current processing technology.

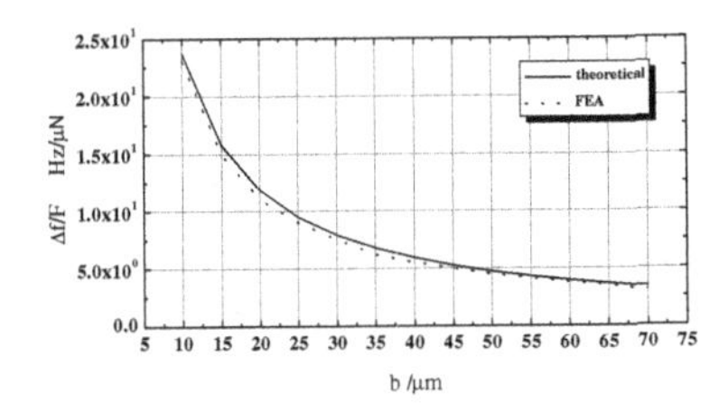

Fig. 2. The axial force is related to thickness *h*

3.2 Resonant Beam Width *b*

From (1), it can be seen that the sensitivity of the resonator to the axial force is inversely proportional to the square of the resonant beam thickness b. The b from 10 to 70 μm is substituted into the formula (1), calculating the $\Delta f/F$. At the same time, use the finite element software to simulate and calculate the corresponding $\Delta f/F$, and show the theoretical results and simulation results in Fig. 3. The results are very close, when the b from 10 μm increase to 70 μm, the sensitivity from 24 Hz/μN reduced quickly to 3 Hz/μN or so. Therefore, in order to ensure that the resonator has a high sensitivity, a smaller b value should be taken. In addition, the selection of b should meet the needs of the exciting capacitor and the pick-up capacitor. The larger the b, the greater the driving force and the detection capacitance can get. That is to say, it is easier to realize the driving on the resonant beam and the detection on vibration state. In summary, b should be taken the middle value of 40 μm.

The length of the resonant beam and the additional mass has an effect on the sensitivity when the additional mass ma cannot be ignored. The two structural parameters were optimized as following, respectively.

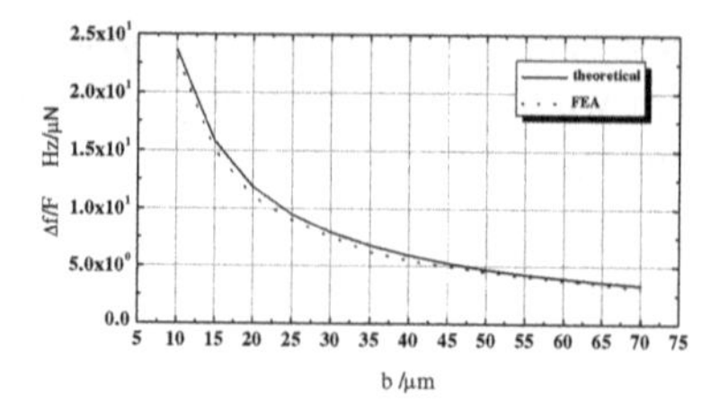

Fig. 3. The axial force is related to width b

3.3 Resonant Beam Length L

The resonant beam length was supposed as 500, 1000, 1500 μm, put them into Eq. (1) respectively, calculating the corresponding axial force sensitivity. At the same time, use the finite element software to simulate and calculate the corresponding $\Delta f/F$, and show the theoretical and simulation results in Figs. 4, 5 and 6.

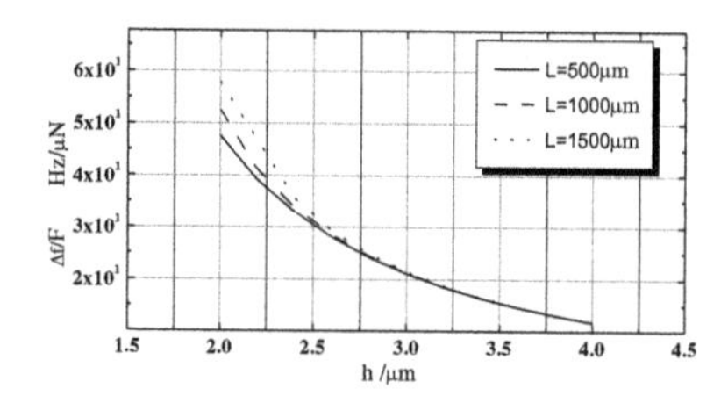

Fig. 4. The axial force is related to thickness h

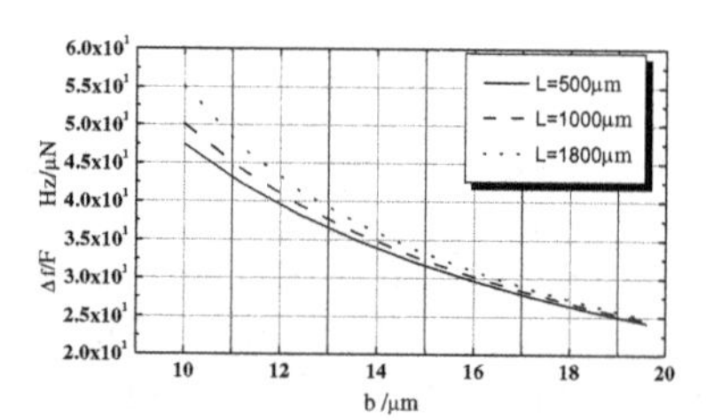

Fig. 5. The axial force is related to width b

As can be seen from Fig. 4, when length L increases from 500 to 1500 μm at $h = 2$ μm, the calculated axial force sensitivity increases from 49 Hz/μN to about 58 Hz/μN, and the rate of change is 18.3%. When $h = 4$ μm, the calculated axial force sensitivity is almost unchanged as the length L increases from 500 to 1500 μm. As a result, the length of the resonant beam has little effect on sensitivity, which is negligible.

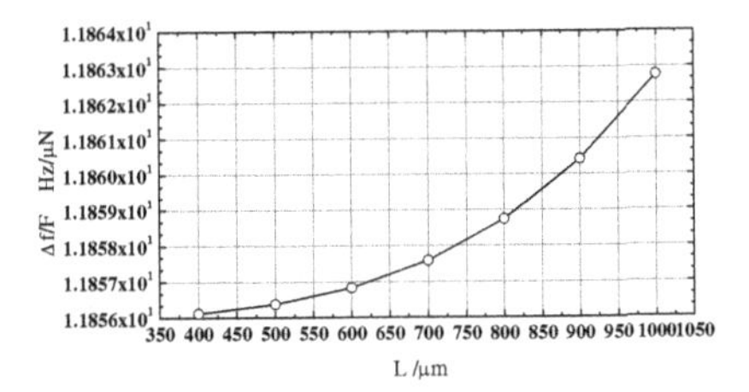

Fig. 6. The axial force is related to length L

As can be seen from Fig. 5, as the width b increases, the change in sensitivity due to the change in L becomes less noticeable. Therefore, when b is 40 μm, the change in L hardly affects the axial force sensitivity.

Figure 6 shows the simulation results of the resonator sensitivity to the axial force with respect to the resonant beam length when h = 4 μm. As the length L increases from 300 to 1000 μm, the axial force sensitivity of the simulation analysis is increased from 11.856 to 11.862 Hz/μN, and the rate of change is 0.05%, which is similar to the theoretical results. Taking the relationship between the length of the resonant beam and the natural frequency into account, while reducing the resonator length can reduce the mechanical vibration nonlinearity, so we choose L is 500 μm.

4 Performance Simulation for Resonator

The resonator structure parameters selected are shown in Table 1. This section uses ANSYS finite element software to simulate the modal frequency and sensitivity of the resonator.

Table 1. Resonator structure parameters

Structure parameters	Length L(μm)	Width b (μm)	Thickness h(μm)
Value	500	40	4

4.1 Modal Frequency Analysis

The resonator driven by the comb teeth has two structural forms as shown in Fig. 7. First, the natural frequency corresponding to the first four modes of the corresponding resonator is calculated, and then the finite element model of the resonator 1 and the resonator 2 as shown in Fig. 7 is established. The natural frequency of the first four order modes is obtained by modal analysis. The second order mode is the inverse mode which is the working mode, and the four-order mode is the inverse mode which is the interference mode. The theoretical results and simulation results are shown in Tables 2 and 3, and the theoretical calculation and simulation results of resonator 1 and resonator 2 are basically consistent. The operating mode frequency of the resonator 2 is much smaller than the operating frequency of the resonator 1. The resonator 1 frequency difference between in-phase and inverse mode is 850 Hz. The resonator 2 frequency difference between in-phase and inverse mode is 2140 Hz. So the resonator 1 is better than the resonator 2.

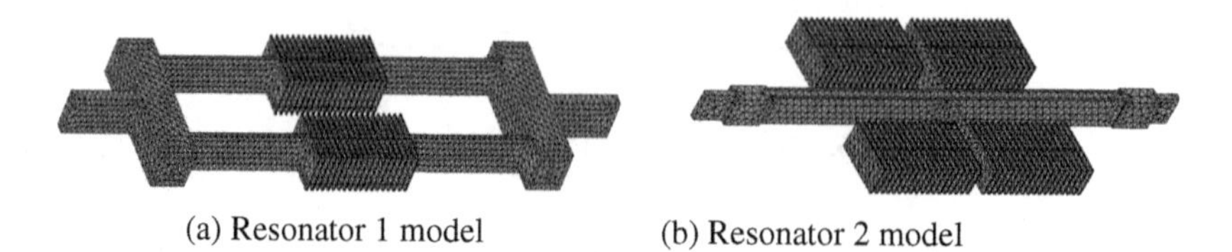

(a) Resonator 1 model (b) Resonator 2 model

Fig. 7. Resonator model

Table 2. Resonator 1 modal frequency

	Order 1	Order 2	Order 3	Order 4
Theoretical (*kHz*)		81.094		271.770
Simulated (kHz)	81.166	82.014	145.221	290.072

Table 3. Resonator 1 modal frequency

	Order 1	Order 2	Order 3	Order 4
Theoretical (kHz)		41.594		50.770
Simulated (kHz)	40.005	42.145	49.105	51.111

4.2 Sensitivity Analysis

Sensitivity analysis determines the resonator sensitivity to the axial force, which is an important performance index of the resonator. The stress distribution obtained by the static analysis of ANSYS finite element software can be used as the pressures in the modal analysis of the resonator. The natural frequency of the resonator under the action of the axial force F is obtained, and the natural frequency changeing is analyzed, and its simulation results are shown in Figs. 8 and 9. The sensitivity of the resonator 1 is slightly higher than that of the resonator 2. In detail, the sensitivity of the resonator 1 is 3.35 Hz/μN, the sensitivity of the resonator 2 is 4.06 Hz/μN. At the same time, the linear degree of sensitivity of the resonator 1 is better than that of the resonator 2.

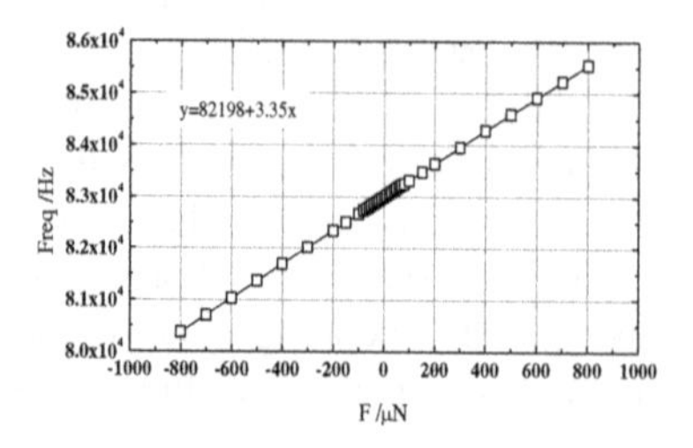

Fig. 8. The resonator 1 sensitivity curve

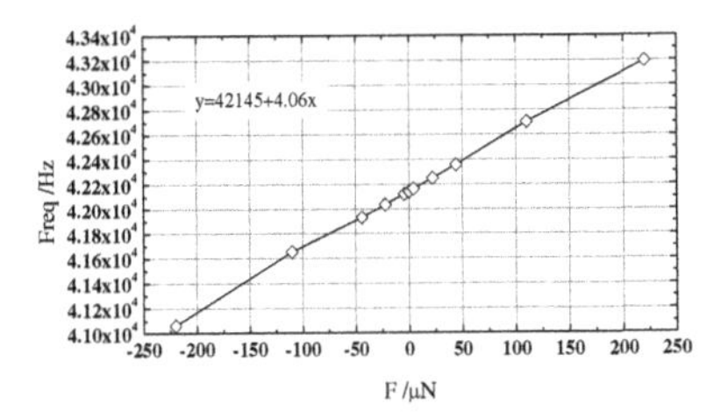

Fig. 9. The resonator 2 sensitivity curve

5 Conclusions

The resonator of resonant accelerometer was designed and analyzed. The sensitivity of the resonator to the axial force is related to width *b,* the thickness *h* and the length *L* of the resonant beam. And the three structural parameters were optimized. The simulation results show that the resonator 1 frequency difference between in-phase and inverse mode is 850 Hz, and the resonator 2 is 2140 Hz, In addition, the sensitivity of the resonator 1 is 3.35 Hz/μN, and resonator 2 is 4.06 Hz/μN. Therefore, the resonator 1 is better than the resonator 2.

Acknowledgments. This study is supported by the National Natural Science Foundation of China under Grant No. 61503018.

References

1. Yazdi, N., Ayazi, F., Najafi, K.: Micromachined inertial sensors. Proc. IEEE **86**, 1640–1659 (1998)
2. Roessig, T.A., Howe, R.T., Pisano, A.P., Smith, J.H.: Surface-micromachined resonant accelerometer. International Conference on Solid-state Sensors and Actuators Chicago, 16–19 June 1997, pp. 859–862
3. Seshia, A.A., Palaniapan, M., Roessig, T.A.: A vacuum packaged surface micromachined resonant accelerometer. J. Microelectromech. Syst. **11**(6), 784–793 (2002)
4. Pedersen, C.B.W., Seshia, A.A.: On the optimization of compliant force amplifier mechanisms for surface micromachined resonant accelerometers. J. Micromech. Microeng. **14**(10), 1281–1293 (2004)
5. Zhang, J., Su, Y., Shi, Q.: Microelectromechanical resonant accelerometer designed with a high sensitivity. Sensors **15**(12), 30293–30310 (2015)

Analysis on Quality and Safety of Toys for Children—Based on the Survey Data of Beijing

Liu Xia(✉), Liu Bisong, Wu Qian, and Li Ya

Quality Management Branch of China National Institute of standardization, Beijing, China
liuxia1010@163.com

Abstract. The damage scenario of toys has a high degree of uncertainty. The parents with children at the age of 0–14 have been investigated by the method of stratified random sampling. The investigation contents include basic information, the possibility of the usage scenarios of children's toys, the possibility of injury under the usage scenarios, the severity of the injury, and recommendations and measures to reduce the safety risk of children's toys. This paper analyzes the factors related to the quality and safety of children's toys, including the correlation of parents' age, education background, occupation and toy purchase channels, and the correlation between the purchase channel and product quality safety level. By using stata12.0, this paper analyzes the relationship between parental heterogeneity and its safety awareness, as well as the relationship between parents' safety consciousness and purchase channel.

Keywords: Toys for children · Uncertainty · Quality · Safety · Relevance

1 Introduction

Due to the unpredictability of the using behaviors of children during use, toy damage scenarios will change with the product, user and environmental characteristics, a feature with a high degree of uncertainty. Based on the subjective evaluation of consumers and the theory of uncertainty, this paper classifies and sums up the 28 kinds of physical hazard factors according to the recalled products, national standard and the network public opinion, and finds 19 kinds of hazard factors. On this basis, the method of random stratified sampling is used to analyze the possibility of the usage scenarios of children's toys and the possibility of injury under the usage scenarios through questionnaires. Then the paper measures the uncertainty of the usage scenarios to lay the foundation for the construction of risk assessment model on the physical hazard of toys for children.

F. Xhafa et al. (eds.), *Advances in Intelligent Systems and Interactive Applications*, Advances in Intelligent Systems and Computing 686,
https://doi.org/10.1007/978-3-319-69096-4_125

2 Questionnaire Survey

In the "National Safety Technical Code on Toys" of National standard GB6675-2014, it is stipulated that those who are in the range of 0–14 years old can be called children. In this paper, the method of the stratified random sampling is used for the investigation. In full consideration of the urban and rural differences, questionnaires were handed out to parents with child or children at the age of 0–14 in a number of kindergartens, large shopping malls and primary schools in Haidian District, Fengtai District, Chaoyang District and Tongzhou District, Beijing. A total of 50 questionnaires were given out and 45 were collected. Among them, 41 were valid questionnaires with the efficiency rate of about 91.1%. The investigation contents include basic information, the possibility of the usage scenarios of children's toys, the possibility of injury under the usage scenarios, the severity of the injury, and recommendations and measures to reduce the safety risk of children's toys.

2.1 Basic Information

The basic information of parents includes age, education background and current occupation. According to the survey results, the respondents are mainly distributed in the age of 20–40 years old, most of them have achieved bachelor's degree or above and are now S&T personnel, educator or general staff. All these indicate that the objects of the survey are mainly senior intellectuals, such as urban white-collar workers or scientific research personnel. The basic information of children includes the age and sex of the child, mainly used to analyze the impact of children's age and gender on the injury cases that occur during the use of children's toys. The survey shows that there are 15 boys and 26 girls in the valid questionnaires, 71% of who involved in the questionnaire are under 36 months (Fig. 1).

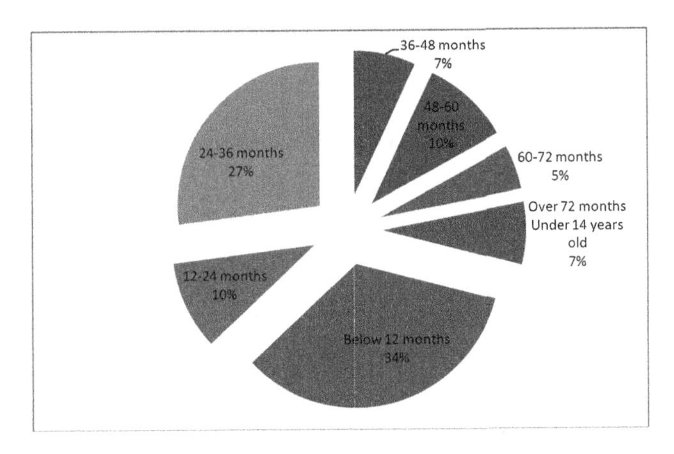

Fig. 1. Distribution statistics of children's ages

2.2 Quality and Safety Level of Common Purchasing Channel of Toys

At present, the common purchasing channels of toys mainly include large stores/shopping malls, small commodity wholesale market, online purchase, large supermarket chains, small supermarkets and roadside stalls.

The common purchasing channels of children's toys mainly cover large stores/shopping malls, small commodity wholesale market, network, large supermarket chains, small supermarkets and roadside stalls. The survey shows that the network, large stores, and large supermarket chains are the most important purchasing channels of toys, and the specific situation is shown in Fig. 2.

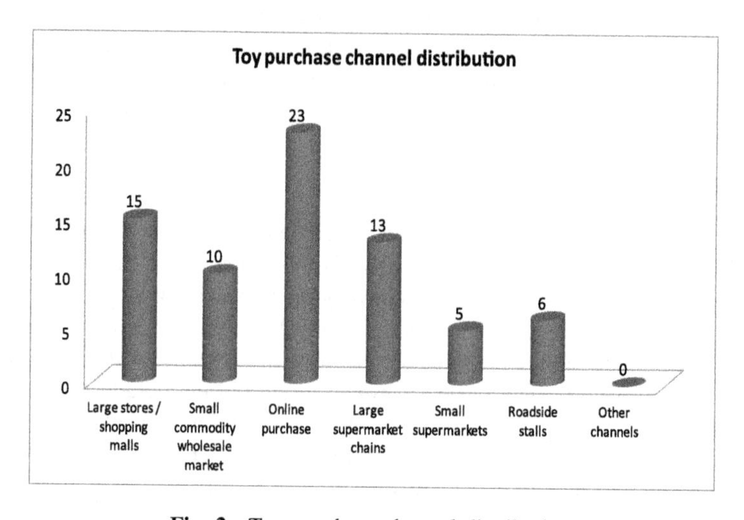

Fig. 2. Toy purchase channel distribution

Table 1 shows the order of the quality and safety levels of toys purchased from different channels based on the questionnaire.

Table 1 Order of the quality and safety levels of toys purchased from different channels

Name	No.						
	Order of the number of toys						
	1	2	3	4	5	6	7
Large stores/shopping malls	19	1	1	5	0	0	0
Small commodity wholesale market	3	0	8	15	0	0	0
Online purchase	2	4	10	2	5	1	0
Large supermarket chains	8	8	3	3	1	2	0
Small supermarkets	0	12	2	0	9	2	0
Roadside stalls	0	0	0	0	2	22	0
Other channels	0	0	0	0	0	0	25

The Borda scores for large stores/shopping malls, small commodity wholesale markets, online purchases, large supermarket chains, small supermarkets, roadside stalls and other channels to buy toys are 138, 95, 89, 113, 88, 26, 0.

According to the above calculation, we can see that the objects of the survey generally agree that the toys purchased from large stores/shopping malls have the best quality, followed by large-scale shopping malls, roadside stalls the worst. The order of their quality is as follows: large stores/shopping malls > large supermarket chains > small commodity wholesale market > online purchase > small supermarkets > roadside stalls > other channels. Among them, the quality of online shopping products, accounting for the largest proportion of the purchase channels, is not as good as the quality of products from small commodity wholesale market, which reminds us to pay attention to the quality and safety issues of online shopping products.

2.3 Concerns of Consumers on Toy Purchasing

Focusing on the concerns of parents on toy purchasing, the questionnaire sets the "3C logo", "use notes", "functional characteristics" and other problems generally concerned by parents. The specific situation is shown in Fig. 3. As can be seen from the figure, when buying the toys, the parents are generally more concerned about the "use notes", "functional characteristics", "3C logo" of the toys.

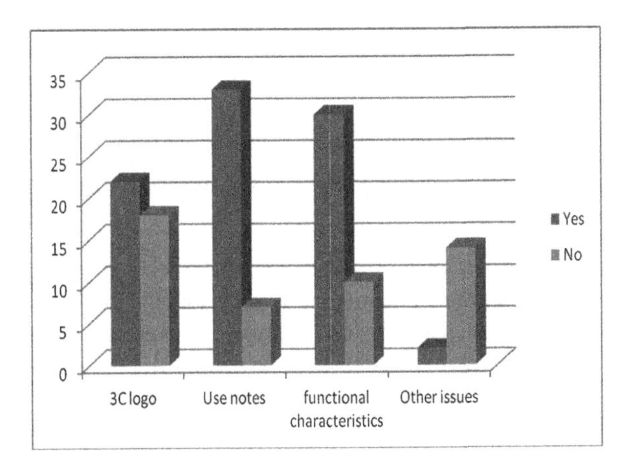

Fig. 3 Statistics for whether parents will be concerned about the quality and safety issues in the purchase of toys

The results of the survey also show that 52% of parents tend to buy domestic brands, while 48% of parents tend to buy foreign brands. Among them, parents who tend to buy domestic brands generally believe that the domestic children's toys with better cost performance are affordable and easy to buy. In the event of quality and safety issues, it is easier for customers to safeguard legal rights if they buy domestic toys.

For the problem that "whether parents will buy toys that are made for children more than 36 months for children at the age of less than 36 months", 55% of the parents said yes, mainly because the toys can be used for a long time with high utilization rate; 45% of the parents believe that, due to the potential safety risks of product quality that may arise, they should not buy toys unsuitable for the children's age.

3 Correlation Analysis on the Quality and Safety Investigation Factors of Children's Toys

3.1 Correlation of Parents' Age, Education Background, Occupation and Toy Purchase Channels

Foreign related research shows that parents' age, education background and occupation may affect the purchase channels of children's toys. In this paper, the above issues are investigated and analyzed in the questionnaire and correlation analysis of the investigation results is carried out by using the stata12.0 software. The results show (See Fig. 4 for detail) that the parents' age and purchase channel are significantly correlated, indicating that the purchase channel is largely affected by age. Since the survey objects at the age of 20–40 account for 92.7% of the total survey objects, the majority of which are young people, the proportion of online purchase is relatively more than that of other channels. In addition, the survey results show that the education background and professional distribution of the parents is not significantly correlative with the purchase channel, indicating that education background and occupation have small impact on the toy purchase channels.

3.2 The Correlation between the Purchase Channel and the Product Quality and Safety Level

From the analysis of the above tables and this table, we can see that the purchase channel has a certain relationship with the product quality and safety level. In this paper, stata12.0 software is used to analyze the correlation between the purchase channel and the toy quality and safety level. The results are shown in Figs. 4 and 5. It can be seen that the quality and safety level of the toy and the purchase channel are significantly correlated with each other at the significance level, indicating that the purchase channel of children's toys has a great influence on the quality and safety level of the toy. So consumers should choose the purchase channel carefully when purchasing the toys (Tables 2 and 3).

Table 2 Correlation between the parents' age, education background, occupation distribution and the purchase channels

		Age of parents	Education background of parents	Occupation distribution of parents	Purchase channel
Age of parents	Pearson correlation	1	0.051	0.070	0.733
	significance (bilateral)		0.914	0.881	0.061
	N	7	7	7	7
Education background of parents	Pearson correlation	0.051	1	0.873[a]	−0.111
	significance (bilateral)	0.914		0.010	0.812
	N	7	7	7	7
Occupation distributionof parents	Pearson correlation	0.070	0.873[a]	1	−0.144
	significance (bilateral)	0.881	0.010		0.758
	N	7	7	7	7
Purchase channel	Pearson correlation	0.733	−0.111	−0.144	1
	significance (bilateral)	0.061	0.812	0.758	
	N	7	7	7	7

[a]Significance correlations at 0.05 level (bilateral)

Table 3 Correlation analysis of purchase channel and product quality and safety level

		Purchase channel	Product quality and safety level
Purchase channel	Pearson correlation	1	0.887[a]
	significance (bilateral)		0.008
	N	7	7
Product quality and safety level	Pearson correlation	0.887[a]	1
	significance (bilateral)	0.008	
	N	7	7

[a]Significant correlations at 0.05 level (bilateral)

3.3 Correlation of the Characteristics and Safety Awareness of Parents and Children and the Purchase Channels

On the basis of research questionnaires, this paper analyzes the relationship between the related information of parents and children and their safety consciousness with pour command by using the method of stata12.0. At the same time, the relationship between parents' safety consciousness and purchase channel is also analyzed in the paper. The selected indexes are shown in Table 4.

Table 4 Source of variable indicators

Problem	The original option (variables)	Purpose
When you buy children's toys, besides brand, what other issues related to quality and safety will you consider?	3C certification mark (q1) Use notes (q2) Suitable age (q3) Functional characteristics (q4)	Safety consciousness
Please choose the toy purchase channel that you usually adopt according to your own life experience	Large stores/shopping malls (c1) Small commodity wholesale market (c2) Online purchase(c3) Large supermarket chain (c4)	Purchase channels
Child's age (months)	0–12; 12–24; 24–36; 36–48 (cage); 48–60; 60–72; >72	——
Parent's age (years)	20–30; 30–40; 40–50 (page); 50–60; >60	——

Table 5 Analysis of relationship between parents' age and safety consciousness

	q1	q2	q3	q4	page
q1	1.000				
q2	−0.055	1.000			
q3	0.104	0.116	1.000		
q4	0.015	0.031	0.271*	1.000	
page	−0.380**	0.103	−0.319*	−0.061	1.000

*p < 0.1; **p < 0.05; ***p < 0.01

(1) Analysis of parents' age and safety consciousness

Table 5 shows the relationship between parents' age and safety consciousness

Through the correlation analysis (Table 5), we can find that parents' age has a significant negative correlation with the concern of 3C certification the degree of concern for the suitable age. While the parents' age has no significant correlation with other variables that represented the parental safety implications, indicating that young parents are more concerned about the toy 3C certification and the suitable use age of toys. The full name of 3C certification is "China Compulsory Certification", a product conformity assessment system implemented by the relevant government in accordance with laws and regulations to protect the personal safety of consumers and national security, and to strengthen product quality management. It can be said that 3C certification is the signal of quality and safety, indicating that the younger parents pay more attention to product quality. At the same time, the correlation analysis also shows that young parents pay more attention to the applicability of toys.

(2) Analysis of relationship between parents' educational level and safety consciousness

Table 6 shows the relationship between parents' educational level and safety consciousness.

Table 6 Analysis of relationship between parents' educational level and safety consciousness

	q1	q2	q3	q4	pedu
q1	1.000				
q2	−0.055	1.000			
q3	0.104	0.116	1.000		
q4	0.015	0.031	0.271*	1.000	
pedu	−0.513**	0.103	0.222*	0.045	1.000

*p < 0.1; **p < 0.05; ***p < 0.01

Through the correlation analysis (Table 6), it can be found that the education background of parents has significantly positive correlation with the concern of the toy 3C. It may be that due to parents with high degree of education will obtain product quality and safety information more initiatively, find more purchase channels and know better about the related standard certification of the product quality and safety.

(3) Correlation of children's age and parents' safety awareness

Table 7 shows the correlation of children's age and parents' safety awareness.

Through the correlation analysis (Table 7), it can be found that the age of the children is significantly negatively correlated with the concern of the use notes of toys, and the parents with younger child are more concerned with the use notes of toys.

Table 7 Analysis of relationship between children's age and parents' safety awareness

	q1	q2	q3	q4	cage
q1	1.000				
q2	−0.055	1.000			
q3	0.104	0.116	1.000		
q4	0.015	0.031	0.271*	1.000	
cage	−0.080**	−0.287*	−0.271	−0.064	1.000

*p < 0.1; **p < 0.05; ***p < 0.01

(4) Correlation of parents' safety awareness and toy purchase channel

Table 8 shows the correlation of parents' safety awareness and toy purchase channel.

Through the correlation analysis (Table 8), we can find that the functional characteristic of toys has a significant positive correlation with the preference of large shopping malls/store, and has a significant negative correlation with the preference for online purchasing. Parents' concern on suitable age group is significantly negatively correlated with the purchase preference for large supermarket chains. All these show that if the parents are concerned about the functional characteristics of toys, they tend to go to large shopping malls/stores to buy toys and personally select the toys to obtain the product properties. Therefore, they will not incline to buy toys from the internet, because the network cannot guarantee parents to have access to the real function information of the toys.

Table 8 Analysis of relationship between security awareness and toy purchase channel

	c1	c2	c3	c4	q1	q2	q3
c1	1.000						
c2	−0.199	1.000					
c3	−0.668***	−0.385**	1.000				
c4	−0.170	−0.098	−0.328**	1.000			
q1	−0.133	0.330**	−0.133	0.089	1.000		
q2	−0.184	−0.062	0.263	−0.116	−0.055	1.000	
q3	0.170	0.098	−0.060	−0.278*	0.104	0.116	1.000
q4	0.345**	0.005	−0.279*	−0.051	0.015	0.031	0.271*

*p < 0.1; **p < 0.05; ***p < 0.01

4 Conclusions

According to the investigation on parents with children aged 0–14 years, we conclude that the Borda scores for catapult toys, electric toys, metal toys, dolls, plastic toys, stroller toys and other types of toys are 48, 179, 115, 167, 180, 120 and 12 respectively. The results of the questionnaire show that plastic toys account for the largest proportion of toys bought by families currently, followed by electric toys, dolls, strollers, metal toys, catapult toys. The objects of the survey generally agree that the toys purchased from large stores/shopping malls have the best quality, followed by large-scale shopping malls, roadside stalls the worst. The order of their quality is as follows: large stores/shopping malls > large supermarket chains > small commodity wholesale market > online purchase > small supermarkets > roadside stalls > other channels.

The correlation analysis on the factors related to the quality and safety of the toys has been carried out. According to the analysis, parents' age and purchase channels are significantly correlative, which indicates that the purchase channels are affected by the age to a large extent; parents' education background and professional distribution have little correlation with the purchase channel, which indicates that education and occupation have small influence on the purchase channel. Toy quality and safety level and the purchase channels are significantly correlative, indicating that the purchase channels of children's toys have great impact on the safety level of toys. So consumers should carefully choose the purchase channels in the purchase of toys. There is a significant positive correlation between the degree of concern about the functional characteristics of the toy and the preference of the large shopping malls/stores, and there is a significant negative correlation between the degree of concern about the functional characteristics of the toy and the preference of online purchase. Parents' concern on toys for children in suitable age group has a significant negative correlation with the purchase preference for large supermarket chains.

Acknowledgements. This paper has been funded by the national key research and development project "Research on key technical standards for quality and safety control of consumer goods" (2016YFF02022600), and the central basic research business project "Research on method for consumer product safety risk assessment based on the simulation technology—taking children's products as an example" (552015Y-3990).

References

1. Chiang, W.C., Pennathur, A., Mitai, A.: Designing and manufacturing consumer products for functionality: a literature review of current function definitions and design support tools. Integr. Manufact. Syst. **12**, 430–448 (2001)
2. Consumer Safety Institute: Manual for risk assessment, report prepared by the dutch consumer safety institute dated February 2005 (2005)
3. EFSA: EFSA responds to Commission's urgent request on dioxins in Irish pork. Press release, 10 December 2008. Available at: accessed 6 May 2011 (2008)
4. European Commission: Guidelines for the notification of dangerous consumer products to the competent authorities in accordance with article 5/3 of directive 2001/95/EC. European Commission, Brussels, Belgium (2004b)
5. Flaherty, E.: Safety first: the consumer product safety improvement act of 2008. Loyola Consum. Law Rev. **21**, 372–384 (2008)
6. Kinnersley, S., Roelen, A.: The contribution of design to accidents. Saf. Sci. **45**, 31–60 (2007)
7. Liu, X., Luo, Hong-qi: Research on the methods of safety hazard identification for the consumer products1 based on product life cycle. In: The International Conference on Computer and Information Science, Safety Engineering (CAISSE), pp. 156–161 (2012)
8. Pyke, D., Tang, C.S.: How to mitigate product safety risks proactively? Process, challenges and opportunities. Int. J. Logistics Res. Appl. Lead. J. Supply Chain Manage. **13**, 243–256 (2010)
9. GB 6675–2003, National Safety Code for Toys [S]
10. Zhu, J.: Research on aggregation method of two uncertain preference information among group decisions. Control Decis. Mak. **21**(8):889–897 (2006)
11. Yu, P., Mu, H.: Research on risk assessment model of quality and safety influence factor of consumer products. China Manage. Sci. **18**, 171–176 (2010)
12. Liu, X., Tang, W., Yang, Y., Wang, L.: Risk assessment—new trends of safety management of China's consumer goods. Stand. Sci. **6**, 76–79 (2009)
13. Fu, Yu., et al.: Information security risk assessment method based on the bayesian network. J. Wuhan Univ. **5**(2006), 631–634 (2006)
14. Zhu, H.: River Environment Risk Model based on Uncertainty Theory and its Early Warning Index System [D]. Hunan University, China (2012)
15. Sun, A.: Urban Rainstorm Waterlogging Risk Assessment Based on the Scene simulation [D]. East China Normal University, China (2011)

Author Index

F. Xhafa et al. (eds.), *Advances in Intelligent Systems and Interactive Applications*, Advances in Intelligent Systems and Computing 686,
https://doi.org/10.1007/978-3-319-69096-4